Fluid Dynamics of Particles, Drops, and Bubbles

This book is a modern presentation of multiphase flow, from basic principles to state-of-the-art research. It explains dispersed fluid dynamics for bubbles, drops, or solid particles, incorporating detailed theory, experiments, simulations, and models while considering applications and recent cutting-edge advances.

The book demonstrates the importance of multiphase flow in engineering and natural systems, considering particle size distributions, shapes, and trajectories as well as deformation of fluid particles and multiphase flow numerical methods. The scope of the book also includes coupling physics between particles and turbulence through dispersion and modulation, and specific phenomena such as gravitational settling and collisions for solid particles, drops, and bubbles. The eight course-based chapters feature over 100 homework problems, including theory-based and engineering application questions. The final three reference-based chapters provide a wide variety of particle point-force theories and models.

The comprehensive coverage will give the reader a solid grounding for multiphase flow research and design, applicable to current and future engineering. This is an ideal resource for graduate students, researchers, and professionals.

Eric Loth is a Rolls-Royce Professor of Mechanical and Aerospace Engineering at the University of Virginia. He is a Fellow of the American Society of Mechanical Engineers (ASME) and the American Institute of Aeronautics and Astronautics (AIAA). He has also received honors and awards from NASA and the National Science Foundation.

Fluid Dynamics of Particles, Drops, and Bubbles

ERIC LOTH
University of Virginia

Shaftesbury Road, Cambridge CB2 8EA, United Kingdom

One Liberty Plaza, 20th Floor, New York, NY 10006, USA

477 Williamstown Road, Port Melbourne, VIC 3207, Australia

314–321, 3rd Floor, Plot 3, Splendor Forum, Jasola District Centre, New Delhi – 110025, India

103 Penang Road, #05–06/07, Visioncrest Commercial, Singapore 238467

Cambridge University Press is part of Cambridge University Press & Assessment,
a department of the University of Cambridge.

We share the University's mission to contribute to society through the pursuit
of education, learning and research at the highest international levels of excellence.

www.cambridge.org
Information on this title: www.cambridge.org/9780521814362

DOI: 10.1017/9781139028806

First published 2023

A catalogue record for this publication is available from the British Library.

Library of Congress Cataloging-in-Publication Data
Names: Loth, Eric, author.
Title: Fluid dynamics of particles, drops, and bubbles / Eric Loth, University of Virginia.
Description: Cambridge, United Kingdom ; New York, NY, USA : Cambridge University Press, 2023. |
Includes bibliographical references and index.
Identifiers: LCCN 2022054292 (print) | LCCN 2022054293 (ebook) | ISBN 9780521814362 (hardback) |
ISBN 9781139028806 (epub)
Subjects: LCSH: Multiphase flow. | Drops. | Granular flow. | Bubbles–Dynamics.
Classification: LCC TA357.5.M84 L68 2023 (print) | LCC TA357.5.M84 (ebook) | DDC 620.1/
064–dc23/eng/20230111
LC record available at https://lccn.loc.gov/2022054292
LC ebook record available at https://lccn.loc.gov/2022054293

ISBN 978-0-521-81436-2 Hardback

Additional resources for this publication at www.cambridge.org/loth.

Contents

Preface

This book provides an overview of dispersed multiphase fluid dynamics of bubbles, drops, and solid particles, whose trajectory is primarily determined by the surrounding flow. The motion of the particles and impact on the fluid is described using governing equations, basic physics, theory, experimental/numerical results, and empirical models.

The text is organized in 11 chapters that provide an introduction, basic equations of motion, particle and coupling classifications, particle–turbulent interactions, an overview of multiphase numerical methods, and reference details for point-force approaches.

The first eight chapters are designed for a course in multiphase flow. The introduction of Chapter 1 describes the importance of multiphase flow to engineering and natural systems. Chapter 2 is a brief review of single-phase fluid dynamic equations, which are used in Chapter 3 to obtain equations of motion for a single spherical particle. Chapter 4 considers particle size distributions, shapes, and trajectories as well as deformation of fluid particles (drops and bubbles). Chapter 5 characterizes multiphase flow-coupling interactions based on particle concentration, along with coupling regime classification. Chapter 6 reviews basic principles of single-phase turbulence, which are used in Chapter 7 to identify coupling physics between particles and turbulence. Chapter 8 overviews numerical methods for multiphase flows and discusses the pros and cons of various approaches. Chapters 1–8 are written for use in a course and have recommended homework problems.

Reference details on the point-force model are provided in Chapters 9–11. Chapter 9 focuses on the fluid dynamic drag on a particle for a wide variety of conditions, while Chapter 10 considers details of lift and other fluid dynamic forces for one-way coupling. Three-way and four-way coupling effects for a point-force description are reviewed in Chapter 11.

Additional information for multiphase flow physics is available in the books of Wallis (1969), Clift et al. (1978), Soo (1990), Drew and Passman (1998), Kleinstreuer (2003), Brennen (2005), Michaelides (2006), Leal (2007), Prosperetti and Tryggvason (2007), Crowe et al. (2011), and Marshall and Li (2014). Heat and mass transfer aspects are also discussed by Williams (1965), Kuo (1986), Oran and Boris (1987), and Sirignano (2010), while dense flow treatments (where particle collisions dominate) are covered by Gidaspow (1994) and Marchisio and Fox (2013).

Acknowledgments

This book endeavors to present dispersed multiphase fluid dynamics within a simple engineering discussion, based on research in the community by my colleagues, whom I thank. I am indebted to Professor G. M. Faeth (1936–2005), who introduced me to the subject, the many great students at the University of Illinois and the University of Virginia who provided valuable feedback (you know who you are!), as well as the team at Cambridge University Press for their superb editorial guidance.

I dedicate this book to my family for all their support, especially my wife Marie.

To honor these people, 100% of the author's proceeds from this book will be donated to the International Committee of the Red Cross (ICRC). The opinions expressed in this text are those of the author (and not necessarily of the ICRC).

Nomenclature

Unless otherwise stated, the numbers in parentheses correspond to the numbered equations presented throughout the text.

Roman Letters

a	Spheroid angle of attack (10.42)
a	Speed of sound (2.42)
A	Area or boundary surface (1.2)
$\mathcal{A}$	Bubble oscillation amplitude (4.59)
A*	Normalized surface area (9.49)
b	Richardson–Zaki exponent (11.17)
B	Spalding transfer number (3.100)
$\mathcal{B}$	Compressibility pressure constant (2.36)
Bo	Bond number (4.43)
c	Coefficient, such as (3.76)
c	Specific heat (2.29)
const.	Arbitrary constant of order unity
C	Force coefficient, such as (2.87)
C	Cumulative distribution function (4.4)
Ca	Capillary number (4.61a)
d	Equivalent volumetric diameter of a particle (1.1b)
$\mathcal{d}$	Particle-path derivative (1.17a) or diameter derivative (4.3b)
D	Macroscopic length scale of physical domain (1.3)
$\mathcal{D}$	Fluid path derivative (1.17b)
e	Internal energy (2.23)
e	Coefficient of restitution (8.32a)
E	Particle aspect ratio (4.38)
E	Turbulent kinetic energy per wavenumber (6.93)
$\mathcal{E}$	Young's modulus (11.39)
f	Stokes drag correction factor (3.31)
f	Ordinary frequency (2.95)
F	Turbulent kinetic energy per frequency (6.97)
$\mathcal{F}$	Marker function (8.53)
$\mathbf{F}$	Force (3.2)

Fr	Froude number (2.71)
$\mathbf{g}$	Gravity acceleration vector (2.13)
$\mathscr{g}$	Gravity drift parameter (7.18)
$\mathcal{G}$	Flow strain (4.67a)
$\mathbf{G}$	Gradient of viscous stresses (2.11)
h	Spread parameter for a size distribution (4.24)
$\mathscr{h}$	Enthalpy (2.38b)
H	History force kernel (10.53a)
$\mathbf{i}$	Unit vector (1.8)
$\mathbf{I}$	Impulse (11.32a)
I	Moment of inertia (4.40)
I^*	Normalized inertia about broadside axis (4.40)
J*	Normalized McLaughlin lift parameter (10.9)
k	Turbulent kinetic energy of the surrounding fluid (6.17)
$\mathscr{k}$	Thermal conductivity (2.25)
$\mathcal{K}$	Compressibility exponent (2.36)
K_{ij}	Viscous stress tensor on face i in direction j (2.15a)
Kn	Knudsen number (1.20)
ℓ	Distance (1.20) or wavelength (2.91)
m	Mass (1.4)
u	Viscosity ratio function (9.68b)
m	Particle mass loading (5.6)
$\dot{m}$	Mass flux per unit time (3.97)
M	Mach number (2.45)
$\mathcal{M}$	Mass fraction (2.30a)
Mo	Morton number (4.44)
MW	Molecular weight of a gas (1.24)
n	Wavenumber (6.92)
$\mathbf{n}$	Outward normal vector (2.1)
n_p	Number of particles per mixed-fluid volume (5.1)
N	Number of particles (5.1), realizations (5.26), nodes (6.106), or parcels (8.10)
$\mathcal{N}$	Number of collisions per unit volume of mixture (5.54)
NISI	No index summation intended
NTP	Normal temperature and pressure (Table A.1)
O	Order of a term
Oh	Ohnesorge number (4.54)
p	Continuous-phase pressure neglecting local flow around particle (1.15e)
P	Pressure around or inside of particle (1.12)
Pr	Prandtl number (2.29)
$\mathcal{P}$	Probability distribution function (4.1)
q	Arbitrary variable (1.7)
$\mathbf{q}$	Arbitrary vector (1.8)
$\mathbf{Q}$	Diffusive (nonconvective) flux (2.1)
$\dot{\mathcal{Q}}$	Heat transfer rate (3.108)

r	Radial distance (1.11)
r	Dimensionless particle roughness (9.43b)
r_p	Volumetric particle radius (1.1b)
R	Acceleration parameter (3.88b)
$\mathcal{R}$	Gas constant (1.24b)
Re	Reynolds number (2.79)
s	Speed ratio (9.32)
s	Particle size ratio (11.45b)
$\mathcal{S}$	Shape oscillation mode (4.70)
Sc	Schmidt number (2.35)
Sh	Sherwood number (3.103)
St	Stokes number (5.13)
t	Time (1.15)
t	Dimensionless turbulence intensity (9.43a)
T	Temperature (1.24)
$\mathcal{T}$	Torque (8.1c)
$\mathbf{u}$	Continuous-phase velocity neglecting local flow around the particle (1.15b)
$\mathbf{U}$	Continuous-phase velocity including local flow around the particle (1.12a)
$\mathbf{v}$	Velocity of the particle centroid (1.15a)
$\mathbf{V}$	Internal particle velocity field (1.12c)
$\mathbf{w}$	Relative velocity of the dispersed phase (1.15d)
$\mathbf{W}$	Faxen-corrected relative velocity (9.3f)
We	Weber number (4.41)
x	Streamwise direction for Cartesian coordinates (1.9)
$\mathbf{X}$	Position vector (1.9)
$\mathcal{X}$	Relative position vector (6.60)
x_i	General Cartesian direction (1.9)
y	Wall-normal direction for Cartesian coordinates (1.9)
Y	Drag-power parameter (4.18)
z	Spanwise direction for Cartesian coordinates (1.9)
$\mathcal{Z}$	Transfer function (8.49)

Greek Letters and Other Symbols

α	Volume fraction (5.2)
β	Collision impact parameter (11.59)
δ	Kronecker delta (2.15b) or boundary-layer thickness (2.85b)
Δ	Discretization increment (space or time)
ε	Turbulent dissipation of continuous phase (6.33)
ϵ	Small perturbation ($\ll 1$)
ϕ	Azimuthal angle coordinate (1.11)
Φ	Velocity potential (2.57)
γ	Gas specific heat ratio (2.39)
Γ	Gamma function (4.27)
η	Kolmogorov length scale of the turbulence (6.86)

ϑ	Particle impact angle (11.59)
κ	Boltzmann constant (5.29b)
λ	Taylor length scale of the turbulence (6.95)
Λ	Integral length scale of the turbulence (6.71)
μ	Dynamic viscosity (1.6)
μ_p^*	Viscosity ratio (1.6)
Π	Multiphase coupling parameter (5.35)
ν	Kinematic viscosity (2.74)
θ	Polar angle for particle-centered coordinates (1.11)
Θ	Mass diffusivity (2.32)
Θ^*	Normalized turbulent particle diffusivity (7.17)
ρ	Density (1.4)
ρ_p^*	Density ratio (1.5)
σ	Surface tension (3.32)
§	Section number
τ	Time scale (3.88a) or temporal shift (6.53)
Υ	Correlation coefficient (6.53)
$\forall$	Volume (1.1)
$\boldsymbol{\omega}$	Continuous-phase fluid vorticity (2.51)
ω^*	Dimensionless continuous-phase vorticity (10.4a)
$\boldsymbol{\Omega}$	Angular rotation rate (Figure 10.1)
Ω^*	Dimensionless particle rotation rate (10.4b)
ψ	Stream function (2.50)
Ψ	Tangential velocity ratio (11.51)
ζ	Gaussian random number (8.64)
χ	Terminal velocity ratio (11.14)

Subscripts

all	All particles
avg	Average
@p	Continuous-fluid property extrapolated to particle center
b	Bin
body	Body force
buoy	Buoyancy effect
Br	Brownian motion
curv	Curvature
clean	Clean conditions where the interface is fully mobile
coll	Collisions
cont	Contaminated conditions
conv	Convective
crit	Critical condition where flow transitions
d	Volumetric diameter or dispersed phase
diss	Dissipation
dyn	Dynamic pressure

D	Drag or overall macroscopic domain scales
E	Particle aspect ratio
$\mathcal{E}$	Eulerian
eff	Effective
eq	Equilibrium
f	Continuous-phase fluid
fm	Free-molecular value
fr	Wall friction
ft	Frozen-turbulence limit
g	Gas
g	Gravity
$\mathcal{G}$	Subgrid filter
gap	Gap between particle surface and wall or another particle
H	History force effect
I	Interface between particle and surrounding fluid
inj	Injection
Kn	Knudsen
∇k	Due to gradients in the turbulent kinetic energy
ℓ	Liquid or wavelength
L	Lift
$\mathcal{L}$	Lagrangian property
lam	Laminar
LN	Log-normal size distribution values
m	Mixed-fluid value
M	Mach number
m-m	Between two molecules
$m\mathcal{E}$	Moving Eulerian property
min	Minimum
n	Normal direction
nat	Natural frequency
o	Initial or reference state
osc	Related to oscillations
p	Particle phase (dispersed phase) or unhindered pressure
P	Parcel (group of particles) or local pressure
p/P	Particles per parcel (8.10b)
p,Δ	Particles that are within a computational cell (8.10a)
p-p	Particle–particle spacing (5.43)
plastic	Onset of plastic deformation
prod	Turbulent production
proj	Projected
pseudo	Pseudoturbulence
r	Radial direction
Re	Reynolds number effect
ref	Normal temperature and pressure conditions

rel	Local difference between particle and surrounding fluid
rms	Root of the mean of the squares (6.16)
RR	Rosin–Rammler size distribution values
S	Fluid-stress value
sd	Strong drift (7.19b)
sep	Separated wake conditions
shear	Linear velocity shear
sphere	For an equivalent spherical particle
stag	Stagnation conditions based on isentropic rest
sub	Subcritical conditions
super	Supercritical conditions
surf	Surface averaged quantity, typically of particle
t	Tangential
∇T	Due to gradients in temperature of continuous phase
term	Terminal characteristic of a particle in quiescent fluid
tot	Total conditions
trans	Transitional flow condition
turb	Turbulent
vapor	Vapor
viscous	Viscous
wall	Wall interaction effect
wd	Weak drift (7.19a)
yield	Yield stress value
α	Finite volume fraction effects
$\forall$	Added mass effect (3.76)
Δ	Cell resolution discretization
η	Kolmogorov scale of turbulence
Λ	Integral scale of turbulence
θ	Polar component
∞	Long-time or far-field property

Superscripts

in	Before interaction with wall
out	After interaction with wall
n	Time-step index
+	Normalized by wall shear-stress values
*	Dimensionless

Functions of an Arbitrary Property q

$\hat{q}$	Lagrangian path average (5.15)
$\bar{q}$	Eulerian time average (6.1)
q'	Instantaneous fluctuation from time average (6.4)
$\langle q \rangle$	Ensemble average for many particles (7.8a)
q''	Deviation from ensemble average (7.8b)

$\widehat{q}$	Spatial averaged value for LES (6.44)
$\underset{\smile}{q}$	Spatial-filtered perturbation for LES (6.44)
∇'^2	Stokes specialized spherical operator (3.15)

Comparison Symbols

$\equiv$	Equal by definition
$\approx$	Approximately equal
$\sim$	Same order of magnitude
$\lesssim$	Approximately less than
$\gtrsim$	Approximately more than
$\ll$	Much smaller
$\gg$	Much larger
$\perp$	Normal component
$\parallel$	Parallel component
O(q)	On the order of q
$q \rightarrow 0$	q can be neglected

1 Introduction to Multiphase Fluid Dynamics

1.1 Multiphase Flow and Book Scope

The term *multiphase flow* denotes fluid dynamics with two or three phases, where each phase can be a solid, liquid, or gas. If one of these phases is distributed in small, unattached elements (such as particles, drops, or bubbles), this phase is a dispersed matter and the combination of phases becomes a "dispersed" multiphase flow, which distinguishes it from a free-surface multiphase flow (such as wind over water). The four primary combinations for the dispersed two-phase flows shown in Figure 1.1 include:

(1) Solid particles in a gas
(2) Liquid drops in a gas
(3) Solid particles in a liquid
(4) Gas bubbles in a liquid

Three-phase flow is an extension of two-phase flow and consists of two primary combination sets: liquid drops and solid particles in a gas (e.g., spray and soot in a combustion chamber) or gas bubbles and solid particles in a liquid (e.g., bubbles and sand in the ocean). Herein, we will use the term "particle" to represent either a solid particle, a liquid drop, or a gas bubble.

This book focuses on flow regimes for which the dynamics of the particles (solid particles, drops, or bubbles) is primarily controlled by the surrounding fluid. Such conditions can include a complex array of particle dynamics phenomena such as acceleration by drag, gravitational settling, turbulent dispersion, collisions, coalescence, breakup, etc. The presence of a large concentration of particles can also lead to coupling effects between the phases that can modify the flow fields of the surrounding fluid, especially when the surrounding flow is turbulent.

Heat and mass transfer as well as dense flows are not a focus of this text but can be important for many multiphase flows. Multiphase heat transfer can stem from conduction, convection, and/or radiation, while mass transfer can be associated with evaporation, condensation, and/or combustion. Dense flow can include granular flow and other regimes where particle motion is dominated by interaction with other particles (instead of by interaction with the surrounding flow). Readers with an interest in heat and mass transfer and/or dense flows should consider the texts recommended in the preface.

In general, analysis of multiphase flow requires careful consideration of the physics, governing equations, models, and numerical schemes that are key to the

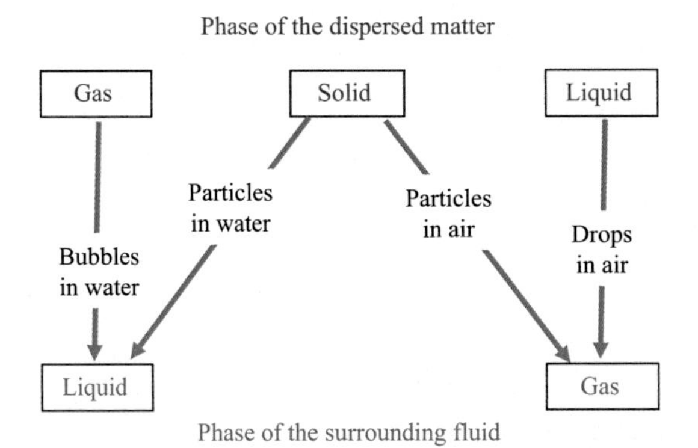

Figure 1.1 Various combinations for the dispersed phase and the surrounding fluid phase for two-phase flows.

characteristics and flow regimes of interest. These aspects are discussed in Chapters 2–11. The first eight chapters include homework sets (e.g., §1.6 for this chapter), and the last three chapters include reference material.

The rest of Chapter 1 introduces and identifies systems with dispersed multiphase flow and then provides key nomenclature, reference frames, and assumptions. The following section (§1.2, where § indicates a section of the text) discusses several engineered and natural systems with respect to multiphase flow. The remaining sections of this chapter set forth key terminology and assumptions for dispersed multiphase flow (§1.3), the key velocity reference frames used for multiphase flow (§1.4), and the assumption of continuum conditions (§1.5).

1.2 Multiphase Flow in Engineered and Natural Systems

Particle-laden flows are important to a wide variety of engineered and natural systems. Engineered systems include aerospace, atmospheric, biological, chemical, civil, mechanical, and nuclear applications. Examples of two-phase flows are listed in Table 1.1a for engineered systems and in Table 1.1b for natural systems. The flows can also be grouped in the following three system types:

(1) Energy and propulsion systems
(2) Manufacturing, processing, and transport systems
(3) Environmental and biological systems

These three groups are considered in the following subsections in terms of the key multiphase flow issues and physics.

1.2.1 Multiphase Flow in Energy and Propulsion Systems

Energy systems often include multiphase flow. For example, wind turbine blades must be designed to withstand particle erosion as well as ice accretion (Figure 1.2). Both

Table 1.1 Two-phase flow combinations for (a) engineered systems and (b) natural systems.

(a)

Two-phase flows	Engineering applications
Solid particles in a gas	Solid rockets, fluidized beds, particle separators, clean rooms
Liquid droplets in a gas	Fuel sprays, printing, coating and plasma sprays, cooling, aerosols
Solid particles in a liquid	Chemical slurries, liquid filters, sedimentation, fluidized beds
Gas bubbles in a liquid	Naval vessels, bubble columns, molten metal baths, boilers

(b)

Two-phase flows	Natural systems
Solid particles in a gas	Snow, sand, volcanic ash, pollen in air
Liquid droplets in a gas	Rain, drizzle, fog, ocean spray
Solid particles in a liquid	Sand or sediment flowing in a river
Gas bubbles in a liquid	Underwater plumes, cresting waves

Figure 1.2 Icing on a wind turbine blade caused by impact of freezing rain or drizzle (Larrsen, 2021).

icing and erosion are complex multiphase flows that involve particle trajectories that are influenced by the local aerodynamics and the impact physics. Another energy system that includes multiphase flow is energy production from combustion, whose products can include soot particles, which are ideally filtered from the exhaust gases. For coal combustion, particle filtering of soot particles and ash for the exhaust are important, as indicated by particulate control devices (see the right-hand side of Figure 1.3). A common approach for such filtering is to employ cyclone filters

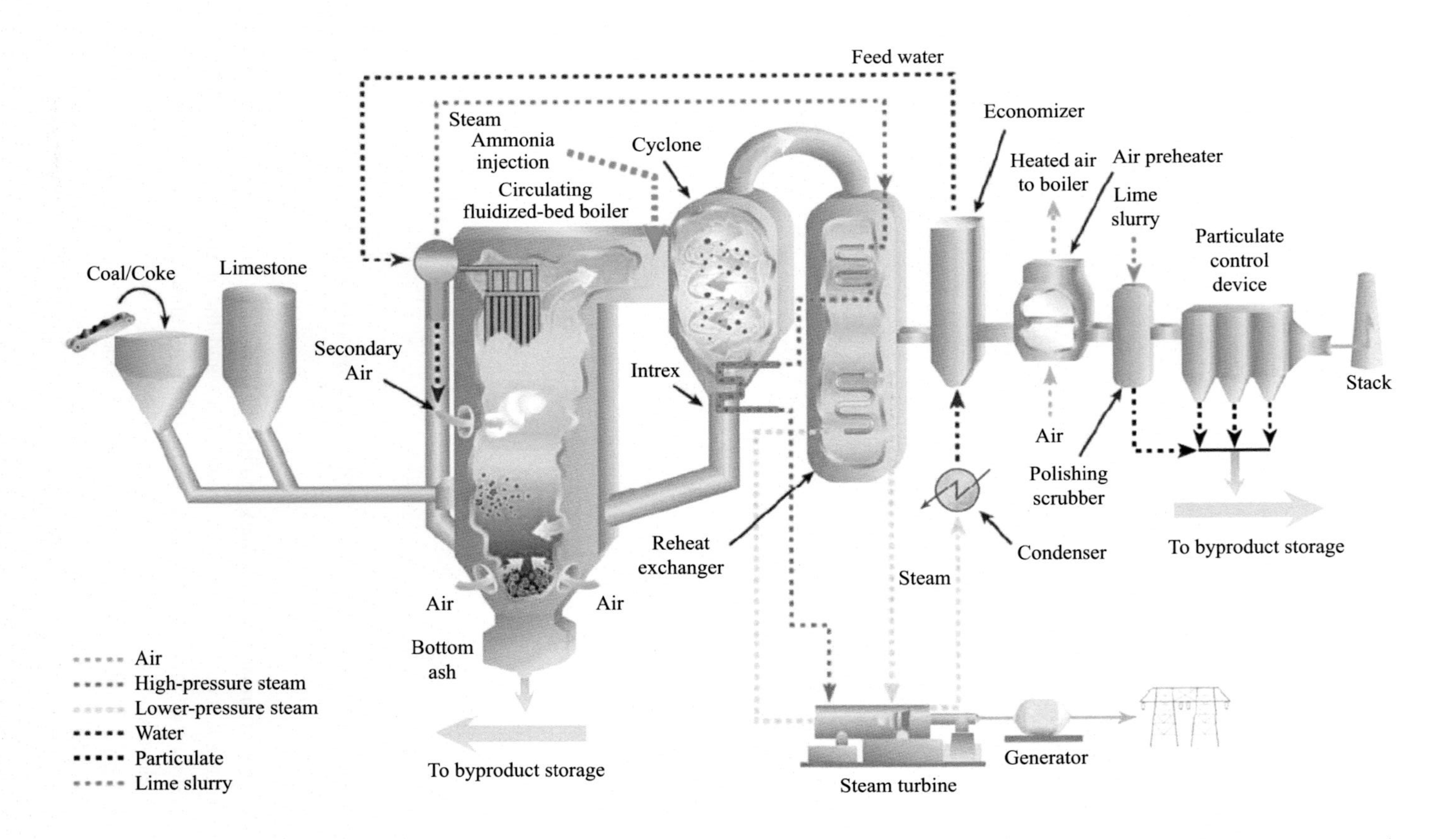

Figure 1.3 Schematic of a coal combustion plant with circulating fluidized bed and downstream particle separation using return channels, cyclones, and eventually cloth filters to help ensure a clean exhaust to atmosphere. Photograph (2020) Alyeska Pipeline Service Company (APSC), used with permission.

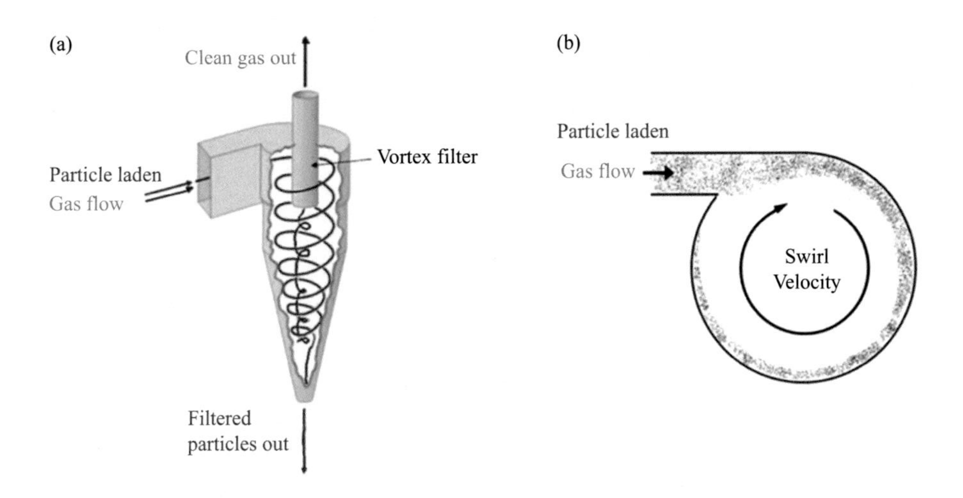

Figure 1.4 Schematic of a dust collector based on a cyclone separator used to allow downward gravitational collection of particles while cleaner gas tends to exhaust upward as clean (particle-free): (a) overall view; and (b) cross-sectional view showing particles centrifuged outward by swirl velocity.

(Figure 1.4), which use centrifugal and gravity forces to achieve separation of the particles from the gas stream. Multiphase flow for these coal-fired plants is also critical for the fuel feed and fluidized bed regions (see the left-hand side of Figure 1.3).

Air-breathing power and propulsion systems also generally involve multiphase flow. For example, a turbojet engine can have particles in the inlet, the combustor, and the exhaust, as shown in Figure 1.5a. Particulate matter from the entering flow can lead to blade damage or problematic blade accretion in the high-speed compressor section. To avoid this, a centrifugal separator can be used to cyclone out the particles. Following the compressor, the combustor also includes multiphase flow stemming from the spray nozzles used to produce fuel droplets. The combustion process is enhanced when the droplets are small (e.g., less than 40 μm in diameter), which can be achieved with rapid drop breakup using a pressurized swirl nozzle (Figure 1.5b). When these drops are uniformly distributed in the combustor, which can be achieved with mixing by turbulent dispersion, the combustion efficiency is higher and the soot is lower, thus minimizing environmental impact.

Multiphase flow can also be important for energy storage systems that support intermittent renewable energy (such as wind and solar energy). For example, compressed air energy storage can be made highly efficient if the process is made nearly isothermal (Qin et al., 2014). This can be achieved by spraying droplets (with a high surface area for enhanced heat transfer) during compression for energy storage (Figure 1.6) and during expansion for energy regeneration.

Another energy-producing system that promotes grid decarbonization and involves multiphase flow is a nuclear power plant. These systems generally use the heat from the nuclear reactor to convert water into steam to drive power-generating turbines. The boiling initially creates a bubbly flow (whose volume is mostly composed of liquid)

(a)

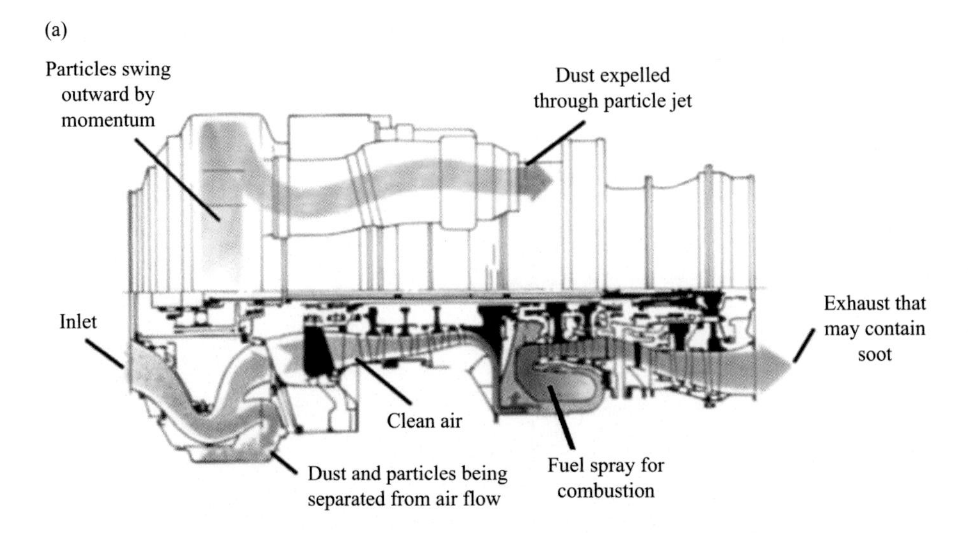

(b)

Figure 1.5 (a) Multiphase aspects of an air-breathing engine that includes a fuel spray in the combustor (Yoon, 2001); and (b) close-up of pressure swirl spray where instabilities transition the conically spreading liquid sheet to ligaments and then into droplets, which can spread through turbulence (Prakash et al., 2014).

and eventually creates a steam flow (a gas vapor flow with drops). This process is associated with high heat and mass transfer rates as well as nonuniform turbulent flow and interfacial dynamics governed by surface tension, all of which are important but cannot be easily predicted. As such, high-fidelity multiphase flow experiments and simulations are critical to effective and efficient plant design.

Multiphase flow is also important to solid rocket propulsion since the combustion processes involve particulate burning. The solid propellant starts off as a packed

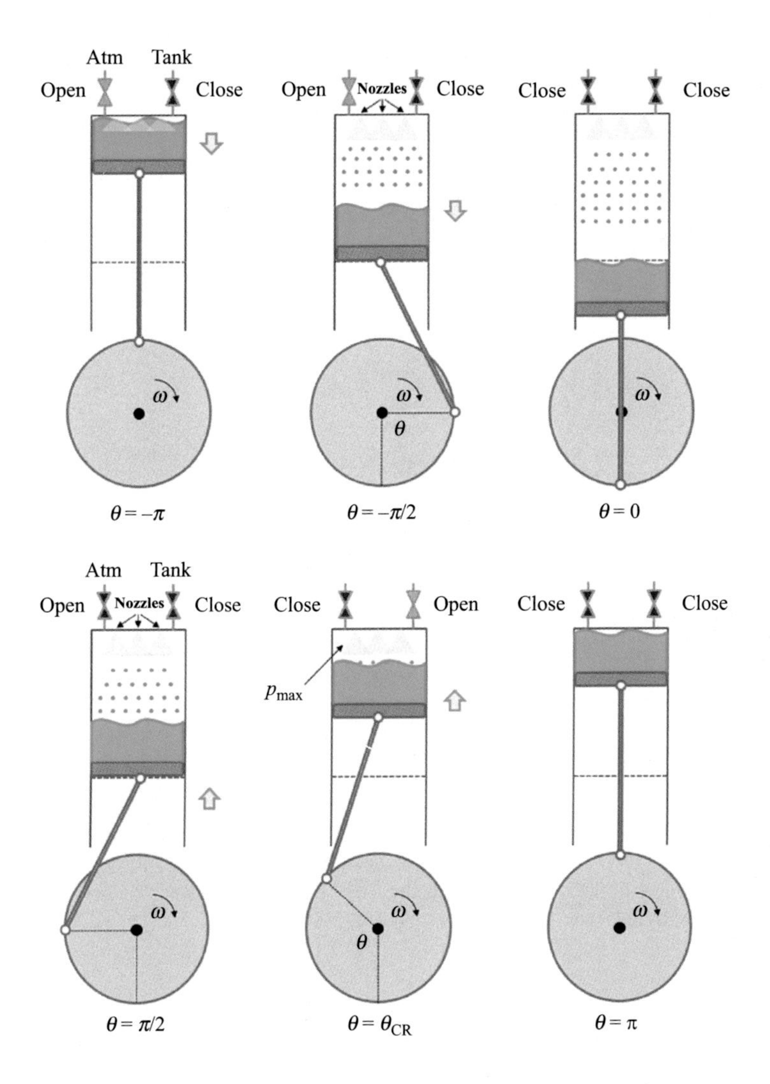

Figure 1.6 Injection of spray droplets for a liquid piston compression system to achieve near-isothermal conditions for compressed air energy storage (Qin et al., 2014).

mixture of powdered fuels and oxidizers integrated with a binder (Figure 1.7a). This mixture can include aluminum particles to achieve very high reaction temperatures (1,500–3,500 K). As shown in Figure 1.7b, many of the particles do not combust immediately and instead break off as the binder around them burns away. As these particles move downstream, they mostly melt, evaporate, and generally combust, yielding high-pressure gas products before they reach the end of the combustion chamber. However, any particles that are still molten are problematic since they can impact and accrete around the nozzle throat (leading to a slag buildup that can cause choking) or can impact the nozzle surface with high-velocity impacts (leading to ablation).

(a)

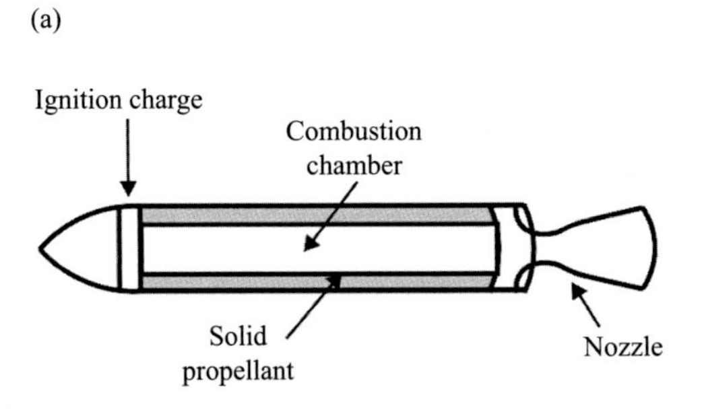

(b)

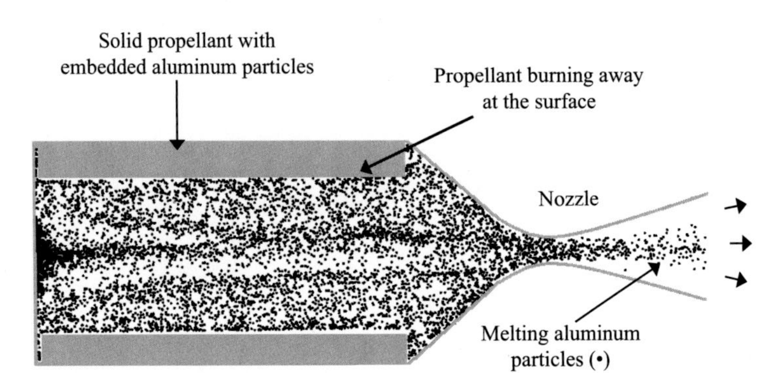

Figure 1.7 (a) Schematic of a solid propellant rocket that uses a combustion chamber to create high-enthalpy gas for propulsion. (b) Particles distributed in the combustion chamber that are released from solid propellant walls as the surface burns away and generally combust as they move downstream, that is, to the right (Najjar, 2005).

1.2.2 Multiphase Flow in Manufacturing, Processing, and Transport Systems

Manufacturing processes often involves multiphase flows. For example, coatings on surfaces are often applied using spray systems that are designed to generate droplets that are small enough to ensure a smooth finish, yet large enough to avoid being carried away from the target zone by surrounding air crossflows.

To create extremely durable metal coatings, one may employ plasma spraying (Figure 1.8a). In this type of coating process, metal powder is injected into a high-temperature plasma arc jet, created by a high-voltage difference (Figure 1.8b). The particles melt and then impinge on the target surface, where they rapidly crystallize to form a single-crystal metallic coating. This coating can survive hostile environments (needed for gas turbine blades) and can have extremely long usage without wear (needed for medical implants). The coating performance is strongly related to the multiphase flow aspects, such that the combination of the flow conditions and temperatures as well as the material and size of the particles are carefully controlled.

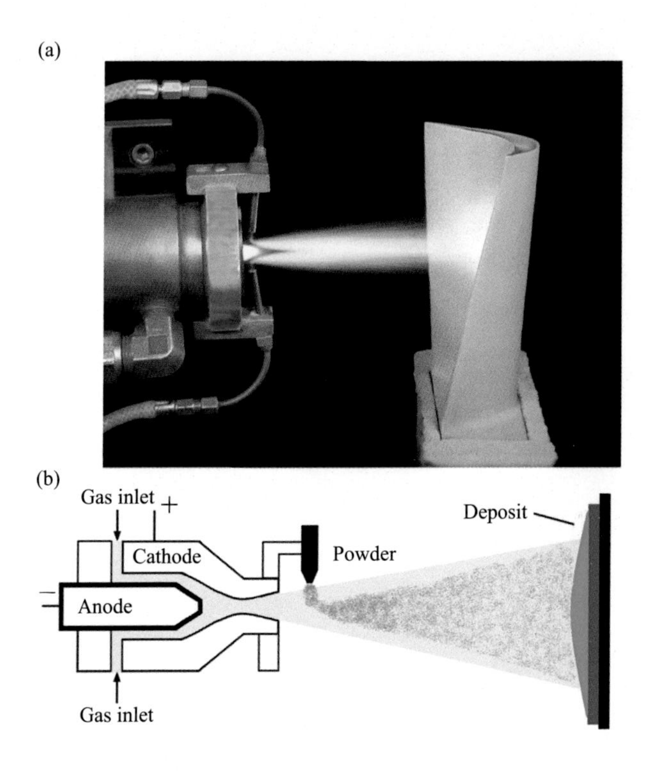

Figure 1.8 Plasma sprays: (a) applied to a turbine blade (JTJ, 2013); and (b) schematic of a spray system where metal particles are entrained into gas field (yellow) that exits as high-temperature plasma causing the particles to melt (or even evaporate) before they deposit on the target surface, where they create a highly durable metallic coating based on single-crystal growth.

Another multiphase manufacturing process is the production of powders. Many foods, detergents, and pharmaceuticals require their powders to be very fine with high consistency of size and shape. These characteristics can be achieved with an industrial spray dryer (Figure 1.9a), where liquid sprays are converted to powders through solidification. For this process, a warm liquid emulsion feed is sprayed downward into a large chamber, where the droplets cool and solidify as they fall to create a fine powder (Figure 1.9b). To create micropowders (Figure 1.9c), some dryers apply high-frequency vibrating spray nozzles to create extremely small droplets.

Multiphase processing is used in the production of many chemical products to promote chemical reactions and mixing. For example, liquid chemical reactors often employ bubble columns and recirculating reactors (Figure 1.10) to increase interfacial surface area and thus the corresponding reaction and mass transfer rates between the phases. Other examples of multiphase reactors include bio-oxygenation, distillation, and chromatography columns as well as fermentation bioreactors.

Multiphase flow can also be used to transport solid particles in pipe systems. This can be achieved by air-driven pneumatic conveyors (generally used for low particle concentrations at high speeds) or by liquid-driven slurries (generally used for high particle concentrations at low speeds). Generally, the flows are turbulent, making the

(a)

(b) (c)

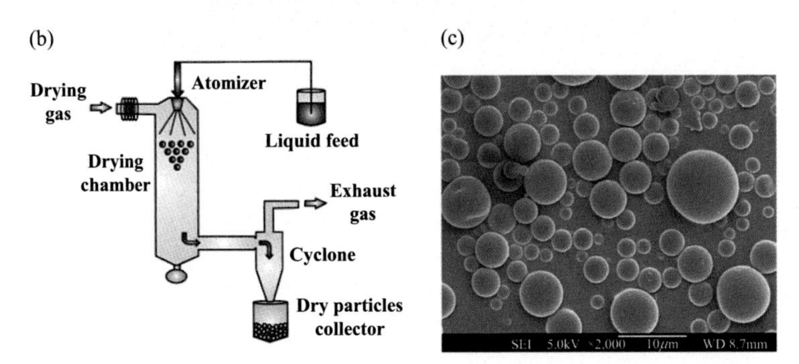

Figure 1.9 Spray dryers used to create powders: (a) a large-scale industrial spray dryer facility (GEA, 2020); (b) schematic of a process that includes an atomizer to create powder in a dry gas and the resulting mixture is fed to a cyclone filter to extract the product downward while clean air exhausts upward (Sosnick and Seremeta, 2015); and (c) microparticles (ranging from 2–10 μm in diameter) generated by a vibrating mesh spray imaged with a scanning electron microscope (Lee et al., 2011).

interactions quite complex. In such cases, understanding the particle trajectory dynamics and their interaction with the driving flow is important to help reduce pipe pressure losses (to minimize energy requirements) while ensuring particles are fully suspended within the flow (to maximize transport efficiency).

Transporting liquids with pumps can also involve multiphase flow, since the impellers can yield problematic cavitation. This occurs when the blade speeds are high enough to cause the local liquid pressure to drop below the vapor pressure. For example, steam bubbles are formed out of water at low enough pressure. These vapor bubbles can reduce impeller efficiency since they can lead to flow separation. In addition, the vapor bubbles entering high-pressure regions will rapidly collapse, and this can lead to significant blade damage (surface pitting) if this occurs near the

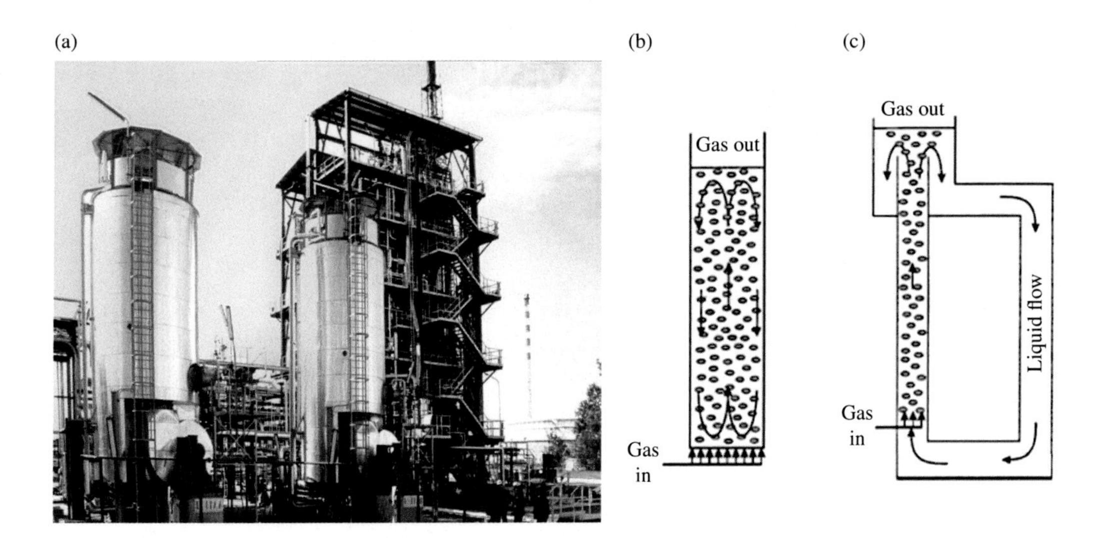

Figure 1.10 Bubble column reactors: (a) an industrial single-column reactor (P. M. Raimundo, ENI, IFP Energies Nouvelles, Etablissement de Lyon, France, personal communication, 2020); (b) schematic of a single-column reactor; and (c) schematic of a recirculating airlift reactor (Mudde and Saito, 2001).

surface. Cavitation can be similarly problematic for propellers of ships and underwater vehicles in terms of propulsive efficiency and blade damage.

However, multiphase flow can be beneficial for transporting liquids in turbulent flows when used for drag reduction. In particular, microbubbles (about 10–100 μm in diameter) and long-chain polymers (with molecular weight greater than 10^5) in turbulent boundary layers have both been found to significantly reduce the shear drag by as much as 75% (Figure 1.11a). In both cases, the injected microbubbles or polymers reduce the near-wall turbulent momentum mixing through dissipative interactions with the boundary-layer vortices. These microbubbles and polymers are active when their response time scales (drag based for the bubbles and coil based for the polymers) are similar to the turbulent microscales (Pal et al., 1988). The polymer approach has been most successful for engineered systems. For example, polymers injected into the Alaskan oil pipeline (Figure 1.11a) allow oil flow rates to be increased by as much as 50% without increasing the pumping power. Polymers can also be mixed with water to reduce liquid skin friction in fire hoses, thereby increasing nozzle water velocity and the resulting distances and heights that can be reached by the exiting water stream. Interestingly, the bubble approach for drag reduction is observed in nature. In particular, penguins have the ability to collect a gas layer under the feathers before entering the water and then later release this gas underwater in the form of microbubbles (Figure 1.12a). This significantly reduces their fluid dynamic skin friction drag, allowing them to make a burst of speed to catch prey or to jump out of the water. Marine vessel designers are investigating injection of microbubbles along their hulls to help reduce fuel consumption using this bio-inspired concept (Figure 1.12b).

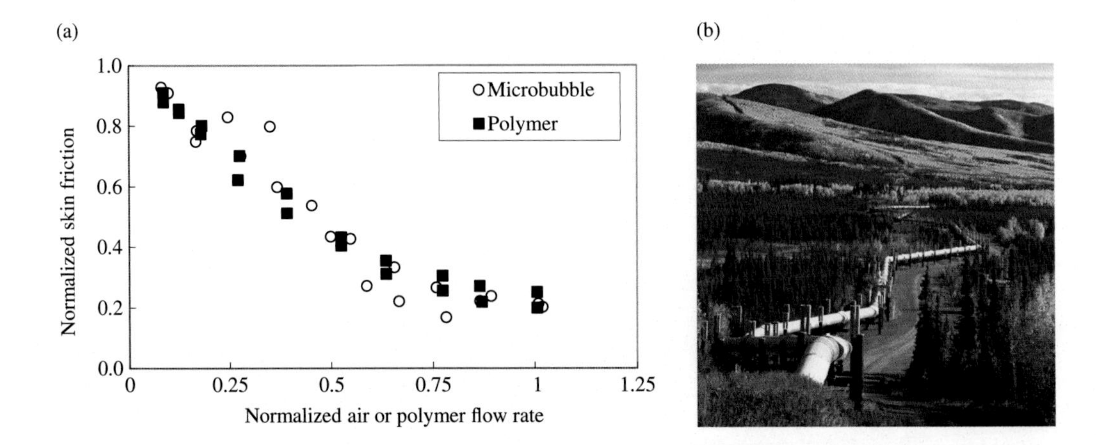

Figure 1.11 Turbulent drag reduction with multiphase flow: (a) measured in a boundary layer (Pal et al., 1988) based on normalized concentrations for microbubbles (relative to 30% by volume loading) or long-chain polymers (relative to 0.5% by mass loading); and (b) the Trans-Alaska Pipeline, which uses polymers that increased the pipe flow from 1.44 billion barrels per day to 2.14 billion barrels per day (Nielsen et al., 1999).

Microbubbles can also be used for specialized manufacturing with a process called *sonoluminescence*, in which sound is converted to light (Figure 1.13). In particular, micron-scaled bubbles can be subjected to strong, highly focused acoustic oscillations at their natural frequency to generate rapid and large oscillations in bubble diameter. The contraction portion of these oscillations can lead to high gas compression with the extreme pressures and temperatures such that the gas inside becomes a plasma (Figure 1.14a). The emitted light when this gas turns into a plasma can even be seen with the unaided eye (Figure 1.14b). The increased local temperatures can allow the production of protein microspheres (Figure 1.14c), whereby a shell of protein is formed over the microbubble through cross-linking.

1.2.3 Multiphase Flow in Biomedical and Environmental Systems

Multiphase flow is also critical to a wide range of biomedical systems and environmental systems. In some case, particles are intentionally introduced by medical procedures for the purpose of diagnostic imaging or drug delivery. For example, microbubbles can be injected into the bloodstream and subjected to sonic waves so that they oscillate and create acoustic signals that can be used for diagnostic purposes. Differences in these acoustic signals can be used to identify regions of good versus bad blood flow circulation, and even identify tumors (Figure 1.13). Blood itself is also multiphase compilation of liquids (plasma) and soft particles (platelets, red blood cells, white blood cells, lymphocytes, etc.). This can create complex rheology in the smaller vessels. In addition, blood can also include small calcium particles, which can problematically deposit and build up on vessel walls. The resulting accretion is called plaque and can cause a narrowing or blockage of the blood vessels, which leads to poor blood circulation,

(a)

(b)

Figure 1.12 Drag reduction using bubbly flow: (a) emperor penguins releasing bubbles from their feathers to reduce hydrodynamic drag (Nolan, 2018); and (b) concept design of microbubbles injected along a ship hull to reduce drag (Kaisha, 2010).

especially if near a constriction or turn. Plaque deposits can be especially dangerous if they break off from a vessel wall and move downstream as an agglomeration (a collection of many smaller accreted particles). This larger particle can be transported downstream to a smaller blood vessel, where it can lodge itself and obstruct the blood flow. If this happens in a brain blood vessel, the blockage may lead to a stroke.

An example of a biomedical system with multiphase air flow is an inhaler, which rapidly delivers a medicinal aerosol spray to the respiratory system (Figure 1.15a). The fine aerosol droplets are intended to evaporate as they reach the lungs. Some of the droplets will instead deposit on mucous surfaces along the complex flow path

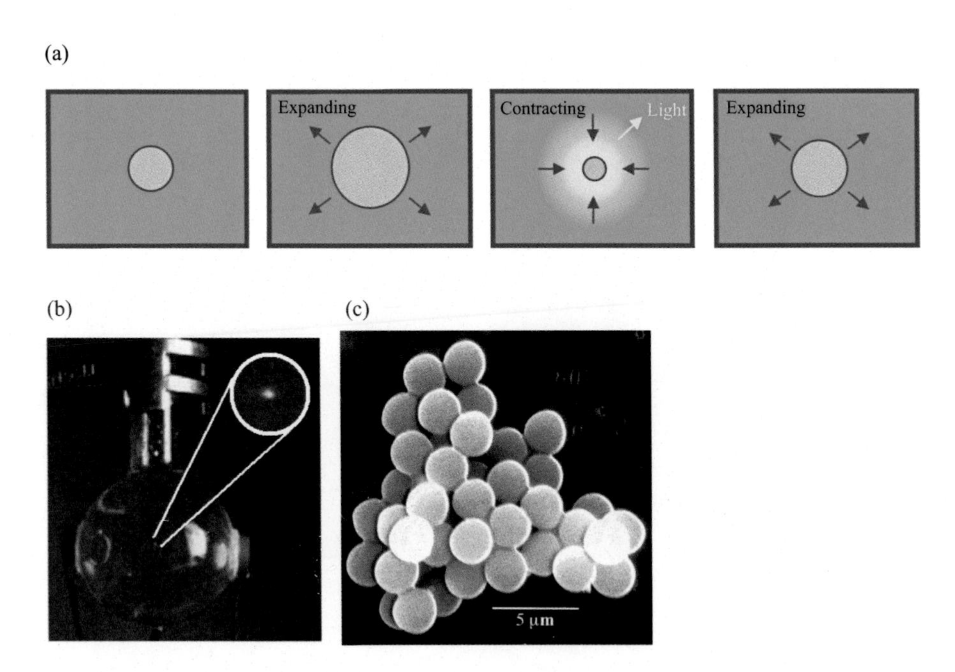

Figure 1.13 Sonoluminescence of microbubbles: (a) diagram of the contraction–expansion process whereby the maximum contraction causes very high pressures and temperatures, yielding light emission; (b) experimental setup to excited microbubble volume dynamics (Geisler, 1999); and (c) protein microspheres produced by this process (Suslick and Price, 1999).

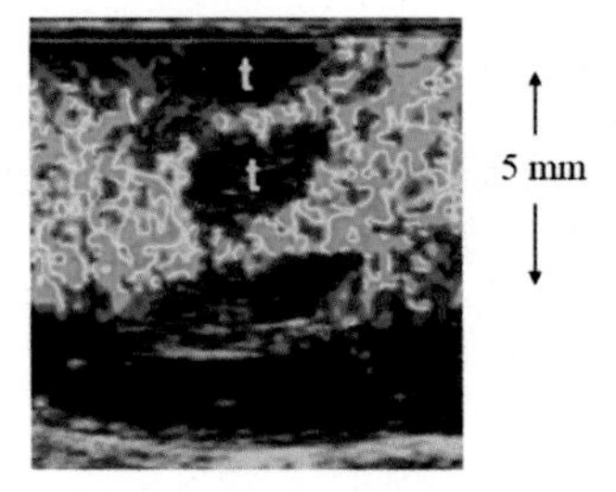

Figure 1.14 Ultrasonic Doppler image of a rabbit liver injected with microbubbles (about 3 μm in diameter) showing acoustic emission signals from the bubbles as colored regions where lack of signals can allow tumor regions to be detected, as marked with "t" (Forsberg et al., 1999).

through the mouth and throat. The deposition mechanism depends on particle size, as it is generally driven by particle inertia and gravity for drops with diameters greater than 1 μm but by stochastic Brownian motion for drop diameters of 0.1 μm or less. As a result of deposition and evaporation, the average drop size changes along the pathway from the mouth to the lungs (Figure 1.15b). One must account for these factors to predict the overall effectiveness of aerosol drug delivery.

However, some airborne droplets are harmful, as they can carry diseases such as the Covid virus. These droplets can be dispersed via breathing, but especially by sneezing and

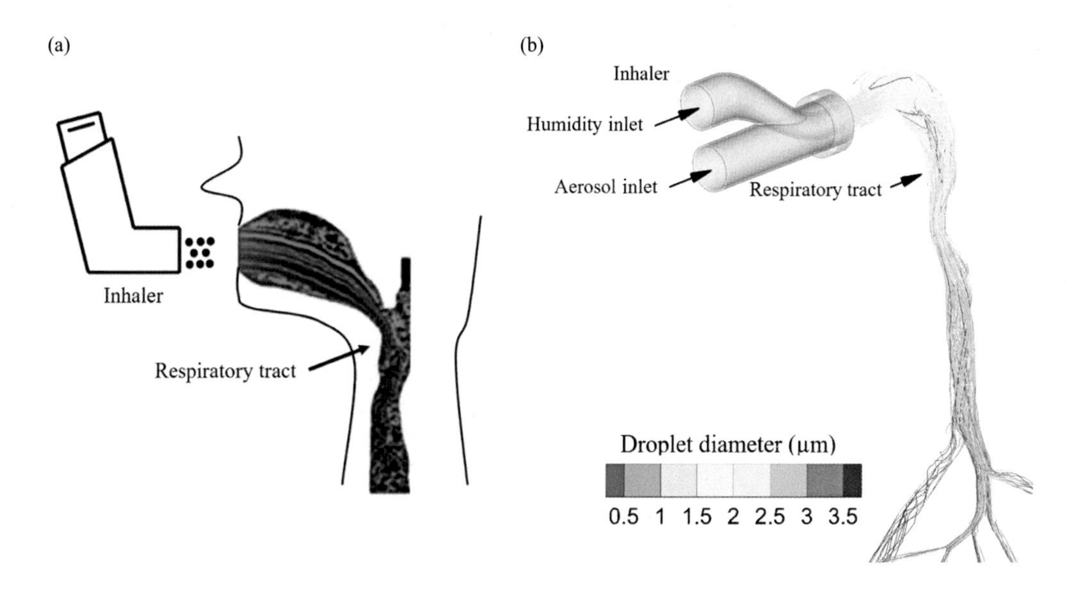

Figure 1.15 Aerosol droplet trajectories: (a) in the simulated geometry of an upper respiratory tract (mouth and throat) to study inhaler aerosol delivery effectiveness (based on a figure by Niven et al., 1994); and (b) change in droplet sizes as the trajectories move to the lungs (Longest, 2019).

Figure 1.16 Aerosol droplets resulting from a human cough can travel significant distances (Butler, 2020).

coughing (Figure 1.16). The larger droplets can have significant inertia, leading to longer distances of travel, and therefore a facial mask or covering can be crucial to help prevent breathing in these droplets. Breathing in solid particles in the air can also be unhealthy. These airborne particles can stem from atmospheric pollution caused by both man-made and natural systems. Our respiratory pathway employs mucous deposition so that most particles don't reach the lungs, where they can do damage to the sensitive lung tissue.

(a)

AQI	AQI category	$PM_{2.5}$ (μg/m³)	PM_{10} (μg/m³)
0–50	Good	0–15	0–54
51–100	Moderate	16–40	55–154
101–200	Unhealthy	41–150	155–254
201–300	Very unhealthy	151–250	255–354
301 or more	Hazardous	251 or more	355 or more

(b)

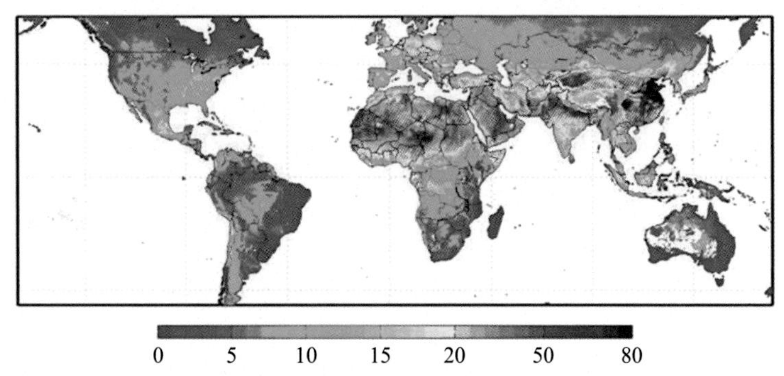

(c)

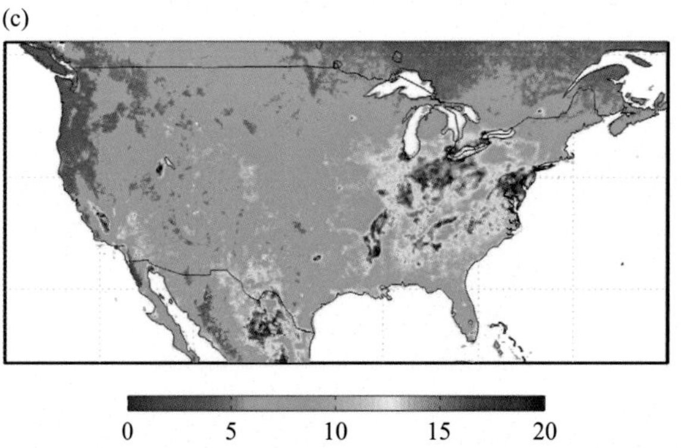

Figure 1.17 Atmospheric particle concentration: (a) Air Quality Index (AQI) as a function of particle matter (PM) concentration and size, where PM_{10} and $PM_{2.5}$ are respectively based on particle diameters of 10 μm or less and based on 2.5 μm or less; (b) global and (c) North American distribution of $PM_{2.5}$ in levels of μg/m³ (van Donkelaar, 2010).

Because of the potential harm of solid particles to human and other animal respiratory systems, the distribution of these particles in the near-surface atmosphere is of special interest. To help assess the potential health impact of airborne particles, a standardized Air Quality Index (AQI) was established by the United States Environmental Protection Agency based on $PM_{2.5}$ and PM_{10}, defined respectively as concentration of particle matter with diameters of 2.5 and 10 μm or less (Figure 1.17a). However, these distributions can be difficult to predict and to control since particle concentrations are influenced by natural environments and human activities that vary

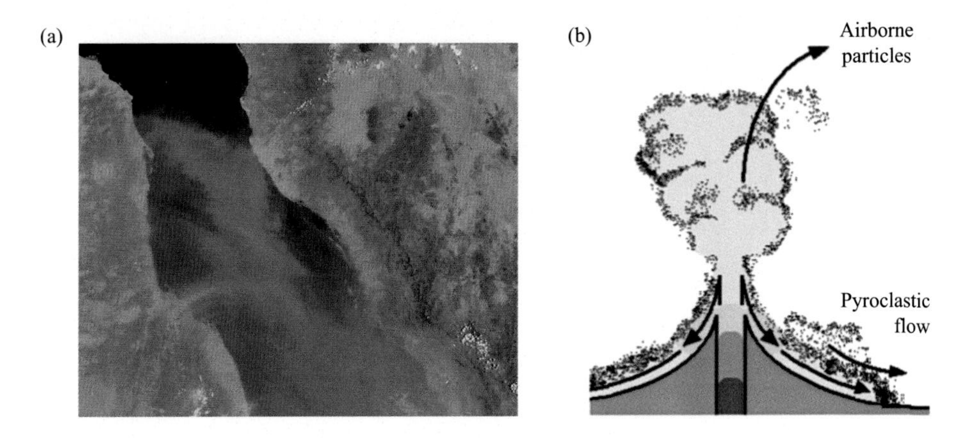

Figure 1.18 Natural events that cause air-laden multiphase flow: (a) atmospheric spread of particles over the Red Sea region by a dust storm (Ahmadi, 2005); and (b) volcanic-induced particle ejection whereby most particles are ejected into the atmosphere, but high-concentration ash-laden pyroclastic flows can also be driven downward.

widely with time and location. For example, the world's highest $PM_{2.5}$ concentrations (Figure 1.17a) tend to be in arid regions and deserts, such as northern Africa, where there are high concentrations of surface particles (e.g., sand) that can be lifted into the air, and there is little ground vegetation to help capture and filter such airborne particles. In addition, regions with heavy transportation, manufacturing, construction, and carbon-based energy production also lead to high $PM_{2.5}$ concentrations, such as regions in the US below the Great Lakes and in the Mid-Atlantic (Figure 1.17b). Furthermore, topography can play an important role for localized regions. For example, weather inversions in low-level areas surrounded by mountain can lead to smog (fog or haze combined with smoke and other atmospheric pollutants) that can persist for long periods. Recently, the PM_1 concentrations (based on particle diameters of 1.0 μm or less) has raised concern since these nanoparticles can more easily reach the lungs and may be medically harmful (though there is not enough understanding to implement within the AQI).

In addition, the transport of larger atmospheric particles is also important since these are more likely to impact natural habitats, surface vegetation, etc. Particles injected into the atmosphere from forest fires or dust storms can be carried hundreds of miles (Figure 1.18a). Volcanic eruptions can also create airborne particles (Figure 1.18b), which can distribute fine ash on a continental scale. Figure 1.18b shows that volcanos can also create *pyroclastic* flows, whereby high particle concentrations create a high mixture density that moves downward along the surface of the volcano and then spreads radially outward. These pyroclastic flows (driven by changes in density) can be particularly dangerous due to their near-zero oxygen content, high temperatures (up to 600°C), fast speeds (up to 100 km/hr), and large ranges (they can travel as far as 60 km). In fact, pyroclastic flows are the main cause of human death from volcanic events (Chester, 1993). The airborne ash can curtail airplane flight

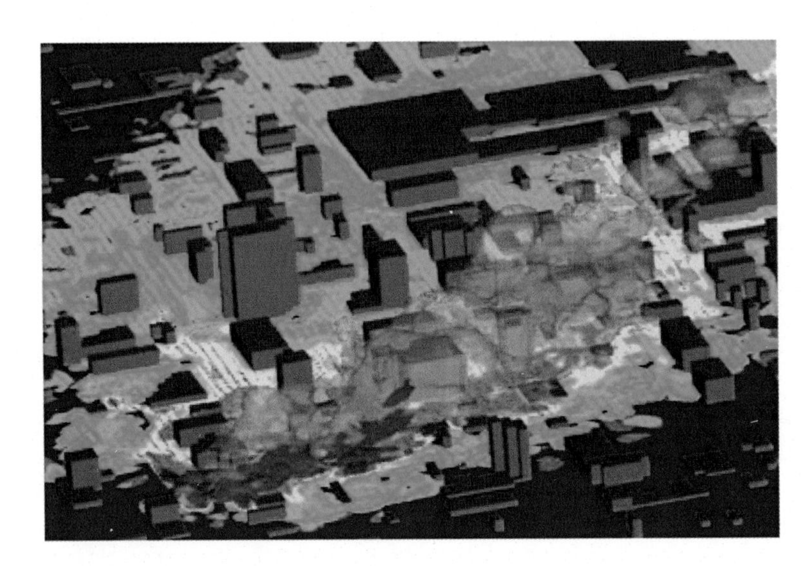

Figure 1.19 Simulated wind-driven dispersal of hazardous airborne particles in an urban area (DeCroix, 2003).

paths in the short term as well as also impact agriculture and surface water quality in the long term. Hazardous airborne particles can also be created when there are fires at chemical or manufacturing facilities. Another concern is the nefarious release of dangerous particles in highly populated urban areas where wind flow interacting with buildings dictates localized dispersal (Figure 1.19).

The transport of sediment in natural bodies of water can also be a concern. The settling and suspension transport is governed by turbulence, topology, and weather events, all of which can be problematic. For example, high sedimentation can reduce river capacity, leading to flooding that negatively impacts agricultural and/or populated regions. On the other hand, erosion of rivers and coastal areas can cause loss of animal habitats and natural hurricane barriers.

In some cases, water streams contain man-made solid particles from manufacturing or waste treatment plants. Such particles should be removed but are difficult to settle, as they are nearly buoyant. One removal method employs floatation though bubble attachment. In this three-phase flow process, microbubbles are added to the solid unwanted matter (Figure 1.20a), so it is carried upward through flotation and then skimmed off before it can proceed downstream (Figure 1.20b).

Transport of bubbles in natural bodies of water is also important to the environment. For example, carbon dioxide and methane bubbles released from ocean beds influence the temperature, salinity, and acidity of the ocean water. Another example is the man-made introduction of bubble plumes into freshwater reservoirs and other nearly static water bodies. The turbulent mixing by these plumes can increase oxygen content and control pH levels to improve the overall ecological health of these freshwater bodies.

Bubble plumes have even been used by humpback whales and dolphins to create "bubble nets," as shown in Figure 1.21. In particular, the animals dive below and then

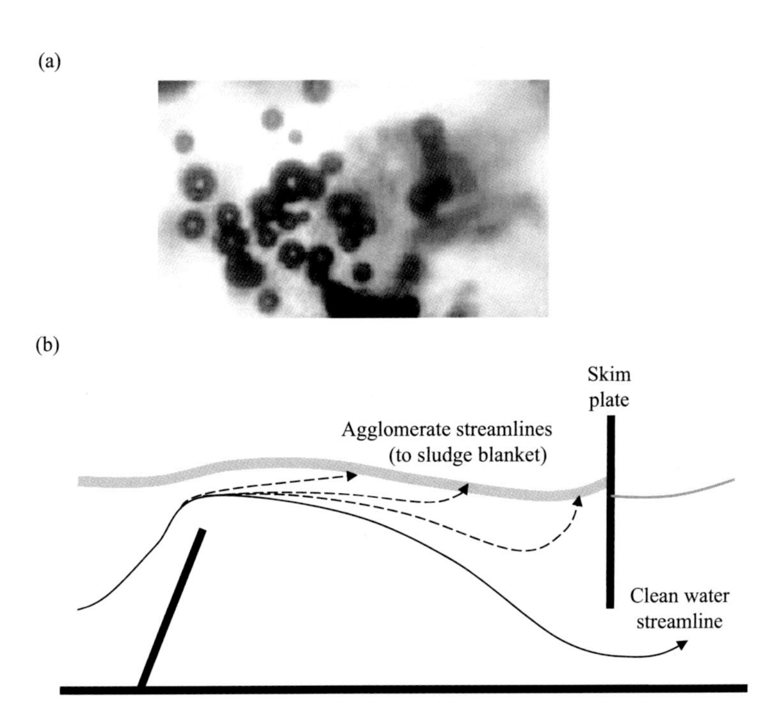

Figure 1.20 Particles filtered from a water stream by attaching small bubbles (ranging from 30–300 μm) to cause unwanted particle matter to float to the surface for skimming. (a) An image of a buoyant agglomeration consisting of microbubbles (dark spheres) attached to a single loosely aggregated flocculation particle (large gray shape) in order to cause enhanced flotation (Leppinen and Dalziel, 2004). (b) Schematic of a water stream with path lines of buoyant agglomerates (particles surrounded by microbubbles) filtered to the surface while clean water passes underneath the skimmer.

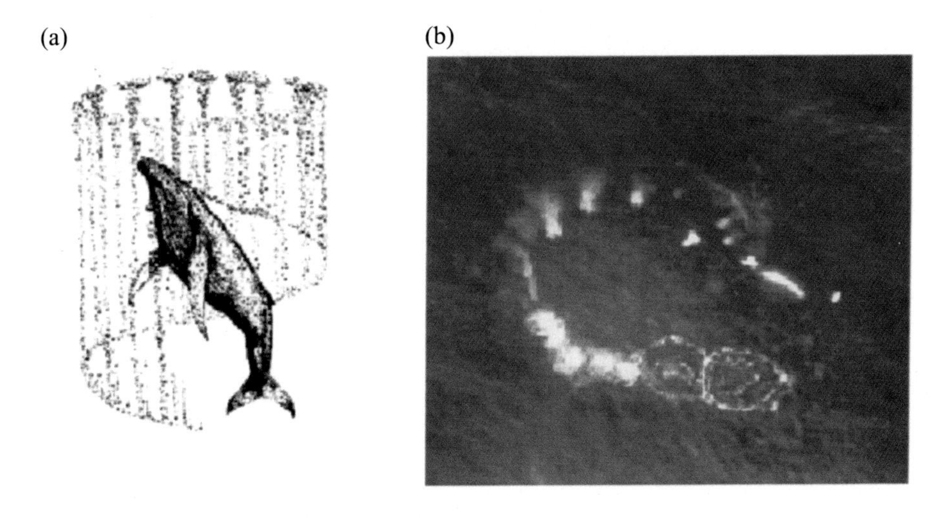

Figure 1.21 A humpback whale making a bubble net using its blowhole: (a) schematic of the bubble release along a helical path; and (b) surface photograph of bubbles reaching the surface (Leighton, 2004).

exhale air through their blowholes as they spiral upward to create a cylindrical wall of bubbles. This bubble wall temporarily "traps" the prey, allowing the predator to feed more easily.

1.3 Basic Terminology and Assumptions for Particle Fluid Dynamics

1.3.1 Basic Multiphase Definitions

A few essential definitions will be introduced to describe the basic features of a dispersed multiphase flow. First, a *particle* is defined as an unattached, freely moving body immersed in a fluid, which is the *surrounding flow*. The particle phase may be a solid, a liquid (i.e., drop), or a gas (i.e., bubble), while the surrounding fluid can be a gas or a liquid. This text specifically focuses on dispersed flow for which particle motion is primarily influenced by that of the surrounding flow rather than by particle collisions. Such flows are called *dilute*. In contrast, *dense flow* occurs when particle collisions dominate particle motion, which is more likely at high particle concentrations. Recommended texts that focus on dense flows are noted in the preface.

For dispersed flow, the collection of all the particles in the flow domain is defined as the *dispersed phase*, while the surrounding fluid is defined as the *continuous phase*. The latter definition assumes that the surrounding flow can be considered as a continuum (an assumption discussed further in §1.5). Thus, a two-phase dispersed flow has a single dispersed phase of particles and a single continuous phase of fluid. However, this concept can be extended to three phases (e.g., a continuous-phase liquid flow that includes both bubbles and solid particles as the dispersed phases).

Dispersed flow can also include the special case of two immiscible liquids, that is, two fluids separated by surface tension and are insoluble (no molecular mixing) so that each fluid's properties (density, viscosity, etc.) remain distinct. For example, oil drops in water generally have a preserved interface between the two liquids. As a result, the dynamics of such oil drops in water are similar to that of solid particles with the same density and size. Notably, this special case only applies to liquids since gases are miscible with other gases.

1.3.2 Basic Particle Nomenclature

In this section, the key variables for the basic properties of a particle (dispersed phase) and the surrounding fluid (continuous phase) are defined. These (and all other variables and symbols used in this text) are listed in the "Nomenclature" section near the beginning of this book.

To generalize the notation, we first define subscripts for the phases. The continuous-phase variables and properties will be designated by the subscript f (e.g., ρ_f is the surrounding fluid's density and μ_f is its dynamic viscosity), while the dispersed-phase variables and properties will be typically designated by the subscript p (e.g., ρ_p is the particle density). For a fluid particle (i.e., a drop or a bubble), one may

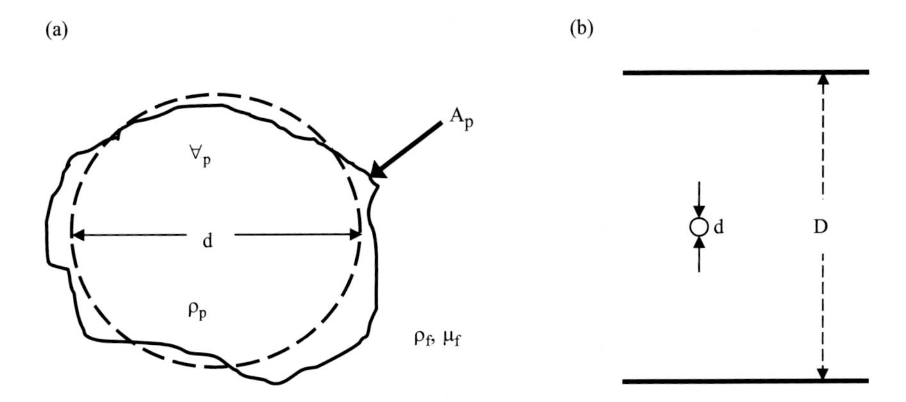

Figure 1.22 Schematic of (a) a nonspherical particle (solid line) with given volume and surface area surrounded by a continuous-phase fluid density and viscosity, along with the equivalent-volume sphere (dashed line), and (b) a particle that is small compared to a channel width, such that d « D.

define the internal particle viscosity as μ_p (Figure 1.22a). In addition, we designate a gas or a liquid property with the respective subscripts *g* and *ℓ*.

In terms of geometry, the particle volume is $\forall_p$ and its surface area is A_p. For an arbitrary shape (Figure 1.22a), we then define the *volumetric diameter* d as the diameter of a sphere that has the same volume as that of the particle, as follows:

$$\forall_p = \pi d^3/6 \qquad \textit{for a sphere with a diameter } d. \tag{1.1a}$$

$$d \equiv 2r_p \equiv \left(6\forall_p/\pi\right)^{1/3} \qquad \textit{for spherical or nonspherical shape.} \tag{1.1b}$$

Note that the use of $\equiv$ denotes the definition of a variable (not just an equality) and that (1.1b) defines the volumetric radius $\left(r_p\right)$ as equal to half the volumetric diameter. If the particle is a sphere, its radius and diameter are equal to the volumetric radius and diameter. If the particle is not a sphere, the particle surface area will always be larger than the surface area of a spherical particle that has the same volume, as follows:

$$A_p \geq A_{p,sphere} = 4\pi r_p^2 = \pi d^2. \tag{1.2}$$

The extent of this inequality increases as the particle shape deviation (from a sphere) increases.

For dispersed flow (the focus of this text), the particle is assumed to be much smaller than the macroscopic length scale of the continuous-flow domain, which is defined as D. This assumption can be expressed as

$$d \ll D. \tag{1.3}$$

Note that D may represent the diameter of a pipe (as in Figure 1.22b), the width of a jet, or another key macroscopic length scale of the flow domain.

Next we consider ratios of the density and viscosity. The particle density (ρ_p) is defined as the ratio of particle mass (m_p) to particle volume:

$$\rho_p \equiv m_p/\forall_p. \tag{1.4}$$

This density relative to that of the fluid (ρ_f) is important for many multiphase flows. For example, the *specific gravity* is the ratio of the particle density to the water density. As such, a particle will float or sink in water depending on whether the specific gravity is less than or greater than unity.

To make a more general comparison for any surrounding fluid density, the particle density ratio is defined as

$$\rho_p^* \equiv \rho_p/\rho_f. \tag{1.5}$$

In this text, a * superscript is used to indicate a dimensionless version of a variable. Similarly, a fluid particle will have a particle viscosity and a surrounding fluid viscosity that can be used to define the particle viscosity ratio as

$$\mu_p^* \equiv \mu_p/\mu_f. \tag{1.6}$$

These ratios for density and viscosity will be shown later to influence particle dynamics, and the limits for very small and very large ratios are often important.

To consider such limits, a set of key comparison symbols is defined for a dimensionless quantity (using the example variable q) as follows:

$$q \to 0 \textit{ indicates q is very small (e.g., < 0.01 or less) and may be neglected.} \tag{1.7a}$$

$$q \ll 1 \quad \textit{indicates q is small (e.g., ~ 0.1 or less) but is considered finite.} \tag{1.7b}$$

$$q \tilde{<} 1 \quad \textit{indicates q is typically less than unity.} \tag{1.7c}$$

$$q \sim 1 \quad \textit{indicates q is of the order unity.} \tag{1.7d}$$

$$q \gtrsim 1 \quad \textit{indicates q is typically more than unity.} \tag{1.7e}$$

$$q \gg 1 \quad \textit{indicates q is large (e.g., 10 or more).} \tag{1.7f}$$

$$q \to \infty \quad \textit{indicates q is very large (e.g., 100 or more).} \tag{1.7g}$$

For example, solid and liquid densities are generally much greater than gas densities, so liquid drops and solid particles in a gas can be characterized by $\rho_p \gg \rho_f$. Such particles are defined as *high-density* particles. In contrast, gas bubbles in a liquid with $\rho_p \ll \rho_f$ are defined *high-buoyancy* particles. Particles with intermediate densities with $\rho_p \sim \rho_f$ (e.g., solid particles with a specific gravity near unity in water) are defined as *near-neutrally buoyant*. If the particle and the surrounding are both fluids, the viscosity ratio can also be defined. Since liquid viscosities are generally much greater than gas viscosities, a drop in a gas corresponds to $\mu_p \gg \mu_f$, while a gas bubble corresponds to $\mu_p \ll \mu_f$.

1.3.3 Vector Notation

In the next section, two types of multiphase velocities are defined. These will employ both vectors and scalars. Herein, a vector (e.g., force, velocity, etc.) is represented by **boldface** (e.g., $\mathbf{q}$), while the scalar magnitude of a vector (a positive value) is represented by a regular typeface (e.g., $q = |\mathbf{q}|$).

Cartesian vector components can be defined using subscripts for the x-, y-, and z-directions or using index notation in the 1, 2, and 3 directions. For example, an arbitrary vector $\mathbf{q}$ can be expressed using x, y, and z and 1, 2, and 3 subscripts as

$$\mathbf{q} \equiv q_x \mathbf{i}_x + q_y \mathbf{i}_y + q_z \mathbf{i}_z = q_1 \mathbf{i}_1 + q_2 \mathbf{i}_2 + q_3 \mathbf{i}_3 = q_i \mathbf{i}_i \tag{1.8a}$$

$$q \equiv |\mathbf{q}| = \sqrt{q_x^2 + q_y^2 + q_z^2} = \sqrt{q_i q_i} \qquad \textit{for } i = 1, 2, \textit{and } 3. \tag{1.8b}$$

The double appearance of the subscript i in the right-hand side (RHS) product indicates a tensor summation operation over all the index values (i = 1, 2, and 3). In this text, this notation will be generally used whenever index summation is intended (otherwise, summation is not intended). The position vector $\mathbf{X}$ is also useful to consider reference frames and is defined based on a position relative to reference point. Following (1.8), it can be expressed in terms of Cartesian components as

$$\mathbf{X} \equiv x\mathbf{i}_x + y\mathbf{i}_y + z\mathbf{i}_z = X_1 \mathbf{i}_1 + X_2 \mathbf{i}_2 + X_3 \mathbf{i}_3 = X_i \mathbf{i}_i \qquad \textit{for } i = 1, 2, \textit{and } 3. \tag{1.9}$$

For a Cartesian coordinate system, the divergence of a vector $\mathbf{q}$ (with components q_x, q_y, and q_z) and the gradient of a scalar q can be expressed in coordinate notation or tensor notation (using the position vector components) as follows:

$$\nabla \cdot \mathbf{q} \equiv \frac{\partial q_x}{\partial x} + \frac{\partial q_y}{\partial y} + \frac{\partial q_z}{\partial z} = \frac{\partial q_1}{\partial X_1} + \frac{\partial q_2}{\partial X_2} + \frac{\partial q_3}{\partial X_3} \equiv \frac{\partial q_i}{\partial X_i}. \tag{1.10a}$$

$$\nabla q \equiv \frac{\partial q}{\partial x}\mathbf{i}_x + \frac{\partial q}{\partial y}\mathbf{i}_y + \frac{\partial q}{\partial z}\mathbf{i}_z = \frac{\partial q}{\partial X_1}\mathbf{i}_1 + \frac{\partial q}{\partial X_2}\mathbf{i}_2 + \frac{\partial q}{\partial X_3}\mathbf{i}_3 \equiv \frac{\partial q}{\partial X_i}\mathbf{i}_i. \tag{1.10b}$$

The left-hand side (LHS) and RHS provide compact forms for divergence of a vector and gradient of a scaler.

For a spherical coordinate system, the coordinates in the radial, polar, and azimuthal direction are r, θ, and ϕ directions as shown in Figure 1.23. As such, a velocity vector $\mathbf{q}$ can be expressed in terms of the components q_r, q_θ, and q_ϕ. In this case, the divergence and gradient for a spherical reference frame can be expressed as

$$\nabla \cdot \mathbf{q} = \frac{1}{r^2}\frac{\partial(r^2 q_r)}{\partial r} + \frac{1}{r\sin\theta}\frac{\partial(q_\theta \sin\theta)}{\partial\theta} + \frac{1}{r\sin\theta}\frac{\partial\left(q_\phi\right)}{\partial\phi}. \tag{1.11a}$$

$$\nabla q \equiv \frac{\partial q}{\partial r}\mathbf{i}_r + \frac{1}{r}\frac{\partial q}{\partial\theta}\mathbf{i}_\theta + \frac{1}{r\sin\theta}\frac{\partial q}{\partial\phi}\mathbf{i}_\phi. \tag{1.11b}$$

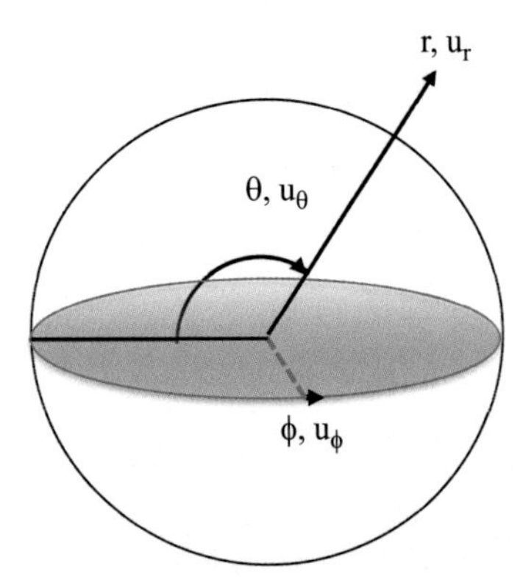

Figure 1.23 Spherical coordinate system for a velocity **u**, where r is the radial coordinate and u_r is the radial velocity component (both in the direction of unit vector $\mathbf{i}_r$), where θ is the polar angle and u_θ is the polar velocity component, and where ϕ is the azimuthal angle about the axis $\theta = 0$ and u_ϕ is the azimuthal velocity component.

This coordinate system is often useful for considering flow around a particle, as will be shown in Chapter 3. For this, azimuthal symmetry will be generally assumed for the flow, which indicates there is no dependence on ϕ and u_ϕ is zero.

1.4 Resolved-Surface and Point-Force Velocity Fields

The velocity fields for multiphase flow are generally categorized as (1) *resolved-surface* velocities, which consider how the flow moves locally around the particle surface; or (2) *point-force* velocities, which neglect the particle surface and instead consider motion based on the particle centroid. These two flow fields are quite different (the first is based on physical geometry, and the second is based on theoretical approximation), but each will be used at different times throughout the book. In the following subsections, these two categories of velocity fields are first discussed individually and then they are compared.

1.4.1 Resolved-Surface Velocity Fields

The resolved-surface velocity fields are defined relative to a particle surface whose interface locations are defined by $\mathbf{X}_I$, as shown in Figure 1.24, where the subscript I indicates the particle surface interface. Relative to this interface, the continuous-phase velocity field outside of the particle surface is denoted by **U**. For this surrounding flow, each point on the surface ($\mathbf{X}_I$) can experience viscous and pressure stresses. The combination of these stresses integrated over the particle surface determines the net fluid dynamic forces acting on a particle. The pressure stresses on the particle are based on P_f and only contribute toward the surface normal direction. As will be discussed in Chapter 2, the viscous stresses are based on the velocity field

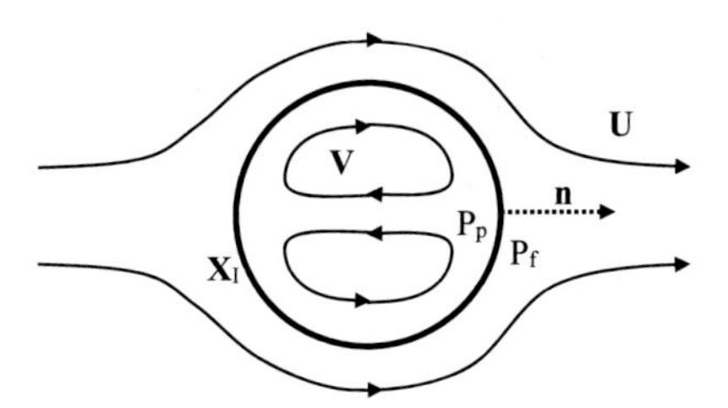

Figure 1.24 Schematic of fluid particle cross-section interface (thick solid line) whose points have coordinates $\mathbf{X}_I$ and whose outward normal is **n**, which separates the two resolved-surface velocity and pressure fields, given as **U** and P_f for the external surrounding fluid and given as **V** and P_p for the internal particle fluid.

gradients and the surrounding fluid viscosity. If the particle is a solid, P_f is often simplified as P.

If the particle is a fluid, there can also be an internal flow defined by **V** (as shown in Figure 1.24). Note that **U** and **V** take into account the local geometry of the particle surface, which need not be spherical. Thus, the two resolved-surface velocity fields and pressure fields for a fluid particle are given as follows:

$$\mathbf{U} \equiv \text{continuous-phase velocity external to the particle surface.} \tag{1.12a}$$

$$P_f \equiv \text{continuous-phase pressure external to the particle surface.} \tag{1.12b}$$

$$\mathbf{V} \equiv \text{dispersed-phase velocity within the particle surface.} \tag{1.12c}$$

$$P_p \equiv \text{dispersed-phase pressure within the particle surface.} \tag{1.12d}$$

The internal velocity field will create internal viscous stresses (based on μ_p) and a field of internal pressure (P_p). There will be a jump condition between P_f and P_p at the interface due to surface tension and surface curvature. Formulations for these internal fluid stresses and the pressure jump for a particle will be discussed in Chapter 3.

If there is no mass transfer across the interface, the normal components of **U** and **V** will be equal at the interface surface. If one further assumes a continuum surrounding flow with finite viscosity (see §1.5), the tangential components of **U** and **V** must also be equal at the interface $(\mathbf{X}_I)$. Therefore, there is no jump in the velocity at the interface, as follows:

$$\mathbf{U}_I = \mathbf{V}_I \text{ at } \mathbf{X}_I \quad \textit{for continuum flow with no mass transfer.} \tag{1.13}$$

This is often described as the "no-slip" condition. If the particle's shape and size are also fixed in time, the external and internal flow field streamlines associated with **U** and **V** will follow along the particle surface geometry (as shown in Figure 1.24).

1.4.2 Point-Force Velocities

In many cases, only the net fluid dynamic force on the particle is of interest (not the details of the local fluid and particle velocity fields near the interface). In such a case,

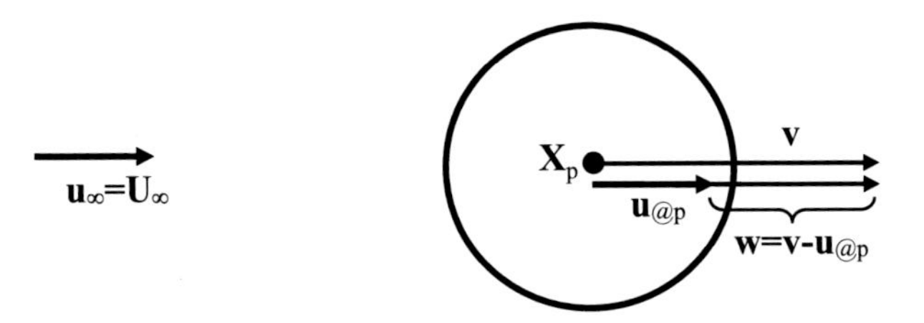

Figure 1.25 Point-force velocity schematics for a nonspinning solid particle with centroid-based velocity ($\mathbf{v}$) exposed to a spatially uniform, unhindered, continuous-phase velocity ($\mathbf{u}_{\infty}$), creating a relative velocity ($\mathbf{w}$) based on the difference when extrapolated to the particle centroid ($\mathbf{u}_{@p}$).

this net force can be assumed to act at a single point located at the centroid of the particle, that is, as a *point force*. The corresponding point-force velocities are also based on the particle centroid $(\mathbf{X}_p)$, as illustrated in Figure 1.25. In particular, the translational velocity $(\mathbf{v})$ is the particle centroid velocity, which can vary with time (t) along the particle path. The particle's translational velocity and centroid position vectors can be expressed using the Cartesian unit vectors of (1.9), as follows:

$$\mathbf{v}(t) \equiv v_x(t)\mathbf{i}_x + v_y(t)\mathbf{i}_y + v_z(t)\mathbf{i}_z. \tag{1.14a}$$

$$\mathbf{X}_p(t) \equiv x_p(t)\mathbf{i}_x + y_p(t)\mathbf{i}_y + z_p(t)\mathbf{i}_z. \tag{1.14b}$$

Since these are referenced to the centroid, these are particle Lagrangian reference frame values.

Next we define the velocity of surrounding fluid in the point-force reference frame $(\mathbf{u})$ as the fluid velocity that neglects the presence and motion of the particle. Thus, the flow defined by $\mathbf{u}$ is "unhindered" by the particle. In particular, $\mathbf{u}$ ignores the local details of the flow path deflection around the particle surface or any other aspect of the particle's presence (and is thus nonphysical). Note that if there are other particles in the domain, $\mathbf{u}$ takes into account their integrated influence on the flow but not the influence of the particle for which it is being referenced.

If one interpolates the unhindered surrounding velocity field $(\mathbf{u})$ to the particle centroid $(\mathbf{X}_p)$, this is termed the surrounding fluid velocity "seen" by the particle and is denoted as $\mathbf{u}_{@p}$. This hypothetical velocity allows one to define a relative particle velocity $(\mathbf{w})$ as the difference between the particle and that of the unhindered surrounding fluid, so that $\mathbf{w} = \mathbf{v} - \mathbf{u}_{@p}$. This relative velocity can be related to various net fluid dynamic forces acting on the particle. For example, the drag force will be found to act in the direction of $-\mathbf{w}$ (drag resists relative velocity motion), and the drag magnitude increases as the relative velocity magnitude increases.

Similarly, the pressure of the unhindered surrounding fluid (which ignores the particle's presence) can be interpolated to the particle centroid as $p_{f@p}$. The definitions for the preceding velocities and this pressure field can be summarized as follows:

$$\mathbf{v} \equiv \text{translation velocity of particle centroid} \equiv \left.\frac{\mathbf{X}_p(t+\Delta t) - \mathbf{X}_p(t)}{\Delta t}\right|_{\Delta t \to 0}. \quad (1.15a)$$

$$\mathbf{u} \equiv \text{surrounding fluid velocity unhindered by the particle's presence.} \quad (1.15b)$$

$$\mathbf{u}_{@p}(t) \equiv \mathbf{u}(\mathbf{X}_p, t) \equiv \text{unhindered flow velocity interpolated to } \mathbf{X}_p. \quad (1.15c)$$

$$\mathbf{w}(t) \equiv \mathbf{v}(t) - \mathbf{u}_{@p}(t) \equiv \text{particle relative velocity to the unhindered flow.} \quad (1.15d)$$

$$p \equiv \text{continuous-phase pressure neglecting local particle disturbances.} \quad (1.15e)$$

It is important to note that $\mathbf{u}_{@p}$ is hypothetical since the particle does indeed affect the surrounding fluid and since the surrounding fluid does not exist at the centroid (inside the particle). However, $\mathbf{u}_{@p}$ is mathematically and computationally convenient since it defines a local relative velocity that is proportional to drag, lift, and other fluid dynamic particle forces.

The preceding point-force approach can be similarly applied to other properties. Particle rotation about the centroid can be characterized by angular velocity $(\mathbf{\Omega}_p)$, and the unhindered surrounding fluid vorticity (which ignores the particle's presence) can be interpolated to the particle centroid as $\omega_{f@p}$. Particle temperature can be characterized by a single temperature (T_p), which reflects the integrated temperature within the particle volume, and the unhindered surrounding fluid temperature can be interpolated to the particle centroid as $T_{f@p}$.

1.4.3 Contrasting Point-Force and Resolved-Surface Velocities

The point-force and resolved-surface velocity fields are contrasted in Figure 1.26, where the surrounding point-force flow field $\mathbf{u}$ is independent of the particle presence (Figure 1.26a), while the resolved surrounding velocity $\mathbf{U}$ around the particle (Figure 1.26b) has increased complexity. In addition, the resolved continuous-fluid velocity $(\mathbf{U})$ is only defined outside the particle interface, while $\mathbf{u}_{@p}$ is defined at the

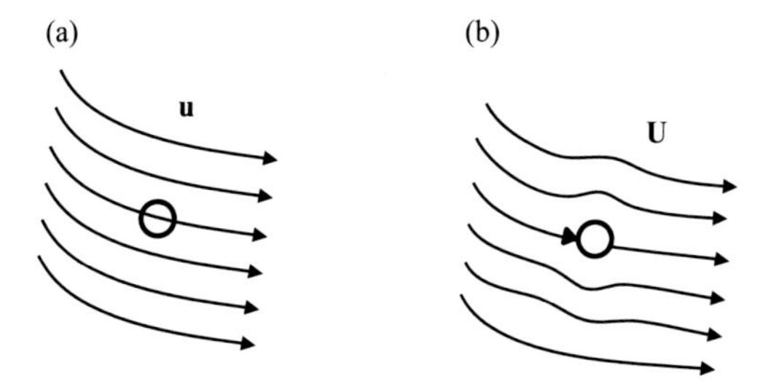

Figure 1.26 Comparing streamlines for different representations of the surrounding velocity field for a stationary particle ($\mathbf{v} = 0$), showing (a) the unhindered flow field ($\mathbf{u}$), which extends hypothetically to the particle centroid where it produces a relative velocity at the centroid, that is, $\mathbf{v} \neq \mathbf{u}$ at $\mathbf{X}_p$, and (b) the resolved-surface flow around same particle showing that $\mathbf{U}$ deviates around the particle surface and matches the surface velocity, that is, $\mathbf{U} = \mathbf{V}$ at $\mathbf{X}_I$.

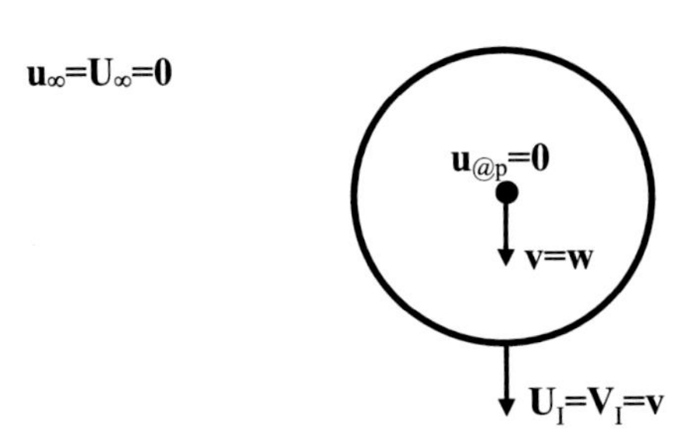

Figure 1.27 A solid nonspinning particle falling in a still fluid. The resolved-surface and point-force surrounding fluid velocities are both far from the particle, but the resolved-surface value tends to the particle velocity as it approaches a point on the particle surface (due to the no-slip condition). The point-force surrounding fluid velocity remains zero up to and even inside the particle.

particle centroid. Furthermore, the point-force velocity fields define a relative velocity of the particle to surrounding fluid (1.15d), while there is no relative velocity between the phases for the resolved-surface fields (1.13).

To further compare the point-force and resolved-surface velocities, consider the simplified case where the particle is solid and not spinning, so $\mathbf{V}$ becomes a constant within the interface. As such, all parts of the particle move at the centroid velocity, that is, $\mathbf{v} = \mathbf{V}$. If we further assume the surrounding fluid is a constant value and steady far from the particle, this far-field velocity is the same for both reference frames, that is, $\mathbf{u}_\infty = \mathbf{U}_\infty$. However, as we get closer to the particle interface, the velocity fields of $\mathbf{u}$ and $\mathbf{U}$ differ, since $\mathbf{u}$ will stay a constant, but $\mathbf{U}$ will be altered by the presence of the particle. For example, consider a particle falling at velocity $\mathbf{v}$ in a large quiescent tank so $\mathbf{u}_\infty = \mathbf{U}_\infty = \mathbf{0}$ as in Figure 1.27. However, the surrounding fluid near the particle will be affected by the local pressure and shear stresses associated with the particle movement. For finite viscosity, the no-slip condition of (1.13) dictates $\mathbf{U}_\mathrm{I} = \mathbf{V}_\mathrm{I} = \mathbf{v}$. In contrast, the unhindered velocity neglects the presence of the particle motion, including at the centroid, so $\mathbf{u}_{@\mathrm{p}} = 0$. As such, $\mathbf{U} \neq \mathbf{u} = \mathbf{0}$ in the vicinity of the particle, and the particle's relative translational velocity in this stagnant medium is given by $\mathbf{w} = \mathbf{v}$.

If one considers multiple particles in a domain, it should be noted that $\mathbf{u}_{@\mathrm{p}}$ for a given particle only neglects the flow disturbances associated with the particle at $\mathbf{X}_\mathrm{p}$. However, the $\mathbf{u}$ velocity field generally can be modified by the combined point forces associated with all other particles in the domain. For example, if we consider a large number of particles falling in one portion of an otherwise static fluid, the downward momentum induced by the falling particles will lead to a regional downdraft of the fluid so $\mathbf{u} \neq 0$. In this way, $\mathbf{u}_{@\mathrm{p}}$ incorporates the net downdraft induced by all the other particles.

1.4.4 Point-Force Time Derivatives

Since flow and particle properties can vary in space ($\mathbf{x}$) and in time (t), it is important to properly define their rate of change with regard to a specific reference frame. The temporal variation can be based on three reference frames, as shown in Figure 1.28:

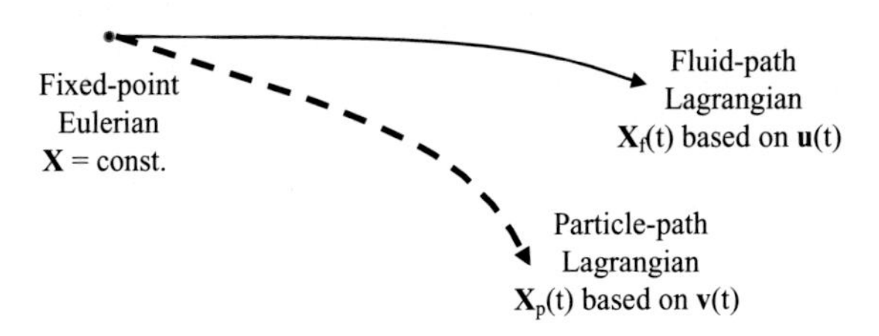

Figure 1.28 Schematic showing the three reference frames based on an Eulerian point ($\mathbf{X}$ = constant), a Lagrangian-particle path (based on $\mathbf{X}_p$ and $\mathbf{v}$ as a function of time) and Lagrangian-fluid path (based on $\mathbf{X}_f$ and $\mathbf{u}$ as a function of time). The particle and fluid element paths differ since the particle is also falling downward.

- Eulerian, referenced to a fixed point in space
- Particle-Lagrangian, referenced to the particle centroid and its velocity $\mathbf{v}$
- Fluid-Lagrangian, referenced to the fluid path and its velocity $\mathbf{u}$

Note that the Lagrangian derivative is also sometimes called the material derivative, as it follows a specific material (the fluid or the particle).

To compare the three reference frames, consider the analogy of a bridge over a river with a duck. The Eulerian reference frame is effectively the stationary (no velocity) vantage point at the bridge. The fluid-Lagrangian reference frame is the vantage of a fluid element (a tracer), which moves exactly with the stream's speed and direction (e.g., a small leaf on the river surface). The particle-Lagrangian reference frame is the vantage of the duck, whose motion is influenced by the river flow but can have a distinct trajectory as it paddles, since the duck velocity $(\mathbf{v})$ will not generally equal the river velocity $(\mathbf{u})$. So the amount and direction of paddling by the duck will dictate its relative velocity $(\mathbf{w})$.

The time derivatives for each of these reference frames are expressed in the following equation using an arbitrary scalar q. The Eulerian time derivative is based on temporal changes at a fixed point within the domain.

$$\frac{\partial q}{\partial t} \equiv \left. \frac{q(\mathbf{X}, t + \Delta t) - q(\mathbf{X}, t)}{\Delta t} \right|_{\Delta t \to 0}. \tag{1.16}$$

In contrast, the particle-Lagrangian and fluid-Lagrangian time derivatives are respectively defined along the particle path (defined with by $\mathbf{v}$) and the fluid path (defined with $\mathbf{u}$) as:

$$\frac{d q}{d t} \equiv \left. \frac{q(\mathbf{X} + \mathbf{v}\Delta t, t + \Delta t) - q(\mathbf{X}, t)}{\Delta t} \right|_{\Delta t \to 0} = \frac{\partial q}{\partial t} + \mathbf{v} \cdot \nabla q. \tag{1.17a}$$

$$\frac{\mathcal{D} q}{\mathcal{D} t} \equiv \left. \frac{q(\mathbf{X} + \mathbf{u}\Delta t, t + \Delta t) - q(\mathbf{X}, t)}{\Delta t} \right|_{\Delta t \to 0} = \frac{\partial q}{\partial t} + \mathbf{u} \cdot \nabla q. \tag{1.17b}$$

These RHS expressions make use of the chain rule and the notation of (1.10b) while assuming that the scalar field is continuously differentiable in space. The two LHS derivatives can be related using the relative velocity of (1.15d) as follows:

$$\frac{d\mathrm{q}}{dt} = \frac{\mathcal{D}\mathrm{q}}{\mathcal{D}t} + \mathbf{v}\cdot\nabla\mathrm{q} - \mathbf{u}\cdot\nabla\mathrm{q} = \frac{\mathcal{D}\mathrm{q}}{\mathcal{D}t} + \mathbf{w}\cdot\nabla\mathrm{q}. \tag{1.18}$$

Referring to our previous river analogy, time changes seen in a Eulerian frame would be like those seen looking straight down from the bridge, where one may observe a floating leaf appear from under the bridge that would then float downstream out of one's field of view. Time changes seen in a particle-Lagrangian frame would be like those seen by the duck, which may observe different small leaves on the river surface due to its relative velocity.

Next we consider the time derivatives for a vector (that varies continuously in space). The changes in the vector as seen by a particle-Lagrangian reference frame can be related to the Eulerian and fluid-Lagrangian time derivatives in vector form ($\mathbf{q}$) and in Cartesian tensor form ($\mathrm{q_i}$) as follows:

$$\frac{d\mathbf{q}}{dt} = \frac{\partial\mathbf{q}}{\partial t} + (\mathbf{v}\cdot\nabla)\mathbf{q} = \frac{\partial\mathbf{q}}{\partial t} + (\mathbf{u}\cdot\nabla)\mathbf{q} + (\mathbf{w}\cdot\nabla)\mathbf{q} = \frac{\mathcal{D}\mathbf{q}}{\mathcal{D}t} + (\mathbf{w}\cdot\nabla)\mathbf{q}. \tag{1.19a}$$

$$\frac{d\mathrm{q_i}}{dt} = \frac{\partial\mathrm{q_i}}{\partial t} + \mathrm{v_j}\frac{\partial\mathrm{q_i}}{\partial\mathrm{X_j}} = \frac{\partial\mathrm{q_i}}{\partial t} + \mathrm{u_j}\frac{\partial\mathrm{q_i}}{\partial\mathrm{X_j}} + \mathrm{w_j}\frac{\partial\mathrm{q_i}}{\partial\mathrm{X_j}} = \frac{\mathcal{D}\mathrm{q_i}}{\mathcal{D}t} + \mathrm{w_j}\frac{\partial\mathrm{q_i}}{\partial\mathrm{X_j}}\ \mathit{for}\ \mathrm{j} = 1,2,3. \tag{1.19b}$$

Note that a key assumption in applying the preceding derivative relations to a flow property is that that the property (density, velocity field, etc.) continuously varies in space.

1.5 Continuum Criteria and Conditions

1.5.1 Normal Temperature and Pressure

A helpful reference condition for properties of a given fluid is Normal Temperature and Pressure (NTP), which is defined by a reference (room) temperature and pressure: $\mathrm{T_{ref}} = 293$ K and $\mathrm{p_{ref}} = 101,320\ \mathrm{N/m^2}$. NTP properties are given in Table A.1 for two gases (air and methane) and two liquids (water and ethanol).

1.5.2 Continuum Criteria

This text generally assumes a "continuum" approximation for the surrounding fluid. In this limit, the drag on a body will be shown to be a function of pressure and viscosity (Chapter 3). This is true if the fluid properties are based on bulk averages of a large number of molecular interactions. However, this assumption is not appropriate if one considers flow length scales similar to the molecular scales on the surrounding fluid.

To clarify the difference between continuum and noncontinuum conditions, consider a particle surrounded by gas molecules (Figure 1.29). If we consider individual molecules, their collisions with the particle surface will be stochastic and discrete. Since these collisions determine the fluid influence on the particle at any given time,

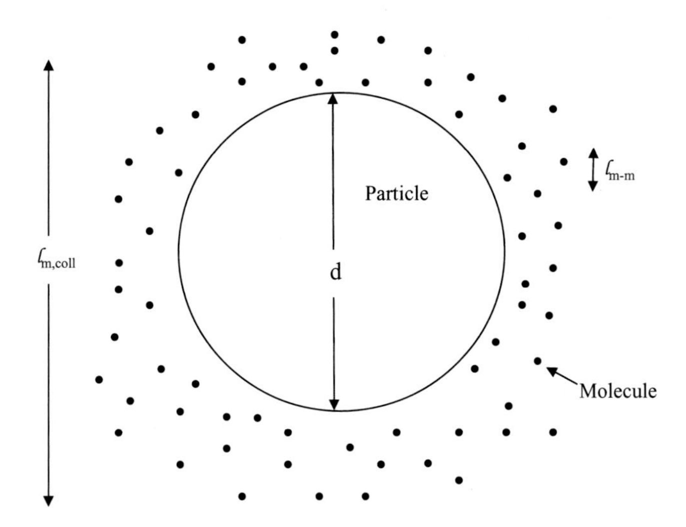

Figure 1.29 Schematic of molecules (small solid spheres) of a surrounding gas near a small particle (e.g., 10 nanometers). The length-scale criterion for density and temperature is based on the molecular spacing ($\ell_{m\text{-}m}$), while the criterion for continuum pressure and viscosity is based on the mean free path between molecular collisions ($\ell_{m,coll}$) where $\ell_{m,coll} \gg \ell_{m\text{-}m}$.

the influence of the fluid will also be stochastic and discrete. If we instead consider continuum conditions, the number of molecular interactions (averaged in both time and space) is large enough such that their integrated influence becomes deterministic and continuous. The net fluid influence can then be expressed in terms of pressure and fluid stresses on the surface. For the example of a fixed particle in a static fluid, the pressure at the continuum scale is a constant while the influence at the molecular scale is based on the individual molecular impacts. In the following, the criterion for continuum conditions is characterized in terms of length scales for a gas and is then considered more generally for a liquid as the surrounding fluid.

For a gas that makes up the surrounding fluid, there are three key length scales associated with the molecules as shown in Figure 1.29:

- The effective diameter of a molecule (d_m), based on its cross-sectional area
- The average spacing between molecules ($\ell_{m\text{-}m}$), which determines the gas density once combined with the molecular mass
- The average distance that a molecule will travel before hitting another molecule ($\ell_{m,coll}$), termed the molecular mean free

Continuum conditions for particles are typically defined in terms of the fluid stresses (since these determine the forces on the particle) and thus require $d \gg \ell_{m,coll}$. To quantify this effect, the particle Knudsen number (Kn_p) is defined as the ratio of the molecular mean free path length to the particle diameter:

$$Kn_p \equiv \frac{\ell_{m,coll}}{d}. \tag{1.20}$$

This ratio can be used to set a criterion for continuum conditions, as follows:

$$\mathrm{Kn_p} \to 0 \qquad \textit{Gas flow around the particle is a continuum.} \tag{1.21}$$

Thus, a very small $\mathrm{Kn_p}$ corresponds to a statistically large number of molecular collisions distributed over the particle surface to ensure that the pressure and viscous stress fields are in continua (i.e., both are smoothly varying in space) and are not a function of the Knudsen number. To illustrate the latter assumption, the particle drag for a sphere at $\mathrm{Kn_p} = 0.01$ is about 98% of the drag at $\mathrm{Kn_p} \to 0$ (where the latter is only a function of viscosity and pressure). As such, $\mathrm{Kn_p} < 0.01$ is often used as a reasonable approximation for (1.21).

Notably, for a gas at typical atmospheric conditions, the average spacing between molecules is typically 10–20 times the size of molecules (such that $\ell_{\mathrm{m-m}} \gg d_{\mathrm{m}}$), while the mean free path is about 30 times larger than the spacing between molecules (such that $\ell_{\mathrm{m,coll}} \gg \ell_{\mathrm{m\text{-}m}}$). This latter difference is illustrated in Figure 1.29, where the molecular spacing is small relative to the particle size (indicating the particle is nearly large enough for the surrounding gas density to be considered as a continuum). However, the mean free path is larger than the particle diameter (indicating that this corresponds to strongly noncontinuum conditions). As such, the continuum conditions based on pressure and viscous stresses (1.21) are even more stringent than those based on density (and temperature).

To consider the continuum assumption for a single-phase flow, one may similarly define a fluid domain-scale Knudsen number in terms of the fluid domain length scale (D of 1.3), as follows:

$$\mathrm{Kn_D} \equiv \frac{\ell_{\mathrm{m,coll}}}{\mathrm{D}}. \tag{1.22}$$

This length scale ratio can be similarly used to set a continuum criterion for single-phase flow, as follows:

$$\mathrm{Kn_D} \to 0 \qquad \textit{Gas flow acts as a continuum at the domain length scales.} \tag{1.23}$$

Since the particle diameters herein are assumed to be much smaller than that of the fluid domain (per 1.3), it follows that $\mathrm{Kn_D} \ll \mathrm{Kn_p}$. Therefore, if continuum conditions are reasonable for flow around the particle, they will automatically also be reasonable for the overall flow field.

To compute the Knudsen number for a particular gas, we first define an *ideal gas* as having the following relationship among pressure, temperature, density, and the specific gas constant $(\mathcal{R}_g)$:

$$\mathrm{p}_g = \rho_g \mathcal{R}_g \mathrm{T}_g. \tag{1.24a}$$

$$\mathcal{R}_g \equiv \mathcal{R}_{\mathrm{univ}}/\mathrm{MW}_g. \tag{1.24b}$$

The latter equation employs the universal gas constant, $\mathcal{R}_{\mathrm{univ}} = 8314.47\,\mathrm{J/(kmol\text{-}K)}$ and the molecular weight of the gas (MW_g). Next we employ the hard sphere model for the molecules (Hirschfelder et al., 1954), so that the mean free path is related to the fluid viscosity (based on the mean molecular collision cross section), the gas density (based on the molecular spacing), and the mean relative molecular velocity magnitude ($v_{m\text{-}m}$, based on the average speed of molecules they approach each another), as follows:

$$\ell_{m,coll} = \frac{2\mu_g}{\rho_g v_{m\text{-}m}} = \frac{\mu_g}{\rho_g}\frac{\sqrt{\pi}}{\sqrt{2\mathcal{R}_g T_g}} \qquad \textit{for an ideal gas.} \tag{1.25a}$$

$$v^2_{m\text{-}m} = 8\mathcal{R}_g T_g/\pi \tag{1.25b}$$

The expression of (1.25b) relates the mean relative kinetic energy of the molecules to the gas temperature (T_g) and the gas constant. While the expression of (1.25a) can be used with (1.20) to determine if the continuum approximation is reasonable for a flow about a particle. For example, the mean free path for air (based on Table A.1 properties) yields $\ell_{m,coll}$ of about 0.07 μm. If one applies a continuum criterion of $Kn_p < 0.01$, this requires $d > 7\,\mu m$ for the flow about the particle. As such, smaller particles (or lower gas densities, e.g., at a high altitude) may result in significant noncontinuum effects.

For a liquid, the continuum criteria based on pressure and viscous stresses are not as well defined. This is because the molecules are more tightly spaced and tend to interact continuously, such that a mean free path description is not appropriate. However, the mean molecular spacing $(\ell_{m\text{-}m})$ is still relevant to define fluid density. For typical liquids, this spacing is less than a nanometer, so that density can be considered a continuum for particle diameters as small as 100 nanometers. This continuum criterion $(d > 100\ell_{m\text{-}m})$ can also be applied for the effects of pressure and viscosity for liquids, that is, the average drag of a particle is unaffected by molecular spacing for this condition.

While these length-scale criteria can be used to determine continuum conditions in an averaged sense, random molecular effects can also be considered in an unsteady sense. In particular, discrete molecular collisions can cause particles as large as 1 μm (in a liquid or a gas) to move irregularly due to Brownian motion. This effect requires an analysis based on kinetic theory (instead of length scales), as will be discussed in §5.3. In addition, Knudsen number effects for a particle in a gas are discussed in §9.2.2–9.2.3. Otherwise, the rest of this text will assume continuum conditions for flow around the particle and for the flow in the domain.

1.6 Chapter 1 Problems

(P1.1) With 100–150 words and one or two figures, identify and discuss an engineering or environmental system that involves a dispersed multiphase flow with solid particles, drops, and/or bubbles that was not presented with a figure in §1.2. Focus your discussion on the importance of this flow system, the relevant multiphase interaction physics, and any research issues. To support the latter, cite two archival research references that are relevant to this flow.

(P1.2) Consider a wind tunnel where the flow is steady and is moving in the positive x-direction (left to right) through a contraction. The unhindered air speed variation with downstream distance for $x > 0$ is given as $u_o(1 + x/L)^{1/2}$, where $u_o = 0.25\,m/s$ and $L = 0.4\,m$. Consider a solid nonrotating particle that is also moving left to right in this flow, but its inertia causes its centroid to move at a nearly fixed speed of 0.24 m/s along the tunnel centerline. When the particle is at $x/L = 0.5$, determine (a) $\mathbf{U}, \mathbf{V}$, and $d\mathbf{U}/dt$ on the particle surface as well as (b) $\mathbf{u}_{@p}, \mathbf{w}, d\mathbf{v}/dt$, and $D\mathbf{u}/Dt$.

(P1.3) A tank of water that sloshes back at forth with a spatially uniform cyclic velocity given by $u = u_D \sin(2\pi t/\tau_D)$, where u_D is the peak velocity magnitude and is given by 0.5 m/s, while τ_D is the oscillation period and is given by 1.5 s. In this flow, consider a neutrally buoyant particle whose centroid has a cyclic velocity given by $v = v_D \sin(2\pi t/\tau_D)$, where $v_D = 0.5u_D$ due to particle inertia (ignoring phase lag). At $t = 0.5$ s, determine the time-varying (a) $\mathbf{U}, \mathbf{V}$, and $d\mathbf{U}/dt$ on the particle surface as well as (b) $\mathbf{u}_{@p}, \mathbf{w}, d\mathbf{v}/dt$, and $D\mathbf{u}/Dt$.

(P1.4) For an altitude of 10 km (where $p = 26{,}400$ Pa, $T = -50°C$, and $\mu_f = 1.447 \times 10^{-5}\,kg/m\text{-}sec$), determine the minimum diameter of a particle for which the surrounding air can considered to be in continuum conditions.

2 Single-Phase Flow Equations and Regimes

The primary assumptions and formulations for single-phase flow regimes are reviewed in this chapter. The governing partial differential equations for general fluid dynamics are developed in §2.1, while equations of state and associated flow regimes are developed in §2.2. Assuming incompressible conditions, flow rotation and the stream function are discussed in §2.3, while inviscid hydrodynamics are presented in §2.4. Viscous effects are reintroduced in §2.5 along with the Reynolds number. The various viscous regimes are characterized via the Reynolds number in §2.6 and flow instability mechanisms are introduced in §2.7. For additional discussion of single-phase fluid dynamics equations, the reader is referred to White (2016) and Schlichting and Gertsen (2017).

It should be noted that the fluid dynamic discussion and governing equations in this chapter assume a continuum flow (§1.5) whereby the flow properties vary smoothly in space and time. For a body in such a continuum surrounding gas, the length scale of any fluid gradients must be much larger than the mean free path of the molecules, that is, the Knudsen number based on a differential length scale must be very small.

In terms of notation, the single-phase fluid velocity and pressure in this chapter will be defined using $\mathbf{u}$ and p, but the equations apply equally with $\mathbf{U}$ and P_f as the variables for single-phase flow. As such, the Chapter 2 single-phase equations can be extended to multiphase flow for either the unhindered or resolved-surface flow fields as discussed in §1.4. These equations can also be used to describe the flow inside a particle using $\mathbf{V}$ and P_p as the variables.

2.1 Conservation Equations and Fluid Properties

The transport partial differential equations (PDEs) that govern single-phase flows are based on the conservation of mass, momentum, energy, and species. These equations can be developed using the Reynolds transport theorem (RTT) and the associated fluid properties. In the following, this transport theorem is defined and then employed to develop transport PDEs for fluid density, velocity, temperature, and species concentration. For use in these transport equations, we will define fluid properties such as bulk and kinematic viscosity, total and internal fluid energy, as well as thermal and molecular diffusivity. In addition, this chapter will present several dimensionless parameters to characterize flow regimes, including the Froude, Mach, Prandtl, Reynolds, and Schmidt numbers.

2.1.1 Reynolds Transport Theorem and Mass Conservation

Reynolds Transport Theorem of a Fluid

The Reynolds transport theorem states that the time rate of change of a transport quantity within a control volume is equal to the sum of the net inward flux through the control volume's surface and the internal generation rate of the same quantity within the volume. In more concise terms, the rate of increase of a quantity in a volume equals the flux rate across the boundary plus the generation rate within the volume.

The RTT concept can determine the time rate of change of a moving quantity for a wide range of scenarios. For example, the rate change of a specific animal population in a game reserve is the sum of the flux through the region's boundary (which will be positive if more animals cross into the region) and the rate of internal generation (which will be positive if the birth rate exceeds the death rate). The RTT concept can be applied to other transported quantities, such as goods, money, diseases, and so on. In some cases, a quantity can move by convection, and in other cases it can be spread by diffusion. For example, a chemical in a still pond can diffuse by molecular motion (spread out) within the pond or it can convect downstream (move with fluid velocity) and be released in a river. In other cases, only convection is important. For example, the mass of water in a still pond does not diffuse (assuming the containing surface is not porous), but it can convect downstream in a river.

The RTT can also be applied in fluid dynamics to determine the relevant equations of motion. To do this, we must first differentiate between an "intensive" quantity and an "extensive" fluid quantity. An extensive property is proportional to the volume considered for uniform conditions. For example, the mass of a fluid is extensive as it will increase as the measuring volume increases. In contrast, an intensive quantity is one that is independent of the measuring volume size for uniform conditions. For example, density (mass of fluid per unit volume) is intensive since it will be the same throughout a uniform flow, regardless of the size of measuring volume employed.

To convert the RTT text into a PDE, consider an intensive quantity (q) that moves locally at a velocity ($\mathbf{u}$) throughout a fixed control volume ($\forall$) with a defined control surface area (A) and outward normal unit vector ($\mathbf{n}$), as shown in Figure 2.1. The RTT volume-based terms can be expressed as volume integrals of the intensive quantity (q), while the flux-based terms can be expressed as surface integrals based on convection ($\mathbf{u}$) and diffusion rate ($\mathbf{Q}$), as follows:

$$\frac{\partial}{\partial t}\oiiint_{\forall} q\, d\forall = -\oiint_{A} q(\mathbf{u}\cdot\mathbf{n})\, dA - \oiint_{A} \mathbf{Q}\cdot\mathbf{n}\, dA + \oiiint_{\forall} \dot{q}\, d\forall. \tag{2.1}$$

Each term in this equation is discussed in the following paragraph.

The left-hand side (LHS) of (2.1) is the time rate of change of q integrated over the control volume based on the Eulerian time derivative (i.e., based on a fixed point in space). As a result of the volume integral, the integrated time rate of change represented by the LHS is an extensive quantity. For example, if q is density, the LHS represents the time rate of change of the control volume mass. The first term on the right-hand side (RHS) is the flux across the boundary due to convection. This term can

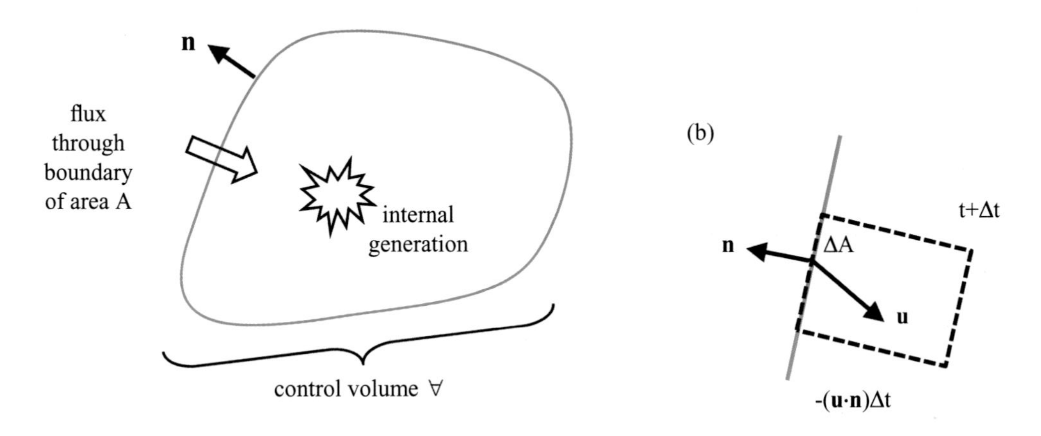

Figure 2.1 (a) Reynolds transport theorem schematic for an arbitrary control volume (∀) where total change in time of a transported intensive quantity is equal to the sum of its (convective and nonconvective) flux through the boundary (which has surface area A and outward normal vector **n**) and of its internal generation, and (b) representation of the convective flux through a local surface element area (ΔA) where the fluid moves at velocity **u** such that the amount of fluid that fluxes into the domain over a time Δt is given by $-(\mathbf{u}\cdot\mathbf{n})(\Delta t)(\Delta A)$.

be obtained by considering a fluid element crossing a differential surface area (ΔA), as shown in Figure 2.1b. Note that only the movement perpendicular to the surface contributes to the flux since tangential movement makes no contribution. Over a time Δt, the fluid element moving perpendicularly into this surface has a length displacement given by $-\mathbf{u}\cdot\mathbf{n}\,\Delta t$, where the negative sign is required to describe inward flux. Taking Δt and ΔA to be infinitesimally small, the flux rate (the first term on the RHS of (2.1) becomes the surface integral of the product of the quantity per unit volume (q) and the penetration volume per unit time $(-\mathbf{u}\cdot\mathbf{n}\,dA)$. The second term on the RHS includes the diffusive flux rate $(\mathbf{Q})$, which is the rate of flux of q through the surface that spreads due to nonconvective effects. This nonconvective flux rate is zero for mass and momentum conservation, but it is nonzero for energy transport (due to thermal conduction) and species transport (due to species diffusion). The third and final term on the RHS of (2.1) is the volume integral of the internal generation rate $(\dot{q})$, which is the net rate at which a quantity is created (or destroyed if a negative value) within the control volume.

The integral-based RTT form shown in (2.1) for fluid dynamics can be transformed into a PDE, which is often more practical to employ, by first converting the surface integrals into volume integrals. For this, we assume that q and **u** are finite and continuously differentiable in space and that the control volume is time invariant (e.g., the volume has fixed shape and a single fixed speed or is stationary). Based on these assumptions, one can employ the Gauss divergence theorem so that the surface integral of the flux terms (using the surface outward normal **n**) can be transformed into a volume integral, as follows:

$$\oiint_A [q(\mathbf{u}\cdot\mathbf{n}) + \mathbf{Q}\cdot\mathbf{n}]\, dA = \oiiint_{\forall} [\nabla\cdot(q\mathbf{u}) + \nabla\cdot\mathbf{Q}]\, d\forall. \tag{2.2}$$

By combining (2.1) and (2.2), all the terms can be consistently expressed as volume integrals. Since the size and location of the volume integral is arbitrary, an infinitesimally small volume with locally uniform flow allows the integrands to be equated, yielding the following:

$$\frac{\partial q}{\partial t} + \nabla\cdot(q\mathbf{u} + \mathbf{Q}) = \dot{q}. \tag{2.3}$$

This is the Eulerian Reynolds transport PDE for an intensive fluid quantity.

The Lagrangian version of this equation can be obtained by employing (1.17b) for the time derivative of q and rewriting the dot product term using the following gradient product vector identity:

$$\frac{\mathcal{D}q}{\mathcal{D}t} \equiv \frac{\partial q}{\partial t} + \mathbf{u}\cdot\nabla q = \frac{\partial q}{\partial t} + \nabla\cdot(q\mathbf{u}) - q(\nabla\cdot\mathbf{u}). \tag{2.4}$$

Combining this result with (2.3) yields the Lagrangian Reynolds transport PDE (along the fluid path):

$$\frac{\mathcal{D}q}{\mathcal{D}t} + q(\nabla\cdot\mathbf{u}) + \nabla\cdot\mathbf{Q} = \dot{q}. \tag{2.5}$$

These results for the Reynolds transport of an intensive fluid quantity can be used to obtain the transport equations for fluid mass, momentum, total energy, and species per unit volume. In cases where diffusion can be neglected (as will be shown for transport of fluid mass or momentum), $\mathbf{Q} = 0$, so this equation is simplified accordingly.

Transport of Mass

For the mass transport equation, the transport PDE is obtained by setting q to be the fluid density (ρ_f). If we assume that mass is neither created nor destroyed (no nuclear reactions), then $\dot{\rho}_f = 0$ for a single-phase flow. Furthermore, we can assume that mass only moves by convection so that the diffusive fluxes can be neglected ($\mathbf{Q} = 0$). As such, application of (2.3) and (2.5) yields the following Eulerian and Lagrangian mass transport PDEs for fluid mass:

$$\frac{\partial \rho_f}{\partial t} + \nabla\cdot(\rho_f\mathbf{u}) = 0. \tag{2.6a}$$

$$\frac{\mathcal{D}\rho_f}{\mathcal{D}t} + \rho_f(\nabla\cdot\mathbf{u}) = 0. \tag{2.6b}$$

These PDEs are referred to as the conservation of mass or the continuity equations. The Eulerian PDE (2.6a) shows that the rate of density change at a point is related to the divergence of momentum per unit volume, while the Lagrangian PDE (2.6b) shows that the rate of density change along a fluid streamline is related to the divergence of the fluid velocity.

The Eulerian transport PDE (2.6a) can be expressed in Cartesian component form or in tensor form as shown in the following:

$$\frac{\partial \rho_f}{\partial t} + \frac{\partial}{\partial x}(\rho_f u_x) + \frac{\partial}{\partial y}(\rho_f u_y) + \frac{\partial}{\partial z}(\rho_f u_z) = 0, \tag{2.7a}$$

$$\frac{\partial \rho_f}{\partial t} + \frac{\partial}{\partial X_i}(\rho_f u_i) = 0 \qquad \textit{for } i = 1, 2, \textit{and } 3. \tag{2.7b}$$

The Lagrangian version can be similarly expanded in Cartesian component or tensor form. Furthermore, spherical versions of the Eulerian and Lagrangian mass transport PDEs can be obtained using the divergence defined in (1.11a).

If the volume of the fluid element can change in shape but does not change in size because it is incompressible and the quantity does not diffuse, it has constant mass and thus constant density along the fluid path This is defined as an *isochoric* flow and its Lagrangian time derivative is zero. Based on (2.6b), this indicates that the divergence of the velocity field must then also be zero:

$$\nabla \cdot \mathbf{u} = \mathbf{0} \qquad \textit{for constant density along fluid path.} \tag{2.8}$$

As such, the "divergence-free" flow of (2.8) is equivalent to stating that the fluid density is constant (and incompressible) along its fluid path. This is a convenient mathematical simplification that will be used later for appropriate flow conditions.

2.1.2 Transport of Momentum

The momentum transport equation can be obtained by setting the intensive transport quantity as the fluid momentum per unit volume ($\mathbf{q} = \rho_f \mathbf{u}$) and applying RTT using (2.1)–(2.5). In doing so, we note that fluid momentum cannot diffuse (it can only be transported by velocity), so $\mathbf{Q} = 0$ for each of the three momentum components. Next, we consider the last term in (2.1), which describes internal generation rate, for fluid momentum. Since momentum can be changed (accelerated or decelerated) via Newton's second law ($\mathbf{F} = m\mathbf{a}$), the rate of this change per unit volume at a fixed location is equal to the sum of the applied forces per unit volume:

$$\dot{\mathbf{q}}_{mom} = \mathbf{F}_{fluid}/\forall_f = \left(\mathbf{F}_{body} + \mathbf{F}_{surf}\right)/\forall_f. \tag{2.9}$$

The RHS shows that the forces acting on a fluid element include body forces (proportional to mass) and surface forces (proportional to surface area), where both are discussed in the following.

Body forces are those forces which are proportional to the mass of an object and can generally include gravitational, electrostatic, and magnetic body forces. Since electrostatic and magnetic forces are generally weak in two-phase flows, they will be neglected in this textbook so gravitational force is the only body force. This force is based on the mass of fluid and gravity (where $\mathbf{g}$ is the gravity acceleration vector), as follows:

$$\mathbf{F}_{body} = m_f \mathbf{g} = \rho_f \forall_f \mathbf{g} \tag{2.10}$$

As shown by the RHS using fluid density, the gravity force per unit volume is $\rho_f\mathbf{g}$.

The surface forces in fluid dynamics are the pressure forces and the viscous forces, where both are a product of stress and an area. The pressure acts perpendicular to a surface area with a stress given by the pressure (p). The viscous stresses can act both perpendicularly and tangentially to a surface and are related to the flow viscosity and the gradients in flow velocity. In the following, we obtain their combined effect in mathematical form.

The pressure and viscous stresses can be obtained by considering an elemental volume $(\Delta x\Delta y\Delta z)$, as shown in Figure 2.2. For example, we can consider the effects of the pressure force in the y-direction and assume there is a pressure difference (Δp) due to a gradient, as shown in Figure 2.2a. Since pressure only acts normal to a surface, the y-direction force is based only on the top and bottom surfaces, whose area is $\Delta x\Delta z$. Since the pressure acts on the volume toward the surface (inward normal),

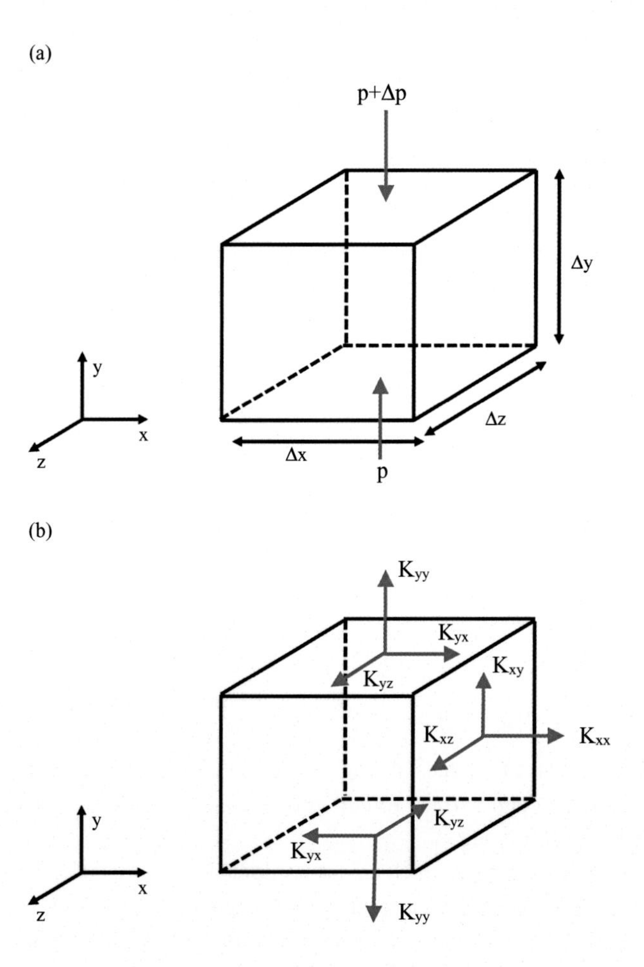

Figure 2.2 Fluid dynamic stresses on a fluid element with a discrete volume $(\Delta x\Delta y\Delta z)$. (a) Pressures on the top and bottom y-faces that act over area $\Delta A = \Delta x\Delta z$ but differ due a pressure gradient. (b) Shear stresses, K_{ij}, where first index is the face where the stress acts (the face outward normal is in the i-direction) and second index is the direction of the resulting force (the force acts in the j-direction).

the net force in the y-direction per unit volume is $-\Delta p/\Delta y$, which approaches $-\partial p/\partial y$ in the limit of an infinitesimally small volume. Applying a similar approach for the x-direction and the y-direction, the net pressure force $(\mathbf{F}_{press})$ in the limit of an infinitesimally small volume is based on the pressure gradient.

A similar result can be obtained for the viscous stresses, except that there are three components of stress applied to each surface face, as shown in Figure 2.2b. The viscous stress tensor, K_{ij}, has two indices, where the first indicates face and the second indicates direction. In particular, the stress acts on the face whose outward normal is in the i-direction and the resulting force is directed in the j-direction. Notably, the viscous stress tensor is symmetric $(K_{ij} = K_{ji})$, as will be shown later.

If the viscous stresses on a given face are approximately uniform, the top face (pointing in the y-direction) with area $\Delta x \Delta z$ would include shear stresses K_{yx} in the x-direction and K_{yz} in the z-direction, as well as a normal viscous stress K_{yy} in the y-direction. Similarly, the bottom face would include shear stresses K_{yx} in the negative x-direction K_{yz} and the negative z-direction as well as a normal viscous stress K_{yy} in the negative y-direction. Therefore (as with pressure gradients), a shear-stress gradient is needed to impart a net force on the fluid volume. Assuming a uniform gradient in the i-direction for each of the stresses and applying the limit of an infinitesimally small volume, the resulting viscous force per unit volume ($\mathbf{G}$) can be written in tensor or vector form as follows:

$$G_i \equiv \frac{F_{visc,i}}{\forall_f} = \frac{\partial K_{ij}}{\partial X_j} \qquad \textit{for}\ j = 1, 2, \textit{and}\ 3. \tag{2.11a}$$

$$\mathbf{G} \equiv \frac{\mathbf{F}_{visc}}{\forall_f}. \tag{2.11b}$$

It should be noted that the result of (2.11a) employs tensor symmetry.

Combining the preceding forces, the momentum generation rate can be written as follows:

$$\dot{\mathbf{q}}_{mom} = \frac{\mathbf{F}_{body} + \mathbf{F}_{surf}}{\forall_f} = \frac{\mathbf{F}_{body} + \mathbf{F}_{press} + \mathbf{F}_{visc}}{\forall_f} = \rho_f \mathbf{g} - \nabla p + \mathbf{G}. \tag{2.12}$$

Including this generation rate into (2.3) and (2.4) with $\mathbf{q} = \rho_f \mathbf{u}$, the resulting momentum transport PDEs in Eulerian and Lagrangian form are given (after some manipulation) as follows:

$$\frac{\partial(\rho_f \mathbf{u})}{\partial t} + \nabla \cdot (\rho_f \mathbf{u} \otimes \mathbf{u}) = \rho_f \mathbf{g} - \nabla p + \mathbf{G}. \tag{2.13a}$$

$$\rho_f \frac{\mathcal{D}\mathbf{u}}{\mathcal{D}t} = \rho_f \mathbf{g} - \nabla p + \mathbf{G}. \tag{2.13b}$$

These are the celebrated Navier–Stokes equations, and these two forms can be related using mass conservation (2.6a) by writing the outer product for (2.13a) as follows:

$$\nabla \cdot (\rho_f \mathbf{u} \otimes \mathbf{u}) = \rho_f (\mathbf{u} \cdot \nabla)\mathbf{u} = \mathbf{u} \nabla \cdot (\rho_f \mathbf{u}). \tag{2.14}$$

Note that (2.13b) could also have been derived by applying Newton's second law along the fluid path of a fluid element, where the LHS is the product of mass and acceleration of a fluid parcel (per unit volume), while the RHS is the total force being applied to the fluid element (per unit volume). In both cases, the viscous stresses are related to the flow field, as discussed later.

The relationship between the fluid viscous stress and the fluid element changes is effectively similar to relating stress and strain for a solid. In particular, the strain on a fluid element can be related to the rates of shape deformation and volumetric change, which can be expressed in terms of the fluid element velocity gradients. For Cartesian coordinates, the resulting linear relationship between viscous stresses and these rates is given as follows (White, 2016):

$$K_{ij} = \mu_f\left(\frac{\partial u_i}{\partial X_j} + \frac{\partial u_j}{\partial X_i}\right) + \mu_{bulk,f}\delta_{ij}\frac{\partial u_k}{\partial X_k} \qquad \textit{for } k = 1, 2, \textit{and } 3. \tag{2.15a}$$

$$\delta_{ij} = \begin{cases} = 1 & \text{if } i = j \\ = 0 & \text{if } i \neq j \end{cases}. \tag{2.15b}$$

The first term on the RHS of (2.15a) is the stress based on the rate of deformation strain, whereas the second term is based on the rate of volumetric strain, which employs the Kronecker delta (δ_{ij}) of (2.15b). The proportionality between stress and strain is based on two viscosities: the shear viscosity (μ_f) for the deformation strain, and the bulk viscosity $(\mu_{bulk,f})$ for the volumetric strain. The bulk viscosity $(\mu_{bulk,f})$, is sometimes called the second viscosity coefficient and is typically modeled with Stokes' hypothesis as follows:

$$\mu_{bulk,f} = -2\mu_f/3. \tag{2.16}$$

This result is formally appropriate for monoatomic gases with moderate compressibility but has been found to be quite reasonable for a wide range of flows, and so will be employed for the rest of this text. However, it may not hold under extreme conditions such as within a shock wave where extreme compressibility occurs over a short distance (Gad-el-Hak, 1995).

If one considers fluids where the density can be considered constant along the fluid path, the velocity divergence will be zero (2.8) so that the last term in (2.15a) is also zero. Such a divergence-free flow (2.8) yields the following:

$$K_{ij} = \mu_f\left(\frac{\partial u_i}{\partial X_j} + \frac{\partial u_j}{\partial X_i}\right) \qquad \textit{for density constant along fluid path.} \tag{2.17}$$

This result shows that the viscous stress tensor is symmetric $(K_{ij} = K_{ji})$, such as $K_{xy} = K_{yx}$.

For spherical coordinates, the corresponding spherical stress tensor components for divergence-free flow with azimuthal symmetry (Figure 1.23) can be expressed as follows:

$$K_{rr} = 2\mu_f\frac{\partial u_r}{\partial r}. \tag{2.18a}$$

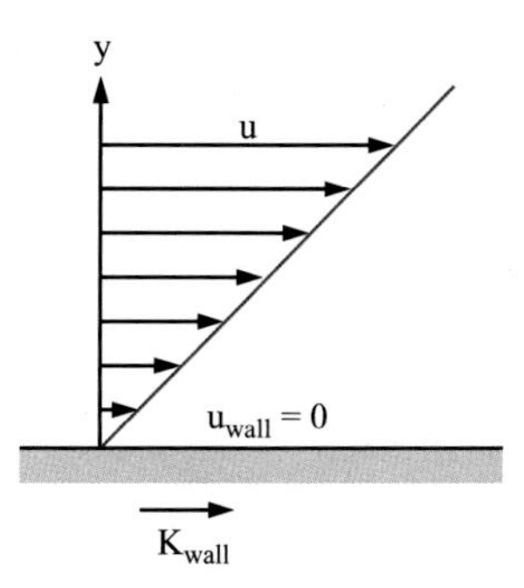

Figure 2.3 Shear stress acting on a stationary wall due to a velocity gradient caused by the no-slip condition, where direction of shear stress on the wall (K_{wall}) is in the direction of the flow. Due to an equal and opposite reaction, the stationary wall applies a shear stress on the fluid that is in the reverse direction.

$$K_{r\theta} = \mu_f \left[r \frac{\partial}{\partial r}\left(\frac{u_\theta}{r}\right) + \frac{1}{r}\frac{\partial u_r}{\partial \theta} \right]. \tag{2.18b}$$

$$K_{\theta\theta} = 2\mu_f \left[\frac{u_r}{r} + \frac{1}{r}\frac{\partial u_\theta}{\partial \theta} \right]. \tag{2.18c}$$

$$u_\phi = 0 \quad \textit{and} \quad \frac{\partial u_\theta}{\partial \phi} = \frac{\partial u_r}{\partial \phi} 0 \qquad \textit{for azimuthal symmetry.} \tag{2.18d}$$

Note that the spherical stress tensor is also symmetric (e.g., $K_{r\theta} = K_{\theta r}$).

If there is a solid surface, the viscous stresses will give rise to a shear force on the surface when there are velocity gradients. These gradients can arise since the fluid at the surface must accomodate the surface velocity (equivalent to 1.13). If we consider a stationary surface relative to our velocity reference frame (the surface velocity is zero), then this condition is given as follows:

$$\mathbf{u} = 0 \text{ at } \mathbf{X}_{surf} \qquad \textit{no-slip condition at a stationary surface.} \tag{2.19}$$

If there is also a finite flow speed some distance away from the wall, this leads to a velocity gradient. For example, consider a flow in the x-direction, which has a linear velocity gradient in the y-direction, and is acting on a wall defined by $y = 0$, as shown in Figure 2.3. In this case, the wall shear stress K_{wall} equals K_{yx} since the face is normal to the y-direction and the shear force is in the x-direction. The no-slip condition in (2.19) yields $u_x = 0$ at $y = 0$, and the near-wall velocity gradient can be used to obtain the wall shear stress from (2.17) as follows:

$$K_{wall} = \mu_f (\partial u_x / \partial y)_{y=0}. \tag{2.20}$$

This stress acts along the wall surface and yields a force on the wall in the direction of the fluid (so the force of the wall on the fluid in the opposite direction).

Returning to our governing equation, substituting either (2.17) or (2.18) into the momentum equation of (2.13a) and, assuming constant density and viscosity, yields the following:

$$\rho_f \frac{\partial \mathbf{u}}{\partial t} + \rho_f(\mathbf{u} \cdot \nabla)\mathbf{u} = \rho_f \mathbf{g} - \nabla p + \mu_f \nabla^2 \mathbf{u} \qquad \textit{for constant density and viscosity.} \tag{2.21}$$

This is the Eulerian momentum transport PDE for constant density and viscosity. The last term of the RHS includes the Laplacian of the fluid velocity, indicating that the viscous stresses effectively diffuse momentum. For example, consider the role of shear stress away from a wall, such as in a shear layer where there is a velocity gradient. In such flows, the higher-speed fluid imparts a shear stress in the direction of the flow and will transfer increased momentum to the lower-speed fluid. Similarly, the lower-speed fluid will impart a shear stress in the opposite direction on the higher-speed fluid and can serve to decrease its momentum. As such, viscous stresses tend to diffuse velocity gradients. Thus, one may consider viscosity as the effective diffusivity of fluid momentum caused by viscous stresses.

The assumption of constant viscosity used in (2.21) will generally be used in this text. However, it is worth noting that the viscosity can change with temperature. Interestingly, a temperature increase generally causes a decrease in liquid viscosity but causes an increase in gas viscosity. White (2016) provides viscosity–temperature relationships for several fluids. For dry air, the sensitivity for temperature in the range of 250–750 K can be approximated as follows:

$$\mu_f = \mu_{f,ref}(T/T_{ref})^{2/3}. \tag{2.22}$$

In this expression, $\mu_{f,ref}$ is the viscosity at $T_{ref} = 293\,K$ based on NTP conditions (per Table A.1). This relationship with a 2/3 exponent and a reference viscosity is also reasonable for several other gases in the range of 250–400 K (White, 2016).

Viscosity can also vary with fluid strain for some fluids, which are referred to as "non-Newtonian fluids." In particular, molten polymers and colloidal suspensions (e.g., paint) can have a nonlinear relationship between the velocity gradients and the viscous stresses, whereby the viscosity varies with the strain. However, this text will generally assume all fluids are "Newtonian" in that the viscosity does not depends on the strain on the fluid.

2.1.3 Transport of Energy and Species

Transport of Energy

For the fluid energy transport equation, we should consider both thermal and kinetic energy. The thermal (or internal) energy is related to the temperature of the fluid (stemming from vibration and random molecular motions of the fluid, which only come to rest when the temperature reaches 0 K). In contrast, the kinetic energy is related to the velocity of the fluid (stemming from mean molecular motion of the fluid). Combining these energies, the thermal energy per unit mass (e) plus the kinetic energy per unit mass ($\frac{1}{2}u^2$) yield the total fluid energy per unit mass (e_{tot}), as follows:

$$e_{tot} \equiv e + \frac{1}{2}u^2 \tag{2.23}$$

Setting the total fluid energy per unit volume ($q = \rho_f e_{tot}$) as the intensive scalar, the transport equation of energy can then be obtained by applying the Reynolds transport theorem. For energy, this is equivalent to using the first law of thermodynamics, whereby the energy change of a system per unit time is equal to the heat fluxing through the surfaces and the internal generation rate (the sum of internal energy release and work done on the fluid per unit time).

The fluid total energy transport equation can be obtained using (2.3) as follows:

$$\frac{\partial(\rho_f e_{tot})}{\partial t} + \nabla \cdot (\rho_f e_{tot} \mathbf{u}) + \nabla \cdot \mathbf{Q}_e = \rho_f \dot{e}_{tot}. \tag{2.24}$$

Special attention must be paid to the diffusive flux (last term on the LHS) and the internal generation (RHS).

The diffusive flux is nonzero because thermal energy can conduct due to molecular diffusion. This conduction can be obtained with Fourier's law, which states that the heat flux per unit area is equal to the product of the fluid thermal conductivity (k_f) and the gradient of fluid temperature, as follows:

$$\mathbf{Q}_e = -k_f \nabla T. \tag{2.25}$$

where the negative sign indicates that the energy moves from high temperature to low temperature.

The rate of internal generation of the total fluid energy inside a control volume can generally include the release of thermal energy by chemical reaction or phase change, as well as the increase of total energy by mechanical work on the fluid (e.g., a pump) or a decrease by fluid dynamic work done by the fluid within the control volume. Herein, we will generally neglect chemical reactions, radiation, phase changes, and mechanical work so that energy generation is only due to the fluid dynamic work done on the fluid. This work can be expressed as the dot product of the fluid dynamic forces and the fluid velocity. For example, the work rate per unit volume due to gravity forces is $\rho_f \mathbf{g} \cdot \mathbf{u}$. The pressure work rate on a fluid volume element can be considered as in Figure 2.2a. By considering pressure forces in all three directions, the pressure work rate per unit volume is $-\nabla \cdot (p\mathbf{u})$. Similarly, the viscous stress work per unit volume can be obtained as $\mathbf{G} \cdot \mathbf{u}$. Combining these elements with (2.24) and (2.25), the total fluid energy transport equation becomes

$$\underbrace{\frac{\partial(\rho_f e_{tot})}{\partial t}}_{\substack{\text{time rate of change}\\ \text{of total energy}}} + \underbrace{\nabla \cdot (\rho_f e_{tot} \mathbf{u})}_{\substack{\text{Convective flux}\\ \text{of total energy}}} = \underbrace{\rho_f \mathbf{g} \cdot \mathbf{u}}_{\text{Body force work}} + \underbrace{\mathbf{G} \cdot \mathbf{u}}_{\substack{\text{Viscous}\\ \text{stress work}}} \underbrace{- \nabla \cdot (p\mathbf{u})}_{\substack{\text{Pressure}\\ \text{stress work}}} + \underbrace{\nabla \cdot (k_f \nabla T)}_{\substack{\text{Heat flux due to}\\ \text{thermal conductivity}}} . \tag{2.26}$$

This PDE includes LHS terms of the time rate of change of total energy and the convective flux with RHS terms for the body force work, the fluid-stress work, and the heat flux due to thermal conductivity. As noted previously, this PDE neglects phase change (such as fluid evaporation or condensation), combustion or chemical reaction (which can lead to energy release), and radiation (which can lead to energy deposition or release) while also neglecting mechanical work and/or external heat transfer applied to the fluid.

In some cases, it is helpful to consider a transport PDE for the fluid temperature changes. In this case, the fluid internal energy per unit volume ($\rho_f e$) can be set as the transport quantity using (2.3) and the internal energy can be related to the fluid temperature and the specific heat at constant volume ($c_{\forall}$). If one further assumes an ideal gas with constant thermal conductivity, the internal energy and the temperature transport PDE can be expressed as follows:

$$e = c_{\forall} T. \tag{2.27a}$$

$$\frac{\partial(\rho_f c_{\forall} T)}{\partial t} + \nabla \cdot (\rho_f c_{\forall} T \, \mathbf{u}) = \mathbf{G} \cdot \mathbf{u} - \nabla \cdot (p\mathbf{u}) + k_f \nabla^2 T. \tag{2.27b}$$

The corresponding Lagrangian form for incompressible flow is then

$$\rho_f \frac{\mathcal{D}(c_{\forall} T)}{\mathcal{D}t} = \mathbf{G} \cdot \mathbf{u} - \nabla \cdot (p\mathbf{u}) + k_f \nabla^2 T. \tag{2.28}$$

The last RHS term in (2.27b) and in (2.28) includes a Laplacian of the temperature and is thus a diffusion term. As such, thermal conductivity acts to diffuse temperature gradients in a fluid (similar to how viscosity diffuses momentum gradients).

It is interesting to compare the rates at which thermal energy and momentum diffuse in a flow, where the effective diffusivity of momentum is proportional to viscosity (μ_f, with units of mass per distance per time), while the diffusivity of thermal energy is proportional to conductivity (k_f, with units of energy per distance per time per degrees Kelvin). To compare these diffusivities with the same units, the thermal conductivity can be divided by the fluid specific heat at constant pressure ($c_{p,f}$, with units of energy per mass per degrees Kelvin). The resulting nondimensional ratio of the momentum to thermal diffusivities is defined as the Prandtl number:

$$Pr_f \equiv \frac{\text{viscous (momentum) diffusivity}}{\text{molecular thermal diffusivity}} \equiv \frac{\mu_f c_{p,f}}{k_f}. \tag{2.29}$$

As such, the Prandtl number of a fluid indicates how fast a velocity gradient is diffused compared to how fast a temperature gradient is diffused. Thus, a high Prandtl number indicates that velocity differences diffuse faster as compared to temperature differences. For steady conditions where diffusion and conduction are balanced, faster rates can be equated to achieving a fixed amount of diffusion over shorter distances. Thus, the Prandtl number will be proportional to the thickness of a thermal diffusion layer relative to a momentum diffusion layer.

For gases, the Prandtl number is on the order of unity (e.g., 0.7 for air at NTP), which indicates that rates of thermal and momentum diffusivity in a gas are similar.

Thus, a thermal boundary layer (caused by the temperature difference between a freestream fluid and the surface) will have roughly the same thickness as that of a momentum boundary layer (caused by the velocity difference between a freestream fluid and the surface). However, the Prandtl number for liquids varies widely depending on the fluid type and temperature. For example, Pr_f ranges from 0.004–0.03 for liquid metals (very good heat conductors), ranges from 1.7–13.7 for water (a moderate conductor), and ranges from 50–100,000 for oils and polymer melts (poor heat conductors). For a fluid with a small Prandtl number (such as that seen in liquid metals), the thermal layers will be much thinner (or shorter lived) compared to the momentum layer so that thermal equilibrium can be expected to be achieved more quickly. The opposite would occur for fluids with a high Prandtl number.

Transport of Molecular Species

The final transport equation considered in this section is for molecular species. A species is a component of the fluid that has a singular molecule type. For example, nitrogen is a component species of air. The component fraction of a species in a fluid can be expressed in terms of the species mass fraction $(\mathcal{M})$ or the species density $(\rho_{\mathcal{M}})$, where the two are related by the overall fluid density (ρ_f) as follows:

$$\mathcal{M} \equiv \frac{\text{mass of a species}}{\text{mass of fluid mixture}}. \tag{2.30a}$$

$$\rho_{\mathcal{M}} \equiv \frac{\text{mass of species}}{\text{volume of fluid mixture}} = \rho_f \mathcal{M}. \tag{2.30b}$$

Since the species density equals the mass fraction of a species per unit volume, it can be summed over all species in the fluid to obtain the overall fluid density:

$$\rho_f \equiv \frac{\text{mass of fluid}}{\text{volume of fluid}} = \sum_{i=1}^{N_{species}} \rho_{\mathcal{M},i}. \tag{2.31}$$

In this expression, $N_{species}$ is the total number of species in a fluid (and i is a summation index).

Since the density of an individual species $(\rho_{\mathcal{M}})$ is an intensive quantity, its rate of change in a control volume is governed by the Reynolds transport theorem. If there are no chemical reactions (molecules do not change), the generation term for species will be zero, like that for mass. Species, such as mass and momentum, can be transported across the surface boundary by convection based on $\mathbf{u}$ (which represents the mean motion of molecules at a point). However, a species, can also be spread by diffusion through the random motion of molecules. For example, consider the spread of dye in water, where the dye represents an individual species (dye has a distinct chemical structure). If a drop of concentrated dye is released at one point in a container of still water $(\mathbf{u} = 0)$, the dye will slowly spread until its concentration is uniform and diluted throughout the container. Its spread is due to molecular diffusivity of one species (in this case, dye molecules) within a fluid (in this case, water molecules), where the mass diffusivity is denoted $\Theta_{\mathcal{M}}$.

The molecular flux of a species in a miscible fluid can be described by Fick's law. This law states that the molecular flux per unit area for a species ($\mathbf{Q}_{\mathcal{M}}$) is the product of the molecular diffusivity ($\Theta_{\mathcal{M}}$) and the magnitude of the gradient of the species mass concentration ($\rho_{\mathcal{M}}$):

$$\mathbf{Q}_{\mathcal{M}} = -\Theta_{\mathcal{M}} \nabla \rho_{\mathcal{M}} = -\Theta_{\mathcal{M}} \nabla (\rho_f \mathcal{M}). \tag{2.32}$$

The negative sign on the RHS indicates that a species will flux in the direction from high-concentration to low-concentration regions. Note that $\Theta_{\mathcal{M}}$ is the diffusivity of species of mass fraction $\mathcal{M}$ relative to the overall fluid mixture, and thus is defined by both the species being considered and that of the surrounding fluid. For example, the diffusivity of oxygen in air is different from that in helium.

Applying the preceding diffusive flux to the RTT of (2.3) yields the Eulerian molecular species transport PDE, which is expressed as follows in terms of the species density and then in terms of the species mass fraction (both for the case where fluid density and molecular diffusivity are constant):

$$\frac{\partial \rho_{\mathcal{M}}}{\partial t} + \nabla \cdot (\rho_{\mathcal{M}} \mathbf{u}) = \nabla \cdot [\Theta_{\mathcal{M}} \nabla \rho_{\mathcal{M}}]. \tag{2.33a}$$

$$\frac{\partial \mathcal{M}}{\partial t} + \nabla \cdot (\mathcal{M} \mathbf{u}) = \Theta_{\mathcal{M}} \nabla^2 \mathcal{M} \qquad \textit{for constant } \rho_f \textit{ and } \Theta_{\mathcal{M}}. \tag{2.33b}$$

Note that the RHS of both equations includes no generation terms, since we have assumed no chemical reactions.

The PDEs of (2.33) state that the time rate of changes of species (the first term on LHS) stems from convective flux (the second term on LHS) and molecular diffusion (the RHS term). To understand the physics better, one may consider two limiting cases: (a) no time dependence and (b) no convection. For these two limits, (2.33b) becomes

$$\mathbf{u} = \Theta_{\mathcal{M}} \left(\frac{\nabla \mathcal{M}}{\mathcal{M}} \right) \qquad \textit{for steady local concentration field}. \tag{2.34a}$$

$$\frac{\partial \mathcal{M}}{\partial t} = \Theta_{\mathcal{M}} \nabla^2 \mathcal{M} \qquad \textit{for diffusion with no convection}. \tag{2.34b}$$

These two PDEs are considered in the following discussion.

For the steady condition with finite concentration gradients, the transport of species by convection in one direction must be balanced by diffusion in the other direction. Therefore, the effective diffusion speed is the RHS of (2.34a). This speed is based on the product of the species diffusivity ($\Theta_{\mathcal{M}}$, with units of length^2/time) and the normalized gradient (the term in in parentheses, with units of length^{-1}). Therefore, diffusion speed increases with the normalized concentration gradient, which scales inversely with the mixing layer thickness. For example, water vapor has a diffusivity of about 0.26 cm^2/s in air, so a gradient region of about 2 cm in thickness yields a local diffusion speed of about 0.13 cm/s.

For the zero-convection limit, (2.34b) indicates that the diffusion rate will be fastest when the Laplacian of the concentration is highest. For example, Figure 2.4a shows a

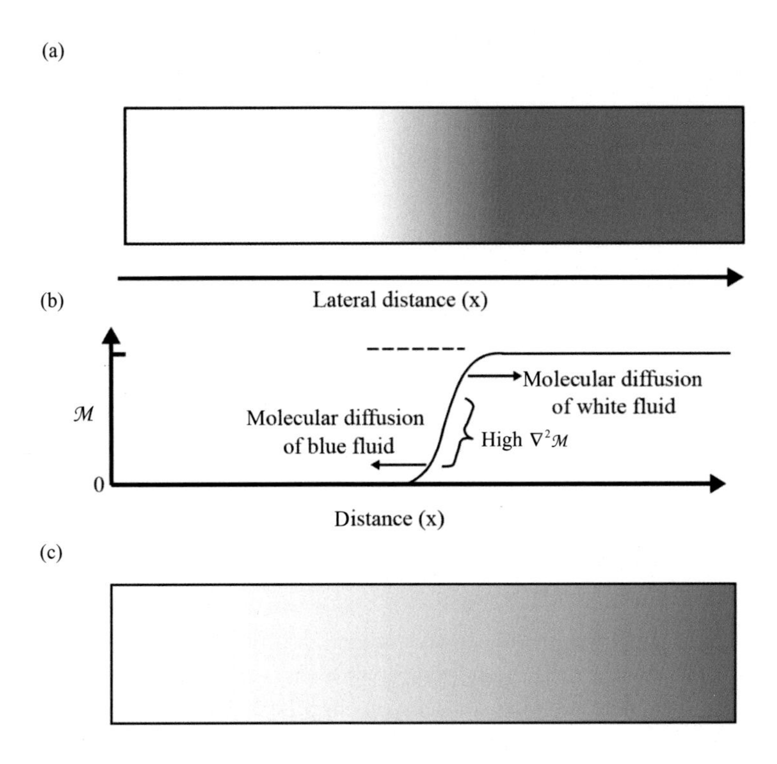

Figure 2.4 Mixing of fluids: (a) a white fluid on the left and a blue fluid on the right, showing an intermediate region where the two fluids are mixed; (b) instantaneous mass concentration of the blue fluid ($\mathcal{M}$), which will diffuse the fastest where the gradient is steepest and the diffusion direction will be the opposite direction of its gradient (blue fluid diffuses to the left); and (c) resulting mixing of both fluids at a later time due to molecular diffusion.

species (indicated by the blue color) that has a higher concentration in the right half of a container with still fluid. There will be a high gradient of this species at the mixing region (Figure 2.4b), causing the species to spread to the left, where the concentration is lower (consistent with negative RHS of 2.32). Over time, this spreading reduces the concentration gradients (Figure 2.4c), so the diffusion rate will lessen. Eventually, the species concentration will be uniformly distributed throughout the container. In this case, there are no more concentration gradients, so this uniform concentration will be the steady-state condition.

For most gas species, the species diffusivity is of the order of $10^{-5} - 10^{-4}\ \mathrm{m}^2/\mathrm{s}$. For most liquids, the species diffusivity is of the order 10^{-10} to $10^{-8}\ \mathrm{m}^2/\mathrm{s}$, which is many times smaller than that for a gas. Thus, the scent of a flower diffuses in a room of still air much faster than dye moves in a still pool. An increase in temperature will increase $\Theta_{\mathcal{M}}$ for both gases and liquids since temperature increases the random motion of molecules. Therefore, a fragrance diffuses faster in warm air. Molecular diffusivity is generally measured and empirically defined (like that for viscosity and thermal conductivity) at a reference temperature. For example, Table A.1 gives the NTP diffusivity of methane in air and of ethanol in water.

To compare the spread rates of species to that for momentum, one may use the Schmidt number, which is defined as the momentum diffusivity normalized by the mass diffusivity:

$$\mathrm{Sc} \equiv \frac{\text{momentum diffusivity}}{\text{mass diffusivity}} \equiv \frac{\mu_f}{\rho_f \Theta_{\mathcal{M}}}. \tag{2.35}$$

For air and most gases, the Schmidt number is of order unity, indicating that the mass and momentum diffusion rates are similar. However, water and most liquids have small diffusivities (as noted earlier), which leads to high Schmidt numbers (e.g., on the order of 10^3). Mass diffusion is therefore much slower than momentum diffusion for a liquid. Thus, a small region of dye in still water will take much longer to spread out (to uniform concentration) compared to the time it takes for a small vortex in water to dissipate its velocity (to a still fluid).

2.2 Thermodynamic Closure and Flow Compressibility

In order to apply the previous conservation equations (for mass, momentum, energy, and species) when density and temperature vary, relations are needed to determine how these changes are related to changes in pressure and fluid energy. The relations are needed to ensure a closed set of equations for solution, and thus can be considered as the thermodynamic "closure" relations. These relationships are discussed in the following for both gases and liquids, followed by a discussion of flow compressibility's impact on velocity.

2.2.1 Thermodynamic Closures

The thermodynamic closure between pressure and density is often termed the *equation of state*. For gases, changes in pressure generally lead to significant changes in density. However, liquid density changes with pressure are often so weak that these changes are generally ignored. To determine when this is a reasonable assumption, one may consider a generalized equation of state that can apply to both liquid and gas compressibility via the empirical Tait equation:

$$\left(\frac{\rho}{\rho_{ref}}\right)^{\mathcal{K}} = \frac{p + \mathcal{B}\, p_{ref}}{(1+\mathcal{B})p_{ref}} \quad \textit{Tait equation of state.} \tag{2.36}$$

This equation employs a compressibility exponent ($\mathcal{K}$), a pressure factor ($\mathcal{B}$), and reference values for density and pressure (ρ_{ref} and p_{ref}) at NTP conditions (Table A.1).

For liquids at isothermal (constant temperature) conditions, the constants $\mathcal{B}$ and $\mathcal{K}$ are much greater than unity; for example, values for water are $\mathcal{B}_\ell = 2{,}955$ and $\mathcal{K}_\ell = 7.15$ (Thompson, 1972, p. 289). These high values indicate that liquid density is a very weak function of pressure. For example, a 1% increase in water density at NTP conditions requires 200 atmospheres of pressure! Thus, water and most liquids are generally considered to be incompressible, so the density can be assumed to be constant.

For gases, there is no pressure factor $\left(\mathcal{B}_g = 0\right)$, and the pressure–density relationship via $\mathcal{K}$ depends on the thermal characteristics of the process. This relationship has two important thermodynamic limits: *isothermal* (constant temperature) and *isentropic* (constant entropy). These limits lead to different values of $\mathcal{K}$, as discussed in the following.

In an isothermal process, the ideal gas assumption (1.24) combined with constant temperature yields a simple linear relationship between pressure and density. The relationship can be expressed in terms of initial values for density and pressure (ρ_o and p_o) as follows:

$$\frac{\rho}{\rho_o} = \frac{p}{p_o} \quad \textit{for gas in an isothermal process.} \tag{2.37}$$

Note that an isothermal process requires either work or external energy transfer to keep the temperature fixed if there is a volume change. For example, isothermal compression can be obtained if the process is very slow, so that high heat conduction to the surroundings ensures a constant internal temperature. Such an isothermal process requires an increase in fluid entropy.

For an isentropic process, it is important to first define the relationships with respect to temperature for changes in fluid internal energy (e) in terms of the specific heat at constant volume $(c_{\forall})$ and for changes in fluid enthalpy $(\hbar = e + p/\rho)$ in terms of the specific heat at constant pressure $\left(c_p\right)$:

$$c_{\forall} \equiv \left.\frac{\partial e}{\partial T}\right|_{\forall=\text{const.}} \tag{2.38a}$$

$$c_p \equiv \left.\frac{\partial \hbar}{\partial T}\right|_{p=\text{const.}} = \left.\frac{\partial (e + p/\rho)}{\partial T}\right|_{p=\text{const.}} \tag{2.38b}$$

The ratio of these two specific heats (γ) is defined as follows:

$$\gamma \equiv \frac{c_p}{c_{\forall}}. \tag{2.39}$$

These relationships can be simplified by defining a "calorically perfect" gas as an ideal gas for which c_p and $c_{\forall}$, and thus γ are independent of temperature. The calorically perfect gas assumption allows a linear relationship between energy and temperature. For a gas, γ is related to the molecular degrees of freedom and for moderate temperatures equals 5/3 for a monatomic gas (like helium), and 7/2 for a diatomic gas (such as oxygen or nitrogen). The latter value $(\gamma = 1.4)$ is generally also used to describe the specific heat ratio for air at moderate temperatures.

To see how specific heats impact gas properties due to a temperature change, consider gas enclosed in a piston chamber, as shown in Figure 2.5. If the piston is locked in place (constant volume), an addition of heat will cause a gas temperature rise that is proportional to $c_{\forall}$, per (2.38a). If the experiment is instead conducted with a

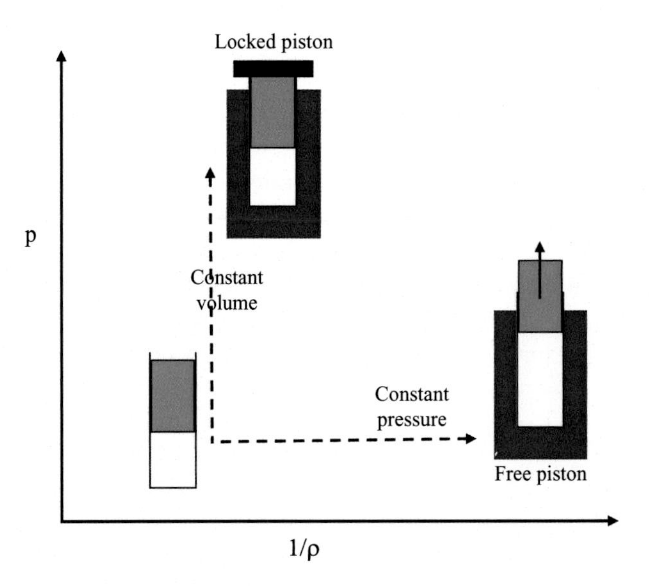

Figure 2.5 Pressure dependence on specific volume (inverse of density) for gas in a piston chamber based on two different heat-addition processes: a constant volume (locked piston) process and a constant pressure (free piston) process. The free piston requires γ-fold heat addition to have the same temperature rise as the locked piston.

free piston (where pressure is constant, so volume expands), an addition of heat causes a gas temperature rise that is proportional to c_p, per (2.38b). If both processes achieve the same temperature rise, the heat addition required for the free piston is γ times larger than that for the locked piston. Conversely, if both processes employ the same amount of heat addition, the temperature rise for the locked piston is γ times larger than that for the free piston.

If there is no energy transfer nor energy lost due to viscous dissipation, the process is reversible since the entropy is constant. This is referred to as an isentropic process. For a calorically perfect gas with isentropic conditions, the specific heat ratio can be used to relate the changes in pressure, density, and temperature (White, 2016), as follows:

$$\frac{p_g}{p_{o,g}} = \left(\frac{\rho_g}{\rho_{o,g}}\right)^{\gamma} = \left(\frac{T_g}{T_{o,g}}\right)^{\frac{\gamma}{\gamma-1}} \quad \textit{for gas in an isentropic process.} \tag{2.40}$$

To see the influence of heat transfer on pressure variations for a volume change, consider piston compression of a gas as shown in Figure 2.6. If the piston is insulated to ensure no heat transfer, the process is *adiabatic*. Thus compressing the chamber will cause an increase in density, temperature, and pressure. If this adiabatic process also has negligible energy losses due to gas viscosity, it will be an isentropic process and the pressure will rise with density per (2.40). If, on the other hand, the temperature is to be held fixed (isothermal), heat must flow outside of the control volume through the surface, such as by a cooling water jacket. In this case, the gas pressure rises linearly with density increase per (2.37).

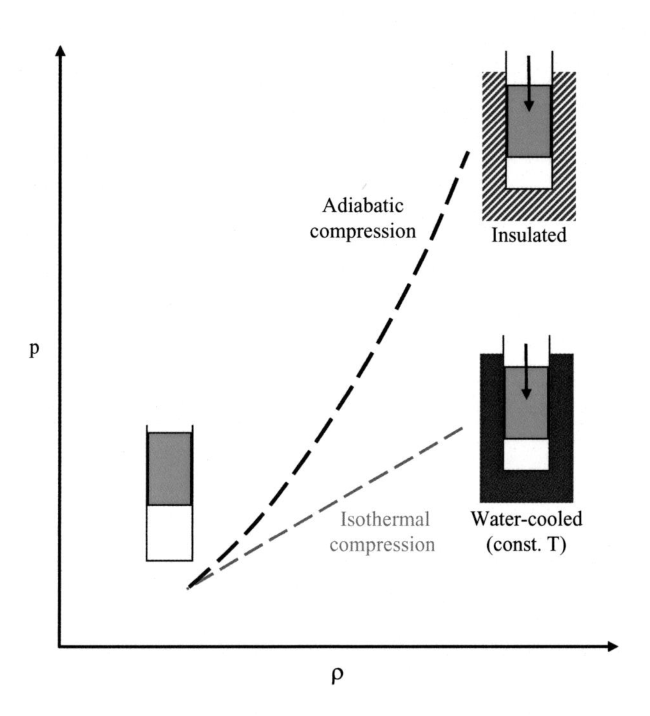

Figure 2.6 Pressure–density relationship for gas in a piston chamber for two compression processes with the same volume (density change) but different heat transfer conditions: adiabatic (no heat transfer through surface) and isothermal (no change in temperature).

One may also compare the pressure–density relationship for a gas compared to that for a liquid. If one converts (2.37) or (2.40) to the form of (2.36), the isothermal process for a gas is consistent with $\mathcal{K}_g = 0$ and $\mathcal{B}_g = 0$, while the isentropic process for a gas is consistent with $\mathcal{K}_g = \gamma$ and $\mathcal{B}_g = 0$. Thus, the Tait equation can be thought of as a generalized relationship that can apply to an isothermal gas, an isentropic gas, or a liquid for different limiting cases. The difference in the pressure–density relationship for an isentropic gas and for a liquid is qualitatively shown by the log–log plot of Figure 2.7. This shows that liquid densities are much higher than gas densities, but the log–log slope is also much higher ($\mathcal{K}_\ell = 7.15$ for water versus for $\mathcal{K}_g = 1.4$ air), since liquids are far less compressible.

Another important state relationship is between energy and velocity. For a calorically perfect gas, (2.38a) provides a linear relationship between temperature and fluid energy ($e = c_\forall T$) such that the total energy of (2.23) can be expressed as follows:

$$e_{tot} = c_\forall T + \frac{1}{2}u^2 \qquad \textit{for a calorically perfect gas.} \qquad (2.41)$$

For a system with no heat transfer nor mechanical work applied to the fluid, the total energy (LHS) will be constant. In this case, the RHS is also constant so an increase in gas velocity will result in a decrease in temperature. These offsetting changes represent an increase in the mean motion of the molecules (related to velocity)

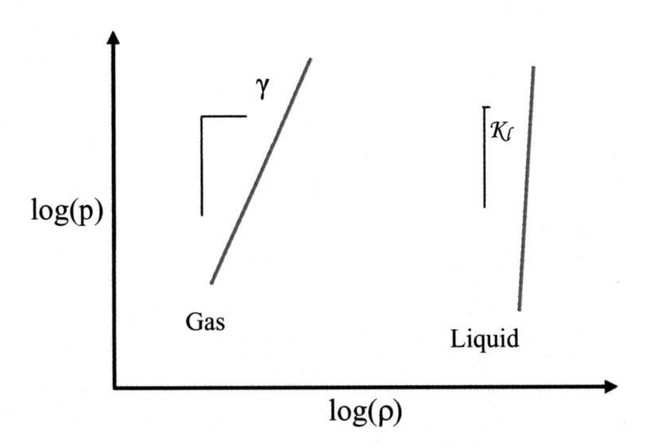

Figure 2.7 Pressure–density relationship for a liquid and an isentropic perfect gas (no work applied nor heat added).

at the expense of the random molecular motion (related to temperature). If the flow is also isentropic (frictionless), then this temperature decrease will be associated with pressure and density decreases based on (2.40).

2.2.2 Speed of Sound and Mach Number

The preceding relationships can also be used to define the acoustic speed of a fluid (a), also known as the speed of sound. This is the speed of pressure waves caused by a small isentropic density disturbance in an otherwise still fluid. As such, it can be expressed as follows:

$$a^2 \equiv \left.\frac{\partial p}{\partial \rho}\right|_{\text{const.entropy}} = \mathcal{K}\frac{(p+\mathcal{B}p_o)}{\rho}. \tag{2.42}$$

The RHS of (2.42) is obtained by assuming a calorically perfect fluid whose specific heats are constant. For a liquid, the speed of sound (a_ℓ) is therefore given by the following:

$$a_\ell = \sqrt{\mathcal{K}_\ell(p+\mathcal{B}_\ell p_o)/\rho_\ell}. \tag{2.43}$$

For a gas in an isentropic process, the combination of (2.40) and (2.42) yields $\mathcal{B}_g = 0$ and $\mathcal{K}_g = \gamma$ so that the speed of sound is simply

$$a_g = \sqrt{\gamma \mathcal{R}_g T_g}. \tag{2.44}$$

The speed of sound for example fluids at NTP are given in Table A.1. The large $\mathcal{B}$ and $\mathcal{K}$ values for a liquid yield a high sound speed, such as 1,482 m/s for water. In comparison, the speed of sound for a gas is much lower, such as 343 m/s for air. The sound waves move in all directions at a speed relative to the local fluid velocity.

To determine the relative significance of acoustics and convection, the Mach number is defined as the ratio of flow velocity magnitude to the speed of sound:

$$M \equiv \frac{\text{convection speed}}{\text{acoustic speed}} \equiv \frac{u}{a}. \tag{2.45}$$

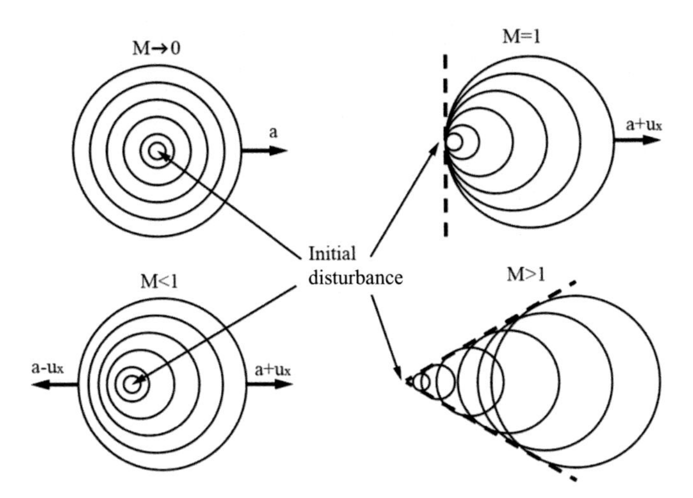

Figure 2.8 Acoustic waves as the Mach number changes from incompressible flow ($M \rightarrow 0$), subsonic flow ($M < 1$), sonic flow ($M = 1$), and supersonic flow ($M > 1$).

Since acoustic waves move at a speed relative to the flow convection, the Mach number gives rise to four different regimes as shown in Figure 2.8 and as listed in the following:

- $M \rightarrow 0$, incompressible flow, so waves travel at the same speed in all directions
- $M < 1$, subsonic flow, so waves travel faster in the downstream direction
- $M \sim 1$, transonic flow, so waves only travel in the downstream direction
- $M > 1$, supersonic flow, so waves travel only downstream in a Mach cone

As an analogy, consider a bridge over a river from which someone drops pebbles and looks at the waves relative to the bridge. If the river is moving very slowly, the waves travel in all directions. However, a faster river limits the upstream wave propagation speed seen from the bridge. In fact, no waves will travel upstream if the current if the river current is fast enough. The subsonic case can also be considered via acoustics, where the change in propagation speed gives rise to the Doppler effect for the sound of a train passing by a stationary listener. The sound frequency is higher as the train approaches, but then will be lower after it passes and is moving away.

To quantify how density varies with the Mach number, one may consider the one-dimensional versions of the momentum and energy PDEs (2.13 and 2.26) along a variable-area stream tube for isentropic flow. Based on (2.40) and (2.41), the pressure, density, and temperature all rise as the velocity slows down and these properties reach their maximum values when the flow stagnates. These values are referred to as the stagnation temperature, density, and pressure (T_{stag}, ρ_{stag}, and p_{stag}). The stagnation values may be combined with the Mach number definition to relate the more general values for a moving flow (Liepmann and Roshko, 1957) as follows:

$$\frac{T_g}{T_{stag,g}} = \left(\frac{\rho_g}{\rho_{stag,g}}\right)^{\gamma-1} = \left(\frac{p_g}{p_{stag,g}}\right)^{\frac{\gamma-1}{\gamma}} = \left(1 + \frac{\gamma-1}{2}M^2\right)^{-1} \quad \textit{if isentropic.} \tag{2.46}$$

This relationship can also be used to determine when it is reasonable to assume the density is approximately constant with respect to velocity changes. For example, an isentropic acceleration from stagnation conditions to $M = 0.2$ yields a density reduction of 2% for $\gamma = 1.4$. As such, flows with $M < 0.2$ are widely considered to be incompressible. Note that $M < 0.2$ corresponds to flow speeds less than 68 m/s in air and less than 296 m/s in water. As such, many gas flows and the vast majority of liquid flows can be considered incompressible.

2.3 Incompressible Flow Characteristics

As discussed in §2.2, $M \to 0$ corresponds to incompressible flow, indicating changes in the flow velocity will not affect the fluid density along its path. If all other mechanisms that can cause changes in density (including combustion, heat transfer, and mechanical compression) are also negligible, then the density for a fluid element stays constant along its path, as follows:

$$\frac{\mathcal{D}\rho_f}{\mathcal{D}t} = 0 \quad \textit{for incompressible flow.} \tag{2.47}$$

However, constant density along a fluid path does not mean density is constant everywhere since a flow may be "stratified" with different densities. For example, atmospheric flows are generally stratified since higher altitudes tend to correspond to lower densities. Similarly, ocean flows can be stratified whereby regions of increased salt concentration are denser than those with less salinity. If stratification can also be neglected, the fluid density can be considered constant and uniform throughout, as follows:

$$\rho_f = \text{const.} \qquad \textit{uniform density flow.} \tag{2.48}$$

Such uniform density conditions significantly simplify flow analysis. In the following, we will consider three field characteristics for a constant density flow: stream function, vorticity, and velocity potential.

2.3.1 Stream Function

Flow streamlines about a body are an intuitive means of understanding a flow field. In two-dimensional flows, these streamlines can be quantified through the stream function (ψ) for incompressible flow. For such a flow, the velocity field has zero divergence (per 2.8), which be expressed in Cartesian or spherical coordinates (per 1.10a or 1.11a), as follows:

$$\nabla \cdot \mathbf{u} = \frac{\partial u_x}{\partial x} + \frac{\partial u_y}{\partial y} + \frac{\partial u_z}{\partial z} = 0. \tag{2.49a}$$

$$\nabla \cdot \mathbf{u} = \frac{1}{r^2} \frac{\partial (r^2 u_r)}{\partial r} + \frac{1}{r \sin\theta} \left[\frac{\partial (u_\theta \sin\theta)}{\partial \theta} + \frac{\partial (u_\phi)}{\partial \phi} \right] = 0. \tag{2.49b}$$

If the flow is further limited to be two-dimensional in Cartesian coordinates ($u_z = 0$ and no gradient in the z-direction) or axisymmetric in spherical coordinates ($u_\phi = 0$ and no gradients in the ϕ-direction), the respective stream functions can be defined such that they satisfy the incompressibility condition:

$$\frac{\partial \psi}{\partial y} \equiv u_x \quad \textit{and} \quad \frac{\partial \psi}{\partial x} \equiv -u_y \quad \textit{for } (x, y) \textit{ Cartesian flow}. \tag{2.50a}$$

$$\frac{\partial \psi}{\partial r} \equiv -r \sin\theta \, u_\theta \quad \textit{and} \quad \frac{\partial \psi}{\partial \theta} \equiv r^2 \sin\theta \, u_r \quad \textit{for } (r, \theta) \textit{ spherical flow}. \tag{2.50b}$$

In particular, one may show that the stream function automatically satisfies incompressibility by substituting (2.50a) or (2.50b) into (2.49a) or (2.49b).

Once obtained, the stream function field can help describe the local flow characteristics. For example, lines of constant ψ help describe the instantaneous flow direction since these "streamlines" are everywhere parallel to the velocity. Furthermore, for steady flow, the volumetric flow between a pair of streamlines becomes constant. This means that the product of the velocity and the cross-sectional area is also constant, so a reduction in the gap between two streamlines indicates a local increase in velocity. Therefore, the flow streamlines give an indication of the velocity direction as well as the changes in velocity magnitude.

2.3.2 Vorticity and Irrotational Flow

While the stream function characterizes the direction of a fluid element relative to a fixed coordinate system, vorticity characterizes the rotationality of a fluid element about itself. In particular, vorticity ($\boldsymbol{\omega}$) is defined as the curl of the velocity field. In the following, we consider Cartesian and cylindrical forms of the vorticity. The vorticity vector for a Cartesian coordinate system and the associated vorticity magnitude are given as follows:

$$\boldsymbol{\omega} \equiv \nabla \times \mathbf{u} = \left(\frac{\partial u_z}{\partial y} - \frac{\partial u_y}{\partial z} \right) \mathbf{i}_x + \left(\frac{\partial u_x}{\partial z} - \frac{\partial u_z}{\partial x} \right) \mathbf{i}_y + \left(\frac{\partial u_y}{\partial x} - \frac{\partial u_x}{\partial y} \right) \mathbf{i}_z. \tag{2.51a}$$

$$\omega = \sqrt{\omega_x^2 + \omega_y^2 + \omega_z^2} \tag{2.51b}$$

The Cartesian vorticity components in (2.51b) are the three terms in parentheses in (2.51a), and the magnitude of the vorticity is a scalar field that shows the degree of rotationality throughout the flow.

The vorticity for a cylindrical coordinate system with the axial symmetry ($u_z = 0$ and no changes along the z-axis) is given as follows:

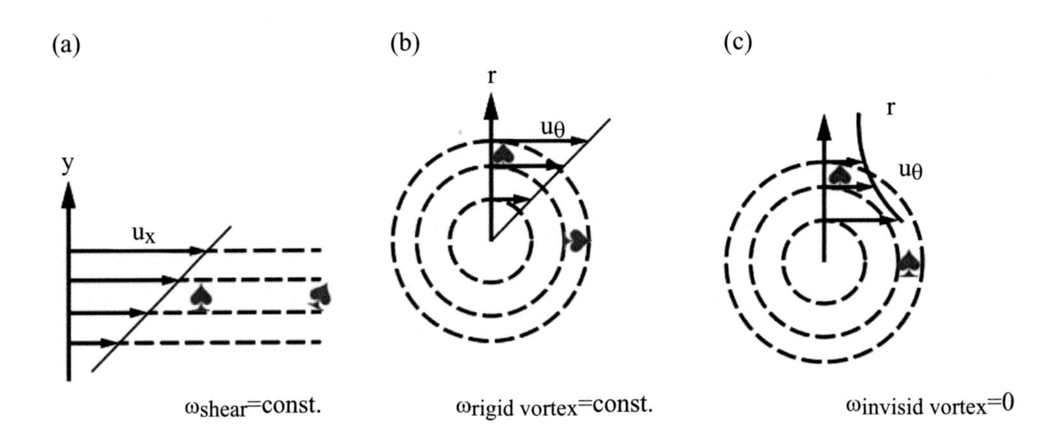

Figure 2.9 Schematic of simple flows, with dashed lines indicating streamlines and arrows indicating local velocity vectors for (a) linear shear flow, (b) rigid vortex, and (c) inviscid vortex. The flows include a marker (♠) that only rotates along a fluid path when there is vorticity, and thus does not rotate for the inviscid vortex.

$$\boldsymbol{\omega} \equiv \nabla \times \mathbf{u} = \frac{1}{r}\left(\frac{\partial(ru_\theta)}{\partial r} - \frac{\partial u_r}{\partial \theta}\right)\mathbf{i}_z \qquad \textit{for axial cylindrical symmetry.} \tag{2.52}$$

In this simplified flow, ω_z is the only vorticity component. A flow with finite (nonzero) vorticity for any of its components is considered a *rotational* flow.

Based on the two preceding forms in Cartesian and cylindrical coordinates, one may define two fundamental flows of uniform vorticity: the *linear shear* and the *rigid vortex* flows, which have only one velocity component. The linear shear flow has parallel streamlines (in one direction), but the velocity magnitude changes linearly in a direction normal to these streamlines. For example, Figure 2.9a shows a flow in the x-direction, where the speed (u_x) varies linearly with y. In this case, $\boldsymbol{\omega} = \omega_z \mathbf{i}_z$, and the vorticity magnitude describes the flow shear:

$$\omega_{shear} = \omega_z = \left|\frac{\partial u_x}{\partial y}\right| = \text{const.} \qquad \textit{linear shear flow.} \tag{2.53}$$

Such a shear flow can occur when a fluid moves along a wall where it must come to rest at the wall due to the no-slip condition (e.g., Figure 2.3) or when two fluids move in parallel but at different velocities (e.g., where a high-speed jet enters into a lower-speed surrounding flow).

The rigid vortex flow has cylindrical streamlines with a tangential velocity (u_θ) that increases linearly with radius (r), as shown in Figure 2.9b. In this case, the vorticity is given by the following:

$$\omega_{vortex} = \left|\frac{2u_\theta}{r}\right| = \text{const.} \qquad \textit{rigid vortex flow.} \tag{2.54}$$

This result shows that vorticity is equal to twice the fluid angular velocity. This is called a rigid vortex flow since it acts as if it is in solid body rotation (where viscosity is effectively infinite).

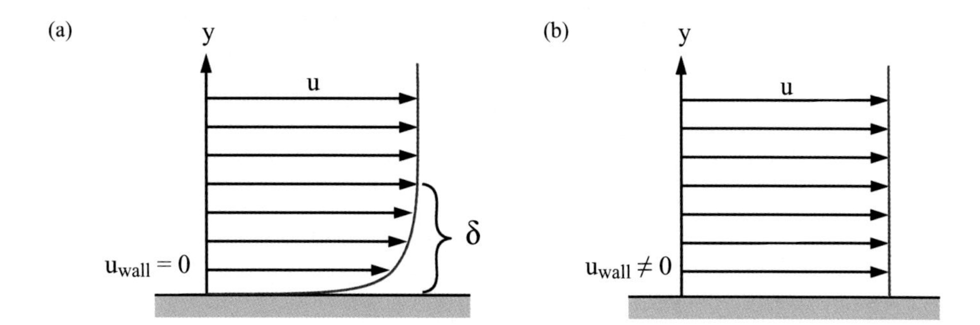

Figure 2.10 Simple velocity fields for flow past a stationary wall showing (a) no-slip conditions where viscosity effects cause the velocity to be zero at the wall ($u = 0$ at $y = 0$) resulting in a boundary layer thickness (δ); and (b) slip conditions where viscosity effects are ignored so that only the normal component of velocity is zero at the wall.

In contrast to the flows of (2.53) and (2.54), a flow with no vorticity is termed *irrotational.* An example of an irrotational cylindrical flow is the inviscid vortex given by the following:

$$u_\theta = \frac{\text{const.}}{r} \qquad \textit{inviscid vortex flow } (\omega = 0). \tag{2.55}$$

As shown in Figure 2.9c, the flow streamlines still circulate about the origin, but a fluid element in this flow does not rotate about itself (as shown by the marker in the figure). Therefore, this flow is irrotational.

In reality, most flows will have some portions that are rotational and other portions that are irrotational. For example, a circular vortex caused by a drain in a sink (or in the eye of a hurricane) will generally have a rotational inner core that follows (2.54) beyond which the vorticity will weaken and can eventually transition to an inviscid outer region that follows (2.55). Similarly, the two-dimensional flow over a wall can have a nearly linear shear near the wall per (2.53), but far from the wall the flow streamlines may be approximately dictated by irrotational flow. For example, the flow in Figure 2.10a shows a viscous rotational region of thickness δ beyond which $(y > \delta)$ the flow is approximately irrotational.

For a two-dimensional flow in the x-y plane, where ω_z is the only vorticity component, one may take the spatial derivatives of (2.50a) and substitute them into (2.51a). The result is a Poisson equation for the stream function and the vorticity given as follows:

$$\nabla^2 \psi = -\omega_z. \tag{2.56}$$

This result shows that the stream function can be used to determine the vorticity field (or vice versa) for a two-dimensional flow. However, a flow that is irrotational throughout allows additional characterization, as discussed in the following section.

2.3.3 Velocity Potential and Superposition

A flow that is irrotational $(\boldsymbol{\omega} = 0)$ gives rise to another flow field variable: the velocity potential, Φ. This field arises for zero vorticity since it corresponds to $\nabla \times \mathbf{u} = \mathbf{0}$ (per

2.51), and since $\nabla \times (\nabla q) = \mathbf{0}$ for any scalar field q, the velocity potential can be set as q, as follows:

$$\nabla \Phi \equiv \mathbf{u}. \tag{2.57}$$

A flow for which Φ can be defined in this way is called a *potential flow*. While this requires the flow to be irrotational, it does not require the flow to be incompressible. However, if incompressibility is additionally assumed, this combination yields the following:

$$\nabla \cdot \mathbf{u} = \mathbf{0} \quad \text{for irrotational incompressible flow} \tag{2.58a}$$

$$\nabla \times \mathbf{u} = \boldsymbol{\omega} = \mathbf{0} \tag{2.58b}$$

If one combines (2.57) and (2.58a), it can be shown that the velocity potential satisfies a Laplacian PDE, and if one combines (2.56) and (2.58b) it can be similarly shown that the two-dimensional stream function also satisfies a Laplacian PDE:

$$\nabla^2 \psi = 0 \quad \text{for irrotational incompressible flow} \tag{2.59a}$$

$$\nabla^2 \Phi = 0 \tag{2.59b}$$

A key advantage of the Laplacian forms of (2.59) is that linear superposition can be used for stream functions and for velocity potentials. For example, the combination of three stream functions results in a new stream function that also satisfied (2.59):

$$\nabla^2(\psi_1 + \psi_2 + \psi_3) = \nabla^2\psi_1 + \nabla^2\psi_2 + \nabla^2\psi_3 = 0 = \nabla^2\psi. \tag{2.60}$$

A similar expression can be given for the combination of the velocity potentials and of the associated velocities. As a result, the combined stream function, velocity potential, and velocity field are the linear sums of the individual fields, as follows:

$$\psi = \psi_1 + \psi_2 + \psi_3 \tag{2.61a}$$

$$\Phi = \Phi_1 + \Phi_2 + \Phi_3 \quad \text{superposition of irrotational incompressible flows.} \tag{2.61b}$$

$$\mathbf{u} = \mathbf{u}_1 + \mathbf{u}_2 + \mathbf{u}_3 \tag{2.61c}$$

This superposition principle allows one to create a wide range of flow fields by combining fields of simple flows.

2.3.4 No-Slip versus Slip Boundary Conditions

As mentioned previously, flows often have rotational (viscous) and irrotational (inviscid) regions. For a stationary wall, viscosity results in a no-slip boundary condition, so the fluid velocity at the wall will be zero (per 1.13), which causes a velocity shear as shown in Figure 2.10a. However, inviscid flow allows a slip boundary condition on the surface so only the normal component will be zero (the tangential component is free), as shown in Figure 2.10b. These two boundary conditions can be expressed as follows:

$$\mathbf{u} = 0 \qquad \textit{no-slip condition on a stationary wall.} \tag{2.62a}$$

$$\mathbf{u}\cdot\mathbf{n} = 0 \quad \textit{slip condition on a stationary wall.} \tag{2.62b}$$

Often, one may employ the slip boundary condition and the irrotational PDEs to help describe the flow far from the wall, whereas the no-slip boundary condition and the irrotational PDEs are needed to describe the flow near the wall and to describe the wall shear stress. These two limits of inviscid flow and viscous flow are considered in the next two sections.

2.4 Inviscid Incompressible Flow and Froude Number

2.4.1 Inviscid Irrotational Bernoulli Equations

If one can neglect the viscous effects, then the flow may be assumed to be irrotational. If one further assumes incompressible flow, then the conditions of (2.58) are satisfied. As shown in the following, the result allows us to directly relate pressure and velocity fields through the fluid momentum equation. To show this, the incompressible momentum transport of (2.21) for no viscosity can be expressed as follows:

$$\frac{\partial \mathbf{u}}{\partial t} + (\mathbf{u}\cdot\nabla)\mathbf{u} = -\frac{1}{\rho_f}\nabla p + \mathbf{g} \qquad \textit{for incompressible inviscid flow.} \tag{2.63}$$

One may use the vector dot product identity to rewrite the second term of this LHS as follows:

$$(\mathbf{u}\cdot\nabla)\mathbf{u} = \frac{1}{2}\nabla(\mathbf{u}\cdot\mathbf{u}) - \mathbf{u}\times(\nabla\times\mathbf{u}) = \frac{1}{2}\nabla u^2 - \mathbf{u}\times\boldsymbol{\omega}. \tag{2.64}$$

For zero vorticity, the second term on this RHS is zero and the remaining term can be substituted into (2.63). If the flow is also steady, (2.63) can be rearranged to express the pressure gradient as the sum of a dynamic component and a gravitational component:

$$\nabla p = -\tfrac{1}{2}\rho_f \nabla u^2 + \rho_f \mathbf{g} = (\nabla p)_{dyn} + (\nabla p)_{grav} \qquad \textit{if also steady.} \tag{2.65}$$

The first term on this RHS is the fluid dynamic pressure gradient (due to changes in velocity), and the second term is the gravitational pressure gradient (due to body forces). If gravity acts downward (in the negative y-direction), this second term can be expressed as follows:

$$(\nabla p)_{grav} \equiv \rho_f \mathbf{g} = -\rho_f g \mathbf{i}_y. \tag{2.66}$$

This term is sometimes called the *hydrostatic effect* because it determines the pressure increase with depth in still water.

Integration of (2.65) using the RHS of (2.66) can therefore be used to yield the "total" pressure field as follows:

$$p_{tot} \equiv p + \tfrac{1}{2}\rho_f u^2 + \rho_f g y = \text{const.} \qquad \textit{steady incompressible inviscid flow.} \tag{2.67}$$

This is the irrotational Bernoulli equation and shows that the total pressure is constant throughout an incompressible irrotational steady flow field (not just along a streamline). This result also allows us to obtain the local flow pressure (p) given the local flow speed (u) and height (y).

For a fixed height, the stagnation pressure is the pressure recovered if the flow is brought to rest, and the corresponding pressure rise for this process is the dynamic pressure (p_{dyn}), as follows:

$$p_{stag} \equiv p + \tfrac{1}{2}\rho_f u^2 = \text{const.} \qquad (2.68a)$$

$$p_{dyn} \equiv \tfrac{1}{2}\rho_f u^2 \qquad \textit{for constant depth} \qquad (2.68b)$$

This relationship shows that the maximum pressure occurs when the flow comes to rest, while the minimum pressure occurs where the local flow speed is highest (note that $p_{stag} = p_{tot}$ at $y = 0$). A similar qualitative relationship between static and stagnation pressures was found for compressible flow as noted by (2.46).

2.4.2 Froude Number

The result of (2.67) can be simplified if the hydrostatic effects are negligible. To determine when this assumption is reasonable, one may compare effects of hydrodynamics to hydrostatics (on the RHS of 2.67) using a dimensionless parameter called the Froude number. To obtain this parameter, we identify the characteristic length in the gravitational direction as D_y and the characteristic speed as u_D, such as the body height and freestream velocity as shown in Figure 2.11a. We then assume that all other characteristic lengths

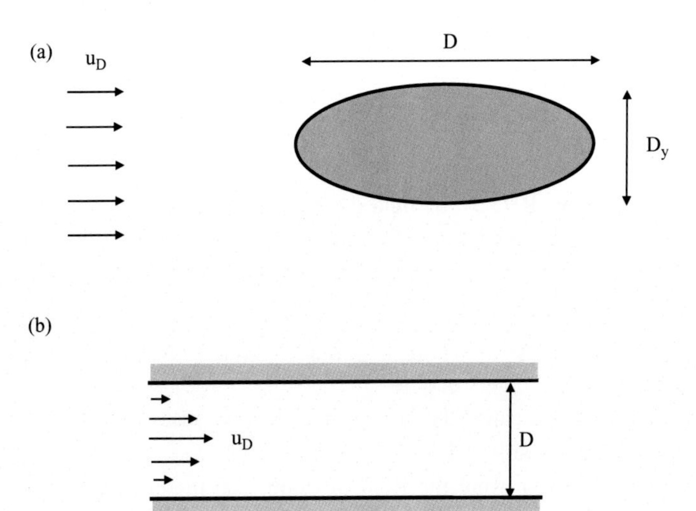

Figure 2.11 Example of characteristic flow scales for velocity and length, for (a) an external uniform flow over a body, which has a length D and a height D_y; and (b) an internal pipe flow that has a mean flow velocity based on u_D and pipe diameter D.

and velocities in the domain are proportional to the characteristic length scale and velocity scale. These two values are then used to normalize the length and velocities in (2.65), while the pressure is normalized with twice the dynamic pressure (2.68b):

$$\mathrm{u}^* = \mathrm{u}/\mathrm{u}_\mathrm{D}. \tag{2.69a}$$

$$\nabla^* = \mathrm{D}_\mathrm{y}\nabla. \tag{2.69b}$$

$$\mathrm{p}^* = \mathrm{p}/\left(\rho_\mathrm{f}\mathrm{u}_\mathrm{D}^2\right). \tag{2.69c}$$

Applying these nondimensional values to the steady-state version of (2.63) yields the following:

$$\nabla^*\mathrm{p}^* = -\tfrac{1}{2}\nabla^*(\mathrm{u}^*)^2 - \mathbf{i}_\mathrm{y}/\mathrm{Fr}_\mathrm{D}. \tag{2.70}$$

The first term on the RHS represents the hydrodynamic effects, while second term represents the hydrostatic effects and is proportional to the inverse of the domain Froude number (Fr_D), which is given as follows:

$$\mathrm{Fr}_\mathrm{D} \equiv \frac{\mathrm{u}_\mathrm{D}^2}{\mathrm{g}\mathrm{D}_\mathrm{y}}. \tag{2.71}$$

Thus, the Froude number represents the ratio of the convective effects (hydrodynamics) to the gravitational effects (hydrostatics).

The Froude number can be used to assess various flow limits. For $\mathrm{Fr}_\mathrm{D} \rightarrow 0$, the velocity effects are negligible so that hydrostatics determine the pressure gradient. For $\mathrm{Fr}_\mathrm{D} \rightarrow \infty$, the effect of hydrostatics is negligible so that the hydrodynamics determine the pressure gradient. Gas flows tend to have a combination of high velocities and smaller scales such that $\mathrm{Fr}_\mathrm{D} > 100$, so hydrostatic effects can be neglected. For liquid flows, the velocities are often smaller, so both effects can be important. For example, a submarine moving at 4 m/s with a body depth of 10 m yields $\mathrm{Fr}_\mathrm{D} = 0.25$, such that the hydrostatics will be roughly four times stronger than hydrodynamics. As another example, a fish at moving at 2 m/s with a body height of 10 cm has $\mathrm{Fr}_\mathrm{D} = 4$, so hydrodynamic effects will instead be roughly four times stronger.

2.5 Viscous Incompressible Flow and Reynolds Number

In this section, we retain our assumption of incompressible flow but reintroduce the effects of rotationality and viscosity in the momentum equation. The relative importance of these viscous effects is then quantified by the nondimensional Reynolds number.

2.5.1 Viscous Incompressible Flow Equations

For incompressible flow with constant viscosity, the continuity and momentum equations are respectively given by the Navier–Stokes equations of (2.8) and (2.21) and are restated as follows:

$$\nabla \cdot \mathbf{u} = 0. \tag{2.72a}$$

$$\rho_f \frac{\partial \mathbf{u}}{\partial t} + \rho_f(\mathbf{u} \cdot \nabla)\mathbf{u} = \rho_f \mathbf{g} - \nabla p + \mu_f \nabla^2 \mathbf{u}. \tag{2.72b}$$

This is a closed set of PDEs since it contains two variables (**u** and p) and two equations. The equations can be expressed in Cartesian tensor form (with a repeated index indicating summation), as follows:

$$\frac{\partial u_i}{\partial X_i} = 0 \qquad \textit{for } i = 1, 2, \textit{and } 3. \tag{2.73a}$$

$$\frac{\partial u_i}{\partial t} + u_j \frac{\partial u_i}{\partial X_j} = g_i - \frac{1}{\rho_f}\frac{\partial p}{\partial X_i} + \frac{\mu_f}{\rho_f}\frac{\partial}{\partial X_j}\left(\frac{\partial u_i}{\partial X_j}\right) \qquad \textit{for } j = 1, 2, \textit{and } 3. \tag{2.73b}$$

The last term on the RHS of (2.73b) suggests the definition of kinematic viscosity:

$$\nu_f \equiv \mu_f / \rho_f. \tag{2.74}$$

The kinematic viscosity is effectively the momentum diffusivity per unit mass (instead of per unit volume as in μ_f) and has metric units of m^2/s.

For spherical coordinates with azimuthal symmetry (2.18d), the incompressible continuity equation, the momentum equations of (2.72) and the spherical Laplacian operator can be written as follows:

$$\frac{1}{r^2}\frac{\partial(r^2 u_r)}{\partial r} + \frac{1}{r \sin\theta}\frac{\partial(u_\theta \sin\theta)}{\partial\theta} = 0. \tag{2.75a}$$

$$\frac{\partial u_r}{\partial t} + u_r \frac{\partial u_r}{\partial r} + \frac{u_\theta}{r}\frac{\partial u_r}{\partial \theta} - \frac{u_\theta^2}{r} = g_r - \frac{1}{\rho_f}\frac{\partial p}{\partial r} + \nu_f\left[\nabla^2 u_r - \frac{2u_r}{r^2} - \frac{2}{r^2}\frac{\partial(u_\theta \sin\theta)}{\sin\theta\, \partial\theta}\right]. \tag{2.75b}$$

$$\frac{\partial u_\theta}{\partial t} + u_r \frac{\partial u_\theta}{\partial r} + \frac{u_\theta}{r}\frac{\partial u_\theta}{\partial \theta} + \frac{u_r u_\theta}{r} = g_\theta - \frac{1}{\rho_f r}\frac{\partial p}{\partial \theta} + \nu_f\left[\nabla^2 u_\theta + \frac{2}{r^2}\frac{\partial u_r}{\partial\theta} - \frac{u_\theta}{r^2 \sin^2\theta}\right]. \tag{2.75c}$$

$$\nabla^2 q = \frac{1}{r^2}\frac{\partial}{\partial r}\left(r^2 \frac{\partial q}{\partial r}\right) + \frac{1}{r^2 \sin\theta}\frac{\partial}{\partial\theta}\left(\sin\theta \frac{\partial q}{\partial\theta}\right). \tag{2.75d}$$

In the momentum equations of (2.75b) and (2.75c), the viscous terms are associated with the square brackets and kinematic viscosity.

2.5.2 Reynolds Number

To identify flow regimes with respect to the influence of viscosity, it is useful to simplify the conditions by neglecting gravitational effects (consistent with $Fr_D \to \infty$) and assuming steady flow. In this limit, the Cartesian momentum equation (2.73b) in vector and tensor form is as follows

$$\rho_f(\mathbf{u}\cdot\nabla)\mathbf{u} = -\nabla p + \mu_f\nabla^2\mathbf{u}. \tag{2.76a}$$

$$u_j\frac{\partial u_i}{\partial X_j} = -\frac{1}{\rho_f}\frac{\partial p}{\partial X_i} + \nu_f\frac{\partial}{\partial X_j}\left(\frac{\partial u_i}{\partial X_j}\right) \qquad \textit{for}\ j = 1, 2, \textit{and}\ 3. \tag{2.76b}$$

The steady-flow assumption is reasonable if the boundary conditions and flow geometry are fixed in time (with no external forcing) and if the effects of viscosity are significant enough to prevent the flow from becoming unstable (as will be discussed later). The resulting PDE of (2.76) includes a convection term (LHS) as well as pressure and viscous terms (RHS). For most flows about a surface, pressure is critical to the surface stresses, regardless of whether the flow is viscous or inviscid. However, the convection and viscous terms are not always as critical. To characterize their relative influence, the momentum PDE can be made nondimensional but done such that pressure terms are always retained.

To make the PDE of (2.76) dimensionless, three characteristic scales are needed. For the first two, we employ a characteristic length scale (D) and a characteristic flow velocity (u_D), as shown in Figure 2.11. As an external flow example, D can be based on body length and u_D on the freestream velocity far upstream of the body. As an internal flow example, D can be based on a pipe diameter and u_D on the area-averaged flow speed through its cross section. The flow density (ρ_f) is used as the third scaling variable (as in 2.69c) in order to normalize pressure. The resulting dimensionless variables then become

$$\mathbf{u}^* = \mathbf{u}/u_D. \tag{2.77a}$$

$$\mathbf{x}^* = \mathbf{x}/D. \tag{2.77b}$$

$$p^* = p/\left(\rho_f u_D^2\right). \tag{2.77c}$$

Applying these dimensionless variables to (2.76a) yields the following:

$$\nabla p^* = -(\mathbf{u}^*\cdot\nabla)\mathbf{u}^* + \frac{1}{Re_D}\nabla^2\mathbf{u}^*. \tag{2.78}$$

The resulting momentum equation includes the domain Reynolds number (Re_D) defined qualitatively and quantitatively as follows:

$$Re_D \equiv \frac{\text{convection effects on pressure}}{\text{viscous effects on pressure}} \equiv \frac{\rho_f D u_D}{\mu_f}. \tag{2.79}$$

The quantitative definition of Re_D results from making the PDE dimensionless, while the qualitative definition stems from comparing the two terms on the RHS (the first is the convection term, and the second is the viscous term) of (2.78). The nondimensional PDE has two important consequences: similarity and characterization.

In terms of similarity, the solution to this dimensionless PDE will be the same for all flows of a given Re_D and domain shape. This flow similitude is often useful when conducting experiments. For example, the nondimensional velocity field obtained for a small model in a wind tunnel will be the same as that for a large model in flight so long as the Re_D and the geometry of the domain (boundaries and body) are the same. As such, the ratio of convection to viscous effects will only be a function of Re_D regardless of any changes in D, u_D, ρ or μ.

In terms of characterization, the PDE also shows that Re_D represents the relative importance of convection and viscous terms, which leads to various Reynolds number regimes, as discussed in the next section.

2.6 Reynolds Number Regimes

To characterize Reynolds number regimes, it is helpful to first define *laminar flow* as dynamically stable viscous flow. As such, a laminar flow has enough viscosity to prevent an unsteady perturbation from causing flow instabilities, so a flow will return to a steady state after it is perturbed. Therefore, laminar flows can only remain unsteady if they are continually driven by unsteady forces. For example, laminar blood flow is only unsteady due to the pulsatile nature of the heart. Since laminar flow is often steady, (2.78) is appropriate. The steady linear shear flow of Figure 2.10a is an example of a laminar flow (which is so named since these flows often have streamlines stacked in layers, i.e., in lamina).

In general, laminar flow is ensured when the viscous effects (relative to the convection effects) are large enough, which occurs when the Reynolds number is small enough ($Re_D \ll 1$). In particular, we define a critical Reynolds number ($Re_{D,crit}$) as the maximum value for laminar flow with a given domain geometry, so the flow will be laminar if $Re_D < Re_{D,crit}$. In the following, we consider several Reynolds number regimes: $Re_D \to 0$, $Re_D \ll 1$, $Re_D \sim 1$, and $1 \ll Re_D < Re_{D,crit}$ (all of which are laminar) as well as $Re_D > Re_{D,crit}$ (which includes transition to turbulence). For the laminar conditions discussed, we will assume there is no unsteady forcing nor effects of initial conditions, so the flow will be steady and (2.76) and (2.78) are generally applicable.

2.6.1 Laminar Flow Regimes

The limit of $Re_D \to 0$ is called *creeping flow*, and such a flow is always laminar. If we apply this limit to the steady dimensionless momentum equation (2.78), the convection terms will be negligible compared to the viscous terms. Retaining the pressure term (since it is important for all Reynolds numbers), the resulting dimensional momentum PDE (assuming there is no unsteady forcing) becomes

$$\nabla p = \mu_f \nabla^2 \mathbf{u} \qquad \textit{for } Re_D \to 0 \textit{ and steady flow.} \tag{2.80}$$

This result indicates that pressure stresses balance the shear stresses at every point in the flow. And because this PDE is linear, it is theoretically tractable for a variety of simple geometries and conditions.

The next regime of $Re_D \ll 1$ occurs when the convection term in (2.78) has a finite but weak effect on the overall flow. In this case, the nonlinear convective term (LHS) can be linearized using the freestream velocity (u_∞) to obtain the following:

$$\rho_f(\mathbf{u}_\infty \cdot \nabla)\mathbf{u} = -\nabla p + \mu_f \nabla^2 \mathbf{u} \qquad \textit{for } Re_D \ll 1 \textit{ and steady flow.} \tag{2.81}$$

This laminar flow regime is called *linearized flow* due to the convection term *linearization*. It has also been called *Oseen flow*, since this linearization was proposed by Oseen (1910) for the theory of flow over surfaces. Since (2.81) is a linear PDE, analytical solutions are available for simple geometries (e.g., parallel flow and small shape changes), and the results are often reasonable for $Re_D < 0.1$. However, obtaining these solutions is often more complicated than for (2.80) due to the added convection term.

For the intermediate condition of $Re_D \sim 1$, both the convective and viscous effects may have a strong influence on the pressure gradient. As such, convection cannot be generally neglected nor linearized, and instead must be retained in the steady momentum equation, as in (2.76), as follows:

$$\rho_f(\mathbf{u} \cdot \nabla)\mathbf{u} = -\nabla p + \mu_f \nabla^2 \mathbf{u} \qquad \textit{for } Re_D \sim 1 \textit{ and steady flow.} \tag{2.82}$$

Some simple geometry cases still allow a theoretical flow solution for this laminar flow regime. In particular, a flow that is unidirectional with velocity gradients only normal to the flow will cause the LHS convection term to be zero, so the solution simply balances pressure stress and viscous stresses. For example, parallel flow between two plates (one fixed and one moving with the flow direction) gives rise to a linear shear flow, as in Figure 2.9a and the vorticity of (2.53). Other examples are the Poiseuille solutions for parallel flow in a circular pipe or between two plates (White, 2016). However, flows that change direction within the domain will cause the LHS convection term to be significant, which makes the PDE nonlinear and generally renders it analytically intractable unless the convection effects are weak.

If one considers the limit of $Re_D \gg 1$, the convection effects may be much stronger than the viscous effects. If one further considers conditions where the flow is stable ($Re_D < Re_{D,crit}$) and is not subjected to unsteady forcing, the flow remains laminar and we retain the same PDE as (2.82):

$$\rho_f(\mathbf{u} \cdot \nabla)\mathbf{u} = -\nabla p + \mu_f \nabla^2 \mathbf{u} \qquad \textit{for } 1 \ll Re_D < Re_{D,crit} \textit{ and steady flow.} \tag{2.83}$$

For laminar flow at very high Reynolds numbers, one is tempted to assume that viscous effects are generally weak since they appear to be negligible when applying $Re_D \to \infty$ for (2.78). However, viscous effects will be generally important for $Re_D \gg 1$ whenever there is a high shear region (e.g., due to the no-slip condition on a stationary surface). In fact, viscous effects will be strong for unidirectional flows with velocity gradients only normal to the flow (e.g., Poiseuille pipe flow) since the convection term will again be zero. Even when the flow is not unidirectional, viscous effects can still be important for $Re_D \gg 1$ but are constrained to small regions where velocity shear is high but then can be negligible everywhere else.

(a)

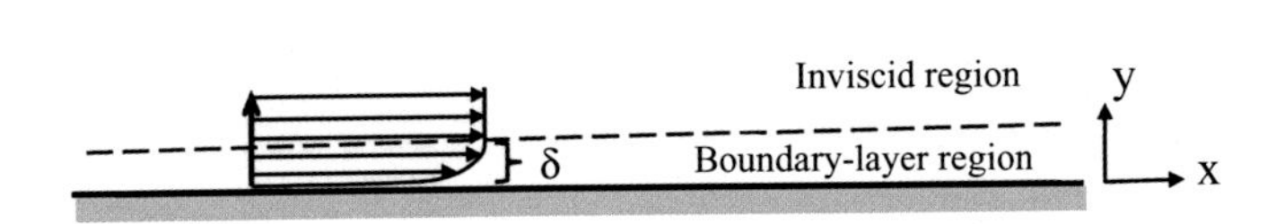

(b)

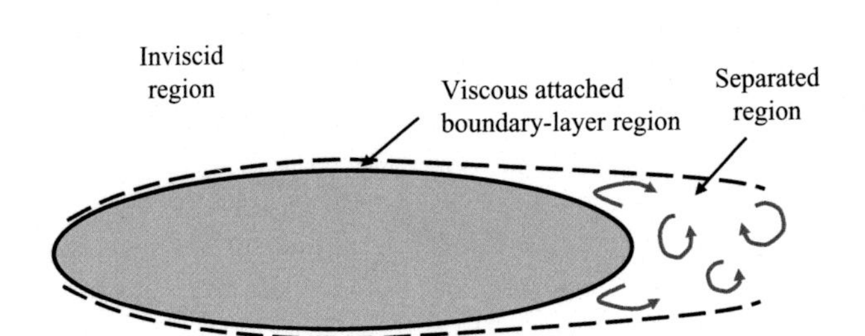

Figure 2.12 Flows with at $Re_D \gg 1$: (a) zoom-in of slowly growing boundary layer of local thickness δ along a flat plate of length D with uniform velocity above the plate; and (b) the boundary layer along an object of length D, where the flow separates of the aft surface (creating recirculating flow) whereby the viscous region is no longer thin and attached.

In particular, viscous effects will be significant very near any solid surfaces due to the no-slip condition, but an inviscid assumption may be reasonable for the rest of the flow. For example, the flow shown in Figure 2.12a has a velocity-gradient region (of height of δ) where viscous effects are important and determine the wall shear stress. This region is known as the *boundary layer* (Prandtl, 1904) and δ is the boundary-layer thickness. Above the boundary layer, the flow may be considered uniform and inviscid, as shown in Figure 2.12b. As may be expected, δ/D is small for a high Re_D (the thin boundary-layer assumption) but can vary greatly in size. For example, δ can be a kilometer for an atmospheric boundary layer or a few millimeters for the flow over a bird's wing.

As previously noted, a flow with $Re_D \gg 1$ with changes in flow direction can often be divided into small viscous regions and large inviscid regions. For flows over a body, the "inner region" is defined as near the surface with significant viscous effects, while the "outer region" is defined as an inviscid field away from the surface. The outer region can be described with the irrotational PDEs of §2.5. Employing the thin boundary-layer approximation ($\delta \ll D$), the inner region flow satisfies the two additional assumptions:

- The streamwise gradients are weak relative to the wall-normal gradients.
- The wall-normal velocities are small compared to the streamwise velocities.

Applying the preceding assumptions to the two-dimensional flow of Figure 2.12a, the steady x- and y-momentum equations are simplified (by neglecting small-order terms) as follows:

$$\rho_f u_x \frac{\partial u_x}{\partial x} + \rho_f u_y \frac{\partial u_x}{\partial y} = -\frac{\partial p}{\partial x} + \mu_f \frac{\partial^2 u_x}{\partial y^2} \tag{2.84a}$$

for a thin laminar boundary layer

$$\frac{\partial p}{\partial y} = 0 \tag{2.84b}$$

The first equation (2.84a) is the x-momentum whereby streamwise convection (LHS terms) is balanced with the streamwise pressure gradient and the tangential shear stress (RHS terms). The second equation (2.84b) is the y-momentum whereby the convection and stress terms are small compared to those in the first equation, so the wall-normal gradient is zero, yielding a constant pressure across the boundary layer thickness. Therefore, the inviscid pressure distribution just above the boundary layer is the same as the pressure distribution on the wall.

For the simplified case of a smooth flat plate in the x-direction with no streamwise pressure gradient, the theoretical velocity profile is given by the Blasius solution (White, 2016). This profile is a function of the wall-normal distance (y), the boundary-layer thickness (δ, defined as the height where the velocity is 99% of the freestream value), and the Reynolds number based on downstream distance (Re_x). The Blasius theoretical profile can be modeled as follows:

$$u/u_\infty \approx \tanh\left[1.65\, y/\delta + (y/\delta)^3\right] \tag{2.85a}$$

$$\delta \approx \left(5.29\, Re_x^{-1/2}\right) x \qquad \textit{for laminar flow } \left(e.g., Re_x < 10^6\right) \tag{2.85b}$$

$$Re_x \equiv \rho_f u_\infty x / \mu_f \tag{2.85c}$$

This result shows that the velocity profile varies smoothly within the boundary layer and that the thickness increases with $x^{1/2}$.

For the more general case of a streamwise pressure gradient and/or wall curvature, the boundary layer may separate, as shown in part of Figure 2.12b. In this case, the thin boundary-layer approximation used in (2.84) is no longer applicable. Furthermore, such separated regions at high Reynolds numbers will generally become unsteady due to instabilities, leading to transitional or turbulent flow, as discussed in the following subsection (which will require a different PDE).

2.6.2 Transitional and Turbulent Flow Regimes

Flow fields are more likely to be inherently unstable as Re_D increases since the relative effect of viscous damping is reduced. This is particularly true if the flow is separated, which can occur at the rear section of a smooth body as shown in Figure 2.12b. This separation is more distinct at a sharp corner, as shown in Figure 2.13a. The boundary layer is attached upstream of the corner, but the flow separates just after the corner. This shear layer caused by this flow separation becomes unstable, yielding growing eddy structures (as discussed in §2.7). As such, this portion of the flow is no longer

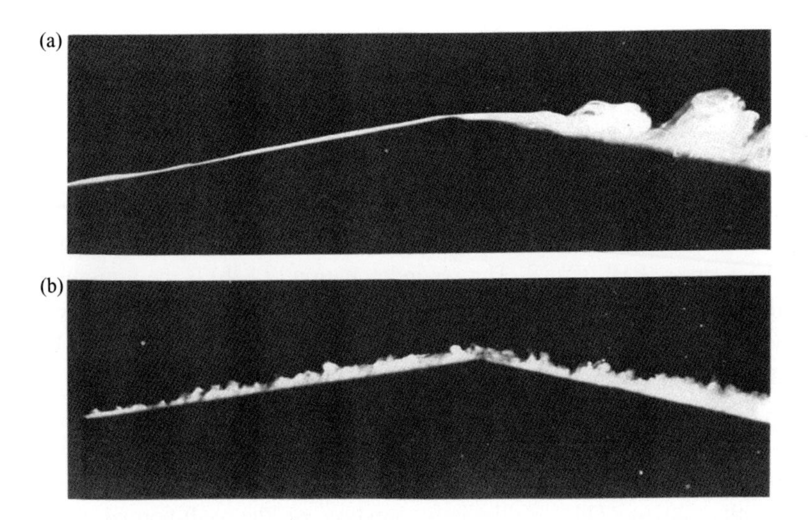

Figure 2.13 Boundary layers visualized with white dye at a high Reynolds number over a convex surface (Head, 1982): (a) laminar flow upstream that separates at the corner leading to an unsteady transition to turbulence; and (b) a turbulent flow upstream of the corner, which remains turbulent but attached after the turn.

laminar and instead includes a "transitional" region where the instabilities are growing though are primarily two-dimensional, followed by a "turbulent" region where the structures are three-dimensional and complex with a wide range of wavelengths.

If one considers the same corner geometry but with much higher Reynolds numbers, the transition to turbulence occurs for the portion upstream of the corner, as shown in Figure 2.13b. For a given geometry, the minimum Reynolds number for the transition to turbulence is designated as $Re_{D,turb}$. Comparing the boundary layers upstream of the corner as shown in Figures 2.13a and 2.13b, it can be seen that the turbulent one is thicker, which is due to the high degree of momentum mixing. This mixing makes a turbulent boundary layer fuller and more resistant to flow separation. This is evident in Figure 2.13b, where the boundary layer remains attached after the corner (whereas the laminar boundary layer of Figure 2.13a becomes separated).

Since transitional flows $(Re_{D,crit} < Re_D < Re_{D,turb})$ and turbulent flows $(Re_D > Re_{D,turb})$ are unstable, they are both governed by the three-dimensional unsteady Navier–Stokes equations:

$$\rho_f \frac{\partial \mathbf{u}}{\partial t} + \rho_f(\mathbf{u} \cdot \nabla)\mathbf{u} = -\nabla p + \mu_f \nabla^2 \mathbf{u} \quad \textit{transitional or turbulent flow.} \tag{2.86}$$

These are the most difficult flow equations to solve (requiring computers) since they describe a complex flow field that is three-dimensional, unsteady, highly nonlinear, with a wide range of length and time scales, However, these equations are needed once the flow is turbulent or transitional.

The Reynolds number range for which transition occurs will vary with domain geometry. Generally, free boundaries, curvature, and corners tend to lower the critical

Table 2.1 Reynolds number flow regimes and incompressible momentum equations in tensor notation (summation with j = 1,2,3).

Re	Regime	Momentum PDE
$Re_D \rightarrow 0$	Creeping flow	$\frac{\partial p}{\partial X_i} = \mu_f \frac{\partial}{\partial X_j}\frac{\partial u_i}{\partial X_j}$
$Re_D \ll 1$	Linearized flow	$\rho_f u_{j\infty}\frac{\partial u_i}{\partial X_j} = -\frac{\partial p}{\partial X_i} + \mu_f \frac{\partial}{\partial X_j}\frac{\partial u_i}{\partial X_j}$
$Re_D \sim 1$	Nonlinear flow	$\rho_f u_j\frac{\partial u_i}{\partial X_j} = -\frac{\partial p}{\partial X_i} + \mu_f \frac{\partial}{\partial X_j}\frac{\partial u_i}{\partial X_j}$
$Re_D \gg 1$, steady	Laminar boundary layer	$\rho_f u_x \frac{\partial u_x}{\partial x} + \rho_f u_y \frac{\partial u_x}{\partial y} = -\frac{\partial p}{\partial x} + \mu_f \frac{\partial^2 u_x}{\partial y^2}$
$Re_{D,crit} < Re_D < Re_{D,turb}$	Transitional flow	$\rho_f \frac{\partial u_i}{\partial t} + \rho_f u_j \frac{\partial u_i}{\partial X_j} = -\frac{\partial p}{\partial X_i} + \mu_f \frac{\partial}{\partial X_j}\frac{\partial u_i}{\partial X_j}$
$Re_D > Re_{D,turb}$	Turbulent flow	$\rho_f \frac{\partial u_i}{\partial t} + \rho_f u_j \frac{\partial u_i}{\partial X_j} = -\frac{\partial p}{\partial X_i} + \mu_f \frac{\partial}{\partial X_j}\frac{\partial u_i}{\partial X_j}$

Reynolds number for transition. In particular, the most unstable flows are those away from walls, such as separation regions, free-shear layers, wakes, and jets. For example, a planar shear layer can transition based on $Re_{\delta,crit} \approx 100$, where δ is the shear-layer thickness. A circular jet (with curvature) can have even greater sensitivity with $Re_{D,crit} \approx 50$, where D is the jet diameter. In contrast, flow in a pipe is stabilized by the presence of the walls. For example, a smooth circular pipe has $Re_{D,crit} \approx 2,300$. However, a pipe with a square cross section (whose corners promote instability) yields $Re_{D,crit} \approx 1,000$, and a curved pipe will also reduce $Re_{D,crit}$. An attached boundary layer along a flat plate (with a stabilizing surface and no curvature nor corners) can be stable up to $Re_{\delta,crit} \approx 3,500$, which corresponds to $Re_{D,crit} \approx 500,000$, where D is the plate length). Roughening the surface for flow in a pipe flow or on a wall will also make the flow less stable (introducing small corner and curvature flow around the roughness elements), reducing $Re_{D,crit}$. Therefore, flow geometry and Reynolds number both play key roles in the transition to turbulence.

2.6.3 Laminar vs. Transitional vs. Turbulent Flow

The various flow regimes and momentum PDEs discussed in the preceding section are summarized in Table 2.1. To see the differences between the flow regimes ranging from creeping flow all the way to turbulent flow, it is helpful to visualize the competing effects of diffusion to convection and the role of instability and unsteadiness.

To understand these effects, Osborne Reynolds in 1883 examined water flow in transparent pipes at various speeds and diameters by releasing a dye in the center of the pipe to visualize the onset of any flow instabilities. In these celebrated experiments, Reynolds found that a certain combination of diameter and flow speed yielded transition and turbulence (consistent with the Reynolds number criteria discussed

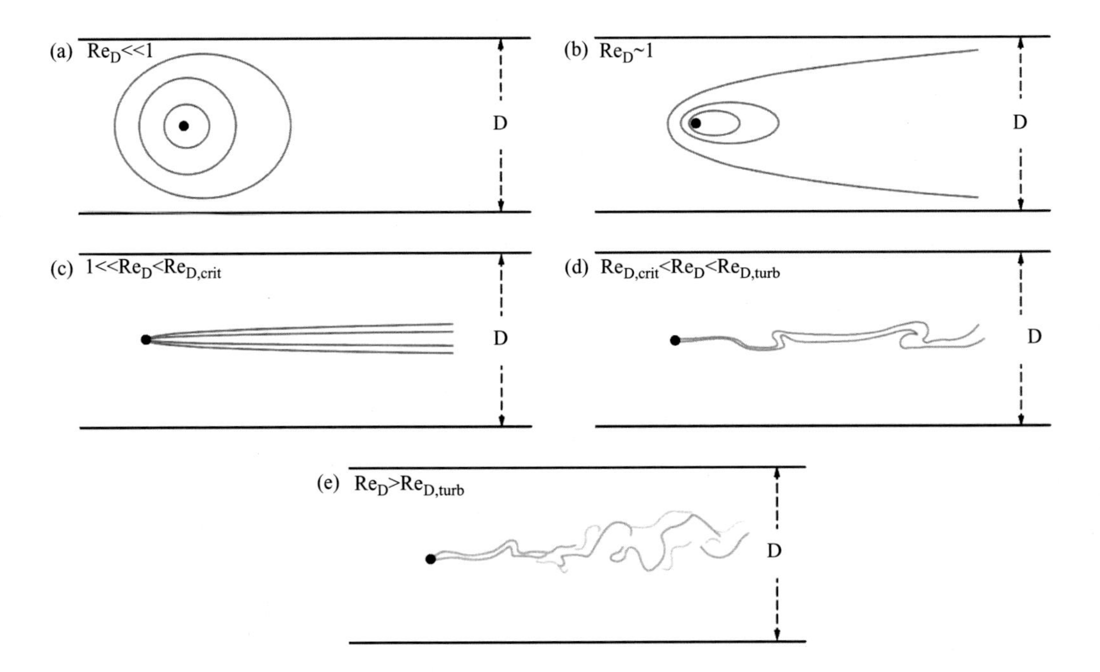

Figure 2.14 Illustration of the diffusion and convection of a fluid tracer species injected in a pipe center (at the black dot) for a left-to-right flow with (a) creeping flow, (b) flow with significant diffusion and convection, (c) high Reynolds number laminar flow, (d) transitional flow, and (e) turbulent flow.

previously). For a smooth circular pipe, the experiment of Reynolds and others are consistent with $\mathrm{Re}_{\mathrm{D,crit}} \approx 2,300$ and $\mathrm{Re}_{\mathrm{D,turb}} \approx 4,000$.

The dye experiment by Reynolds visualized the transition from laminar to turbulent flow. However, it cannot be used to readily investigate $\mathrm{Re}_\mathrm{D}(= \rho_\mathrm{f} U_\mathrm{D} D/\mu_\mathrm{f})$ effects within the laminar flow regime since the viscous diffusion rate will scale with $\mu_\mathrm{f}/\rho_\mathrm{f}$ while the dye diffusion rate will scale with $\Theta_{\mathcal{M}}$, and these two diffusivities are quite different for a liquid. In particular, the ratio of these two diffusivities is the Schmidt number $(\mathrm{Sc} = \mu_\mathrm{f}/\rho_\mathrm{f}\Theta_{\mathcal{M}})$ and is of the order 10^3 for liquids. However, gases have a Schmidt number of order unity, so their momentum and mass diffusion rates are about the same. Therefore, visualizing the competing effects of convection to mass diffusion for a gas species will also approximately illustrate the competing effects of convection to viscous diffusion (which scale with Re_D). Based on this, we can construct a hypothetical experiment to investigate the Re_D regimes by considering the release of a tracer gas in a transparent pipe with an air flow and imagining that we can "see" the species concentration of the tracer gas within this airflow.

Using this approach, Figure 2.14 illustrates the flow conditions ranging from creeping flow to turbulent flow. For the creeping flow ($\mathrm{Re}_\mathrm{D} \ll 1$) of Figure 2.14a, molecular diffusion (and thus viscous effects) dominate convection, so the tracer diffuses nearly radially outward from the point of injection. For $\mathrm{Re}_\mathrm{D} \sim 1$ of Figure 2.10b, downstream convection and upstream diffusion are of the same order (e.g., the balance indicated by 2.34a), so it much easier for the tracer to spread downstream. For the steady high Reynolds number laminar flow ($1 \ll Re_D < Re_{D,crit}$) of Figure 2.14c, convection will dominate diffusion, so the tracer moves primarily in the flow direction with only weak transverse diffusion. This is akin to a thin boundary layer where momentum diffusion is confined to a thin streamwise region.

If the Reynolds number is increased further such that $Re_D > Re_{D,crit}$ and the flow goes unstable as in Figure 2.14d, the flow is subjected to unsteady flow structures that serve to increase local mixing with lateral deviations (normal to the mean flow direction). At yet higher Reynolds numbers where the flow is turbulent ($Re_D > Re_{D,turb}$) as in Figure 2.14e, the flow contains complicated unsteady three-dimensional structures that substantially increase lateral mixing. This increased mixing causes the lateral distribution of the species to spread faster for turbulence than for laminar flow.

Since the turbulent regime leads to high mixing of momentum, this causes the profiles of the time-averaged velocity in the pipe to be much fuller than the velocity profile for laminar flow, as shown qualitatively by the profile insets in Figure 2.15. This is due to high-speed fluid in the pipe center region being moved toward the walls while low-speed fluid near the walls is transported toward the pipe center. Since the wall shear stress is proportional to the velocity gradient at the wall (recall 2.20), the fuller velocity profile for turbulent flow yields a larger average fluid shear stress on the

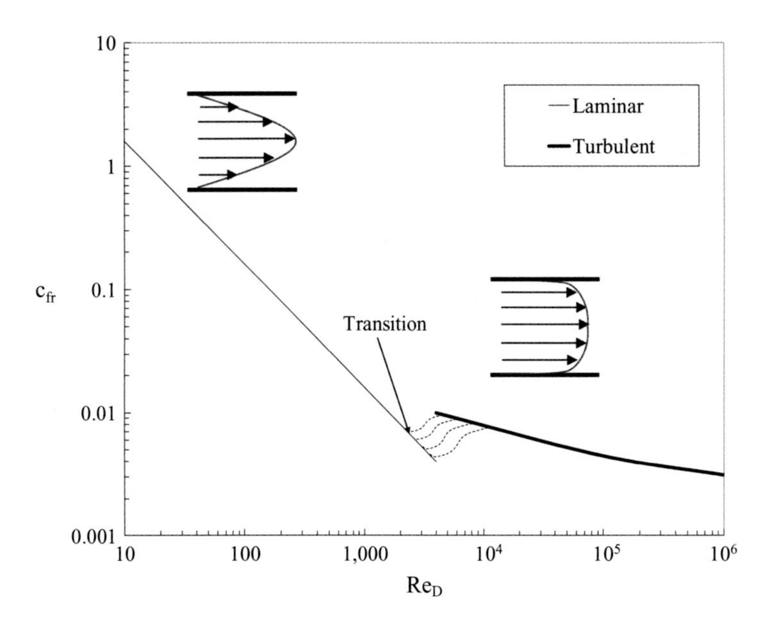

Figure 2.15 Fluid skin friction coefficient for a smooth fully developed pipe flow as a function of the pipe's Reynolds number for laminar flow ($Re_D < Re_{D,crit}$), transitional flow ($Re_{D,crit} < Re_D < Re_{D,turb}$), and turbulent flow ($Re_D > Re_{D,turb}$). The insets qualitatively indicate the shape of the time-averaged velocity profiles for laminar and mean turbulent flow.

wall compared to that for laminar flow. These changes can be quantified using the fluid friction coefficient (C_{fr}) defined as the ratio of the time-averaged wall shear stress to the mean dynamic pressure, which can be related to the axial pressure gradient:

$$C_{fr} \equiv \frac{\overline{K}_{wall}}{\frac{1}{2}\rho_f u_D^2} = \frac{\partial \overline{p}}{\partial x}\left(\frac{D}{2\rho_f u_D^2}\right). \tag{2.87}$$

The LHS dimensionless parameter is also called the Fanning friction factor (and is one-fourth of the Darcy friction factor, which uses a different normalization).

Based on this definition, the friction coefficient for a circular straight pipe is plotted in Figure 2.15 for laminar, transitional, and turbulent flow. Laminar flow allows a theoretical Poiseuille flow solution (White, 2016), but a turbulent regime has no theoretical solution, so one must rely on empirical models based on experimental results, such as the Blasius empirical fit (White, 2016). These laminar and turbulent fluid friction values are both a function of the pipe Reynolds number:

$$C_{fr} = 16\,Re_D^{-1} \qquad \textit{for laminar pipe flow.} \tag{2.88a}$$

$$C_{fr} \approx 0.08\,Re_D^{-1/4} \qquad \textit{for turbulent pipe flow.} \tag{2.88b}$$

As shown in Figure 2.15, turbulence causes an increase in the skin friction coefficient, consistent with its fuller velocity profile. In between, there is a broad range of transition paths from laminar to turbulent flow. This is due to the high sensitivity of instability growth rate to small deviations in flow spatial or temporal uniformity. For example, slight changes in the entrance to the pipe or surface finish can cause significant changes in $Re_{D,crit}$ and $Re_{D,turb}$. Consequently, C_{fr} in the transitional regime is difficult to quantify, other than being bounded by the laminar and turbulent limits (2.88a and 2.88b).

For flat plates, the boundary layers have much fuller profiles when they are turbulent instead of laminar (like that for a pipe), as shown in Figure 2.16. The turbulent profile is "fuller" since most of the velocity distribution is close to the

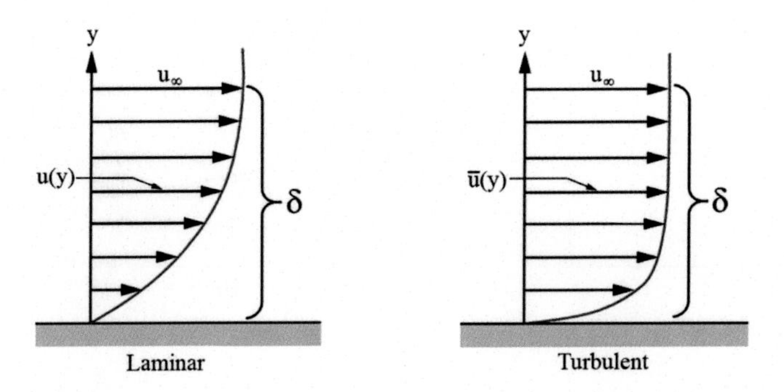

Figure 2.16 Velocity profiles for (a) laminar and (b) turbulent boundary layers of the same thickness, where the profile of the turbulent time-averaged velocity is much fuller near the wall, resulting in higher wall shear stress compared to that of the laminar boundary layer.

freestream velocity. This fuller profile creates a higher-velocity gradient at the wall, thereby increasing fluid skin friction. For an airfoil, this leads to increased drag, which is generally detrimental to aerodynamic performance. However, the fuller velocity profile also increases the boundary layer's resilience to flow separation. In particular, turbulent boundary layers are more likely to stay attached on a convex surface compared to laminar boundary layers (Figure 2.13). For an airfoil, resilience to flow separation is generally favorable. As such, turbulence can lead to both positive and negative flow performance characteristics.

2.7 Flow Instability Mechanisms

The root cause for a flow to transform from laminar to transitional to turbulent conditions is flow instability. In particular, a flow can be considered unstable, stable, or neutrally stable. A flow is said to be unstable if an initial small perturbation grows larger over time as it convects. On the contrary, a flow is stable if the perturbations (with no further forcing) decays in amplitude over time. Energy dissipation by viscous stresses can cause such a decay. A neutrally stable flow will result in perturbations that neither grow nor decay.

In general, flow instabilities are related to the flow geometry and Reynolds number as well as the boundary conditions and any imposed forces. For example, flat plate boundary layers at high Reynolds numbers are subject to Tollmien–Schlichting instabilities stemming from streamwise disturbances that grow as they move downstream. These same types of instabilities can be found in pipe flow (Lessen et al., 1968) and other wall-bounded flows. Such wall-bounded instabilities require consideration of viscous effects and can be complex to analyze. However, two of the most common instabilities, Rayleigh–Taylor and Kelvin–Helmholtz, are inviscid and more readily analyzable. Their physics and frequencies are discussed in the following subsection.

2.7.1 Inviscid Flow Instabilities

One may investigate Rayleigh–Taylor and Kelvin–Helmholtz instabilities by considering two parallel-flowing incompressible immiscible inviscid fluids that have different densities and velocities, as shown in Figure 2.17. Since they are immiscible, the interface allows a density discontinuity where Fluid 1 (Figure 2.17a) has density ρ_1 and Fluid 2 (Figure 2.17b) has a density ρ_2. Since they are inviscid, the interface allows a velocity discontinuity where Fluid 1 has a velocity field $\mathbf{u}_1$ while Fluid 2 has a velocity field $\mathbf{u}_2$. The initial velocities for each fluid field are set as constants $(u_{1,o}, u_{2,o})$ with a flat interface at $y_I = 0$:

$$u_1 = u_{1,o} \quad \textit{with} \quad \rho_1 \quad \textit{for}\ y > 0 \qquad \textit{at}\ t = 0 \tag{2.89a}$$

$$u_2 = u_{2,o} \quad \textit{with} \quad \rho_2 \quad \textit{for}\ y < 0 \tag{2.89b}$$

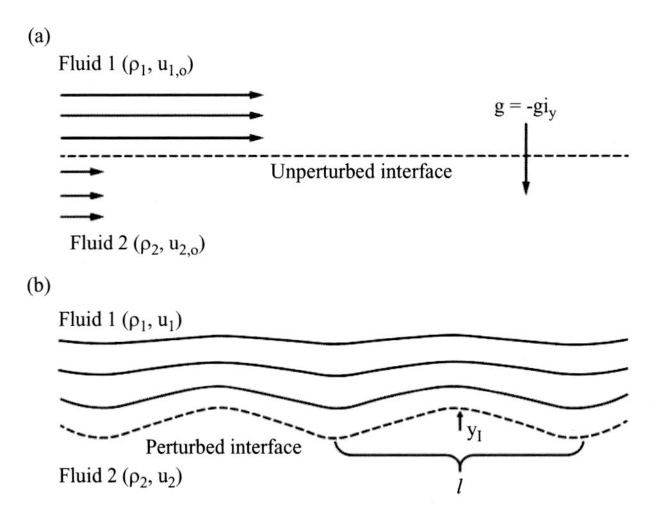

Figure 2.17 An irrotational incompressible flow with an interface that allows a velocity difference and/or a density difference: (a) initial conditions (no perturbations) with two fluids moving horizontally and with gravity vector as vertical; and (b) a later time with a perturbed interface (y_I) that has a wavelength (l) that impacts streamlines in the above fluid.

Since both fluids are incompressible and inviscid, their unsteady flow fields are each irrotational and so satisfy a Laplacian PDE for their velocity potentials (2.59b) as follows:

$$\nabla^2\Phi_1 = 0 \quad \textit{for flow in Stream 1.} \tag{2.90a}$$

$$\nabla^2\Phi_2 = 0 \quad \textit{for flow in Stream 2.} \tag{2.90b}$$

Notably, this irrotationality does not apply to the interface since it contains a discontinuous velocity difference, which is an infinitely thin vorticity layer. As such, the interface is hypothetical since viscosity would cause lead to a smoothly varying velocity change in a shear layer of finite thickness. Similarly, mass diffusivity would cause a smoothly varying density change.

Once the discontinuous interface is perturbed, it may become unstable depending on gravity and/or hydrodynamic effects. The perturbed interface (with height of y_I) can be modeled as a traveling sinusoidal wave that extends infinitely in the x-direction with perturbations in the y-direction (Figure 2.17b). This sine wave can be described in terms of an initial amplitude ($y_{I,o}$, which is a constant) as well as a wavelength (ℓ) and a frequency (f) as follows:

$$y_I = y_{I,o}e^{i(x/\ell - tf)}. \tag{2.91}$$

The wave thus moves in the x-direction at a speed of $f\ell$.

Since the interface joins the two fluids with no thickness, the velocities normal to the interface must be same on both sides (the tangential velocities can have a difference). If the curvatures are weak, the normal velocities can be approximated as the vertical velocities such that $u_{1y} = u_{2y}$ at the interface. This boundary condition can be expressed in terms of the velocity potentials (using 2.57) as follows:

$$\frac{\partial \Phi_1}{\partial y} = \frac{\partial \Phi_2}{\partial y} \quad a\ y = y_I. \tag{2.92}$$

If we neglect surface tension between the fluids, the pressure on either side of the interface will be equal. Based on this assumption, the pressure gradient in the unsteady Bernoulli equation (2.63) can be neglected at the interface, providing a second boundary condition:

$$\rho_1\left(\frac{\partial \Phi_1}{\partial t} + u_1\frac{\partial \Phi_1}{\partial x} + y_I g\right) = \rho_2\left(\frac{\partial \Phi_2}{\partial t} + u_2\frac{\partial \Phi_2}{\partial x} + y_I g\right) \quad at\ y = y_I. \tag{2.93}$$

Note that the gravity force acts downward, so $\mathbf{g} = -g\mathbf{i}_y$, where g is a positive scalar. Far from the interface, the interface perturbations will have no effect, so

$$u_1 = u_{1,o} \quad for\ y \to \infty. \tag{2.94a}$$

$$u_2 = u_{2,o} \quad for\ y \to -\infty. \tag{2.94b}$$

This provides the final flow boundary condition.

Since the flow is irrotational and incompressible, the linear superposition principle of (2.61b) can used to decompose the unsteady potential field as a sum of the initial unperturbed flow field (2.89 and Figure 2.17a) and the unsteady perturbations that travel with a wavelength and frequency (Figure 2.17b). Applying the boundary conditions of (2.92)–(2.94) for a given wavelength yields an ODE for the interface position (y_I) as a function of time and space (Drazin and Reid, 1981). Assuming weak perturbations, the two eigenvalues of the ODE are frequencies that can be expressed in terms of the fluid densities, far-field velocities, and the wavelength, as follows:

$$f = \frac{(\rho_1 u_{1,o} + \rho_2 u_{2,o}) \pm \sqrt{-\rho_1\rho_2(u_{1,o} - u_{2,o})^2 - (\rho_1 + \rho_2)(\rho_1 - \rho_2)g\ell}}{\ell(\rho_1 + \rho_2)}. \tag{2.95}$$

The interface perturbation will grow in time if this frequency has an imaginary component, that is, if the term within the square root is negative. This instability criterion is next considered for two special limiting conditions: (a) a density difference with initially uniform velocity; and (b) a velocity difference with initially uniform density.

The case of a density difference with no initial velocity ($u_{1,o} = u_{2,o} = 0$) is a stratified flow for which (2.95) can be used to define the Rayleigh–Taylor instability frequency as follows:

$$f = \pm\sqrt{\frac{-(\rho_1 - \rho_2)g}{\ell(\rho_1 + \rho_2)}} \quad for\ Rayleigh\text{–}Taylor\ instability. \tag{2.96}$$

If the above fluid is lighter than the below fluid ($\rho_1 < \rho_2$), the term in the square root is positive, so the flow is theoretically stable to perturbations for incompressible and

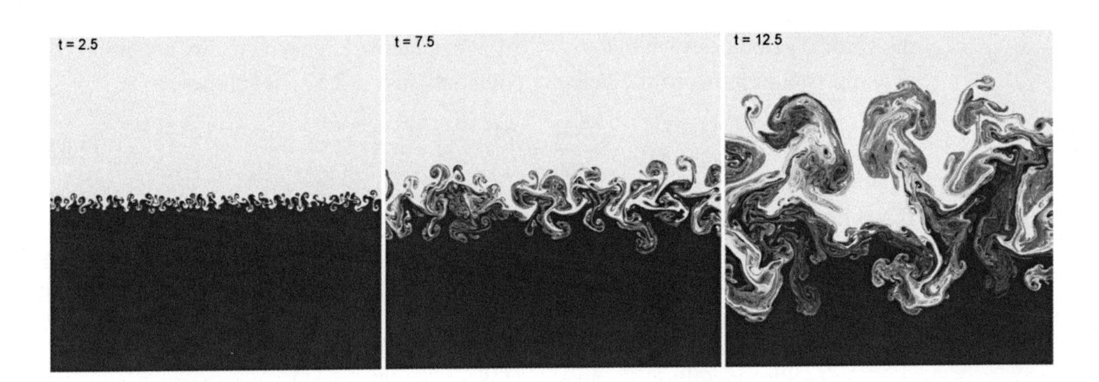

Figure 2.18 A heavier fluid over a lighter fluid that is perturbed and results in an evolving Rayleigh–Taylor instability (three different instances in times are shown), leading to larger and more complicated interface instabilities and flow structures at later times (Springel and Dullemond, 2011).

inviscid conditions. In particular, a perturbation to the interface will stably oscillate without growth (though any viscous damping would cause these oscillations to decay in amplitude). However, if the above fluid is denser than the below fluid ($\rho_1 > \rho_2$), the flow is dynamically unstable, so any perturbations in the interface will grow exponentially with time. This is called the Rayleigh–Taylor instability. An example of this is shown in Figure 2.18, where a lighter fluid is placed below a heavier fluid. The smallest wavelengths grow quickest since these have the highest frequencies, per (2.96). However, larger and larger wavelengths occur over time, which cause the interface to become more convoluted, complicated, and distributed over a larger vertical extent. This generally leads to transition into turbulent flow if viscosity effects are small. And since this instability is due to the gravitational acceleration acting normal to the interface, a Rayleigh–Taylor instability can also grow whenever there is a fluid acceleration in the direction of the lighter fluid.

In the other limit of uniform density ($\rho_1 = \rho_2$) with a velocity difference, a perturbation can lead to a Kelvin–Helmholtz instability. For the conditions shown in Figure 2.17a, the result of (2.95) for uniform density yields the following:

$$f = \frac{u_{1,o} + u_{2,o}}{2\ell} \pm \frac{\sqrt{-(u_{1,o} - u_{2,o})^2}}{2\ell} \quad \textit{for Kelvin–Helmholtz instability.} \tag{2.97}$$

The first term on the RHS is the real part of the frequency and indicates that the convective speed of this perturbation ($f\ell$) is equal to the average speeds of the two fluids, so $u_I = ½(u_{1,o} + u_{2,o})$ The second term on the RHS is the imaginary part that of the frequency and indicates that any velocity difference across the shear layer (regardless of how small or which fluid is moving faster) will create instability.

While the Rayleigh–Taylor instability (heavy over a light fluid) is intuitive, the Kelvin–Helmholtz instability is more complex. To understand how this instability

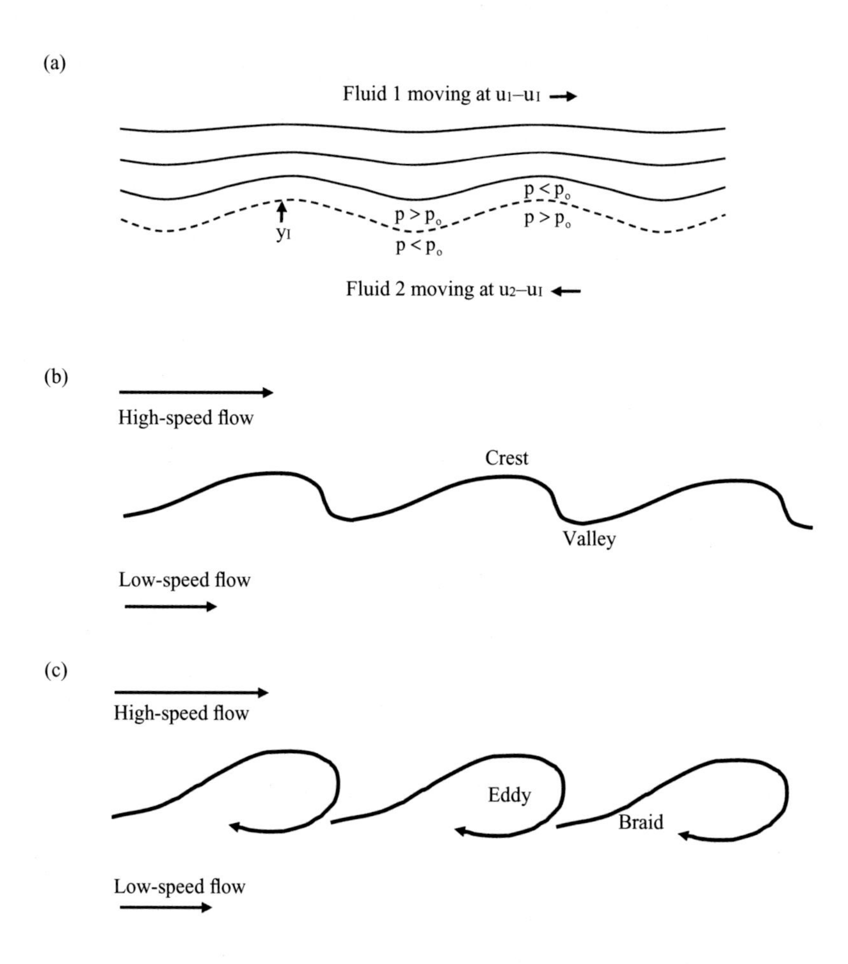

Figure 2.19 Kelvin–Helmholtz instability for two flows moving past each other with different speeds (and potentially different densities) separated by an inviscid interface: (a) a reference frame moving at speed of the sinusoidal interface (dashed line) creating variations in the high-speed fluid streamlines and the pressure field; (b) the local pressure decrease at a crest pulls it even farther upward while the local velocity increase pulls it faster to the right (and the opposite occurs for fluid in the valleys); and (c) a schematic of the eddy-and-braid flow after the amplified instabilities yield interacting vortex structures.

mechanism arises, we consider the local pressure relative to the initial pressure (p_o) in a reference frame moving with the interface speed (u_I). As shown in Fig. 2.19a for this reference frame, the high-speed fluid moves to the right and the low-speed fluid moves to the left. Since the flow is irrotational, the stagnation pressure within a given fluid is approximately constant (2.68a). As such, the local static pressure will be higher when the velocity is low ($p > p_o$ when $u < u_o$) and lower when the velocity is high ($p < p_o$ when $u > u_o$). Next we consider how the velocity varies due to the perturbation valleys and hills shown in Figure 2.19a. Since the flow is incompressible and since the high-speed flow is moving faster than the interface perturbation, continuity

(a)

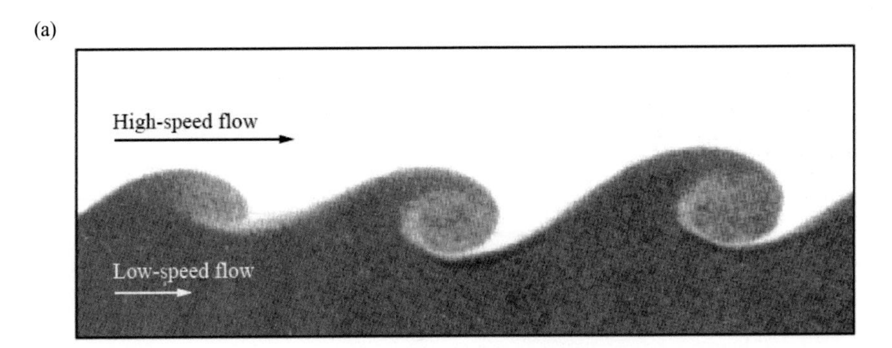

(b)

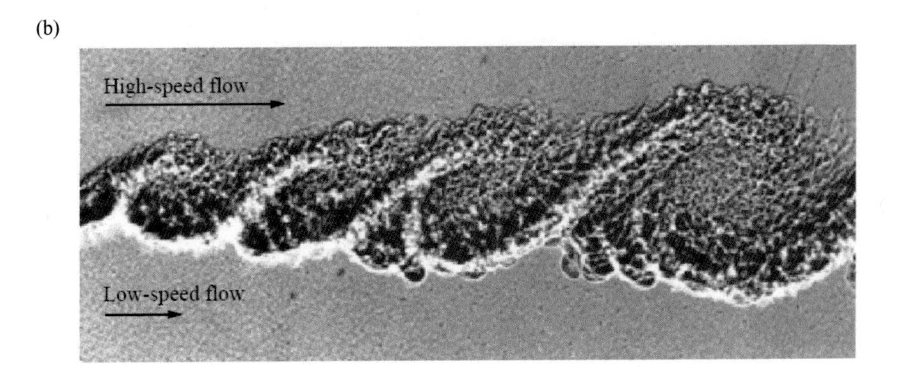

Figure 2.20 Images for parallel streams with different velocities where the shear-layer creates eddy-and-braid structures: (a) transitional conditions with moderate Reynolds numbers yielding two-dimensional flow (Thorpe, 1971); and (b) fully turbulent conditions at high Reynolds numbers yielding three-dimensional flow with a large range of length scales (Brown and Roshko, 1974).

(as discussed in §2.4) will cause Fluid 1 to decelerate in the valleys and accelerate at the crests. As such, the static pressure of the high-speed fluid will tend to be higher in the valleys and lower in the crests. If we consider the pressure field of Fluid 2, we find the opposite effect. Thus, the combination is a pressure difference in the direction of the surface perturbations. With no counteracting surface tension forces, this will cause the perturbations to grow. As such, this inviscid flow is always unstable. As a result, the Kelvin–Helmholtz instability is the most common instability mode in unbounded flows with velocity gradients (jets, wakes, free-shear flows).

Once the instabilities became large, the interface will no longer be represented by a simple sinusoidal perturbation. Instead, two-dimensional vortices will form whereby the fluid at the crest is moving faster on average and therefore pushes to the right and up, while the fluid in the valleys gets pushed to the left and downward (as in Figure 2.19b). As a result, the crest fluid eventually overtakes the valley fluid and the roll-up vortices lead to the "eddy-and-braid" features of Figure 2.19c.

Note that the Kelvin–Helmholtz instability can be eliminated if there are competing forces that stabilize the interface or that dampen the perturbation energy. In particular, a stably stratified flow (lighter fluids moving above heavier fluids) can ensure the square root of (2.95) is positive for large enough wavelengths. Surface tension can add another stabilizing force that tends to be strongest at smaller wavelengths (higher curvatures). The addition of viscosity can diffuse the velocity gradients and dampen their energy, thereby reducing the growth rate of these instabilities. Therefore, the Kelvin–Helmholtz instability is more pronounced at higher Reynolds numbers (where viscous effects tend to be weaker), where it is more likely to induce turbulent flow.

For example, experimental results for parallel streams with velocity differences are shown in Figure 2.20, where the eddy-and-braid dynamics of Figure 2.19c can be observed. At the moderate Reynolds number for Figure 2.20a, the roll-up vortices are nearly two-dimensional, indicating transitional flow. For the higher Reynolds number of Figure 2.20b, the flow is fully turbulent, with a wide range of three-dimensional, unsteady, and effectively stochastic structures (as will be discussed in Chapter 6) superposed over the large-scale eddy-and-braid structures.

2.8 Chapter 2 Problems

As you work through the following problems, show all steps and list any needed assumptions.

(P2.1) A pond with a volume of 940 m^3 has an average fish birth rate of 2,500 per year and an average death rate of 3,800 per year. For two months a year, the pond is fed by a creek, which enters the pond with an average flow speed of 0.17 m/s, a cross-sectional area of 0.15 m^2 and a fish concentration of 0.02 fish/m^3. The pond has a spillover outlet to keep its volume constant, but a screen ensures no fish exit the pond. Starting from (2.1), apply the Reynolds transport theorem by drawing a control volume for the pond, define q and evaluate all terms to find the annual rate of change of the fish population for the pond.

(P2.2) Starting from the integral form of the Reynolds transport theorem (2.1), derive the Eulerian transport equation (2.3).

(P2.3) Consider a wind tunnel test with a 10 cm $\times$ 10 cm square cross section that is 1 m long, with steady air flow of 25 m/s. Upstream of the test section, there is a spray nozzle that injects droplets at the same speed as the air flow with an average drop diameter of 50 μm and a total mass flow rate of 10 grams per second. Within the test section, evaporation and wall impacts continuously remove 10% of the drop mass from the airstream. Apply the Reynolds transport theorem by drawing a control volume for the test section, defining an appropriate q variable for the transport of water drop mass in the test section, and quantitively evaluating all four terms (two are volume integrals and two are area integrals) in (2.1) based on this q.

(P2.4) Starting with (2.3), (a) obtain the Eulerian x-momentum transport equation in component form (with u_x, u_y, u_z, K_{xx}, etc.) by setting the transport quantity $q = \rho u_x$, and by considering all the forces that act in the x-direction using a control volume diagram for pressure stresses and another for viscous stresses; and (b) then derive the PDE for $\mathcal{D}u_x/\mathcal{D}t$ assuming incompressible flow.

(P2.5) Consider a wind tunnel with incompressible inviscid steady flow moving in the positive x-direction (left to right) through a contraction with an air speed variation for $x > 0$ given as $u_o(1 + x/L)^{1/2}$, where $u_o = 0.25\,m/s$ and $L = 0.4\,m$. To analyze this flow, first decompose (2.21) into x- and y-momentum equations, where y is in the vertical direction. Then obtain the pressure gradients for air at normal density in the x- and y-directions at $x = 1\,m$.

(P2.6) (a) Starting from the control volume form of (2.1), derive the Eulerian species transport PDE using Fick's law for the diffusive flux, assuming constant molecular diffusivity. Then identify the physical interpretation of each term in the resulting transport equation. (b) Using the results from (a), and assuming incompressible flow, obtain the Lagrangian time derivative for the species mass fraction ($\mathcal{D}\mathcal{M}/\mathcal{D}t$).

(P2.7) Consider steady quasi-one-dimensional air flow moving at 100 m/s and at NTP in a duct (Station 1) that then expands to twice the cross-sectional area at a downstream location (Station 2) with no change in altitude nor entropy. Employing (2.1) and (2.46), write two equations for M_1/M_2 and T_1/T_2 and use these together to solve for M_2, and then for pressure change (p_2-p_1). (b) For an incompressible inviscid process for the same area change, find the resulting pressure change (p_2-p_1). (c) Quantify and discuss the differences between the incompressible and isentropic pressure changes as related to the average flow Mach number. (d) For an isothermal process with the same velocity changes as in (a), is energy added to or extracted from the flow?

(P2.8) Consider a lake exposed to still dry air where the water vapor mass fraction in the air is 3% just above the water surface but reduces to 0% far away. At a height of 10 cm above the water surface, the vapor mass fraction gradient is 0.1%/cm and the diffusivity of the water vapor in air is $0.26\,cm^2/s$. Based on Fick's law, determine the mass flux per unit area of water vapor. Use this result to determine how much the lake level would change in one month.

(P2.9) Make a rough approximation of a horizontal slice of a hurricane as a solid body vortex for the interior (2.54 for $r < r_1$) and as an inviscid vortex for the exterior (2.55 for $r < r_1$), and assume both regions are incompressible with negligible radial and vertical velocities. At the interface ($r_1 = 100\,m$), these two regions have the same pressure and tangential velocity ($u_1 = 40\,m/s$) and the far-field ($r \to \infty$) conditions are based on standard conditions for an altitude of 2,000 m (pressure of 79.5 kPa and a density of $1.01\,kg/m^3$). (a) Determine the radial distribution of the tangential velocity for both the exterior and interior regions. (b) Use these velocity distributions to determine the pressure at the interface (p_1) and at the vortex center (p_0).

(P2.10) Starting from (2.7b) and (2.13a), derive the mass and momentum PDEs of (2.72) in Cartesian coordinates, assuming a constant viscosity.

(P2.11) To define a nondimensional pressure, one may normalize by viscosity, $p^* = pD/(\mu_f u_D)$, instead of by density. (a) Starting from the PDE of (2.76b) and normalizing pressure by viscosity, derive a steady dimensionless form of the PDE in terms of Reynolds number. (b) Use this result to obtain the creeping-flow dimensional PDE for $Re_D \to 0$.

(P2.12) Consider a thin, two-dimensional, attached, laminar boundary layer along a wall in the x-direction with $Re_D \gg 1$. Starting from (2.83), obtain the relationships of (2.84).

(P2.13) Consider NTP flow of water in a pipe with an inner diameter of 4 cm. Determine the flow rate ranges (in liters per minute) corresponding to laminar, transitional, and turbulent flow regimes, then compute the axial pressure gradient at the start and at the end of the transitional flow range.

(P2.14) Consider a 5 m/s wind over initially stagnant water (e.g., a pond) at NTP. Determine which wavelength(s) can cause a small perturbation to grow into waves, assuming inviscid flow without surface tension. Discuss how surface tension and viscosity would affect these waves.

3 Governing Equations for an Isolated Spherical Particle

This chapter develops the point-force equations of motion for a single spherical particle moving in an unbounded fluid (which is treated as a continuum). In §3.1, the particle equations of motion are considered as a sum of pointwise forces. The quasisteady drag force is then described for solid and fluid particles in §3.2 for Reynolds numbers ranging from creeping flow to turbulent flow. Three different acceleration forces are discussed in §3.3, and these are combined with the drag force in §3.4 to provide various equations of motion. Finally, heat and mass transfer effects on the particle are discussed in §3.5.

3.1 Equations of Motion and Force Decomposition

The primary equations of motion for a particle are the ordinary differential equations (ODEs) for position and velocity. The Lagrangian time derivative of the particle centroid position $(\mathbf{X}_\mathrm{p})$ is defined by the particle centroid velocity $(\mathbf{v})$ as

$$\frac{d\mathbf{X}_\mathrm{p}}{dt} \equiv \mathbf{v}. \tag{3.1}$$

The Lagrangian time derivative of the particle momentum from Newton's Second Law of Motion equals the sum of the forces acting on the particle. These forces include the mass-based body forces $(\mathbf{F}_\mathrm{body})$, the area-based surface forces $(\mathbf{F}_\mathrm{surf})$, and the contact-based collision forces $(\mathbf{F}_\mathrm{coll})$:

$$\frac{d}{dt}\left(\mathrm{m_p}\mathbf{v}\right) = \mathbf{F}_\mathrm{body} + \mathbf{F}_\mathrm{surf} + \mathbf{F}_\mathrm{coll}. \tag{3.2}$$

For constant particle mass, the equation can be restated as follows: The product of particle mass and Lagrangian acceleration is equal to the sum of the forces acting on the particle. Integration of (3.1) and (3.2) together yields the particle trajectory as a function of time for a reference frame that is either fixed at a single location or linearly translating at a constant speed. If the reference frame is rotating, Coriolis and centrifugal relative forces must also be added.

Let us now consider each of the forces on the RHS of (3.2). For the remainder of this chapter, we assume that there is only one particle, and it does not contact any walls. As such, the collision forces can be neglected. This leaves only the body force (proportional to particle mass) and the surface force (proportional to particle surface area). As noted in §2.1.2, the body force is assumed to be dominated by the gravity since electromagnetic forces are considered negligible. The gravitational force $(\mathbf{F}_g)$ based on the gravitational acceleration $(\mathbf{g})$ and particle mass (m_p) is

$$\mathbf{F}_{body} \approx \mathbf{F}_g = m_p\mathbf{g}. \tag{3.3}$$

The fluid dynamic surface force on a particle is the sum of the pressure and viscous stresses integrated over the particle surface area and can be written (based on §2.1.2) as

$$\mathbf{F}_{surf} = \iint \left(-P\mathbf{n} + K_{ij}n_j\right) dA_p \qquad \textit{for } j = 1, 2, \textit{and } 3. \tag{3.4}$$

The pressure and viscous stresses in this integral are based on the resolved-surface pressure and velocity fields (P and $\mathbf{U}$) as defined in (1.12). If the particle is a sphere and azimuthal symmetry is assumed (2.18d), the fluid dynamic surface force can be expressed as

$$\mathbf{F}_{surf} = \iint (-P\mathbf{i}_r + K_{rr}\mathbf{i}_r + K_{r\theta}\mathbf{i}_\theta) dA_p. \tag{3.5}$$

This equation uses the coordinate system identified in Figure 1.23, where $\mathbf{i}_r$ is the unit vector in the outward radial direction, $\mathbf{i}_\theta$ is the unit vector in the polar (θ) direction, and dA_p is the differential particle surface area. Notably, the surface force of the fluid acting on the particle $(\mathbf{F}_{surf})$ imparts an equal but opposite surface force by the particle on the fluid.

Under certain idealized conditions (as will be presented in §3.4), $\mathbf{F}_{surf}$ can be decomposed into a linear sum of various point forces:

$$\mathbf{F}_{surf} = \mathbf{F}_D + \mathbf{F}_L + \mathbf{F}_\forall + \mathbf{F}_H + \mathbf{F}_S + \mathbf{F}_{Br} + \mathbf{F}_{\nabla T}. \tag{3.6}$$

These individual point-force components (each described later in this book) include the following:

(a) Drag force $(\mathbf{F}_D)$, which resists the relative velocity.
(b) Lift force $(\mathbf{F}_L)$ due to particle spin or unhindered fluid shear.
(c) Virtual-mass force $(\mathbf{F}_\forall)$ due to acceleration of the fluid around the particle.
(d) History force $(\mathbf{F}_H)$ due to unsteadiness of the viscous stress over the particle.
(e) Fluid-stress force $(\mathbf{F}_S)$ due to the fluid stresses neglecting the particle.
(f) Brownian motion $(\mathbf{F}_{Br})$ due to discrete molecular interactions around the particle.
(g) Thermophoresis force $(\mathbf{F}_{\nabla T})$ via a fluid temperature gradient.

While the point-force linear decomposition of (3.6) is widely used, this decoupling is only provable for certain theoretical limits and only for specific subsets of these forces. In cases where a theoretical description is not available, empirical models can be used for a validated range of conditions, but the linear decomposition may not be guaranteed. For complex conditions where neither theoretical nor empirical models are appropriate and available and/or there is coupling between forces, the surface forces can instead be determined using (3.5) as discussed in Chapters 9–11.

Keeping in mind these limitations, the rest of this chapter will employ the point-force approach of (3.6) assuming a continuum surrounding flow (e.g., $Kn_p \ll 1$ for a gas) such that $\mathbf{F}_{Br}$ and $\mathbf{F}_{\nabla T}$ (discussed in §5.3 and §9.2.3) are negligible. In addition, we assume viscous incompressible flow (so equations in §2.5 can be utilized) and no particle spin nor unhindered fluid vorticity, such that $\mathbf{F}_L$ (discussed in §10.1) is also negligible. Among the remaining surface forces to be considered in this chapter ($\mathbf{F}_D, \mathbf{F}_\forall, \mathbf{F}_H,$ and $\mathbf{F}_S$), the drag force is typically the most dominant and is the focus of the following section.

3.2 Drag for a Spherical Particle in Steady Flow

In this section, we consider conditions where the drag force not only dominates but is the only surface force. In later sections, we will consider other surface forces and then integrate these all together.

For steady uniform flow past a spherical particle moving at a constant speed, all the acceleration forces can be neglected ($\mathbf{F}_\forall = \mathbf{F}_H = \mathbf{F}_S = 0$). If the freestream flow gradients are negligible and the particle is not spinning, lift forces can also be ignored ($\mathbf{F}_L = 0$). If we further restrict ourselves to continuum conditions, then the Brownian and thermophoresies forces can be ignored ($\mathbf{F}_{Br} = \mathbf{F}_{\nabla T} = 0$). In this case, the only surface force remaining is the quasisteady drag force ($\mathbf{F}_{surf} = \mathbf{F}_D$). This force can then be related to the steady-state stresses acting over the surface of the particle, based on the particle outward normal $(\mathbf{n}_p)$.

Based on (3.5), the drag force is the sum of a pressure-based *form drag* and a viscous-based *friction drag* (with $\mathbf{n} = \mathbf{n}_p$), as follows:

$$\mathbf{F}_D = \underbrace{\mathbf{F}_{pressure}}_{\text{form drag}} + \underbrace{\mathbf{F}_{viscous}}_{\text{friction drag}} = \underbrace{\iint(-P\mathbf{n})dA_p}_{\text{form drag}} + \underbrace{\iint(K_{ij}n_j)dA_p}_{\text{friction drag}} \quad \textit{for } j = 1, 2, \textit{and } 3. \tag{3.7}$$

An example where only one of these occurs is an infinitely thin plate. If the plate is parallel to the flow, only friction (viscous) drag will be present. However, if the plate is normal to an oncoming flow, only form (pressure) drag will be present. For the more general case, particles will have both form drag and friction drag.

The particle drag force will generally depend on the relative velocity of the particle. In particular, the direction of the drag force $(\mathbf{F}_D)$ will oppose that of the relative velocity $(\mathbf{w})$, as shown in Figure 3.1a. Furthermore, the drag force magnitude (F_D) increases with the relative speed (w) and with the particle size, which can be characterized by the volumetric diameter (d). To quantify this dependence, it is important to consider the flow Reynolds number (as discussed in §2.6). For a particle, the characteristic flow speed and length scale can be taken as the relative velocity and

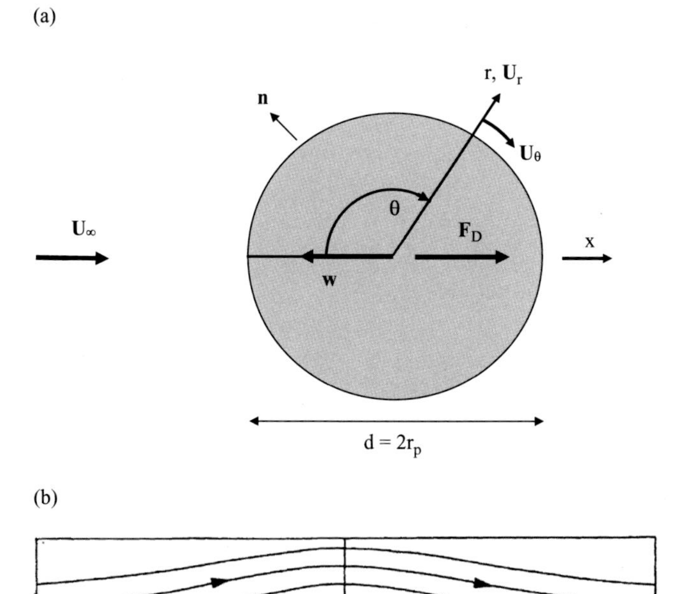

Figure 3.1 Uniform flow past a stationary spherical particle ($\mathbf{v} = 0$, such that $\mathrm{w} = -\mathrm{U}_\infty$) showing (a) the spherical coordinate system based on particle centroid with directions of particle surface normal, velocities, and drag force and the polar angle (where the upstream stagnation point is at $\theta = 0$ and $\mathrm{r} = \mathrm{r}_\mathrm{p}$); and (b) theoretical streamlines for steady flow in creeping-flow conditions (Clift et al., 1978).

the diameter, so the particle Reynolds number $\left(\mathrm{Re}_\mathrm{p}\right)$ can be defined using the surrounding fluid viscosity as follows:

$$\mathrm{Re}_\mathrm{p} \equiv \frac{\rho_\mathrm{f}\mathrm{wd}}{\mu_\mathrm{f}} \qquad \textit{where } \mathrm{w} \equiv |\mathbf{w}| = \left|\mathbf{v} - \mathbf{u}_{@\mathrm{p}}\right|. \tag{3.8}$$

Similar to the definition of Re_D of (2.79), the particle Reynolds number represents the ratio of convection to viscous effects for the flow around the particle.

In the following, we consider various regimes to investigate the drag force, including theory for a solid particle in creeping flow, theory for a fluid particle in creeping flow $\left(\mathrm{Re}_\mathrm{p} \to 0\right)$, theory for a particle in inviscid flow $(\mu_\mathrm{f} \to 0)$, and finally empirical models for a solid particle at finite Reynolds numbers.

3.2.1 Solid Sphere in Creeping Flow

Consistent with the steady creeping-flow limit for single-phase flows discussed in §2.6, the creeping flow over a particle is defined as $\mathrm{Re}_\mathrm{p} \to 0$. In this limit, the convection terms in the fluid momentum equation can be neglected, yielding a balance of the pressure and viscous stresses (as in 2.80):

$$\nabla P = \mu_f \nabla^2 \mathbf{U} \qquad \textit{for } Re_p \to 0. \tag{3.9}$$

Note that this PDE employs the resolved-surface pressure and velocity fields. Using this equation, Stokes (1851) derived the drag force for a spherical solid particle subjected to a flow that is otherwise uniform. It is perhaps the most important theoretical result for particulate fluid dynamics. The basic steps of the Stokes derivation are outlined in this subsection.

For the Stokes derivation of drag, we assume the particle centroid is fixed $(\mathbf{v} = 0)$ with constant size $(r_p = \text{const.})$ and there is no spin so the particle surface velocity is also stationary, whereby $\mathbf{V} = \mathbf{0}$. Since viscosity effects are important, the flow on the particle surface must obey the no-slip condition. As such, the flow velocity must come to a rest on the surface, so $\mathbf{U} = \mathbf{0}$ (as in 1.13). The far-field flow velocity (far away from the particle) is set with speed U_∞, moving left to right as shown in Figure 3.1a. Therefore, the resolved-surface fluid velocity very far from the surface and at the surface is given as follows:

$$\mathbf{U} = U_\infty \mathbf{i}_x \quad \textit{as } r \to \infty. \tag{3.10a}$$

$$\mathbf{U} = 0 \qquad \textit{at } r = r_p. \tag{3.10b}$$

Since this flow is azimuthally symmetric (per Figure 1.23), it contains only radial and tangential velocity components (U_r and U_θ), which can be used to transform (3.10), as follows:

$$U_r = -U_\infty \cos\theta \qquad \textit{as } r \to \infty. \tag{3.11a}$$

$$U_\theta = U_\infty \sin\theta \qquad \textit{as } r \to \infty. \tag{3.11b}$$

$$U_r = U_\theta = 0 \qquad \textit{for } r = r_p. \tag{3.11c}$$

These are the far-field and surface boundary conditions for steady uniform viscous flow past a fixed spherical particle.

The radial and tangential momentum equations in spherical coordinates can be based on (2.75b) and (2.75c) but simplified for steady flow with no convection terms nor gravitational acceleration to yield:

$$\frac{1}{\mu_f}\frac{\partial P}{\partial r} = \frac{\partial^2 U_r}{\partial r^2} + \frac{2}{r}\frac{\partial U_r}{\partial r} - \frac{2U_r}{r^2} + \frac{1}{r^2}\frac{\partial^2 U_r}{\partial \theta^2} + \frac{\cot\theta}{r^2}\frac{\partial U_r}{\partial \theta} - \frac{2}{r^2}\frac{\partial U_\theta}{\partial \theta} - \frac{2U_\theta \cot\theta}{r^2}. \tag{3.12a}$$

$$\frac{1}{r\mu_f}\frac{\partial P}{\partial \theta} = \frac{\partial^2 U_\theta}{\partial r^2} + \frac{2}{r}\frac{\partial U_\theta}{\partial r} - \frac{U_\theta}{r^2 \sin^2\theta} + \frac{1}{r^2}\frac{\partial^2 U_\theta}{\partial \theta^2} + \frac{\cot\theta}{r^2}\frac{\partial U_\theta}{\partial \theta} + \frac{2}{r^2}\frac{\partial U_r}{\partial \theta}. \tag{3.12b}$$

To solve these PDEs with the boundary conditions of (3.11), Stokes used the stream function (ψ) for the spherical coordinate system:

$$\frac{\partial\psi}{\partial\theta} \equiv U_r r^2 \sin\theta. \qquad (3.13a)$$

$$\frac{\partial\psi}{\partial r} \equiv -U_\theta r \sin\theta. \qquad (3.13b)$$

Rearranging (3.13) as equations for U_r and U_θ and then substituting these into the PDEs of (3.12) yields stream function radial and tangential momentum equations written as follows:

$$\frac{\partial P}{\partial r} = \frac{\mu_f}{r^2 \sin\theta}\frac{\partial}{\partial\theta}\left(\nabla'^2\psi\right). \qquad (3.14a)$$

$$\frac{\partial P}{\partial\theta} = \frac{-\mu_f}{\sin\theta}\frac{\partial}{\partial r}\left(\nabla'^2\psi\right). \qquad (3.14b)$$

The RHS of these two equations includes the Stokes specialized spherical differential operator for a variable q, which is defined as follows:

$$\nabla'^2 q \equiv \frac{\partial^2 q}{\partial r^2} + \frac{1}{r^2}\frac{\partial^2 q}{\partial\theta^2} - \frac{\cot\theta}{r^2}\frac{\partial q}{\partial\theta}. \qquad (3.15)$$

Note that this operator is different from the spherical Laplacian operator of (2.75d).

Next, (3.14a) is differentiated with respect to θ and (3.14b) with respect to r so that the resulting equations have the same LHS pressure terms. Then equating the RHS terms yields a single PDE for the stream function (using again the Stokes specialized operator) as

$$\nabla'^2\left(\nabla'^2\psi\right) = 0. \qquad (3.16)$$

To solve this fourth-order PDE, one must rewrite the boundary conditions in terms of ψ. The surface boundary conditions given by (3.11c) can be transformed using (3.13) to yield

$$\frac{\partial\psi}{\partial\theta} = \frac{\partial\psi}{\partial r} = 0 \qquad \textit{at } r = r_p. \qquad (3.17)$$

Similarly, the far-field boundary conditions of (3.11a) and (3.11b) are transformed as

$$\psi = -\tfrac{1}{2}U_\infty r^2 \sin^2\theta + \text{const.} \qquad \textit{as } r \to \infty. \qquad (3.18)$$

The constant in (3.18) is arbitrary (it does not affect the velocity field), and therefore can be set to zero.

The solution for (3.16)–(3.18) may be obtained by performing separation of variables by assuming ψ to be the product of $\sin^2(\theta)$ and a function of radius (White, 2016). This yields the Stokes solution for the stream function about a solid sphere in creeping-flow conditions:

$$\psi = -\frac{U_\infty r^2}{2}\left(1 - \frac{3r_p}{2r} + \frac{r_p^3}{2r^3}\right)\sin^2\theta. \qquad (3.19)$$

The resulting streamlines are symmetric both front to back (as seen in Figure 3.1b) and top to bottom. They follow the particle surface at $r = r_p$ and tend to straight lines at $r \to \infty$. The streamline directly upstream of the particle comes to rest at a stagnation point ($\theta = 0$ and $r = r_p$), while the flow in the rear stays attached (no flow separation).

The portion within the parentheses of (3.19) has three terms that represent three different effects. The first term represents a base flow consistent with the far-field condition (3.18). The second term corresponds to a "Stokeslet" (named for Stokes) that represents a viscous correction whose influence decays as r^{-1}. The third term corresponds to a "doublet," which represents an inviscid correction whose influence decays more rapidly from the particle surface, with r^{-3}. In Figure 3.1a, the deviations of the streamlines due to the particle presence are a combination of the Stokeslet and doublet effects.

The Stokes solution for the stream function can also be used to obtain the local velocity and pressure fields (needed to obtain particle drag). Substituting the stream function solution into (3.13), the radial and tangential velocities can be obtained as

$$U_r = -U_\infty \cos\theta \left(1 - \frac{3r_p}{2r} + \frac{r_p^3}{2r^3}\right). \tag{3.20a}$$

$$U_\theta = U_\infty \sin\theta \left(1 - \frac{3r_p}{4r} - \frac{r_p^3}{4r^3}\right). \tag{3.20b}$$

As with (3.19), the three terms in the parentheses correspond to far-field, viscous, and inviscid components.

To obtain the pressure field for the form drag, one may relate the differential pressure at a point in the fluid in terms of radial and tangential pressure gradients and then relate these gradients to the stream function by using (3.14):

$$dP = \frac{\partial P}{\partial r} dr + \frac{\partial P}{\partial \theta} d\theta = \left[\frac{\mu_f}{r^2 \sin\theta} \frac{\partial}{\partial \theta}\left(\nabla'^2 \psi\right)\right] dr + \left[\frac{-\mu_f}{\sin\theta} \frac{\partial}{\partial r}\left(\nabla'^2 \psi\right)\right] d\theta. \tag{3.21}$$

The RHS terms in parentheses (specialized operators of the stream function) can be simplified using (3.15) and (3.19) as follows:

$$\nabla'^2 \psi = -\frac{3U_\infty r_p}{2} \frac{\sin^2\theta}{r}. \tag{3.22}$$

By combining (3.21) and (3.22), integrating this equation with respect to r and θ, and applying the far-field pressure boundary condition, the pressure in the flow domain becomes

$$P = P_\infty + \frac{3U_\infty r_p \mu_f}{2r^2} \cos\theta. \tag{3.23}$$

This result shows that the fluid pressure far upstream ($r \to \infty$) equals P_∞, while the pressure on the surface $\left(r = r_p\right)$ varies with θ. The surface pressure variation is symmetric top to bottom but is asymmetric front to back since the pressure on the surface decreases left to right. In particular, the surface pressure peaks at the front stagnation point ($r = r_p$ and $\theta = 0$) but drops to a minimum at the rear stagnation point ($r = r_p$ and $\theta = \pi$). This net reduction in pressure (due to viscous losses) causes the pressure-based drag to be in the streamwise direction (see Figure 3.1a).

The preceding pressure field can be integrated over the particle surface to obtain the form drag based on (3.7):

$$\mathbf{F}_{\text{pressure}} = -\iint P\mathbf{i}_r dA_p. \tag{3.24}$$

Because of top–bottom symmetry for (3.23), the net form drag is equal to twice the force acting on the upper half of the sphere ($\theta = 0$ to π). Using this result with the relation $\mathbf{i}_r = -\mathbf{i}_x \cos\theta$ and considering a differential surface area as $dA_p = \pi r_p^2 \sin(\theta)d\theta$, the pressure-based drag with azimuthal symmetry becomes

$$\mathbf{F}_{\text{pressure}} = 2\pi r_p^2 \mathbf{i}_x \int_0^{\pi} P\cos\theta\sin\theta d\theta = 2\pi r_p \mu_f \mathbf{U}_\infty. \tag{3.25}$$

The RHS indicates that form drag is linearly proportional to particle radius, fluid viscosity, and the far-field velocity.

To determine the viscous component of drag (i.e., friction drag), the spherical form of the viscous stresses can be employed. In particular, the viscous stress component in the tangential direction due to the resolved-surface velocity gradients (2.18b) is given by

$$K_{r\theta} \equiv \mu_f \left[r\frac{\partial}{\partial r}\left(\frac{U_\theta}{r}\right) + \frac{1}{r}\left(\frac{\partial U_r}{\partial \theta}\right)\right]. \tag{3.26}$$

This stress can be obtained on the sphere by using the fluid velocity of (3.20). Note that K_{rr} can be similarly obtained with (2.18a) and (3.20) but yields zero on the surface and so is neglected. Integrating $K_{r\theta}$ on the particle surface, the viscous shear force contribution to drag is given by

$$\mathbf{F}_{\text{viscous}} = 4\pi r_p \mu_f \mathbf{U}_\infty. \tag{3.27}$$

This is the friction drag, which turns out to be twice the form drag (3.25). As such, the ratio of form drag to friction drag for a solid particle in creeping flow is 0.5.

The total drag force is then the sum of the form and friction drags:

$$\mathbf{F}_D = \mathbf{F}_{\text{pressure}} + \mathbf{F}_{\text{viscous}} = 6\pi r_p \mu_f \mathbf{U}_\infty = 3\pi d\mu_f \mathbf{U}_\infty. \tag{3.28}$$

This is the creeping-flow drag for a fixed spherical particle subjected to a far-field velocity. It is often referred to as the *Stokes drag*, owing to the author of the derivation.

The preceding result can be generalized to include the case where the particle is moving at fixed velocity ($\mathbf{v}$). The far-field velocity is still $\mathbf{U}_\infty$ but at the surface, the fluid must move at the centroid speed ($\mathbf{U}_I = \mathbf{v}$). To analyze this case, it is helpful to instead consider a reference frame moving with a velocity $\mathbf{v}$ so that $r = 0$ is at the particle centroid and the spherical steady Navier–Stokes equations can be centered on the particle to solve for the relative velocity field $\mathbf{U}_{rel} = \mathbf{U}_\infty - \mathbf{v}$. Since $\nabla P = \mu_f \nabla^2 \mathbf{U}_{rel} = \mu_f \nabla^2 \mathbf{U}$, the PDE is unchanged. However, the boundary conditions for the moving reference frame must be modified as follows:

$$\begin{matrix} \mathbf{U} = \mathbf{U}_\infty - \mathbf{v} & \textit{as } r \to \infty \\ \mathbf{U} = 0 & \textit{as } r = r_p \end{matrix} \qquad \textit{for particle and frame moving at } \mathbf{v}. \tag{3.29}$$

Comparing this to the boundary conditions of (3.10), one may simply replace the freestream velocity ($\mathbf{U}_\infty$) in the preceding fixed particle equations with the relative freestream velocity ($\mathbf{U}_\infty - \mathbf{v}$) for a moving particle, so (3.28) becomes

$$\mathbf{F}_D = 3\pi d\mu_f(\mathbf{U}_\infty - \mathbf{v}) = -3\pi d\mu_f \mathbf{w} \qquad \textit{for } \mathrm{Re}_p \to 0. \tag{3.30}$$

As expected, the drag force resists the relative velocity, so it acts in the opposite direction. Note that the RHS of (3.30) employs the point-force relative velocity (of 1.15d) and thus assumes that any far-field nonuniformities have a weak effect on drag. Note also that the Stokes drag force is linearly proportional to the relative velocity, so it is a “linear” drag (an aspect that will be employed later to simply analysis of particle motion).

While the Stokes drag is only exact for the limit of $\mathrm{Re}_p \to 0$, it agrees well with experimental data (within 2%) for Re_p values as large as 0.1 for solid spheres (Clift et al., 1978). To quantify the variation from the Stokes drag for other conditions (e.g., different Re_p values or fluid particles), we define the dimensionless “Stokes correction factor” f, based on the force magnitude as

$$f \equiv \frac{F_D}{3\pi d\mu_f w}. \tag{3.31}$$

As such, $f = 1$ for a solid spherical particle in creeping-flow conditions (per 3.30) In particular, we define f_q as the Stokes correction factor when the parameter q causes a change the drag force. For example, the next section considers, creeping flow for a fluid particle to obtain $f_{\mu p}$, which characterizes the effects of finite particle viscosity (μ_p). Later sections in this chapter will describe models for f_{Re} to characterize effects of finite particle Reynolds numbers (Re_p). Stokes correction factors for other parameters (e.g., f_M to account for the effects of relative particle Mach number) are discussed in Chapter 9.

3.2.2 Fluid Sphere in Creeping Flow

A fluid particle (such as a drop in a gas or a bubble in a liquid) can have motion inside the particle. For example, Figure 3.2 shows internal streamlines for a drop falling slowly $(\mathrm{Re}_p \ll 1)$ in a fluid for a reference frame fixed to the particle centroid. The surrounding fluid (indicated by blue arrows) moves upward relative to this centroid, which causes the interior fluid near the drop surface to be pulled upward, leading to internal vortices. The corresponding internal recirculating flow is called *Hill’s vortex*. This can reduce the drag force as compared to a solid nonrotating particle, which has

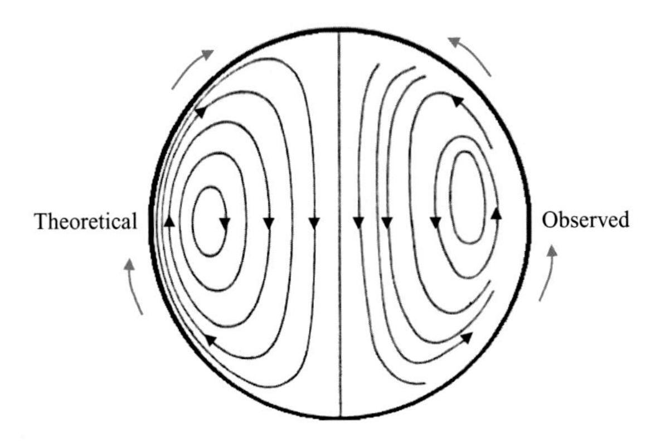

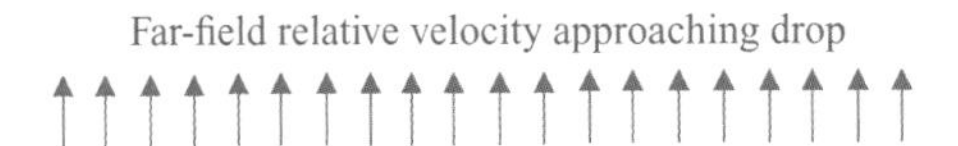

Figure 3.2 Recirculating internal flow for a drop of glycerin falling through castor oil at a near-creeping-flow condition, with theoretical streamlines on the left half and experimentally observed streamlines on the right half (Batchelor, 1967). The particle is falling downward, so the relative motion of the approaching fluid is upward, as shown by blue arrows.

no internal circulation that is consistent with zero tangential velocity at the interface in this reference frame.

To quantify the recirculating tangential velocity for a fluid particle, one must consider the pressure and viscous stresses at the interface between two fluids, as shown in Figure 3.3a. Just as the pressure must be considered both outside and inside of this interface (P_f and P_p), the viscous stresses acting on the interface surface are both outside ($K_{nn,f}$ and $K_{nt,f}$) and inside ($K_{nn,p}$ and $K_{nt,p}$). In addition, an immiscible interface has a surface tension (σ, which is defined relative to the two fluids) that creates a tangential stress based on the local surface curvature. The two radii of curvature ($r_{curv,1}$ and $r_{curv,2}$) are defined by the two orthogonal planes that intersect along the surface normal ($\mathbf{n}$), as shown in Figure 3.3b.

Across a fluid particle interface, the sum of the pressure, viscous, and surface tension stresses must be balanced. The resulting boundary conditions that relate the external to the internal properties in the normal and tangential directions are given as

$$\left(P_p - P_f\right)_I - \left(K_{nn,p} - K_{nn,f}\right)_I = \sigma\left(\frac{1}{r_{curv,1}} + \frac{1}{r_{curv,2}}\right) \quad \textit{for fixed shape} \tag{3.32a}$$

$$\left(K_{nt,p} - K_{nt,f}\right)_I = \mathbf{i}_t \cdot \nabla\sigma \tag{3.32b}$$

The first equation balances the pressure and viscous stresses in the normal direction (LHS) with the surface tension stresses (RHS). The second equation balances the

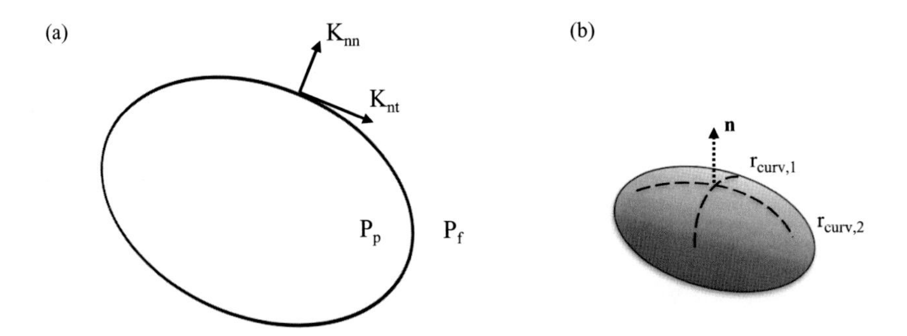

Figure 3.3 Schematics of fluid particle: (a) cross section showing the external and internal pressure fields (P_f and P_p) and the normal and tangential viscous stress components (K_{nn} and K_{nt}) acting at a point on the surface; and (b) perspective view of a particle showing surface curvature defined by the two local radii of curvature: $r_{curv,1}$ in one plane and $r_{curv,2}$ in an orthogonal plane.

tangential stresses with the surface tension gradient (where $\mathbf{i}_t$ is the unit vector in the tangential direction).

These equations are simplified if we consider a spherical shape and a clean interface. A spherical shape causes both radii of curvature to equal to the volumetric radius, r_p, which simplifies the RHS of (3.32a). A clean interface is devoid of surface contaminants, which yields a uniform surface tension ($\sigma = \text{const.}$) that eliminates the RHS of (3.32b). If we further assume azimuthal symmetry, the viscous stresses are given by (2.18d), so the normal and tangential stress balances become

$$\left[P_p - P_f\right]_I + \left[2\mu_f \frac{\partial U_r}{\partial r} - 2\mu_p \frac{\partial V_r}{\partial r}\right]_I = \frac{2\sigma}{r_p} = \frac{4\sigma}{d} \qquad (3.33a)$$

$$\mu_p \left[r \frac{\partial}{\partial r}\left(\frac{V_\theta}{r}\right)\right]_I = \mu_f \left[r \frac{\partial}{\partial r}\left(\frac{U_\theta}{r}\right)\right]_I \qquad (3.33b)$$

at $r = r_p$.

Equation (3.33a) balances the interface difference in pressure and radial viscous stresses with those created by surface tension. Since the radial viscous stresses are generally weak, the pressure increase across the interface is primarily due to surface tension, which causes the pressure inside the particle to be higher than that of the surrounding fluid (and this difference will increase as the particle radius decreases). Equation (3.33b) indicates that the tangential shear stress for the internal and external fluids must match.

In addition to the interface force balance, a velocity boundary condition can be prescribed at the interface. For a spherical particle of constant size, the no-slip velocity boundary conditions given by (1.13) on the fluid interface can be expressed as follows:

$$U_r = V_r = 0 \qquad (3.34a)$$

$$U_\theta = V_\theta \qquad (3.34b)$$

Interface velocities at $r = r_p$.

The latter result indicates that the tangential velocities of the particle fluid and the surrounding fluid are equal at the interface. Thus, (3.33a,b) and (3.34a,b) describe the interface boundary conditions between the internal and external flow fields for a clean spherical fluid particle of fixed radius.

Next, the external and internal flow fields are considered, where we designate the internal stream function as Ψ_p and the external stream function as Ψ_f. The external flow stream function is still defined by the external velocity field per (3.13) and governed by the fourth-order creeping-flow PDE of (3.16). In a similar manner, the internal stream function is based on the internal velocity components and is governed by the same creeping-flow PDE:

$$V_r = \frac{1}{r^2 \sin\theta} \frac{\partial \psi_p}{\partial \theta}. \tag{3.35a}$$

$$V_\theta = -\frac{1}{r \sin\theta} \frac{\partial \psi_p}{\partial r}. \tag{3.35b}$$

$$\nabla'^2 \left(\nabla'^2 \psi_p \right) = 0. \tag{3.35c}$$

In addition, Ψ_p and Ψ_f are linked by the velocity and tangential stress boundary conditions on the interface of (3.33a,b) and (3.34a,b), which can be expressed in terms of the stream functions as follows:

$$\psi_f = \psi_p \tag{3.36a}$$

$$\frac{\partial \psi_f}{\partial r} = \frac{\partial \psi_p}{\partial r} \qquad \textit{at } r = r_p. \tag{3.36b}$$

$$\mu_f \frac{\partial}{\partial r}\left(\frac{1}{r^2} \frac{\partial \psi_f}{\partial r} \right) = \mu_p \frac{\partial}{\partial r}\left(\frac{1}{r^2} \frac{\partial \psi_p}{\partial r} \right) \tag{3.36c}$$

These are sufficient to determine the velocity field, but the normal stress boundary condition of (3.33a) is needed to determine the pressure field.

The analytical solution that satisfies the internal and external flow PDEs and the preceding boundary conditions is described by Clift et al. (1978) and is named the Hadamard–Rybczynski solution (as it was independently derived by these two people in 1911). The resulting stream functions generalized for a relative particle velocity (w) at steady conditions are given as a function of the particle–fluid viscosity ratio, μ_p^* of (1.6), as follows:

$$\psi_f = -\frac{w r^2 \sin^2\theta}{2} \left[1 - \frac{3 r_p \left(2 + 3\mu_p^* \right)}{2r \left(3 + 3\mu_p^* \right)} + \frac{3 r_p^3 \mu_p^*}{2r^3 \left(3 + 3\mu_p^* \right)} \right]. \tag{3.37a}$$

$$\psi_p = -\frac{w r^2 \sin^2\theta}{4\left(1 + \mu_p^* \right)} \left[1 - \frac{r^2}{r_p^2} \right]. \tag{3.37b}$$

In these expressions, w is the magnitude of the relative velocity (a positive number), that is, $w = |\mathbf{w}|$, per (1.8b). The internal recirculating flow of (3.37b) is Hill's vortex, and Figure 3.2 shows good comparison for the theoretical and experimental streamlines for a falling drop.

To compute drag for the spherical fluid particle with recirculation, one must integrate the external surface pressure and viscous stresses. The distribution of the external fluid pressure (P_f) can be obtained in a manner similar to that used for the solid particle, as in (3.21)–(3.23), to yield

$$P_f = P_\infty + \frac{wr_p\mu_f}{2r^2}\left(\frac{2+3\mu_p^*}{1+\mu_p^*}\right)\cos\theta. \tag{3.38}$$

As with the solid particle, the pressure at the fore section $(\theta < 90^\circ)$ is higher than the aft section $(\theta > 90^\circ)$ so there is a net pressure drag in the x-direction (which opposes the relative velocity). However, the pressure variation for the fluid particle is modified by the term in the parentheses, which is reduced when there is finite particle viscosity, which will reduce the form drag.

The results of (3.33a) and (3.38) can also be combined to determine the internal fluid pressure (P_p) as well as the pressure difference at the interface, as follows:

$$P_p = P_\infty + \frac{2\sigma}{r_p} + \frac{5wr\mu_p}{r_p^2\left(1+\mu_p^*\right)}\cos\theta. \tag{3.39a}$$

$$\left(P_p - P_f\right)_I = \frac{2\sigma}{r_p} + \frac{w\mu_p}{r_p}\left(\frac{2+13\mu_p^*}{2+2\mu_p^*}\right)\cos\theta. \tag{3.39b}$$

This latter equation indicates that the internal pressure increases across the interface due to surface tension (σ), while viscous stresses cause a tangential variation.

The steady drag force can be obtained by integrating both the external pressure and viscous stresses distribution over the particle surface (3.5) to yield

$$\mathbf{F}_D = -3\pi d\mu_f \mathbf{w}\left(\frac{2+3\mu_p^*}{3+3\mu_p^*}\right) \qquad \textit{for } Re_p \to 0. \tag{3.40}$$

This drag for a fluid particle in creeping flow has the same linear dependence on particle radius and relative velocity as noted for the Stokes drag, but it includes an additional dependency on the particle–fluid viscosity ratio per the term in the parentheses. The associated correction factor based on the ratio of fluid particle drag to solid particle drag is then equal to this term:

$$f_{\mu_p} \equiv \frac{F_D\left(\mu_p, Re_p \to 0\right)}{F_D\left(\mu_p \to \infty, Re_p \to 0\right)} = \frac{2+3\mu_p^*}{3+3\mu_p^*}. \tag{3.41}$$

This Hadamard–Rybczynski theoretical result has been confirmed by experiments with various clean fluids (Loth, 2008a) and is considered for various example limits in the following.

For a drop in a gas, the particle viscosity ratio will be very high so the drag approaches that of a solid particle:

$$f_{\mu_p \gg \mu_f} \approx 1 \qquad \textit{for a drop in a gas at } Re_p \rightarrow 0. \tag{3.42}$$

Similarly, the external velocity and pressure fields also approach that of the solid particle. If the viscosity ratio is more moderate, the drag will be reduced per (3.41). For example, one may consider the Stokes corrections for viscosity ratios of unity and much less than unity:

$$f_{\mu_p = \mu_f} = 5/6 \qquad \textit{for a clean immiscible drop at } Re_p \rightarrow 0. \tag{3.43a}$$

$$f_{\mu_p \ll \mu_f} \approx 2/3 \qquad \textit{for a clean bubble at } Re_p \rightarrow 0. \tag{3.43b}$$

The reduction in drag as the viscosity ratio decreases is due to the internal recirculation that reduces both the surface shear stress and the surface pressure variation. However, this drag reduction can be counteracted by the influence of surface contaminants, as discussed in the following subsection.

Influence of Contaminants

Some liquid systems contain contaminants, very small solid particles that can collect at the interface of a bubble or a drop. These contaminants are generally composed of organic molecules or other microscale particles that can adhere to the particle surface. This adherence can be specially strong if they have asymmetric chemistry. For example, particles with a hydrophobic head and a hydrophilic tail will lodge themselves at the interface, with the tail pointing toward the water side. Such contaminants are called surface-active agents or *surfactants* for short. On the interface, surfactants can reduce surface mobility and surface tension.

The effect of surfactants on surface tension can be seen with the example of a needle placed on a water surface. If placed gently on the surface of tap water, the needle will float as surface tension counteracts the gravitational force. However, if one adds a small amount of dish detergent (which has a high concentration of surfactants), the needle will sink because the surface tension has been reduced (the needle is more easily wetted). Since surfactants can affect surface tension, a nonuniform surfactant concentration on the interface will lead to nonuniform surface tension. These tangential gradients of surface tension lead to viscous stresses that drive fluid to regions of higher surface tension per (3.32b). These stresses are known as Marangoni stresses, and the "tears of wine" phenomenon is an example of the Marangoni effect (Venerus and Simavilla, 2015). This effect can alter the shear stress distribution on a fluid particle and the associated drag.

At high concentrations, surfactants can also impact fluid particle drag due to their influence on surface mobility. A fluid particle moving relative to the surrounding fluid will tend to collect more surfactants, which will be swept toward the rear of the particle, where they will concentrate. As the surfactant concentration increases, this will create an immobile surface that locally hinders relative motion of the fluid particle surface. When the concentration is sufficiently high so that the surface is fully covered by surfactants, the surface is said to be "fully contaminated." This yields a fully immobile surface consistent with a zero-velocity boundary condition on the interface

Table 3.1 Typical surfactant concentrations and bubble contamination conditions for bubbles in various levels of water purity (Clift et al., 1978; McLaughlin, 1996).

Water type	Surfactant concentration (g/L)	Conditions for $d < 0.7$ mm	Conditions for 0.7 mm $< d < 5$ mm
Tap water	$\sim 2 \times 10^{-1}$	Fully contaminated	Fully contaminated
Distilled water	$\sim 1 \times 10^{-4}$	Fully contaminated	Partially contaminated
Hyperclean	$\sim 1 \times 10^{-5}$	No data	Clean

(3.11c). This immobility eliminates internal recirculation so that the external flow field velocity and pressure will be the same as that for a solid particle. As a result, the drag for a fully contaminated fluid particle in creeping flow will be the same as that for a solid particle so that

$$f \approx 1 \qquad \textit{for fully contaminated conditions at } Re_p \to 0. \tag{3.44}$$

If the surfactant concentration only partially inhibits surface motion, this condition is defined as a "partially contaminated" surface. In this case, the creeping-flow drag correction will have a lower bound given by the clean Hadamard–Rybczynski condition (3.41) and an upper bound set by the Stokes drag (3.44). More details on the influence of surfactant concentration on drag are given in §9.3.1, where reduced surface mobility tends to increase drag.

The quantity of surfactants varies with the type of liquid. Surfactants are most likely to appear in water because many organic substances are easily dissolved or dispersed in water (and are thus notoriously difficult to remove). They naturally occur in high concentrations in most environmental water systems (rivers, lakes, and oceans), so that bubbles in these systems can be considered fully contaminated. Even tap water has enough contaminants to generally yield fully contaminated conditions. As shown in Table 3.1, distilled water (water obtained by boiling water and condensing the steam into another container) reduces the surfactant concentration levels by three orders of magnitude. This mitigates contamination for bubbles with diameters greater than 0.7 mm, but smaller bubbles (which have a higher surface-to-volume ratio) will generally still be fully contaminated. Additional purification methods (deionization, reverse osmosis, UV light, etc.) can produce "hyperclean" water, which has an order of magnitude decrease in surfactant concentration. This allows smaller bubble sizes to be partially contaminated and larger bubbles to be nearly clean so that (3.43b) applies for creeping flow. For nonaqueous liquids, contaminants are more easily removed so that clean conditions can be possible, so that (3.41) can apply for creeping flow.

3.2.3 Sphere in Inviscid Flow and the Paradox

The creeping-flow condition discussed in the preceding section assumes that convection effects are negligible compared to viscous effects. While there are theoretical extensions to include weak convection effects for small but finite values of Re_p (see §9.2.1), there is no closed-form analytical viscous flow solution for $Re_p > 1$ because of the nonlinearities of the Navier–Stokes equation. However, if the viscous effects are ignored completely, one may obtain an analytical solution for the flow around a sphere based on the potential flow theory of §2.4. As will be shown, this inviscid flow solution can help describe the pressure distribution over the front of the particle at high Re_p conditions. In addition, added-mass effects at high Reynolds numbers are reasonably described with the inviscid flow solution. Thus, the inviscid flow limit can serve as a helpful building block at high Reynolds numbers, and the basic derivation steps are outlined in the following.

For incompressible inviscid flow over a stationary sphere, the flow field is governed by the potential flow PDE of (2.59):

$$\nabla^2 \psi = 0 \qquad \textit{for}\ r \geq r_p. \tag{3.45}$$

The surface boundary condition for inviscid flow is given by the slip condition of (2.62b) and the far-field condition is the same as in (3.10a), yielding:

$$\mathbf{U} \cdot \mathbf{n} = 0 \ \ i.e.\ \ U_r = 0 \qquad \textit{for}\ r = r_p. \tag{3.46a}$$

$$\mathbf{U} = \mathbf{U}_\infty = U_\infty \mathbf{i}_x \qquad \textit{for}\ r \to \infty. \tag{3.46b}$$

For the inviscid PDE and these boundary conditions, the stream function solution and the corresponding radial and tangential velocities can be obtained (White, 2016) as follows:

$$\psi = -\frac{1}{2} U_\infty r_p^2 \sin^2\theta \left(\frac{r^2}{r_p^2} - \frac{r_p}{r} \right). \tag{3.47a}$$

$$U_r = -U_\infty \cos\theta \left(1 - \frac{r_p^3}{r^3} \right). \tag{3.47b}$$

$$U_\theta = U_\infty \sin\theta \left(1 + \frac{r_p^3}{2r^3} \right). \tag{3.47c}$$

For these results, the first term in the parentheses corresponds to the far-field solution, while the second term corresponds to an inviscid doublet that satisfies the surface boundary condition.

To determine the form drag, one must consider the pressure stresses. The inviscid surface pressure can be obtained with the surface velocity of (3.47c) evaluated at $r = r_p$, combined with the Bernoulli relation (2.68a) to give

$$P_I = P_\infty + \frac{1}{2} \rho_f U_\infty^2 \left(1 - \frac{9}{4} \sin^2\theta \right) \quad \textit{for the surface of a stationary particle.} \tag{3.48}$$

This result can be generalized to include the case where the particle is moving at fixed velocity ($\mathbf{v}$). The far-field velocity is still $\mathbf{U}_\infty$ but the fluid velocity at the particle surface must consider the particle velocity and the slip condition. In particular, the normal component for the fluid and the particle are equal, that is, $\mathbf{U} \cdot \mathbf{n} = \mathbf{v} \cdot \mathbf{n}$. If we instead apply a moving a reference frame at $\mathbf{v}$ (so that $r = 0$ is at the particle centroid and the spherical steady Navier–Stokes equations can be used), the boundary conditions then become

$$\mathbf{U} \cdot \mathbf{n} = 0 \qquad \textit{at } r = r_p. \tag{3.49a}$$

$$\mathbf{U} = \mathbf{U}_\infty - \mathbf{v} \qquad \textit{as } r \to \infty. \tag{3.49b}$$

Comparing this to the boundary conditions of (3.46), one may replace the freestream velocity ($\mathbf{U}_\infty$) in the preceding fixed particle equations with the relative freestream velocity ($\mathbf{U}_\infty - \mathbf{v}$) for a moving particle, so (3.48) becomes

$$P = P_\infty + \frac{1}{2}\rho_f w^2\left(1 - \frac{9}{4}\sin^2\theta\right) \quad \textit{for the surface of a moving particle.} \tag{3.50}$$

The pressure distribution can be made dimensionless by defining the pressure coefficient (C_P) based on the difference between the local and far-field pressures normalized by the dynamic pressure (stemming from the relative particle velocity):

$$C_P \equiv \frac{P - P_\infty}{\frac{1}{2}\rho_f w^2}. \tag{3.51}$$

Note that a pressure with $C_P = 1$ corresponds to the stagnation pressure (2.68a), while $C_P = 0$ corresponds to the freestream pressure. Employing this coefficient and using (3.50) yields the following inviscid result

$$C_P = \left(1 - \frac{9}{4}\sin^2\theta\right) \qquad \textit{for the surface of a moving particle} \tag{3.52}$$

This surface pressure distribution is thus symmetric both top to bottom and front to back, where the latter is shown by Figure 3.4. The pressure maximum is equal to the stagnation pressure at both $\theta = 0$ and $\theta = \pi$ (where the velocity is zero) and drops to at a minimum at $\theta = \pi/2$ (where the velocity is highest). Because of the front-to-back pressure symmetry, there will be zero pressure drag $\left(\mathbf{F}_{\text{pressure}} = 0\right)$. Furthermore, the absence of viscosity (and any viscous stresses) also results in zero friction drag ($\mathbf{F}_{\text{viscous}} = \mathbf{0}$). Therefore, the potential flow solution yields no drag force:

$$\mathbf{F}_D = \mathbf{F}_{\text{pressure}} + \mathbf{F}_{\text{viscous}} = 0 \quad \textit{for inviscid flow past a sphere.} \tag{3.53}$$

This theoretical result contradicts observations of substantial drag on bodies at high Reynolds numbers. This contradiction is known as d'Alembert's paradox, as posed by the French mathematician in 1752. He noted that the "theory, in all rigor," cannot explain the "resistance of fluids." The theory, at that time, was based on potential flow owing to the pioneering work of Euler.

The paradox posed by D'Alembert caused a split between mathematicians and fluid engineers since it was well understood that relative motion causes fluid dynamic drag.

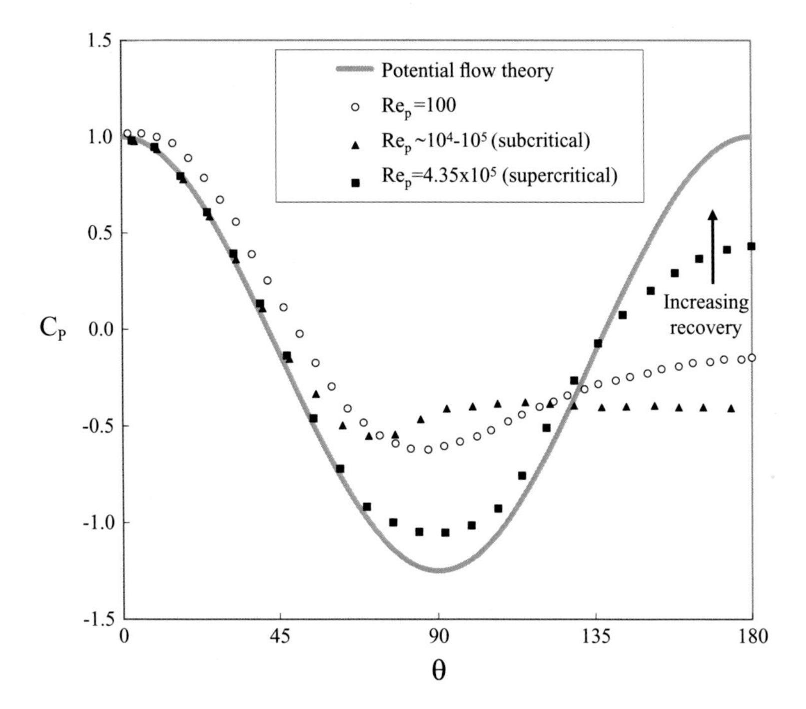

Figure 3.4 Experimental and theoretical pressure coefficient distributions over a solid sphere for incompressible flow (Clift et al., 1978, and Schlichting and Gersten, 2017). In all cases, the front stagnation point ($\theta = 0°$) is based on the freestream stagnation pressure (static pressure plus dynamic pressure). However, viscous effects lead to a loss in stagnation pressure at the rear stagnation point ($\theta = 180°$) and can cause flow separation further upstream. For $\text{Re}_\text{p} = 100$ and subcritical conditions, the pressure at this point does not even recover the freestream static pressure, which leads to large pressure drag. However, supercritical flow leads to an increasing pressure recovery in the rear (per the arrow in the figure) due to a reduction in flow separation.

Interestingly, Sir Isaac Newton had earlier sided with the engineers when he analyzed drag resistance for spherical projectiles (cannonballs) in 1710. In particular, Newton proposed that the drag linearly scales with the dynamic pressure ($\frac{1}{2}\rho_\text{f}\text{w}^2$) and the projected cross-sectional area ($\pi\text{d}^2/4$ for a sphere). This reasoning can be used to define the drag coefficient (C_D) for an object in flight, as follows:

$$\text{C}_\text{D} \equiv \frac{\text{F}_\text{D}}{\left(\frac{\pi}{4}\text{d}^2\right)\left(\frac{1}{2}\rho_\text{f}\text{w}^2\right)} = \frac{\text{F}_\text{D}}{\frac{\pi}{8}\text{d}^2\rho_\text{f}\text{w}^2}. \tag{3.54}$$

Newton empirically proposed that the drag coefficient for a sphere at high speeds $\left(\text{Re}_\text{p} \gg 1\right)$ was a constant of about 0.5 and that the drag force acts in the direction opposite to the relative velocity. As seen in Figure 3.5, this proposal is reasonable for a significant range of Reynolds numbers but is inconsistent with the d'Alembert's inviscid flow theory.

This paradox (the contradiction of theory and observation) was solved for creeping flow by Stokes (1851). However, Stokes drag yields $\text{C}_\text{D} = 24/\text{Re}_\text{p}$, that is, $\text{C}_\text{D} \to 0$

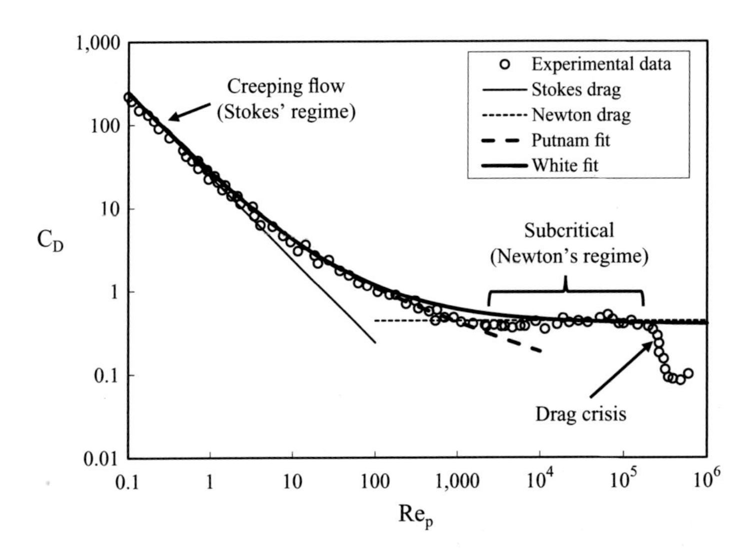

Figure 3.5 Drag coefficient for a smooth solid sphere in incompressible flow conditions at various Reynolds numbers (White, 2016) showing experimental data and various theoretical and empirical drag correlations along with Stokes' and Newton's regimes.

as $Re_p \to \infty$ (see Figure 3.5), so the high Reynolds number limit was still not understood. This part of the paradox was finally solved in 1904 by Prandtl, who recognized that a thin boundary layer can form when the flow is attached, and that viscous effects occur in this small region close to the surface, which results in viscous drag. Furthermore, flow separation of this boundary layer (caused by the adverse pressure gradient in the rear of a body) hinders pressure recovery, which can yield significant form drag. Since then, detailed experimental measurements have shown that the drag coefficient of a sphere varies with Re_p, as shown in Figure 3.5. This variation is discussed in the next section.

3.2.4 Solid Sphere at Finite Reynolds Numbers

Finite Reynolds Number Flow and Pressure

To bridge a model for drag between Stokes' regime (creeping flow) and the Newton's regime (nearly constant drag coefficient), it is helpful to consider how the flow characteristics (on the surface and in the wake) change due to variations in Re_p. For a smooth solid sphere in incompressible flow, one may identify a set of seven Reynolds number regimes, as follows:

- $Re_p < 0.1$: symmetric front-to-back attached flow and a laminar wake.
- $0.1 \leq Re_p < 22$: attached flow over the sphere surface and a laminar wake.
- $22 \leq Re_p < 130$: flow separated at the rear but with a steady laminar wake.
- $130 \leq Re_p < 2,000$: laminar boundary layer separates and a transitional wake.

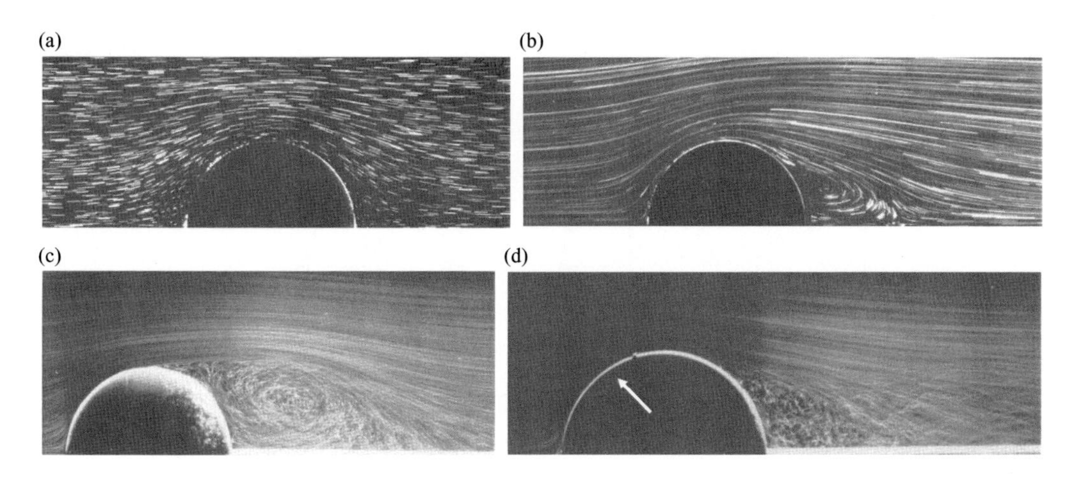

Figure 3.6 Flow over a sphere for various Reynolds number regimes (Werlé, 1980): (a) near-creeping flow at $Re_p = 0.1$, where flow is symmetric from top to bottom, and nearly symmetric from left to right with no flow separation (Stokes' regime); (b) intermediate flow at $Re_p = 56.5$, where separation occurs at the rear, yielding a steady symmetric laminar wake (slight unsteadiness will occur at about $Re_p = 130$); (c) time-averaged subcritical flow at $Re_p = 15{,}000$ ($<Re_{p,crit}$) with laminar boundary layer separation resulting in a fully turbulent wake (Newton's regime); and (d) time-averaged supercritical flow at $Re_p = 30{,}000$ with a trip wire (shown by the white arrow) placed to prematurely transition the boundary layer to turbulent flow before the separation point (the trip wire causes $Re_{p,turb} < 30{,}000$), yielding less flow separation.

- $2{,}000 \le Re_p < 200{,}000$: laminar boundary layer and turbulent wake.
- $200{,}000 \le Re_p < 500{,}000$: boundary layer transitions before separating.
- $Re_p > 500{,}000$: boundary layer is fully turbulent before separating.

The last three regimes are denoted subcritical, critical, and supercritical based on the boundary layer being laminar, transitional, and turbulent before separating, and all have a turbulent wake. These seven regimes also occur for a nonspherical particle shape but with lower Reynolds number values for the ranges. In the following, the regimes are discussed in terms of their flow characteristics (using Figures 3.6 and 3.7) and their drag behavior (using Figures 3.4 and 3.5), with more details in Clift et al. (1978), White (2016), and Schlichting and Gersten (2017).

For $Re_p < 0.1$ (near-creeping), the flow is fully attached, and the streamlines are symmetric as shown in Figure 3.6a. There is an extensive viscous region that extends several radii away from the particle surface. The drag is close to Stokes drag, which has a 0.5 form/friction drag ratio.

For $0.1 \le Re_p < 22$, the viscous region reduces in radial extent and the flow at the rear becomes more stagnant but is still attached. By $Re_p = 1$, the total drag is about 14% larger than that predicted by Stokes drag, but the form/friction drag ratio is still within 1% of that for creeping flow (le Claire et al., 1970). By $Re_p = 20$, the viscous

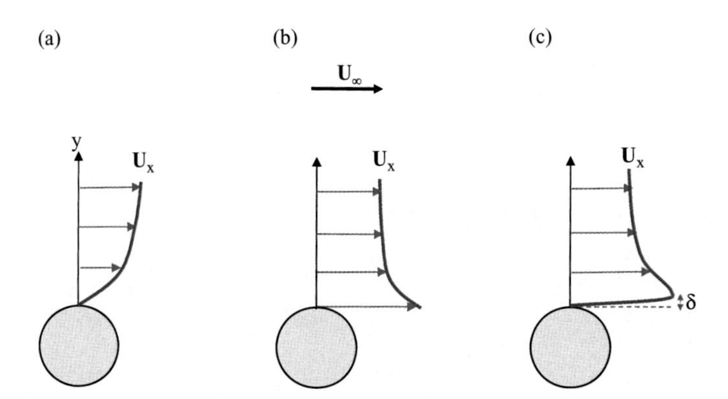

Figure 3.7 Qualitative sketches of velocity profiles on a stationary sphere at $\theta = \pi/2$ for flow moving left to right: (a) $Re_p \sim 1$, where the flow is slower nearer the surface; (b) inviscid conditions, where the flow is faster nearer the surface; and (c) an intermediate Reynolds number condition with $Re_p \gg 1$ (and flow separation at $\theta > \pi/2$), which represents a combination of inviscid and viscous regions, where the latter includes a thin boundary layer of thickness δ, with nearly constant pressure.

region has reduced to only about one particle radius thickness from the surface (Figure 3.7a) and the form/friction drag ratio increases to 0.58.

For $22 \leq Re_p < 130$, flow separation at the rear of the particle first appears around $Re_p > 22$, and the separation point moves upstream from $\theta \approx 180°$ as the Reynolds number increases, for example, to $\theta \approx 180°$ at $Re_p = 100$. The wake yields a steady recirculation pattern (Figure 3.6b). In the front part of the sphere, a thin boundary layer develops. Since pressure is nearly constant across the boundary layer thickness (2.84b), the surface pressure is approximately equal to that above the boundary layer, where the flow is largely inviscid. By $Re_p = 100$, the pressure coefficient (shown in Figure 3.4) yields negative values near $\theta \approx 90°$, indicating that flow above the boundary layer has accelerated to speeds greater than the freestream velocity based on (2.68a) and (3.51). The pressure stays negative for the rear section (it does not recover to the freestream value), yielding significant form drag (due to its difference with the pressure in the front). By $Re_p = 120$, the form/friction drag ratio nearly reaches unity (due to increased pressure effects), and the overall drag is more than threefold larger than that predicted by Stokes drag.

For $130 \leq Re_p < 2,000$ the boundary layer on the front of the sphere becomes thinner as Re_p increases. The flow above the boundary layer tends to the inviscid flow solution with high velocity near $\theta \approx 90°$ (Figure 3.7b). However, inside the boundary layer, the flow comes to rest on the sphere surface (Figure 3.7c). The separation point continues to move forward and reaches $\theta \approx 80°$ at $Re_p = 2,000$, with an increasingly unsteady wake as Re_p increases. The drag at $Re_p = 1,000$ is about 17 times that of Stokes drag and is consistent with $C_D \sim 0.4$.

For $2,000 \leq Re_p < 200,000$ (subcritical flow), the boundary layer on the front becomes thinner but stays laminar and the separation point stays nearly constant at $\theta \approx 80°$ with a highly turbulent wake (Figure 3.6c). As shown in Figure 3.4, C_P for the portion upstream of $\theta \approx 45°$ is similar to that for potential flow over a sphere (3.52), and C_P reaches a minimum at $\theta \approx 70°$, indicating that the peak velocity above the boundary layer occurs just upstream of separation. Downstream of the flow separation, the pressure coefficient is negative (indicating the surface pressure is below that of the freestream pressure) and nearly uniform (consistent with large flow separation). The low aft pressure in the subcritical regime causes the form drag to be high and dominate the drag force; for example, the form/friction drag ratio is about 9:1 for Re_p of 100,000 (Hoerner, 1965).

For $200,000 \leq Re_p < 500,000$ (critical flow), the boundary layer on the front becomes transitional before $\theta \approx 80°$ and the separation point moves rearward due to a fuller boundary layer caused by increased mixing. The higher pressure in the aft section of the sphere reduces the form drag. Since form drag accounts for more than 90% of the drag just prior to this transition, the large drop in form drag due to reduced separation more than counteracts the increase in friction drag on the foreward side due to change from a laminar to a turbulent boundary layer. As such, the overall drag coefficient drops rapidly as shown in Figure 3.5. The phenomenon, termed the "drag crisis," was first observed by the famous French architect Gustav Eiffel, who also conducted wind tunnel research. This can cause C_D to reach as low as ~ 0.12 at $Re_p = 500,000$.

For $Re_p > 500,000$ (supercritical flow), the upstream attached boundary layer becomes fully turbulent before separation, so this regime can be classified as $Re_p > Re_{p,turb}$. In this regime, the separation point is moved to $\theta \sim 130°$, thereby reducing the separation zone (Figure 3.6d). The pressure coefficient in Figure 3.4 reflects the delayed flow separation, whereby the pressure coefficient follows the potential flow solution up to $\theta \sim 130°$ (just before separation). The smaller wake also yields $C_P > 0$ in the aft section, indicating the pressure recovers to values greater than the freestream pressure in this region. This is consistent with a large reduction in pressure drag compared to subcritical flow (Figure 3.4). In the supercritical regime, the boundary layer further thins and there is a gradual upstream movement of the separation point. This combination slowly increases drag coefficient as Reynolds number increases.

In the following, empirical models for the drag coefficient are described for the regimes varying from near-creeping to subcritical, along with some comments about variations in the supercritical Reynolds number.

Finite Reynolds Number Drag Coefficient

In general, the drag force can be normalized with either the drag coefficient (3.54) or the Stokes correction (3.31):

$$\mathbf{F}_D = -\frac{1}{8}\pi\rho_f|\mathbf{w}|\mathbf{w}d^2C_D \quad (3.55a)$$

$$f_{Re} \equiv \frac{F_D}{3\pi d\mu_f w} = \frac{Re_p C_D}{24} \quad (3.55b)$$

for any Re_p.

The RHS of (3.55b) provides the relationship between these two drag normalizations.

As shown in Figure 3.5, Stokes' regime (based on $f = 1$) and Newton's regime (based on a constant C_D) represent the following two important limits for a smooth solid particle in uniform steady flow:

$$C_D = \frac{24}{Re_p} \quad \textit{for a solid sphere at } Re_p \to 0. \tag{3.56a}$$

$$C_{D,crit} \approx 0.42 \quad \textit{for a solid sphere at } 2{,}000 < Re_p < 200{,}000. \tag{3.56b}$$

For other conditions (such as a nonspherical and/or nonsmooth particle), $C_{D,crit}$ is the "critical" drag coefficient that is nearly constant in the subcritical regime, $Re_{p,sub}$ is the subcritical Reynolds number where the wake first becomes fully turbulent, and $Re_{p,crit}$ is the Reynolds number where the attached boundary layer first becomes transitional, as follows:

$$C_D = C_{D,crit} \quad \textit{for } Re_{p,sub} < Re_p < Re_{p,crit}. \tag{3.57}$$

In addition, the minimum Reynolds number for which the drag coefficient first reduces to 0.3 is defined as the transcritical Reynolds number $\left(Re_{p,trans}\right)$ and is about 350,000 for a solid sphere (Figure 3.5).

For a smooth solid sphere as seen in Figure 3.5, the drag at intermediate Reynolds numbers is approximately bounded by the Stokes and Newton regimes of (3.56), with a smooth transition in between these limits. There are many empirical formulae that have been proposed to handle this nonlinear drag regime. For example, the model of Schiller-Naumann (1933), written in terms of the drag coefficient or the Stokes correction factor using (3.55b) is

$$C_D = \frac{24}{Re_p}\left(1 + 0.15\,Re_p^{0.687}\right) \tag{3.58a}$$

$$f_{Re} = 1 + 0.15\,Re_p^{0.687} \quad \textit{for a solid sphere at } Re_p < 1{,}000. \tag{3.58b}$$

For (3.58a), the first term on the RHS is the Stokes limit, while the second term in parentheses is an empirical correction for the intermediate regime. A similar empirical correlation for moderate Reynolds numbers given by Putnam (1961) can be written as follows:

$$C_D = \frac{24}{Re_p}\left(1 + \frac{Re_p^{2/3}}{6}\right) \tag{3.59a}$$

$$f_{Re} = 1 + Re_p^{2/3}/6 \quad \textit{for a solid sphere at } Re_p < 1{,}000. \tag{3.59b}$$

To extend the applicability to higher Reynolds numbers, White (2016) proposed a three-term empirical correlation that can be written as follows:

$$C_D = \frac{24}{Re_p} + \frac{6}{1 + \sqrt{Re_p}} + 0.4 \tag{3.60a}$$

$$f_{Re} = 1 + \frac{Re_p/4}{1 + \sqrt{Re_p}} + \frac{Re_p}{60} \quad \textit{for a solid sphere at } Re_p < 2 \times 10^5. \tag{3.60b}$$

As can be seen with these expressions, it is often more convenient to use the drag coefficient for high Re_p (where C_D tends to a constant) and to use the Stokes correction factor for low Re_p (where f tends to a constant).

Comparing the different models in Figure 3.5, White's fit gives reasonable agreement for a large Reynolds number range, whereas Putnam's model is more accurate for intermediate Reynolds numbers (e.g., 200–6,000). Additional models for C_D are given in Chapter 8.

Notably, $Re_{p,crit}$ and $Re_{p,tran}$ can be reduced by surface changes that can hasten the transition to turbulence. This can include tripping the boundary layer with a wire as shown in Figure 3.6d. More generally, the addition of freestream turbulence or particle roughness can cause the drag crisis to occur at smaller Reynolds numbers. As an example, a baseball with a diameter of 75 mm (3″) and a speed of 36 m/s (80 mph) has a Reynolds number of about 180,000. If the ball were perfectly smooth, this would correspond to subcritical conditions. However, the stitching on the baseball "trips" the attached boundary layer to turbulent, resulting in a lower C_D so that the baseball can better retain faster speeds. The addition of dimples on a golf ball similarly reduces pressure drag by more than 80% by tripping the boundary layer. However, most particles have length and velocity scales much smaller than a baseball or golf ball in flight, so the Reynolds numbers will generally be limited to the subcritical regime of (3.57).

3.3 Surface Forces Due to Accelerations

While drag is the generally the most important steady-state force, there are surface forces associated with the accelerations of the unhindered surrounding fluid and/or the particle that can also be important. These acceleration-based forces include the fluid-stress force, the added-mass force, and the history force. In the following, these three forces are considered for an isolated spherical particle, where more generalized versions that account for nonspherical shapes, Reynolds number effects, and walls are discussed in §10.2–§10.4, while effects of surrounding particles are discussed in §11.2.

3.3.1 Fluid-Stress Force

To derive the fluid-stress force acting on a particle, we first consider the fluid-stress force acting on an element of fluid ($\mathbf{F}_{surf,f}$) for an incompressible single-phase flow. For a fluid element of volume $\forall_f$ and surface area A_f, the fluid-stress force is based on the pressure and viscous stresses acting on that element. As such, this force can be obtained by the area integral of pressure and shear stresses (3.4), which can be converted to a volume integral by using the Gauss divergence theorem (2.2) as follows:

$$\mathbf{F}_{surf,f} \equiv \iint \left(-p\mathbf{n} + K_{ij}n_j\right)dA_f = \iiint \left(-\nabla p + \nabla \cdot K_{ij}\right)d\forall_f \quad \textit{for}\ j = 1, 2, \textit{and}\ 3. \tag{3.61}$$

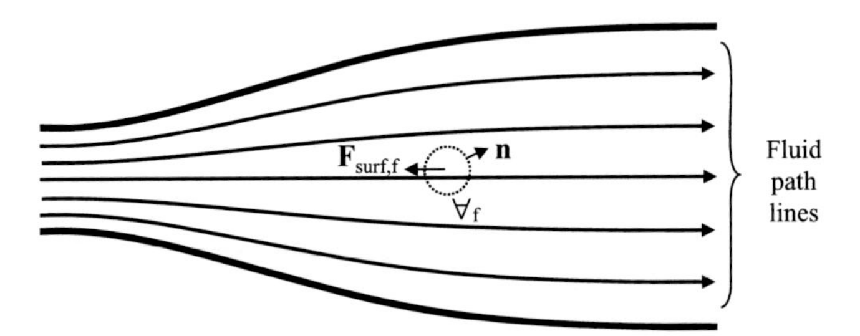

Figure 3.8 Schematic of steady single-phase flow streamlines (fluid path lines) through an expanding geometry with a spherical fluid element shown by dashed circle. This element has a volume $\forall_f$ and a surface outward normal unit vector $\mathbf{n}$. The fluid (including the fluid element) slows down as it moves through the expansion, such that it has a negative acceleration along the fluid path (even though the flow is steady from a Eulerian point-of-view). As such, the fluid stress force ($\mathbf{F}_{surf,f}$) causes the fluid element to decelerate, which is consistent with the velocity reduction as it moves downstream. If the fluid element is replaced by a solid particle with same velocity, acceleration and volume (such that the particle does not hinder the flow), then $\mathbf{F}_S = \mathbf{F}_{surf,f}$.

In this expression, the outward surface unit normal is denoted as $\mathbf{n}$ in vector notation and n_j in tensor notation. Using (2.13b), the fluid-stress force on a fluid element can then be expressed in terms of the fluid and gravitational accelerations. If one assumes a small integration volume (with weak flow gradients), the volume integral can be removed, yielding

$$\mathbf{F}_{surf,f} = \forall_f(-\nabla p + \mathbf{G}) = \forall_f \rho_f \left(\frac{\mathcal{D}\mathbf{u}}{\mathcal{D}t} - \mathbf{g} \right). \tag{3.62}$$

This relationship occurs regardless of the shape of the fluid element so long as it is small. Comparing the LHS and the RHS, the fluid stress is based on gravitational and Lagrangian fluid acceleration along a flow streamline.

Note that Lagrangian acceleration can be significant even for steady flows so long as the fluid element "sees" a change in fluid velocity along its path. For example, let Figure 3.8 represent an inviscid steady air flow through an expanding duct. The flow reduces in velocity as it moves downstream based on the increase in cross-sectional area and mass conservation. The fluid element undergoes this same velocity reduction along its path, yielding nonzero Lagrangian deceleration. In the limiting case of inviscid flow with negligible gravity effects ($\mathbf{G} = \mathbf{g} = 0$), this Lagrangian deceleration is driven solely by the streamwise pressure gradient per (3.62).

Returning to the definition of (3.61), we note that the fluid-stress force is based on a surface integral and is not impacted by what is inside the volume. Therefore, if the volume occupied by the fluid element volume is replaced by that of a particle with the same volume, the fluid stress on the particle ($\mathbf{F}_S$) will be the same as that on the fluid element ($\mathbf{F}_{surf,f}$), as follows:

$$\mathbf{F}_S = \mathbf{F}_{surf,f} \quad \textit{if} \quad \forall_p = \forall_f. \tag{3.63}$$

Therefore, one may employ the unhindered pressure and velocity fields of the surrounding flow (i.e., the values theoretically extrapolated to the particle centroid) to get the fluid-stress force acting on a particle with the same volume:

$$\mathbf{F}_S = \forall_p(-\nabla p + \mathbf{G})_{@p} = \rho_f \forall_p \left(\frac{\mathcal{D}\mathbf{u}}{\mathcal{D}t} - \mathbf{g}\right)_{@p}. \tag{3.64}$$

While this result assumes incompressible flow, it makes no assumptions regarding the particle shape, interface condition, or Reynolds number.

As mentioned previously, the fluid-stress force on a particle can be important even when the flow is steady if the velocity field is nonuniform. Even if the flow is steady and uniform, fluid stress can be nonzero due to the gravity term. In particular, (3.64) for an unhindered flow that is steady and uniform (so the viscous stress and flow acceleration terms are zero) yields the hydrostatic pressure gradient of (2.66):

$$\mathbf{F}_S = -\forall_p(\nabla p)_{grav} = -\forall_p(\rho_f \mathbf{g}) \equiv \mathbf{F}_{buoy} \qquad \textit{for steady uniform flow.} \tag{3.65}$$

The RHS defines the buoyancy force as the fluid stress for steady uniform flow, which acts in the opposite direction of the gravitational force (3.3). In addition, the buoyancy force is associated with the fluid density and particle volume:

$$m_{f@p} \equiv \rho_f \forall_p. \tag{3.66}$$

This is the "displaced mass" and is defined as the mass of a fluid element that has the same volume as that of the particle. This displaced mass is the basis for Archimedes' principle of buoyancy. Thus, buoyancy is a surface force owing to hydrostatic pressure gradients and will balance with the gravitational force if the particle density equals the surrounding fluid density.

To summarize, the fluid-stress force acting on a particle is equal to the fluid dynamic surface forces that would act on a fluid element at the same location with the same volume. This force incorporates the effects of both Lagrangian fluid acceleration and gravitational acceleration on the fluid. However, it does not consider effects of relative acceleration between the fluid and the particle, which are important for the next two surface forces.

3.3.2 Added-Mass Force

In the preceding analysis of the fluid-stress force, it is assumed that both the particle and surrounding fluid move together with the same velocity and acceleration. However, this is not generally true. If the particle has a relative acceleration with respect to the surrounding fluid, there will also be a surface force termed the *added-mass force*. This force arises because any *relative* particle acceleration will cause some of the local fluid mass to be carried along with the particle and thus be accelerated to some degree. For example, imagine the particle has the shape of a cup in a still inviscid flow (so there is no drag). If the cup is accelerated in the direction of the open end, the fluid caught in this portion will accelerate as well (as will some of the fluid at the sides and the back).

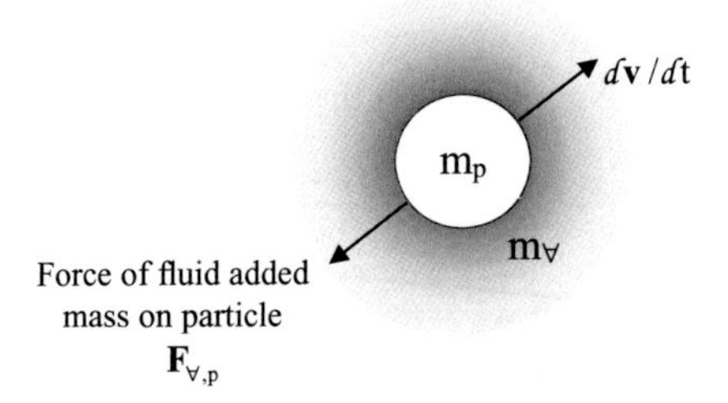

Figure 3.9 Schematic of a particle accelerating in an otherwise still unbounded fluid. The particle mass (m_p) occupies the white circular interior, while the virtual mass portions of the surrounding fluid ($m_\forall$) is shown as the shaded distributed portion outside of the particle. Since this added mass is forced to accelerate with the particle, the fluid exerts an added-mass force on the particle ($\mathbf{F}_{\forall,p}$), which is opposite to the direction of particle acceleration.

For a particle shaped as a sphere, the amount of surrounding fluid that is accelerated with the particle will be less but will still be finite. The fluid very close to the sphere will be accelerated at nearly the particle acceleration $(d\mathbf{v}/dt)$, while the fluid that is farther away will have less acceleration. As qualitatively depicted in Figure 3.9, the shaded regions indicate that the particle acceleration influence on the surrounding fluid varies radially. By considering the integrated effect, the surrounding fluid can be hypothetically divided into two masses: one that is not affected at all by the particle acceleration, and one that is fully affected and accelerates the exact same amount as the particle. The latter mass is called the "added mass" or the "virtual mass" of the particle and is denoted $m_\forall$.

For the example of the cup accelerating, this requires a force on the fluid based on the combined cup mass and added mass with the cup acceleration, that is, $(m_{cup} + m_\forall)(d\mathbf{v}_{cup}/dt)$. Based on the equal and opposite reaction force (Newton's third law), the force of the cup on the fluid is equal and opposite to the fluid force on the cup. Therefore, the force on the cup that is associated with the added mass is $-m_\forall(d\mathbf{v}_{cup}/dt)$. Similarly, the added mass force applied by the surrounding fluid on an accelerating particle $(\mathbf{F}_{\forall,p})$ is based on the particle added mass and particle acceleration $-(d\mathbf{v}/dt)m_\forall$. Thus, as shown in Figure 3.9, the added-mass force is in the opposite direction of the particle acceleration.

To quantify the added-mass force, one may employ a work principle to relate the force on the particle to the kinetic energy change. In particular, the added-mass work done by the fluid on the particle is the product of the force on the particle and the particle velocity. To solve for the added mass, this work can also be equated to the time rate of change of the fluid kinetic energy. Using potential flow theory for inviscid incompressible flow, Lamb (1945) applied this work principle to obtain the added-mass force of a spherical particle for two limits:

- $\mathbf{F}_{\forall,p}$ for particle Lagrangian acceleration $(d\mathbf{v}/dt)$ in a fluid otherwise at rest $(\mathbf{u} = 0)$.
- $\mathbf{F}_{\forall,f}$ for fluid Lagrangian acceleration $(\mathcal{D}\mathbf{u}/\mathcal{D}t)$ with a particle held at rest $(\mathbf{v} = 0)$.

Using superposition for potential flow (2.61), one may linearly combine the flow solutions for these two conditions to obtain the added-mass force for the general condition as

$$\mathbf{F}_{\forall} \equiv \mathbf{F}_{\forall,f} + \mathbf{F}_{\forall,p}. \tag{3.67}$$

In the following, the two limits are considered independently and then combined.

In the first limit of a particle accelerating in an otherwise still fluid, $\mathbf{v}$ changes in time but there is no unhindered velocity ($\mathbf{u} = 0$ throughout the domain). As such, $\mathbf{U}$ represents the local flow velocity due to the particle–induced disturbance, which tends to zero far from the particle ($\mathbf{U}_{\infty} = 0$). The fluid kinetic energy associated with this disturbance can be expressed in terms of the fluid density and fluid velocity integrated throughout the fluid volume as follows:

$$\text{K.E.} \equiv \frac{1}{2}\iiint \rho_f U^2 d\forall_D = \frac{1}{2}\iiint \rho_f(\nabla\Phi \cdot \nabla\Phi) d\forall_D. \tag{3.68}$$

The RHS integral is based the fluid volume in the domain (it does not include the particle volume) and uses (2.57) to write this result in terms of the velocity potential. This volume integral can then be converted to an area integral based on the Gauss divergence theorem, yielding the following:

$$\text{K.E.} = \frac{1}{2}\rho_f \iiint \nabla \cdot (\Phi\nabla\Phi) d\forall_D = \frac{1}{2}\rho_f \iint (\Phi\nabla\Phi \cdot \mathbf{n}) dA_D. \tag{3.69}$$

The RHS is an integral of the fluid surface areas, which are composed of particle surface (A_p) and far-field surface (A_{∞}). For the far field, the particle disturbance is zero $(\mathbf{U}_{\infty} = 0)$, so the kinetic energy contribution on A_{∞} is also zero. On the particle surface, $\mathbf{U}$ is nonzero (since $\mathbf{U} \cdot \mathbf{n} = \mathbf{v} \cdot \mathbf{n}$) so this surface integral must be retained. Notably, the fluid surface for the domain integral has an outward normal which points toward the particle and so opposes the particle outward normal the particle (i.e., $-\mathbf{n}_D = \mathbf{n}_p$).

If one expresses the integration kernel on the RHS of (3.69) as $\Phi\mathbf{U} \cdot \mathbf{n}$, this can be evaluated with the potential flow solution for a sphere moving in the x-direction $(\mathbf{v} = v\mathbf{i}_x)$ using a reference frame moving at $\mathbf{v}$ by replacing U_{∞} with $-v$ in (3.47b) and (3.47c). These velocities can then be returned to a static reference frame by adding back in $\mathbf{v}$, yielding the radial and tangential velocities (and their relation to the velocity potential) as follows:

$$U_r = -v\left(r_p^3/r^3\right)\cos\theta = \frac{\partial\Phi}{\partial r}. \tag{3.70a}$$

$$U_\theta = -\frac{1}{2}v\left(r_p^3/r^3\right)\sin\theta = \frac{1}{r}\frac{\partial\Phi}{\partial\theta}. \tag{3.70b}$$

Thus, the inviscid velocity disturbance in the static reference frame is zero in the far field ($\mathbf{U} = 0$ at $r \to \infty$) and satisfies the slip boundary condition on the surface (e.g., $\mathbf{U} = \mathbf{v}$ at $\theta = 0$ and π). The integrated effect is a local inducement of the surrounding flow in the particle direction.

To quantify the flow energy induced by the particle movement, the velocity potential can be obtained from (3.70) and combined with the velocity to obtain the RHS kernel of (3.69). Integrating over the particle surface, the net kinetic energy of the surrounding fluid is found to be

$$\text{K.E.} = \frac{1}{4}\rho_f \forall_p v^2. \tag{3.71}$$

Recalling that product of the added-mass force applied to the fluid and the particle velocity equals the rate of change of this kinetic energy. Taking this to be equal to the Lagrangian the time derivative of (3.71) yields the following:

$$-\mathbf{F}_{\forall,p} \cdot \mathbf{v} = d(\text{K.E.})/dt = \frac{1}{4}\rho_f \forall_p d\left(v^2\right)/dt = \frac{1}{2}\rho_f \forall_p (d\mathbf{v}/dt) \cdot \mathbf{v}. \tag{3.72}$$

The negative sign on the LHS arises because the added-mass force applied to the particle is equal and opposite to the added-mass force applied to the fluid. The resulting added-mass force on the particle can then be written in terms of particle acceleration and the added mass as follows:

$$\mathbf{F}_{\forall,p} = -\frac{1}{2}\rho_f \forall_p d\mathbf{v}/dt = -m_\forall d\mathbf{v}/dt. \tag{3.73a}$$

$$m_\forall = \frac{1}{2}\rho_f \forall_p. \tag{3.73b}$$

The first equation shows that the direction of the added-mass force is opposite to the particle acceleration, which is consistent with Figure 3.9. The second equation quantifies the added mass, that is, the amount of surrounding fluid that is accelerated when the particle is accelerated.

In the second limit, the particle is stationary ($\mathbf{v} = 0$) but all the surrounding fluid is uniformly accelerating (i.e., $\mathbf{u}$ changes in time, but not in space). In this case, $\mathbf{F}_{\forall,f}$ is the added-mass force on the particle required to keep it stationary while being subjected to a surrounding fluid acceleration. Since this force is equal in magnitude and opposite in direction to the force applied on the fluid, the change in kinetic energy for inviscid conditions can be obtained using a similar approach as in (3.73) to give the following:

$$\mathbf{F}_{\forall,f} \cdot \mathbf{u}_{@p} = \mathcal{D}(\text{K.E.})/\mathcal{D}t = \frac{1}{2}\rho_f \forall_p \left(\mathcal{D}\mathbf{u}_{@p}/\mathcal{D}t\right) \cdot \mathbf{u}_{@p}. \tag{3.74a}$$

$$\mathbf{F}_{\forall,f} = \frac{1}{2}\rho_f \forall_p \left(\mathcal{D}\mathbf{u}_{@p}/\mathcal{D}t\right) = m_\forall \left(\mathcal{D}\mathbf{u}_{@p}/\mathcal{D}t\right). \tag{3.74b}$$

This result shows that the limit of fluid acceleration past a static particle yields the same added mass as that for particle acceleration past an otherwise stationary fluid.

For the general case where both the particle and the fluid are accelerating, the velocity potentials can be linearly combined (3.67), so the net added-mass force is

$$\mathbf{F}_\forall \equiv \mathbf{F}_{\forall,f} + \mathbf{F}_{\forall,p} = \frac{1}{2}\rho_f \forall_p \left(\frac{\mathcal{D}\mathbf{u}_{@p}}{\mathcal{D}t} - \frac{d\mathbf{v}}{dt}\right) \quad \textit{for inviscid flow over a sphere.} \tag{3.75}$$

As such, the added-mass force is based on to the particle's relative acceleration to the fluid.

To compare the effect of added mass to that of fluid stress, the added mass of (3.73b) can be normalized by the displaced fluid mass of (3.66) to define an added-mass coefficient ($c_\forall$):

$$c_\forall \equiv \frac{m_\forall}{m_{f@p}} = \frac{1}{2}. \tag{3.76}$$

The RHS value is based on (3.73b) for a sphere in potential flow.

While the preceding result was derived for inviscid flow, the added-mass force can also be derived in the creeping-flow limit (where viscous effects dominate). The result by Maxey and Riley (1983) can be expressed as follows:

$$\mathbf{F}_\forall = \frac{1}{2}\rho_f \forall_p \left(\frac{d\mathbf{u}_{@p}}{dt} - \frac{d\mathbf{v}}{dt}\right) = -\frac{1}{2}\rho_f \forall_p \frac{d\mathbf{w}}{dt} \qquad \textit{for } \mathrm{Re}_p \to 0. \tag{3.77}$$

The RHS simplification employs the definition of relative particle velocity. The added mass for this force is the same as that obtained for inviscid flow in (3.73b), indicating that $c_\forall = ½$ is appropriate for both the inviscid and creeping-flow extremes. However, the acceleration for the creeping flow (3.77) is based on the particle path, while that for the inviscid flow (3.75) is based on the fluid path. Maxey (1993) noted that this difference is negligible in creeping flow with a uniform free-stream, so one may use (3.75) for the creeping-flow conditions. Furthermore, this equation has been found to be appropriate for spheres with intermediate and high (Newton-regime) Reynolds numbers (as discussed in §10.2). As such, (3.75) is widely used for a spherical particle.

3.3.3 History Force

The last fluid-surface acceleration force is the history force ($\mathbf{F_H}$), which arises due to unsteady viscous stresses on the particle's surface during relative acceleration. Like quasisteady drag, this force acts in the direction opposite to particle relative velocity, and so it is sometimes labeled the unsteady drag force. Like quasisteady drag, the history force is based on viscous stresses and can be derived analytically for creeping flow $(\mathrm{Re}_p \to 0)$. However, unlike quasisteady drag, the history force becomes negligible as $\mathrm{Re}_p \gg 1$.

To understand the genesis of the history force, consider a fluid with high viscosity over a flat plate where initially the plate is stationary ($\mathbf{V} = 0$) and flow is still ($\mathbf{U} = 0$). The plate is then impulsively accelerated to a fixed velocity $\mathbf{V}$ in a direction parallel to the surface, as shown in Figure 3.10a. Just after the plate is impulsively started, only the fluid immediately near the wall is pulled along, creating a small region of high shear near the wall. Over a period of time, fluid shear and viscosity will induce fluid velocities farther and farther above the wall. This thickening viscous layer will cause the wall shear velocity gradient (and associated shear force on the plate) to slowly

(a)

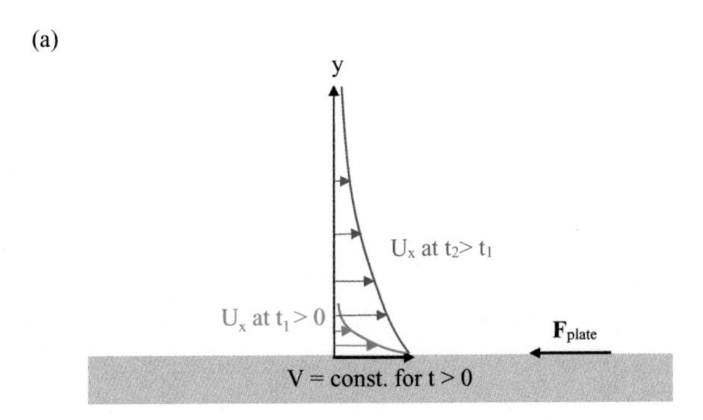

(b)

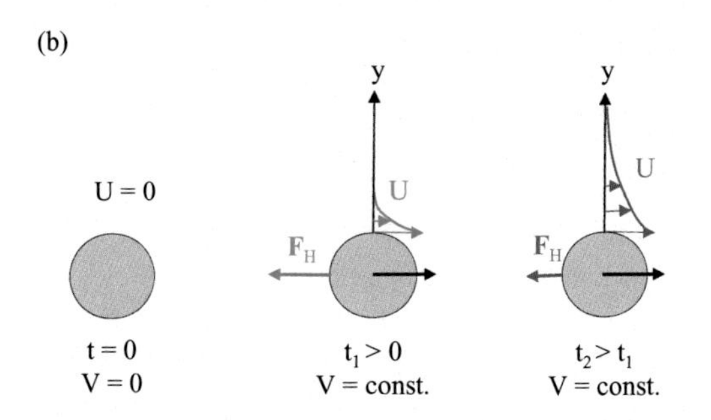

Figure 3.10 History force effect: (a) a horizontal plate initially at rest in a still fluid is instantly accelerated to a new and constant velocity V, inducing a horizontal velocity (U_x) in the surrounding fluid that extends farther upward with increasing time and thus reduces the force of the fluid on the plate ($\mathbf{F}_{plate}$) over time; (b) a particle initially at rest is instantly accelerated to a new and constant velocity V, inducing a flow around the particle (U). In both cases, the velocity profile extends farther away so that the shear at the surface and the net viscous force gradually reduce with time (even though V is a constant after the initial impulse).

decay in time. Note that the plate pulls the fluid to the right while the fluid pulls the plate to the left with an equal and opposite force (shown as $\mathbf{F}_{plate}$ in Figure 3.10a).

In the creeping-flow limit $\left(Re_p \to 0\right)$, the impulsively started plate flow is known as Stokes' First Problem (as it came before the Stokes' solution for drag of a sphere). In this case, the velocity field is governed by the unsteady flow equation (2.72b), without the convective terms, as follows:

$$\rho_f \frac{\partial \mathbf{U}}{\partial t} = \rho_f \mathbf{g} - \nabla P + \mu_f \nabla^2 \mathbf{U}. \tag{3.78}$$

As there is no external pressure gradient and the flow is constrained to horizontal motion only, this equation can be further simplified for the x-momentum equation as follows:

$$\rho_f \frac{\partial U_x}{\partial t} = \mu_f \frac{\partial^2 U_x}{\partial y^2}. \tag{3.79}$$

The solution to this PDE (with the initial and boundary conditions described previously) is a time-dependent fluid velocity field (Crowe et al., 2011) consistent with the evolution shown in Figure 3.10a. The instantaneous shear stress acting at the wall $(y = 0)$ can then be used to obtain the unsteady fluid force on the plate $(\mathbf{F}_{plate})$ based on the plate surface area (A_{plate}), as follows:

$$\mathbf{F}_{plate} = A_{plate}\left(\mu_f \frac{\partial U_x}{\partial y}\right)_{y=0} \mathbf{i}_x = -A_{plate}\sqrt{\frac{\rho_f \mu_f}{\pi t}}\mathbf{V} \qquad \textit{for } V_{t>0} = \text{const.} \qquad (3.80)$$

This viscous fluid force is initially infinite (where the theory is ill-posed) but decays in time and tends to zero when all the fluid above the plate is moving at speed $\mathbf{U} = \mathbf{V}$. This decrease over time is caused by the continued extension of the induced velocity from the plate, thereby reducing the velocity gradient at the wall.

These same creeping conditions may be generalized for plate that accelerates smoothly in time. In this case, the fluid shear force (again found by Stokes) is represented by a time integral that includes the history of all the previous accelerations as follows:

$$\mathbf{F}_{plate} = -\sqrt{\frac{\rho_f \mu_f}{\pi}} A_{plate} \int_0^t \left(\frac{d\mathbf{V}}{d\tau} \frac{1}{\sqrt{t-\tau}}\right) d\tau \qquad \textit{for } V \textit{ continuously varying.} \qquad (3.81)$$

Note that the integral on the RHS employs a dummy variable τ to represent intermediate time periods. In both (3.80) and (3.81), the shear force is proportional to the fluid viscosity and decays in time as $t^{-1/2}$.

Similar to the impulsively started plate of Figure 3.10a, we can consider a spherical particle initially at rest but is then impulsively accelerated to a constant speed (Figure 3.10b). As with the plate, the particle will be subjected to a shear velocity gradient on its surface that creates a friction force. This viscous fluid force is initially infinite (where the theory is ill-posed) but the transient portion decays in time and vanishes when the flow around the particle becomes steady. At that point, only the steady-state drag would remain (i.e., 3.30). The transient portion of this viscous fluid force (beyond the steady-state drag value) is the particle history force.

For the more generalized case of particle acceleration, Basset (1888) obtained an analytical creeping-flow solution $(Re_p \to 0)$ for a spherical particle. The result can be expressed via the particle relative velocity as follows (Crowe et al. 2011):

$$\mathbf{F}_H = -\frac{3}{2} d^2 \sqrt{\pi \rho_f \mu_f} \left[\int_0^t \left(\frac{d\mathbf{w}}{d\tau} \frac{1}{\sqrt{t-\tau}}\right) d\tau + \frac{\mathbf{w}_{t=0}}{\sqrt{t}}\right] \qquad \textit{for } Re_p \to 0. \qquad (3.82)$$

As such, the history force acts in the opposite direction of the relative particle acceleration. In the square brackets, there is an integral for all previous accelerations and a term for the discontinuous initial impulse. Both of these contributions have a decay rate of $t^{-1/2}$, as was found for a flat plate. Given the dependence on accelerations that happened in the past, this force is often called the particle history force. For the creeping-flow condition, it is sometimes called the Basset force.

At higher Reynolds numbers, convection effects complicate the flow, and there is no analytical solution for the history force for $Re_p > 1$. Instead, only semi-empirical expressions (based on simulations and experiments) are available (see §10.2). Fortunately, these results show that convection effects significantly reduce the relative importance of the history force. As a result, the history force can be neglected for $\mathrm{Re_p} > 100$ if $\rho_p/\rho_f > 10$, or if the particle relative acceleration is small compared to gravity.

In summary of §3.3, the fluid-stress, the added-mass, and the history forces are based on a combination of the gravitational acceleration (via buoyancy), the Lagrangian fluid acceleration, and the Lagrangian particle acceleration. All three of these acceleration forces are proportional to the fluid density, and so tend to be greater for particles in a liquid. The history force is the most complicated of them all (as it is an integrodifferential force), but (thankfully) it can often be neglected due to either high particle density ratios or high particle Reynolds numbers.

3.4 Simplified Equation of Motion

3.4.1 Translational Equations of Motion

While the previous section considered various surface force components individually, one may also derive the Equation of Motion for the more general case of particle and fluid acceleration in creeping flow $(\mathrm{Re_p} \to 0)$ to directly determine the combined effects. In this unified approach, all the surface forces in §3.2 and §3.3 are considered simultaneously and (with lengthy derivation) yield a single equation of momentum for a spherical solid particle of fixed mass in a fluid, where relative acceleration is present. This was first obtained as the Basset–Boussinesq–Oseen (BBO) equation (named for the authors), who assumed no particle spin nor freestream vorticity so that lift force can be neglected. This creeping-flow unified equation was rederived by Maxey and Riley (1983), who fixed some minor errors to yield the current form:

$$\left(\rho_p + c_\forall \rho_f\right)\forall_p \frac{d\mathbf{v}}{dt} = -3\pi\mu_f d\mathbf{w} + \forall_p\left(\rho_p - \rho_f\right)\mathbf{g} + \rho_f\forall_p\left(\frac{\mathcal{D}\mathbf{u}_{@p}}{\mathcal{D}t} + c_\forall \frac{d\mathbf{u}_{@p}}{dt}\right)$$
$$-\frac{3}{2}d^2\sqrt{\pi\rho_f\mu_f}\left[\int_0^t \left(\frac{d\mathbf{w}}{d\tau}\frac{1}{\sqrt{t-\tau}}\right)d\tau + \frac{\mathbf{w}_{t=0}}{\sqrt{t}}\right]. \tag{3.83}$$

This is known as the Maxey–Riley equation and was obtained using the unsteady incompressible flow equations of (3.78). Notably, it also consistent with a linear combination of the surface and body force expressions of (3.2), (3.3), (3.30), (3.64), (3.77), and (3.82). This indicates that the linear combination postulated in (3.6) is appropriate for creeping flow. Note that this momentum equation (and the others employed in this section) assumes constant particle mass. However, if the mass changes rapidly with time, then (3.2) should be employed to compare the change in momentum to all the applied forces.

To extend this equation to finite Reynolds numbers (e.g., $\mathrm{Re_p} > 1$), this equation is often modified by (a) employing the Stokes correction factor defined by (3.31), (b) utilizing the inviscid added mass force of (3.75), and (c) neglecting the history force of (3.82) by assuming either high Reynolds numbers, high density ratios, and/or low accelerations. The result is the following:

$$\left(\rho_p + c_\forall \rho_f\right)\forall_p \frac{d\mathbf{v}}{dt} = -3\pi d\mu_f f\mathbf{w} + \forall_p\left(\rho_p - \rho_f\right)\mathbf{g} + \rho_f \forall_p\left(\frac{\mathcal{D}\mathbf{u}_{@p}}{\mathcal{D}t} + c_\forall \frac{\mathcal{D}\mathbf{u}_{@p}}{\mathcal{D}t}\right). \tag{3.84}$$

Even though this linear combination is not based on a unified theory, this Equation of Motion has been found to be reasonably accurate for solid spherical particle motion for a variety of conditions for incompressible flow (Loth, 2000; Crowe et al., 2011). Corrections for nonspherical particle shape, lift, flow compressibility, history forces at finite Re_p, interactions with other particles or walls, etc., are given in Chapters 9–11.

In the following, the Equation of Motion is modified to help interpret particle motion for some limiting cases. First, the LHS of (3.84) is used to define an "effective mass" as the sum of particle mass and the added mass (which accelerate together):

$$m_{eff} = m_p + m_\forall = \left(\rho_p + c_\forall \rho_f\right)\forall_p. \tag{3.85}$$

As noted on the RHS, the particle mass is proportional to particle density, while the added mass is proportional to fluid density. As such, the added mass is generally negligible for a solid particle in a gas $\left(\rho_p \gg \rho_f\right)$, while the particle mass is generally negligible for a gas bubble in a liquid $\left(\rho_p \ll \rho_f\right)$.

The terms associated with gravity on the RHS of (3.84) can be similarly grouped by defining the "effective" gravitational force as follows:

$$\mathbf{F}_{g,eff} = \left(m_p - m_{f@p}\right)\mathbf{g} = \forall_p\left(\rho_p - \rho_f\right)\mathbf{g} = \left(\pi d^3/6\right)\left(\rho_p - \rho_f\right)\mathbf{g}. \tag{3.86}$$

The RHS employs the particle diameter (1.1b) and shows that this force is based on the difference between particle and fluid density. As such, the effective gravitational force becomes the body force $\left(\mathbf{F}_{g,eff} \approx \mathbf{F}_g\right)$ for a solid particle in a gas $\left(\rho_p \gg \rho_f\right)$ but becomes the upward buoyancy force $\left(\mathbf{F}_{g,eff} \approx \mathbf{F}_{buoy}\right)$ for a gas bubble in a liquid $\left(\rho_p \gg \rho_f\right)$.

Combining (3.84)–(3.86) yields a concise Equation of Motion for a particle of constant mass:

$$m_{eff}\frac{d\mathbf{v}}{dt} = -3\pi d\mu_f f\mathbf{w} + \mathbf{F}_{g,eff} + \left[\rho_f \forall_p (1 + c_\forall)\right]\frac{\mathcal{D}\mathbf{u}_{@p}}{\mathcal{D}t}. \tag{3.87}$$

By defining two new parameters as the particle response time $\left(\tau_p\right)$ and the acceleration parameter (R), the equation can be expressed in terms of the particle acceleration:

$$\tau_p \equiv \frac{m_{eff} w}{F_D} = \frac{\left(\rho_p + c_\forall \rho_f\right)d^2}{18\mu_f f} = \frac{4\left(\rho_p + c_\forall \rho_f\right)d}{3\rho_f w C_D}. \tag{3.88a}$$

$$R \equiv \frac{m_{f@p} + m_\forall}{m_{eff}} = \frac{\rho_f + \rho_f c_\forall}{\rho_p + \rho_f c_\forall} = \frac{1 + c_\forall}{\rho_p^* + c_\forall}. \tag{3.88b}$$

$$\frac{d\mathbf{v}}{dt} = -\frac{\mathbf{v} - \mathbf{u}_{@p}}{\tau_p} + (1 - R)\mathbf{g} + R\frac{\mathcal{D}\,\mathbf{u}_{@p}}{\mathcal{D}\,t}. \tag{3.88c}$$

Notably, a high-density particle in a gas $\left(\rho_p \gg \rho_f\right)$ yields $R \to 0$ while a spherical bubble in a liquid $\left(\rho_p \gg \rho_f\right)$ yields $R \to 3$ and $c_{\forall} = 1/2$:

$$\frac{d\mathbf{v}}{dt} = -\frac{\mathbf{v} - \mathbf{u}_{@p}}{\tau_p} + \mathbf{g} \qquad \textit{for } \rho_p \gg \rho_f. \tag{3.89a}$$

$$\frac{d\mathbf{v}}{dt} = -\frac{\mathbf{v} - \mathbf{u}_{@p}}{\tau_p} + -2\mathbf{g} + 3\frac{\mathcal{D}\mathbf{u}_{@p}}{\mathcal{D}t} \qquad \textit{for } \rho_p \ll \rho_f. \tag{3.89b}$$

For Newton-based drag with C_D as a constant per (3.56b), τ_p varies with w. For linear drag with f as a constant (3.60b), τ_p is a constant. This linear drag limit is mathematically convenient since it allows the Equation of Motion to be decomposed into uncoupled scalar equations for different directions. For example, the x-momentum equation (for v_x) can be solved separately from the y-momentum equation. Employing this linear drag, the following considers the influence of the particle response time and the acceleration parameter in terms of particle motion.

To understand the influence of the particle response time (τ_p) we can neglect gravity and fluid acceleration effects (e.g., a neutrally buoyant particle in steady flow). If we set an initial relative velocity $(\mathbf{w}_o \equiv \mathbf{v}_o - \mathbf{u}_{@p})$, (3.88c) becomes

$$\tau_p d\mathbf{w} = -\mathbf{w}dt \tag{3.90a}$$

for negligible gravity and fluid acceleration effects.

$$\mathbf{w} = \mathbf{w}_o e^{-t/\tau_p} \tag{3.90b}$$

Based on (3.90b), τ_p is the time scale that controls how fast the particle relative velocity is reduced by drag, as shown in Figure 3.11. For example, the relative velocity reduces to 36.8% of its original value by $t = \tau_p$. At long times relative to the response time, the velocity is negligible. As such, $t \gg \tau_p$ yields $\mathbf{w} \to 0$ $(\mathbf{v} \approx \mathbf{u}_{@p})$.

To interpret the influence of the acceleration parameter (R), we consider two limits. First, we assume drag and gravity-based forces can be neglected (e.g., a neutrally buoyant particle moving at the same speed as the fluid). In this case, (3.88c) becomes

$$\frac{d\mathbf{v}}{dt} = R\frac{\mathcal{D}\mathbf{u}_{@p}}{\mathcal{D}t} \qquad \textit{for negligible drag and gravity effects.} \tag{3.91}$$

Thus, the acceleration parameter is the ratio of the particle acceleration to the fluid acceleration (hence its name). For a solid particle in a gas $\left(\rho_p \gg \rho_f\right)$, $R \to 0$ so the fluid acceleration dependence can be neglected. For a spherical bubble in a liquid $\left(\rho_p \gg \rho_f\right)$, $R \to 3$ so the bubble acceleration will be three times greater than that of the fluid (if drag and gravity-based forces are neglected).

For the second limit of R, we neglect drag and fluid acceleration, so (3.88c) becomes

$$\frac{d\mathbf{v}}{dt} = (1 - R)\mathbf{g} \qquad \textit{for negligible drag and fluid acceleration effects.} \tag{3.92}$$

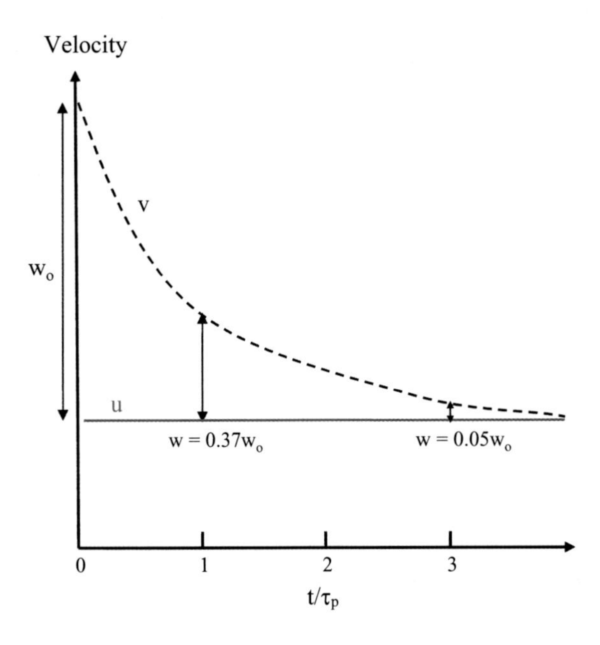

Figure 3.11 Particle velocity change with time (normalized by the particle response time) where the initial relative velocity ($v > u$ at $t = 0$) reduces exponentially for linear drag until equilibrium is reached with the fluid velocity (if gravitational effects are neglected).

As such, a very dense particle ($R \rightarrow 0$) will initially accelerate at $\mathbf{g}$ (downward) due to gravity, while a spherical bubble ($R \rightarrow 3$) will initially accelerate at $-2\mathbf{g}$ (upward) due to buoyancy. The higher magnitude of acceleration for the buoyant particle is due to the combined effects of added-mass and fluid-stress force.

3.4.2 Terminal Velocity

Over a long time in a uniform steady flow, the particle accelerations will cease as the effective gravitational force will be balanced by the drag force. This balance can be used to determine the "terminal" velocity as follows:

$$\mathbf{w}_{\mathrm{term}} = \tau_p(1 - R)\mathbf{g} \quad \textit{for steady-state balance of drag and gravity effects.} \tag{3.93}$$

Based on (3.88a), the terminal velocity can be expressed in terms of either the Stokes correction factor or the drag coefficient, respectively, as follows:

$$\mathbf{w}_{\mathrm{term}} = \frac{\left(\rho_p - \rho_f\right)\mathbf{g}d^2}{18\mu_f f_{\mathrm{term}}} \tag{3.94a}$$

$$\mathbf{w}_{\mathrm{term}} = \left(\rho_p - \rho_f\right)\mathbf{g}\sqrt{\frac{4d}{3\rho_f C_{D,\mathrm{term}}}\frac{1}{|\rho_p - \rho_f||\mathbf{g}|}} \tag{3.94b}$$

$$\mathrm{Re}_{p,term} \equiv \frac{\rho_f w_{term} d}{\mu_f} \quad \textit{for any } \mathrm{Re}_p. \tag{3.94c}$$

These results demonstrate that w_{term} increases with d^2 in the Stokes drag regime ($f = \text{const.}$) but increases with $d^{1/2}$ in the Newton drag regime ($C_D = \text{const.}$).The intermediate drag regime will be bounded by these limits but will depend on the terminal Reynolds number, $\mathrm{Re}_{p,term}$. For example, using the White drag coefficient model of (3.60a) yields the following:

$$C_{D,term} = \frac{24}{\mathrm{Re}_{p,term}} + \frac{6}{1+\sqrt{\mathrm{Re}_{p,term}}} + 0.4 \quad \textit{for a solid sphere at } \mathrm{Re}_p < 2 \times 10^5. \tag{3.95}$$

Starting with an initial guess for $C_{D,term}$, one may iterate using (3.94b,c) and (3.95) to obtain a terminal velocity. As another option, Ferry and Balachandar (2001) proposed a two-step explicit method to find the terminal Stokes correction for a sphere as follows:

$$\mathrm{Re}_{Stokes} = \rho_f |\rho_p - \rho_f| g d^3 / (18\mu_f^2). \tag{3.96a}$$

$$f_{term} \approx [1 + 0.76\mathrm{Re}_{Stokes}(1 + 0.025\mathrm{Re}_{Stokes})]^{0.18} \quad \textit{for } \mathrm{Re}_p < 1{,}000. \tag{3.96b}$$

In the first step, Re_{Stokes} is the particle Reynolds number for creeping-flow conditions $(\mathrm{Re}_p \to 0)$. In the second step, f_{term} is a model for finite Re_p effects that can be used in (3.94a) to compute w_{term} (from which the $\mathrm{Re}_{p,term}$ can also be obtained). This method gives a drag coefficient which is accurate to within 2% of the experimental data of Figure 3.5.

The preceding relations can be used to find the size of water drops and contaminated bubbles for a given terminal Reynolds numbers based on White's drag model. For example, $\mathrm{Re}_{p,term}$ of 0.1 (an upper limit of Stokes drag reasonability) corresponds to terminal velocity conditions for a 40 μm diameter water drop in air or a 60 μm diameter bubble in water. An intermediate $\mathrm{Re}_{p,term}$ of 100 corresponds to a 600 μm diameter water drop in air or a 950 μm diameter contaminated bubble in water. Drops and bubbles with diameters of 1 mm or more will generally deform at terminal conditions (§4.3 and §4.4), so their terminal velocities must instead use a nonspherical drag model that accounts for deformation (§9.3.2). As such, terminal velocities for spheres at subcrtical Reynolds numbers are typically associated with solid particles in air. For example, the Newton regime with Re_p ranging from 2,000–200,000 corresponds to spherical ice particles (hail) ranging from 3.1–40 mm in diameter.

3.5 Mass and Heat Transfer for an Isolated Spherical Particle

The basic relations for heat and mass transfer are next considered for an isolated spherical particle. This section assumes uniform temperature within the particle interior while neglecting radiation and chemical reactions. For a detailed and comprehensive discussion on multiphase heat and mass transfer (including effects of internal temperature variations, radiation, and combustion), the reader is referred to texts of Clift et al. (1978), Soo (1990), Brennen (2005), Michaelides (2006), and Sirignano (2010).

3.5.1 Mass Transfer

For mass transfer, the particle mass can change due to interactions with the surrounding fluid; for example, a drop can condense or evaporate. In the following, we identify these interaction mechanisms and develop quantitative expressions for the rate of transfer. To start, we define the particle mass transfer rate, $\dot{m}_p$, as the change of particle mass per unit time. The rate is thus positive when particle mass increases (e.g., for a condensing drop in a gas, or for a solidifying particle or a cavitating bubble in a liquid) and negative when the particle mass reduces (e.g., for an evaporating drop in a gas, or for a melting particle or a condensing bubble in a liquid). Assuming a constant particle density, the mass transfer rate can then be related to the change in particle diameter, as follows:

$$\dot{m}_p \equiv \frac{d m_p}{d t} = \frac{\pi}{6}\frac{d\left(\rho_p d^3\right)}{d t} = \frac{\pi \rho_p d^2}{2}\frac{d d}{d t}. \tag{3.97}$$

Thus, an increasing particle diameter is consistent with a positive particle mass transfer.

When particle mass transfer occurs, the transferred matter changes phase as it crosses the particle surface, but it retains the same molecular composition. For the example of a condensing drop of water, the transferred matter changes from water vapor to liquid water (both H_2O) when crossing the interface. Therefore, the mass transfer rate will depend on the species mass fraction ($\mathcal{M}$) of the water vapor that exists in the surrounding gas. In the following, $\mathcal{M}$ is more generally denoted as the mass fraction in the surrounding fluid for the species that has the same molecular composition of the transferred matter. For the example case of a salt particle in water, $\mathcal{M}$ is the mass fraction of liquid salt in the water.

Based on Fick's law (2.32), mass transport of a species occurs when there is a gradient in the species mass fraction. For mass transfer to a particle, this gradient is caused by the difference between the mass fraction in the far-field ($\mathcal{M}_\infty$) relative to that in the fluid just above the particle surface ($\mathcal{M}_{surf}$). If $\mathcal{M}_\infty > \mathcal{M}_{surf}$, then mass will be transferred from the surrounding fluid to the particle (and vice versa if $\mathcal{M}_\infty < \mathcal{M}_{surf}$). The actual species mass fraction difference normalized by the differnce for the case of 100% species in the far field is quantified as the Spalding number (B) defined as follows:

$$B \equiv -\left(\frac{\mathcal{M}_\infty - \mathcal{M}_{surf}}{1 - \mathcal{M}_{surf}}\right). \tag{3.98}$$

Note that this definition includes a negative sign. As such, $B = 0$ indicates no mass transfer (since there is no concentration gradient), $B < 0$ represents a positive mass transfer (increasing particle mass), and $B > 0$ represents a negative mass transfer (decreasing particle mass).

To determine $\mathcal{M}_{surf}$, the fluid temperature just above the particle surface can be assumed to be equal to the temperature of the particle $\left(T_{surf} = T_p\right)$, and the mass fraction of the species can be based on saturated conditions at that temperature. For water vapor in air, the mass fraction for saturated conditions (100% relative humidity) can be empirically related (Monteith and Unsworth, 1990) to the local gas temperature (in °C) and air pressure (in N/m^2) as follows:

$$\mathcal{M} = \frac{380\ N/m^2}{p_g} \exp\left[17.27 \frac{T}{T + 237.3}\right] \quad \textit{for saturated conditions.} \tag{3.99}$$

For example, the surface saturated mass fraction of water vapor just above a 20°C water drop in air is 1.44%.

The mass fraction in the far-field ($\mathcal{M}_\infty$) can range anywhere between saturated value (based on the far-field temperature, T_∞) and zero. For example, water vapor at 50% relative humidity in 20°C far-field air has a mass fraction of 0.72%. As such, a drop with the same temperature in this air yields $B = 0.073$ per (3.98), which would cause evaporation. Since the saturated mass fraction increases with temperature (3.99), a warmer droplet would increase $\mathcal{M}_{surf}$, resulting in a larger B and therefore faster evaporation. When the surrounding fluid has spatial variations in $\mathcal{M}$ that exist in the absence of the particle, the far-field value can be replaced by the unhindered mass fraction hypothetically extrapolated to the particle centroid $\left(\mathcal{M}_{@p}\right)$, as follows:

$$B \equiv -\left(\frac{\mathcal{M}_{@p} - \mathcal{M}_{surf}}{1 - \mathcal{M}_{surf}}\right). \tag{3.100}$$

In general, the mass transfer rate from the particle will be proportional to B.

In order to quantify the mass transfer as a function of the Spalding number, we first consider the creeping-flow regime $\left(Re_p \to 0\right)$. In this regime, the convection effects can be neglected and the species transport in the surrounding flow is only by molecular diffusion. In particular, Fick's law (2.32) for incompressible flow indicates that the mass transfer rate is related to the particle surface area $\left(A_p\right)$, the species diffusivity of the particle matter after phase change relative to the surrounding fluid matter ($\Theta_\mathcal{M}$), and the mass fraction gradient, as follows:

$$\dot{m}_p = -A_p \rho_f \Theta_\mathcal{M} \nabla \mathcal{M} \quad \textit{for } Re_p \to 0. \tag{3.101}$$

For the example of a water drop in air, $\Theta_\mathcal{M}$ is the diffusivity of water vapor in the surrounding air and is generally based on the unhindered fluid temperature $\left(T_{f@p}\right)$. For a spherical particle, the surface area is πd^2 and the mass fraction gradient will scale with B/r_p. If one further assumes $|B| \ll 1$, a theoretical result can be obtained by solving the species PDE with an approach similar to that used for Stokes drag (Clift et al., 1978), yielding the following:

$$\dot{m}_p = -2\pi d \rho_f \Theta_\mathcal{M} B \quad \textit{for } Re_p \to 0\ \&\ |B| \ll 1. \tag{3.102}$$

This result can be used to define a dimensionless mass transfer, the Sherwood number (Sh), as follows:

$$\mathrm{Sh} \equiv -\frac{\dot{m}_p}{\pi d \rho_f \Theta_{\mathcal{M}} B}. \tag{3.103}$$

Note that Sh = 2 for the limits of creeping flow and weak gradient limits based on (3.102). One may also use this expression with (3.97) to obtain the particle diameter rate of change as follows:

$$\frac{d\mathrm{d}}{dt} = -\frac{2\rho_f \mathrm{Sh}\, \Theta_{\mathcal{M}} B}{\rho_p d}. \tag{3.104}$$

If the Sherwood number is fixed as a constant, this can be integrated in time to relate the instantaneous particle to its initial value ($d_o = d_{t=0}$) and to define a mass transfer time scale ($\tau_{\dot{m}}$) as follows:

$$d^2 = d_o^2 - \left(\frac{4\rho_f B \Theta_{\mathcal{M}} \mathrm{Sh}}{\rho_p}\right) t = d_o^2\left(1 - \frac{t}{\tau_{\dot{m}}}\frac{B}{|B|}\right) \tag{3.105a}$$

$$\tau_{\dot{m}} \equiv \frac{\rho_p d_o^2}{4\rho_f |B| \Theta_{\mathcal{M}} \mathrm{Sh}} \qquad \textit{for}\ \mathrm{Sh} = \mathrm{const.} \tag{3.105b}$$

This result of (3.105a) is often referred to as the "d-squared law," which indicates particle surface area varies linearly with time at a rate that is inversely proportion to the mass transfer time scale. Furthermore, this time scale of (3.105b) also equals the time required for the particle to disappear if $B > 0$ or for the particle surface area to double if $B < 0$.

If the Reynolds number and finite Spalding number are no longer small, one may employ the semi-empirical Sherwood number model of Ranz and Marshall (1952) for a spherical particle, as follows:

$$\mathrm{Sh} = \left[2 + 0.6 \cdot \mathrm{Re}_p^{1/2} \cdot \mathrm{Sc}_f^{1/3}\right] \frac{\ln(1+B)}{B}. \tag{3.106}$$

This model reduces to Sh = 2 for the creeping flow and small gradients but otherwise includes the effects of enhanced mass transfer due to convection (which occurs at finite Re_p values) as well as the nonlinear effect for Spalding numbers that are no longer small.

3.5.2 Heat Transfer

Whenever there is particle mass transfer (as discussed previously), the phase change can lead to particle heat transfer. Even when there is no mass transfer, heat transfer can occur when the particle temperature is different from that of the surrounding fluid. Based on the Rayleigh transport theorem, the Lagrangian rate of change of the particle's internal energy is equal to the heat transfer to the particle (flux) plus the energy absorbed from the surrounding fluid due to any phase change of the particle matter (generation). This equation can be converted to an enthalpy equation for a constant pressure process (2.38b) as:

$$m_p \frac{d(h_p)}{dt} = m_p \frac{d(c_{p,p} T_p)}{dt} = \dot{Q}_p + h_{phase} \dot{m}_p. \tag{3.107}$$

On the LHS, h_p is the particle enthalpy, which is then written in terms of the particle specific heat at constant pressure $(c_{p,p})$ and the particle temperature (T_p). On the RHS, $\dot{Q}_p$ is the heat transfer to the particle and h_{phase} is the enthalpy required for any particle matter to undergo a phase change to be consistent with the phase of the fluid. For the example of a drop of water in a gas, h_{phase} is equal to the heat of vaporization of water (a positive quantity) from a liquid to a vapor.

During mass transfer, the particle surface temperature generally stays constant, so the LHS of (3.107) is approximately zero. As such, the two RHS terms must be equal in magnitude and opposite in sign. Thus, negative mass transfer to the particle $(\dot{m}_p < 0)$ is associated with a positive heat transfer to the particle $(\dot{Q}_p > 0)$. For example, an evaporating drop must absorb heat from the surrounding fluid to phase change the liquid to a gas.

To solve for the heat transfer rate in (3.107), one may consider the theoretical limit of $Re_p \to 0$ (negligible convection) and no mass transfer $(B = 0)$. In this case, heat transfer is based on Fourier's conduction law (2.25) and is thus proportional to the particle surface area, the surrounding fluid thermal conductivity (k_f), and the temperature gradient. This gradient is proportional to the difference between the unhindered fluid temperature and the particle surface temperature $(T_{f@p} - T_p)$ and the exact solution for heat transfer (Clift et al., 1978) is given as follows:

$$\dot{Q}_p = 2\pi d k_f (T_{f@p} - T_p) \qquad \textit{for } Re_p \to 0 \textit{ and } B = 0. \tag{3.108}$$

As such, a fluid temperature higher than the particle temperature corresponds to positive heat transfer, that is, heat flowing into the particle. In addition, this limiting result can be used to define the dimensionless heat transfer rate as the particle Nusselt number:

$$Nu \equiv \frac{\dot{Q}_p}{\pi d k_f (T_{f@p} - T_p)}. \tag{3.109}$$

Based on this definition, heat transfer to a spherical solid particle in creeping-flow conditions is given by $Nu = 2$. Furthermore, a constant Nusselt number indicates that the heat transfer rate is linearly proportional to the temperature difference and particle diameter.

If the particle specific heat is constant, (3.107) and (3.109) can be combined as follows:

$$\frac{dT_p}{dt} = \frac{6 Nu k_f}{\rho_p c_{p,p} d^2}(T_{f@p} - T_p) + \frac{3 h_{phase}}{c_{p,p} d} \frac{dd}{dt}. \tag{3.110}$$

If the surrounding fluid temperature and the Nusselt number are also constant and if mass transfer is negligible, this ODE can be integrated to obtain the relative temperature difference as a function of time. Denoting the initial particle temperature (at t = 0) as $T_{p,o}$, the normalized temperature difference can be characterized in terms of a thermal response time (τ_T) as follows:

$$\frac{T_p(t) - T_{f@p}}{T_{p,o} - T_{f@p}} = \exp\left(-\frac{6Nu\mathcal{k}_f t}{\rho_p c_{p,p} d^2}\right) = \exp\left(-\frac{t}{\tau_T}\right) \qquad \textit{for } Nu = \text{const.} \tag{3.111a}$$

$$\tau_T \equiv \frac{\rho_p c_{p,p} d^2}{6Nu\mathcal{k}_f} \tag{3.111b}$$

As such, the temperature difference decays exponentially where the rate depends on the thermal response time. At $t/\tau_T = 1$, this difference is reduced to 36.8% of its original value. By $t/\tau_T = 4.6$, it is only 1% of its original value.

As with mass transfer, the effects of convection will serve to increase heat transfer rates relative. The Ranz–Marshall model (1952) for spherical particles incorporates these effects via the surrounding fluid Prandtl number (2.29) as follows:

$$Nu = \left(2 + 0.6 \cdot Re_p^{1/2} \cdot Pr_f^{1/3}\right) \frac{\ln(1+B)}{B}. \tag{3.112}$$

This model reverts to Nu = 2 as $Re_p \to 0$ and B = 0 per (3.108) and (3.109). While the Ranz–Marshall relationships for Nusselt number and Sherwood number are widely employed, additional models have been developed to account for temperature variations within the particle, high mass transfer rates, and particle spacing (Sirignano, 2010 and Michaelides, 2006).

3.5.3 Comparison of Transfer Time Scales

The momentum, heat, and mass transfer mechanisms discussed in this chapter can be summarized in terms of the key ODEs using unhindered fluid values hypothetically extracted to the particle centroid as follows (where the momentum equation assumes constant mass, which is shown later to be often reasonable):

$$\frac{d\mathbf{v}}{dt} = -\frac{\mathbf{v} - \mathbf{u}_{@p}}{\tau_p} + (1-R)\mathbf{g} + R\frac{\mathcal{D}\mathbf{u}_{@p}}{\mathcal{D}t}. \tag{3.113a}$$

$$\frac{dT_p}{dt} = -\frac{1}{\tau_T}\left(T_p - T_{f@p}\right) + \frac{3\mathcal{h}_{phase}}{c_{p,p}d}\frac{dd}{dt}. \tag{3.113b}$$

$$\frac{dd}{dt} = -\frac{1}{\tau_{\dot{m}}}\frac{d_o^2}{2d}\frac{\mathcal{M}_{surf} - \mathcal{M}_{@p}}{\left|\mathcal{M}_{surf} - \mathcal{M}_{@p}\right|}. \tag{3.113c}$$

The associated time scales for the ODEs are given as follows:

$$\tau_p = \frac{\left(\rho_p + c_{\forall}\rho_f\right)d^2}{18\mu_f f}. \tag{3.114a}$$

Table 3.2 Characteristics of momentum, mass, and heat transfer from an incompressible fluid to a spherical particle in terms of the differences that drive these rates, the fluid diffusion properties that characterize the physics, the dimensionless numbers that quantify these rates, the theoretical values of these numbers for creeping flow (and weak gradient for heat and mass transfer), and the associated rate time scales.

Transfer characteristic	Momentum transfer	Mass transfer	Heat transfer
Transferred to the particle	$\mathbf{F}_D$	$\dot{m}_p$	$\dot{Q}_p$
Driving difference	$\mathbf{w} = \mathbf{v} - \mathbf{u}_{@p}$	$\mathcal{M}_{surf} - \mathcal{M}_{@p}$	$T_p - T_{f@p}$
Fluid diffusion property (@p)	μ_f	Θ_p	k_f
Dimensionless numbers	f & C_D	Sh	Nu
Limit for creeping flow (& B→0)	f = 1 & $C_D = 24/Re_p$	Sh = 2	Nu = 2
Transfer time scale	τ_p	$\tau_{\dot{m}}$	τ_T

$$\tau_T \equiv \frac{\rho_p d^2 c_{p,p}}{6 Nu k_f}. \tag{3.114b}$$

$$\tau_{\dot{m}} \equiv \frac{\rho_p d_o^2}{4\rho_f Sh|B|\Theta_{\mathcal{M}}}. \tag{3.114c}$$

As shown in Table 3.2, the transfer rates are driven by differences between fluid and particle properties in a direction that tends to equilibrate that difference. Additionally, the rates of the momentum, mass, and heat transfer are based on fluid diffusion properties, and each has a theoretical solution for the creeping-flow regime.

It is interesting to also compare the time scales listed in the last row of Table 3.2. If one of the time scales is much smaller than the others, it indicates that this characteristic will change faster. If we compare the drag-based particle response time to the temperature-based thermal response time scale for creeping-flow conditions, the ratio can be expressed as follows:

$$\frac{\tau_p}{\tau_T} = \frac{2k_f}{3\mu_f c_{p,p}}\left(\frac{\rho_p + c_{\forall}\rho_f}{\rho_p}\right) \quad \textit{for } Re_p \to 0. \tag{3.115}$$

This ratio depends on the fluid conductivity and viscosity as well as the particle specific heat. However, it is independent of particle size, and the term in the parentheses can be ignored for $\rho_p \gg \rho_f$. For a water drop in air with creeping flow, $\tau_p/\tau_T = 0.22$ (based on Table A.1), indicating that velocity differences decay about five times faster than temperature. However, an air bubble in water for creeping flow yields $\tau_p/\tau_T = 164$ so that the temperature differences decay much faster. It is therefore often reasonable to assume that bubbles are in thermal equilibrium $(T_p \approx T_f)$.

When the drag-based time scale is compared to the mass transfer time scale for creeping-flow conditions, the ratio can be expressed as follows:

$$\frac{\tau_p}{\tau_{\dot{m}}} = \frac{4|B|\Theta_{\mathcal{M}}}{9\nu_f}\left(\frac{\rho_p + c_{\forall}\rho_f}{\rho_p}\right). \tag{3.116}$$

As such, the ratio of these time scales for heavy particles depends only on the Spalding number and the surrounding fluid molecular diffusivity to viscosity (and again is independent of particle size). For water drops in air, the mass diffusivity of water vapor in air is $0.24\,\text{cm}^2/\text{s}$ so that the time scale ratio is $0.71\ |\text{B}|$. Since B is generally much less than unity, the time scale ratio of (3.116) is very small. As such, it is reasonable to assume constant particle mass for the momentum equation, as was done in (3.113a), where the assumption is as follows:

$$\frac{d}{dt}\left(m_p \mathbf{v}\right) \approx m_p \frac{d\mathbf{v}}{dt} \quad \textit{for } \tau_p \ll \tau_{\dot{m}}. \tag{3.117}$$

For this condition, the relative velocity will tend to drive mass transfer effects (and not the other way around). This is true for most multiphase flow conditions.

3.6 Chapter 3 Problems

As you work through the following problems, show all steps and list any needed assumptions.

(P3.1) Starting with the velocity field of (3.20) for a stationary particle, derive the friction drag force (viscous effects) for a moving particle as a function of the relative velocity **w**.

(P3.2) (a) For a clean spherical gas bubble $\left(\mu_p^* \rightarrow 0\right)$ in the creeping-flow limit subjected to a far-field velocity of $U_\infty \mathbf{i}_x$, obtain the pressure-based form drag using the pressure of (3.38). (b) Determine and discuss how this form drag compares to that for creeping flow past a solid sphere and that for inviscid flow past a sphere. (c) Using the stream function of (3.37), determine the maximum surface velocity on the clean bubble in terms of U_∞ and compare this to that for a solid particle in creeping flow and that for a particle in inviscid flow.

(P3.3) Show that the stream function of (3.47a) for inviscid flow over a fixed sphere is consistent with (a) the velocity fields of (3.47b,c, and (b) continuity based on (2.49b). Using these velocity fields, (c) show that they satisfy the far-field and surface velocity boundary conditions, (d) derive the associated surface pressure coefficient pressure as a function of θ, and (e) discuss the flow regions for which this C_P can be used for a particle in the subcritical regime and explain your reasoning.

(P3.4) Consider a 20 micron diameter spherical sand particle with a density of $2,600\,\text{kg/m}^3$ with no initial velocity $(v = 0\,\text{m/s})$ released into a horizontal air cross flow of $0.05\,\text{m/s}$ in the x-direction at NTP, where gravity acts in the y-direction. (a) Explain why it is reasonable to assume the only surface force is quasisteady drag for this case, and then determine the instantaneous particle acceleration using Stokes drag. (b) With the same assumptions as in (a), determine the steady particle velocity when drag and gravity forces are balanced. (c) Determine the particle Reynolds numbers for (a) and (b), then comment on whether the Stokes drag assumption is reasonable for (a) and (b).

(P3.5) For a solid particle moving at a velocity $\mathbf{v}$ in a still fluid ($\mathbf{u} = 0$), (a) obtain expressions for the creeping flow and potential flow surface pressure coefficient as a function of θ and Re_p in the referernce frame of the particle and discuss the qualitative differences; (b) obtain expressions for the radial velocity (U_r) in this same frame for creeping and potential flow; and (c) determine the normalized radial distance from the particle center (r/r_p), where U_r along the x-axis becomes 98% of the far-field value for creeping and potential flow, and discuss the qualitative difference.

(P3.6) Consider a golf ball of 42.677 mm in diameter traveling at 40 m/s in NTP. a) Assuming a smooth sphere, determine the magnitude of the drag force as well as the friction drag and form drag (based on their estimated ratio). b) Repeat a) but include effects of the dimples by assuming they act as a trip wire that reduces the form drag by four-fold but increases the friction drag by two-fold (compared to that of a solid sphere). c) Discuss the flow physics associated with the drag changes from a) to b).

(P3.7) For a particle at a Reynolds number of about 10^5, how would you expect surface roughness or freestream turbulence intensity to qualitatively affect the flow field around the particle and the drag coefficient?

(P3.8) Starting from (3.68), derive the added-mass force and the added mass of (3.73) for a particle accelerating in the x-direction at a velocity $\mathbf{v}$ in an otherwise still inviscid fluid.

(P3.9) Starting from (3.84), obtain (3.88c) and (3.94a,b) and note all assumptions.

(P3.10) Consider a 1 mm diameter solid particle held in place and surrounded by an oil whose velocity is zero at $t = 0$ and then linearly ramps up by $t = 0.005\,\text{s}$ to $0.02\,\text{m/s}$ in the x-direction. The oil has a viscosity of $0.1\,\text{kg/m-s}$ and a density of $910\,\text{kg/m}^3$. Determine the fluid-stress force, the added-mass force, and the history force acting on the particle at $t = 0.002\,\text{s}$. Determine which of these forces has the largest magnitude and which has the smallest; then discuss the qualitative reason for these differences.

(P3.11) Consider a 3 mm diameter copper particle starting from rest falling in a steady horizontal NTP air flow of $4\,\text{m/s}$. Obtain the terminal velocity and terminal repsonse time based on White's drag coefficient, using iteration. b) Using numerical integration, integrate the particle velocity up until one terminal response time, and determine the fraction of the vertical equilibrium velocity achieved at that time. c) Repeat b) for the fraction of horizontal velocity and comment on the difference.

(P3.12) For a 200 micron diameter water drop in air at NTP, (a) obtain w_{term} and Re_p using the non-iterative technique of (3.96) and using the Schiller–Nauman drag coefficient model (with iteration).

(P3.13) Consider a solid spherical particle released with an initial velocity (v_o) into a stagnant gas ($u = 0$). (a) Starting from (3.88c), obtain an analytical expression for $v(t, \tau_p, v_o, g)$ for Stokesian drag $(f = 1)$. (b) Repeat (a) except use Newton drag $(C_D = 0.42)$ with $\tau_p = \tau_{p,term}$. (Hint: Use the tanh function). (c) Compare the analytical solution from (a) and (b) by plotting v/w_{term} as a function of $t/\tau_{p,term}$ and discuss the difference.

(P3.14) A helium balloon of 26 cm in diameter whose latex skin weighs 3.1 grams is inflated to a pressure of 1.05 atmospheres and then is released in the field where the wind is blowing horizontally at 3 m/s. Calculate all the forces at the time of release and its acceleration at release.

(P3.15) Consider the helium balloon of (P3.14) but now inside a car and floating attached to a lightweight string. If the car is accelerating in the x-direction at $3\ m/s^2$, determine the equilibrium angle of the string (relative to the car) during this acceleration.

(P3.16) (a) Starting from (3.107), apply the necessary assumptions and integrate to obtain (3.111).

(P3.17) Consider a water drop initially stationary at 20°C with an initial diameter of 20 μm. This drop is exposed to still air at one atmosphere pressure with 50% humidity and a temperature of 40°C. Assume the water vapor mass diffusivity in air is $3.0 \times 10^{-6} m^2/s$. (a) Use the creeping-flow approximations to determine the particle drag response time, the particle thermal response time, and the mass transfer time scale at the initial conditions. (b) Explain whether the droplet will evaporate or condense.

4 Particle Sizes, Shapes, and Trajectories

Unlike the previous chapter, where we considered a single spherical particle, this chapter will consider nonspherical particles as well as groups of particles. In §4.1, large groups of particles with sizes that vary significantly are described using effective averages. Next, §4.2 discusses nonspherical solid particles and how these shapes affect the motion in free fall. Nonsphericity effects are discussed for drops in free fall in §4.3 and for bubbles in free rise in §4.4. Finally, quasisteady shape deformation due to shear is considered for drops and bubbles in §4.5, while unsteady shape and size deformation dynamics are considered for fluid particles in §4.6.

4.1 Particle Size Distribution

As discussed in Chapter 3, particle size plays a key role in the forces that determine its trajectory and for the heat and mass transfer rates that determine its temperature and size. Thus, it is important to characterize the size distribution for a given multiphase flow system. Particle sizes can be broadly classified by their volumetric diameter (d) as ultrafine (ranging from 10 nm to 1 μm), fine (from 1 μm to 1 mm), or coarse (from 1–100 mm). Examples sizes of manufactured and naturally occurring particles are identified in Figure 4.1. In some cases, the particles are named according to a size range; for example, water drops in air are called "aerosols" if their diameter (d) is less than 1 μm (not visible by the human eye), "droplets" for $1\,\mu m < d < 500\,\mu m$ (e.g., clouds and fog), and "drops" for $d > 500\,\mu m$ (e.g., rain).

The degree of variation of particle sizes in a given multiphase system is also important. A group of particles whose diameters vary significantly is defined as *polydisperse* (e.g., the broad range of particle sizes shown in Figures 1.9c or 4.2). Conversely, a group of particles whose diameters are nearly uniform is defined as *monodisperse* (e.g., Figure 1.13c). The following three sections discuss particle size distributions in terms of basic definitions (§4.1.1), weighted and effective averages (§4.1.2), and analytical size distributions (§4.1.3).

4.1.1 Size Probability Distribution Functions

Measured size distributions for solid particles can be obtained by separating the particles into a series of binned size ranges with sieves (screens), where each bin

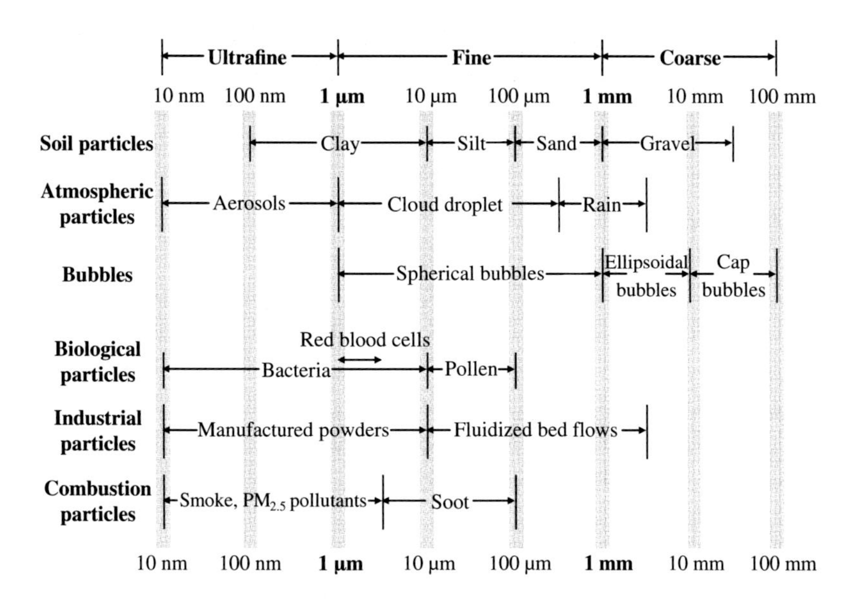

Figure 4.1 Classification of different particles based on based on volumetric diameter as shown by horizontal axis and based on origination as shown by vertical axis.

Figure 4.2 Ceramic particles with spherical shapes and a wide range of sizes (Sun et al., 2019).

contains particles within a specific size range. These bins can then be weighed to determine the mass fraction for each of these size ranges. These mass fractions can be used to establish the mass-based probability distribution function (PDF) as a function of particle diameter, denoted as $\mathcal{P}_\forall(d)$. One may also define a number-based PDF, denoted as $\mathcal{P}_N(d)$, based on the number fraction of particles in each bin. Number-based distributions are often obtained using optical techniques and image counting. While $\mathcal{P}_N$ and $\mathcal{P}_\forall$ are generally different, both are dimensional with units of 1/length (based on the bin width) and can be expressed in the form of discrete or continuous probabilities.

For a discrete number-based probability, $\mathcal{P}_{N,i}$ is the probability per bin width Δd_i that a particle diameter will be in the range of $d_i - ½\Delta d_i$ to $d_i + ½\Delta d_i$ (with average diameter d_i) for a bin of index i. As such, $\mathcal{P}_{N,i}$ is equal to the number of particles in that bin ($N_{p,i}$) relative to the bin width (Δd_i) and the total number of particles in all bins ($N_{p,all}$):

$$\mathcal{P}_{N,i} \equiv \frac{N_{p,i}}{N_{p,all}\ \Delta d_i} \qquad \text{No Index Summation Intended (NISI)} \tag{4.1a}$$

$$N_{p,all} = \sum_{i=1}^{N_b} N_{p,i} \tag{4.1b}$$

In the second expression, N_b is the total number of all bins.

If the bin width is uniform, the bin range can be expressed as $d_i - ½\Delta d$ to $d_i + ½\Delta d$ and the product of this range and the sum of all the size probabilities is unity:

$$\Delta d \sum_{i=1}^{N_b} \mathcal{P}_{N,i} = 1. \tag{4.2}$$

As such, $\mathcal{P}_N$ has units of probability per length. In the rest of this section, we will assume uniform bin widths for simplicity.

An example discrete number-based PDF for spray droplets is shown in Figure 4.3a for a uniform bin width of 10 μm. The probability of occurance in a bin is $\mathcal{P}_N \Delta d$ so changing the bin width changes this probability. For example, the probability of particle being between sizes 45–55 μm is 16.6%, whereas the probability between sizes 49.5–50.5 μm is 1.66% (assuming $\mathcal{P}_N$ is the same). The discrete PDF is sometimes called the rectangular skyline function owing to its discontinuous profile.

The statistical uncertainty of the probability in each bin is equal to $(N_{p,i})^{-1/2}$, assuming the sampling error has a Gaussian distribution. For example, 100 particles in a bin corresponds to a 10% uncertainty for $\mathcal{P}_N$ of that bin, while 10,000 particles are required to reduce the uncertainty to 1%.

To avoid the discontinuous aspects of the discrete PDF, one may instead employ a smooth curve based on a statistically large number of particles (or as a curve fit), as shown in Figure 4.3a for a continuous probability distribution function. In this case, $\mathcal{P}_N(d)$ is the probability per differential bin width and the integral of the PDF over all differential bin widths is unity:

$$\mathcal{P}_N(d) \equiv \frac{N_{<d+\mathrm{d}d/2} - N_{<d-\mathrm{d}d/2}}{N_{p,all}\ \mathrm{d}d}. \tag{4.3a}$$

$$\int_0^\infty \mathcal{P}_N(d)\ \mathrm{d}d = \frac{N_{<\infty}}{N_{p,all}} = 1. \tag{4.3b}$$

In these expressions, $N_{<q}$ represents the number of particles whose diameter is less than q, and $\mathcal{P}_N$ again has units of probability per length. Note that the continuous and discrete PDFs approach each other in the limit of $\Delta d \rightarrow 0$ and $N_p \rightarrow \infty$.

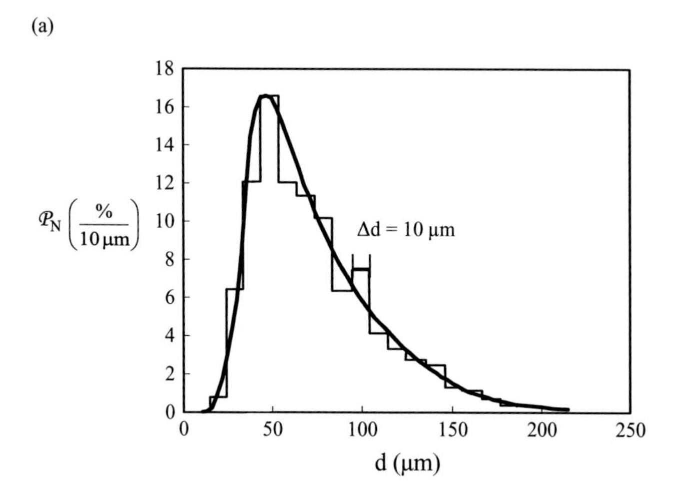

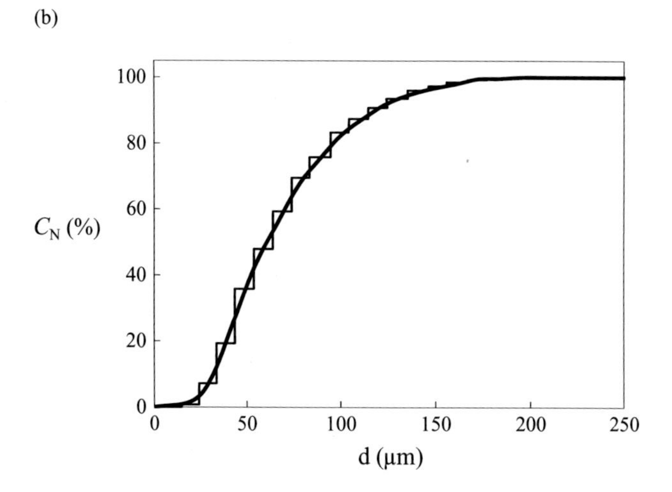

Figure 4.3 Example droplet spray number-based size statistics (Kuo, 1986) showing discrete and continuous forms of (a) the distribution function and (b) the corresponding cumulative distribution function.

Another way to characterize a given particle size distribution is through the cumulative distribution function (CDF). For number-based distributions, the continuous CDF is the integral of the continuous PDF from the smallest diameter to the diameter d, while the discrete CDF is the sum of the discrete PDFs for all the bins with particles less than the diameter d:

$$\mathcal{C}_N(d) \equiv \int_0^d \mathcal{P}_N(d)\, dd. \tag{4.4a}$$

$$\mathcal{C}_{N,i} \equiv \sum_{j=1}^{N_i} \mathcal{P}_{N,j} \Delta d_j. \tag{4.4b}$$

As such, the number-based CDF represents the number fraction of particles that have a diameter of d or less. For example, if the CDF of 30 μm is 0.34, then 34% of the particles have diameters of 30 μm or less. Based on the PDFs of Figure 4.3a, the corresponding discrete and continuous CDFs are given in Figure 4.3b. Note that $\mathcal{C}_N(d)$ monotonically increases with particle size and equals unity when d becomes the largest particle diameter for a group. Note also that the size PDF has units of probability/length, while the size CDF has units of probability. And, as with the PDFs, the discrete and the continuous CDFs approach each other as $\Delta d \rightarrow 0$ and $N_p \rightarrow \infty$.

In addition to the aforementioned *number-based* distributions, one may also characterize a given particle group using *volume-based* distributions. In particular, the discrete volume-based PDF ($\mathcal{P}_{\forall,i}$) is the volumetric fraction per bin width Δd (for an average particle volume in the bin) Also, the product of the bin width and the sum of all the bin PDFs is unity if Δd is a constant, as follows:

$$\mathcal{P}_{\forall,i}(d) \equiv \frac{\forall_{<d+\Delta d/2} - \forall_{<d-\Delta d/2}}{\forall_{p,all}\, \Delta d_i} = \frac{N_{p,i}\left(\pi d^3/6\right)}{\forall_{p,all}\, \Delta d_i}. \tag{4.5a}$$

$$\Delta d \sum_{i=1}^{N_b} \mathcal{P}_{\forall,i} = \frac{\forall_{<\infty}}{\forall_{p,all}} = 1 \quad \textit{for uniform bin width.} \tag{4.5b}$$

In these expressions, $\forall_{<q}$ is the volume of particles whose diameter is less than q, $\forall_{p,all}$ is the sum volume of all the particles, and $\mathcal{P}_\forall$ again has units of probability per length. Correspondingly, the continuous volume-based probability distribution function ($\mathcal{P}_\forall$) can be expressed as follows:

$$\mathcal{P}_\forall(d) \equiv \frac{\forall_{<d+dd/2} - \forall_{<d-dd/2}}{\forall_{p,all}\, dd} = \mathcal{P}_N(d)\frac{N_{p,all}\left(\pi d^3/6\right)}{\forall_{p,all}}. \tag{4.6a}$$

$$\int_0^\infty \mathcal{P}_\forall(d)\, dd = 1. \tag{4.6b}$$

The first expression relates the number-based PDF to the volume-based PDF using the particle volume. In a similar fashion, the continuous volume-based CDF is the integral of the PDF and can be related to the ratio of number-based integrals as follows:

$$\mathcal{C}_\forall(d) \equiv \int_0^d \mathcal{P}_\forall(d)\, dd = \frac{\int_0^d \mathcal{P}_N(d)\, d^3 dd}{\int_0^\infty \mathcal{P}_N(d)\, d^3 dd}. \tag{4.7a, 4.7b}$$

An example a volume-based cumulative distribution function is shown in Figure 4.4. One may similarly define the mass-based PDF and CDF, $\mathcal{P}_m$, and $\mathcal{C}_m$, which are equal

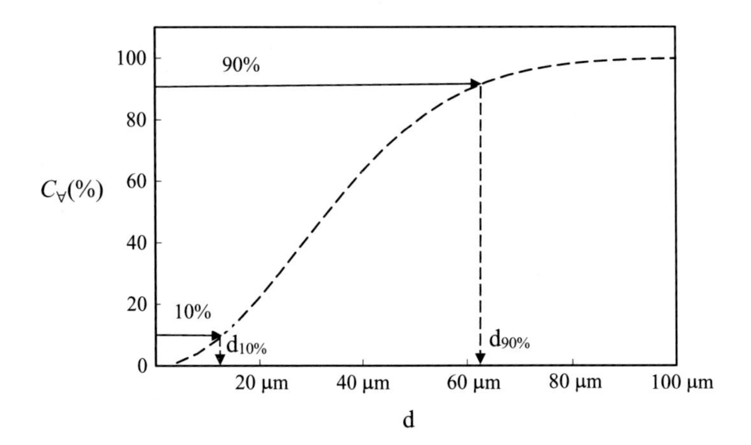

Figure 4.4 Schematic of a volume-based cumulative distribution function indicating the diameters for 10% and 90% mass fractions assuming constant particle density, where $d_{10\%}$ is 13 μm and a $d_{90\%}$ is 62 μm.

to $\mathcal{P}_\forall$ and $C_\forall$ if ρ_p is constant, and where the "mass fraction diameter" $d_{q\%}$ can be defined via C_m as follows:

$$C_m(d) \equiv \int_0^d \mathcal{P}_m(d)\, dd = C_\forall(d) \quad \textit{for uniform particle density.} \tag{4.8a}$$

$$C_m(d_{q\%}) = q\%. \tag{4.8b}$$

As such, $d_{q\%}$ is denoted as the particle diameter that has a CDF mass fraction of q%. For example, the particles of diameter $d_{60\%}$ or smaller will constitute 60% of the total particle mass.

To estimate the relative range of particle diameters in a cloud, it is common to compare the diameters for $C_\forall = 90\%$ and $C_\forall = 10\%$ as shown in Figure 4.4. The ratio between these diameters ($d_{90\%}/d_{10\%}$) indicates the breadth of the particle size distribution. The example in Figure 4.4 has a rather wide distribution with $d_{90\%}/d_{10\%}$ equal to 4.77. Such a wide size distribution is termed a *polydisperse* distribution. In contrast, a *monodisperse* distribution will have $d_{90\%}/d_{10\%}$ of nearly unity. To quantify the diameter ratio for a monodisperse criterion, one must consider distribution of particle velocities, as discussed in the next section.

4.1.2 Weighted Averages and Effective Diameters

Weighted-Average Diameters

For a given size distribution for a group of particles, one may define a "weighted-average diameter," d_{ij}, based on the ratio of the number-based PDF weighted by d^i relative to that weighted by d^j:

$$d_{ij} \equiv \left[\int_0^\infty \mathcal{P}_N(d) d^i \, dd \Big/ \int_0^\infty \mathcal{P}_N(d) d^j \, dd \right]^{\frac{1}{(i-j)}} \quad NISI. \tag{4.9}$$

Note that the exponent 1/(i–j) ensures that d_{ij} has units of length. Using this general notation, one may define the number-averaged diameter (d_{10}), the area-averaged diameter (d_{20}), and the volume-averaged diameter (d_{30}) as follows:

$$d_{10} \equiv \int_0^\infty \mathcal{P}_N(d) d \, dd. \tag{4.10a}$$

$$d_{20} \equiv \left[\int_0^\infty \mathcal{P}_N(d) d^2 \, dd \right]^{\frac{1}{2}}. \tag{4.10b}$$

$$d_{30} \equiv \left[\int_0^\infty \mathcal{P}_N(d) d^3 \, dd \right]^{\frac{1}{3}}. \tag{4.10c}$$

For spherical particles, d_{20} can be used to determine the average particle surface area and d_{30} can be used to determine the average particle volume.

One may also use d_{30} to convert between the number-based PDF of (4.1) and the volume-based PDFs of (4.5a) and (4.6a) in a discrete and continuous sense as follows:

$$\mathcal{P}_{\forall,i} = \frac{\mathcal{P}_{N,i} d_i^3}{d_{30}^3}. \tag{4.11a}$$

$$\mathcal{P}_\forall(d) = \mathcal{P}_N(d) \left(\frac{d}{d_{30}} \right)^3. \tag{4.11b}$$

Based on such relationships, it can be shown that diameter weighting results in $d_{30} \geq d_{20} \geq d_{10}$. For a particle group, the most relevant weighting depends on whether the characteristics of interests are proportional to the particle diameter, area, or volume. In some cases, it is a combination of these averages.

Effective Terminal Velocity

When a multiphase system contains a large number of particles, it may not be feasible to measure or simulate the motion of each particle. However, when they move together at roughly the same velocity, one may define an average group velocity as the *effective velocity*. Such a group velocity is reasonable if the particle size distribution is relatively narrow, that is, if the ratio $d_{90\%}/d_{10\%}$ is near unity (the criteria for this condition will be given later). In this case, the particles moving together can be defined as a *particle cloud*, and the particle diameter that describes their behavior is the *effective* diameter.

To characterize a group-averaged velocity, we will assume the particle motion is dominated by drag and gravitational forces. For the particle cloud, the sum of the effective particle gravitational forces (3.86) and the sum of the drag forces (3.55) can be based on the number-based PDF assuming constant particle density but variable drag coefficient (C_D) as follows:

$$\sum_{i=1}^{N_{p,all}} F_{g,eff,i} = g|\rho_p - \rho_f| \sum_{i=1}^{N_{p,all}} \forall_{p,i} = \frac{\pi g|\rho_p - \rho_f|}{6} N_{p,all} \int_0^\infty \mathcal{P}_N(d) d^3 \, \mathcal{d}d. \tag{4.12a}$$

$$\sum_{i=1}^{N_{p,all}} F_{D,i} = \frac{\pi}{8} \rho_f w_{eff}^2 \sum_{i=1}^{N_{p,all}} C_{D,i} d_i^2 = \frac{\pi}{8} \rho_f w_{eff}^2 N_{p,all} \int_0^\infty \mathcal{P}_N(d) C_D(d) d^2 \, \mathcal{d}d. \tag{4.12b}$$

In these expressions, the summations are over all particles, $\forall_{p,i}$ is the volume of each particle, and w_{eff} is the effective relative velocity of the particle cloud. If we further assume steady-state conditions, the net drag and gravity forces will balance resulting in an effective terminal velocity $(w_{term,eff})$ for the cloud, which can be obtained with the RHS integrals of (4.12a) and (4.12b) if C_D is given as a function of diameter. In the following, this force balance is evaluated for: (1) Newton-based drag, (2) Stokes-based drag, and (3) generalized drag for intermediate Reynolds numbers.

Effective Diameter for Newton Regime

For particles in the Newton-drag regime (e.g., $2{,}000 < Re_p < 200{,}000$ for a sphere), the drag coefficient is approximately constant and can be set as the critical drag coefficient, $C_{D,crit}$. As such, the balance of the RHS integrals of (4.12a) and (4.12b) at terminal velocity yields the following:

$$\frac{3}{4} \rho_f w_{term,eff}^2 C_{D,crit} \int_0^\infty \mathcal{P}_N(d) d^2 \, \mathcal{d}d = g|\rho_p - \rho_f| \int_0^\infty \mathcal{P}_N(d) d^3 \, \mathcal{d}d. \tag{4.13}$$

Based on (4.10b) and (4.10c), the LHS integral is proportional to the area-averaged diameter, d_{20}, while the RHS is proportional to the volume-averaged diameter, d_{30}. Solving for the effective particle diameter and the effective terminal velocity for Newton-based drag yields

$$d_{eff} \equiv \int_0^\infty \mathcal{P}_N(d) d^3 \, \mathcal{d}d \Big/ \int_0^\infty \mathcal{P}_N(d) d^2 \, \mathcal{d}d = \frac{d_{30}^3}{d_{20}^2} = d_{32} \qquad \textit{for } C_D = C_{D,crit}. \tag{4.14a}$$

$$w_{term,eff} = \sqrt{4|\rho_p - \rho_f| d_{32} g / (3\rho_f C_{D,crit})} \tag{4.14b}$$

As such, the effective diameter is d_{32}, which is commonly known as the Sauter mean diameter (SMD). In fact, the SMD is perhaps the most commonly referenced average particle diameter because it represents the combination of gravity and drag effects at high particle Reynolds numbers.

Effective Diameter for the Stokes Regime

If one instead considers small particle Reynolds numbers consistent with creeping flow $(Re_p \to 0)$, the drag force is linearly proportional to the particle diameter, as in

(3.30) for a solid sphere. In this case, the balance of the drag and gravitational forces from (4.12) yields

$$18 w_{term,eff} \mu_f \int_0^\infty \mathcal{P}_N(d) d\, dd = g\left|\rho_p - \rho_f\right| \int_0^\infty \mathcal{P}_N(d) d^3\, dd. \quad (4.15)$$

As a result, the effective diameter and terminal velocity are as follows:

$$d_{eff} \equiv \sqrt{\int_0^\infty \mathcal{P}_N(d) d^3\, dd \Big/ \int_0^\infty \mathcal{P}_N(d) d\, dd} = \sqrt{\frac{d_{30}^3}{d_{10}}} = d_{31} \quad \textit{for } Re_p \to 0. \quad (4.16a)$$

$$\mathbf{w}_{term,eff} = d_{31}^2 \left|\rho_p - \rho_f\right| \mathbf{g}/(18\mu_f) \quad (4.16b)$$

Since this effective diameter (d_{31}) is a ratio of the volume-weighted integral and the diameter-weighted integral, it is known as the *volume-width diameter*. This average diameter applies to the group velocity so long as the drag force is linearly proportional to the relative velocity as in (3.30).

Effective Diameter for an Intermediate Reynolds Number Regime

In the intermediate Re_p range, the relationship between the effective diameter and velocity is more complicated due to the nonlinear drag behavior. To simplify this relationship, one may assume a relatively narrow range of particle Reynolds numbers (e.g., a factor of two change in Re_p) such that the drag force can be approximated as follows (Maude and Whitmore, 1958):

$$F_D \approx \text{const. } \rho_f \nu_f^2 (w\, d/\nu_f)^Y \quad \textit{for moderate changes in } Re_p. \quad (4.17)$$

The exponent on the RHS is defined as the drag-power parameter (Y), which is related to the drag coefficient as follows:

$$C_D \approx c_Y (Re_p)^{Y-2} \quad \textit{for moderate changes in } Re_p. \quad (4.18)$$

Note that Y and c_Y can only be approximated as constants within a moderate range of Reynolds numbers (Loth et al., 2004). These values will approach $Y = 1$ and $c_Y = 24$ for $Re_p \ll 1$ and will approach $Y = 2$ and $c_Y = C_{D,crit}$ for $Re_p \gg 1$. To obtain Y at intermediate Reynolds numbers, one may take the log of both sides of (4.18) and differentiate to yield the following:

$$\frac{\partial[\ln(C_D)]}{\partial[\ln(Re_p)]} \approx Y - 2. \quad (4.19)$$

For solid spherical particles, one may substitute the White drag coefficient (3.60a) into (4.19) to obtain the drag-power parameter as follows:

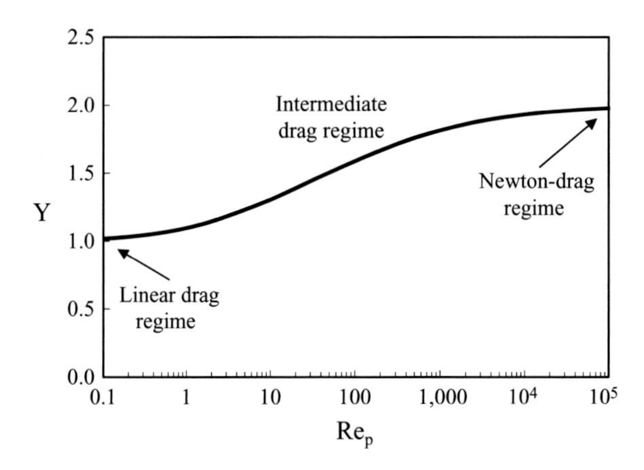

Figure 4.5 Drag-power parameter (Y) as a function of the particle Reynolds number (Re_p) based on the White drag coefficient for solid spherical particles in an incompressible fluid.

$$Y = 2 - \left[\frac{24}{Re_p} + \frac{3\sqrt{Re_p}}{\left(1+\sqrt{Re_p}\right)^2}\right] \Big/ \left[\frac{24}{Re_p} + \frac{6}{1+\sqrt{Re_p}} + 0.4\right]. \tag{4.20}$$

As shown in Figure 4.5, the parameter transitions smoothly from the linear drag limit ($Y = 1$) to the Newton-drag limit ($Y = 2$).

For a nearly monodisperse set of particles, for which c_Y and Y are approximately constant and a group velocity is appropriate, (4.18) can be substituted into with (4.12a) to obtain the effective diameter using the weighted diameter definition of (4.9) and the corresponding effective group terminal velocity as follows:

$$d_{eff} = d_{3Y} \equiv \left[\int_0^\infty \mathcal{P}_N(d) d^3 \, dd \Big/ \int_0^\infty \mathcal{P}_N(d) d^Y \, dd\right]^{\frac{1}{(3-Y)}}. \tag{4.21a}$$

$$w_{term,eff} = \sqrt{4|\rho_p - \rho_f| d_{3Y} g/(3\rho_f C_{D,eff})}. \tag{4.21b}$$

As such, d_{3Y} is the effective diameter for particles moving as a group at an intermediate effective Reynolds number ($Re_{p,eff}$). This diameter can be obtained with an iterative solution with C_D of (3.60a) and Y of (4.20) at this Reynolds number. These expressions revert to (4.16) for the creeping-flow limit (with $Y = 1$ and d_{31}) and to (4.14) for the Newton-drag limit (with $Y = 2$ and d_{32}). Next we consider which size distributions are appropriate for an effective group velocity.

Appropriateness of Effective Diameter

Group-averaged particle motion occurs when the individual velocities do not vary far from the group effective velocity. Such a group of particles can be considered monodisperse. For sparse flow conditions (where particle collisions are rare),

Gauthier et al. (1999) proposed an effective velocity criterion based on the mass fraction diameter ratio as follows:

$$w_{term} \approx w_{term,eff} \quad \textit{for sparse systems} \quad \textit{if } d_{90\%}/d_{10\%} < 1.2. \tag{4.22}$$

This is quite similar to the Crowe et al. (2011) criterion for monodisperse systems. For creeping flow, (4.22) corresponds to $\pm 10\%$ variation in relative velocity for 80% of the particles by mass. This criterion is generally reasonable for a wide range of Reynolds numbers but restricts the particles to a narrow size range. As such, many multiphase flows do not meet this criterion and should be considered polydisperse, which underscores the importance of characterizing the particle size distribution.

The strict criterion of (4.22) can be relaxed if there are many particle collisions, since these collisions tend to equilibrate the particle velocities. One example of this is fluidized beds where a high concentration of particles is levitated with an upward flowing gas. For fluidized beds, group motion is observed at much broader particle sizes but is a strong function of the effective Reynolds number (Loth et al., 2004) and can be estimated as follows:

$$w_{term} \approx w_{term,eff} \quad \textit{for fluidized beds} \quad \textit{if } d_{90\%}/d_{10\%} < 1.4Y. \tag{4.23}$$

The criterion for group behavior is more stringent at lower Reynolds numbers because the terminal velocity is more sensitive to particle diameter, as seen in comparing (4.16b) to (4.14b).

4.1.3 Analytical Probability Distribution Functions

Despite the wide variety of mechanisms that can influence particle size distributions, the shapes of many size distribution PDFs are often similar. Because of this, several standardized analytic functions have been proposed to describe PDF distributions, each requiring only a few fit parameters. This section discusses the two most common analytical PDF models: the Rosin–Rammler distribution and the log-normal distribution.

Rosin–Rammler Distribution

The Rosin–Rammler distribution was originally developed to characterize the cumulative mass PDF ($\mathcal{C}_m$) of pulverized coal (Rosin and Rammler, 1933). In particular, this distribution was formulated based on the mass fraction obtained from a series of sieves, where each sieve would only allow particles that are smaller than the sieve's pore size to pass through. By using a series of finer and finer screens, the $\mathcal{C}_m$ can be obtained for a range of discrete particle diameters. Rosin and Rammler found that a log–log plot involving $(1 - \mathcal{C}_m)$ and particle diameter often yielded a nearly straight line, as shown in Figure 4.6a. A fit to this straight line is the basis of the Rosin–Rammler distribution:

$$\mathcal{C}_m(d) = 1 - \exp\left[-(d/d_{RR})^{\frac{1}{h_{RR}}}\right]. \tag{4.24}$$

This formulation employs two empirical parameters for a particular size distribution: the Rosin–Rammler reference diameter (d_{RR}) and spread parameter (h_{RR}). This

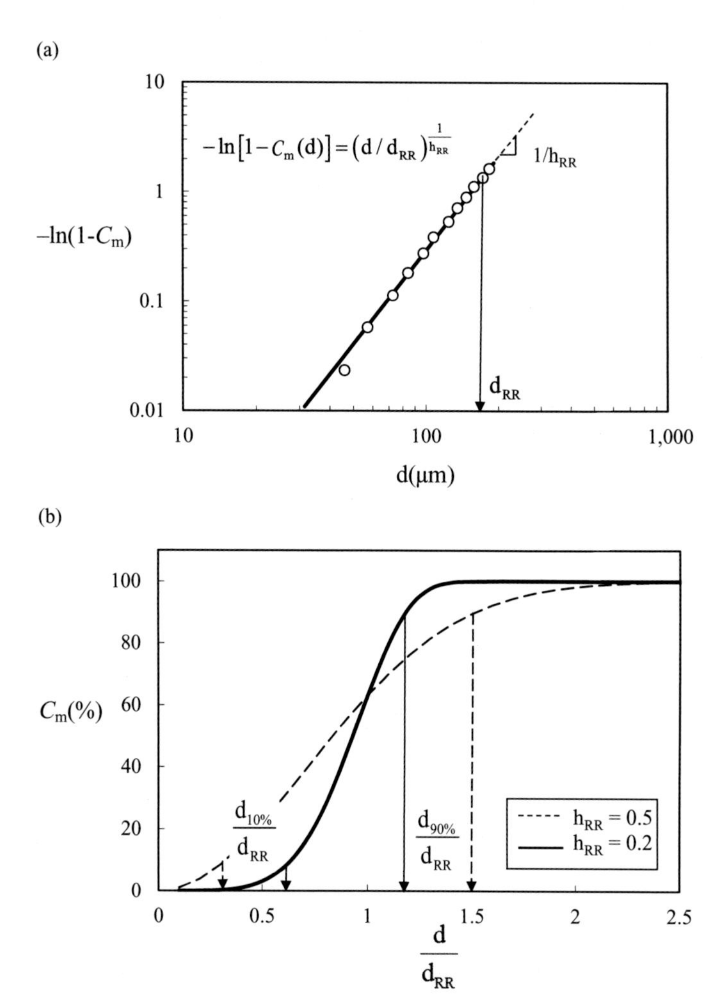

Figure 4.6 Rosin–Rammler mass-based CDF distributions as a function of particle size: (a) comparison of experimental spray data to a fit with $h_{RR} = 0.33$ (Mugele and Evans, 1951); and (b) influence of the Rosin–Rammler spread parameter on diameter ratios, where a larger h_{RR} increases $d_{90\%}/d_{10\%}$ (and distribution broadness).

reference diameter corresponds to $d_{RR} \approx d_{63\%}$ since setting $d = d_{RR}$ in (4.24) yields $C_m = 63\%$. The "spread parameter" (h_{RR}) is the inverse of the log–log slope as shown in Figure 4.6a. While the Rosin–Rammler distribution was developed for pulverized materials, it has been found to be quite robust in representing mass distributions for powders, bubbles, and drops (e.g., Figure 4.6 is for a spray). As a result, the Rosin–Rammler distribution is perhaps the most commonly used PDF for particle mass distribution. To understand the influence of d_{RR} and h_{RR}, we consider various relationship with diameter ratios and weighted diameters.

The spread parameter quantifies the breadth of the size distribution as illustrated in Figure 4.6b, where an increase in h_{RR} yields a net increase in $d_{90\%}/d_{10\%}$. This relationship can be obtained by combining (4.8b) and (4.24) as follows:

$$h_{RR} = 0.324 \ln(d_{90\%}/d_{10\%}). \tag{4.25}$$

The (4.22) criterion for a monodisperse system thus corresponds to $h_{RR} < 0.06$.

While the Rosin–Rammler distribution was developed for mass-based distributions, it can be converted to number-based distributions assuming constant particle density. For example, the number-based Rosin–Rammler PDF is given as follows:

$$\mathcal{P}_N(d) = \frac{(d/d_{RR})^{\frac{1-4h_{RR}}{h_{RR}}}}{h_{RR} d_{RR} \Gamma(1 - 3h_{RR})} \exp\left[-(d/d_{RR})^{\frac{1}{h_{RR}}}\right]. \tag{4.26}$$

This PDF employs the statistical Gamma function, $\Gamma(q)$, with key properties as follows:

$$\Gamma(q) \equiv \int_0^\infty x^{q-1} e^{-x} dx. \tag{4.27a}$$

$$\Gamma(q+1) = q\Gamma(q). \tag{4.27b}$$

$$\Gamma(1/2) = \sqrt{\pi}. \tag{4.27c}$$

$$\Gamma(1) = \Gamma(2) = 1. \tag{4.27d}$$

The first expression uses a dummy variable x, and the Gamma function is plotted in Figure 4.7a.

Based on (4.9), (4.26), and (4.27), the weighted d_{ij} average diameter for Rosin–Rammler distributions with j = 0 is given as follows:

$$d_{i0} \equiv \left[\int_0^\infty \mathcal{P}_N(d) d^i dd\right]^{\frac{1}{i}} = \left[\frac{\Gamma[1 + (i-3)h_{RR}]}{\Gamma(1 - 3h_{RR})}\right]^{\frac{1}{i}} d_{RR}. \tag{4.28}$$

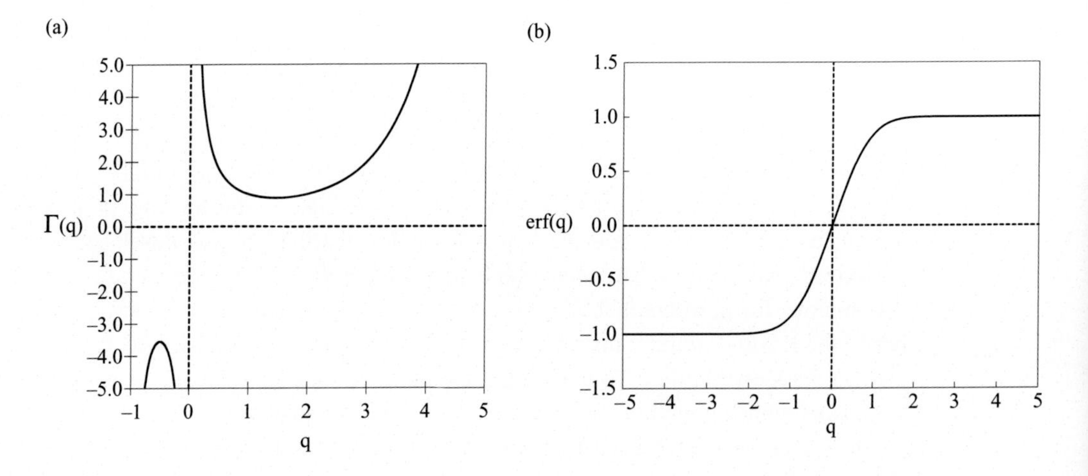

Figure 4.7 The variations of (a) the Gamma function and (b) the error function.

Using this result with (4.14a) and (4.16a), the Sauter mean diameter and the volume-width diameter can each be expressed as a function of the Rosin–Rammler reference diameter and spread parameter as follows:

$$d_{32} = d_{RR} \frac{1}{\Gamma(1 - h_{RR})}. \tag{4.29a}$$

$$d_{31} = d_{RR} \frac{1}{\sqrt{\Gamma(1 - 2h_{RR})}}. \tag{4.29b}$$

Since $\Gamma(0)$ is undefined (Figure 4.7a), this description of d_{32} will also be undefined when $h_{RR} = 1$, while that for d_{31} will be undefined when $h_{RR} = ½$. These h_{RR} values are thus upper limits for which the Rosin–Rammler distribution can be used to obtain these weighted diameters. As an example, the limit of $h_{RR} < ½$ to evaluate d_{31} corresponds to $d_{90\%}/d_{10\%} < 4.68$ based on (4.25). Thus, number-based characteristics derived from the Rosin–Rammler distribution can be problematic when the size distributions are too broad (Mugele and Evans, 1951). To allow for improved performance for number-based characteristics, the log-normal distribution, discussed in the following is often used.

Log-Normal Distribution

The log-normal size distribution employs a Gaussian variation for the log of a normalized diameter to define the number-based PDF as follows:

$$\mathcal{P}_N(d) = \frac{1}{\sqrt{2\pi}\, h_{LN} d} \exp\left\{ -\frac{1}{2} \left[\frac{\ln(d/d_{LN})}{h_{LN}} \right]^2 \right\}. \tag{4.30}$$

In this equation, h_{LN} is the log-normal spread parameter and d_{LN} is the log-normal reference diameter. This PDF can also be integrated based on (4.4a) to give the corresponding number-based CDF as follows:

$$\mathcal{C}_N(d) = \frac{1}{2} \left\{ 1 + \text{erf}\left[\frac{\ln(d/d_{LN})}{h_{LN}\sqrt{2}} \right] \right\}. \tag{4.31}$$

This equation employs the Error function (erf), with key properties as follows:

$$\text{erf}(q) = \frac{2}{\sqrt{\pi}} \int_0^q e^{-x^2} dx. \tag{4.32a}$$

$$\text{erf}(-q) = -\text{erf}(q). \tag{4.32b}$$

$$\text{erf}(0) = 0. \tag{4.32c}$$

$$\text{erf}(\infty) = 1. \tag{4.32d}$$

The first expression again uses a dummy variable x, and the Error function is plotted in Figure 4.7b. Combining (4.31) and (4.32c), it can be shown that $d = d_{LN}$ corresponds to $\mathcal{C}_N = 50\%$, that is, d_{LN} is the diameter that equally divides the particle distribution by number.

(a)

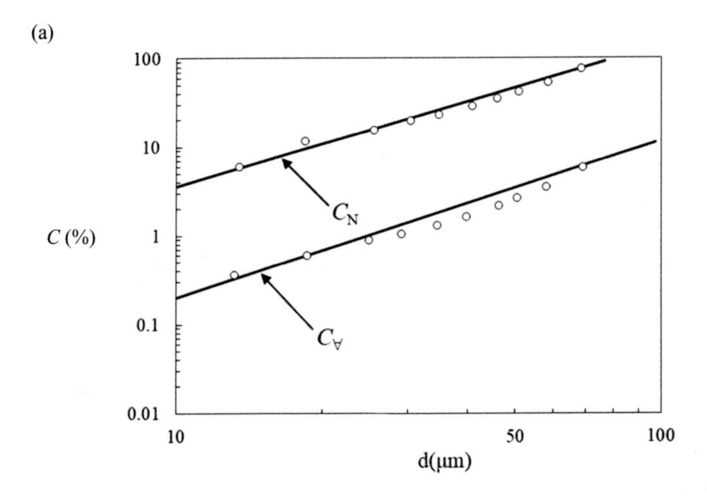

(b)

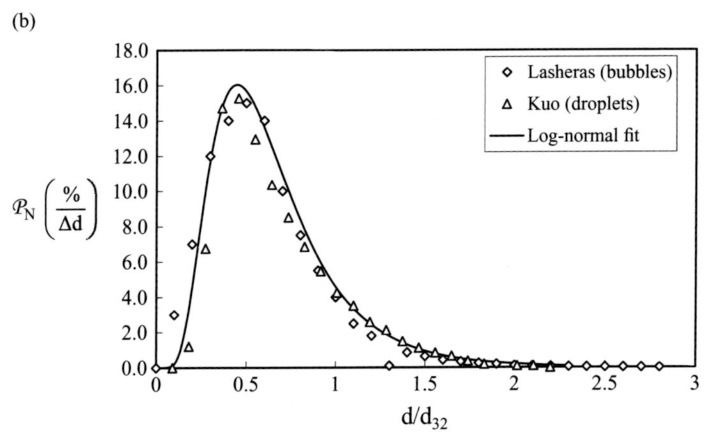

Figure 4.8 (a) An experimental and log-normal CDF (with $h_{LN} = 0.71$ and $d_{90\%}/d_{10\%} \approx 6.2$) for number-based and volume-based size distributions where both will reach 100% for the largest particle size (Mugele and Evans, 1951); and (b) a number-based PDF based on $\Delta d = 0.1d_{32}$ with a log-normal distribution using $h_{LN} \approx 0.22$, for which the number-based peak probability occurs at $d/d_{32} \approx 0.5$ and reasonably matches data for bubbles (Martinez-Bazan et al., 1999b) and droplets (Kuo, 1986).

To obtain the volume-based CDF using the log-normal distribution, one may integrate the volume-weighted PDF per (4.7) to yield

$$C_\forall(d) = \frac{1}{2}\left\{1 + \text{erf}\left[\frac{\ln(d/d_{LN}) - 3h_{LN}^2}{h_{LN}\sqrt{2}}\right]\right\}. \tag{4.33}$$

The difference in the number-based and volume-based CDFs of Figure 4.8a shows that $C_\forall < C_N$ for given particle diameter (a general result). For this example distribution, particles with a diameter of 50 μm or less constitute 50% of the particles by number, but only constitute about 3% of the total particle volume. This difference is a

Table 4.1 Relationship between various broadness ratios ($d_{90\%}/d_{10\%}$ and d_{32}/d_{31}) and the spread parameters for Rosin–Rammler and log-normal distributions.

	Rosin–Rammler		Log-normal	
$\frac{d_{90\%}}{d_{10\%}}$	h_{RR}	$\frac{d_{32}}{d_{31}}$	h_{LN}	$\frac{d_{32}}{d_{31}}$
1	0.00	1.00	0.00	1.00
2	0.23	1.07	0.27	1.04
3	0.36	1.28	0.43	1.10
4	0.45	1.91	0.54	1.16
5	0.52	Undefined	0.62	1.21

result of the d^3 weighting used for $C_\forall$ (4.7), which increases the contributions for the larger particles.

This log-normal distribution can be used to obtain the mass fraction diameter defined by (4.8b). In particular, $d_{50\%}$ (the diameter that divides the particles into sets of equal volume and of equal mass for constant particle density) is given as follows:

$$d_{50\%} = d_{LN} \exp\left(3h_{LN}^2\right). \tag{4.34}$$

Similarly, the log-normal spread parameter can be related to the size ratio by (4.8b) and (4.33):

$$h_{LN} \approx 0.39 \ln\left(d_{90\%}/d_{10\%}\right). \tag{4.35}$$

From this relationship, the monodisperse criterion of (4.22) can be expressed as $h_{LN} < 0.07$ The log-normal mean diameter and spread parameter can also be used to obtain the general d_{ij} weighted diameter using (4.9) and (4.33), as follows:

$$d_{ij} = d_{LN} \exp\left[\frac{1}{2}(i+j){h_{LN}}^2\right]. \tag{4.36}$$

This relationship has no upper limits on h_{LN} for computing d_{32} and d_{32} and is more straightforward than that for the Rosin–Rammler distribution (4.28). Moreover, the log-normal distribution is generally better for experimental number-based distributions (as in C_N of Figure 4.8a), whereas Rosin–Rammler is generally better for mass-based distributions (as in C_m of Figure 4.6a). Notably, the differences between the log-normal and Rosin–Rammler forms will lead to different ratios of d_{32}/d_{31} for the same $d_{90\%}/d_{10\%}$ (especially as the size distribution becomes broader), as shown in Table 4.1. These differences indicate that care should be taken in deciding which distribution model to use.

However, some data are difficult to fit with either of these models as shown in Figure 4.9. While the Rosin–Rammler distribution is a better fit than the log-normal distribution (as expected for a mass-based distribution), the upper limit distribution is even better. The upper limit distribution combines a conventional log-probability distribution with a prescribed maximum particle size to prevent excessive tails at large diameters (Kuo, 1986). In general, the upper limit model improves predictions of

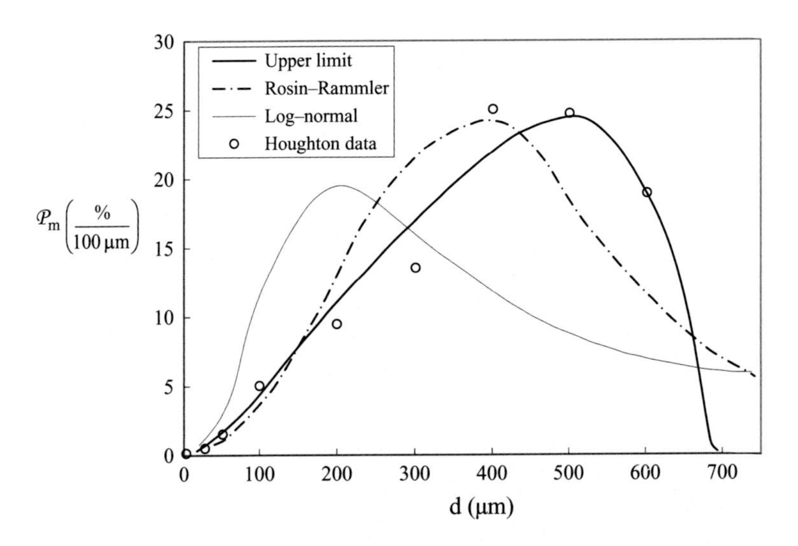

Figure 4.9 Various analytical mass-based PDFs compared to that of an experimental spray distribution with $d_{90\%}/d_{10\%} \approx 6$ (Houghton, 1950; Muegle and Evans, 1951).

mass-based or volume-based distributions for the largest diameters (e.g., Figure 4.9) but requires three parameters (instead of two) and is not as widely used. Other analytical PDF models include the Nukiyama–Tanasaw function (Kuo, 1986), the log-hyperbolic function (Crowe et al., 2011), and evolution distributions (Lasheras, 1998). These models can give improved agreement for specific particle distributions but are not as generally robust nor as widely used as the Rosin–Rammler and log-normal models. In summary, no single analytical model for the PDF size distribution is best suited for all data in terms of both number-based and volume-based characteristics. Instead, a model is best chosen based on the characteristics of primary interest and the general shape of the experimental data.

4.2 Solid Particle Shapes and Motion in Free Fall

4.2.1 Shape Classifications

While solid particle shapes are often approximated as a sphere, nonspherical shapes can occur due to natural or manufacturing processes. Such particles can be classified as either regularly shaped or irregularly shaped. Irregularly shaped solid particles have no geometric symmetry and often occur through the fracture of a particle (Figure 4.10a) or coalescence of particles (Figure 4.10b). In contrast, regularly shaped particles are generally symmetric and often have rounded features or edges (Figure 4.10c,d). Rounded features are often a result of solidification of liquid particles or of weathering and/or smoothing of fragmented particles.

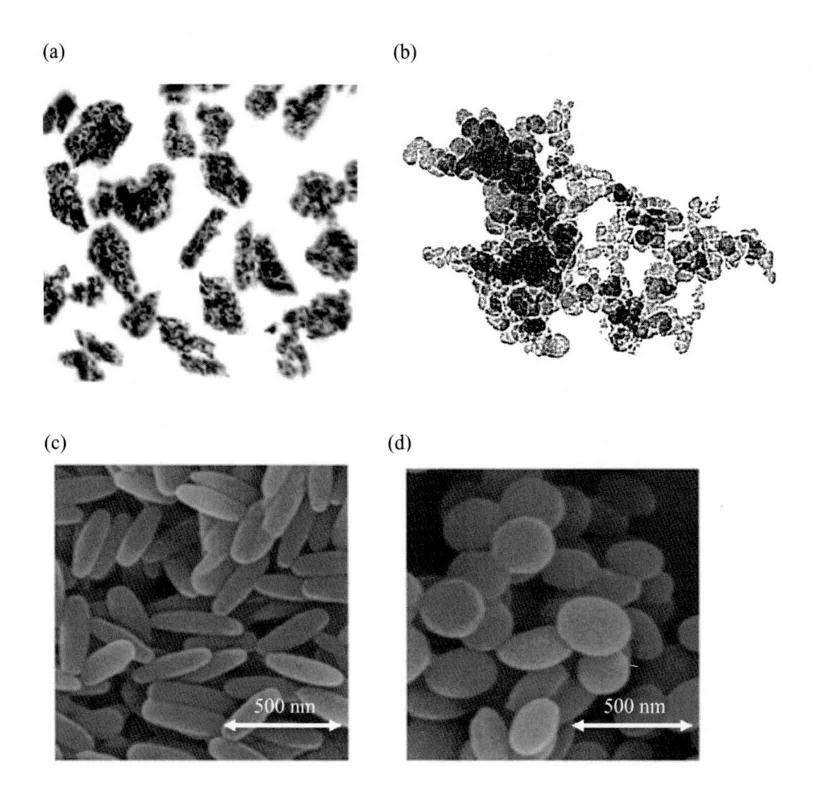

Figure 4.10 Examples of irregularly shaped micron-scale particles: (a) metal dust particles formed by fragmentation (Wostenberg, 2005); and (b) a single (micron-sized) soot particle from a Diesel engine formed by agglomeration of many spherules, which are approximately 30 nm in scale (Roessler, 1982). Examples of regularly shaped polystyrene nanoparticles (Barua et al., 2013) include (c) cylinder-like ellipsoids and (d) disk-like ellipsoids.

In many cases, nonspherical regularly shaped particles can be approximated as *ellipsoids*. Ellipsoids are defined to have elliptical projected areas when viewed from three perpendicular axes. The lengths along on each axis (d_1, d_2, and d_3 as shown in Figure 4.11a) can be called the particle's *diameters*. The product of these diameters is equal to the cube of the volumetric diameter, as follows:

$$\forall_p = \frac{\pi}{6} d_1 d_2 d_3 = \frac{\pi}{6} d^3. \tag{4.37}$$

As such, the sphere is the special case of an ellipsoid when the three lengths are equal ($d_1 = d_2 = d_3$).

Another type of ellipsoid is the spheroid, for which only two of the diameters are equal. In this case, one cross-sectional shape is a circle and the other two are equal ellipses, as in Figure 4.11b,c. This leads to an axis of symmetry along the unique diameter that is perpendicular to the circular cross section formed by the other two equal diameters. As shown in Figure 4.11, the length along the axis of symmetry is denoted $d_{\parallel}$ and the two equal lengths about the axis of symmetry (which form the

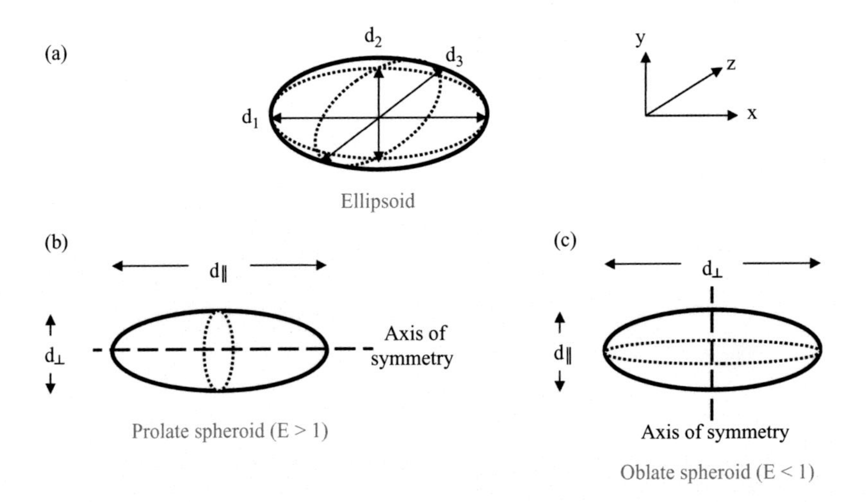

Figure 4.11 Representations of ellipsoidal particles showing (a) general geometry with cross sections represented by dashed lines and diameters defined by arrows; (b) a prolate spheroid; and (c) an oblate spheroid. For both spheroids, the long-dashed line indicates the axis of symmetry, while the short-dashed line is a cross section about this axis. In addition, both spheroids are shown in an orientation with the maximum cross-sectional area facing the y-axis.

circular cross section) are denoted $d_\perp$. The ratio between the parallel and perpendicular lengths defines the particle's aspect ratio (E) as follows:

$$E \equiv d_\parallel / d_\perp. \tag{4.38}$$

Based on (4.37) and (4.38), the volumetric diameter of a spheroid can be expressed in terms of these lengths and the aspect ratio as

$$d = \left(d_\perp^2 d_\parallel\right)^{1/3} = d_\perp E^{1/3} = d_\parallel E^{-2/3}. \tag{4.39}$$

The aspect ratio also defines whether a spheroid is *prolate* (when $E > 1$ as in Figure 4.11b) or *oblate* ($E < 1$ as in Figure 4.11c). The prolate case approaches a needle shape for $E \rightarrow \infty$, while the oblate case approaches a thin circular disk for $E \rightarrow 0$. The limit of $E = 1$ is a sphere ($d_1 = d_2 = d_3$).

However, regular surfaces can also have edges. For example, solid particles formed through crystallization tend to be in the form of cylinders, cones, and cuboids (Figure 4.12). Cuboids have surfaces that are orthogonal rectangular faces and yield a cube if these faces are squares. Regularly shaped particles can also have higher complexity. For example, ice particle shapes can include a hexagonal disk (Figure 4.13a), a cylinder with a hexagonal cross section (Figure 4.13b), and a dendritic snowflake (Figure 4.13c). When modeling such shapes, a typical approach is to ignore the higher complexity and approximate the shapes as oblate spheroids (especially when the aspect ratio is very small) or prolate spheroids (especially when the aspect ratio is very large). This simplification is helpful since spheroids allows simple approximations for their volume based on only two dimensions. For example,

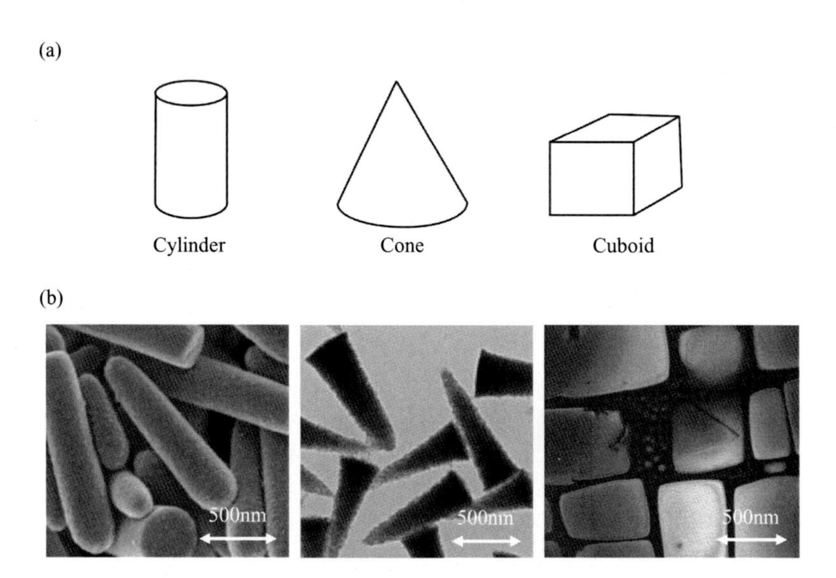

Figure 4.12 Examples of regularly shaped geometries with edges: (a) schematic of cylinder, cone, and cuboid; and (b) particles with nearly geometric shapes (Hagemans, 2020).

Figure 4.13a can be approximated as a very oblate spheroid, and Figure 4.13b as a very prolate spheroid. The particle dynamics in free fall, as described in the following section, are often reasonably predicted using such spheroidal approximations.

4.2.2 Orientation and Trajectory Dynamics

Free-fall (or free-rise) conditions are defined when the surrounding fluid is static (or at least uniform), so that the particle reaches an equilibrium terminal velocity, based on the balance of drag and effective gravitational forces. This terminal velocity will be parallel to the direction of gravity, and if there are no instabilities about this direction, this yields *rectilinear* motion. In addition, a nonspherical particle will have a specific orientation with respect to the fall direction relative to its shape. This orientation is controlled by the particle geometry and the fluid dynamics around its surface.

At a low enough particle Reynolds number to ensure a steady wake (e.g., $Re_p <$ 50 for a cylinder), the initial orientation for ellipsoidal particles depends on the initial condition. However, over long times it will reach a steady "broadside" orientation, where the particle's maximum cross-section faces the direction of relative motion. As such, the particle tends to an orientation where the drag is maximized. This orientation is based on dynamic stability considerations, as shown in Figure 4.14. If one considers static stability, a spheroid with either exactly streamlined or broadside orientation will have no net torque, since the surface pressure and viscous stresses are symmetric (e.g., Figure 4.14a), yielding static stability.

However, dynamic stability considers whether a small orientation perturbation will grow or decay. For a spheroidal particle falling in a nearly streamlined orientation with

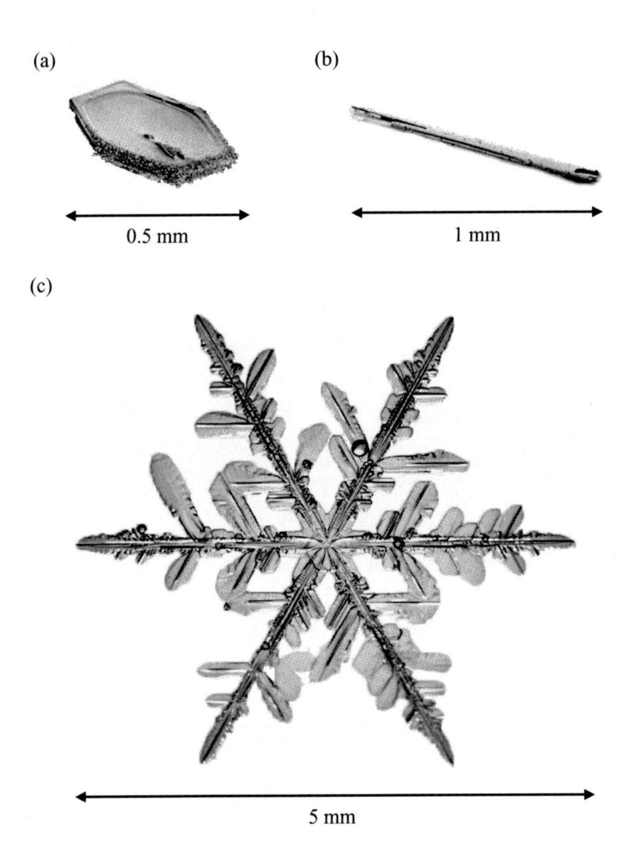

Figure 4.13 Examples of water crystals formed in the atmosphere whereby vapor condenses directly into ice (i.e., not by freezing water drops), including (a) hexagonal disk, (b) snow needle, and (c) a snowflake (Erbe, 2005; Libbrecht, 2005). The large size and complexity of a snowflake's dendritic (frilly) shape is due to long-time formation at high humidity and high altitudes.

a small angular perturbation (Figure 4.14b), the angle of attack relative to the oncoming flow results in a fluid dynamic torque that will cause the perturbation angle to grow, resulting in dynamic instability. This instability is the reason why an arrow or a rocket must be stabilized with countertorqueing fins to retain a streamwise orientation. In contrast, a particle falling in a broadside orientation (Figure 4.14c) with a small perturbation angle produces a counteracting torque that tends to return the particle to its broadside orientation. For example, the cylinder shown in Figure 4.15a has a steady wake and falls broadside. As such, the broadside orientation is both statically and dynamically stable. Thus, broadside is the commonly observed steady-state orientation for spheroidal particles in a free fall at low Reynolds numbers.

However, there are four exceptions at low Reynolds numbers for which a steady broadside orientation is not realized for a nonspherical shape:

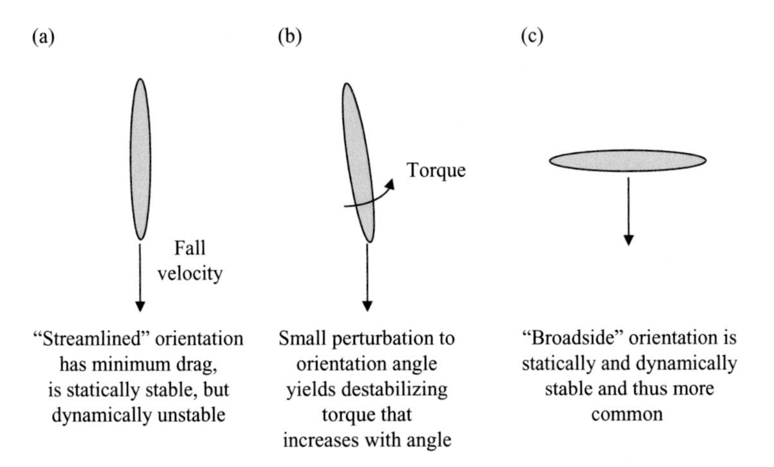

Figure 4.14 Stability (based on trajectory orientation) of spheroids in free fall for creeping flow, where only broadside orientation is statically and dynamically stable.

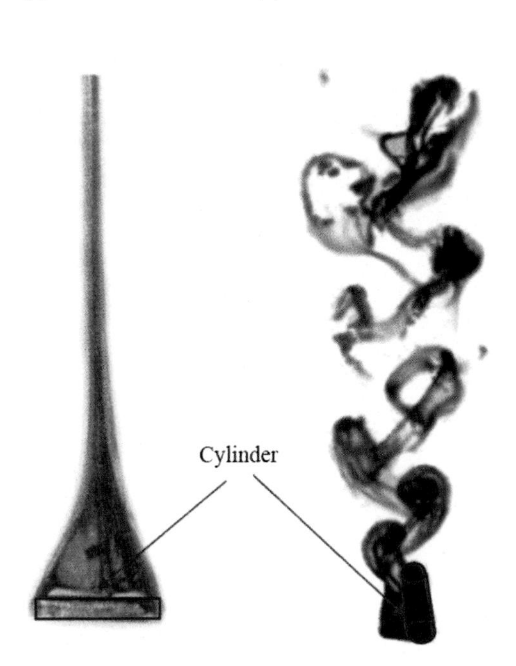

Figure 4.15 Cylinders falling broadside at terminal velocity with wake visualization: (a) cylinder viewed "lengthwise" at $Re_p = 40$, where the wake is steady; and (b) cylinder viewed "endwise" at $Re_p = 70$, where the unsteady wake includes shed vortices (Jayaweera and Mason, 1965). Red outlines show cylinder trajectory orientations for (a) and (b).

- Influence of neighboring particles (or nearby walls) that can cause a particle to orient itself streamwise due to collisions or fluid dynamic interactions.
- Fall in a non-Newtonian fluid (where viscosity varies with fluid shear), for which streamlined orientations can be dynamically stable (Galdi et al., 2002).
- Noncontinuum effect for very small particles whereby random molecular collisions lead to random orientations (discussed further in §5.3).
- Irregular particle shape (nonspheroidal) that prevents a symmetric torque-free equilibrium orientation and causes rotation and/or tumble (Gogus et al., 2001).

These exceptions apply whether or not the particle is at terminal velocity.

At higher Reynolds numbers where wake unsteadiness appears (e.g., $Re_p > 50$ for a cylinder), nonspherical particles will still fall in a broadside orientation. However, this will be accompanied by "secondary motion" due to the wake oscillations (Figure 4.15b). For prolate particles (e.g., cylinders and spheroids with $E > 1$), this secondary motion typically leads to mild rocking of the particle orientation about the broadside orientation for Reynolds numbers as high as the subcritical regime (Stringham et al., 1969, Clift et al., 1978).

For oblate particles (e.g., disks and spheroids with $E < 1$), the secondary motion at higher Reynolds numbers can be stronger and more complex, as shown in Figure 4.16. In particular, disk free-fall trajectories have four primary types of motion:

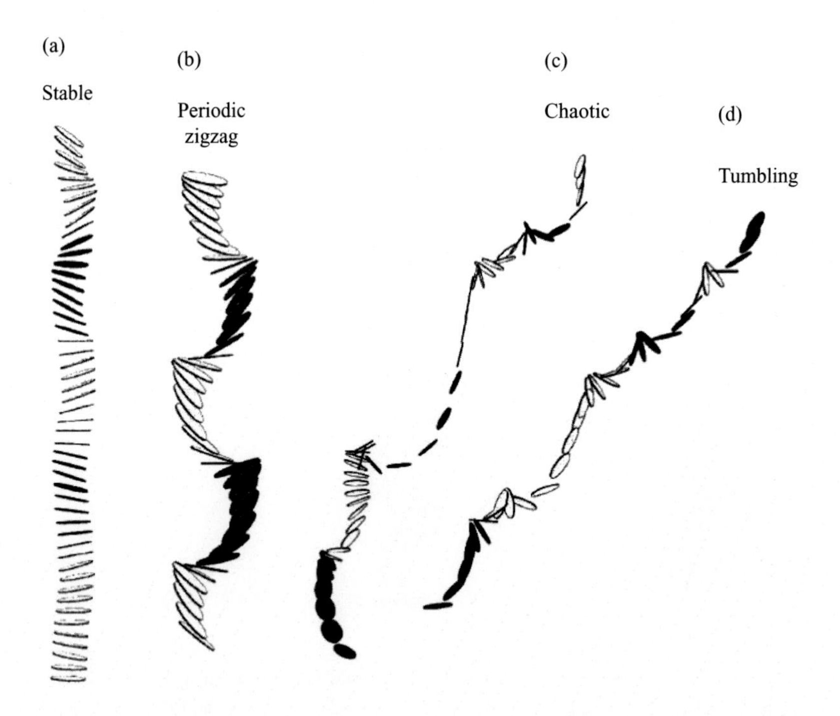

Figure 4.16 Illustrations of different solid disk motions in free fall where one side of the disk is black and the other side is white to clarify when the disk flips over (Field et al., 1997). The four regimes shown include (a) mild rocking settling into to steady vertical motion (nearly rectilinear with no flipping), (b) periodic zigzag (side-to-side periodic oscillation with no flipping), (c) chaotic (alternates between tumbling and periodic oscillation with occasional flipping), and (d) tumbling (flips continuously end-over-end with significant lateral movement).

- Stable motion (Figure 4.16a), where any initial secondary motion decays so the disk eventually falls in a stable broadside configuration.
- Periodic motion (sometimes also called fluttering motion) with three subcategories: (i) zigzag motion (Figure 4.16b); (ii) spiral motion in a helical path; and (iii) hybrid motion which transitions between zigzag and spiral (but never flips over).
- Chaotic motion (Figure 4.16c), where the particle randomly transitions between the periodic and tumbling motions (occasional flipping) with horizontal drift.
- Tumbling motion (Figure 4.16d), where the particle turns end over end (continuous flipping) with substantial horizontal drift.

These motion types depend on particle Reynolds numbers and a dimensionless particle inertia (I^*) defined as follows:

$$I^* \equiv \frac{I_{p\perp}}{\rho_f d^5} = \frac{\pi \rho_p d_{\parallel}}{64 \rho_f d_{\perp}}. \tag{4.40}$$

In this relationship, $d_{\perp}$ is the disk diameter and $d_{\parallel}$ is the disk thickness. As shown by the regime map of Figure 4.17, $Re_p < 100$ generally yields steady trajectories for all inertias, while $Re_p > 100$ often results in significant secondary motion, Generally, this motion is tumbling for $I^* > 0.04$, chaotic for $0.04 < I^* < 0.01$ and periodic (zigzag, spiral, or hybrid) for $I^* < 0.01$.

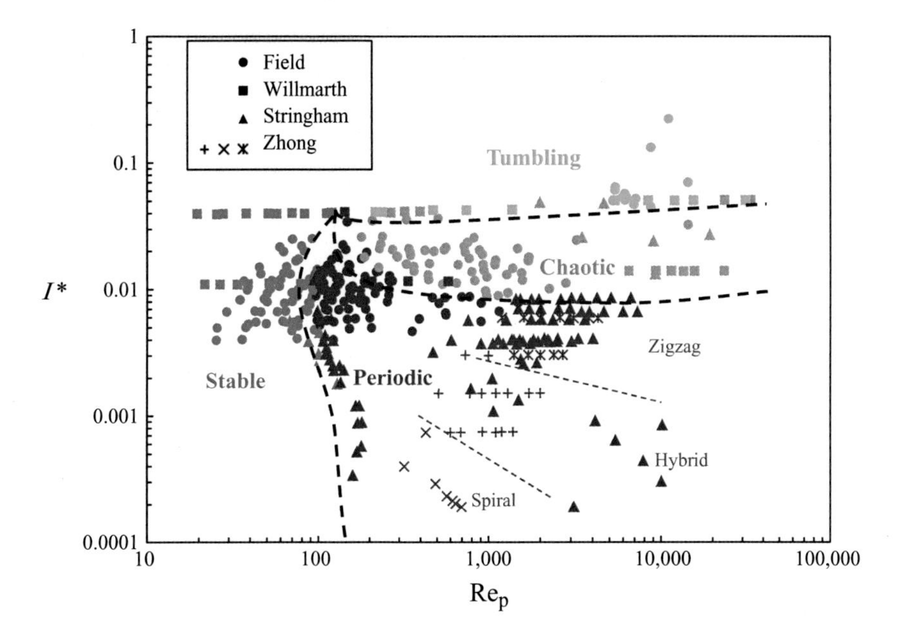

Figure 4.17 Disk motion regimes in free fall based on experimental data, based on data reported by Willmarth et al. (1964), Stringham et al. (1969), Field et al. (1997), Zhong et al. (2011), and Estaban et al. (2019), where symbol shape indicates the data source. Symbol color indicates whether the observed motion was steady, tumbling, chaotic, or periodic (where the periodic motion is broken into three categories of spiral, transitional, or zigzag).

4.3 Drop Shapes and Breakup in Free Fall

4.3.1 Weber Number and Deformation

Drops can deform in free fall due to their relative velocity, especially as their size increases and/or surface tension decreases. There is a common misconception that a deformed drop in free fall has a teardrop shape that includes a tail at the rear (💧). For example, this tear shape is commonly used as a graphical symbol for rain in weather maps and reports. In reality, the tear shape *never* occurs in free fall. Instead, the teardrop shape only occurs when a drop is released from a solid surface, such as from the tip of an injection needle. Once the drop is released, surface tension will quickly eliminate the tail and the drop shape will tend to a sphere if surface tension effects dominate. Otherwise, it will deform as an oblate spheroid, flattened in the vertical direction (⬬). As such, the heavy rain graphic on weather maps would be more accurately designated with such ellipses (so long as they are not confused as flying saucers!). In the following, we explain why large raindrops tend to oblate shapes by first defining the dimensionless parameters that control this deformation.

Dimensionless Parameters

Drops free falling in a gas deform based on the interplay of surface tension and pressure stresses. Variation of the static pressure on the surface tends to cause a deviation from a sphere. Surface tension tends to force the shape back toward a sphere since a reduction in local radius of curvature will lead to an increase in surface tension to counter this effect. In general, pressure variations scale with the dynamic pressure, which scales with $\rho_f w^2$ per (3.51), while the surface tension stresses scale with σ/d per (3.33a). To characterize these competing effects, the nondimensional Weber number (We) is defined as follows:

$$\text{We} \equiv \frac{\rho_f w^2 d}{\sigma} = \frac{\text{dynamic pressure effects}}{\text{surface tension effects}}. \tag{4.41}$$

As such, the Weber number is a ratio of effects that cause deformation (numerator) relative to those that prevent deformation (denominator), so an increase in Weber number corresponds to an increase in deformation. In particular, a wide variety of experiments for $\text{Re}_p > 1$ (Clift et al., 1978) yield the following regimes for drops and bubbles:

$$\text{We} \ll 1 \quad \textit{approximately spherical.} \tag{4.42a}$$

$$\text{We} \sim 1 \quad \textit{moderate deviations from a sphere.} \tag{4.42b}$$

$$\text{We} \gg 1 \quad \textit{large deviations from a sphere.} \tag{4.42c}$$

In general, the Weber number determines the onset and degree of deformation whereas the Reynolds number determines the shape of the deformation once the Weber number is high. Both the Weber and Reynolds numbers depend on the instantaneous relative velocity of the particle (w), regardless of whether flow is steady.

For a steady flow, the relative velocity will tend to the terminal velocity condition (w_{term}), where the drag force balances the effective gravitational forces. This equilibrium condition allows the Weber number and Reynolds number to be replaced by the Bond number and the Morton number. The Bond number (also called the Eötvös number) is based on the effective gravitational force per unit volume, $g|\rho_p - \rho_f|$ from (3.86), relative to the surface tension stresses for a spherical shape, which scale with σ/d from (3.33a), normalized by the drop diameter (d) to convert these stresses into a force per unit volume. The nondimensional parameter that characterizes the competing effects of these forces can thus be defined as follows:

$$\mathrm{Bo} \equiv \frac{gd^2|\rho_p - \rho_f|}{\sigma} = \frac{\text{gravitational effects}}{\text{surface tension effects}}. \tag{4.43}$$

Compared to the Weber number, the Bond number has the advantage that it can be obtained without knowledge of the terminal velocity. A disadvantage of the Bond number is that it is only appropriate for characterizing terminal velocity conditions.

In order to integrate viscous effects with an appropriate dimensionless parameter for terminal velocity conditions, one may apply dimensional reasoning and the Buckingham-π theorem (White, 2016) with the surrounding fluid viscosity (μ_f) to obtain a second dimensionless parameter as the Morton number (Mo):

$$\mathrm{Mo} \equiv \frac{g{\mu_f}^4|\rho_p - \rho_f|}{\rho_f^2\sigma^3}. \tag{4.44}$$

The Morton number has no direct physical interpretation in terms of competing effects but incorporates effects of viscosity. Generally, shapes have increased deformation as the Bond number increases and more complexity as Morton number decreases (similar to the influence of Weber number and Reynolds number).

These four dimensionless parameters can be related at the terminal velocity condition as follows:

$$\frac{\mathrm{We}_{term}^2}{\mathrm{Re}_{p,term}^4} = \frac{\mathrm{Mo}}{\mathrm{Bo}}. \tag{4.45}$$

Furthermore, the terminal Weber number can be related to the Bond number and the terminal drag coefficient by combining (3.94b), (4.41), and (4.43), as follows:

$$\mathrm{We}_{term} = \frac{4\mathrm{Bo}}{3C_{D,term}}. \tag{4.46}$$

For fluid particles in the Newton-based drag regime (where C_D is approximately constant), this yields a linear relationship between We_{term} and Bo. As such, increasing the Bond number is similar to increasing the terminal Weber number. And based on (4.45), increasing the Morton number for a fixed Bond number effectively decreases the terminal Reynolds number for a fixed Weber number.

Historically, empirical correlations for drop deformation have been based on Bo and Mo (rather than on We and Re_p). This is because the Bond and Morton numbers can be controlled experimentally a priori, and it is straightforward to vary one of these parameters while holding the other one constant. For example, one may vary the Bond number by changing the drop volumetric diameter while fixing the Morton number by keeping the same fluid viscosity. Similarly, one may vary the Morton number by changing the fluid viscosity while fixing the Bond number with a constant drop size. In comparison, it is more difficult to conduct parametric experiments that vary We while keeping Re_p fixed, and vice versa.

However, drop deformation characterized by Bo and Mo are only appropriate at terminal velocity. As such, the deformation at any other speed *cannot* be characterized with Bo and Mo, and instead relies on We and Re_p. For example, a drop accelerating to terminal conditions will have an evolving relative velocity with $w < w_{term}$. On the other hand, a drop injected into a gas at high relative speed (as occurs in a spray system) often has $w > w_{term}$. In these examples, the relative velocity will be accompanied by changes in Re_p and We and thus in deformation. In contrast, Bo and Mo will not change with time in these examples, and so will fail to characterize changes in the deformation. As such, Re_p and We are preferred as a more general set of parameters for quasisteady prediction. In the following subsection, we outline the general influence of the Weber and Reynolds number on the shape of deformed drops in free fall as well as empirical relationships for their aspect ratios and drags.

Terminal Shapes for Drops in a Gas

For a drop falling in a gas at the terminal velocity, there is a steady-state balance between gravity and the drag forces. Increasing the Weber number leads to increasing deformation, as shown in Figure 4.18 for water drops falling in air. A key nondimensional parameter used to quantify the degree of nonspherical deformation is the aspect ratio, E of (4.38). For negligible deformation, the drop shape is a sphere with $E = 1$. If the terminal Weber number is low, the drops are nearly spherical based on (4.42a). As an example, the 0.5 mm drop in Figure 4.18a has $E \approx 0.99$ (about 1% deviation from a sphere).

Moderate deformation in free-fall conditions is generally associated with an oblate shape with $E < 1$, yielding an ellipse as viewed from the side (●) and a circular cross section as viewed from above. The deformation increases (as aspect ratio decreases) with the Weber number. For example, a 1 mm diameter water drop at terminal velocity in air falls at about at 4 m/s with $Re_p \approx 320$ and $We \approx 0.27$ so the drop is nearly spherical (Clift et al., 1978). If the drop size is increased to 2 mm, $We \approx 1$ and the drop becomes more deformed with $E \approx 0.9$.

To understand why the shape flattens for a free-falling drop with small to moderate deformation, consider an oblate spheroidal drop, as shown in Figure 4.19. For a small Weber number at the onset of weak deformation, the dynamic pressure variations along the surface can be roughly approximated based on the flow over a sphere at $Re_p = 100$, as given in Figure 3.4. At the bottom of the drop (which first sees the oncoming relative velocity), the surface pressure will rise to its highest value by recovering the stagnation pressure ($C_P = 1$). At the sides, the relative surrounding gas velocity (just outside of the boundary layer) speeds up to its maximum so that the surface pressure falls to a minimum ($C_P \approx -0.6$). At the top (rear) the surface pressure recovery is hampered

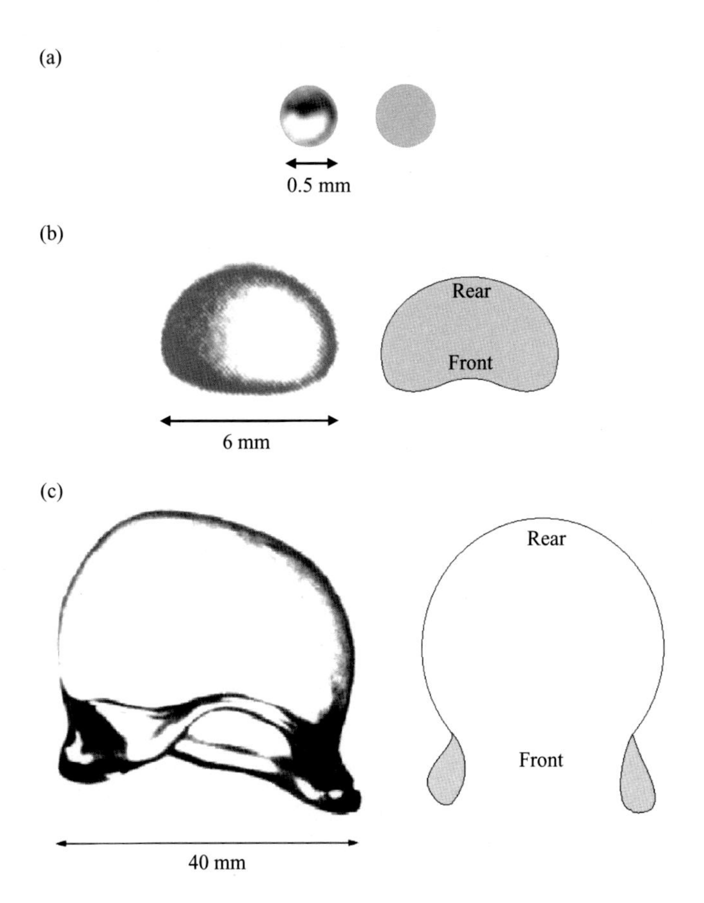

Figure 4.18 Photographs of water drops falling in air (with overall width indicated) and corresponding cross-sectional shapes along the axis of symmetric: (a) spherical shape for d $\approx$ 0.5 mm at $Re_p \approx 70$ and $We \approx 0.03$ (Garg and Nayar, 2004); (b) flattened shape for d $\approx$ 5 mm at $Re_p \approx 4{,}500$ and $We \approx 11$ (Clift et al., 1978); and (c) bag drop for d $\approx$ 9 mm at $Re_p \approx 6{,}600$ and $We \approx 16$ (Mason, 1978).

by flow separation and only rises to about the freestream pressure ($C_P \approx 0$). Since the internal liquid pressure is approximately constant, these external pressure variations cause the drop to expand more at the sides where this pressure is lower, as shown in Figure 4.19. The smaller radius of curvature at the sides creates an increase in surface tension stress to match the increased difference between internal pressure and surface pressure. This causes the drop to tend toward an oblate spheroid with $E < 1$, and the higher pressure at the front causes this surface to be more flattened.

As the drop diameter and the Weber number increase further, the stagnation pressure on the front (bottom) surface becomes significantly larger than at the rear (top) surface, since the rear portion has a separated flow (Figure 3.4). As a result, the drop shape develops significant front-to-back asymmetry, becoming more flattened on the bottom. For $We > 8$, this can even cause a dimple, as seen by front surface concavity in the cross-sectional view of Figure 4.18b, for which $E \approx 0.73$. As the

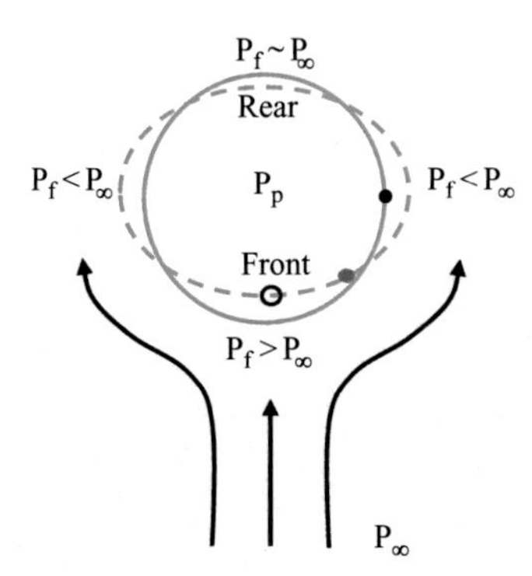

Figure 4.19 Illustration of the aerodynamic surface pressures over a nearly spherical drop shown by the solid blue line (as in Figure 4.18a) that causes deformation into an oblate spheroid at significant Weber numbers shown by the dashed blue line (as in Figure 4.18b). The streamlines caused by a relative velocity (as in falling at terminal conditions) yield a bottom stagnation point (○) where the surface aerodynamic pressure is the highest and where a Rayleigh–Taylor instability is most likely to occur. The initial location most likely for a Kelvin–Helmholtz instability to occur is shown by a black solid circle (•) for a spherical drop and by a blue ellipse (⬬) for an oblate spheroidal drop.

droplet deforms, it increases its cross-sectional area (compared to that of a sphere) seen by the relative velocity. This higher area (as well as the increased flow separation that often occurs) generally increases form drag.

At very high Weber numbers, the aspect ratio tends to a minimum (E_{min}) and the intermediate aspect ratio can be empirically described (Loth, 2008a) as follows:

$$E_{min} \approx 0.25 \qquad \textit{for}\ \mathrm{We} \gg 1 \qquad \textit{at}\ \mathrm{Re_p} > 100. \tag{4.47a}$$

$$E = 1 - (1 - E_{min})\tanh(0.07\mathrm{We}) \tag{4.47b}$$

The prediction of (4.47b) is compared to experimental data for drops in a gas in Figure 4.20 indicating good agreement, where $E_{min} \approx 0.25$ is consistent with the drop inset image during a breakup. However, drops are generally only stable for We > 12 (as discussed in the next section).

Similarly, high quasisteady deformation leads to a maximum drag coefficient ($C_{D,max}$) that can be used to model the drag change relative to that of a sphere (3.58) for a range of Reynolds and Weber numbers (Loth, 2008a), as follows:

$$C_{D,\max} = 8/3 \tag{4.48a}$$

$$C_{D,\mathrm{sphere}} = \frac{24}{\mathrm{Re_p}}\left(1 + 0.15\,\mathrm{Re_p^{0.687}}\right) \tag{4.48b}$$

$$C_D = C_{D,\mathrm{sphere}}\left(1 - \Delta C_D^*\right) + C_{D,\max}\,\Delta C_D^* \qquad \textit{at}\ \mathrm{Re_p} > 100. \tag{4.48c}$$

$$\Delta C_D^* \text{ » } \tanh\left[0.0008\left(\mathrm{We}\mathrm{Re_p^{0.2}}\right)^{1.45}\right] \tag{4.48d}$$

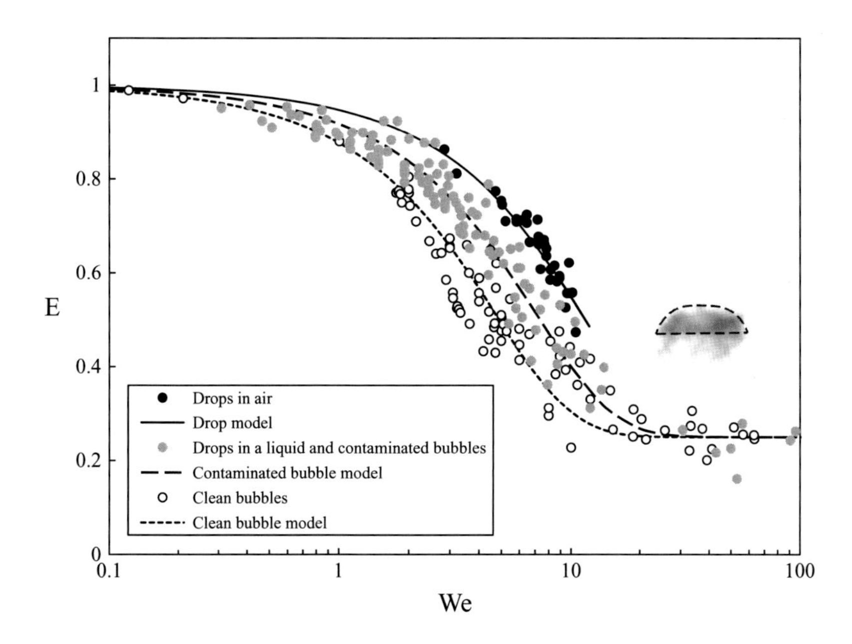

Figure 4.20 Aspect ratios at $Re_p > 100$ for drops in air (•), for contaminated drops and bubbles in a liquid (•), and for clean bubbles (○), based on experimental data from various sources (Loth, 2008a) compared to empirical models, where the inset shows a drop being broken up by a shock wave (Joseph et al., 1999), which reveals a lenticular shape with $E \sim 0.25$, as shown by dashed-line shape.

The model compares well with data for $100 < Re_p < 2{,}000$, as shown in Figure 4.21. Further details for the drag and aspect ratio are given in §9.3.2 for a variety of Reynolds and Weber numbers.

Both Figures 4.20 and 4.21 show a limit of $We < 12$ for the experimental data. This is due to unsteady deformation leading to breakup. For example, the shape of Figure 4.19 subjected to an increased Weber number will increase stagnation pressure in the front. This causes the dimple to grow and rapidly penetrate inward, creating the bag shape of Figure 4.18c, for which a steady aspect ratio is no longer appropriate. This parachute-like shape is highly sensitive to instabilities and can quickly lead to drop breakup, as shown in Figure 4.22. For diameters greater than 6 mm, a water drop in a gas often undergoes complex unsteady deformation and breakup (corresponding to Weber numbers greater than 12). Predicting the drop breakup is discussed in the next section.

4.3.2 Drop Breakup

Drops can break up when they are subjected to a fluid dynamic instability strong enough to overcome the surface tension effects. Breakup of a single isolated drop generally occurs when the Weber number is high and the drop is nonspherical. In particular, an instability can cause a locally concave interface (e.g., the dimpled or bag

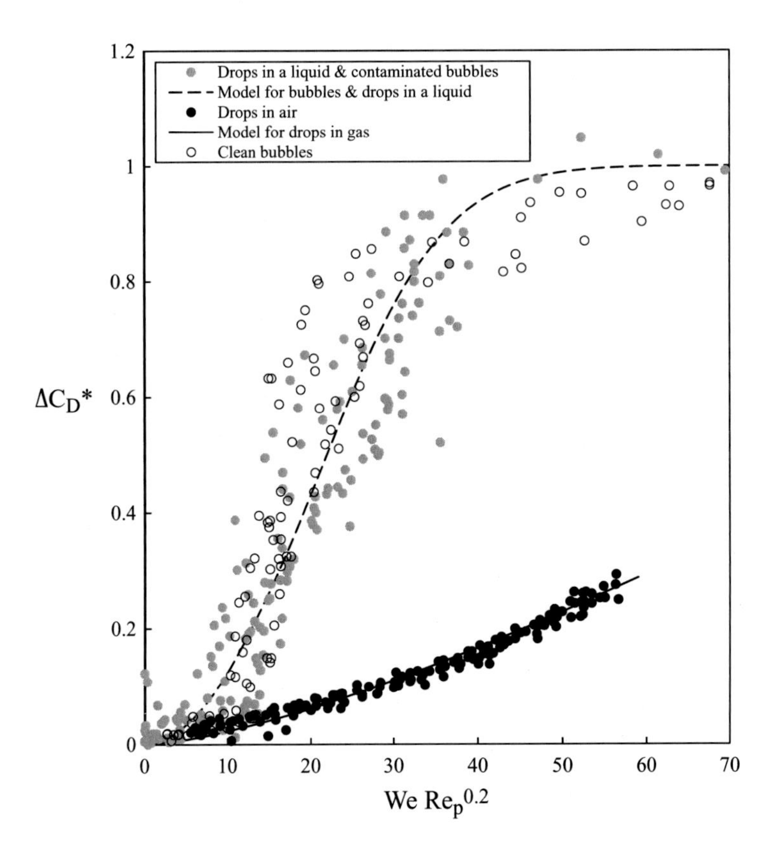

Figure 4.21 Incremental drag ratios of fluid particle at $Re_p > 100$, including clean bubbles (○), contaminated drops and bubbles in a liquid (●) and drops in air (●), based on experimental data from various sources (Loth, 2008a) compared to two empirical models.

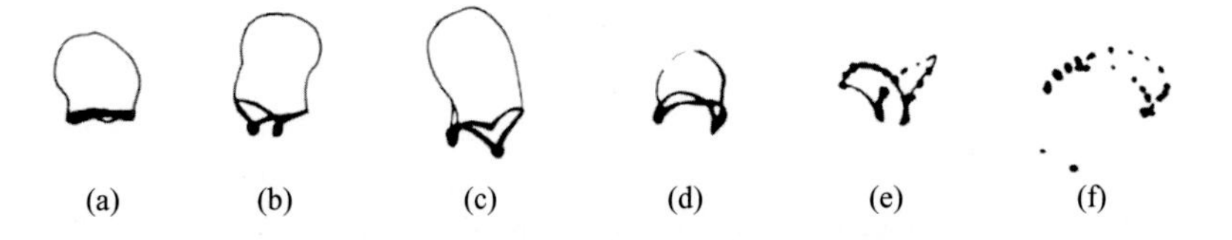

Figure 4.22 Bag-drop breakup process illustrated by time sequences for a 10 mm diameter water drop in air (Clift et al., 1978). In (a), a bag-drop shape is formed with a thin liquid shell and a toroidal ring at the bottom. In (b) and (c), the toroidal ring becomes unstable, leading to two pendant drops. In (d), the liquid shell has become too thin and breaks. In (e), the toroid also breaks up and gives rise to several small drops.

shapes of Figure 4.18), which can grow and sever the fluid body into two or more smaller bodies (Figure 4.22). However, drop breakup tends to occur in other modes since bag-mode breakup is only observed when there are little disturbances (very still air). A variety of conditions lead to other breakup processes, especially when the surrounding flow has spatial or temporal gradients. These include vibrational modes, stripping modes, lenticular breakup, shattering, and so on (Lefebvre et al., 1988). Fortunately, the onset of breakup for all these processes are all primarily related to the Weber or Bond number, as discussed in the following.

Breakup caused by an interfacial fluid density difference is related to the Rayleigh–Taylor instability of Figure 2.18, for which a denser fluid is placed over a lighter fluid and subjected to acceleration (e.g., by gravity). Such a condition occurs at the bottom of a falling drop where liquid is above the gas ("o" on Figure 4.19). Because of this density difference, small local perturbations can grow due to the Rayleigh–Taylor instability. Since these instabilities are caused by gravitational acceleration and are countered by surface tension, the Bond number (4.43) can be used to determine instability onset. Indeed, such a criterion is reasonable for terminal velocity conditions. However, a criterion based on Weber number is more general as it can handle the conditions when the drop velocity is unsteady (not yet a terminal velocity) and when gravity is not significant (breakup by a shock wave or an initial injection velocity). As such, breakup criteria based on the Weber number are typically more common and useful, especially when the relative velocity is known.

The breakup criterion based on Weber number stems from a velocity difference that causes a pressure difference and thus can be related to the inviscid Kelvin–Helmholtz instability of Figure 2.19. This type of instability occurs across an interface when one side is exposed to a higher velocity than the other side. Such a condition effectively occurs when considering the inviscid flow past a drop (ignoring viscous effects in the boundary layer), since the surrounding fluid above this boundary layer is moving at a significant velocity relative to that of the fluid inside the drop. The resulting Kelvin–Helmholtz instabilities can cause small perturbations to grow on the surface and lead to increasing interface distortion, which can ultimately result in drop breakup.

The two-dimensional Kelvin–Helmholtz instability of Figure 2.19 can be modified to include surface tension by assuming a flat interface and neglecting effects of gravity and viscosity (Drazin and Reid, 1981), as follows:

$$f = \frac{(\rho_1 U_{1,o} + \rho_2 U_{2,o}) \pm \sqrt{-\rho_1\rho_2(U_{1,o} - U_{2,o})^2 + \frac{(\rho_1+\rho_2)2\pi\sigma}{\ell}}}{\ell(\rho_1 + \rho_2)}. \tag{4.49}$$

At the critical condition where the interface is neutrally stable, the square root term in (4.49) will be zero, and one may define an associated critical wavelength as follows:

$$\ell_{crit} = \frac{2\pi\sigma(\rho_1 + \rho_2)}{(U_{1,o} - U_{2,o})^2 \rho_1\rho_2}. \tag{4.50}$$

This is the minimum wavelength for which the instability will grow, that is, the interface will be stable due to surface tension if $\ell < \ell_{crit}$ but unstable due to the velocity difference if $\ell > \ell_{crit}$.

To apply the preceding inviscid theory to a drop interface, one must identify the two fluid densities, the velocity difference and the wavelength. If we consider Stream 1 as the surrounding fluid and Stream 2 as the drop internal fluid, then ρ_1 is set as the gas density and ρ_2 as the liquid density. Along this vein, $U_{1,o}$ is approximated as the inviscid airflow velocity above the unperturbed interface (U_{surf}). This approximation is appropriate for a thin attached boundary layer, which is reasonable for $Re_p \gg 1$ and the front portion of the drop ($\theta < 90°$). Similarly, $U_{2,o}$ is the inviscid liquid velocity below the interface (V), but this will generally be small since liquid viscosity is much greater than the gas viscosity so that $U_{2,o} \approx 0$. Therefore, the next step is to determine the values for U_{surf} and ℓ.

For small deformations, the inviscid velocity can be approximated by that for a sphere (3.47), which can be expressed in terms of the relative velocity and polar angle, as follows:

$$U_{surf} = \frac{3}{2} w \sin\theta. \tag{4.51}$$

As such, the velocity is initially zero at the stagnation point ($\theta = 0$, indicated by "o" in Figure 4.19), and then rises to its highest velocity at the sides ($\theta = 90°$, indicated by •). For a spherical shape, this is the most likely location for a Kelvin–Helmholtz instability to occur. However, the curvature is highest here for an oblate-shaped drop, indicating the stabilizing surface tension stresses are also highest at this point. Since the instability is most likely to first form when there is a combination of both low curvature (closer to the front) and high relative velocity (closer to the sides), the instability initiation location is generally between these points. Experimental observations have indicated that this occurs at $\theta \approx 38°$ from the leading edge as illustrated by the blue ellipse (⬬) in Figure 4.19. At this location, $U_{surf} \approx 0.92w$. To determine an approximate maximum wavelength for this region, one may use the distance between the points on either side of the front surface, that is, $\ell \approx d\sin(38°)$. Substituting these values into (4.50) and (4.51), the resulting critical particle diameter, d_{crit} (above which breakup will occur) is

$$d_{crit} \approx 12 \frac{\sigma\left(\rho_p + \rho_f\right)}{w_{crit}^2 \rho_p \rho_f} \quad \textit{at } Re_p \gg 1 \cdot \tag{4.52}$$

Based on (4.41), the corresponding critical Weber number can be defined by using this critical diameter and the associated critical relative velocity as

$$We_{crit} \equiv \frac{\rho_f w_{crit}^2 d_{crit}}{\sigma} \approx 12 \frac{\rho_p + \rho_f}{\rho_p} \quad \textit{for inviscid conditions.} \tag{4.53}$$

Thus, instabilities leading to breakup will occur if $We > We_{crit}$, which simplifies to $We > 12$ for a drop in a gas ($\rho_p \gg \rho_f$). This result is consistent with experiments of droplets with low viscosity (Clift et al., 1978; Lefebvre et al., 1988).

As mentioned before, a Rayleigh–Taylor instability can also occur since there is a heavy fluid (liquid) suspended over a lighter fluid (air) with a gravitational acceleration at the bottom of the drop. Taylor (1949) investigated a spherical drop in this manner and obtained $Bo_{crit} = 16$. Since C_D of a deformed drop just before breakup corresponds to about 2, this Bond number criterion combined with (4.46) yields

$We_{crit,term} \approx 12$. Thus, the inviscid Kelvin–Helmholtz and Rayleigh–Taylor theoretical approaches are both quantitatively consistent with experimental breakup of low-viscosity drops.

However, the preceding analyses neglect the damping that can occur when either the drop viscosity (μ_p) or surroundng fluid viscosity (μ_f) are high. Such viscous effects can stabilize the perturbations, causing an increase in the droplet size and/or velocity needed for breakup (and thus an increase in We_{crit}). This viscous damping can be characterized by the dimensionless Ohnesorge number (Oh), which compares the viscous stresses to the surface tension stresses. The Ohnesorge number based on the particle viscosity is denoted as Oh_p, while that based on the surrounding fluid viscosity is denoted as Oh_f:

$$Oh_p \equiv \frac{\mu_p}{\sqrt{\rho_p \sigma d}} = Oh_f \frac{\mu_p^*}{\sqrt{\rho_p^*}}. \tag{4.54a}$$

$$Oh_f \equiv \frac{\mu_f}{\sqrt{\rho_f \sigma d}} = \frac{\sqrt{We}}{Re_p} = \left(\frac{Mo}{Bo}\right)^{1/4}. \tag{4.54b}$$

The RHS of (4.54a) shows that the differences in the particle and fluid Ohnesorge numbers are related to the viscosity and density ratios. The RHS of (4.54b) shows the fluid Ohnesorge number can be related to the Weber and Reynolds numbers or to the Morton and Bond numbers.

For drops in a gas, generally $Oh_f \ll Oh_p$ such that Oh_p (with μ_p) will have the greatest impact on viscous damping of breakup. As shown in Figure 4.23, higher Oh_p values leads to higher critical Weber numbers (due to viscous damping), which is reflected by the empirical fit:

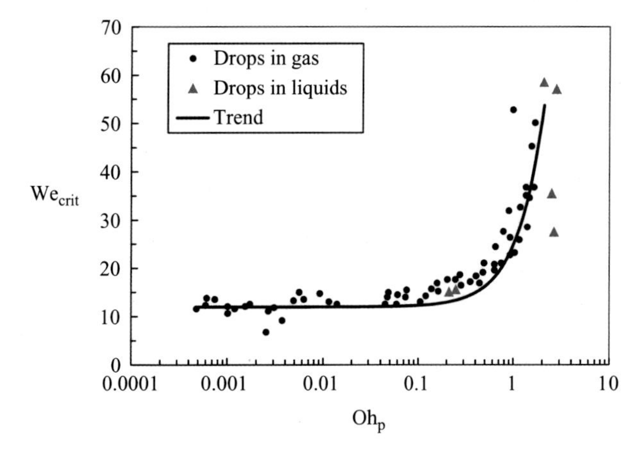

Figure 4.23 Critical Weber number as a function of the particle Ohnesorge number for drops in gas and in immiscible liquids based on Pilch and Erdman (1987).

$$\mathrm{We}_{\mathrm{crit}} \approx 12\left(1 + 1.07\mathrm{Oh}_{\mathrm{p}}^{1.6}\right) \qquad \textit{for drops in a gas.} \tag{4.55}$$

This correlation should be used with caution for $\mathrm{Oh}_p > 1$ where there is significant scatter. However, this correlation and the data in Figure 4.23 show that $\mathrm{Oh}_p < 0.1$ allows viscous effects to be ignored such that $\mathrm{We}_{crit} \approx 12$ is appropriate.

It is interesting that Figure 4.23 shows that the breakup of drops in immiscible liquids can also reasonably represented by (4.55) so long as $\mathrm{Oh}_f \ll 1$. This is somewhat surprising since (4.53) suggests a greater threshold for stability when the two fluid densities are similar corresponding to $\mathrm{We}_{crit} \approx 8\pi$. However, drops in a liquid (as compared to drops in a gas) tend to have a high-amplitude secondary motion (unstable trajectories), which may serve to enhance their shape instabilities (Loth, 2008a) and explain why the observed breakup is closer to $\mathrm{We}_{crit} \approx 4\pi$. This secondary motion also occurs for bubbles, as discussed in the following section.

4.4 Bubble Shapes and Trajectories in Free Rise

4.4.1 Free-Rise Deformation

As with a drop, the shape of a gas bubble depends on its interaction with the surrounding fluid. Examples of bubble shapes in free rise are shown in Figure 4.24 and are qualitatively consistent with the guidelines of (4.42). For example, a 0.5 mm bubble in a glass of champagne (Figure 4.24a) rises at a terminal velocity such that

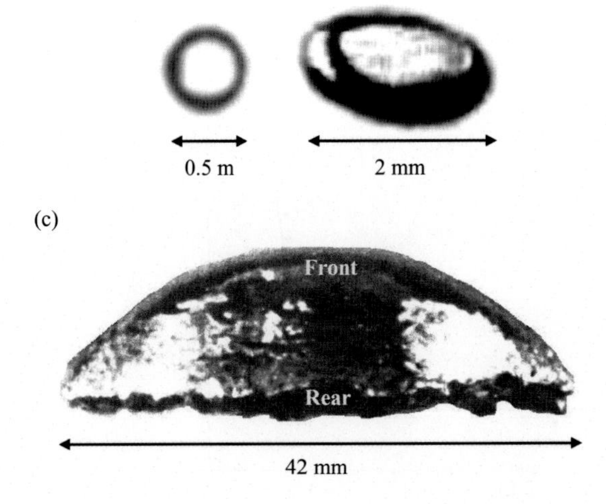

Figure 4.24 Photographs of bubbles rising in a liquid: (a) nearly spherical bubble rising in champagne at $\mathrm{Re}_p \approx 25$ and $\mathrm{We} \approx 0.02$ (Liger-Belair and Jeandet, 2002); (b) ellipsoidal air bubble rising in tap water at $\mathrm{Re}_p \approx 880$ and $\mathrm{We} \approx 2.5$ (Tomiyama et al., 2002a); and (c) spherical-cap bubble in water at $\mathrm{Re}_p \approx 2{,}500$ and $\mathrm{We} \approx 50$ (Clift et al., 1978).

We « 1 and so is nearly spherical with E ≈ 1. A similar result occurs for this size of a bubble in tap water.

As the Weber number increases, a bubble deforms to an oblate spheroid for the same reasons as described by Figure 4.19. An example of an oblate shape at a Weber number of about 2 for air in tap water (where the surface is fully contaminated) is shown in Figure 4.24b, for which E ≈ 0.6. If this same size bubble were placed in hyperclean water so that the surface is clean (Table 3.1), the drag would be lower, so the velocity and Weber number would be higher, thus yielding even more deformation and a lower E (Tomiyama et al., 2002a). Note that the Bond number (4.43) is the same whether a bubble is clean or contaminated and thus cannot explain this difference. This is another reason why the Weber number is more qualified as a fundamental predictor of deformation than the Bond number.

Large bubbles rising with a high Weber number (e.g., We > 20) tend to form cap shapes as shown in Figure 4.24c. This shape is defined by a rounded top and a flat bottom. The flat surface of the bottom (rear) is caused by flow separation, which yields a constant pressure along this surface. Since the density is small inside the bubble and surface tension effects are weak, this aft liquid pressure is the same as the pressure throughout the bubble. The semisharp bubble edges (high curvature) at the bubble circumference are also possible because surface tension effects are weak (We » 1). For deformation (We > 20), the bubble shape will depend on the Reynolds number, with the following approximate regimes (Figure 4.25):

- $0.5 < Re_p < 40$, ellipsoidal cap that may be dimpled.
- $40 < Re_p < 110$, a spherical cap with laminar steady closed wake.
- $110 < Re_p < 200$, a spherical cap with a wake that may be unsteady.
- $200 < Re_p < 600$, a spherical cap with an unsteady shedding wake.
- $600 < Re_p < 1{,}500$, a spherical cap with a highly unsteady wake, causing breakup.
- $Re_p > 1{,}500$, a spherical cap with a turbulent open wake and fluctuating aft surface.

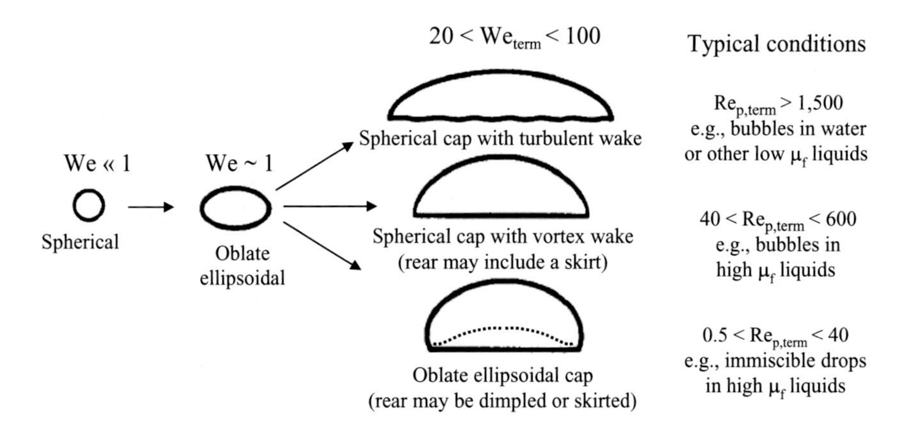

Figure 4.25 Steady shape categories of bubbles or immiscible drops in uncontaminated liquids (Bhaga and Weber, 1981; Loth, 2008a).

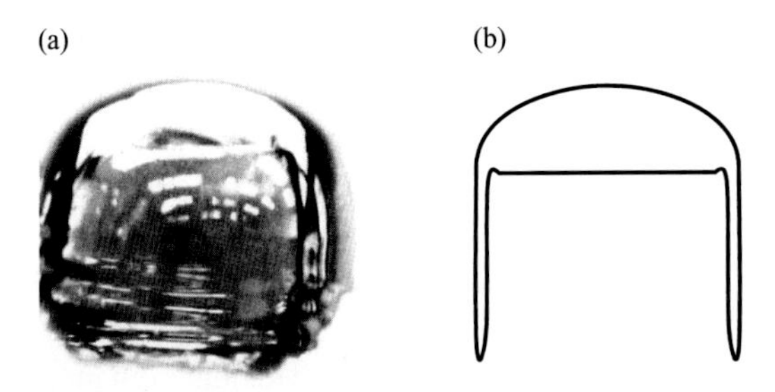

Figure 4.26 Skirted bubble: (a) photograph of a bubble rising in mineral oil at $Re_p \approx 100$ (Van Dyke, 1982); and (b) cross-sectional schematic of the interface showing a skirt below the spherical cap.

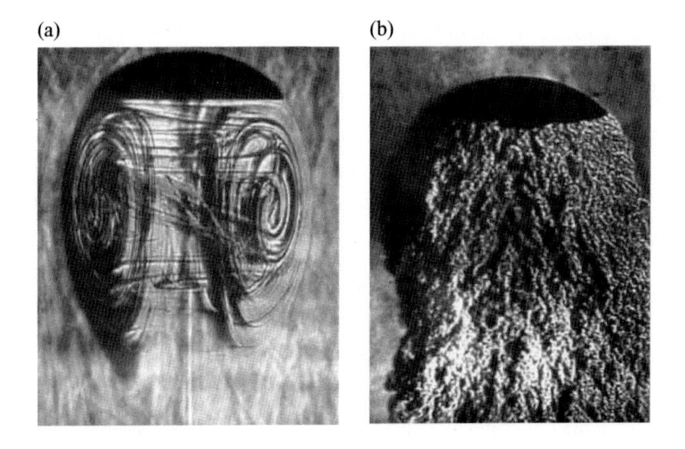

Figure 4.27 Shadowgraphs of spherical-cap bubbles (Wegener et al., 1971): (a) with a steady wake that is laminar and closed ($Re_p \approx 180$) and (b) with an unsteady wake that is turbulent and open ($Re_p \approx 17{,}000$).

The shapes are overviewed in the text that follows, and additional details are given by Wegener et al. (1971), Bhaga and Weber (1981), and Loth (2008a).

For high deformation at $0.5 < Re_p < 40$ (extreme-viscosity liquids), an oblate ellipsoidal cap shape can occur. The separation region and aft portion are very small at $Re_p < 5$, but both grow as Re_p increases. For $40 < Re_p < 110$, the flat aft portion and the wake size continues to grow, and the top is a spherical cap. However, other shapes are also possible at moderate Reynolds numbers, including thin disk-like shapes for $10 < We < 20$ and $30 < Re_p < 100$ and a skirted bubble (Figure 4.26) for $8 < Re_p < 600$ and $We > 40$. For the skirted bubble, viscous stresses pull down the gas at the outer edges (enclosing the recirculation region). The flows around these bubbles and in the wake are steady and laminar.

For high deformation at $110 < Re_p < 200$ (high-viscosity liquids), the wake for a spherical cap bubble is still an external toroidal vortex, as shown in Figure 4.27a. When closed, the wake resembles the internal Hill's vortex of Figure 3.2. However, the wake can be unsteady and even lead to vortex shedding due to the higher Reynolds numbers. For $200 < Re_p < 600$, the vortex becomes highly unstable with strong shedding. The wake becomes very unstable at about Re_p of 600, generally leading to breakup. In fact, stable spherical cap bubbles have not been observed for $600 < Re_p < 1{,}500$.

For high deformation at $Re_p > 1{,}500$ (water-like viscosities), a spherical-cap shape is found (Figures 4.24c and 4.27b) where the rounded top (front) surface is characterized by nearly potential flow like that over the front of a sphere since viscosity effects are weak. The wake of the spherical-cap bubble in Figure 4.27b is turbulent (complex and highly unsteady). The wake is also "open" in the sense that turbulent structures shed far downstream and there is not a well-defined recirculation region. These wake characteristics are similar to that for a solid sphere at these Reynolds numbers (§3.2.4). On close inspection of Figure 4.24c (in water), the bottom surface has fluctuations. These are due to unsteady pressure fluctuations in the turbulent wake.

In general, bubble shapes at very high deformation tend to spherical caps in Figure 4.24c and 4.27 for which the minimum aspect ratio is nearly a constant and given as follows:

$$E_{min} \approx 0.25 \quad \textit{for spherical cap bubbles at} \quad We > 40 \textit{ and } Re_p > 40. \tag{4.56}$$

This result is for clean, partially contaminated, and fully contaminated bubbles, and happens to be the same minimum aspect ratio as for drops in a gas (4.47a).

The aspect ratio for bubbles with an intermediate Weber number depends on the level of contamination. In particular, there is less deformation for contaminated (versus clean) conditions, as this reduces surface velocity and thus surface pressure variations. As shown in Figure 4.20, the influence of Weber number for $Re_p > 100$ can be empirically modeled (Loth, 2008a) as follows:

$$E = 1 - (1 - E_{min}) \tanh(0.165We) \quad \textit{for clean bubbles.} \tag{4.57a}$$

$$E = 1 - (1 - E_{min}) \tanh(0.11We) \quad \textit{for contaminated bubbles.} \tag{4.57b}$$

This is similar to the trends for drops. However, bubbles can be stable at $We > 12$, at which drops in a gas generally break up. This is consistent with (4.53), which predicts that We_{crit} based on Kelvin–Helmholtz instabilities is much higher for bubbles than for drops in a gas. In fact, careful creation of a bubble in a still liquid can allow stable Weber numbers as large as 100. However, Rayleigh–Taylor instabilities cause breakup at yet higher Weber numbers.

As can be expected, increased deformation of a bubble at higher Weber numbers causes an increase in its drag coefficient. Interestingly, the incremental increases are nearly the same for clean bubbles, contaminated bubbles, and contaminated drops in a liquid for $Re_p > 100$, as shown in Figure 4.21. The generalized drag increase for contaminated conditions in a liquid can be modeled using the theoretical maximum drag coefficient ($C_{D,max}$) for spherical cap bubbles (Joseph, 2006) and the Schiller–Naumann spherical drag model as follows:

$$C_{D,max} = \frac{8}{3} + \frac{14.24}{Re_p} \tag{4.58a}$$

$$C_{D,sphere} = \frac{24}{Re_p}\left(1 + 0.15Re_p^{0.687}\right) \quad \textit{for contaminated bubbles.} \tag{4.58b}$$

$$C_D = C_{D,sphere}(1 - \Delta C_D^*) + C_{D,max}\Delta C_D^* \tag{4.58c}$$

$$\Delta C_D^* \approx \tanh\left[0.0016\left(WeRe_p^{0.2}\right)^{1.9}\right] \tag{4.58d}$$

As shown in Figure 4.21, this drag model gives reasonable performance, though there is significant scatter (much more than for drops in air), which can be attributed to secondary motion effects. Note that (4.58b) is only for contaminated bubbles (and contaminated immiscible drops in a liquid) since it uses the solid sphere model of (3.58) for $C_{D,sphere}$. For clean bubbles, the drag coefficients of (4.58) should be replaced as discussed in §9.3.1 and §9.3.2.

4.4.2 Free-Rise Trajectory Dynamics

For small bubbles in a liquid with spherical shapes, the free-rise trajectory in a still fluid (quiescent conditions) is a rectilinear (straight line) path upward. However, the onset of deformation with an oblate spheroidal shape is often accompanied by secondary (side-to-side) trajectory motion, as shown in Figure 4.28. In the following, the characteristics of this motion are first described, then quantified, and finally explained.

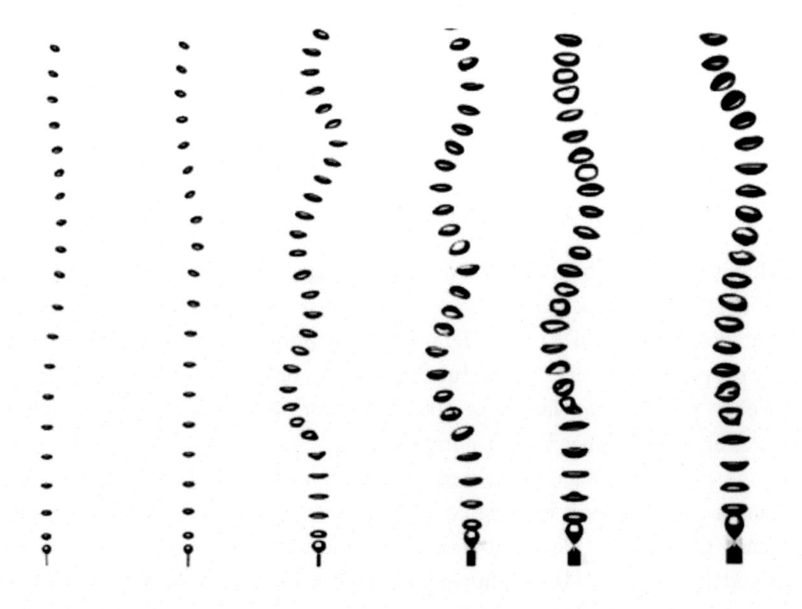

Figure 4.28 Sequential images over time of single bubble trajectories rising in deionized water for volumetric diameters (left to right) of 1.86 mm, 2.23 mm, 2.93 mm, 3.65 mm, 4.32 mm, and 4.4 mm for which the approximate Reynolds numbers range from 700–1,200 (Yan et al., 2017).

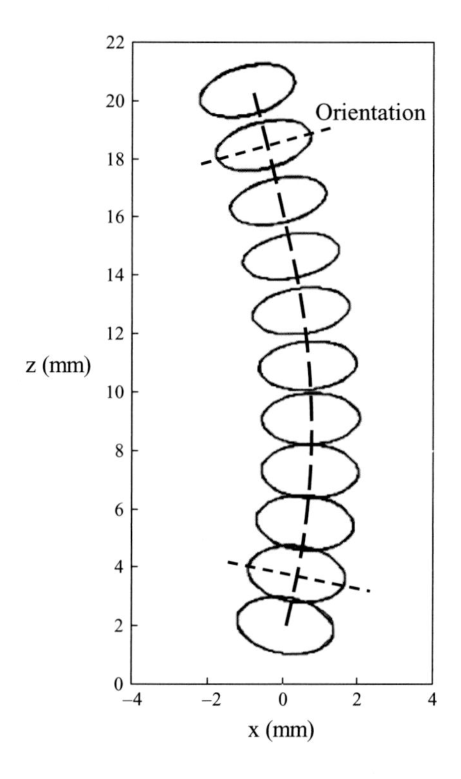

Figure 4.29 Successive images of a 2.5 mm bubble with E ~ 0.5 rising in water over time (Ellingsen and Risso, 2001) exhibiting a trajectory with secondary motion (based on the centroid path indicated by long dashes) and changes in orientation based on the shape axis of symmetry relative to the horizon (as indicated by short dashes).

As shown in Figure 4.29, secondary motion for an oblate bubble can include both trajectory oscillations (deviations in the centroid from the vertical axis) and orientation oscillations (tilting of the spheroid axis of symmetry from the vertical axis). Furthermore, the trajectory oscillations have two primary modes, as shown in Figure 4.30: *zigzag* (sinusoidal motion within a plane) and *spiral* (3D helicoidal motion). While zigzag motion tends to be preferred for larger bubbles (as noted in the caption of Figure 4.30), both can occur for all diameters as they can depend on slight differences in the initial release. In addition, a bubble in one mode can sometimes transfer into the other mode while maintaining similar frequencies. For example, air bubbles between 1.5–6.0 mm in diameter in water have a zigzag frequency (f_{osc}) of about 7 Hz but a spiraling frequency of about 5 Hz (Mercier et al., 1973).

To quantify the spatial deviations associated with these trajectory oscillations, one may define the lateral (horizontal) amplitude of the perturbations about the mean as $\mathcal{A}_{osc}$ (as shown in Figure 4.30 for both zigzag and helicoidal motion). For zigzag motion in the y-plane, the lateral perturbations relative to some initial time (t_o) can be modeled as follows:

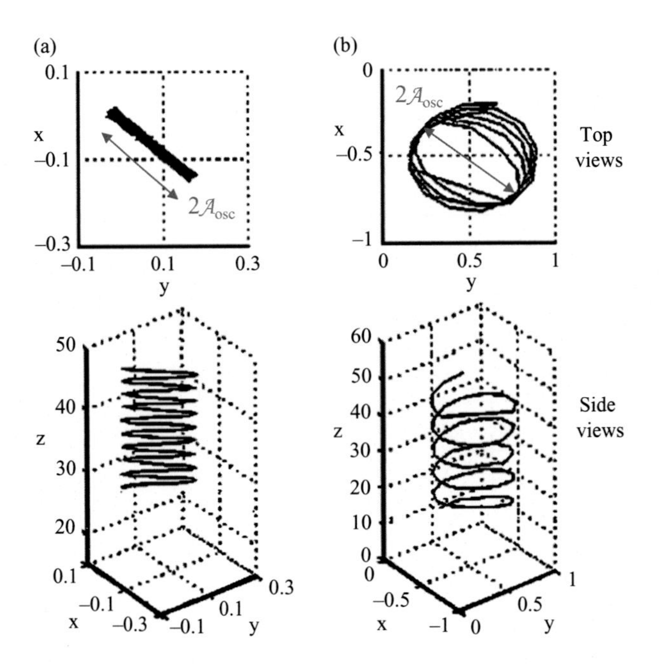

Figure 4.30 Bubble trajectories for a 1.9 mm air bubble in water shown in the horizontal plane (above) and in the three-dimensional volume (below), where axes units are in centimeters (but are highly amplified in the lateral direction to make secondary motion appear more pronounced) with two types of motion: (a) zigzag motion associated with a gentle push-off an a nearly spherical shape; and (b) helicoidal motion associated with a pinch-off and an ellipsoidal shape (Wu and Gharib, 2002). The lateral amplitude of the oscillations ($\mathcal{A}_{osc}$) is shown from the x-y plane views.

$$y_{osc} \approx \mathcal{A}_{osc} \sin\left[2\pi f_{osc}(t - t_o)\right]. \tag{4.59}$$

The model can be extended to spiraling conditions by superimposing a similar expression for x_{osc}, but one that is $\pi/2$ out of phase with y_{osc} to create a helical path.

The lateral amplitudes tend to vary with bubble size. For example, measurements of $\mathcal{A}_{osc}$ for air bubbles in water are shown in Figure 4.31 for both clean conditions (open symbols) and contaminated conditions (solid symbols). In general, secondary motion starts at $d \approx 1.3$ mm for both cases and the amplitude peaks at $d \approx 2$ mm for clean conditions, but at $d \approx 4$ mm (as seen in Figure 4.28) for contaminated bubbles. As a rough approximation, the amplitude is about 3.5 mm as shown by the dashed line empirical fit of Figure 4.31. In this case, the average trajectory of a bubble in water can be roughly approximated as follows:

$$y_{osc} \approx (3.5\,\text{mm}) \sin[2\pi(6\text{ Hz})(t - t_o)] \quad \textit{for}\ 1.5\,\text{mm} < d < 6\,\text{mm}. \tag{4.60}$$

The frequency and magnitude of the oscillations are even preserved in turbulent flow conditions if the trajectory is considered in a reference frame that moves with the local and instantaneous water velocity (Ford and Loth, 1998).

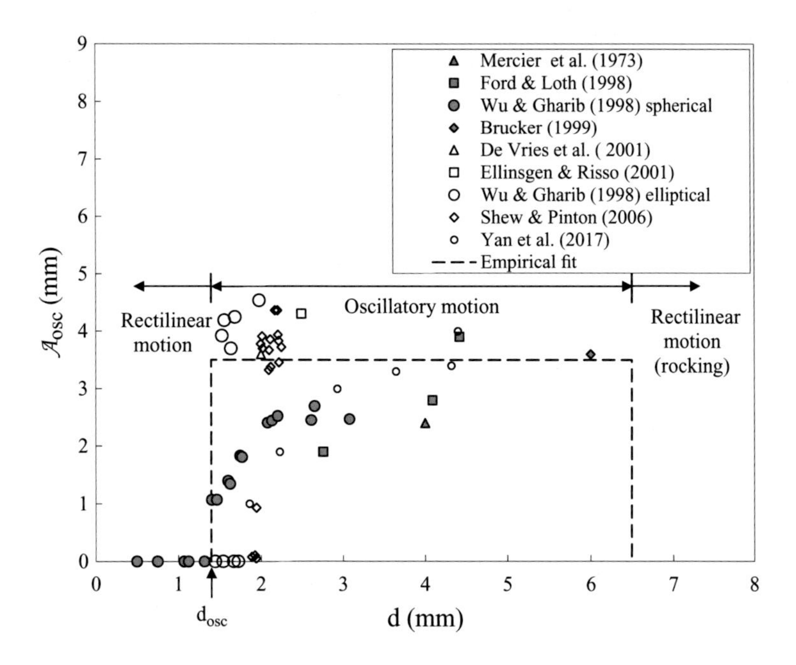

Figure 4.31 Amplitudes of lateral path oscillations for bubbles in water, where solid symbols represent fully contaminated conditions (the surface is completely immobilized), open symbols represent clean conditions, and the dashed line is an approximate average fit.

For both clean and contaminated conditions, the lateral amplitude generally subsides once sizes reach 6.5 mm for air bubbles in water. Larger bubbles of $7\,\text{mm} \leq d \leq 15\,\text{mm}$ have nearly vertical paths with little lateral amplitude but are accompanied by rocking of the bubble orientation (Aybers and Tapucu, 1969). Spherical-cap bubbles ($d > 15\,\text{mm}$) also have a nearly rectilinear motion though weak rocking can still occur for turbulent wake conditions (Fan and Tsuchiya, 1993). Thus, only the oblate shape causes significant secondary trajectories (zigzag and/or spiral). The reported minimum (onset) bubble diameters for which this trajectory oscillation occurs (d_{osc}) are listed Table 4.2 for several fluids. While there is significant scatter, damping due to cleaner conditions and larger Morton numbers (increased viscosity) often results in larger oscillation onset diameters.

Finally, we explain the cause for this secondary trajectory motion. Originally, this motion was thought to be due to unsteady wake shedding from the rear of the bubble. However, wake shedding frequency changes with bubble size (ranging from about 10–40 Hz for the above bubble size range), whereas the trajectory oscillation frequency is nearly independent of bubble size and is lower in magnitude. Furthermore, secondary motion can occur even when the bubble wake is attached with no flow separation (as can occur with a clean bubble at moderate Reynolds numbers). For example, the bubble shown in Figure 4.32a has an attached laminar wake (no vortex shedding) but undergoes significant trajectory oscillation. Based on such experiments, it was found that the primary driving mechanism for trajectory oscillation is an

Table 4.2 Critical bubble diameter for transition from rectilinear path to oscillatory path (either sinusoidal or helical trajectories) in order of decreasing Morton number. The fluid temperature is 21–23°C, unless specified.

Continuous phase	Mo	d_{osc} (mm)	Reference
Ethyl alcohol	1.2×10^{-8}	1.7	Hartunian and Sears (1957)
Methyl alcohol	8.9×10^{-9}	1.3	Hartunian and Sears (1957)
40% (wt.) glycerol	5.7×10^{-9}	2.2	Hartunian and Sears (1957)
30% (wt.) glycerol	1.1×10^{-9}	1.8	Hartunian and Sears (1957)
20% (wt.) glycerol	2.6×10^{-10}	1.5	Hartunian and Sears (1957)
Pure water	2.5×10^{-11}	1.8	Duineveld (1995)
		1.7	Wu and Gharib (2002)
Distilled water	2.6×10^{-11}	1.7	Hartunian and Sears (1957)
Filtered tap water	2.7×10^{-11}	1.4	Saffman (1956)
Tap water	2.6×10^{-11}	1.3	Hartunian and Sears (1957)
		1.3	Haberman and Morton (1954)
Hot tap water (43°C)	5.0×10^{-12}	1.0	Hartunian and Sears (1957)
Hot tap water (49°C)	3.1×10^{-12}	1.0	Haberman and Morton (1954)
Benzene	4.5×10^{-12}	1.2	Hartunian and Sears (1957)

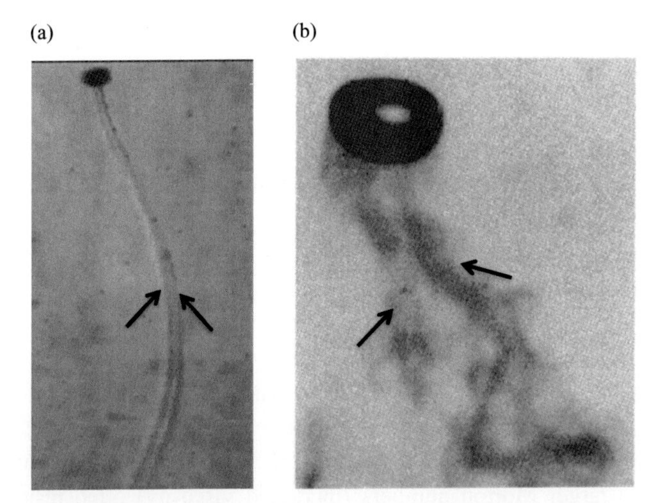

Figure 4.32 Visualization of bubbles undergoing spiral trajectory: (a) shadowgraph of a 2 mm bubble ($Re_p \sim 630$) that leads to a laminar double-thread wake (de Vries et al., 2001); and (b) a dye visualization of a 6 mm ellipsoidal bubble ($Re_p \sim 1,800$), where the wake is effectively turbulent, though remnants of the double-thread vorticity tubes can be seen near the bubble surface (Brucker, 1999). In both cases, a pair of arrows are used to show each of the two threads.

instability of the "double-threaded" wake, as shown by the arrows on Figure 4.32. This wake is composed of two counterrotating vortex filaments that can become unstable (Crow, 1970). In particular, the Crow instability occurs when two inviscid counterrotating vortices are parallel and close to each other. A small perturbation of one of the filaments induces a reinforcing perturbation in the other, until both will sinusoidally undulate over time. For a bubble, these filaments extend along the wake, and their ensuing curvature induces a side force on the bubble, causing lateral motion. This lateral movement, in turn, leads to a change in bubble orientation that causes an asymmetric production of vorticity, resulting in the filament deforming in the other direction, thus completing the cycle of instability (de Vries et al., 2001). Once Re_p exceeds 1,000 (or less at higher surfactant levels), unsteady wake shedding occurs (as in solid particles) and the trajectory oscillation amplitudes tend to increase, though there is still ample evidence of the two counterrotating filaments in the bubble wake (Figure 4.32b). Once Re_p exceeds 2,000, the trajectory oscillations tend to subside. This is attributed to increasing unsteadiness and complexity of the separated wake, which interferes with the Crow instability.

4.5 Shapes and Breakup in Shear Flow

In the previous sections, the shapes of drops and bubbles were shown to be influenced by relative velocity (e.g., in free-fall or free-rise conditions). However, flow shearing can also drive a change of shape. In particular, a fluid particle subjected to linear shear tends to stretch and tilt at an angle θ_ω, as shown in Figure 4.33. If the flow is steady, the drop or bubble will tend to reach an equilibrium shape and angle based on the balance of viscous shear stresses (which drive deformation) and surface tension stresses (which oppose deformation). The shear stresses are proportional to the product of the surrounding flow shear rate (ω_{shear}) and the surrounding fluid viscosity (μ_f). The surface tension stresses are proportional to the ratio of interface surface tension (σ) and particle diameter (d). The ratio of these competing stresses is used to define the shear-based capillary number (Ca_ω), while the ratio of the inertial stress (due to convection) to surface tension is used to define the shear-based Weber number (We_ω), and these two dimensionless parameters are related by the shear-based Reynolds numbers (Re_ω):

$$Ca_\omega \equiv \frac{\mu_f d\omega_{shear}}{\sigma}. \tag{4.61a}$$

$$We_\omega \equiv \frac{\rho_f d^3 \omega_{shear}^2}{\sigma}. \tag{4.61b}$$

$$Re_\omega \equiv \frac{\rho_f d^2 \omega_{shear}}{\mu_f} = \frac{We_\omega}{Ca_\omega}. \tag{4.61c}$$

In general, deformation is primarily controlled by We_ω for $Re_\omega \gg 1$ (where convective effects dominate) and by Ca_ω for $Re_\omega \ll 1$ (when creeping-flow stresses dominate). Also, the RHS of (4.61c) shows how the three dimensionless numbers are related. If the shear effects are weak ($Ca_\omega \ll 1$ and $We_\omega \ll 1$), one may expect nearly spherical shapes. If the shear effects are strong ($Ca_\omega \gg 1$ or $We_\omega \gg 1$), one may expect high deformation and possibly breakup.

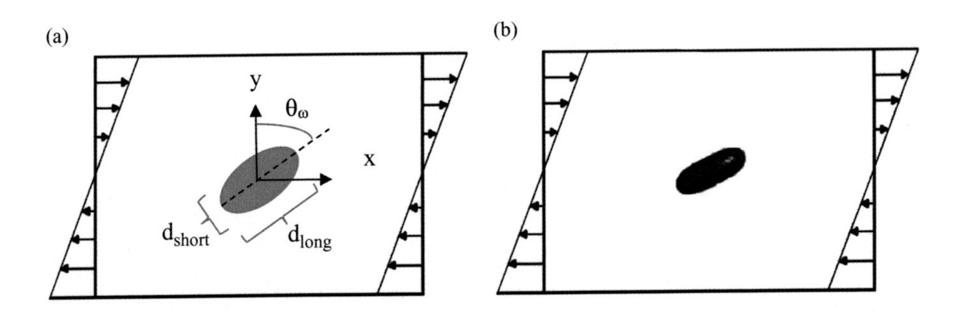

Figure 4.33 Deformation of a fluid particle in a linear shear in the y-direction for a velocity in the x-direction: (a) schematic ellipsoid stretched in the direction of fluid shear (the top portion is pulled to the right and the bottom portion is pulled to the left) with shear-plane diameters shown; and (b) observed shape for $\mu_p = 5\mu_f$, $Re_\omega = 10$, and $Ca_\omega = 0.14$ (Khismatullin et al., 2003).

For small deformations, the shape of the particle tends to an ellipsoid with three different lengths: d_{short}, d_{medium}, and d_{long}. The lengths of d_{short} and d_{long} are in the shear (x-y) plane as shown in Figure 4.33, while d_{medium} is perpendicular to these (in the z-plane). The short and long lengths define the aspect ratio (E) and the square of the medium diameter is equal to the product of the other two diameters as follows

$$d_{short} = E\, d_{long}. \tag{4.62a}$$

$$d_{medium} = E^{1/2} d_{long}. \tag{4.62b}$$

$$d_{short} d_{medium} d_{long} = d^3. \tag{4.62c}$$

Based on these expressions and (4.37), d_{medium} is equal to the volumetric diameter (d).

The creeping-flow limit allows an analytical solution for small deformations in the absence of any relative velocity (w = 0). Using the PDEs of §3.2.2 for fluid particles under shear, the theoretical aspect ratio for clean conditions (Taylor, 1932, 1934; Cox, 1969) is as follows:

$$\frac{1-E}{1+E} = \frac{5\left(19\mu_p^* + 16\right)}{4\left(\mu_p^* + 1\right)\sqrt{\left(19\mu_p^*\right)^2 + (40/Ca_\omega)^2}} \quad \textit{for } Re_\omega \to 0. \tag{4.63}$$

The LHS represents the degree of deformation since it increases from 0 to 1 as E reduces from 1 to 0. Based on the RHS, the deformation limits for very high and low particle viscosity ratios are as follows:

$$E \to 1 \quad \textit{as} \quad \mu_p^* \to \infty. \tag{4.64a}$$

$$\frac{1-E}{1+E} = \frac{Ca_\omega}{2} \quad \textit{as} \quad \mu_p^* \to 0. \tag{4.64b}$$

The first limit (4.64a) indicates that a liquid drop sheared in a gas ($\mu_p \gg \mu_f$) typically has weak distortion in creeping flow. As such, drops in a gas tend to only deform when both

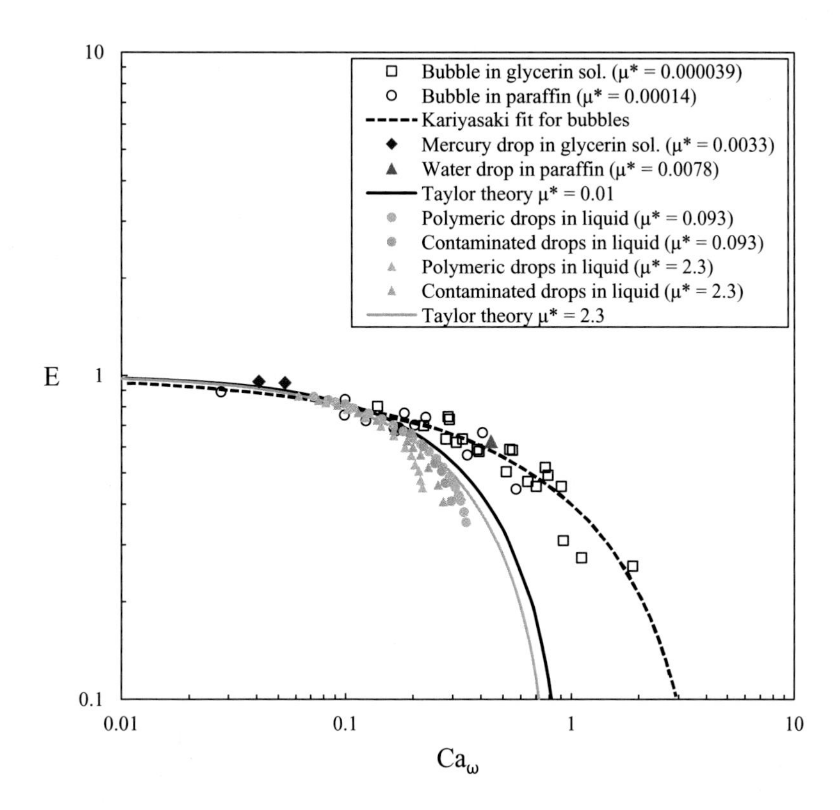

Figure 4.34 Bubble and droplet aspect ratio as a function of capillary number for $0.005 < \mathrm{Re_p} < 5$ for experiments (Taylor, 1934; Kariyasaki, 1987; Hu and Lips, 2003) compared to creeping-flow theory (Taylor, 1932, 1934; Cox, 1969) and the empirical fit of Kariyasaki (1987) for (a) low-viscosity drops and (b) high-viscosity drops (relative to the surrounding liquid).

the Reynolds number and Weber number are significant. In contrast, the second limit (4.64b) indicates that a bubble in a liquid ($\mu_p \ll \mu_f$) can undergo significant deformation for creeping flow. Experimental deformations for gas and liquid particles in other liquids are shown in Figure 4.34 in comparison to the creeping-flow theory. For $\mathrm{Ca}_\omega < 0.1$, deformation increases (E decreases) with the capillary number consistent with (4.63). For $\mathrm{Ca}_\omega > 0.1$, drops in a liquid tend to deform more than the theoretical value, whereas bubbles tend to deform less but are reasonably predicted by a fit proposed by Kariyasaki (1987). These differences can be attributed to large deformation effects.

For creeping flow, the equilibrium (steady-state) orientation angle under shear is as follows:

$$\theta_\omega = \frac{\pi}{4} + \frac{1}{2}\tan^{-1}\left(\frac{19\mu_p \mathrm{Ca}_\omega}{40\mu_f}\right) \quad \textit{for } \mathrm{Re}_\omega \to 0. \tag{4.65}$$

Weak shear ($\mathrm{Ca}_\omega \to 0$) or weak particle viscosity (e.g., a bubble with $\mu_p \ll \mu_f$) yields $\theta_\omega \to 45°$. Strong shear ($\mathrm{Ca}_\omega \to \infty$) or high particle viscosity yields $\theta_\omega \to 90°$, whereby the long diameter of the drop fully aligns in the flow direction and stretches with the flow streamlines. For finite Reynolds numbers, the orientation angles are larger than

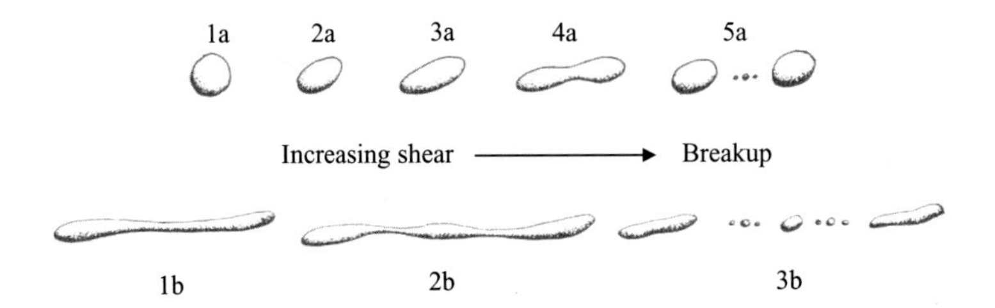

Figure 4.35 Sequential time images of immiscible drops in a liquid (with $\mu_p \approx \mu_f$) subjected to a continually increasing shear level (from left to right, with $Re_\omega \ll 1$) that stretches and deforms the drop, eventually leading to breakup (Clift et al., 1978): (a) in silicon oil yielding $Ca_{\omega,crit} \approx 0.4$ and (b) in corn syrup yielding $Ca_{\omega,crit} > 2$.

that predicted by (4.65) since the increased effect of convection aligns the shapes with streamwise direction. For example, the distorted drop in Figure 4.33 reaches $\theta_\omega \approx 74°$ at $Re_\omega = 10$, while creeping flow predicts $\theta_\omega = 54°$.

The trend of increased orientation angle and increased deformation with increased shear is shown in Figure 4.35a for Time 1a–3a, and is qualitatively consistent with (4.63) and (4.65). By Time 4a, the drop has stretched into two primary masses (on the left and the right) connected by cylindrical ligament, indicating high shear and deformation. At Time 5a, breakup has occurred due to an instability on the interface, yielding two primary drops as well as several smaller drops from the connecting ligament. To characterize this condition, one may define a critical shear capillary number (beyond which breakup occurs) as $Ca_{\omega,crit}$. For creeping flow with clean particles, there are two first-order theoretical solutions: small deformation (from a sphere) for viscosity ratios of order unity (Taylor, 1932), and slender body for small viscosity ratios (Hinch and Acrivos 1979, 1980) as follows:

$$Ca_{\omega,crit} = \frac{8\mu_p^* + 8}{19\mu_p^* + 16} \quad \textit{for } \mu_p^* \sim 1 \tag{4.66a}$$

$$\textit{for shear at } Re_\omega \to 0.$$

$$Ca_{\omega,crit} = 0.05\left(\frac{\mu_p}{\mu_f}\right)^{-2/3} \quad \textit{for } \mu_p^* \ll 1 \tag{4.66b}$$

For elongation flow with a strain $(\mathcal{G})$, a similar capillary number can be defined, and the creeping-flow theories for small-deformation and slender-body breakup are

$$Ca_{\mathcal{G}} \equiv \frac{\mu_f d\mathcal{G}}{\sigma} = \frac{We_{\mathcal{G}}}{Re_{\mathcal{G}}} \tag{4.67a}$$

$$Ca_{\mathcal{G},crit} = \frac{4\mu_p^* + 4}{19\mu_p^* + 16} \quad \textit{for } \mu_p^* \sim 1 \tag{4.67b}$$

$$\textit{for elongation at } Re_\omega \to 0.$$

$$Ca_{\mathcal{G},crit} = 0.15\left(\frac{\mu_p}{\mu_f}\right)^{-1/6} \quad \textit{for } \mu_p^* \ll 1 \tag{4.67c}$$

In this elongation flow, the fluid particle deforms in the strain direction, that is, $\theta_\omega \approx 90°$.

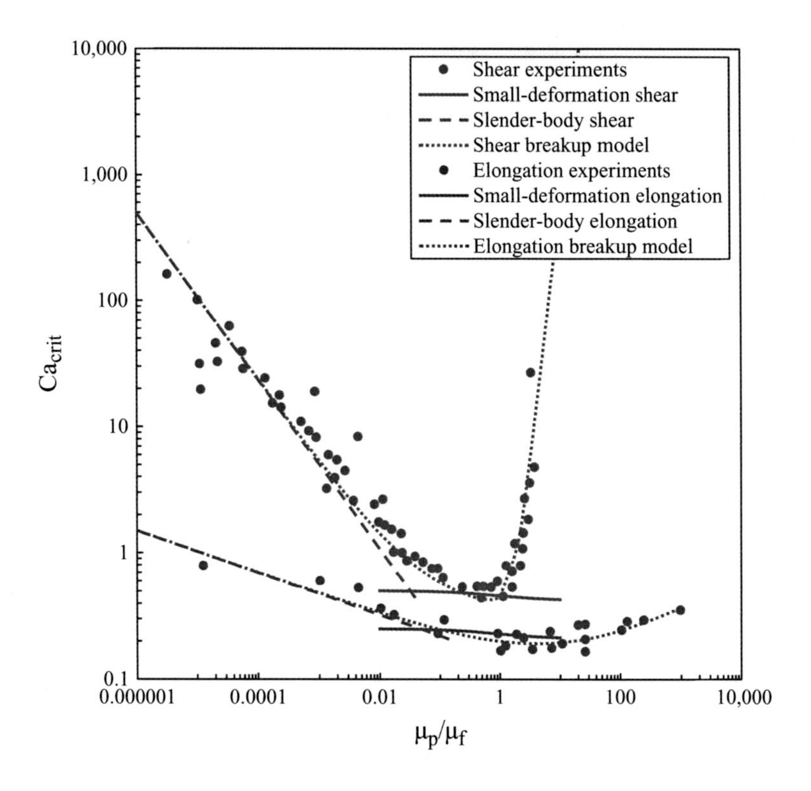

Figure 4.36 Critical capillary numbers as a function of viscosity ratio for clean drops in creeping shear and elongational flows based on experiments (Grace, 1982), first-order theories (Taylor, 1932), slender-body theories (Hinch and Acrivos, 1979, 1980), and semi-empirical models.

The trends for a range of particle viscosities in creeping-flow conditions are shown in Figure 4.36, based on results surveyed by Stegeman (2002). The theories are consistent with the experiments for their appropriate regimes, but higher critical capillary numbers are found at high viscosity ratios. A generalized semi-empirical expression for clean drops can be obtained using slender-body theory combined with empirical terms at higher viscosity ratios, as follows:

$$\mathrm{Ca}_{\omega,\mathrm{crit}} = 0.05\left(\frac{\mu_\mathrm{p}}{\mu_\mathrm{f}}\right)^{-2/3} + 0.35 + 0.05\left(\frac{\mu_\mathrm{p}}{\mu_\mathrm{f}}\right)^{4} \quad \textit{for drops in shear.} \tag{4.68a}$$

$$\mathrm{Ca}_{\mathcal{G},\mathrm{crit}} = 0.15\left(\frac{\mu_\mathrm{p}}{\mu_\mathrm{f}}\right)^{-1/6} + 0.05\left(\frac{\mu_\mathrm{p}}{\mu_\mathrm{f}}\right)^{1/4} \quad \textit{for drops in elongation.} \tag{4.68b}$$

These breakup models are robust for a wide range of viscosity ratios, as shown in Figure 4.36.

It should be noted that the drop of Figure 4.35b encountered much greater shear and deformation before breakup, even though the viscosity ratio was about the same as in Figure 4.35a. This can be attributed to surfactants in the surrounding corn syrup,

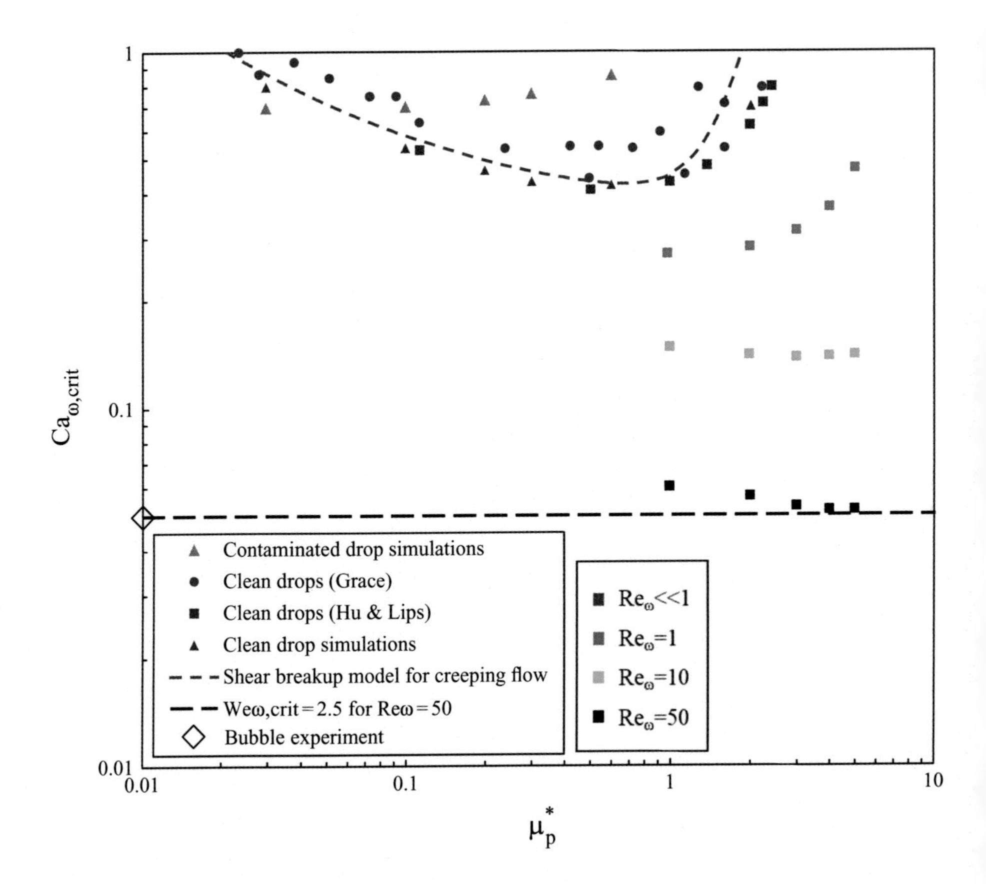

Figure 4.37 Critical capillary number as a function of viscosity ratio and shear Reynolds numbers, with creeping-flow data (Grace, 1982), clean drops simulations at creeping flow (Khismatullin et al., 2003), experiments at various Reynolds numbers as square symbols (Hu and Lips, 2003), and bubble at $Re_\omega = 50$ (Revuelta et al., 2006) along with the model for $Re_{\omega,crit} \ll 1$ (solid line) and the $We_{\omega,crit}$ model for $Re_{\omega,crit} \gg 1$ (dashed line).

which immobilized at least part of the surface, thus reducing internal motion and damping instabilities. As such contamination generally delays breakup and increases $Ca_{\omega,crit}$, as shown in Figure 4.37. This figure also shows that increasing Re_ω causes $Ca_{\omega,crit}$ to be reduced due to the additional presence of convective instabilities. This is consistent with the deformation and breakup being driven by the Weber number at higher Reynolds numbers (Martinez-Bazan et al., 1999a). In fact, the breakup criterion for a drop in an immiscible liquid at high Re_ω is nearly independent of the viscosity ratio and is instead based primarily on a critical Weber number, as follows:

$$We_{\omega,crit} \sim 2.5 \quad \textit{when} \quad Re_\omega > 50. \tag{4.69}$$

This fit is shown as the dashed line in Figure 4.37 and is reasonable for both drops and bubbles.

4.6 Oscillating Shape and Size Modes of Fluid Particles

In addition to the deformation mechanisms associated with relative velocity, shear, and elongation, a fluid particle can also deform due to flow unsteadiness (e.g., turbulent fluctuations) or initial conditions (e.g., release from a needle). The deformation dynamics can be related to various modes called shape *harmonics*, which are classified according to their integer mode values ($\mathcal{S}$). The first harmonic ($\mathcal{S} = 1$) is defined when the shape is fixed as a sphere, but the volume expands and contracts. This mode involves compressibility of the particle and so is generally limited to gas bubbles, as will be discussed in §4.6.2 and §4.6.3. The higher modes ($\mathcal{S} = 2, 3, 4$, etc.) are oscillations between various nonspherical shapes at constant volume and so can occur for gas bubbles as well as liquid drops, as will be discussed in §4.6.1.

4.6.1 Oscillating Shape Modes

The shape modes for constant volume are based on geometric modes with associated natural frequencies. The simplest of these is the second mode ($\mathcal{S} = 2$), which flattens and elongates on two axes. In particular, the mode flucuates from an oblate spheroid on one axis, to a sphere, to a prolate spheroid on a perpendicular axis, back to a sphere, and thereafter the process repeats (Figure 4.38a). Higher modes have more complex shapes, such as the rounded triangular ($\mathcal{S} = 3$) in Figure 4.38b, which flattens and elongates on

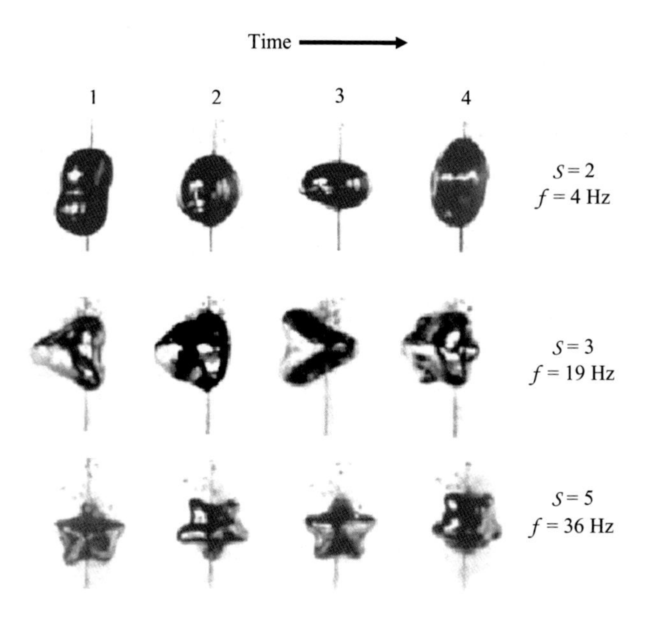

Figure 4.38 Various modes of incompressible shape oscillations of a sessile mercury drop in a liquid excited at various frequencies (Azuma and Yoshihara, 1999). Note that $\mathcal{S} = 1$ corresponds to a volumetric mode (not a shape mode) whereby the sphere diameter changes with time, and thus requires that the fluid particle be compressible.

two axes, and star-like ($\mathcal{S}$ = 5) configurations that oscillate within a plane as shown in 4.38c. Such higher mode harmonics tend to require carefully controlled isolated conditions to avoid any weak asymmetries; for example, flow gradients or the presence of walls or neighboring particles can make it difficult for these higher modes to appear.

To analyze these modes theoretically, one may note that the shape dynamics and their natural frequencies are governed by the inertia of the internal and external fluids and by surface tension. We next assume that the surrounding fluid is stagnant in the far field and there is no relative velocity nor shear ($w = \omega = 0$). The dynamics can be obtained by applying potential flow PDEs for both the interior and exterior fluids (i.e., assuming incompressible inviscid flow) and further assuming the shape oscillations are small perturbations to a spherical shape. This leads to a second-order ODE of the perturbation amplitude. Using this approach, Lamb (1945) found that the natural frequency ($f_{\mathcal{S}}$) of the oscillating shape modes is based on the mode number ($\mathcal{S}$), as follows:

$$(2\pi f_{\mathcal{S}})^2 = \frac{(\mathcal{S}-1)\mathcal{S}(\mathcal{S}+1)(\mathcal{S}+2)\sigma}{\left[(\mathcal{S}+1)\rho_p + \mathcal{S}\rho_f\right] r_p^3} \quad \textit{for } \mathcal{S} = 2, 3, 4, \ldots \; (\textit{NISI}). \tag{4.70}$$

The term in the parentheses on the LHS is the natural *angular* frequency (with units of radians/time), which is the product of 2π and the natural frequency, $f_{\mathcal{S}}$ (with units of cycles/time, e.g., Hz). The relationship of (4.70) is like that of a conventional spring system, for which the square of the angular frequency (LHS) is equal to the ratio of stiffness to mass (RHS). The RHS stiffness (numerator) is based on surface tension, while the mass (denominator) includes the particle mass and the fluid added mass. As such, the frequency decreases as the particle radius increases. In addition, (4.70) indicates that the frequencies increase as $\mathcal{S}$ increases, as shown by Figure 4.38 for large mercury drops. As other examples, the second harmonic ($\mathcal{S} = 2$) leads to a frequency of 120 Hz for a 1 mm radius water drop in air and 148 Hz for a 1 mm radius gas bubble in water.

These frequencies of (4.70) are in good agreement with measured frequencies for millimeter-sized bubbles (Feng and Leal, 1997) and drops (Azuma and Yoshihara, 1999). This agreement is attributed to the weak influence of viscosity for these sizes. Viscosity can play a larger role at smaller particle sizes, but primarily serves to dampen the oscillations rather than changing the natural frequencies.

These shape dynamic modes will be most excited when driven at a frequency close to the natural frequency. These excitations often come in the form of pressure oscillations. It is noted that the shape oscillation frequencies are high compared to the frequencies noted in §4.4.2. As such, their effects are generally decoupled, that is, bubble wake vortex shedding occurs at too low of a frequency to induce shape oscillations. In the following, we consider size oscillations, which will generally occur at even higher frequencies than those of shape oscillations.

4.6.2 Rayleigh–Plesset Equation for Bubble Radius

As noted previously, a fluid particle can undergo size oscillations if the particle is compressible, as in a gas bubble in a liquid. For a spherical shape, these oscillations

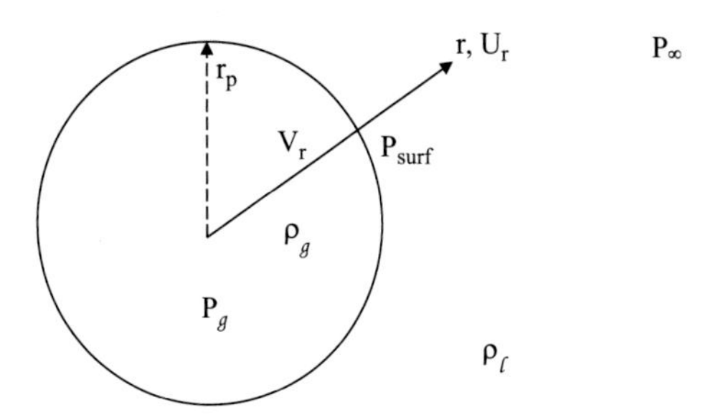

Figure 4.39 Schematic of spherical bubble of radius r_p with uniform gas pressure inside the bubble (P_g) and liquid pressure that varies from surface (P_{surf} at $r = r_p$) to far field (P_∞ at $r \to \infty$).

are termed mode one ($\mathcal{S} = 1$) and can be driven by liquid pressure fluctuations or by an initial impulse that drives the bubble size out of equilibrium, resulting in an oscillatory response.

For a spherical shape, the volume change is related to the change in radius and can be investigated theoretically. The analysis for a spherical compressible bubble surrounded by an incompressible inviscid liquid was first obtained as an ODE by Lord Rayleigh (1917). The effects of surface tension and liquid viscosity were added later by Plesset (1949), resulting in the well-known Rayleigh–Plesset equation for the dynamics of the bubble radius as a function of far-field liquid pressure. The derivation of this ODE is described by Batchelor (1967), and the basic elements are reviewed below for a stagnant fluid, followed by an extension to include a relative velocity and weak liquid compressibility.

As shown in Figure 4.39, consider a spherical gas bubble whose centroid is stationary ($v = 0$) but has a radius (r_p) that varies with time. If the spatial variations are in the radial direction only, spherical coordinates can be used by setting the origin ($r = 0$) at the bubble centroid. The gas density (ρ_g) and gas pressure (P_g) will also vary with time but are spatially uniform within the bubble volume. The surrounding fluid is an incompressible liquid with constant density (ρ_ℓ) but has a pressure that can vary with time and radial distance, ranging from a surface pressure (P_{surf}) to a far-field pressure (P_∞). Resolved-surface values are used since this pressure considers the effects of the bubble surface. In the far field, the fluid is assumed to be stagnant but with a pressure that varies with time:

$$U_\infty = 0 \quad \textit{and} \quad P = P_\infty(t) \quad \textit{as}\ r \to \infty. \tag{4.71}$$

Due to spherical symmetry, the velocity field outside of the bubble will have no angular components ($U_\theta = U_\phi = 0$) and the radial velocity is only a function of radius and time, defined by $U_r(r,t)$. Similarly, the velocity field inside the bubble is given by $V_\theta = V_\phi = 0$ and $V_r(r,t)$. At the surface, these liquid and gas radial velocities are equal (by continuity) and are also equal to the time rate of change of the particle radius (interface velocity) as follows:

$$\dot{r}_p \equiv \frac{\partial r_p}{\partial t} = U_r(r_p, t) = V_r(r_p, t). \tag{4.72}$$

As the bubble radius increases, it will push liquid outward in the radial direction. Since the liquid is incompressible and has spherical symmetry, conservation of mass necessitates that the liquid mass fluxing through any spherical control surface (of fixed radius r) must be the same at any given time. Therefore, we can relate the surface-integrated radial mass flux at the surface to that at some general radius in the surrounding fluid as follows:

$$\rho_\ell\left(4\pi r_p^2\right)\dot{r}_p = \rho_\ell\left(4\pi r^2\right)U_r. \tag{4.73}$$

Since the density is constant, the local liquid radial velocity can be expressed as follows:

$$U_r = \dot{r}_p\left(r_p^2/r^2\right). \tag{4.74}$$

As such, the liquid radial velocity decreases as $1/r^2$ in space and tends to zero as $r \to \infty$ per (4.71) to satisfy conservation of mass.

Conservation of radial momentum is next considered by neglecting gravitational forces (consistent with neglecting relative velocity). Applying the spherical incompressible Navier–Stokes equations (2.75b) for the resolved-surface flow field yields

$$\rho_\ell\frac{\partial U_r}{\partial t} + \rho_\ell U_r\frac{\partial U_r}{\partial r} = -\frac{\partial P}{\partial r} + \mu_\ell\left(\frac{1}{r^2}\right)\left[\frac{\partial}{\partial r}\left(r^2\frac{\partial U_r}{\partial r}\right) - 2U_r\right]. \tag{4.75}$$

Substituting (4.73) into this equation and solving for the radial pressure gradient in the surrounding fluid yields

$$\frac{\partial P}{\partial r} = -\frac{\rho_\ell}{r^2}\frac{\partial}{\partial t}\left(r_p^2\dot{r}_p\right) + \frac{2\rho_\ell}{r^5}\left(r_p^2\dot{r}_p\right)^2. \tag{4.76}$$

Thus, the liquid pressure gradient depends on the radial dynamics of the bubble surface. Integrating this equation from the bubble surface to the far field yields the surrounding fluid pressure difference just above the interface (P_{surf}) relative to that at the far field (P_∞) as

$$P_{surf} = P_\infty + \frac{\rho_\ell}{r_p}\frac{\partial}{\partial t}\left(r_p^2\dot{r}_p\right) - \frac{1}{2}\rho_\ell\dot{r}_p^2. \tag{4.77}$$

Next we consider the stress balance across the interface using (3.33a) to relate the surface liquid pressure to the bubble pressure. We can neglect the gas viscosity (since $\mu_g \ll \mu_\ell$) to yield the following:

$$P_g - P_{surf} = \frac{2\sigma}{r_p} - 2\mu_\ell\left(\frac{\partial U_r}{\partial r}\right)_{surf} = \frac{2\sigma}{r_p} + \frac{4\mu_\ell\dot{r}_p}{r_p}. \tag{4.78}$$

The RHS is obtained via (4.74). Combining this result with (4.77) yields

$$P_g - P_\infty = \rho_\ell r_p\ddot{r}_p + \frac{3}{2}\rho_\ell\left(\dot{r}_p\right)^2 + \frac{1}{r_p}\left(2\sigma + 4\mu_\ell\dot{r}_p\right). \tag{4.79}$$

This equation relates the difference between the bubble gas pressure and far-field liquid pressure to the surface radial dynamics. To get this in the form of an ODE for r_p, the bubble pressure (which varies) must also be expressed as a function of this radius.

The bubble pressure is related to r_p through the compressibility of the gas via the equation of state. To obtain this relationship, we first consider the equilibrium steady-state condition where there is no more forcing and oscillations have subsided ($U_r = 0$). In this case, the surrounding liquid pressure is constant everywhere ($P_{surf,eq} = P_{\infty,eq}$) and can be related to the equilibrium gas pressure ($P_{g,eq}$) and equilibrium radius ($r_{p,eq}$) via a steady version of (4.79), as follows:

$$P_{g,eq} = P_{\infty,eq} + \frac{2\sigma}{r_{p,eq}} = \rho_{g,eq} \mathcal{R}_g T_{g,eq}. \tag{4.80}$$

The RHS employs the ideal gas law (1.24) and the equilibrium gas density ($\rho_{g,eq}$) and temperature ($T_{g,eq}$). Returning to the unsteady condition, the instantaneous gas pressure, density, and radius can be related to their equilibrium values via the polytropic exponent of (2.36) as follows:

$$P_g/P_{g,eq} = \left(\rho_g/\rho_{g,eq}\right)^{\mathcal{K}_g} = \left(r_{p,eq}/r_p\right)^{3\mathcal{K}_g}. \tag{4.81}$$

The RHS of this expression assumes a bubble with constant mass, while the polytropic exponent ($\mathcal{K}_g$) depends on the type of thermodynamic process during expansion and contraction.

The two common limits for the polytropic exponent are: $\mathcal{K}_g = 1$ for an isothermal process, based on (2.37), and $\mathcal{K}_g = \gamma$ for an isentropic process, based on (2.40). The isothermal limit is reasonable when heat transfer is fast relative to bubble dynamics. This occurs when the bubble thermal response time, τ_T of (3.111b), is small compared to the radial oscillation time scale (τ_r), so that the surrounding liquid dictates the bubble temperature. At the other extreme, the isentropic limit is appropriate when $\tau_T \gg \tau_r$. Intermediate conditions ($1 < \mathcal{K}_g < \gamma$) are more complex as they require consideration of finite-rate heat transfer to relate bubble pressure and density (Brennen, 2005).

If we set aside specifying the value of $\mathcal{K}_g$ and substitute (4.81) into (4.79), the result is the celebrated Rayleigh–Plesset equation:

$$r_p \ddot{r}_p + \frac{3}{2}\left(\dot{r}_p\right)^2 + \frac{4\mu_\ell \dot{r}_p}{\rho_\ell r_p} = \frac{P_{g,eq}}{\rho_\ell}\left(\frac{r_{p,eq}}{r_p}\right)^{3\mathcal{K}_g} - \frac{2\sigma}{\rho_\ell r_p} - \frac{P_\infty}{\rho_\ell}. \tag{4.82}$$

On the LHS, the first term is related to the acceleration of the bubble surface, the second term is related to the dynamic pressure, and the third term is based on viscous damping. On the RHS, the first two terms tend the radius to the equilibrium value (like a spring), while the last term is the driving effect by surrounding fluid pressure, which can drive the radius away from the equilibrium via forcing. If P_∞ only drives a single initial impulse and is thereafter steady, the result will be an oscillating response that will dampen over time due to viscosity.

It should be noted that the preceding analysis assumes that the gas inside the bubble does not include any condensable gas. However, a bubble in water will often contain some water vapor inside the bubble (in fact, it is 100% vapor if it is a cavitation bubble), which is condensable at high enough pressure (or low enough temperature). Any condensation will change the equation of state since this vapor will be converted to liquid such that the mass of gas is not preserved. Brennen (1995) noted that one may account for the condensation effect by replacing P_∞ on the RHS of (4.82) with $P_\infty - P_{vapor}$, which reduces the overall system stiffness.

Additional effects include compressibility of the surrounding liquid (which allows for energy to be lost to the far field) and relative motion of the bubble (which increases the average surface pressure on the bubble). To address this, Prosperetti (1987) extended the Rayleigh–Plesset ODE to include the speed of sound in the liquid (a_ℓ) and the bubble relative velocity (w) as follows:

$$\begin{aligned} \rho_\ell r_p \ddot{r}_p + \frac{3}{2}\rho_\ell(\dot{r}_p)^2 + 4\mu_\ell \dot{r}_p/r_p - \frac{\rho_\ell}{a_\ell}\left[r_p^2\dddot{r}_p + 6r_p\dot{r}_p\ddot{r}_p + 2\dot{r}_p^{\,3}\right] \\ = P_{g,eq}(r_{p,eq}/r_p)^{3\mathcal{K}_g} - P_\infty - 2\sigma/r_p - \frac{1}{4}\rho_\ell w^2. \end{aligned} \tag{4.83}$$

The third term on the LHS is the liquid compressibility term, which causes the equation to become a third-order ODE but this effect is negligible when $U_r \ll a_\ell$. The last term on the RHS is an added dynamic pressure due to bubble velocity as used in (3.51) and is negligible when $w \ll U_r$. As with the Rayleigh–Plesset ODE, solving this nonlinear ODE for the time evolution of the bubble radius requires initial conditions for the bubble and the liquid as well as boundary conditions for the far-field pressure. Moreover, the ODEs can also be used to obtain the natural frequency of the bubble volume oscillations, as discussed in the next section.

4.6.3 Bubble Natural Frequency and Size Dynamics

The natural frequency for bubble size dynamics, f_{nat}, may be obtained analytically by considering small perturbations of the second-order ODE as spring system. In particular, the effective ODE stiffness divided by effective mass is equal to $(2\pi f_{nat})^2$. The natural frequency of (4.82) for an incompressible liquid and no relative bubble velocity can then be expressed using (4.80) with the modification to include effects of condensation, as follows:

$$2\pi f_{nat} = \sqrt{\frac{3\mathcal{K}_g(P_\infty - P_{vapor})r_{p,eq} + 2(3\mathcal{K}_g - 1)\sigma}{\rho_\ell r_{p,eq}^3}}. \tag{4.84}$$

As previously noted, $\mathcal{K}_g$ depends on the heat transfer aspects during compression and expansion. The isentropic limit ($\mathcal{K}_g = 1.4$) for natural frequency oscillation is appropriate when $f_{nat}\tau_T \gg 1$. This is reasonable for a pure air bubble at NTP with a radius greater than 5 mm based on (3.114b) and (4.84). On the other hand, the isothermal limit ($\mathcal{K}_g = 1.0$) appropriate for $f_{nat}\tau_T \ll 1$ would require this bubble be smaller than 40 μm. As such, heat transfer can be important for the wide range of sizes between these two limits, but the polytropic effects are not large. For a pure air bubble with a radius of 100 μm, the isentropic and isothermal frequencies are 32 kHz and 27 kHz. However, vapor effects can be large as a cavitation bubble of this size can have a natural frequency as low as 1 kHz. In practice, the measured frequencies for bubbles with sizes from 0.1 to 1.0 mm range from 5–25 kHz (Brennen, 2005).

For a large bubble with isentropic conditions ($\mathcal{K}_g = \gamma$), the impact of surface tension on f_{nat} can be neglected. This assumption yields the well-known Minnaert (1933) frequency of the bubble (neglecting vapor pressure), as follows:

$$f_{nat} = \sqrt{3\gamma P_\infty / \left(\pi^2 \rho_\ell d_{eq}^2\right)}. \tag{4.85}$$

As an example, a 6 mm diameter air bubble at 10 m water depth has a natural frequency of only 1.5 kHz. However, this frequency is still much higher than those associated with shape dynamics (§4.6.1), which are, in turn, much higher than those associated with trajectory oscillations (§4.4.2). As such, these three types of dynamics are typically decoupled, allowing them to be analyzed separately.

A bubble impacted by a single sharp pressure impulse will tend to oscillate at the natural frequency (f_{nat}), but the radial fluctuations will decrease in magnitude as the energy is damped by the liquid viscosity and/or radiated away to the far field. A bubble instead impacted by a continuous driving far-field liquid pressure frequency (f_∞) will have a frequency response closer to that driving frequency. Therefore, the bubble dynamic response is sensitive to the ratio of these frequencies (f_∞/f_{nat}). To investigate this sensitivity, consider a far-field pressure (P_∞) oscillating sinusoidally at a frequency (f_∞) with an amplitude (ϵ) about a mean value $(P_{\infty,eq})$, as follows:

$$P_\infty = P_{\infty,eq}[1 + \epsilon \sin(2\pi f t)] = P_{surf}. \tag{4.86}$$

For a slow driving frequency $(f_\infty \ll f_{nat})$, it is reasonable to assume an isothermal condition during contract and expansion, that is, $T_g \approx T_\ell \approx$ const. In this case, the gas density is linearly proportional to pressure per (2.37) so the bubble radius oscillates about $r_{p,eq}$, as defined by (4.80) and (4.81). For first-order perturbations (small ϵ), the radius is given by the following:

$$r_p = r_{p,eq}\left[1 + \epsilon \sin\left(2\pi f_\infty t\right)\right]^{-1/3} \quad for\, f_\infty \ll f_{nat}. \tag{4.87}$$

Considering (4.86) and (4.87), the bubble radius reaches its minimum when the surface pressure reaches its maximum, and the dimensionless radial oscillations are a cube root smaller than that for the pressure oscillations. As an example of these trends, the pressure and bubble radius dynamics for $f_\infty/f_{nat} = 1/20$ and $\epsilon = 1/20$ are shown in Figure 4.40.

However, when the far-field driving frequency is close to the natural frequency$(f_\infty \sim f_{nat})$, resonance can occur, and the results become more complex. In particular, the nonlinear RHS of (4.82) requires numerical integration for a given set of initial and boundary conditions. For example, Figure 4.41a for $f_\infty = 0.8 f_{nat}$ with $\epsilon = 0.33$ yields highly amplified radial oscillation amplitudes approaching $\pm 50\%$ of the mean value.

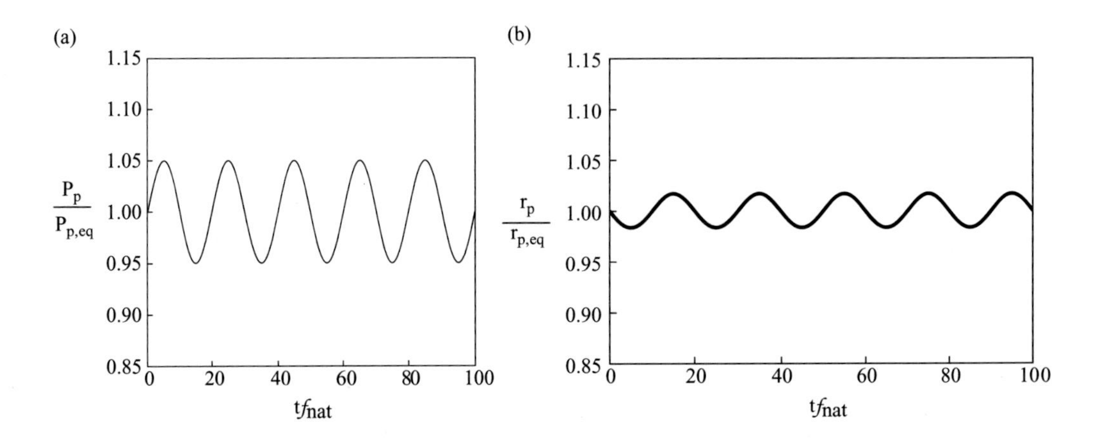

Figure 4.40 Oscillations of a bubble (normalized with the equilibrium values) as a function of time (normalized with the bubble natural frequency) for a pressure excitation of amplitude given by $\epsilon = 1/20$ and a frequency given by $f_\infty = f_{nat}/20$ for (a) bubble pressure and (b) responding bubble radius.

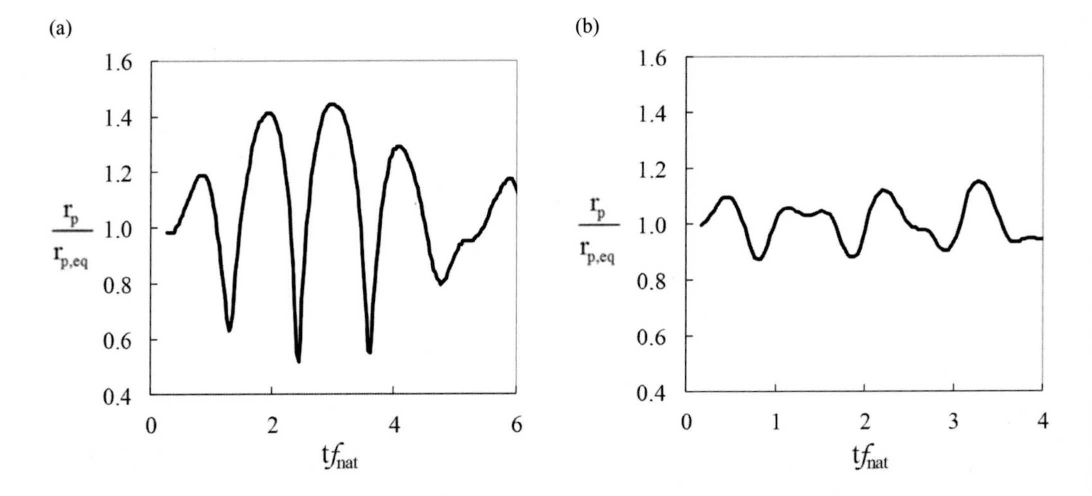

Figure 4.41 Oscillations of the bubble radius as a function of time (using normalization of Figure 4.40) for a 26 μm diameter bubble excited with $\epsilon = 0.33$ at (a) $f_\infty = 0.8\,f_{nat}$ and (b) $f_\infty = 1.8\,f_{nat}$ (Brennen, 1995).

When the bubble dynamics at this frequency ratio cause the radius to reduce by half $\left(r_p/r_{p,eq} = 0.5\right)$, the gas density will increase eightfold, leading to an 18-fold increase in pressure for isentropic conditions (2.40). The dynamics when this bubble is near the minimum radius also differ from a sinusoidal shape due to the more complex physics. In particular, the high inertia liquid moving inward causes continued contraction beyond the quasisteady point. This contraction finally reaches a threshold minimum radius where gas compressibility effects are strong enough to overcome this inertia and drive bubble expansion. As a result, the bubble undergoes a very rapid expansion process, yielding a sharp change (as compared to the smoother sinusoidal change seen in Figure 4.40b).

A second nonlinear example is shown in Figure 4.41b for which the bubble is driven at a frequency significantly above the bubble's natural frequency ($f_\infty = 1.8 f_{nat}$ with $\epsilon = 0.33$). In this case, the range of the bubble radius (maximum to minimum) is not as exaggerated as when the driving frequency was closer to the natural frequency. In fact, the range of radius for Figure 4.41b is closer to the quasisteady amplitudes (about $\pm 10\%$). However, the dynamic response is quite complex and nonlinear due to strong influence of both f_∞ and f_{nat}.

If one carefully and symmetrically excites a bubble precisely at its natural frequency $(f_\infty = f_{nat})$, the amplitudes will be the most amplified, with even greater reduction in diameter due to high inertia liquid moving inward. This can lead to a case where there is a nearly 10-fold reduction in diameter compared to the equilibrium state, yielding density increases approaching 1,000-fold. As a result, the gas pressures can exceed 10,000 atmospheres with interior gas temperatures as high as 10,000 K (Suslick and Flannigan, 2005). This can cause the gas to be converted into a plasma yielding the sonoluminescence phenomenon discussed in §1.2.2. The radial dynamics in these extreme conditions require additional physics beyond those included in (4.82). This may include chemical reactions as well as shock and expansion waves within the bubble due to the high speeds of the radial interface (Kameda and Matsumoto, 1996; Feng and Leal, 1997). At such violent conditions, any asymmetry in the excitation can cause these oscillations to induce nonspherical shape dynamics (as discussed in §4.6.1). These spatial nonuniformities can lead to surface instabilities that cause bubble breakup, as shown in Figure 4.42.

Another example with gas bubble oscillations is an underwater explosion. Such an explosion produces a gas cloud that initially overexpands and then contracts and oscillates at the natural frequency, as shown in Figure 4.43. This figure also shows nonlinear numerical predictions that include radiated energy loss due to finite liquid compressibility (4.83). These predictions reasonably describe the bubble dynamics at the maximum radii and the sharp discontinuity at the minimum radius (as in Figure 4.41a). However, after the first few cycles, breakup and turbulence (not included in the simulations) cause the bubble cloud dynamics to decay more rapidly than predicted (Brennen, 1995; Feng and Leal, 1997).

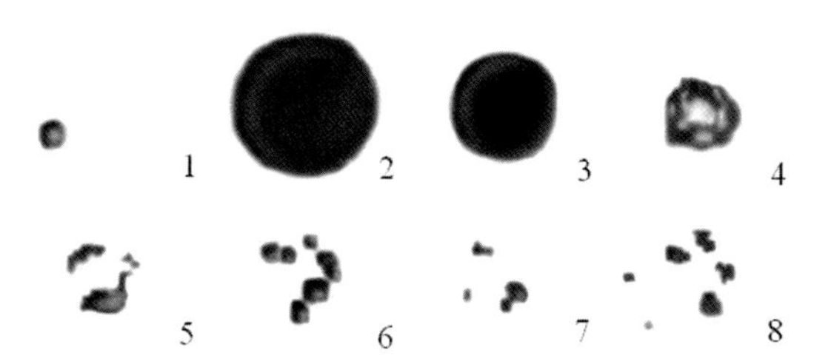

Figure 4.42 Microbubble undergoing acoustic excitation (0.4 microseconds between images). Bubble breakup occurs due to asymmetric contraction and results in several smaller bubbles (Wolfrum et al., 2003).

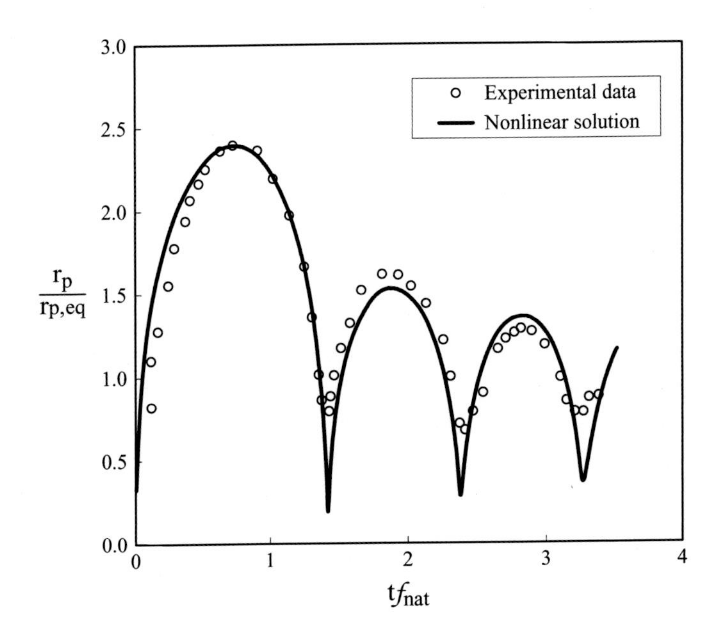

Figure 4.43 Oscillations of the gas cloud radius resulting from an underwater explosion at a depth of 100 m. The data are compared against the compressible nonlinear predictions assuming $r_{p,eq} = 15$ cm (Keller and Kolodner, 1956).

4.7 Chapter 4 Problems

Show all steps and list any needed assumptions.

(P4.1) Consider a cloud of droplets whose size distribution that has a triangularly shaped PDF that peaks at a diameter of 10 μm as shown in the following diagram:

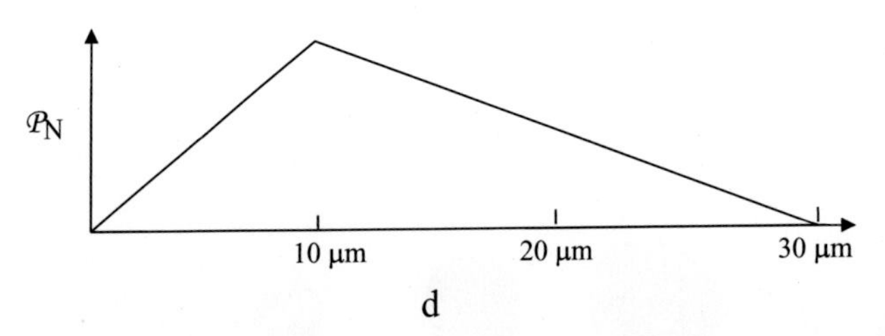

(a) Determine the Sauter mean diameter (d_{32}) and the volume-width mean diameter (d_{31}) and comment on which of these two is more appropriate for the terminal velocity of these particles. (b) Obtain an analytical expression for the corresponding volume-based CDF for these particles and plot this function. (c) Compute both the volume fraction and number fraction of particles with diameters of 10 μm or less. (d) Determine the $d_{90\%}/d_{10\%}$ for this distribution.

(P4.2) (a) Derive an analytical expression for Y as a function of Re_p beginning with (4.18) but using the Putnam fit (3.59) instead of the White fit for the drag. (b) Determine the limiting value for this Y function at very high Reynolds numbers and whether this limit is consistent with Newton-based drag.

(P4.3) Consider a bidisperse water drop cloud with just two sizes, $d_{large} = 600\,\mu m$ and $d_{small} = 400\,\mu m$, where each constitutes 50% of the total particles by number. (a) Using the White drag model, compute the effective terminal velocity of the cloud (for group behavior) in air at NTP (iteration can be used). (b) Compare this result to the individual terminal velocities for 400 μm and 600 μm diameter particles, respectively. (c) Comment on whether it is reasonable to assume group motion for this cloud.

(P4.4) (a) Obtain the relation between the Rosin–Rammler spread parameter and the diameter ratio ($d_{90\%}/d_{10\%}$) given by (4.25) starting with the $\mathcal{C}_\forall$ distribution function based on (4.24). (b) Obtain the log-normal relation given by (4.35) starting with the $\mathcal{C}_\forall$ distribution function of (4.33). (c) To prove that $d_{10} < d_{20} < d_{30}$, show analytically that a log-normal distribution with finite spread ($h_{LN} > 0$) yields $d_{i0} < d_{j0}$, where $j = I + 1$.

(P4.5) Consider the below discrete data set of particle mass by bin (where bin width is 50 μm). a) Convert the results into a discrete $\mathcal{C}_m$ versus bin diameter. b) Based on a least squares fit, find the best Rosin–Rammler diameter and spread parameter (h_{RR} and d_{RR}) and the best log-normal diameter and spread parameter (h_{LN} and d_{LN}) to within 1 μm accuracy for the diameter and within 1% for the spread parameter. Plot the Rosin–Rammler, the log-normal and the discrete $\mathcal{C}_m$ distributions and discuss which of the analytical PDF provides better agreement.

d_i	25 μm	75 μm	125 μm	175 μm	225 μm	275 μm	325 μm	375 μm
$\mathcal{P}_m\,\Delta d$	2%	7%	15%	20%	24%	19%	9%	4%

(P4.6) Based on the definitions of Re_p, We, Mo, and Bo, confirm the relationships given in (4.45), (4.46), and (4.54b).

(P4.7) (a) Determine (with iteration) the terminal velocity and Weber number for a fully contaminated bubble in water with a volumetric diameter of 3 mm assuming a drag coefficient based on a spherical shape. (b) If the bubble is clean, note whether the terminal velocity would qualitatively increase, stay the same, or decrease and explain the reasoning based on fluid dynamics. (c) Redo (a) but take into account deformation effects on drag then explain the qualitative difference.

(P4.8) Consider a drop at a relative speed whose shape is an oblate spheroid, whose internal pressure is constant, whose external surface pressure is based on inviscid flow over a sphere, and whose interface pressure jump is based on the local curvature at $\theta = 0$ and at $\theta = \pi/2$. (a) Use these pressure relationships to develop a theoretical equation for Weber number (We) as a function of only the aspect ratio E; (b) For We = 1, compare E values from this theoretical result to the empirical relation of (4.47) and then discuss the qualitative difference.

(P4.9) Consider a spherical-cap bubble of air rising in still water where the top shape can be approximated with a constant radius of curvature (r_{cap}), while the bottom can be approximated as a flat horizontal interface (see the following diagram). a) In the moving reference frame of the bubble moving at v, obtain the liquid velocity distribution over the bubble surface for potential flow theory, and do the same for pressure (referenced to the pressure at the stagnation point). b) Revise the results in a) for a fixed frame where the far-field velocity is zero and where pressure includes hydrostatic effects and is referenced to the pressure at y = 0.

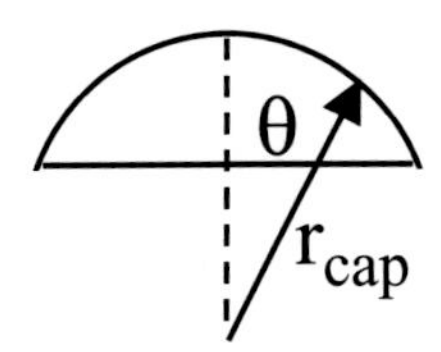

(P4.10) For a spherical-cap bubble with shape as in (P4.9) determine whether the liquid pressure along the surface based on the Bernoulli equations is consistent with the pressure based on surface tension and curvature (assuming a constant air pressure inside the bubble).

(P4.11) Determine the minimum volumetric diameter for a bubble in water for a spherical cap geometry with E = 0.25 at terminal velocity conditions.

(P4.12) Consider a 1 mm diameter ethanol drop in quiescent air. Compute the aspect ratio and drag coefficient of this drop for (a) instantaneous conditions when injected at 10 m/s, (b) terminal velocity conditions, and (c) discuss the difference between these two conditions.

(P4.13) (a) Determine (with iteration) the critical breakup diameter for a water drop falling at terminal velocity in air at NTP (the biggest rain drop) using a drag coefficient based on a spherical shape. (b) Repeat as in (a) except use a drag coefficient with deformation effects and discuss the resulting differences.

(P4.14) Consider an oil lubrication layer between two parallel sliding surfaces that are 4 mm apart with a relative velocity of 2 m/s and includes a 0.5 mm diameter drop of water in this oil layer. The surface tension between the oil and water is 0.024 N/m and the oil has a dynamic viscosity of 0.012 kg/m-s and a specific gravity of 90%. (a) Determine the equilibrium tilt angle and aspect ratio of the drop assuming creeping flow. (b) Do you expect breakup at this shear rate assuming creeping flow? (c) If finite Reynolds number effects are included (no longer creeping flow), how would the results in (a) and (b) qualitatively change?

(P4.15) Starting from (4.74)–(4.75), derive the Rayleigh–Plesset ODE in the isothermal limit.

(P4.16) Consider the isothermal Rayleigh–Plesset ODE for the case of a bubble with equilibrium radius and maximum radii of 1.0 mm and 1.2 mm, respectively, that is being driven at 1 kHz. Estimate the magnitude of each of the ODE terms to determine which terms can be reasonably neglected for these conditions.

(P4.17) (a) Obtain an expression for the natural Mode 1 frequency of a bubble from the isothermal Rayleigh–Plesset ODE by assuming a constant far-field pressure, negligible viscosity, and weak bubble oscillations of the form $r_p = r_{p,eq}\left[1 + \epsilon \sin\left(2\pi f_\infty t\right)\right]$ with $\epsilon \ll 1$. To do this, neglect higher-order ϵ terms and use the remaining sinusoidal terms to solve for f_{nat}. (b) For a 1 mm diameter bubble in water with a far-field liquid pressure of one atmosphere, determine and compare the Mode 1 (radial oscillation) and Mode 2 (shape oscillation) natural frequencies.

(P4.18) Consider the dynamics of a bubble at a depth of 1 m in water at normal temperature. The bubble has an equilibrium diameter of 2 mm, but at time $t = 0$, the diameter is set as 2.5 mm. Write a code to solve the Rayleigh–Plesset ODE neglecting both surface tension as well as viscosity effects. (a) Plot the bubble radius as a function of time over 5 ms for both isothermal and isentropic bubble dynamics. (b) Discuss the results and attach your code.

5 Coupling Regimes for Multiphase Flow

Multiphase flow can include a variety of coupling physics, such as the impact of the surrounding fluid on a particle, the impact of particles on the surrounding fluid, the impact of particles on each other, and so on. The conditions that control these interactions can be used to define specific coupling regimes beyond the isolated particle conditions of Chapters 3 and 4. Identifying these regimes is critical to the analysis of the multiphase system, whether it is from a theoretical, experimental, or computational point of view. This chapter first overviews types of coupling and particle concentration descriptors in §5.1 and then considers aspects of one-way coupling in §5.2, with the special case of Brownian motion in §5.3. The remainder of the chapter considers two-way coupling (§5.4), three-way coupling (§5.5), and four-way coupling (§5.6). Turbulent coupling aspects are discussed later in Chapter 7.

5.1 Coupling Regimes and Particle Concentration

5.1.1 Multiphase Coupling Regimes

As shown in Figure 5.1, multiphase flow can be grouped into four coupling regimes that are defined (and roughly ordered in terms of increasing particle concentration) as follows:

(a) *One-way coupling*, where particle motion is only affected by the surrounding fluid and the particles do not affect the fluid macroscopically (e.g., fall of a single isolated particle).
(b) *Two-way coupling*, where a significant portion of the surrounding fluid is substantially affected by the particles (e.g., entrainment of liquid induced by a bubble plume).
(c) *Three-way coupling*, where particles affect each other by fluid dynamic interactions (e.g., a particle's drag is reduced when it drafts behind another particle).
(d) *Four-way coupling*, where particles affect each other by direct surface contact (e.g., cloud droplets collide and coalesce to form larger raindrops).

The first three coupling regimes are based on fluid dynamic interactions, whereas the fourth coupling regime is based on particle–particle collisions. A flow with any of

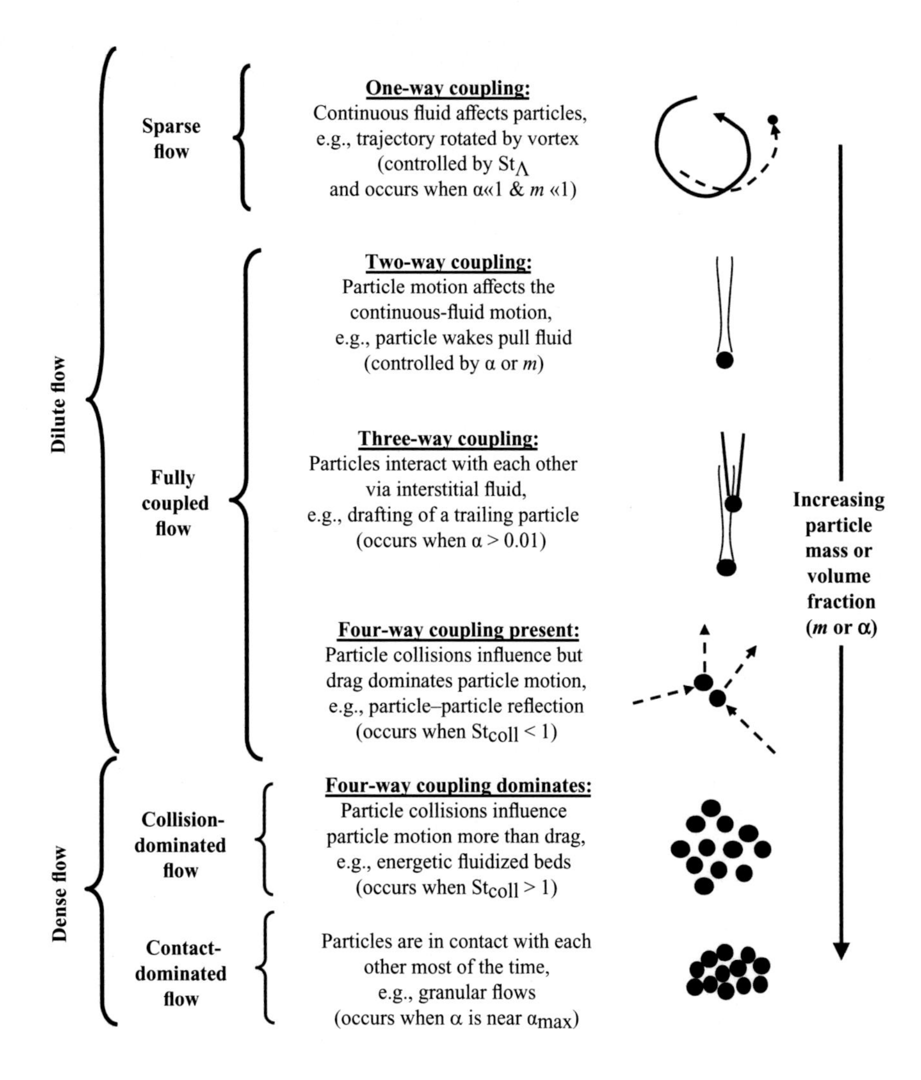

Figure 5.1 Regimes for multiphase coupling including sparse, dilute, and dense flow.

the last three regimes can be termed "strongly coupled,". Notably, increasing particle concentration does not always escalate a flow from two-way to three-way to four-way. For example, solid particles in a gas may exhibit four-way coupling but not necessarily three-way coupling, while a bubble flow may have three-way coupling but not necessarily two-way coupling.

A multiphase flow can also be defined as either "dilute" or "dense" based on the strength of the four-way coupling. Dilute flow (which is the focus of this book) occurs when particle movement is dominated by fluid dynamic interactions. Such flow can also include four-way coupling effects so long as those are secondary (particle motion is mostly influenced by the surrounding fluid). If the particle motion is dominated by collisions (via four-way coupling), this is considered a dense flow. The dense flow regime can be further divided into two regimes: *collision dominated* and *contact dominated*. For collision-dominated regimes, the particles are in contact only for short periods, and their free-flight motions are primarily controlled by rebound velocities. For contact-dominated regimes, the particle concentration is so high that the particles tend to be in continual rolling or sliding contact with each other with negligible fluid dynamics, such as when shaking a container filled with marbles and air.

As shown in Figure 5.1, the dense regime is more likely to occur at high particle concentrations. As such, large gradients in particle concentration can yield some dilute flow portions and some dense flow portions. For example, sediment transport of sand by wind tends to have high particle concentrations near the ground surface (consistent with dense flow) and lower concentrations at higher altitudes (consistent with dilute flow). To establish criteria for these regimes, quantitative descriptors of particle concentrations are needed, as discussed in the following section.

5.1.2 Particle Concentration Parameters

Quantitative particle concentration (used to identify and describe the four coupling regimes) can be expressed using number-based, volume-based, or mass-based parameters.

The number-based particle concentration is generally characterized with the *particle number density* (n_p), defined as the number of particles per unit mixed-fluid volume. The *mixed-fluid volume* ($\forall_m$) includes both the surrounding fluid volume ($\forall_f$) and that of all the particles, where each particle has a given volume ($\forall_p$). Denoting the number of particles in a given mixed-fluid volume as N_p, the number density and mixed-fluid volume are as follows:

$$n_p \equiv \frac{\text{number of particles}}{\text{volume of mixture}} = \frac{N_p}{\forall_m}. \tag{5.1a}$$

$$\forall_m = \forall_f + \sum_{k=1}^{N_p} \forall_{p,k}. \tag{5.1b}$$

The first equation shows that the number density is dimensional; for example, n_p can have units of particles/mm^3. The second equation shows that mixed-fluid volume includes the continuous phase and the sum of all the individual particle volumes (where k is a particle count index).

The most common volume-based particle concentration parameter is the *particle volume fraction* (α), which is defined as the ratio of particle volumes to the mixed-fluid volume and can be related to the number density and the average particle volume using (4.10c), as follows:

$$\alpha \equiv \frac{\text{volume of particles}}{\text{volume of mixture}} = \frac{\sum_{k=1}^{N_p} \forall_{p,k}}{\forall_m}. \tag{5.2a}$$

$$\alpha = n_p \frac{\sum_{k=1}^{N_p} \forall_{p,k}}{N_p} = n_p \frac{\pi}{6} d_{30}^3 = n_p \forall_{p,avg}. \tag{5.2b}$$

As such, the volume fraction increases with both the number density and size of the particles. Notably, the fraction of volume occupied by the continuous (surrounding) phase relative to the mixed-fluid volume is 1-α.

Two more parameters include the particle concentration density and the convective mass flux of particles. In particular, the mass flux of particles through a surface (with outward normal **n**) can be obtained with (2.1) and the transport quantity $\alpha\rho_p$ (the particle mass concentration per unit volume of mixed fluid) as follows:

$$\alpha\rho_p(\mathbf{v}\cdot\mathbf{n}) = \frac{\text{mass flux of particles}}{\text{mixed fluid area}}. \tag{5.3a}$$

$$\alpha\rho_p \equiv \frac{\text{mass of particles}}{\text{volume of mixture}}. \tag{5.3b}$$

The transport quantity of (5.3b) is the particle concentration density.

There are other mass-based particle concentration parameters. The first is the *particle mass fraction* $(\mathcal{M}_p)$, which is defined as the ratio of particle mass to the mixed-fluid mass within a given volume:

$$\mathcal{M}_p \equiv \frac{\text{mass of particles}}{\text{mass of mixture}} = \frac{\sum_{k=1}^{N_p} m_{p,k}}{\left[\alpha\rho_p + (1-\alpha)\rho_f\right]\forall_m} = \frac{\alpha\rho_p}{\alpha\rho_p + (1-\alpha)\rho_f}. \tag{5.4}$$

This definition is similar to that for the species mass fraction (2.30a). The denominator on the RHS of (5.4) can also be used to define a *mixed-fluid density* (also called the *effective density* or *modified density*) as follows:

$$\rho_m \equiv \frac{\text{mass of mixture}}{\text{volume of mixture}} = \frac{\rho_p\sum_{k=1}^{N_p}\forall_{p,k} + \rho_f\forall_f}{\forall_m} = \alpha\rho_p + (1-\alpha)\rho_f. \tag{5.5}$$

A final mass-based particle concentration parameter is the *mass loading* (m), which is defined as the ratio of particle mass to surrounding fluid mass within a given volume:

$$m \equiv \frac{\text{mass of particles}}{\text{mass of surrounding fluid}} = \frac{\sum_{k=1}^{N_p} m_{p,k}}{[(1-\alpha)\rho_f]\forall_m} = \frac{\alpha\rho_p}{(1-\alpha)\rho_f}. \tag{5.6}$$

When the particle density ratio is very high (e.g., solid or liquid particles in a gas), the mass loading can be linearly related to the volume and mass fractions as follows:

$$m \approx \alpha\rho_p/\rho_f \quad \textit{for } \alpha \ll 1 \tag{5.7a}$$

$$m \approx \mathcal{M}_p \quad \textit{for } \mathcal{M}_p \ll 1 \tag{5.7b}$$

$$\textit{and} \quad \rho_p \gg \rho_f.$$

On the other hand, a very low-density ratio (e.g., bubbles in a liquid) causes the mass contribution to be negligible, so generally only volume-based concentrations are used.

The concept of a mixed fluid can be extended to include the mixed-fluid viscosity (μ_m) and mixed-fluid thermal conductivity (k_m), as follows:

$$\mu_m = \mu_f(1+2.5\alpha) \tag{5.8a}$$

$$k_m = (1-\alpha)k_f + \alpha k_p \tag{5.8b}$$

$$\textit{for } \alpha \ll 1.$$

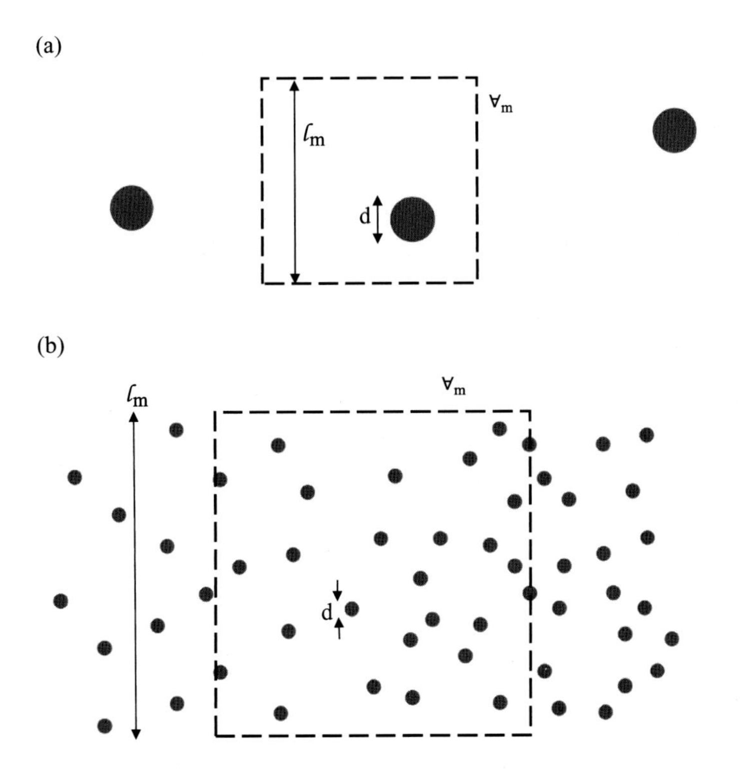

Figure 5.2 Representation of a mixed-fluid control volume (dashed lines) as a cube that allows a (a) well-posed particle concentration, as it can contain at least one particle ($d < \ell_m$); and (b) continuum particle concentration, as it contains many particles ($N_{pm} \gg 1$).

The mixed-fluid viscosity is based on Einstein's theory for a dilute suspension of particles in Stokesian conditions (Soo, 1990), while the mixed-fluid thermal conductivity (k_m) is based on the volumetric fractions and thermal conductivities of each phase (Crowe et al., 2011).

Since the preceding concentration parameters are based on the number of particles in the mixed-fluid volume, the mixed-fluid volume should be large enough to encapsulate at least one particle, as shown in Figure 5.2a. In this case, the particle concentration is defined to be "well-posed" and satisfies volume/size constraints relative to that of the particle as follows:

$$\forall_m > \forall_p \qquad \textit{for well-posed particle concentration.} \tag{5.9a}$$

$$\forall_m^{1/3} = \ell_m > d \tag{5.9b}$$

The second expression defines ℓ_m as the length scale of the mixed-fluid volume assuming a cube.

In addition, a particle concentration at a given time can be considered to vary as a "continuum" if the number of particles in the mixture volume is statistically significant such that small changes in its size or location causes only small changes in the concentration value. Since particle positions are typically randomly distributed (not spaced uniformly as in a lattice structure), this requires that the mixed-fluid volume encompasses many particles, as shown in Figure 5.2b, and is expressed as

$$N_p \gg 1 \quad \textit{in } \forall_m \textit{ for a continuum particle concentration.} \tag{5.10}$$

Typically, 100 particles would satisfy (5.10), and and allow the instantaneous particle concentration to be continuously differentiable in space, so the *gradients* in n_p, α, η, and M are well posed. Comparing (5.9) and (5.10) for particle concentration, the mixed-volume size is larger for continuum conditions compared to that for well-posed conditions. For example, a cloud of drops with diameters of 25 μm and a bulk concentration of 10^2 drops/(mm)^3 satisfies (5.10) by requiring $\ell_m > 1$ mm but only needs $\ell_m > 0.025$ mm to satisfy (5.9). If one considers a bubbly flow with $d = 1$ mm and 10 bubbles/(cm)^3, then continuum particle concentrations require $\ell_m > 30$ mm while well-posed conditions require $\ell_m > 1$ mm.

There are two notable features of (5.10). First, the continuum criterion for particle concertation is conceptually similar to that for fluid density discussed in §1.5, but the scales are much smaller for the latter. For example, a continuum description of air density at NTP requires lengths of at least 16 microns, while liquid water would require lengths of at least 3 microns (where these lengths correspond to about 100 molecules for a cubic volume). Secondly, the criterion of (5.10) can be replaced with that of (5.9) for continuum of the *time-averaged* particle concentration. For example, a mixed-fluid volume with ℓ_m of 0.1 mm for the previous spray case has a 10% chance that a particle is in the volume at a given time (since $N_{p,avg} = 0.1$).

The rest of this chapter discusses the controlling physics and the criteria for various coupling regimes, often making use of the preceding particle concentration parameters. Such qualitative relationships can be critical in determining both the appropriate equations of motion and numerical strategies.

5.2 Macroscopic One-Way Coupling and Domain Stokes Numbers

For one-way coupling, the continuous-phase flow motion is not affected by the presence of the particles. This is also called "sparse" flow, which requires very low particle concentration by both volume and mass, that is, $\alpha \ll 1$ and $\eta \ll 1$. This regime is important to recognize since it allows the continuous-phase governing equations to be considered independent of the particles, greatly simplifying multiphase analysis. For example, the surrounding fluid can be obtained in advance by using a single-phase flow approach (as described in Chapter 2) to consider the movement for a variety of particle situations (sizes, locations, trajectories, etc.). As discussed in the following subsections, the particle motion relative to the surrounding fluid can be characterized via the particle response time.

5.2.1 Domain Stokes Numbers

As discussed in §3.3, particles have a drag-based response time (3.88a), given as follows:

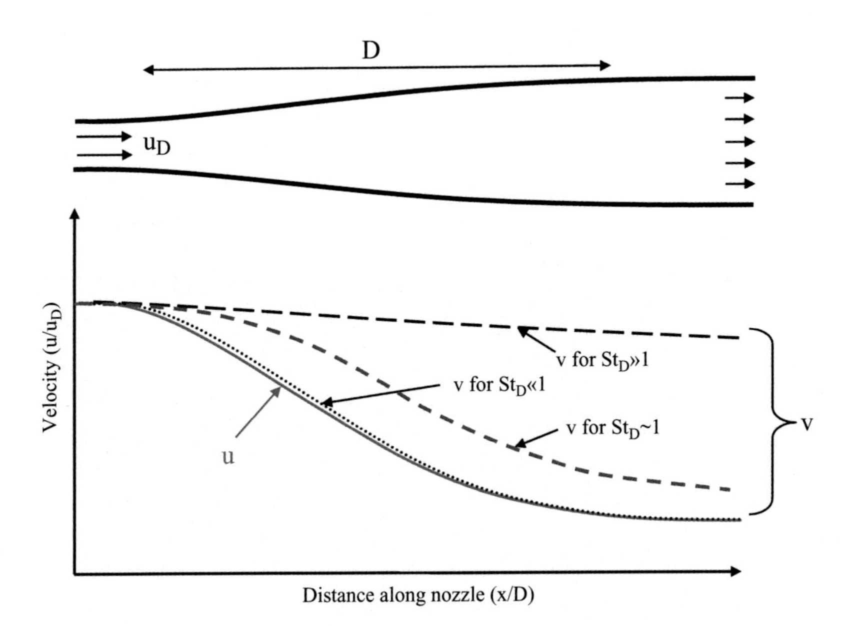

Figure 5.3 Steady flow in a diverging nozzle of length D and with inflow velocity u_D for which the continuous-phase velocity (solid blue line) decelerates along a Lagrangian fluid path and the velocities of a particle-phase (dashed lines) depend on the Stokes number (gravitational effects are neglected).

$$\tau_p \equiv \frac{m_{eff} w}{F_D} = \frac{(\rho_p + c_{\forall}\rho_f)d^2}{18\mu_f f} = \frac{4(\rho_p + c_{\forall}\rho_f)d}{3\rho_f w C_D}. \tag{5.11}$$

This particle response time can be compared to that of the fluid time scale to characterize one-way coupling. In particular, one may define a surrounding fluid time scale for the domain (τ_D) and use it to represent the time it takes for significant changes in the fluid velocity as seen along a particle's path. These changes can be due to temporal gradients for unsteady flow and/or spatial gradients for nonuniform flow. For an unsteady flow that is uniform in space, τ_D can represent the overall period of the flow, such as the pulsation period if there is an oscillating velocity. For a steady flow that is nonuniform in space, τ_D can represent the time associated with fluid to traverse the domain. Generally, we define the domain timescale based on the time for a significant change in fluid velocity seen by the particle using the a macroscopic surrounding fluid velocity scale (u_D) and length scale (D) per §2.4, such that

$$\tau_D \equiv \text{characteristic time scale for fluid velocity changes} = \frac{D}{u_D}. \tag{5.12}$$

For the diverging nozzle flow example in Figure 5.3, u_D may be set as the inflow velocity and D as the diffuser length such that a particle traveling in the general direction of the flow will see significant fluid velocity changes over a time scale given by τ_D.

To consider how fast a particle will respond to the domain scales of the surrounding flow, the ratio of the particle drag-based response time (5.11) to the domain time scale (5.12) can be used to define a *domain Stokes number*:

$$St_D \equiv \frac{\text{particle response time}}{\text{domain time scale}} \equiv \frac{\tau_p}{\tau_D} = \frac{\left(\rho_p + c_\forall \rho_f\right) d^2}{18\mu_f f} \frac{u_D}{D}. \tag{5.13}$$

This dimensionless parameter indicates how fast the particle velocity responds to changes in the surrounding flow velocity.

A particle with $St_D \ll 1$ corresponds to the *drag-dominated* regime since drag causes the particle to be nearly equal to the velocity of the surrounding fluid, as shown, for example, in Figure 5.3. In fact, the limit of $St_D \to 0$ corresponds to an infinitely small particle with no inertia that perfectly follows the flow (e.g., a microscopic dye particle in water so $\mathbf{v} \approx \mathbf{u}_{@p}$).

At the other extreme, $St_D \gg 1$ corresponds to the *inertia-dominated* regime since the particle motion is chiefly governed by its own momentum and initial conditions. As such, the limit of $St_D \to \infty$ corresponds to a particle with such high inertia that it is not affected by the flow at all (e.g., a bullet traversing a short domain where $\mathbf{v} \approx \mathbf{v}_o$).

For the intermediate regime of $St_D \sim 1$, the continuous-phase velocity significantly influences but does not dominate the particle velocity. Therefore, the various Stokes number regimes may be summarized in order of increasing particle inertia as follows:

$St_D \ll 1$ *particle motion closely follows the macroscopic flow.* (5.14a)

$St_D \sim 1$ *particle motion partly follows the macroscopic flow.* (5.14b)

$St_D \gg 1$ *particle motion is weakly affected by the macroscopic flow.* (5.14c)

The influence of the particle Stokes number for a nonuniform flow and for an unsteady flow is demonstrated in the following using two examples.

5.2.2 Particle Response to a Nonuniform Flow

Consider an airfoil moving through a cloud of drops. For a coordinate system based on the airfoil, the drop trajectories approaching as in Figure 5.4a and the domain time scale can be defined based on the chord length (D) and flight speed (u_D). Consistent with (5.14a), drag-dominated drops with negligible inertia ($St_D \to 0$) will move around the airfoil so no trajectories will impact the surface. In contrast, large inertia-dominated drops ($St_D \to \infty$) will have no trajectory deflections and upstream drops will impact the airfoil directly. For intermediate drop sizes, the impact likelihood will also be intermediate with only some of the drops impacting, as they are only partially deflected. This influence of Stokes number on droplet impact is quantified in Figure 5.4b in terms of the droplet *impact efficiency*, defined as the fraction of drops in the projected freestream area that will impact the airfoil. The impact efficiency tends to 0% as $St_D \to 0$ and to 100% $St_D \to \infty$, while intermediate values are found for intermediate St_D. The Stokes number in this plot also takes into account nonlinear drag effects, as discussed next.

(a)

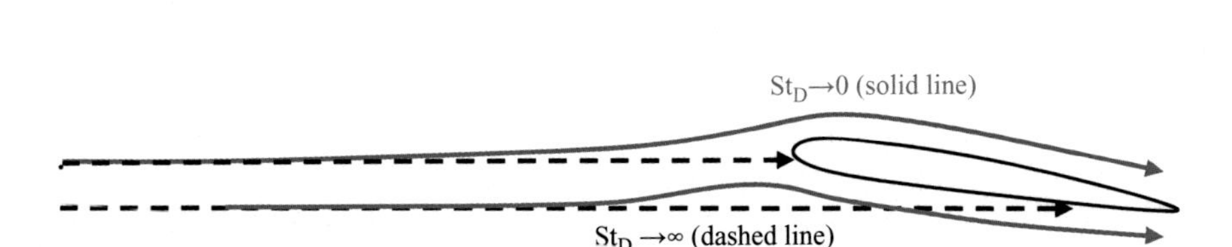

(b)

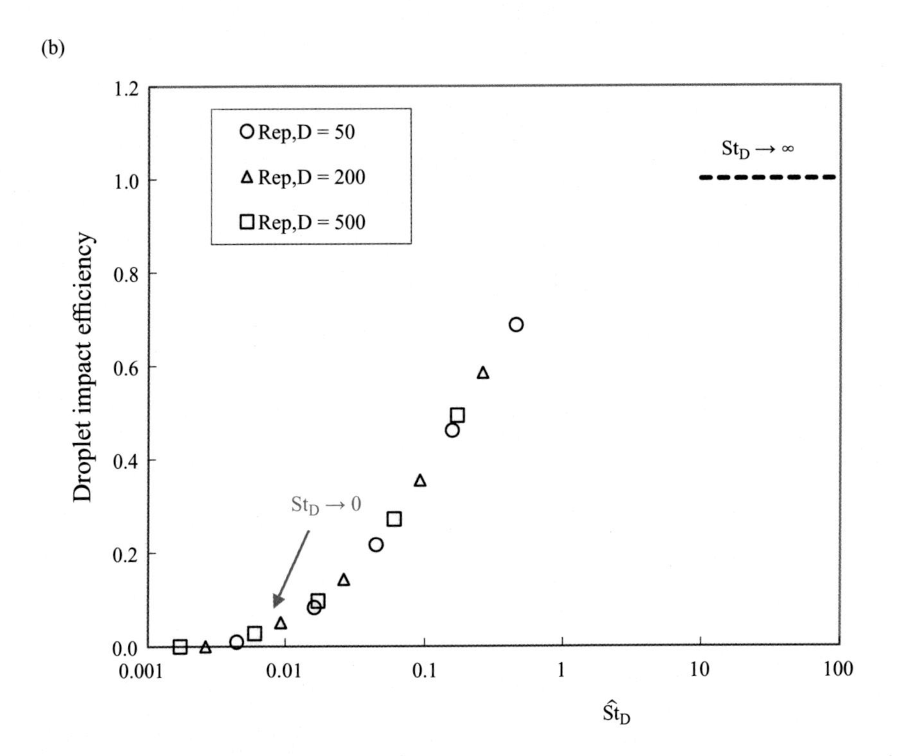

Figure 5.4 Droplets in the air subjected to a moving airfoil: (a) schematic of trajectories, where low inertia drops (solid blue lines) tend to follow the air flow streamlines and go around the airfoil, while high inertia drops (dashed lines) impact directly; and (b) predicted fraction of upstream droplets that impact an airfoil as a function of a path-averaged Stokes number for a different particle impact Reynolds numbers (Bragg, 1982).

The Stokes number given by (5.13) includes the Stokes drag correction (f). If the particle Reynolds number is not small, f depends on Re_p, as in (3.60b). In this case, changes in the particle relative velocity along the particle path will result in changes in Re_p and f as well. To account for these changes, the "path average" of a particle quantity, q, can be defined as follows:

$$\hat{q} \equiv \frac{1}{t_{final} - t_o} \int_{t_o}^{t_{final}} q\, dt. \tag{5.15}$$

This integral is based on the initial time (t_o) and final time (t_{final}) along a particle path. Applying this path average to (5.13) and assuming $\rho_f \ll \rho_p$ (for drops in air), the path-averaged Stokes number can be expressed as follows:

$$\widehat{St}_D = \frac{\widehat{\tau_p}}{\tau_D} = \frac{\rho_p d^2}{18\mu_f \tau_D}\widehat{(1/f)}. \tag{5.16}$$

This path-averaged Stokes number is also called the *modified inertia parameter* (Langmuir and Blodgett, 1946). To evaluate the RHS, one may employ the Putnam fit (3.59b) and integrate per (5.15) as follows:

$$\widehat{(1/f)} = \frac{1}{Re_{p,final} - Re_{p,o}} \int_{Re_{p,o}}^{Re_{p,final}} \frac{d(Re_p)}{1 + Re_p^{2/3}/6}. \tag{5.17}$$

The integral limits include the final Reynolds number ($Re_{p,final}$) and initial Reynolds number ($Re_{p,o}$), which are based on the initial and final relative velocities (w_o and w_{final}). For the airfoil flow of Figure 5.4, the initial relative velocity is that far upstream and can be set as the terminal velocity ($w_o \approx w_{term}$). To characterize the path-averaged Stokes number for droplets that impact, the final relative velocity is that just before impact. A drop with enough inertia to impact the airfoil will tend be only weakly affected by the flow around the airfoils so $w_{final} \approx u_D$. However, if one considers a flight speed on the order of 100 m/s and drop diameters less than a millimeter, the terminal velocity will be much smaller than the flight speed ($w_{term} \ll u_D$). As a consequence, the initial Reynolds number is much smaller than the final Reynolds number ($Re_{p,o} \ll Re_{p,final}$), so that the integration of (5.17) and insertion into (5.16) yields the following:

$$\widehat{St}_D = \frac{\rho_p d^2}{\mu_f \tau_D}\left[\frac{1}{Re_{p,D}^{2/3}} - \frac{\sqrt{6}}{Re_{p,D}}\tan^{-1}\left(\frac{Re_{p,D}^{1/3}}{\sqrt{6}}\right)\right]. \tag{5.18a}$$

$$Re_{p,D} \equiv \frac{\rho_f u_D d}{\mu_f} \approx Re_{p,final}. \tag{5.18b}$$

The second expression defines a particle Reynold number based on u_D and approximates this as the final value. For simplicity, the RHS of (5.17) can be approximated empirically as $[1 + 0.0967\, Re_{p,d}(0.6367)]^{-1}$ to evaluate (5.16). The path-average Stokes number can therefore characterize the particle motion. This is shown in Figure 5.4b for the impact efficiency as a function of the Stokes number (which scales with droplet size, flight speed, chord length, etc.), where the data collapse into a single curve for a variety of Reynolds numbers, demonstrating that path averaging properly accounts for nonlinear drag effects.

5.2.3 Particle Response to an Oscillating Flow

While the preceding example considers a steady nonuniform flow, the domain Stokes number can also be used to characterize particle behavior for an unsteady uniform flow. Consider a horizontal flow that is spatially uniform but has a back-and-forth cyclic velocity given by the following expression:

$$u = u_D \sin(2\pi t/\tau_D) \quad \textit{laminar cyclic flow.} \tag{5.19}$$

In this expression, u_D is the peak velocity magnitude and τ_D is the oscillation period. If we consider particles that are small enough relative to the fluid viscosity, such that $Re_p \ll 1$, one can use (3.83) for the particle equation of motion. For this linear drag equation, the particle response in the horizontal direction is only a function of the unsteady flow in the horizontal direction. In addition, one may consider the long-time behavior (after the initial condition effects have subsided) by considering $t \gg \tau_D$ and $t \gg \tau_p$. In this case, the particle horizontal equation of motion and velocity are given as follows:

$$\begin{aligned}(\rho_p + c_\forall \rho_f)\forall_p \frac{dv}{dt} &= -3\pi\mu_f d(v-u) + \rho_f \forall_p (1+c_\forall)\frac{\partial u}{\partial t} \\ &\quad - \frac{3}{2} d^2 \sqrt{\pi \rho_f \mu_f} \int_0^t \frac{d(v-u)}{d\tau} \frac{1}{\sqrt{t-\tau}} d\tau.\end{aligned} \tag{5.20a}$$

$$v = v_D \sin(2\pi t/\tau_D - \phi). \tag{5.20b}$$

In the second expression, v_D is the magnitude of the particle velocity fluctuations and φ is the phase shift between the particle and the fluid velocities (due to a particle velocity time lag from that of the fluid). One may then substitute (5.19) and (5.20b) into (5.20a) and then integrate in time and compare terms with the same frequency to characterize the particle velocity oscillations (L'Esperance et al., 2006). For a spherical particle ($c_\forall = 1/2$), these oscillations can be expressed in terms of the domain Stokes number of (5.13) and the acceleration parameter of (3.88b), as follows:

$$\left(\frac{v_D}{u_D}\right) e^{i\varphi} = 1 + \frac{iSt_D(\rho_p - \rho_f)(1-R)}{iSt_D(\rho_p - \rho_f) + (1-R)\left(\rho_p + 3e^{i\pi/4}\sqrt{\tfrac{1}{2}\rho_p \rho_f St_D}\right)} \quad \textit{for } Re_p \ll 1. \tag{5.21}$$

The added mass effect causes an inclusion of the fluid density in the particle response while the history force effect is associated with the square root term in the RHS denominator.

If one considers high-density particles ($\rho_p \gg \rho_f$), the equation of motion for the particle velocity vector can be simplified based on (3.89a), as follows:

$$\frac{d\mathbf{v}}{dt} = -\frac{\mathbf{v} - \mathbf{u}_{@p}}{\tau_p} + \mathbf{g} \quad \textit{for } \rho_p \gg \rho_f. \tag{5.22}$$

Substituting (5.20a) and (5.20b) into this equation of motion, the ratio of the particle velocity fluctuations to that of the fluid was obtained by Hinze (1975) as follows:

$$\left(\frac{v_D}{u_D}\right)^2 \approx \frac{1}{1+St_D^2} \quad \textit{for cyclic flow with } Re_p \ll 1 \textit{ and } \rho_p \gg \rho_f. \tag{5.23}$$

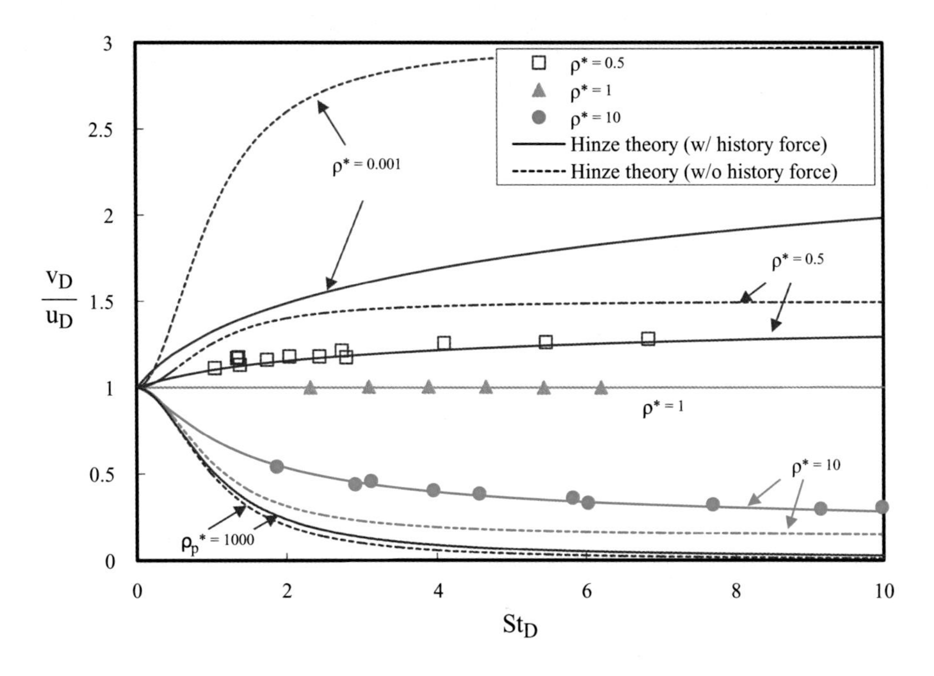

Figure 5.5 Measurements and theoretical predictions (with and without the history forces) for the ratio of particle to fluid velocity magnitudes for various density ratios in a laminar flow oscillating at a single frequency with no gravity effects (L'Esperance *et al.* 2006).

As such, the particle velocity oscillations tend to that of the fluid for small Stokes numbers but tend to zero as the Stokes number (particle inertia) increases.

This ratio for the particle to fluid velocity magnitudes from Hinze's theory is shown in Figure 5.5 for various density ratios, with and without the history force. For high density ratios, the history force has little effect. For smaller density ratios, the history force effect becomes significant (as shown by the difference between the dashed and solid lines). In addition, the effects of the fluid-stress and added-mass forces (associated with finite R) cause lower density particles to have greater accelerations than that of the fluid. In fact, a particle to fluid density ratio of 0.001 combines with a high particle Stokes number can yield a particle velocity magnitude nearly three times greater than that of the surrounding fluid. However, as the Stokes number approaches zero, the particle velocity reverts to that of the fluid velocity (and φ tends to zero) regardless of particle density ratio or history force effects. This is consistent with the regimes of (5.14).

5.3 Microscopic One-Way Coupling by Brownian Diffusion

In the preceding section, we considered the particle response to the largest scales of the surrounding fluid. In this section, we consider the other extreme: particle response to molecular dynamics of the surrounding fluid. In particular, these dynamics can

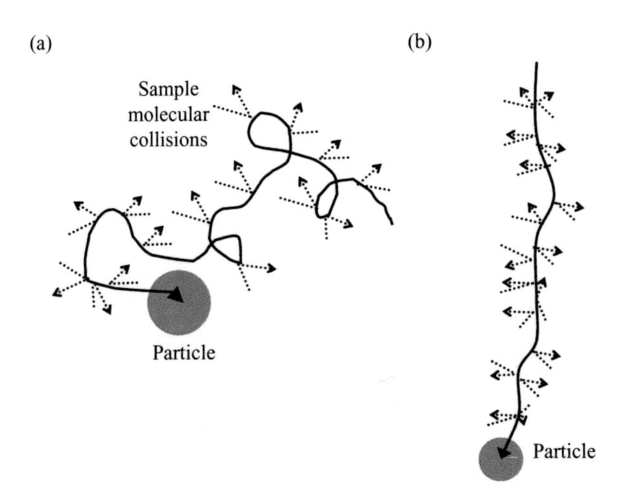

Figure 5.6 A schematic of a small particle subjected to Brownian motion, in which the particle trajectory (solid line) is deflected by many random collisions with molecules (dashed lines) for (a) neutrally buoyant conditions in quiescent fluid (u = 0) where Brownian motion dominates, and (b) particle falling downward due to gravity such that Brownian motion is secondary to terminal velocity motion. Note that this schematic is highly qualitative since particle motion over just 1 μm in distance is typically the result of many millions (or even billions) of molecular interactions.

cause random particle motion due to random collisions of the surrounding fluid molecules on the particle surface (Figure 5.6a). This phenomenon is known as *Brownian motion* since it was first reported by botanist Robert Brown in 1827. In particular, Brown was examining small organic particles (about 1 μm in diameter) in a liquid using a microscope and observed a strange wandering particle trajectory. This motion was not fully understood nor accepted until Einstein's 1905 statistical molecular theory of heat in liquids outlined the laws governing particle movements due to molecular interactions. Einstein's explanation was so compelling that it played a major role in transforming molecules from a hypothetical and controversial postulate to a practical and generally accepted fact. This stochastic motion tends to occur for small particles (diameters on the order of microns) with small movements (displacements on the order of microns), so it is often secondary to larger motion associated with the fluid continuum. For example, Brownian motion can cause small perturbations about the particle terminal velocity, as shown in Figure 5.6b, or about the Stokes-based trajectories discussed in the previous section. In the following, we follow Einstein's analysis to quantify the average long-time particle diffusion due to Brownian motion.

To analyze the motion of a particle under Brownian motion, a set of key assumptions is needed. First, we consider a small, neutrally buoyant particle ($\rho_p = \rho_f$) in an otherwise still fluid, so that there are no effective gravitational forces, and any particle velocity is only caused by Brownian motion. Second, we assume that the fluid surface force on the particle can be decomposed into a random Brownian force based on

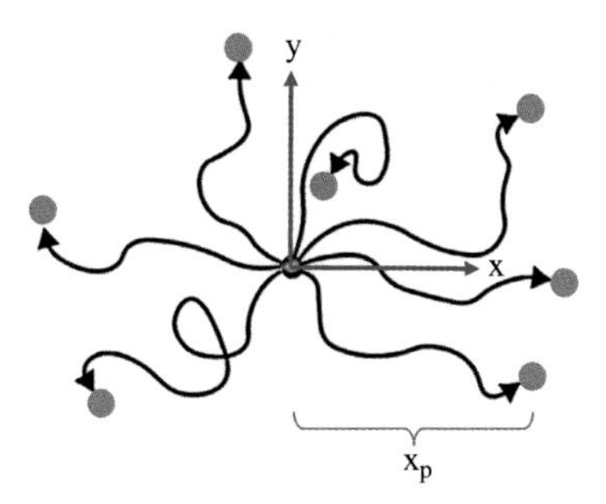

Figure 5.7 Schematic of Brownian motion in an otherwise still fluid for individual particles that start at the same point but are released at different times and move along random paths in all directions. The particle displacement in the x-direction for one particle is shown as x_p, where the ensemble-averaged position of many such particles is zero (no preferred direction), but the ensemble-averaged square of the positions grows with time.

molecular collisions ($\mathbf{F}_{Br}$) and a continuum-based drag force ($\mathbf{F}_D$) based on the continuum-based relative velocity, which can be equated to $\mathbf{v}$ since $\mathbf{u} = 0$. Modifying (3.87) to include these conditions and the extra random Brownian force as a function of time yields the following:

$$m_{eff}\frac{d\mathbf{v}}{dt} = -3\pi d\mu_f \mathbf{v} + \mathbf{F}_{Br}(t). \tag{5.24}$$

This expression neglects the particle history force since this effect is weak for long-time diffusion but should be included for short-time diffusion if the particle density is similar to or lower than that of the fluid (Mainardi and Pironi, 1996). In addition, we assume $f = 1$, but noncontinuum effects on drag may be important for particles in gases, as will be discussed in §9.2.2.

Since the drag of (5.24) is linear, the motion in each Cartesian direction can be considered separately. If we describe the particle velocity in the x-direction using (3.1) and its x-position (x_p), as shown in Figure 5.7, the governing ODE for this motion obtained from (5.24) is as follows:

$$m_{eff}\frac{d}{dt}\left(\frac{dx_p}{dt}\right) = -3\pi d\mu_f \frac{dx_p}{dt} + F_{Br,x}(t). \tag{5.25}$$

Next we consider this motion as a function of time from a statistical point of view for many particle realizations. In particular, we define the *ensemble average* as the group average of a variable q for a statistically large number of particle path realizations released from the same point at different times (t_o) but tracked over the same time duration since release ($\tau = t - t_o$):

$$\langle q(\tau)\rangle \equiv \frac{1}{N_p}\sum_{i=k}^{N_p} q_k(\tau) \qquad \textit{ensemble average for given time shift.} \tag{5.26}$$

On the RHS, k is an index referring to each particle path considered. Multiplying each term in (5.25) by particle position (x_p), using the chain rule for derivatives for the LHS, and taking the ensemble average yields

$$m_{eff}\left[\frac{d}{dt}\left\langle x_p\frac{dx_p}{dt}\right\rangle-\left\langle\left(\frac{dx_p}{dt}\right)^2\right\rangle\right]=-3\pi d\mu_f\left\langle x_p\frac{dx_p}{dt}\right\rangle+\left\langle x_pF_{Br,x}(t)\right\rangle. \quad (5.27)$$

The LHS of this ODE includes terms associated with the particle acceleration, while the RHS includes the effect of drag and the position-weighted average of the Brownian force. If the molecular concentration and kinetic energy of the molecules is homogeneous (fluid density and temperature are statistically uniform in space), then Brownian motion has no preferred direction (Figure 5.7) so long as a statistically large number of particles are considered. Thus there is an equal probability that a particle may drift to the left or to the right so the group average drift is zero. As such, the ensemble-averaged particle position and velocity will be zero, as will the position-weighted average of the Brownian force:

$$\left\langle x_p\right\rangle_{Br}=\left\langle v_x\right\rangle_{Br}=0. \quad (5.28a)$$

$$\left\langle x_pF_{Br,x}(t)\right\rangle=0. \quad (5.28b)$$

Based on (5.28b), the last RHS term of (5.27) will be zero.

Next we apply Einstein's assumption of *equipartition of kinetic energy*, which assumes that the random kinetic energy of particles due to Brownian motion will be equal to the random kinetic energy of the surrounding fluid molecules. This equilibrium of system energy effectively assumes many elastic molecular collisions averaged over many particle trajectories as well as long enough times such that any effects of initial conditions can be neglected ($t \gg \tau_p$). The important equitation principle allows us to relate the particle velocity fluctuations to the molecular velocity fluctuations. In particular, the second term on the LHS of (5.27) reflects the ensemble-average of the kinetic energy of the particle x-velocity fluctuations. The other two directions can be similarly described so that the total particle kinetic energy can then be set equal to that for the fluid molecules, which can be related to the fluid temperature using the Boltzmann constant, κ, and then used to rewrite (5.27) using (5.28a,b) as

$$\frac{1}{2}m_{eff}\left(\left\langle v_x^2\right\rangle+\left\langle v_y^2\right\rangle+\left\langle v_z^2\right\rangle\right)=\frac{3}{2}m_{eff}\left\langle v_x^2\right\rangle=\frac{3}{2}\kappa T_f. \quad (5.29a)$$

$$\kappa\equiv 1.38\times 10^{-23}\,m^2kg/\left(s^2K\right). \quad (5.29b)$$

$$m_{eff}\frac{d\left\langle x_pv_x\right\rangle}{dt}=\kappa T_f-3\pi d\mu_f\left\langle x_pv_x\right\rangle. \quad (5.29c)$$

If one considers long times, the LHS unsteady term of (5.29c) will tend to zero. In this case, the ordinary differential equation can be integrated to obtain the average of the square of the particle positional variation as a function of path time (taking $t_o = 0$ for simplicity) as:

$$\langle x_p x_p \rangle_{Br} = \frac{2\kappa T_f t}{3\pi d \mu_f}. \tag{5.30}$$

The LHS is the one-dimensional spread of particles due to Brownian diffusion. Therefore, the square root of the LHS indicates the average x-distance a particle has moved for a given amount of time. For example, particles that are 1 μm in diameter and diffuse over a 1 second interval at room temperature will have mean lateral spread of about 0.9 μm in water and about 6.8 μm in air based on (5.30) and properties in Table A.1. Since the diffusion is inversely proportional to d and μ_f, the spread will be reduced for larger particles and higher-viscosity fluids.

Since the ensemble-averaged movement is zero (5.28a), similar results can be obtained for the other two Cartesian directions:

$$\langle x_p x_p \rangle_{Br} = \langle y_p y_p \rangle_{Br} = \langle z_p z_p \rangle_{Br} = \frac{2\kappa T_f t}{3\pi d \mu_f}. \tag{5.31}$$

As such, the spread in the x-, y-, and z-directions will all be equal, so the ensemble-averaged Brownian motion yields a spherical cloud of particles (like that for species diffusion) if the fluid is still and there are no effective gravitational forces.

One may also use this result to obtain the Brownian spread rate (Θ_{Br}) at long times (after initial conditions have subsided) in the x-direction, as follows:

$$\Theta_{Br} \equiv \left\langle x_p \frac{d x_p}{d t} \right\rangle_{Br} = \frac{1}{2} \frac{d \langle x_p x_p \rangle_{Br}}{d t} = \frac{\kappa T_f}{3\pi d \mu_f}. \tag{5.32}$$

As a result, the diffusivity is linearly proportional to the fluid temperature, which is consistent with molecular diffusivity of a species in a fluid. Note that (5.32) describes the Brownian diffusivity (which has units of length^2/time), while (5.31) describes the Brownian diffusion (which has units of length^2). However, both aspects are interestingly independent of fluid density, particle density, and particle mass. Furthermore, F_{Br} does not appear on the RHS, which indicates that the detailed dynamics of the Brownian force (e.g., frequency of the molecular collisions) are not important.

Since (5.24) is a linear ODE, one may linearly combine effects of Brownian motion with movement of particles by convection, particle inertia, or gravitational forces. For the example of a particle falling at terminal velocity in a still fluid as in Figure 5.6b, the ensemble-averaged movement (in vector notation) becomes

$$\langle \mathbf{v}_p \rangle = \langle \mathbf{v}_p \rangle_{term} + \langle \mathbf{v}_p \rangle_{Br} = \mathbf{w}_{term}. \tag{5.33}$$

Therefore, Brownian motion generally does not influence the mean motion if one considers a single release point. Instead, it provides a potential spread in position about this mean point for a fixed release location. This spread is important for many microfluidic systems, including the deposition of micron-sized aerosol particles in respiratory tracts (Figure 1.15).

If one considers particles distributed in a domain with a gradient in the particle concentration, Brownian motion can lead to a net migration in the opposite direction.

In particular, the average migration velocity of the particles due to Brownian motion is proportional to the particle concentration gradient, as

$$\Delta \mathbf{w}_{Br} \equiv -\Theta_{Br}\frac{\nabla\alpha}{\alpha} = -\frac{\kappa T_f}{3\pi d\mu_f f_{Kn}}\frac{\nabla\alpha}{\alpha}. \tag{5.34}$$

As such, particles will spread from regions of high concentration to regions of low concentration like that for a species concentration gradient in a single-phase flow. In fact, as the particles become smaller and tend toward the size of molecules, Brownian diffusivity will tend toward the species molecular diffusivity of (2.32), that is, $\Theta_{Br} \rightarrow \Theta_{\mathcal{M}}$, so the speed of (5.34) will tend to that for a species (2.34a).

5.4 Two-Way Coupling Criteria

Two-way coupling occurs when the presence of the particles significantly impacts the surrounding fluid in a bulk sense (not just around the local flow around the particle). This can occur due to density coupling and/or momentum coupling.

The simplest coupling is by density, whereby the effective density of the overall flow is changed by the presence of the particles. This can be characterized in terms of the density coupling parameter (Π_ρ) by using the mixed-fluid density of (5.5), as follows:

$$\Pi_\rho \equiv \frac{\text{change in effective fluid density}}{\text{fluid density in absence of particles}} = \frac{|\rho_m - \rho_f|}{\rho_f} = \alpha\frac{|\rho_p - \rho_f|}{\rho_f}. \tag{5.35}$$

This coupling parameter increases with particle concentration and with the density difference. In terms of fluid physics, Π_ρ is most important when there is a large variation of the mixed-fluid density. For example, if one portion of the flow has a higher effective density (due to gradients in α), it will tend to be pulled downward by gravitational effects relative to the surrounding regions. An example of such "density-driven" flows is shown in Figure 1.18, where a region of high particle concentration in a gas increases the mixed-fluid density, causing the flow to be driven downward. Another example is that of an underwater plume where the reduced density of the mixture causes the fluid mixture to be pulled upward by buoyancy. In these two cases with extreme density ratios, the density coupling parameter simplifies to the mass loading or the volume fraction via (5.5)–(5.7):

$$\Pi_\rho \approx m \qquad \textit{for } \rho_p \gg \rho_f. \tag{5.36a}$$

$$\Pi_\rho \approx \alpha \qquad \textit{for } \rho_f \gg \rho_p. \tag{5.36b}$$

To quantify an approximate criterion for the possibility of significant two-way coupling, one may employ a 1% change in the effective density as follows:

$$\Pi_\rho > 0.01 \qquad \textit{for significant two-way density coupling.} \tag{5.37}$$

The criterion of (5.37) indicates when such coupling *may* occur.

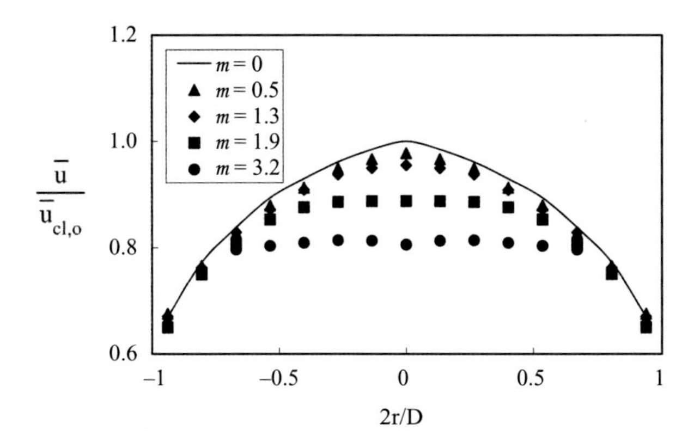

Figure 5.8 Time-averaged air velocity (normalized by the time-averaged centerline velocity for no particles) in an upward-flowing vertical channel flow subjected to downward-falling 243 μm diameter solid particles at various mass loadings (Tsuji et al., 1984).

In addition, particles can impact the surrounding fluid velocity due to momentum effects. This generally occurs when the net particle drag is significant relative to the unladen fluid momentum. For example, an upward-driven pipe flow of air will have a reduced velocity profile if it contains falling particles, as shown in Figure 5.8. The impact on the fluid velocity is amplified (especially near the pipe centerline) as the mass loading of the particles increases.

To characterize such two-way coupling, Crowe et al. (2011) defined the momentum coupling parameter (Π_u) based on the net particle drag in the domain relative to the momentum based on the fluid velocity in the absence of particles ($u_{D,o}$) as follows:

$$\Pi_u \equiv \frac{\text{net drag due to particles on the domain}}{\text{fluid momentum in absence of particles}} = \frac{F_D n_p D^3}{\rho_f u_{D,o}^2 D^2}. \tag{5.38}$$

The RHS numerator includes the force for a single particle (F_D), the number of particles per volume (n_p), and the domain volume (D^3), while the denominator is the momentum flux of the surrounding fluid based on a domain cross-sectional area (D^2). Using (5.3) and (5.11)–(5.13), the momentum coupling parameter for uniform particle size can be expressed as follows:

$$\Pi_u = \frac{\alpha(\rho_p + \rho_f c_\forall)}{\rho_f St_D} \frac{w}{u_{D,o}}. \tag{5.39}$$

As such, the momentum coupling is proportional to particle volume fraction and particle relative velocity and includes the added-mass effect.

If the relative velocity of the particle can be approximated as the terminal velocity, then one may employ the balance of (4.12a) and (4.12b) with (2.71) to obtain the following:

$$\Pi_u = \frac{g|\rho_p - \rho_f|\alpha D^3}{\rho_f u_{D,o}^2 D^2} = \frac{\alpha|\rho_p - \rho_f|}{\rho_f Fr_D} \qquad \textit{for } w \approx w_{term}. \tag{5.40}$$

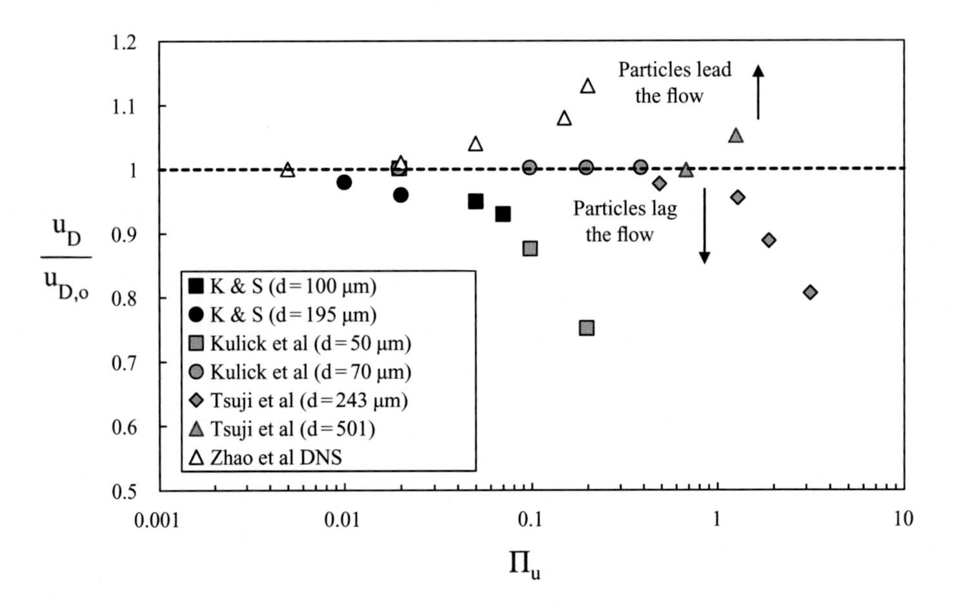

Figure 5.9 Influence of the momentum coupling parameter on the fluid centerline velocity (normalized by that for no particles) for pipe and channel flows driven by a fixed pressure gradient (Tsuji et al., 1984; Kulick et al., 1994; Kussin and Sommerfeld, 2001; Zhao et al., 2013).

The RHS shows that the coupling increases when the density difference becomes larger, as for (5.35), but also when the flow Froude number becomes smaller. The momentum coupling parameter can be approximated in the limit of very high or very low particle density ratios as follows:

$$\Pi_u \approx m/Fr_D \qquad \textit{for } \rho_p \gg \rho_f. \tag{5.41a}$$

$$\Pi_u \approx \alpha c_\forall/Fr_D \qquad \textit{for } \rho_f \gg \rho_p. \tag{5.41b}$$

The latter two expressions show that momentum coupling is related to the mass loading for solid or liquid particles in a gas and related to the volume fraction for bubbles.

While Π_u determines when drag-based two-way coupling may occur, the direction of the relative velocity determines how this flow will change. For example, particles that lead the flow ($v > u$) will tend to increase the flow momentum ($u_D > u_{D,o}$), while particles that lag the flow ($v < u$) tend to reduce the flow momentum ($u_D < u_{D,o}$). These trends are shown in Figure 5.9 in terms of changes to the channel centerline velocity, where the effects begin at coupling values of about 1% and are magnified as the coupling parameter increases.

If the overall fluid momentum is fixed (e.g., by a constant flow rate in a channel), the primary particle influence may instead be on the pressure gradient required to maintain this momentum flux. The variations in pressure gradient for constant mass flow experiments are shown in Figure 5.10, where particles that lead the flow ($v > u$) reduce the

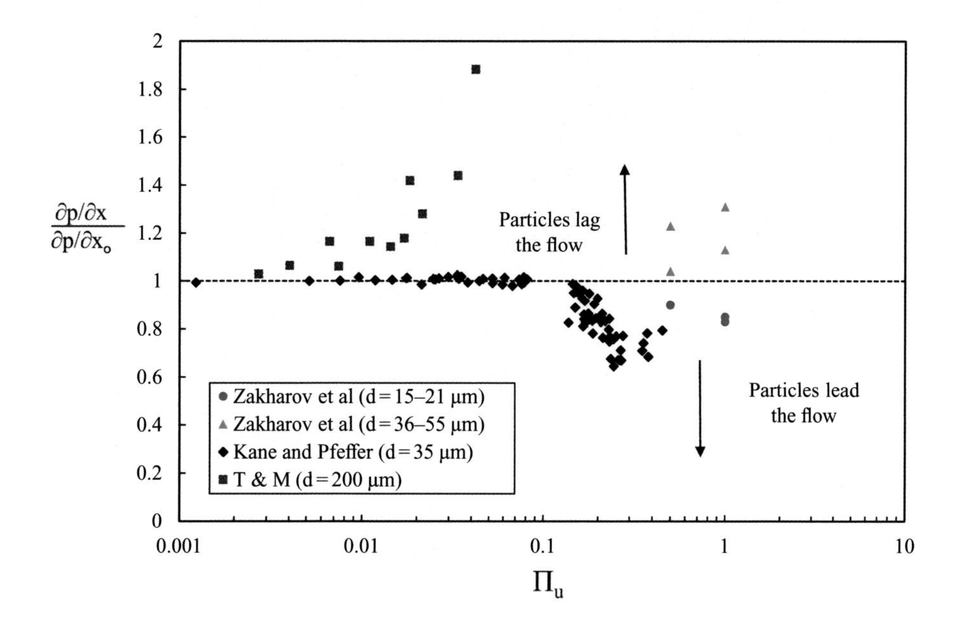

Figure 5.10 Pressure gradient ratios in a pipe and channel flow based on a fixed mass flow rate (Kane and Pfeffer, 1973; Tsuji and Morikawa, 1982; Zakharov et al., 1993).

required pressure gradient, whereas those that lag the flow ($v < u$) increase the required pressure gradient. This figure also shows that the magnitude of the changes becomes more profound as the momentum coupling parameter increases. Based on results shown in Figures 5.9 and 5.10, the onset criterion for continuous-phase momentum change due to particles (for $u_{D,o} > w$) can be approximately given as follows:

$$\Pi_u > 0.01 \quad \textit{for significant two-way momentum coupling.} \tag{5.42}$$

Note that this only considers the impact of particles on the bulk (mean) fluid flow and does not consider the potential impact of particles on turbulence, which will be discussed in Chapter 7.

In summary, two-way coupling can be expected if either (5.37) or (5.42) are satisfied. It should be noted these criteria only indicate if coupling may occur, since sometimes the effects cancel out and since the magnitude of the effects depends on details of flow geometry, particle trajectories, and concentrations variations. In addition, interphase transfer coupling can arise through mass and heat transfer using the relations of §3.5 (Crowe et al. 2011).

5.5 Three-Way Coupling: Particle–Particle Fluid Dynamic Interactions

Three-way coupling refers to the fluid dynamic interactions between local particle fluid fields (Figure 5.1). Such interactions occur when the respective added-mass or viscous

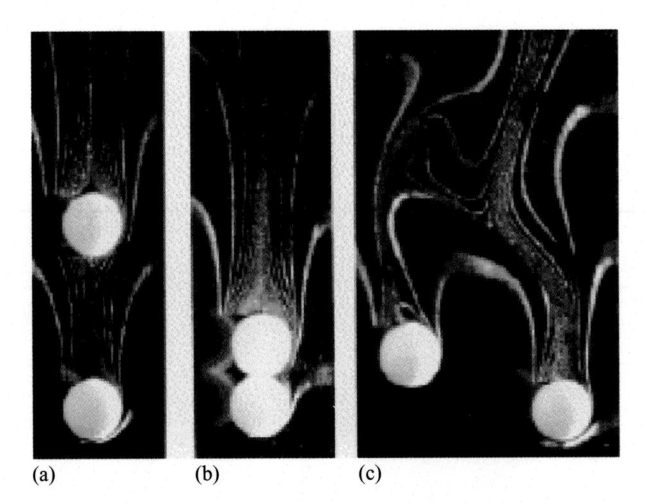

Figure 5.11 Experimental flow visualization of various interactions between falling particles in a liquid, which include (a) drafting, (b) kissing, and (c) pair tumbling (Fortes et al., 1987).

volumes of two or more particles intersect so that they interact in a fluid dynamical sense modifying the surface forces seen by particles. In contrast, four-way coupling is based on direct particle-to-particle contact. In some cases, an interaction can have both three-way coupling and four-way coupling. For example, consider the two particles falling at nearly their terminal velocity in an otherwise stagnant fluid shown in Figure 5.11. In Figure 5.11a, the upper particle is in the wake of the lower particle, so the upper particle sees a local downward flow induced by the lower particle (a three-way coupling effect). This condition is termed *drafting* and allows the trailing particle to travel faster relative to the lower particle. This relative motion often leads to a collision between the particles (a four-way coupling effect), as seen in Figure 5.11b. In this case, the particle Reynolds numbers are small, so significant fluid viscous effects in the gap between the particle cause the trailing particle to slow down before gently contacting the lower particle. This is referred to as *kissing* and may result in both particles traveling vertically together for a short period. As they travel, their combined shape resembles a prolate geometry that is generally unstable (Figure 4.14), leading the particles to rotate about each other. This is referred to as *pair tumbling*. Three-way coupling then causes the particles to repulse, which leads to a lateral distance between the particles (Figure 5.11c).

In this section, we seek to determine the conditions for which three-way coupling is significant to the overall particle motion. It is important to establish such criteria since inclusion of the three-way coupling mechanisms will substantially complicate the analysis (to be discussed in §11.2).

Three-way coupling is principally influenced by the minimum distance between the surfaces of neighboring particles, often known as the *gap distance* (ℓ_{gap}), and the *interparticle spacing* ($\ell_{p\text{-}p}$), defined as the distance between centroids of neighboring particles. For spherical particles (Figure 5.12), the gap distance is the interparticle spacing minus the average of the particle diameters:

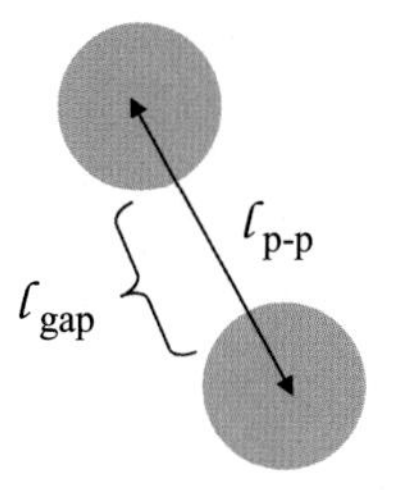

Figure 5.12 Two spherical particles with centroids separated by the interparticle spacing while the surfaces separation is given by the gap spacing (which controls three-way coupling effects).

$$\ell_{p\text{-}p} \equiv \left|\boldsymbol{\ell}_{p-p}\right| = \left|\mathbf{X}_{p,1} - \mathbf{X}_{p,2}\right|. \tag{5.43a}$$

$$\ell_{gap} \equiv \ell_{p-p} - \left(r_{p,1} + r_{p,2}\right) = \ell_{p\text{-}p} - \frac{1}{2}(d_1 + d_2). \tag{5.43b}$$

The ensemble-averaged spacing and gap distance for spherical particles of uniform diameter can be related to the volume fraction via (5.2b) as

$$\langle \ell_{p\text{-}p} \rangle \approx \langle d \rangle \left(\frac{\pi}{6\alpha}\right)^{1/3} \approx \langle \ell_{gap} \rangle + \langle d \rangle. \tag{5.44}$$

For reference, $\alpha = 1\%$ corresponds to an average gap of 2.7 diameters, which is close enough for particles to influence each other fluid dynamically. This average gap will decrease (more three-way coupling) as the particle concentration increases. Local gaps will be even less when there is random particle spacing and/or the shapes are nonspherical.

For a given gap distance, the fluid dynamic influence is most often quantified in terms of changes in the net particle drag. If one considers solid spherical particles, the drag correction due to finite volume effects can be empirically described by the Richardson–Zaki model:

$$f_\alpha = (1 - \alpha)^{-3.5} \quad \textit{for solid spheres.} \tag{5.45}$$

As such, the average particle drag increase for $\alpha = 1\%$ is nearly 4%, and for $\alpha = 10\%$ is nearly 45%, while these increases are even stronger for nonspherical particles (§11.2). On the other hand, three-way coupling reduces the fluid-stress and added-mass forces, and these changes scale directly with volume fraction, so that $\alpha = 1\%$ yields ~1% decrease in F_S and $F_\forall$. As such, three-way coupling has the largest effect on the drag force.

Based on the preceding results, a rough criterion for when particle–particle fluid dynamic interactions become significant and when they dominate can be given as follows:

$$\alpha \gtrsim 0.01 \quad \textit{for significant three-way coupling.} \tag{5.46a}$$

$$\alpha \gtrsim 0.1 \quad \textit{for strong three-way coupling (group movement).} \tag{5.46b}$$

These criteria are more likely to occur in liquid flows, where higher volume fractions are more common.

5.6 Four-Way Coupling: Particle–Particle Collisions

Collision interactions occur when there is surface-to-surface contact between particles. Such four-way coupling leads to changes in each particle's trajectory depending on the collision outcome (which can be a bounce, an adhesion, or a fracture). This coupling is more likely to occur when the particle concentration and the relative velocities between neighboring trajectories are high. The following discusses various mechanisms for these collisions.

For particles falling in a flow at nearly terminal conditions, drafting can cause collisions as shown in Figure 5.11. In addition, differences in particle size or shape can cause different terminal velocities resulting in collisions (Figure 5.13a). Particle collisions can also occur when there are differences in inertia coupled with a change of the surrounding fluid flow speed and/or direction (Figure 5.13b).

Another collision mechanism is the Brownian motion of very small particles (§5.3). This random movement phenomenon, shown in Figure 5.13c, is driven by Brownian diffusivity (Θ_{Br}). Finally, turbulence can also cause particles to collide when they originate from different vortex structures (Figure 5.13d). Collisions induced by either of these two mechanisms are primarily related to the particle velocity fluctuations, so they are more likely when the temperature is high for Brownian collisions or the turbulence intensity is high for turbulent collisions.

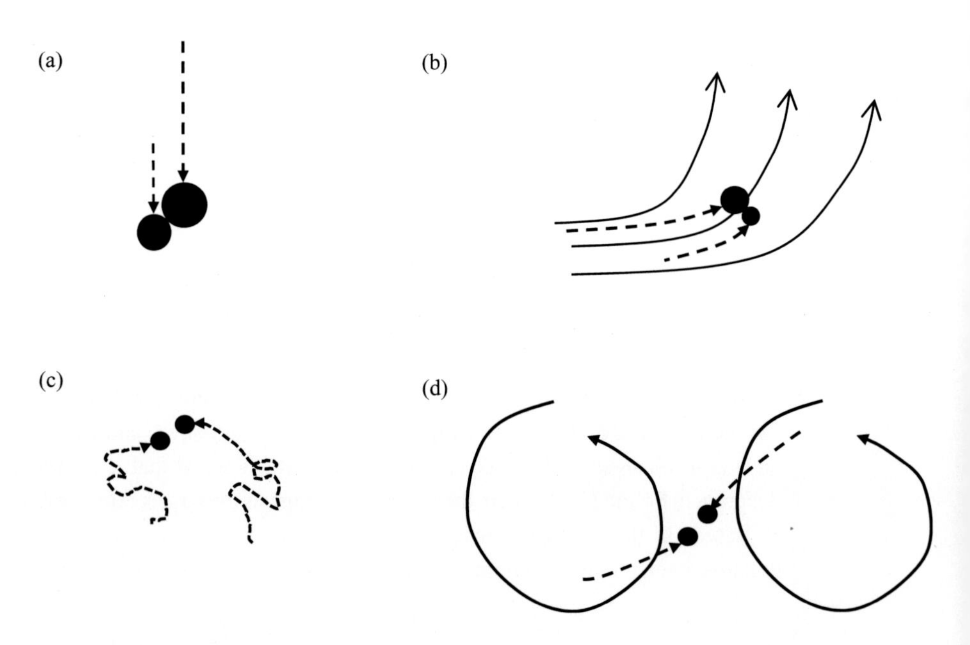

Figure 5.13 Schematic of various particle collision mechanisms: (a) gravitational collision resulting from a large particle having a larger terminal velocity; (b) flow variation collision due to differences in particle response times; (c) Brownian collision due to random motion; and (d) turbulent collisions due to originating in different eddies.

Turbulent four-way coupling will be discussed in §7.7, and the other three mechanisms are discussed in the following section.

5.6.1 Collision Frequencies and Collision Stokes Number

To quantify the likelihood of collision for a given particle, one may define the average time between collisions along its path as τ_{coll}, which is inversely proportional to the *collision frequency* (f_{coll}), which is the average number of collisions per unit time for a single particle:

$$f_{coll} \equiv \frac{\text{collisions experinced by a particle}}{\text{time}} \equiv 1/\tau_{coll}. \tag{5.47}$$

To help characterize the importance of collisions, this time scale (or the frequency) can be compared to the drag-based time scale (τ_p) by defining a collision Stokes number as follows:

$$St_{coll} \equiv \tau_p/\tau_{coll} = \tau_p f_{coll}. \tag{5.48}$$

As illustrated in Figure 5.14a, small collision Stokes numbers ($\tau_p \ll \tau_{coll}$) indicate that the particle responds quickly to the continuous-fluid velocities and undergoes only occasional collisions. As such, collision effects (and four-way coupling) are weak. On the other hand, the high collision Stokes numbers of Figure 5.14b ($\tau_p \gg \tau_{coll}$) yield many collisions during the particle's fluid dynamic response time. As such, collisions are primarily responsible for the particle motion. Intermediate collision Stokes numbers ($\tau_p \sim \tau_{coll}$) indicate that both effects are important and should be considered,

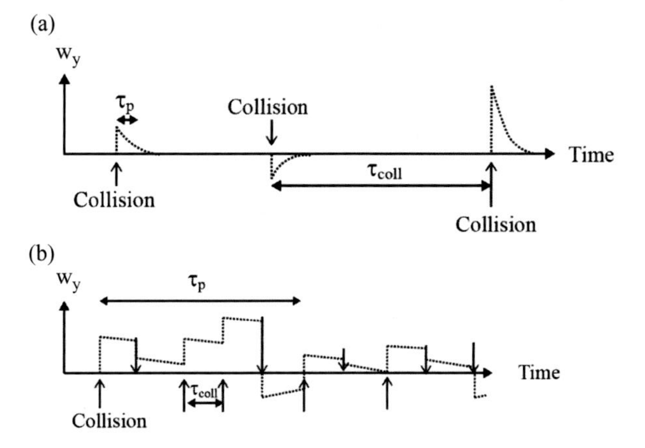

Figure 5.14 Particles in a uniform flow whose relative velocity in vertical direction (w_y) changes due to random collisions for the cases of (a) infrequent collisions ($St_{coll} \ll 1$) such that the particle velocity is mostly determined by drag via the continuous-fluid velocity; and (b) collision-dominated conditions ($St_{coll} \gg 1$) where the particle velocity is mostly determined by collisions with other particles.

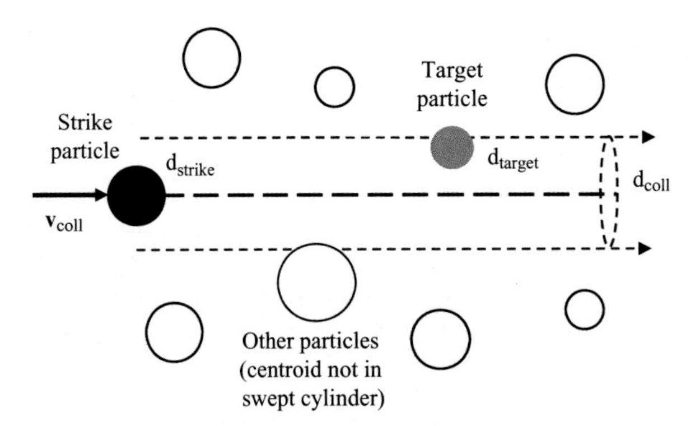

Figure 5.15 Schematic of strike and target particles and the collision swept cylinder with a base centered at the strike particle, with a forward direction based on the collision velocity ($\mathbf{v}_{coll}$), and with a diameter (d_{coll}) equal to the sum of the two particle diameters.

with drag dominating when $\tau_p < \tau_{coll}$ and with collisions dominating when $\tau_p > \tau_{coll}$. Therefore, the four-way coupling regime for collisions can be summarized as follows:

$$St_{coll} \to 0 \quad \textit{particle collisions are negligible (e.g., sparse flow)}. \tag{5.49a}$$

$$St_{coll} < 1 \quad \textit{drag dominates particle motion (dilute flow)}. \tag{5.49b}$$

$$St_{coll} > 1 \quad \textit{collisions dominate particle motion (dense flow)}. \tag{5.49c}$$

The regimes between dilute flow ($St_{coll} < 1$) and dense flow ($St_{coll} > 1$) are noted in Figure 5.1.

To apply the preceding criteria, an estimate of the particle collision frequency is needed. This can be determined by defining a strike particle velocity and diameter ($\mathbf{v}_{strike}$ and d_{strike}) as well as a target particle velocity and diameter ($\mathbf{v}_{target}$ and d_{target}), as shown in Figure 5.15. In this case, the collision velocity (v_{coll}) is defined as the velocity of the strike particle relative to the target particle, that is, in a reference frame where the target particle is stationary.

$$\mathbf{v}_{coll} \equiv \mathbf{v}_{strike} - \mathbf{v}_{target}. \tag{5.50}$$

As also shown in this figure, the swept cylinder interaction volume stems from the collision velocity and a cross-sectional area (A_{coll}) based on the collision diameter (d_{coll}), which is the sum of the two particle diameters:

$$A_{coll} \equiv \frac{\pi}{4} d_{coll}^2. \tag{5.51a}$$

$$d_{coll} \equiv d_{strike} + d_{target}. \tag{5.51b}$$

Any particle with a centroid within this forward-facing cylinder is a target particle, and the one closest to the strike particle is expected to yield a collision.

To quantify the collision frequency (f), the volume of fluid swept out per unit time for this cylinder ($v_{coll}A_{coll}$) can be multiplied by the number of target particles per unit volume ($n_{p,target}$ based on 5.1) to provide the number of expected collisions per unit time for a strike particle:

$$f_{coll} = A_{coll} v_{coll} n_{p,target} \qquad \textit{collision frequency for a strike particle.} \tag{5.52}$$

For a monodisperse distribution with uniform particle diameter, (5.52) can be combined with (5.2b), (5.11), (5.48), and (5.51) to yield

$$St_{coll} = \frac{\alpha d\left(\rho_p + c_{\forall}\rho_f\right) v_{coll}}{3\mu_f f}. \tag{5.53}$$

As such, collision coupling becomes more important with increases in the particle volume fraction, the particle diameter, and the particle collision velocity.

One may also define the net frequency of collisions among all particles occurring per unit volume ($\mathcal{N}$), as follows:

$$\mathcal{N} \equiv \frac{\text{all particle collisions}}{(\text{mixed-fluid volume})(\text{time})} \equiv n_p f_{coll}. \tag{5.54}$$

For a monodisperse size distribution, this net frequency per volume can be related to the collisional velocity by (5.51) and (5.52) as follows:

$$\mathcal{N} = \pi d^2 n_p^2 v_{coll}. \tag{5.55}$$

For a polydisperse mixture with particles binned by size, the net frequency per unit volume for particles in a bin with diameter d_i colliding with target particles within a bin with diameter d_j can be expressed as follows:

$$\mathcal{N}_{ij} \equiv \frac{\pi}{4} d_{coll}^2 n_{p,i} n_{p,j} v_{coll}. \tag{5.56}$$

The number density of particles in a bin (e.g., $n_{p,i}$) can be obtained using the number-based probability size distribution with (4.1) and (5.1), where the sum of these number densities for all the bins equals the overall number density:

$$n_{p,i} = \mathcal{P}_{N,i} n_p / \Delta d_{i.} \tag{5.57a}$$

$$n_p = \sum_{i=1}^{N_b} n_{p,i.} \tag{5.57b}$$

In the second expression, N_b is the number of discrete bins and i is the bin index. To obtain the total collision rate among all particles for a binned distribution, one may sum over all target and strike bins (counting each combination only once), as follows:

$$\mathcal{N} = \sum_{i=1}^{N_b}\left(\sum_{j=1}^{i} \mathcal{N}_{ij}\right). \tag{5.58}$$

In the following, the bin-based collision rates per unit volume are estimated based on terminal velocity differences $\left(\mathcal{N}_{ij,term}\right)$, on fluid domain time scales $\left(\mathcal{N}_{ij,D}\right)$, and on Brownian motion $\left(\mathcal{N}_{ij,Br}\right)$.

5.6.2 Collisions Due to Terminal Velocity Variations

For collisions resulting from bin differences in terminal velocity, the collision velocity is as follows:

$$v_{coll,ij} = \left| w_{term,i} - w_{term,j} \right| \qquad \textit{for} \;\; w = w_{term}. \tag{5.59}$$

For polydisperse particles, the frequency of collision for a single-strike particle of diameter d_i relative to target particles with a diameter d_j and a bin number density $n_{p,j}$ can be expressed as follows:

$$f_{term,j} = A_{coll,ij} v_{coll,ij} n_{p,j} = \frac{\pi}{4}(d_i + d_j)^2 \left| w_{term,i} - w_{term,j} \right| n_{p,j}. \tag{5.60}$$

If one considers solid spherical particles at small Re_p, then $f_{term} = 1$ and w_{term} is proportional to d^2 per (3.94a). From this, the net frequency of all collisions between strike and target particles with a bin number density $n_{p,i}$ and $n_{p,j}$ is as follows:

$$\mathcal{N}_{ij,term} \equiv f_{term,j} n_{p,i} = \frac{\pi g}{72\mu_f}(d_i + d_j)^2 \left| (\rho_p - \rho_f)\left(d_i^2 - d_j^2\right) \right| n_{p,i} n_{p,j}. \tag{5.61}$$

The overall collision frequency for all bins can be obtained by employing (5.58). Note that this collision frequency becomes zero for a monodisperse size distribution $(d = d_i = d_j)$.

5.6.3 Collisions Due to Flow Path Changes

The collision rates due to surrounding fluid velocity changes based on a domain time scale (τ_D) can be obtained by considering the differences between the strike and target particle inertias. In particular, consider horizontal motion of a solid particle in a surrounding gas (such that gravity, added mass, fluid stress, and history forces can be neglected), where $u_x = v_x = 0$ initially. Then consider a "gust" of the surrounding fluid that yields a linear increase in velocity with time as $u_x = u_D(t/\tau_D)$. In this case, the differential equation given by (3.88c) and its solution (dropping the "x" subscripts for conciseness) are given as follows:

$$\frac{dv}{dt} = -\frac{v - u_{@p}(t)}{\tau_p} = -\frac{1}{\tau_p}\left(v + \frac{u_D t}{\tau_D}\right). \tag{5.62a}$$

$$v = \frac{u_D}{\tau_D}\left[(t - \tau_p) + \tau_p e^{-t/\tau_p}\right]. \tag{5.62b}$$

As the particle inertia goes to zero, the particle velocity tends to the fluid velocity (as $\tau_D \rightarrow 0$, $v \rightarrow u$). Considering small times and expanding the exponential as a Taylor series, the particle velocity can be approximated as follows:

$$v \approx \frac{u_D t^2}{2\tau_D \tau_p} \qquad \textit{for} \;\; t \ll \tau_p. \tag{5.63}$$

If we now consider a strike particle and a target particle with different response times ($\tau_{p,strike}$ and $\tau_{p,target}$), their relative velocity difference provides the collisional velocity, which can be used for the net collisional frequency between bins as follows:

$$v_{coll} = \left|v_{strike} - v_{target}\right| \approx \frac{u_D t^2}{2\tau_D}\left|\frac{1}{\tau_{p,strike}} - \frac{1}{\tau_{p,target}}\right|. \tag{5.64a}$$

$$\mathcal{N}_{D,ij} = \frac{\pi}{4}\left(d_i + d_j\right)^2 \frac{u_D t^2}{2\tau_D}\left|\frac{1}{\tau_{p,i}} - \frac{1}{\tau_{p,j}}\right| n_{p,j} n_{p,i}. \tag{5.64b}$$

As such, the collision velocity increases with differences in the particle response times.

5.6.4 Collisions Due to Brownian Motion

For very small particles, collisions can occur due to Brownian motion stemming from their individual random trajectories. For such motion, the random velocity of the particles is related to the fluid temperature by (5.29a). Integrating over the probability distribution function of these random relative velocities, the net collision frequency obtained by Smoluchowski (1916) in bin-based form is given as follows:

$$\mathcal{N}_{ij,Br} = f_{ij,Br} n_{p,j} = \frac{2\kappa T_f}{3\mu_f}\frac{\left(d_i + d_j\right)^2}{d_i d_j} n_{p,i} n_{p,j}. \tag{5.65}$$

For a monodisperse distribution ($d_j = d_j$) and Stokes drag ($f = 1$, often reasonable for small particles), the collision frequency can be converted into a collision Stokes number as follows:

$$St_{coll,Br} \equiv \tau_p f_{Br} = \frac{8}{9\pi}\frac{\kappa T_f \alpha}{d\mu_f^2}\left(\rho_p + c_{\forall}\rho_f\right). \tag{5.66}$$

As such, Brownian-based collisions become more likely as particle concentration and temperature increase, and as particle size decreases. Typically, collisions by Brownian diffusion at NTP conditions are only significant for micron-scaled or smaller particles, whereas the collision frequencies for larger particles are dominated by differences in terminal velocity or changes in the surrounding fluid velocity.

5.7 Chapter 5 Problems

Show all steps and list any needed assumptions.

(P5.1) Consider a 10 cm diameter pipe with NTP air traveling upwards at an average velocity of 10 m/s and 220 μm glass particles with an average upward particle velocity of 8.5 m/s, a density of 2,500 kg/m^3, and a mass loading of 1.3. (a) Determine the mass flux (kg/s) for the air and for the particles across a pipe with a diameter of 6 cm. (b) Determine α, n_p, $\mathcal{M}$, and

ρ_m. (c) How accurate is it to assume the viscosity and thermal conductivity of the mixed fluid are equal to their air values?

(P5.2) Consider a spray of ethanol drops in air at NTP, where the average droplet diameter is 30 μm and the liquid mass flux through the nozzle is 4 g/s. Then consider the axial downstream location where the spray has a cross-sectional diameter of 15 cm and the flow mixture is moving axially at 3 m/s assuming no evaporation and no sidewall impact for the droplets. At this location, determine the following: (a) α, n_p, $\mathcal{M}$, and m; (b) the accuracy of the approximations of (5.7a) and (5.7b); and (c) the minimum ℓ_m values for well-posed and for continuum conditions of the particle concentration.

(P5.3) Consider a stationary wind turbine blade with a chord length (D) of 2 m subjected to cold wind at –10°C and 15 m/s that carries supercooled water droplets (which tend to freeze on impact). Determine the droplet size range for $St_D < 0.1$ (unlikely to impact) and for $St_D > 10$ (very likely to impact). Approximate the RHS of (5.17) as $[1 + 0.0967\ Re_{p,D}(0.6367)]^{-1}$.

(P5.4) Consider a steady two-dimensional corner gas flow in a domain given by $0 \leq x \leq D$ and $0 \leq y \leq 1$ where the velocity field is given as $u_x = x/\tau_D$ and $u_y = -y/\tau_D$. In this flow, consider a solid particle with Stokesian drag ($f = 1$) and a particle response time τ_p, released at $(x,y) = (0.5D,D)$ with no initial relative velocity ($w_o = 0$) and no gravity ($g = 0$). (a) Obtain the velocity and position ODEs that govern the particle trajectory in terms of Stokes number. (b) Use a numerical solver to obtain the particle trajectories until they exit the domain as a function of time for St_D values of 0.01, 0.1, 1, and 10. (c) On one plot, graph the four trajectories and discuss the influence of the Stokes number.

(P5.5) Consider a high-density particle in simple harmonic flow with $u = u_D \sin(2\pi t/\tau_D)$ for which linear drag is the only surface force and gravitational effects are neglected. After long times, the particle velocity will be periodic with but will differ from that of the flow in amplitude and phase angle, $v = v_D \sin(2\pi t/\tau_D - \phi)$. Starting with (5.20a), obtain u_D/v_D as a function of the Stokes number for these conditions given as (5.23), by using the complex form for the fluid and particle velocities such as $u = A\exp(i\omega t)$ and applying Euler's identity. Discuss the trends and fluid physics for this result.

(P5.6) Starting from (5.24), analytically obtain (5.30), showing all the intermediate steps.

(P5.7) Thermal flow currents aided by Brownian motion can help steep tea. However, if there are no flow currents and the tea spreads only by Brownian motion of 0.1 μm diameter particles, determine the time for the particles to spread an average of 2 cm in water at 60°C with $\mu_f = 5 \times 10^{-4}$ kg/m-s.

(P5.8) Consider a plume of bubbles released with a gas flux of 360 cm^3/s at the bottom of a 4 meter deep pond where the plume rises upward as a column. The bubbles have a diameter of 1 mm and an upward velocity of 21 cm/s once they have risen 2 m from the bottom. (a) Assuming sparse flow conditions, determine the volume fraction of bubbles where the average diameter of the plume column is 20 cm. (b) Based on the results from (a), determine whether the bubbles are likely to result in two-way and/or three-way coupling within the plume. (c) If the bubble rise velocities vary due to flow unsteadiness such that the variations on average are ±10% of the terminal velocity, compute the collisional Stokes number and determine if this corresponds to dense four-way coupling conditions.

(P5.9) Starting with (5.59), derive (5.61).

(P5.10) Consider horizontal flow at NTP in the converging section which is a mixture of air and solid spherical particles where the air velocity is described by $u_o(1 + x/L)$ for $x \geq 0$ where $u_o = 10$ m/s and $L = 0.5$ m and the air density may be assumed constant at 1.23 kg/m^3. Over this range, the solid particles have an average mass loading of 0.5. The particles have a density of 1,500 kg/m^3 and a bidisperse size distribution, where 60% (by mass) have a diameter of 50 μm and 40% have a diameter of 100 μm. At $x = 0$, the particles can be assumed to be moving at same horizontal speed as the gas but with a vertical speed based on terminal velocity. (a) Determine the average frequency of collisions for all particles between $x = 0$ and $x = L$ based only on terminal velocity differences. (b) Write a code and numerically integrate the horizontal particle velocities for each droplet size (neglecting gravitational effects) to determine the average frequency of collisions for all particles based on average horizontal velocity difference between $x = 0$ and $x = L$ (include a copy of your code). (c) Compare a) and b) results to determine whether the particle terminal velocity or the fluid velocity changes have a bigger effect on the collision rates.

6 Single-Phase Turbulent Flow

The power of turbulence to rapidly mix and transport mass, momentum, and energy is critical to a wide variety of engineering and environmental systems, as discussed in §1.2. This chapter provides an overview of the key elements of single-phase turbulent flow to serve as a basis for multiphase turbulent flow in Chapter 7. In the following, turbulence decomposition and averaging are described in §6.1 along with examples of turbulent flow. Then the time-averaged transport equations for mass, momentum, and species with closure models are given in §6.2, while various approaches to numerically describe turbulent flow are discussed in §6.3. Turbulent time and length scales as well as the kinetic energy cascade are overviewed in §6.4, and then theoretical turbulent species diffusion is treated in §6.5. For more details of single-phase turbulence, the reader is referred to Pope (2000) and Schlichting and Gertsen (2017).

6.1 Time-Averaged Turbulent Flow

As noted in Chapter 2, turbulence is highly three-dimensional, unsteady, and complex with a wide range of length and time scales. It is also an effectively random process with respect to long times, since the detailed instantaneous structures at one time are not generally correlated with those at a much later time. However, as with many chaotic processes, the mean (time-averaged) statistics of turbulent flows are generally reproducible for a given a set of geometric and deterministic flow-boundary conditions.

6.1.1 Time Averaging and Reynolds Decomposition

The most common turbulent statistic is the time average. For a property, q, at a fixed point in space over a time period τ_{avg} starting at time t_o, the time average is

$$\bar{q}(\mathbf{X}, t_o) = \frac{1}{\tau_{avg}} \int_{t_o}^{t_o+\tau_{avg}} q(\mathbf{X}, t)dt. \tag{6.1}$$

The overbar used on the LHS indicates a long time average, which is also termed a *Reynolds average*. Generally, the averaging time period is much longer than the

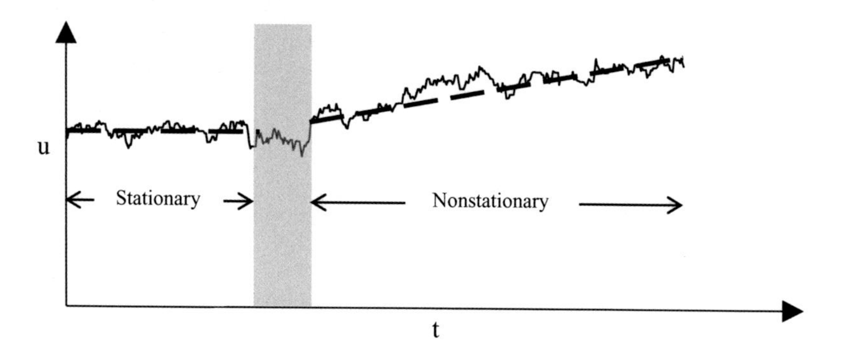

Figure 6.1 Sample time history of instantaneous velocity magnitude (u, shown by solid line) at a fixed point in space for turbulent flow in a pipe as a function of time (t), where the stationary region is based on a constant bulk flow (e.g., the flow rate valve is fixed) and nonstationary region occurs as bulk flow velocity increases (e.g., the valve is slowly opened). Also shown is a rolling average ($\bar{u}$, shown by the dashed line) based on an averaging time.

longest fluid time scale, which is generally the domain time scale (τ_D), yielding the following constraint:

$$\tau_{avg} \gg \tau_D. \tag{6.2}$$

If the resulting time average is independent of the initial time (t_o), then the flow can be regarded as *ergodically stationary* and a function of only space:

$$\bar{q}(\mathbf{X}) = \bar{q}(\mathbf{X}, t_o) \quad \textit{for ergodically stationary turbulence}. \tag{6.3}$$

For example, turbulent flow through a pipe at a constant mass flow rate is ergodically stationary, since the mean velocity at any point will be a fixed value so long as the time averaging is long enough (6.2). If a flow does not meet this criterion, it is designated as nonstationary. Thus, a turbulent pipe flow where the mass flow rate increases over time (e.g., the valve is slowly opened) will be nonstationary, since a time average will depend on when the averaging occurred. This difference between stationary and nonstationary flows is illustrated in Figure 6.1.

For a stationary flow, the fluctuating component is defined as the difference between the instantaneous and mean values of a quantity (where the latter is not a function of time):

$$q'(\mathbf{X}, t) \equiv q(\mathbf{X}, t) - \bar{q}(\mathbf{X}). \tag{6.4}$$

This is sometimes referred to as the Reynolds decomposition. Applying this decomposition to the instantaneous velocity yields mean and fluctuation components expressed in vector or tensor form as follows:

$$\mathbf{u}(\mathbf{X}, t) = \bar{\mathbf{u}}(\mathbf{X}) + \mathbf{u}'(\mathbf{X}, t). \tag{6.5a}$$

$$u_i(X_i, t) = \bar{u}_i(X_i) + u_i'(X_i, t). \tag{6.5b}$$

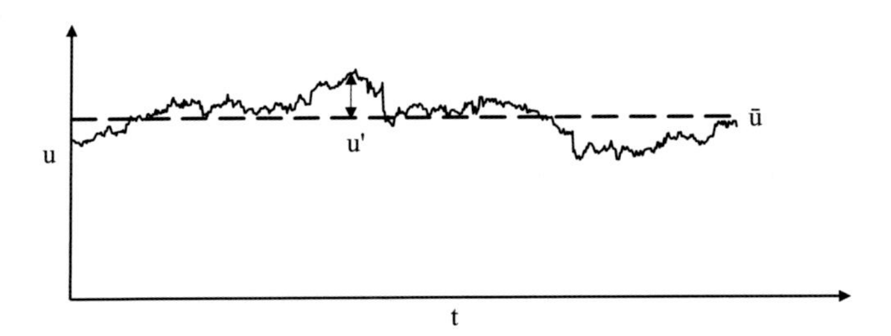

Figure 6.2 Example of instantaneous fluid velocity magnitude (solid line) as a function of time at a fixed location in an ergodically stationary turbulent flow, allowing decomposition into a mean (dashed line) and an instantaneous fluctuation relative to this mean (as indicated by the arrow).

Such a decomposition for a single component of the velocity is illustrated in Figure 6.2 based on measurement at a fixed location.

The decomposition of the streamwise velocity for a complex turbulent flow (between periodic hills) is shown in Figure 6.3, where the instantaneous realization includes complex structures as a function of three dimensions (x, y, z), as shown in Figure 6.3a, while the mean velocity yields a smooth variation in the streamwise and transverse (x-y) plane, as shown in Figure 6.3b, with no variations in the spanwise (z) direction.

6.1.2 Time-Averaged Boundary Layer

One of the most studied turbulent flows with a simple geometry is a boundary layer on a smooth flat wall driven by a uniform steady velocity (u_∞) with zero streamwise pressure gradient. Unlike the laminar flow case, there is no exact solution from the Navier–Stokes equations for the turbulent flow velocity profile. However, the mean velocity profile for a smooth flat plate varies from zero at the wall to the freestream velocity in a repeatable manner for a given Reynolds number and geometry in a thin boundary layer. The boundary-layer thickness (δ) is defined as the distance from the wall where the mean velocity reaches 99% of the freestream velocity, that is, $\bar{u} = 0.99u_\infty$ at $y = \delta$, where y is the wall-normal direction. Experimental results have been used to obtain empirical approximations for the velocity profile and the boundary-layer thickness. For example, the velocity profile for $0.3 < y/\delta < 1.0$ is the outer portion of the flow, and the associated boundary-layer thickness for a fully developed turbulent boundary layer can be empirically approximated with Prandtl's power-law model (White, 2016) using the Reynolds number based on downstream distance (Re_x) as follows:

$$\bar{u}/u_\infty \approx (y/\delta)^{1/7} \qquad \textit{for } y < 0.3\delta. \tag{6.6a}$$

$$\delta \approx \left(0.16\,Re_x^{-1/7}\right)x \qquad \textit{for turbulent flow } \left(\textit{e.g.}, Re_x > 10^7\right). \tag{6.6b}$$

$$Re_x \equiv \rho_f u_\infty x/\mu_f. \tag{6.6c}$$

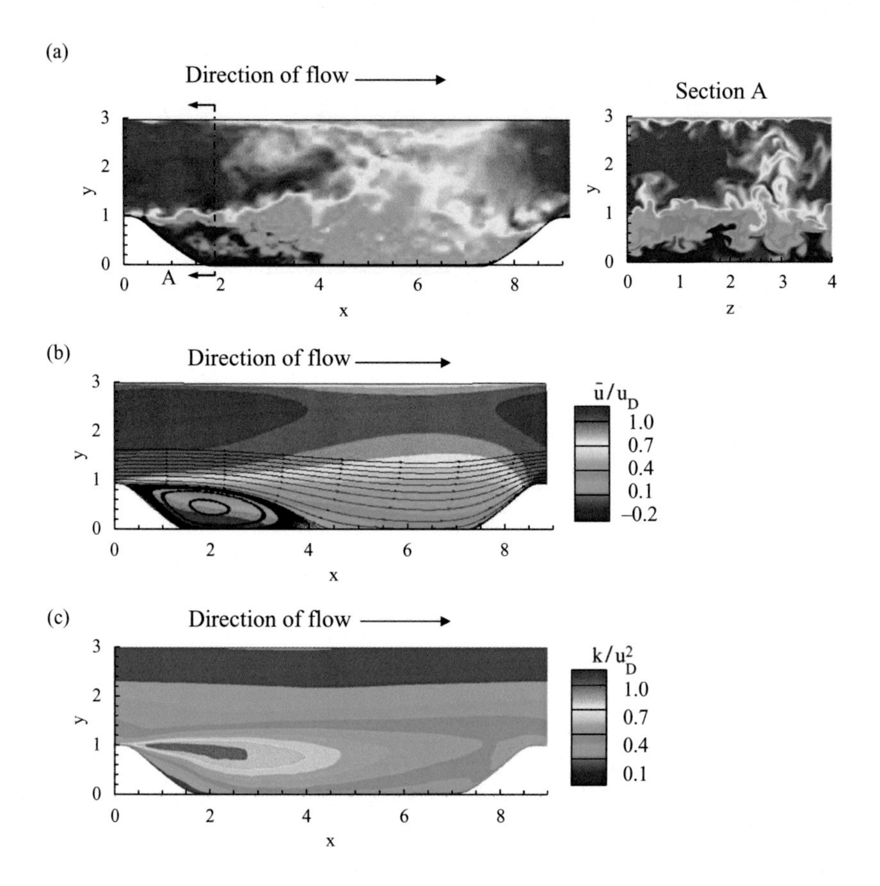

Figure 6.3 Turbulent flow over periodic set of hills with $Re_H = 10{,}595$ based on hill height (H) and inflow velocity (u_D) with flow separation (Balakumar, 2015; Gloerfelt and Cinnella, 2015) showing (a) contours of instantaneous velocity magnitude (red is maximum and blue is minimum) for a streamwise slice at $z/H = 2$ and a spanwise slice at $x/H = 2$; (b) contours of mean velocity magnitude (with average streamlines); and (c) turbulent kinetic energy normalized by inflow velocity.

As shown in Figure 6.4a, this yields a very full velocity profile compared to that of a laminar boundary layer (2.85a). This is because the turbulent case has high-momentum mixing that pulls high-speed fluid closer to the wall. The region associated with $y > 0.3\delta$ is defined as the *outer flow*. As such, the *outer variables* include δ and u_∞, and these can be used to describe an outer time scale as $\tau_\delta = \delta/u_\infty$.

In the other extreme of flow very near the wall, the wall shear stress per (2.20) dominates the turbulent flow velocity profile. The mean shear stress can be expressed in terms of the mean velocity gradient and can be used to define the *wall-friction velocity* (u_{fr}) as follows:

$$\bar{K}_{wall} \equiv \mu_f \left(\frac{\partial \bar{u}}{\partial y}\right)_{wall}. \tag{6.7a}$$

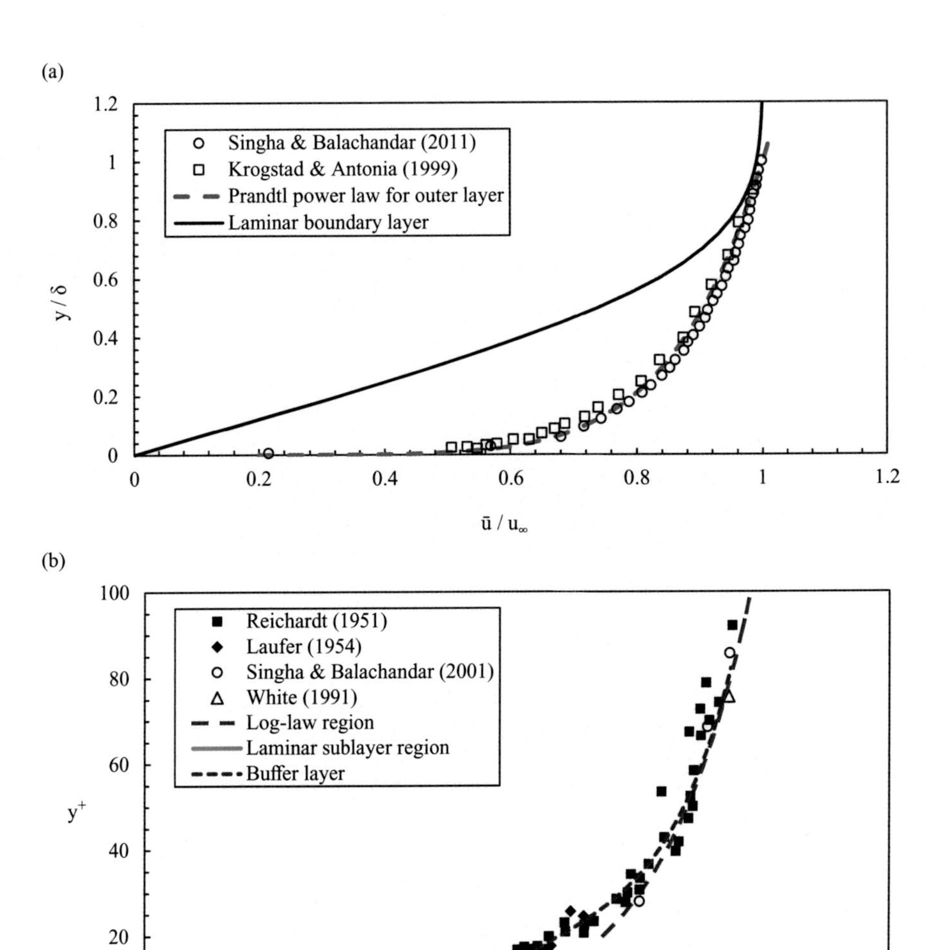

Figure 6.4 A fully developed turbulent boundary-layer mean velocity profile on a flat plate: (a) normalized with boundary-layer thickness and freestream velocity where the slope is steepest near the wall (and much greater than that for a laminar boundary layer); and (b) coordinates normalized with inner region scales (friction velocity and wall unit length scale) where the velocity profile very close the wall is nearly linear, with $u^+ \sim y^+$.

$$u_{fr} \equiv \sqrt{\bar{K}_{wall}/\rho_f}. \tag{6.7b}$$

Prandtl empirically approximated the wall-friction velocity (White, 2016) as follows:

$$u_{fr} \approx \left(0.083\,\text{Re}_x^{-1/14}\right)\bar{u}_\infty. \tag{6.8}$$

The friction velocity can be considered an "inner" variable and can be used to define two more inner variables, the wall-friction length scale (y_{fr}) and time scale (τ_{fr}), as follows:

$$y_{fr} \equiv \frac{\nu_f}{u_{fr}}, \quad \tau_{fr} \equiv \frac{\nu_f}{u_{fr}^2}. \tag{6.9}$$

Using these inner variables, the velocity, distance, and time within the boundary layer can be made dimensionless in terms of wall units, as follows:

$$u^+ \equiv \frac{u}{u_{fr}}, \quad y^+ \equiv \frac{y}{y_{fr}}, \quad t^+ \equiv \frac{t}{\tau_{fr}}. \tag{6.10}$$

Note that this is different from the dimensionless velocity of (6.6a) for outer units.

For the region closest to the wall, the mean shear stress and the velocity gradient are nearly constant. As such, this region is called the *laminar sublayer* and can be approximated in wall units as follows:

$$\bar{u}^+ \approx y^+ \quad \textit{for } y^+ < 5. \tag{6.11}$$

Above this region, the mean velocity follows a "log-law" behavior that depends on the inner scaling and can be empirically modeled with the von Karman constant (White, 2016) as follows:

$$\bar{u}^+ \approx 2.44 \ln y^+ + 5.0 \quad \textit{for } y^+ > 30 \textit{ and } y/\delta < 0.3. \tag{6.12}$$

As shown in Figure 6.4b, the laminar sublayer and log-law regions are reasonably approximated by the aforementioned models for a turbulent boundary layer. The portion between log-law region and the laminar sublayer ($5 < y^+ < 30$) is called the *buffer layer* and can be modeled by interpolating between (6.11) and (6.12).

As may be expected, the inner scales are generally much smaller than the outer scales, that is, $u_\infty \gg u_{fr}$, $\delta \gg y_{fr}$, and $\tau_\delta \gg \tau_f$. For example, a boundary layer of water over a surface with u_∞ of 5 m/s, δ of 40 mm, and τ_δ of 8 ms for the outer variables will have inner variables of u_{fr} of 0.13 m/s, y_{fr} of 7.7 μm, and τ_{fr} of 59 μs. Other simple wall-bounded flows can be similarly described by empirical velocity profiles based on a combination of inner and outer scales. For example, the velocity profile for a circular pipe flow employs inner scales based on the average wall shear stress, and outer scales based on the centerline velocity and pipe radius (White, 2016). However, describing a flow with three-dimensional geometry (e.g., no longer a simple flat plate) requires numerical solution of turbulent transport, as discussed in §6.2 and §6.3.

6.1.3 Fluctuation Averaging and Turbulent Kinetic Energy

In order to obtain the time-averaged equations, it is helpful to consider some additional characteristics of time averaging for an ergodically stationary flow (6.3), especially in terms of the fluid fluctuations. First, we note that the time average of a mean value is the mean itself, while the time average of the fluctuation is zero, as follows:

$$\bar{\bar{q}} = \bar{q}. \tag{6.13a}$$

$$\overline{q'} = 0. \tag{6.13b}$$

Applying (6.13b) to the velocity in vector, tensor, and Cartesian notations yields the following:

$$\overline{\mathbf{u}'} = \overline{u_i'} \equiv 0. \tag{6.14a}$$

$$\overline{u_x'} = \overline{u_y'} = \overline{u_z'} \equiv 0. \tag{6.14b}$$

While the average of the fluctuations is zero, we note that the average of the square of an unsteady quantity includes contributions from both the mean and fluctuating quantities, as follows:

$$\overline{qq} = \bar{q}\,\bar{q} + \overline{q'q'} = \bar{q}\,\bar{q} + \left(q'_{rms}\right)^2. \tag{6.15}$$

The RHS introduces the "rms" subscript as the root mean square of an arbitrary fluctuating quantity q', which is formally defined as follows:

$$q'_{rms}(\mathbf{X}) \equiv \left[\frac{1}{\tau_{avg}} \int_{t_o}^{t_o+\tau_{avg}} [q'(\mathbf{X},t)]^2 dt\right]^{1/2}. \tag{6.16}$$

The rms (also called the standard deviation) quantifies the strength of the fluctuations about the mean. As an example, Figure 6.5a shows a snapshot of the velocity variations in a turbulent boundary layer, and the rms of the velocity fluctuations in the x-, y-, and z-directions are shown in Figure 6.5b. In general, these fluctuation strengths are small compared to the freestream velocity and vary strongly with distance from the wall as well as in terms of the velocity direction.

In addition, the sum of fluctuation strengths in all three directions defines the turbulent kinetic energy per unit mass of fluid in component or tensor form, as follows:

$$k(\mathbf{X}) = \frac{1}{2}\left[\overline{u_x'u_x'} + \overline{u_y'u_y'} + \overline{u_z'u_z'}\right] = \frac{1}{2}\overline{u_i'u_i'}. \tag{6.17}$$

For conciseness, k is often referred to as the *turbulent kinetic energy* (TKE). The TKE is generally highest where the velocity gradients are greatest. For example, the highest TKE is seen in the separated free-shear layer in Figure 6.3c and is seen close to the wall for a boundary layer in Figure 6.5b. The kinetic energy can also be used to define the turbulent strength (u_Λ) as follows:

$$u_\Lambda \equiv u'_{rms} \equiv \sqrt{\frac{2}{3}k} = \sqrt{\frac{1}{3}\left[\overline{u_x'u_x'} + \overline{u_y'u_y'} + \overline{u_z'u_z'}\right]}. \tag{6.18}$$

This velocity represents the average velocity fluctuation strength.

Next we consider two different flow variables (q_1 and q_2), whereby the time average of their sum and the time average of their product can be expressed as follows:

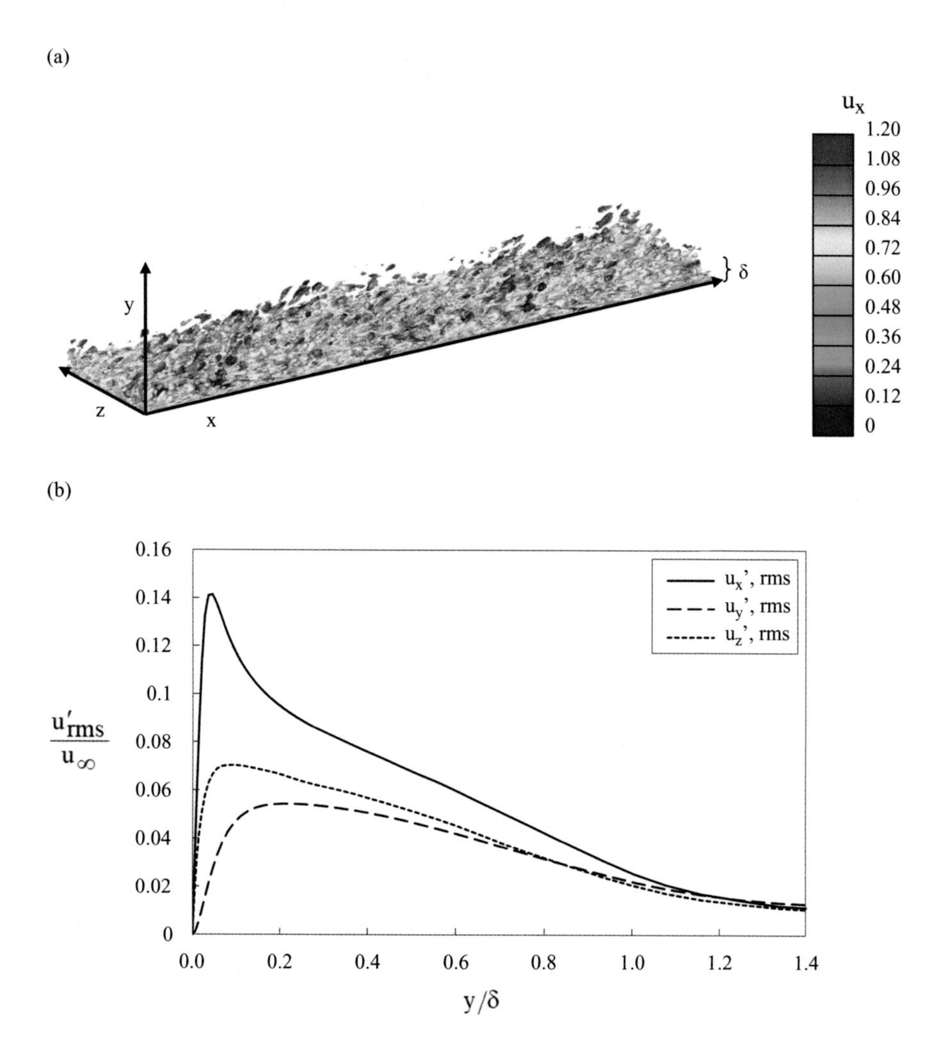

Figure 6.5 Turbulent boundary layer on a flat plate: (a) isosurfaces of vortex structures colored by the magnitude of the streamwise velocity (Arolla and Durbin, 2014); and (b) velocity fluctuation distributions in the turbulent boundary layer at $Re_\delta = 4{,}500$ (Dorgan and Loth, 2004), where y-direction is normal to the wall, and δ is the boundary-layer thickness.

$$\overline{q_1 + q_2} = \bar{q}_1 + \bar{q}_2. \tag{6.19a}$$

$$\overline{q_1 q_2} = \bar{q}_1 \bar{q}_2 + \overline{q'_1 q'_2}. \tag{6.19b}$$

The last term on the RHS of (6.19b) is correlation term and will be zero only if the fluctuations are independent of each other (e.g., if they are separate random processes). As will be seen in the next section, fluctuation correlations can play a major role in turbulent transport.

Finally, we note a few properties of the time and spatial derivatives:

$$\overline{\frac{\partial q}{\partial t}} = \frac{\partial \bar{q}}{\partial t} = 0 \quad since \quad \overline{\frac{\partial q'}{\partial t}} = 0. \tag{6.20a}$$

$$\overline{\frac{\partial q}{\partial x}} = \frac{\partial \bar{q}}{\partial x} \quad since \quad \overline{\frac{\partial q'}{\partial x}} = 0. \tag{6.20b}$$

These properties are helpful when applying time averaging to the transport PDEs.

6.2 Turbulent Transport and Closure Equations (RANS)

6.2.1 Incompressible Time-Averaged Transport Equations

As discussed in §2.6, the unsteady three-dimensional Navier–Stokes PDEs fully describe turbulent flow (2.86). Unfortunately, these equations do not have an analytical solution due to the complexity of turbulence. A fully-resolved numerical solution is possible, but the wide range of length and time scales in turbulence at high Reynolds numbers requires extremely large computational requirements so that solutions are only feasible for modest Reynolds numbers. Therefore, a less computationally intensive time-averaged approach is much more common.

Time-Averaged Mass Transport

For the time-averaged conservation of mass for incompressible flow, we start with the continuity equation, which can be written in vector form (2.72a) or tensor form (2.73a) for the instantaneous velocity as follows:

$$\nabla \cdot \mathbf{u} = \frac{\partial u_i}{\partial X_i} = 0. \tag{6.21}$$

The tensor form includes a summation based on the index $i = 1, 2, 3$. Substituting (6.5) into (6.21) and time averaging yields the mean continuity equation, and subtracting this result from (6.21) yields the fluctuating continuity equation:

$$\nabla \cdot \bar{\mathbf{u}} = 0. \tag{6.22a}$$

$$\nabla \cdot \mathbf{u}' = 0. \tag{6.22b}$$

These PDEs demonstrate that both the mean and fluctuating velocity fields are divergence-free for incompressible flow.

Time-Averaged Momentum Transport

The unsteady incompressible momentum transport given by (2.73b) is known as the Navier–Stokes equation. As in (6.5b) for the velocity, the pressure can be decomposed as follows:

$$p(X_i, t) = \bar{p}(X_i) + p'(X_i, t). \tag{6.23}$$

Substituting the velocity and pressure decompositions into the Navier–Stokes equation, time averaging, and applying (6.13), (6.14), (6.19), and (6.20) yields the tensor-based PDE for the i-direction momentum (with a summation of j = 1, 2, and 3) as follows:

$$\underbrace{\rho_f \frac{\partial \bar{u}_i}{\partial t}}_{\substack{\text{Mean accleration} \\ =0 \text{ for stationary}}} + \underbrace{\rho_f \bar{u}_j \frac{\partial \bar{u}_i}{\partial X_j}}_{\text{Mean convection}} = \underbrace{\rho_f g_i}_{\substack{\text{Gravity} \\ \text{forces}}} - \underbrace{\frac{\partial \bar{p}}{\partial X_i}}_{\substack{\text{Pressure} \\ \text{gradient}}} + \underbrace{\frac{\partial}{\partial X_j}\mu_f\left(\frac{\partial \bar{u}_i}{\partial X_j} + \frac{\partial \bar{u}_j}{\partial X_i}\right)}_{\text{Mean viscous stress}} - \frac{\partial}{\partial X_j}\underbrace{\left[\rho_f \overline{u_i' u_j'}\right]}_{\text{Turbulent stress}}. \tag{6.24a}$$

$$\frac{\partial \bar{u}_i}{\partial t} = 0 \quad \textit{for stationary turbulence}. \tag{6.24b}$$

Since time averaging of flow variables is also known as Reynolds averaging, (6.24a) is often referred to as the Reynolds-Averaged Navier–Stokes (RANS) equation.

It is helpful to identify the individual terms of the momentum PDE of (6.24a). The first term on the LHS is the acceleration of the mean velocity. This term should be retained for a flow that is not ergodically stationary. For stationary turbulence, this term is zero (6.24b), but is generally retained to allow numerical solution with a time-marching scheme (whereby the equations are integrated in time until the unsteady term become negligible, indicating the flow field has converged to the stationary solution).

The second term on the LHS is the mean convection term, which transports fluid momentum with the time-averaged velocity. The first term on the RHS reflects the gravitational forces on a fluid element (unchanged from the original unsteady equation). The next two terms are the mean pressure and viscous stresses (where the average fluctuating stresses are zero). The last term includes a velocity correlation due to averaging the nonlinear convection (per 6.19b). This correlation term has units of force per unit area and so is generally referred to as the *turbulent stress* or the *Reynolds stress*. The sum of the diagonal elements of this velocity tensor (i = j) is twice the turbulent kinetic energy of (6.17). The off-diagonal velocity components (i ≠ j) are also important to provide turbulent momentum transport, as will be discussed in §6.2.2.

Time-Averaged Species Transport

For the mass fraction of a species ($\mathcal{M}$) in an incompressible flow (with no chemical reactions), the unsteady transport PDE is governed by (2.33b). As with the mass and momentum equation, the species is decomposed to mean and fluctuating quantities and then each term in the PDE is time averaged. The resulting mean species transport PDE can then be expressed as follows:

$$\underbrace{\frac{\partial \bar{\mathcal{M}}}{\partial t}}_{\substack{=0 \text{ for} \\ \text{stationary}}} + \underbrace{\frac{\partial(\bar{\mathcal{M}}\,\bar{u}_i)}{\partial X_i}}_{\text{Mean convection}} = \underbrace{\frac{\partial}{\partial X_i}\left(\Theta_{\mathcal{M}}\frac{\partial \bar{\mathcal{M}}}{\partial X_i}\right)}_{\text{Molecular diffusion}} \underbrace{-\frac{\partial\left(\overline{\mathcal{M}' u_i'}\right)}{\partial X_i}}_{\text{Turbulent diffusion}} \quad \textit{for } i = 1, 2, \textit{and } 3. \tag{6.25a}$$

$$\frac{\partial \bar{\mathcal{M}}}{\partial t} = 0 \qquad \textit{for stationary flows.} \tag{6.25b}$$

It is helpful to identify the individual terms of the mean transport PDE (6.25a). As with the RANS equation, the first term on the LHS is zero, but is often retained to allow for a flow that is not ergodically stationary or to help with numerical solution. The second term on the LHS is the mean convection of the species by the mean velocity. The first term on the RHS is the species molecular diffusion due to mean species gradients and the species molecular diffusivity. The second term on the RHS is the species turbulent diffusion due to a correlation of the species and velocity fluctuations, which results from averaging the nonlinear convection. This term requires an empirical closure model, as does the last term on the RHS of (6.24a).

The preceding transport equations can be extended to describe compressible flow by allowing variable density and adding an energy transport PDE (Schlichting and Gersten, 2017; Pope, 2000). In the next sections, the physics of turbulent momentum diffusion is discussed (§6.2.2), followed by its closure models (§6.2.3), and then turbulent species diffusion (§6.2.4).

6.2.2 Turbulent Momentum Diffusion Principles

Turbulent Diffusion Mechanisms and Principles

To understand how the velocity correlation $\overline{u_i'u_j'}$ of (6.24a) is related to turbulent diffusion, consider the turbulent boundary layer of Figure 6.6, which has a mean flow in the x-direction and a velocity profile that increases in the y-direction. Based on the mean velocity, the mean momentum transport is only in the x-direction. However, the velocity fluctuations cause an additional unsteady transport. In particular, consider the transport across an imaginary horizontal transport interface (a plane for which y is a constant) where the mean flow above is at a higher speed than the mean flow below. On both sides of the interface, the y-velocity is fluctuating, with both positive and negative value events

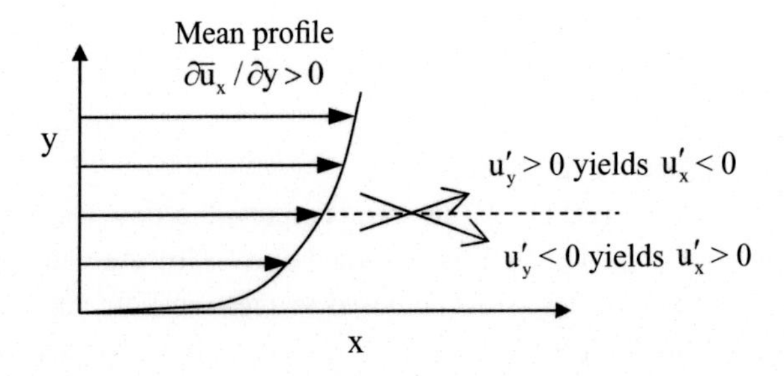

Figure 6.6 A turbulent boundary layer in the streamwise x-direction with increasing velocity in the y-direction. The dashed line is an imaginary transport interface, where fluid below this line tends to have a lower streamwise velocity relative to the fluid above this line. In this case, an upward turbulent fluctuation transports this low-speed fluid above this line, while a downward turbulent fluctuation will move high-speed fluid below this line. Thus, both transport processes result in correlations of $\overline{u_x'u_y'} < 0$ for a positive mean velocity gradient.

occurring: $u'_y > 0$ and $u'_y < 0$. If this fluctuation is positive, fluid below this interface will be transported upward across this plane. Since this fluid arises from the lower speed region, this transported fluid is likely to have a lower streamwise velocity than the fluid above the plane, such that $u'_x < 0$ is likely to occur after this event. This will be especially true as the velocity gradient increases. If one instead considers a downward velocity with $u'_y < 0$, which moves fluid downward across this interface from the high-speed side, this will likely to lead to $u'_x > 0$ relative to the lower region. In both cases, the sign of the x-direction velocity fluctuation tends to be opposite of that for the y-velocity. Therefore, the average of many such events will generally yield $\overline{u'_x u'_y} < 0$ for a postive mean velocity gradient. In this case, momentum is transported downward by turbulence.

If we consider a different flow for which flow speed decreases as y increases, the correlation would tend to be postive since $u'_y > 0$ likely leads to $u'_x > 0$. Therefore, the sign of the velocity gradient (which is negative in this case) is again the opposite of the sign of the correlation. As such, turbulence moves low-speed fluid to high-speed regions (and vice versa) via velocity fluctuations. This momentum mixing causes turbulent boundary layers to have a fuller velocity profile than laminar boundary layers (Figure 6.4a). Finally, if we consider a flow with no mean velocity gradient or no turbulence, the velocity correlation would be zero, and no momentum transport would occur.

Based on the preceding and Figure 6.6, the turbulent transport of momentum is governed by three important principles:

(1) The mixing increases as the velocity gradient increases since the gradient yields a velocity difference (and the mixing goes to zero if there is no velocity gradient).
(2) The direction of turbulent momentum transport is opposite to the direction of the mean velocity gradient since a local increase in fluid momentum stems from fluid in faster regions moving into slower regions.
(3) The magnitude of this mixing increases as the lateral velocity fluctuations increase since they are needed to move momentum laterally (and this mixing goes to zero if there are no fluctuations, as in laminar flow).

These three principles can be summarized in terms of the velocity correlation as follows:

Principle 1: $\overline{u'_i u'_j}$ is proportional to the magnitude of the mean velocity gradient.
Principle 2: $\overline{u'_i u'_j}$ has a sign opposite to that of the mean velocity gradient.
Principle 3: $\overline{u'_i u'_j}$ is proportional to the strength of the turbulent fluctuations.

Finally, it should be noted that the correlation with $i = j$ does not lead to turbulent transport.

Turbulent Diffusion Model

The preceding principles are also the foundation for the commonly used Boussinesq (1997) approximation, for which the Reynolds stress due to mixing is related to the mean velocity gradient and turbulent viscosity (μ_{turb}), as follows:

$$\rho_f \overline{u'_i u'_j} \approx -\mu_{turb}\left(\frac{\partial \bar{u}_i}{\partial X_j} + \frac{\partial \bar{u}_j}{\partial X_i}\right) \qquad \textit{for} \quad i \neq j. \tag{6.26}$$

Principle 1 is reflected by the mean velocity gradient on the RHS, while Principle 2 is satisfied by the negative sign on the RHS. It should be noted that the special case of $i = j$ is explicitly excluded in (6.26), since turbulent mixing stems from lateral transport of streamwise momentum. Principle 3 can be incorporated by making the turbulent viscosity proportional to the turbulent kinetic energy, as will be discussed in §6.2.3.

It is important to note that the Boussinesq model given by (6.26) is based only on qualitative physical reasoning and *not* based on a theoretical or exact solution of the Navier–Stokes equations. Thus, the Boussinesq approximation is empirical and is not necessarily quantitatively correct. In fact, Tennekes and Lumley (1972) referred to this concept as the "gradient-transport fallacy," because it has no theoretical basis and can be significantly inaccurate in many turbulent flows. Despite these shortcomings, this simple approximation is the most common model for the turbulent stress velocity correlation of (6.24a) since no other empirical approximations have been shown to be fundamentally superior. As such, the widely used Boussinesq approximation can be viewed as a necessary (empirical) evil for solving the mean momentum transport (if the solution to the full three-dimensional unsteady flow equations is not available).

If one employs the Boussinesq approximation of (6.26), the generalized turbulent stress tensor can be defined and approximated as follows:

$$\overline{K'_{ij}} \equiv -\rho_f \overline{u'_i u'_j} \approx \mu_{turb}\left(\frac{\partial \bar{u}_i}{\partial X_j} + \frac{\partial \bar{u}_j}{\partial X_i}\right)_{i \neq j} - \frac{2}{3}\rho_f k \delta_{ij}. \tag{6.27}$$

The special case of $i = j$ is handled by the Kronecker delta (2.15b) so that the result is consistent with the definition of turbulent kinetic energy (6.17). Substituting this stress tensor into (6.24a), combining similar terms, and applying the assumption of incompressibility yields the momentum transport equation in tensor form as follows:

$$\rho_f \frac{\partial \bar{u}_i}{\partial t} + \rho_f \frac{\partial \bar{u}_i \bar{u}_j}{\partial X_j} = -\frac{\partial \bar{p}}{\partial X_i} + \rho_f g_i + \frac{\partial}{\partial X_j}\left[(\mu_f + \mu_{turb})\left(\frac{\partial \bar{u}_i}{\partial X_j} + \frac{\partial \bar{u}_j}{\partial X_i}\right) - \frac{2}{3}\rho_f k \delta_{ij}\right]. \tag{6.28}$$

The RHS shows that both laminar and turbulent diffusion of momentum depend on the mean velocity gradients (due to the Boussinesq assumption), where the laminar case has a proportionality based on fluid (laminar) viscosity, while the proportionality between the mean velocity and turbulent stress of (6.26) is defined as the turbulent viscosity (μ_{turb}). In order to solve this transport equation, a model for the turbulent viscosity as a function of turbulence kinetic energy (based on Principle 3) is discussed in the next section.

6.2.3 Models for Turbulent Viscosity and Momentum Closure

Mixing-Length Model

There have been a wide variety of models for turbulent viscosity that have been developed over the last 50 years to "close" the RANS equation. The simplest of these is the algebraic formula based on Prandtl's *mixing-length model*. This model assumes that the μ_{turb} is linearly proportional to the local velocity gradients and the square of a mixing length (Λ_{mix}), which follows from dimensional analysis. For a 2D flow, the mixing-length model for turbulent viscosity and turbulent kinetic energy can be expressed as follows:

$$\mu_{turb} = \rho_f \nu_{turb} = \rho_f \Lambda_{mix}^2 \left| \frac{\partial \bar{u}_x}{\partial y} - \frac{\partial \bar{u}_y}{\partial x} \right|. \tag{6.29a}$$

$$k = \left(\frac{\nu_{turb}}{c_{mix} \Lambda_{mix}} \right)^2. \tag{6.29b}$$

Note that Principle 3 requires that the turbulent viscosity be proportional to the turbulent velocity fluctuations. This dependency is implicitly integrated in (6.29a, since mean velocity gradients lead to instabilities and turbulence (§2.7) that have high-velocity fluctuations (e.g., the near-wall region of Figure 6.5). This is explicitly shown by (6.29b), where ν_{turb} increases as k increases for a given mixing length.

To complete the mixing-length model of μ_{turb} for (6.28), a model of Λ_{mix} is needed. In a free-shear flow or a jet, this length is typically approximated as a fraction of the jet or shear-layer thickness. For a boundary layer, the outer region mixing length is often modeled as 9% of the boundary-layer thickness, but reduces in the near-wall region which can be empirically described as a function of the wall-normal distance with the damping factor of Van Driest (1956):

$$\Lambda_{mix} = 0.09\delta \qquad \textit{for } y/\delta > 0.22. \tag{6.30a}$$

$$\Lambda_{mix} = 0.41y[1 - \exp(-y^+/25)] \qquad \textit{for } y/\delta < 0.22. \tag{6.30b}$$

While this algebraic mixing-length approach for turbulent viscosity is reasonable for simple attached boundary layers and simple free-shear flows, it often provides poor performance for more complex flows and therefore is not commonly used.

One- and Two-Equation Models

To improve on the mixing-length models, most RANS turbulence models instead employ one or more transport PDEs of the turbulent statistics. The most popular of these are the *one-equation* and *two-equation* turbulence models, which refer to the use of either one or two transport PDEs to obtain μ_{turb}. The most common one-equation formulation is the Spalart–Allmaras method (1994), which considers a transport PDE for the turbulent viscosity itself. The most common two-equation formulations include the k-ε model (good for simple free-shear flows); the k-ω model (good for simple boundary-layer flows); and the Shear-Stress Transport (SST) model, which blends the k-ω approach close to the wall with the k-ε far from the wall (Menter, 1994). In general, these two-equation models are very similar in framework but use different coefficients and algebraic relations to achieve closure (Chung, 2002). In the following, we discuss the details of the k-ε approach (Launder and Spalding, 1972), since this model is relatively straightforward and since its basic characteristics are shared by the other models.

The k-ε approach employs transport PDEs for the turbulent kinetic energy and its dissipation. The incompressible transport PDE for the TKE can be obtained by assuming stationary turbulence with the momentum equation of (2.73b). In particular, if this entire PDE is denoted as q_i, the PDE for the turbulent kinetic energy can be obtained by employing Reynolds decompositions for all the values, applying $\overline{u_i q_i} - \bar{u}_i \bar{q}_i$, and then taking a time average of the result. After some manipulation, the resulting transport equation for k is as follows:

$$\underbrace{\frac{\partial k}{\partial t}}_{=0} + \underbrace{\bar{u}_i \frac{\partial k}{\partial X_i}}_{\text{Mean convection}} = \underbrace{-\overline{u_i' u_j'} \frac{\partial \bar{u}_j}{\partial X_i}}_{\text{Turbulence production}} + \underbrace{\frac{\partial}{\partial X_i}\left[\nu_f \frac{\partial k}{\partial X_i} - \frac{1}{2}\overline{u_i' u_j' u_j'} - \frac{1}{\rho_f}\overline{u_i' p'}\right]}_{\text{Viscous diffusion, Turbulent diffusion, Pressure diffusion}} - \underbrace{\nu_f \overline{\frac{\partial u_i'}{\partial X_j}\frac{\partial u_i'}{\partial X_j}}}_{\text{Turbulent dissipation}}. \tag{6.31}$$

Note that this PDE does not include any empiricism and employs two tensor summations ($i = 1, 2, 3$ and $j = 1, 2, 3$). The LHS includes the unsteady term (which is zero for a stationary flow) and a mean flow convection term. The first term on the RHS term is TKE production, which arises in regions of mean shear and is overall positive (since the product of the velocity correlation and the gradients is negative). The second term of the RHS is the diffusion (spreading) of the kinetic energy due to viscous diffusion, turbulent diffusion, and pressure diffusion. The last RHS term is the turbulent dissipation, which causes decay of the turbulent kinetic energy and is based on a correlation of the fluctuating velocity gradients. While not included in (6.31), some formulations also include an additional viscous dissipation associated with the mean velocity gradients (as occurs in laminar flow). However, this extra term is generally small compared to turbulent dissipation and so is generally neglected.

Since the turbulent kinetic energy is a fluid property, it can also be considered in terms of the Reynolds transport theorem of (2.3), as follows:

$$\underbrace{\frac{\partial k}{\partial t}}_{\substack{=0 \text{ for} \\ \text{stationary}}} + \underbrace{\nabla \cdot (k\bar{\mathbf{u}})}_{\substack{\text{Convection} \\ \text{of TKE}}} + \underbrace{\nabla \cdot (\mathbf{Q}_k)}_{\substack{\text{Diffusion} \\ \text{of TKE}}} = \underbrace{\dot{k}}_{\substack{\text{Generation} \\ \text{of TKE}}} = \underbrace{\dot{k}_{prod}}_{\substack{\text{Production} \\ \text{of TKE}}} + \underbrace{\dot{k}_{diss}}_{\substack{\text{Dissipation} \\ \text{of TKE}}}. \tag{6.32}$$

The LHS again includes an Eulerian time derivative (which will be zero for an ergodically stationary flow), the convection of turbulent energy based on the mean velocity, and the kinetic energy diffusion due to velocity fluctuations. The RHS includes the generation of turbulent kinetic energy divided into a production term and a dissipation term. The production is a result of velocity gradients, as noted in (6.31), since these cause flow instabilities (§2.6 and §2.7). The dissipation term is a result of viscosity, which converts the kinetic energy of local velocity fluctuation gradients (RHS of 6.31) into thermal energy. The magnitude of this dissipation term is formally defined as the turbulent energy dissipation (ε). With 6.31 and 6.32, this term is given as follows:

$$\varepsilon \equiv -\dot{k}_{diss} = \nu_f \overline{\frac{\partial u'_i}{\partial X_j} \frac{\partial u'_i}{\partial X_j}}. \tag{6.33}$$

This definition uses a negative sign on the RHS such that ε is always positive.

Note that the TKE transport PDE of (6.31) does not itself close the equations since it introduces additional correlations that require further modeling. To achieve this, the velocity correlation in the production terms can be modeled with the Boussinesq approximation (6.27). The triple velocity correlation in the turbulent diffusion term can be treated with a Fickian model of the Reynolds stress. The pressure diffusion term can be reasonably neglected for incompressible flow. Employing these assumptions, the resulting TKE transport PDE (with the time derivative removed) can be expressed as follows (Shyy et al., 1997):

$$\bar{u}_i \frac{\partial k}{\partial X_i} = \frac{\overline{K'_{ij}}}{\rho_f} \frac{\partial \bar{u}_j}{\partial X_i} + \frac{\partial}{\partial X_i}\left[(\nu_f + \nu_{turb}) \frac{\partial k}{\partial X_i}\right] - \varepsilon \qquad \textit{for}\, j = 1, 2, \textit{and}\, 3. \tag{6.34}$$

This result still requires a model for the turbulent viscosity (as was also needed for the mixing-length model). The k-ε approach next models the turbulent dynamic and kinetic viscosities as follows:

$$\frac{\mu_{turb}}{\rho_f} \equiv \nu_{turb} = \frac{c_\mu k^2}{\varepsilon}. \tag{6.35}$$

The exponents used for k and ε on the RHS are obtained by dimensional analysis, while the nondimensional constant in this model (c_μ) is an empirical parameter. This parameter is typically set as $c_\mu = 0.09$, as this has been found to gives the best predictions of TKE and mean fluid transport for simple free-shear turbulent flows (such as jets, shear layers, and wakes). The model of (6.35) is also qualitatively consistent with Principle 3, whereby the turbulent viscosity increases with the turbulent kinetic energy. The model's inverse dependence on dissipation is qualitatively consistent with a reduction in mixing if TKE is being lost.

However, both (6.34) and (6.35) require knowledge of ε for their solution. This can be obtained by formulating a transport PDE for turbulent dissipation by taking the time derivative of the turbulent kinetic energy PDE, employing Reynolds decompositions for all terms, and then applying a time average with some additional modeling yielding (Shyy et al., 1997)

$$\bar{u}_i \frac{\partial \varepsilon}{\partial X_i} = \frac{1.44\varepsilon}{\rho_f k} \overline{K'_{ij}} \frac{\partial \bar{u}_j}{\partial X_i} + \frac{\partial}{\partial X_i}\left[\left(\nu_f + \frac{\nu_{turb}}{1.30}\right) \frac{\partial \varepsilon}{\partial X_i}\right] - 1.92 \frac{\varepsilon^2}{k}. \tag{6.36}$$

The first term on the RHS is the *production of dissipation*, the second is the *turbulent diffusion of dissipation*, whereas the last term is the *dissipation of dissipation*. These terms are handled through the use of three empirical constants (which include 1.44, 1.30, and 1.92) to finally close the equations.

In summary, the ergodically stationary time-averaged mass and momentum PDEs (6.22a and 6.28) and the turbulent viscosity model (6.35) in tensor form are given as follows:

$$\frac{\partial \bar{u}_i}{\partial X_i} = 0. \tag{6.37a}$$

$$\rho_f \frac{\partial \bar{u}_i \bar{u}_j}{\partial X_j} = -\frac{\partial \bar{p}}{\partial X_i} + \rho_f g_i + \frac{\partial}{\partial X_j}\left[(\mu_f + \mu_{turb})\left(\frac{\partial \bar{u}_i}{\partial X_j} + \frac{\partial \bar{u}_j}{\partial X_i}\right) - \frac{2}{3}\rho_f k \delta_{ij}\right]. \tag{6.37b}$$

$$\mu_{turb} = 0.09 \frac{\rho_f k^2}{\varepsilon}. \tag{6.37c}$$

These equations combined with the PDEs of (6.34) and (6.36) represent a complete set of equations for the k-ε treatment of turbulent flow.

Notably, the laminar and turbulent viscosities both appear in the diffusion terms of (6.34), (6.36), and (6.37b), but their relative importance can vary significantly. In most regions of the flow, turbulence will dominate this diffusion, that is, $\mu_{turb} \gg \mu_f$, so it is reasonable to ignore laminar viscosity. However, regions very close to a wall ($y^+ < 5$) will have very low turbulence such that fluid viscosity will be more important ($\mu_f \gg \mu_{turb}$) and will importantly dictate the wall-shear stress (6.7a). Thus, both laminar and turbulent viscosities must be generally retained for wall flows.

RANS Predictive Performance

A key issue of the k-ε turbulence modeling (and other two-equation models) is the inclusion of several empirical constants (which include 1.44, 1.30, 1.92, and 0.09). These constants have been historically obtained by "tuning," that is, by optimizing their values so that predictions best match experimental data for a select set of validation test cases. These cases have included round and planar jets, grid-generated wakes, and attached boundary layers so these cases can be reasonably predicted. However, turbulence modeling cannot generally predict flows that differ significantly from the validation cases with the same level of fidelity. As such, turbulence models generally have reduced accuracy for complex flows, such as separated flows over curved surfaces, as shown in Figure 6.7a. The mean velocity field profiles for this flow are shown in Figure 6.7b, where the RANS comparison with experiment is only qualitative at x/H = 2 (within the separation bubble) and at x/H = 6 (just downstream of the separation bubble). The k-ε performance is similarly problematic when predicting the velocity correlation, as shown in Figure 6.7c.

Other RANS approaches have been developed to improve the fidelity beyond the two-equation models. For example, the Reynolds stress model (RSM) employs six transport PDEs, for each unique component of the Reynolds stress tensor $\overline{u_i' u_j'}$ (Wilcox, 2006). However, the increase in the number of PDEs results in proportionally greater computational cost for solution, several more empirical constants to be set, and not necessarily an increase in accuracy (e.g., Figure 6.7). As such, these more complex turbulence models are not commonly adopted. The shortcoming of all such time-averaged (RANS) approaches is that they do not resolve any of the turbulent structures, and this lack of fluid physics limits their performance fidelity.

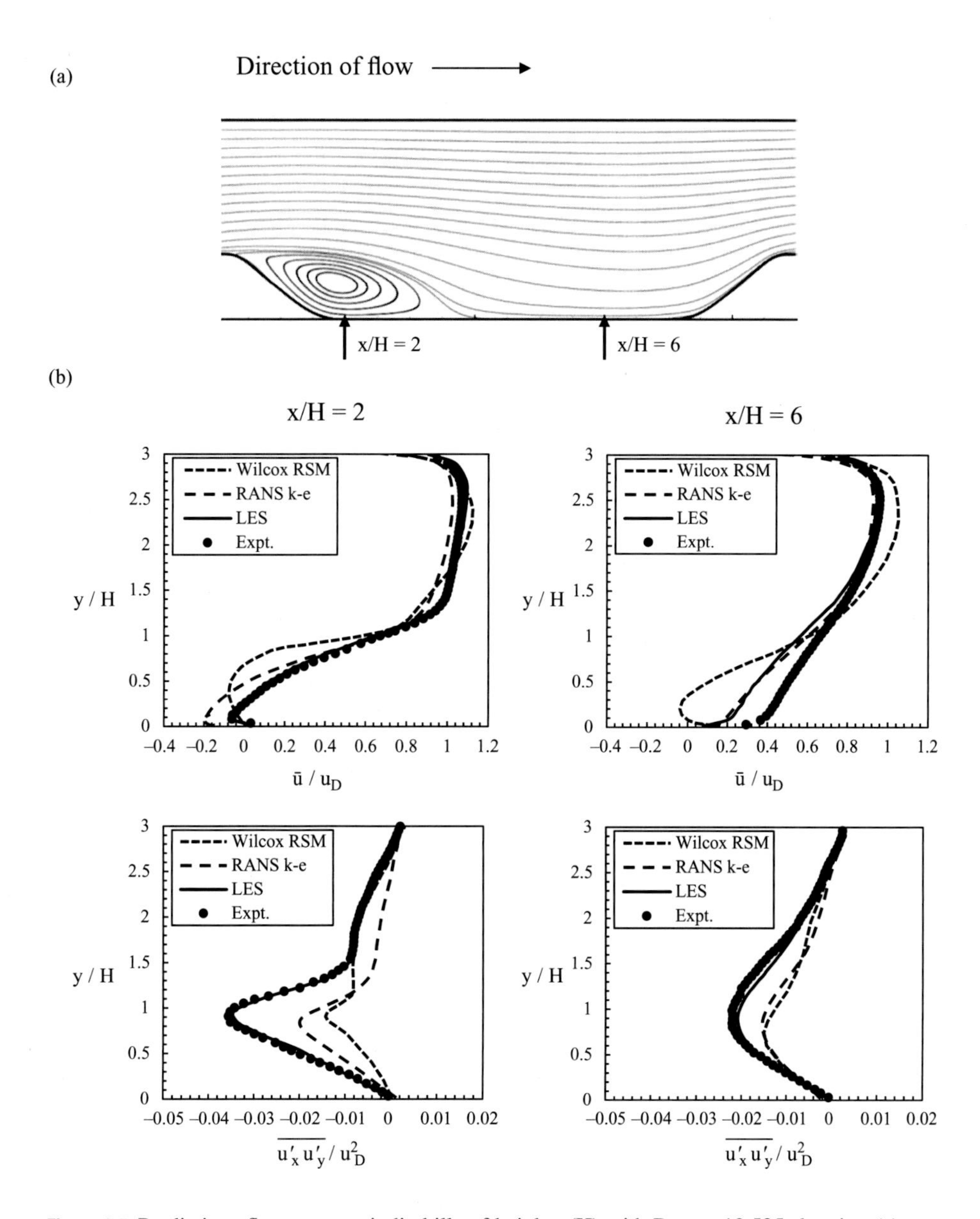

Figure 6.7 Predictions flow over periodic hills of heights (H) with $Re_H = 10{,}595$ showing (a) streamlines for time-averaged flow; and (b) profiles of mean streamwise velocity and turbulent Reynolds stress at x/H = 2 and x/H = 6 for RANS with a k-ε model and a Wilcox ω-based Reynolds stress model, large eddy simulation (LES), and experimental data (Breuer et al., 2009; Maduta and Jakirlic, 2012).

Approaches to improve accuracy by incorporating some or all of the three-dimensional unsteady structures are discussed later in §6.3. However, the preceding turbulence model limitations for the time-averaged momentum transport should be kept in mind for the following discussion of time-averaged species transport.

6.2.4 Turbulent Species Diffusion: Physics and Models

Turbulent Species Diffusion Mechanisms and Principles

As described in §6.2.1, the transport equation for the mean species fraction (6.25a) includes a second-order correlation based on velocity and species fraction fluctuations, $\overline{\mathcal{M}'u_i'}$. To show how this correlation is related to the species turbulent diffusion, consider a simple channel flow with uniform mean velocity in the x-direction with an average mixture of two fluids (one is blue, and one is white), as shown in Figure 6.8a. In a time-averaged sense, there is a positive gradient of the blue fluid mean species in the y-direction with $\partial\bar{\mathcal{M}}/\partial y > 0$. However, we assume that the two fluids are immiscible (no molecular mixing) so the instantaneous blue mass fraction is either 0% or 100% at all points, as shown in Figure 6.8b.

For this flow, vertical velocity fluctuations due to turbulence will move the species across an imaginary horizontal interface (as shown by the dashed line). In particular, upward velocity fluctuations with $u_y' > 0$ will tend to move $\mathcal{M} = 0$ fluid into regions where $\bar{\mathcal{M}} > 0$ (likely yielding $\mathcal{M}' < 0$), while $u_y' < 0$ will tend to move $\mathcal{M} = 1$ fluid into regions where $\bar{\mathcal{M}} < 1$ (likely yielding $\mathcal{M}' > 0$). In both cases, the average lateral transport of species due to turbulence is associated with $\overline{\mathcal{M}'u_y'} < 0$.

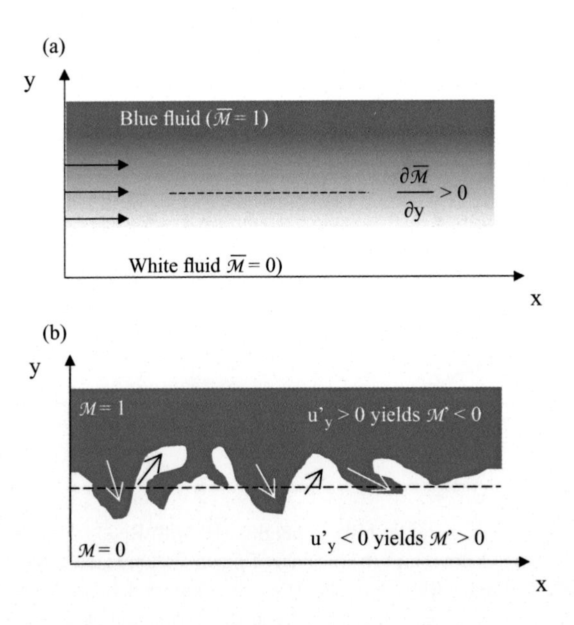

Figure 6.8 A schematic of turbulent channel flow. (a) A uniform mean flow in the x-direction with a vertical (y-direction) gradient of the mean species fraction, where the dashed line is an imaginary transport interface with $\bar{X} = 0.5$. (b) A two-dimensional slice of an instantaneous species fraction. The slice has a convoluted interface due to turbulent structures (in which molecular mixing is ignored), where an upward turbulent fluctuation tends to transport white fluid above this line, while a downward fluctuation velocity will tend to move blue fluid below this line.

If one considers the opposite mass fraction gradient so that the blue fluid is now below the interface (with $\partial\bar{\mathcal{M}}/\partial y < 0$), this would lead to $\overline{\mathcal{M}'u'_y} > 0$. Therefore, the sign of the fluctuating correlation is again opposite to the sign of the mean species gradient. Furthermore, the correlation would go to zero if there is no turbulence or no mean species gradient.

The preceding characteristics can be used to establish three principles for turbulent species mixing (analogous to those for turbulent momentum mixing), as follows:

Principle 1: $\overline{\mathcal{M}'u'_i}$ is proportional to the mean species gradient in the i-direction.
Principle 2: $\overline{\mathcal{M}'u'_i}$ transport is in the opposite direction of this species gradient.
Principle 3: $\overline{\mathcal{M}'u'_i}$ is proportional to the strength of the turbulent fluctuations.

As such, turbulent mixing of a species is driven by velocity fluctuations and a mean concentration gradient, and this mixing tends to diffuse these gradients (not sharpen them).

Turbulent Species Diffusion Model

To model the fluctuation correlation, one may employ a *turbulent Fickian diffusion* approximation (similar to the Boussinesq assumption of 6.26) by introducing the turbulent species diffusivity (Θ_{turb}) such that

$$\overline{\mathcal{M}'u'_i} \approx -\Theta_{turb}\frac{\partial\bar{\mathcal{M}}}{\partial X_i}. \tag{6.38}$$

The RHS includes the mean species gradient per Principle 1, and it has a negative sign consistent with Principle 2. To achieve Principle 3, Θ_{turb} should be related to the turbulent kinetic energy (as will be discussed later). Note that the Fickian diffusion approximation, like that of Boussinesq, is based on qualitative physical arguments and is thus *empirical*. With this important limitation firmly in mind, (6.25a) and (6.38) can be combined to yield the following:

$$\frac{\partial\bar{\mathcal{M}}}{\partial t} + \frac{\partial(\bar{\mathcal{M}}\,\bar{u}_i)}{\partial X_i} = \frac{\partial}{\partial X_i}\left[(\Theta_{\mathcal{M}} + \Theta_{turb})\frac{\partial\bar{\mathcal{M}}}{\partial X_i}\right] \qquad \textit{for}\ i = 1, 2, \textit{and}\ 3. \tag{6.39}$$

The first term on the LHS is zero for an ergodically stationary flow and the second term is mean convection. The RHS contains both molecular and turbulent species diffusion, which play separate but critical roles in overall mixing and have different rates. For example, if cream is placed gently in a coffee in a cup without stirring, it may take a long time for the two fluids to be mixed via molecular diffusion. But if one stirs even gently, this will generally cause flow instabilities and turbulence which allow the fluids to intermingle and then rapidly mix.

To explain the different roles of molecular and turbulent diffusion for the coffee cup example, consider a blue fluid ($\mathcal{M} = 1$) on top of white fluid ($\mathcal{M} = 0$) in a cylindrical chamber, as shown in Figure 6.9. Next, assume the two fluids have the same density so there are no gravitational instabilities. Further, assume that the two fluids are initially separated by a horizontal interface and are stirred by a

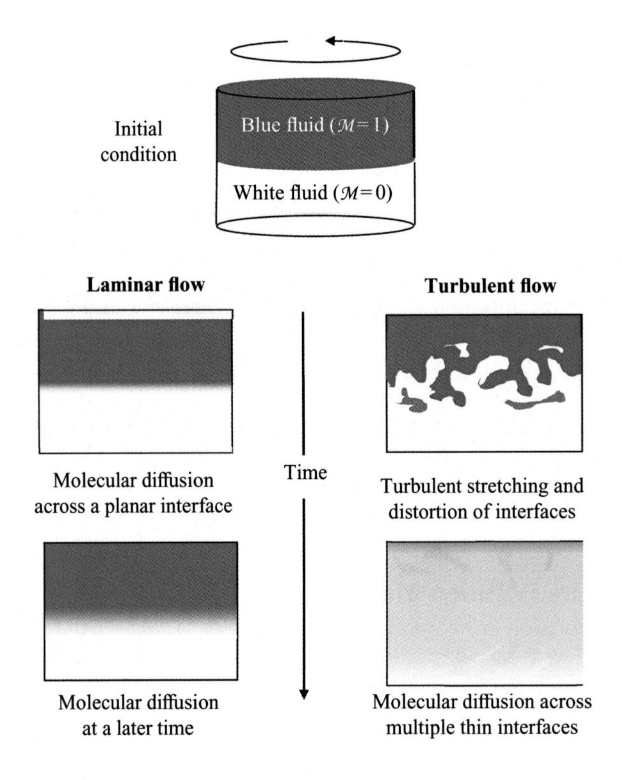

Figure 6.9 Illustration of overall mass diffusion when stirring two miscible fluids (e.g., blue and white paint) with a vortex to rotate the flow in a horizontal plane. The top is the initial condition, while the left and right columns are instantaneous realizations for a vertical slice comparing mixing in laminar flow (with only molecular diffusivity) versus turbulent flow (where flow structures cause the interface to be convoluted and complex, leading to small islands and thin regions coupled with molecular diffusivity that acts over short distances). The RHS will achieve homogeneous species fraction much faster.

two-dimensional vortex centered along the cylinder axis. Finally, assume the two fluids are miscible so they mix over time according to the molecular diffusivity ($\Theta_{\mathcal{M}}$). Because they mix over time, the species distribution is not ergodically stationary so that the LHS unsteady term of (6.39) will be nonzero (as is the case for Figure 6.8).

If the viscosity is high (as with paint) and the stirring is slow (or is zero) such that the flow is laminar with no instabilities (per the LHS of Figure 6.9), there will be no velocity fluctuations and thus no turbulent diffusion of the species. In this case, molecular diffusion (associated with Θ_f) will act in the direction normal to the horizontal interface to diffuse the two fluids into each other. Since molecular diffusion is slow, a later time may only yield a limited vertical extent of mixed fluid. It will generally require a long time before the two liquids are homogeneously mixed from top to bottom ($\mathcal{M} = 0.5$ throughout).

However, if the viscosity is low and the stirring is fast enough such that the flow is turbulent (per the RHS of Figure 6.9), then a cascade of vortices will be created. As a result of the turbulent mixing, the interface between the two liquids will be rapidly stretched, convoluted, and distributed throughout the domain. This leads to blue and white regions broken up into many small and thin regions throughout the domain (instead of confined to a single planer interface). This increases the vertical extent over which both fluids are found and creates a much larger interfacial area (relative to the laminar flow case) upon which molecular diffusion can affect. As a result, $\Theta_{\mathcal{M}}$ only has to operate over very short distances to complete the mixing for a large amount of fluid material. Therefore, one may view species mixing in turbulence as a two-step process:

(1) Turbulent mixing due to Θ_{turb}, whereby a wide range of vortex structures yields a large amount of interfacial area with high concentration gradients.
(2) Molecular diffusion due to $\Theta_{\mathcal{M}}$, which acts on the high concentration gradients and large interfacial areas, yielding rapid local molecular mixing at small scales.

This two-step combination yields a fully mixed state ($\mathcal{M} = 0.5$ throughout) much faster than would occur by molecular mixing alone. In reality, both steps occur simultaneously in a turbulent flow, but they are notionally separated on the RHS of Figure 6.9 to see their individual roles.

It may not be intuitive that both turbulent mixing and molecular mixing are critical to local mixing since $\Theta_{turb} \gg \Theta_{\mathcal{M}}$. However, if there is no molecular diffusion, the two fluids will never mix in a local instantaneous sense (i.e., the mass fraction will either $\mathcal{M} = 0$ or $\mathcal{M} = 1$ throughout if $\Theta_{\mathcal{M}} = 0$). As such, turbulent mixing provides many small interface regions distributed throughout the flow, which greatly enables (but does not replace) the local molecular diffusion process.

To close the species transport (6.39), a model is needed for turbulent diffusivity (Θ_{turb}). This can be related to turbulent viscosity by defining the turbulent Schmidt number (similar to the laminar Schmidt number of 2.35) as follows:

$$\mathrm{Sc}_{turb} \equiv \frac{\mu_{turb}}{\rho_f \Theta_{turb}} \equiv \frac{\nu_{turb}}{\Theta_{turb}} \approx 0.7. \tag{6.40}$$

Since mass and momentum turbulent mixing are driven by the same key principles (as discussed earlier), the turbulent Schmidt number is expected to be of order unity. This assumption is borne out by measurements that indicate $\mathrm{Sc}_{turb} \approx 0.7$ for a wide range of conditions (Faeth, 1987). Therefore, the turbulent momentum diffusivity can be expressed using (6.37c) and (6.40) and used in the species turbulent correlation term of (6.38):

$$\Theta_{turb} = \frac{c_\mu k^2}{\mathrm{Sc}_{turb}\varepsilon} = 0.13\frac{k^2}{\varepsilon}. \tag{6.41}$$

As such, the turbulent species diffusivity can be obtained from solution of the k-ε transport PDEs. However, as noted with models for μ_{turb} as shown in Figure 6.7, predictions of turbulent species transport using (6.39) and (6.41) are often inaccurate for complex flows. Approaches that provide improved accuracy, though at much larger computational cost, are described in the following section.

6.3 Fully Resolved and Partially Resolved Turbulence

To remove the empiricism and potential errors associated with the RANS approaches (which describe the mean flow properties only), flow-field numerical predictions can instead include unsteady three-dimensional features that directly resolve the turbulent structures. Such approaches are denoted as *fully resolved* or *partially resolved*, depending on whether all or some of these turbulent features are resolved. In general, increased resolution is associated with increased predictive accuracy for partially or fully resolved approaches, but at higher computational cost.

Based on the previously described categories, RANS is an "unresolved" approach, whereby only the time-averaged flow quantities are predicted. An example time-averaged flow is shown in Figure 6.3b, where no turbulent structures can be seen. As a result, RANS is the most computationally convenient approach but is generally the least accurate. In the following, fully resolved and partially resolved approaches are discussed.

6.3.1 Fully Resolved Approaches (DNS)

The most accurate approach is to employ the unsteady three-dimensional Navier–Stokes equations to directly capture all length and time scales of the turbulence. This approach is referred to as Direct Numerical Simulation (DNS), and examples of flows predicted by DNS are shown in Figure 6.5a, where high detail can be seen.

For incompressible flow, the tensor forms of the unsteady PDEs for Navier–Stokes (2.73b) and species (2.33b) for turbulant flow with DNS are

$$\rho_f \frac{\partial u_i}{\partial t} + \rho_f \frac{\partial u_i u_j}{\partial X_j} = -\frac{\partial p}{\partial X_i} + \rho_f g_i + \frac{\partial}{\partial X_j}\left[\mu_f\left(\frac{\partial u_i}{\partial X_j} + \frac{\partial u_j}{\partial X_i}\right)\right] \tag{6.42a}$$

for DNS.

$$\frac{\partial \mathcal{M}}{\partial t} + \frac{\partial}{\partial X_i}(\mathcal{M}u_i) = \Theta_{\mathcal{M}} \frac{\partial^2}{\partial X_i^2}\mathcal{M} \tag{6.42b}$$

By incorporating and resolving all the flow structures, DNS eliminates the empiricism associated with turbulence modeling.

The downside of DNS is that computational resources for predicting all the details of turbulence are large and increase rapidly with the flow Reynolds number (see §6.4.8). For example, full resolution of a turbulent boundary layer on a flat plate with no spanwise geometry variation requires more than 10^8 computational nodes for the minimum Reynolds number that provides fully developed turbulence. In addition, the number of computational nodes will increase by another 10-fold for only a threefold increase in Reynolds number. As such, it is generally computationally impractical to employ DNS for most fully developed turbulent flows, and partially resolved methods are often a more practical alternative.

6.3.2 Partially Resolved Approaches (LES)

A partially resolved approach will capture some of the unsteady three-dimensional flow structures. The resolved structures tend to be the ones responsible for most of the turbulence and/or in regions of significant flow separation.

Large Eddy Simulation Methods

The most popular approach in this category is the large eddy simulation (LES), for which the largest structures are captured and the smallest scales are empirically modeled in an average sense. An example of a flow predicted by LES is shown in Figure 6.3a, where all the large structures can be seen but the microstructures are not resolved.

The LES approach is based on filtering at a computational (nonphysical) length scale of $\Delta_{\mathcal{G}}$ and smaller. This user-defined length scale is generally set to be small enough to ensure that the majority of the turbulent kinetic energy is resolved. For the unresolved turbulence (at scales smaller than $\Delta_{\mathcal{G}}$), a turbulence model is instead applied. This strategy is advantageous since it avoids resolving the higher-frequency fluctuations, which represent only a minor fraction of the total kinetic energy and yet are the most computationally demanding portion to capture. In the following, we discuss how LES is applied to the fluid momentum (6.42a), but the same approach can be used for the fluid species (6.42b).

To develop the LES transport equations, a low-pass spatial filter with $\Delta_{\mathcal{G}}$ is applied to the Navier–Stokes equations such that all the velocity components are separated into resolved (unfiltered) and unresolved (filtered) components. While many spatial filter functions are possible (Chung, 2002), the simplest and most common is the Box filter, which simply sets the filter width as equal to the discretization length scale of the cell volume (Garnier et al., 2002) as follows:

$$\Delta_{\mathcal{G}} = (\Delta x \Delta y \Delta z)^{1/3}. \tag{6.43}$$

The RHS is a volume-based computational grid size that may vary in space. Use of the Box filter allows the grid resolution to effectively serve as the low-pass filter whereby the turbulent structures larger than the grid scale are resolved while those that are smaller (and could not be resolved computationally anyway) are filtered.

If the Box filtering is applied, the fluid velocity field (LHS) can be decomposed into a spatially-averaged value ($\widehat{u}$) and a filtered velocity ($\underset{\smile}{u}$), for which only the spatially-averaged value is directly computed and resolved, and thus used for a time-averaged velocity and a perturbation as

$$u_i(X_i, t) = \widehat{u}_i(X_i, t) + \underset{\smile}{u}_i(X_i, t). \tag{6.44a}$$

$$\bar{u}_i(X_i) = \overline{\widehat{u}_i(X_i, t)}. \tag{6.44b}$$

$$\widehat{u}_i'(X_i) = \overline{\widehat{u}_i} - \widehat{u}_i. \tag{6.44c}$$

The second term on the RHS of (6.44a) represents the subgrid fluctuations, which are not resolved and thus must be modeled. Therefore, refining the grid to smaller sizes will reduce the amount of velocity fluctuations to be empirically modeled, thus allowing more of the turbulence to be captured by the resolved velocity fluctuations of (6.44c). However, refining the grid will also increase the number of computational nodes and thus the required computational resources. As such, grid refinement is chosen as a balance of computational resources and predictive fidelity. In general, LES requires much higher grid resolution than RANS to achieve this balance.

Application of Box-filtered spatial averaging to the incompressible Navier–Stokes equations (6.42a) yields the LES momentum PDE in tensor form (index summation included) as follows:

$$\rho_f \frac{\partial \widehat{u}_i}{\partial t} + \rho_f \widehat{u}_j \frac{\partial \widehat{u}_i}{\partial X_j} = \rho_f g_i - \frac{\partial \widehat{p}}{\partial X_i} + \frac{\partial}{\partial X_j}\left[\mu_f\left(\frac{\partial \widehat{u}_i}{\partial X_j} + \frac{\partial \widehat{u}_j}{\partial X_i}\right) - \rho_f \, \underset{\smile}{u_i u_j}\right]. \tag{6.45}$$

This is similar to the form of the "pseudo-unsteady" RANS formulation (6.24a). However, the RANS equations are integrated to converge to a state-state velocity field with no turbulent structures, while the LES velocity field is always unsteady and resolves the large eddies.

It should be noted that LES still requires an empirical model for the subgrid velocity correlation on the RHS of (6.45). For free-shear flows, this correlation can be described with a Smagorinsky subgrid stress model using the resolved (known) velocity gradients and a subgrid turbulent viscosity ($\nu_{\mathcal{G}}$), which can be used to obtain the filtered turbulent kinetic energy ($k_{\mathcal{G}}$) as well as the total kinetic energy (k) and dissipation (ε) as follows:

$$\underset{\smile}{u_i u_j}\,(X_i, t) = -\nu_{\mathcal{G}}\left[\frac{\partial \widehat{u}_i}{\partial X_j} + \frac{\partial \widehat{u}_j}{\partial X_i}\right] + \frac{\delta_{ij}}{3}\,\underset{\smile}{u_i u_i}\,. \tag{6.46a}$$

$$\nu_{\mathcal{G}}(X_i, t) = c_{\mathcal{G}}^2 \Delta_{\mathcal{G}}^2 \sqrt{\frac{1}{2}\left(\frac{\partial \widehat{u}_i}{\partial X_j} + \frac{\partial \widehat{u}_j}{\partial X_i}\right)\left(\frac{\partial \widehat{u}_i}{\partial X_j} + \frac{\partial \widehat{u}_j}{\partial X_i}\right)}. \tag{6.46b}$$

$$k_{\mathcal{G}}(X_i, t) \equiv \frac{1}{2}\,\underset{\smile}{u_i u_i} = \left(\frac{\nu_\Delta}{c_{k\mathcal{G}}\Delta_{\mathcal{G}}}\right)^2. \tag{6.46c}$$

$$k(X_i) = \widehat{k} + \underset{\smile}{k} = \frac{1}{2}\overline{\widehat{u'_i}\widehat{u'_i}} + \overline{k_{\mathcal{G}}}. \tag{6.46d}$$

$$\varepsilon(X_i, t) = k_{\mathcal{G}}^{3/2}/\Delta_{\mathcal{G}}. \tag{6.46e}$$

The LES version of (6.46a) is similar to the RANS version of (6.27). Furthermore, (6.46b) is similar to (6.29a), and (6.46c) is similar to the combination of (6.17) and (6.29b). The total turbulent kinetic energy of (6.46d) includes the resolved TKE and the subgrid. Finally, (6.46e) assumes that all the turbulent dissipation takes place at the subgrid scale and is based on the subgrid TKE and filter-length scale.

For free-shear flows, the subgrid expressions include two empirical constants, $c_{\mathcal{G}}$ for (6.46b) and $c_{k\mathcal{G}}$ for (6.46c). Recommended values of $c_{\mathcal{G}}$ range from 0.065 (Urbin and Knight, 2001) to 0.2 (Piomelli, 1997), while $c_{k\mathcal{G}}$ is often set as 0.05 (Yoshizawa and Horiuti, 1985). More sophistication can be achieved by employing dynamic subgrid models where these empirical subgrid values become a function of the unsteady resolved velocity gradients (Piomelli, 1997; Chung, 2002). Fortunately, $c_{\mathcal{G}}$ and $c_{k\mathcal{G}}$ generally have a small impact on the predictions if the grid resolution is sufficiently high, so further complexity is often not critical.

For wall-bounded flows with LES, wall damping can be included by modifying $c_{\mathcal{G}}$ of (6.46) so it reduces near the wall as follows (Balaras et al., 1996):

$$c_{\mathcal{G},\text{wall}} = \min\left\{c_{\mathcal{G}}, \left(\frac{0.41y_{\text{wall}}}{\Delta_{\mathcal{G}}}\right)^2\left[1 - \exp\left(-y^+/25\right)^3\right]\right\}. \tag{6.47}$$

This is similar to the RANS-type wall damping of (6.30b), but with $c_{\mathcal{G}}\Delta_{\mathcal{G}}$ instead of Λ_{mix}. Far from the wall, this equation reverts to the free-shear constant ($c_{\mathcal{G}}$).

Hybrid RANS/LES Methods

While the preceding LES approaches help reduce inaccuracies associated with the use of empirical constants as compared to RANS approaches, the computational grid requirements for LES at high Reynolds numbers, especially for wall-bounded flows, can be quite high (see §6.4.8). This can render many flows impractical to compute even with LES. This problem has given rise to hybrid approaches that are more computationally efficient than LES but are generally more accurate than a RANS approach.

The hybrid RANS/LES approaches generally employ RANS-like equations in regions with an attached boundary layer (where RANS is generally accurate and LES would be very expensive) and apply LES-like equations in regions where the flow is separated (where RANS tends to be inaccurate and LES can be efficient).

As an example, hybrid RANS/LES is used to simulate the flow past a cylinder in Figure 6.10, where the upstream attached boundary layer over the surface is treated as a RANS region while the downstream separated flow with complex three-dimensional unsteady structures is treated as an LES region. As such, the hybrid approach can be a good alternative. In fact, simulating this particular flow with RANS is generally not possible, as the strong separation can prevent numerical convergence to a steady solution of the RANS PDE. On the other hand, resolving the detailed turbulent structures within the thin boundary layer upstream with LES at this Reynolds number is beyond the capability of most computers.

Given that LES and RANS already have a wide range of approaches, there are plethora of approaches for hybrid RANS/LES. However, these all generally stem from the detached eddy simulation method (Spalart et al., 1997). Often, the key aspect that distinguishes the various approaches is the technique used to identify the LES and RANS regions as well as the model that blends these two regions at their interface (Nichols and Nelson, 2003). However, all these approaches are three dimensional and unsteady whereby some of the kinetic energy is resolved (which can be obtained by time averaging the simulation results) and some is modeled (e.g., at the subgrid scales

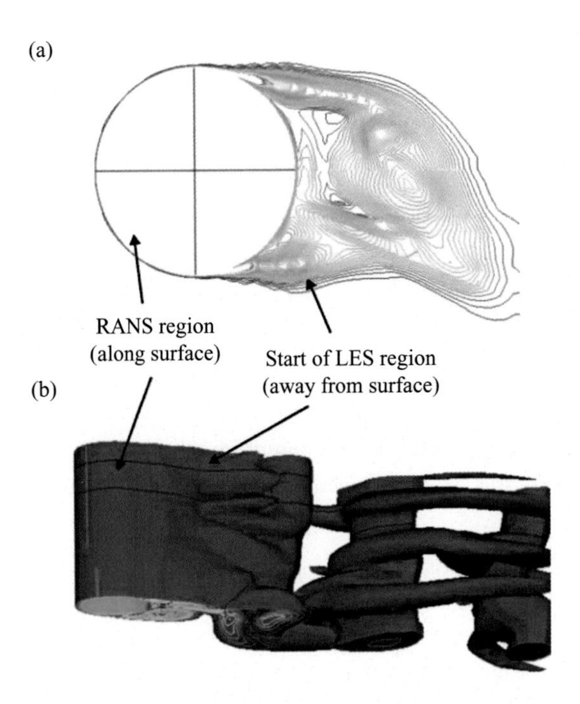

Figure 6.10 Hybrid RANS/LES simulation for flow over a cylinder with a diameter-based Reynolds number of 3×10^6 showing instantaneous (a) a cross-sectional view of cylinder, zoomed-in to show boundary-layer thickness; and (b) a span-wise view zoomed out to show large eddy structures in the wake.

in the LES region or by a kinetic energy PDE in the RANS regions). The resolved and modeled kinetic energy contributions can be added together to give the full kinetic energy. An example of this is shown in Figure 6.11 for a cylinder wake, where the resolved portion is larger than the subgrid portion (indicating that LES captures most of the TKE) and the combination correctly approaches the DNS TKE.

6.4 Time and Length Scales for Turbulent Flow

As noted previously, turbulence is characterized by a wide range of time and length scales, as shown in Figures 6.3a, and 6.5a. In the following section, characterization of turbulence is considered as well as the most energetic turbulent length and time scales. This is followed by a description of the smallest scales in the turbulence, and then the spectrum properties that relate the full range of scales.

6.4.1 Homogeneous Isotropic Stationary Turbulence

To help characterize the turbulent scales, we consider two types of turbulence: homogeneous isotropic stationary turbulence (HIST) and nonhomogeneous anisotropic

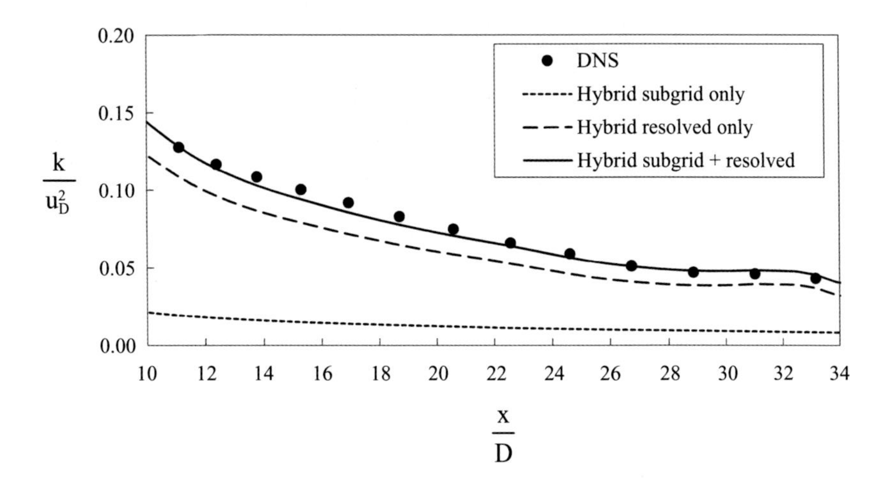

Figure 6.11 Predictions of turbulent kinetic energy along the centerline of a cylinder wake based on DNS and the Nichols–Nelson hybrid model for $Re_D = 800$ (Rybalko et al., 2008). This shows that the hybrid model predictions of kinetic energy (the sum of the subgrid and resolved portions) reasonably represent the kinetic energy predicted by the most accurate DNS technique.

turbulence (NAsT). HIST is an idealized condition based on three assumptions. First, *homogeneous* turbulence is defined as having no spatial gradients for the mean statistics such that the kinetic energy and dissipation are independent of *location*:

$$\nabla k = 0 \textit{ and } \nabla \varepsilon = 0 \qquad \textit{for homogeneous turbulence.} \tag{6.48}$$

This is equivalent to stating that the turbulent statistics do not vary if we translate the coordinate system. Second, *isotropic* turbulence is defined as having equal magnitude of the velocity fluctuations in *all directions*:

$$\overline{u'_x u'_x} = \overline{u'_y u'_y} = \overline{u'_z u'_z} = u_\Lambda^2 \qquad \textit{for isotropic turbulence.} \tag{6.49}$$

This is equivalent to stating that the turbulent statistics do not vary if we rotate the coordinate system. Third, HIST is stationary per (6.3). As a result, the root mean square (rms) values of the velocity fluctuations for HIST are equal at all locations, in all directions, and for all times. HIST has been an important "historical" foundation since it greatly simplifies the analysis of turbulence.

In contrast, NAsT is the other extreme of turbulence, whereby the statistical properties vary substantially in space and by direction. These aspects greatly complicate the theoretical analysis of turbulence, so the flows can be considered "nasty" in this sense. The separated flow turbulence shown in Figure 6.3c has NAsT characteristics since there are high gradients in turbulent kinetic energy, indicating nonhomogenous TKE, which violates (6.48). Similarly, the turbulent boundary layer of Figure 6.5b has rms velocity fluctuations that vary greatly with wall normal distance. In addition, there are differences between the u_x, u_y, and u_z fluctuation strengths, which indicates anisotropic turbulence, which violates (6.49). The NAsT aspects are especially profound in the near-wall region ($y/\delta < 0.1$), where the fluctuation

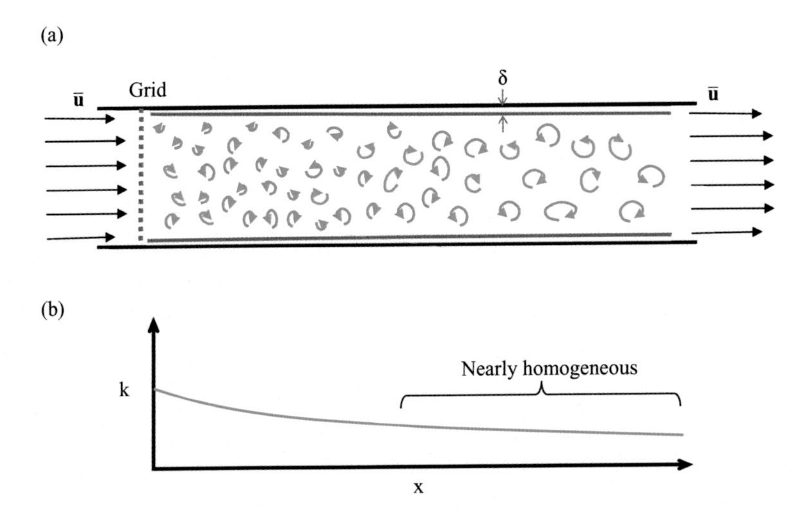

Figure 6.12 A wake flow due to an upstream grid placed in a wind tunnel of a constant cross-sectional area and thin boundary layers (of thickness δ) along the walls showing the following: (a) A schematic of turbulent structures in blue with mean flow subtracted (i.e., showing $\mathbf{u}'$). The structures increase in size as they move downstream. (b) The change in time-averaged turbulent kinetic energy along the centerline of the tunnel, but the decay is slow far downstream such that the kinetic energy is stationary and nearly homogeneous.

intensities differ greatly and are a strong function of location. This is because the near-wall region has the strongest mean velocity gradients, and these are responsible for turbulence production, but u_y fluctuations are limited in the wall-normal direction by the zero-velocity condition at $y = 0$.

Fortunately for the sake of analysis, many flows tend to have at least some regions that can be considered as "nearly" HIST. These regions are generally far from any solid surfaces and are associated with near-uniform mean velocity. For example, the turbulence strengths of Figure 6.5b at $y/\delta > 1.1$ are nearly the same in all three directions and have weak spatial variations. In particular, the rms of all three the velocity fluctuations are within 10% and have spatial variations less than 20%. As such, this region is *nearly* HIST.

Another flow that includes nearly HIST regions is an ergodically stationary grid-generated wake in a duct of constant cross section as illustrated in Figure 6.12. In particular, the regions far downstream of the grid and away from the walls generally have a uniform mean velocity with velocity fluctuations that are nearly the same in all three directions (Snyder and Lumley, 1971). Along the tunnel centerline, the turbulent kinetic far downstream can be nearly constant (nearly homogeneous), as illustrated in Figure 6.12b, indicating nearly homogeneous turbulence. In addition, downstream of the grid, there are no mechanisms for producing TKE, so its transport equation in the x-direction using (6.32) becomes

$$\frac{\partial k}{\partial t} + \bar{u}_x \frac{\partial k}{\partial x} = \dot{k}_{diss} = -\varepsilon \qquad \textit{for uniform mean flow in x-direction.} \tag{6.50}$$

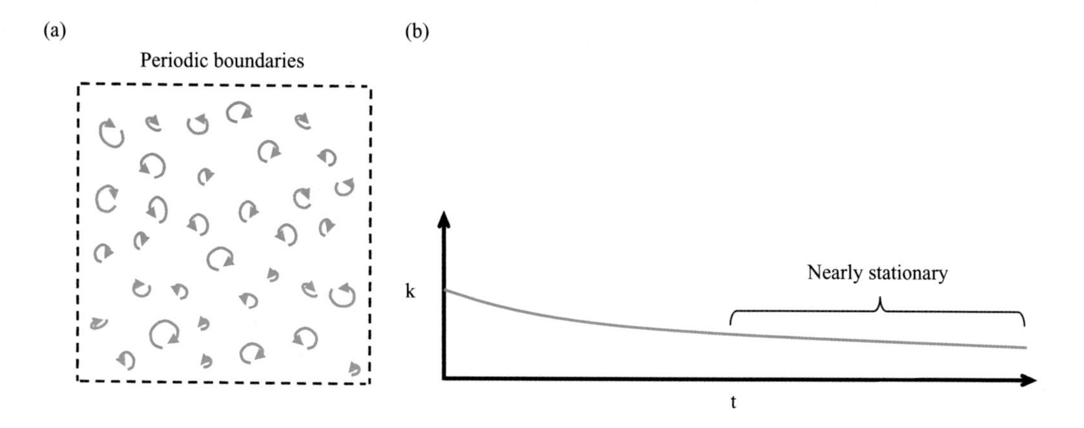

Figure 6.13 Turbulence in a box due to random velocity fluctuations in a cube with no mean flow: (a) periodic boundary conditions with turbulent structures at a given time; and (b) turbulent kinetic energy averaged over space (instead of time), which initially drops quickly but then decays slowly such that the kinetic energy is homogeneous and nearly stationary after longer times.

Since viscous effects are always present in turbulence, the dissipation rate is always negative. As such, kinetic energy may decay slowly in the downstream direction, resulting in weakly nonhomogenous flow so the flow is only nearly HIST.

A third nearly HIST flow is called *turbulence in a box*, a numerical creation where periodic boundary conditions are used to ensure homogeneous conditions and where an initial random velocity field with equal strength in all directions (with no walls) is used to ensure isotropic conditions (Figure 6.13a). To further ensure fully developed turbulence, the Navier–Stokes equations must employ a suitably low viscosity (i.e., a high enough Reynolds number). However, the lack of turbulence production for this flow (after the initial conditions) due to an absence of mean velocity gradients causes the TKE to decay over time, as shown in Figure 6.13b, yielding nonstationary turbulence. As such, HIST is not fully satisfied for this flow.

As shown by the preceding examples, a turbulent flow can either be fully stationary but only nearly homogeneous and isotropic (Figure 6.12) or be fully homogeneous and isotropic but only nearly stationary (Figure 6.13). Thus, a flow can never be exactly HIST in actual flow conditions. To determine whether a flow can be considered nearly HIST, one may propose that the variations are quantitatively limited as follows:

$$\frac{\left|\overline{u'_x u'_x} - u_\Lambda^2\right|}{u_\Lambda^2}, \frac{\left|\overline{u'_y u'_y} - u_\Lambda^2\right|}{u_\Lambda^2}, \& \frac{\left|\overline{u'_z u'_z} - u_\Lambda^2\right|}{u_\Lambda^2} < 0.1 \qquad \textit{for nearly isotropic}. \tag{6.51a}$$

$$\frac{|\nabla k|}{k} D < 0.2 \qquad \textit{for nearly homogeneous}. \tag{6.51b}$$

$$\left|\frac{\partial k}{\partial t}\right| \frac{\tau_D}{k} < 0.2 \qquad \textit{for nearly stationary}. \tag{6.51c}$$

The choice for the domain scales (D and τ_D) for the second and third expressions can be subjective but is often based on as the length scale over which the overall velocity gradient occurs and the time scale based on the length and a mean flow velocity (u_D) via (5.12). For a pipe flow, D and u_D can be the cross-sectional diameter and the area-averaged velocity. For a boundary layer, they can be the thickness and the freestream velocity. For a jet flow, they can be the jet diameter and centerline velocity.

In the following, we will generally assume nearly HIST flow to simplify the analysis. If one also assumes incompressible flow, the dissipation of (6.33) can be simplified as follows (Hinze, 1975):

$$\varepsilon = \nu_f \overline{\frac{\partial u_i'}{\partial X_j}\frac{\partial u_i'}{\partial X_j}} \approx 15\nu_f \overline{\left(\frac{\partial u_1'}{\partial X_1}\right)^2} \quad \textit{for HIST}. \tag{6.52}$$

This tensor form includes summation over i and j indices (for 1, 2, *and* 3).

6.4.2 Velocity Correlation Coefficient

To characterize the degree of spatial and temporal correlation of turbulent structures, one may define the dimensionless velocity *correlation coefficient* Υ based on the time average of two velocity fluctuations in the same direction (u_i) but at different times (with a time shift τ) and/or different locations (with a spatial shift vector χ) as follows:

$$\Upsilon_i(\mathbf{X}, t, \chi, \tau) \equiv \frac{\overline{u_i'(\mathbf{X}, t)u_i'(\mathbf{X}+\chi, t+\tau)}}{\overline{u_i'(\mathbf{X}, t)u_i'(\mathbf{X}, t)}} \quad \textit{NISI}. \tag{6.53}$$

Note that this correlation is dimensionless, and (6.53) has No Index Summation Intended (NISI): i is 1, 2, *or* 3. If we consider HIST, the correlation coefficient will be a function of space shift and time shift only (independent of **X** and t) and thus may be written (again with NISI) as follows:

$$\Upsilon_i(\chi, \tau) \equiv \frac{\overline{u_i'(\mathbf{X}, t)u_i'(\mathbf{X}+\chi, t+\tau)}}{\overline{u_i'(\mathbf{X}, t)u_i'(\mathbf{X}, t)}} \quad \textit{for HIST}. \tag{6.54}$$

Next, we consider the correlation characteristics for various shifts in time and space.

For the first limit, when both temporal and spatial shifts are zero, the correlation coefficient simply reverts to unity by its definition. At the other two limits of infinitely long temporal shifts ($\tau \to \pm\infty$) or infinitely large spatial shifts ($\chi \to \pm\infty$), the two fluctuations will be statistically uncorrelated, since turbulence is stochastic (not periodic) for long times and distances, so the correlation coefficient tends to zero. These three limits can be summarized as follows:

$$\Upsilon_i(\chi = 0, \tau = 0) = 1. \tag{6.55a}$$

$$\Upsilon_i(\chi = 0, \tau = \pm\infty) = 0. \tag{6.55b}$$

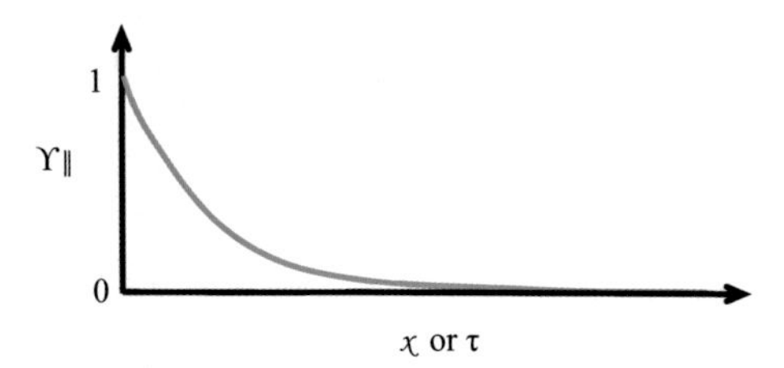

Figure 6.14 Correlation coefficient for streamwise velocity fluctuations as a function of a shift in space (x) or time (τ), where the coefficient equals unity for zero shift and approaches zero for an infinite shift.

$$\Upsilon_{\mathrm{i}}(\chi = \pm\infty, \tau = 0) = 0. \tag{6.55c}$$

Intermediate temporal and/or spatial shifts bridge these limits, whereby the correlation coefficient decays from unity to zero as qualitatively shown in Figure 6.14. The decorrelation of the velocity fluctuations is a result of changes in space and time as the turbulent structure convect, break up, merge, distort, and dissipate.

Fluctuations due to weather can serve as an analogy to explain temporal and spatial shift dependencies. For example, consider a correlation based on the average amount of rain minus the monthly mean as a function of space and time (Υ_{rain}). For temporal and spatial shifts that are both nearly zero, the correlation will be nearly unity; for example, the local and instantaneous rain fluctuation from the mean measured only 1 meter away and only 1 second later will be nearly the same, that is, $\Upsilon_{\mathrm{rain}} \approx 1$ for $\chi \approx 0$ and $\tau \approx 0$. However, if we consider a long temporal shift we would expect zero correlation. For example, the rain fluctuation at a fixed location at noon on March 20, 2022, is not expected to be correlated to the rain fluctuation on March 20, 2023, since $\Upsilon_{\mathrm{rain}} \approx 0$ for very long times. A large spatial shift (χ) with no time shift ($\tau = 0$) will also result in a zero correlation. For example, a very rainy day in New York does not mean it will be very rainy in Beijing at the same time, so $\Upsilon_{\mathrm{rain}} \approx 0$ for very far distances. For intermediate time or space differences, such as a few hours or a few kilometers, some correlation is likely but is not 100%, that is, $\Upsilon_{\mathrm{rain}} < 1$ for $\tau > 0$ and/or $\chi > 0$.

Returning to the velocity fluctuations, we consider three combinations of time and space shift based on the three reference frames shown in Figure 6.15: Eulerian, Lagrangian, and moving-Eulerian. The Eulerian frame indicates a shift in time but no change in location ($\chi_{\mathrm{E}} = 0$). The Lagrangian frame includes a shift in time and a simultaneous shift in space along the fluid path (χ_{L} based on an integration of local $\mathbf{u}$ and τ). The moving-Eulerian frame includes a shift in time and a simultaneous shift in space along the mean fluid path (χ_{mE} based on $\bar{\mathbf{u}}$ and τ), where the mean fluid velocity is defined as follows:

$$\mathbf{u} = \bar{\mathbf{u}} + \mathbf{u}' = \left(\bar{\mathrm{u}}_{\parallel} + \mathrm{u}'_{\parallel}\right)\mathbf{i}_{\parallel} + \mathrm{u}'_{\perp,1}\mathbf{i}_{\perp,1} + \mathrm{u}'_{\perp,2}\mathbf{i}_{\perp,2}. \tag{6.56}$$

The spatial shifts for these three reference frames are thus defined as

$$\chi_{E} \equiv 0 \qquad \textit{Eulerian reference frame.} \tag{6.57a}$$

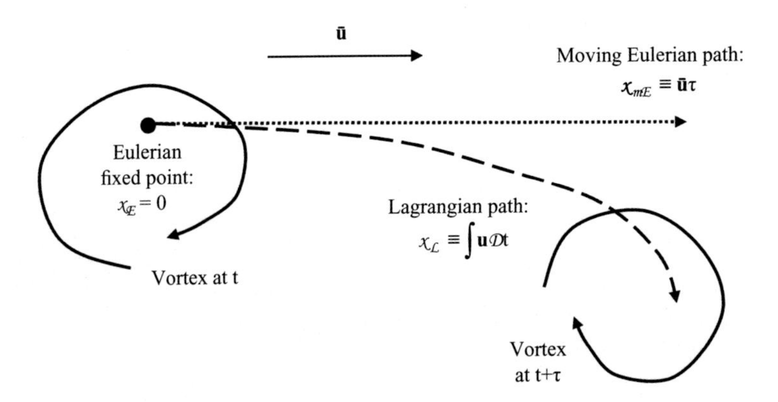

Figure 6.15 Reference frames for a flow with a horizontal mean velocity but an instantaneous velocity that varies with time and space: (a) Eulerian based on a fixed point in time (solid circle); (b) Lagrangian path (long-dash) based on instantaneous fluid velocity; and c) moving-Eulerian path (short-dash) based on the mean velocity.

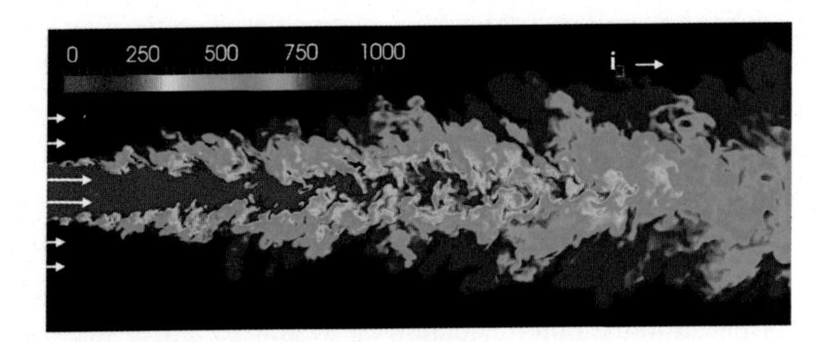

Figure 6.16 Plane cut of a turbulent jet where the streamwise direction (left to right) is indicated by unit vector in the upper right and color contours indicate the Lagrangian positions for various time shifts (τ) since injection. For example, purple indicates the fluid that was most recently injected (Shin et al., 2017).

$$x_{\mathcal{L}} \equiv \int_{t}^{t+\tau} \mathbf{u} d\tau \qquad \textit{Lagrangian reference frame.} \tag{6.57b}$$

$$x_{mE} \equiv \bar{\mathbf{u}}\tau = \bar{u}\mathbf{i}_{\parallel}\tau \qquad \textit{moving Eulerian reference frame.} \tag{6.57c}$$

If we consider the jet flow of Figure 6.16, the Eulerian reference frame corresponds to a fixed point in space while the moving-Eulerian trajectory corresponds to two points along the mean streamwise direction (left to right). The Lagrangian trajectory undergoes a complex three-dimensional path as it interacts with the vortex structures. The flow-field color contours are based on the Lagrangian positions for different time shifts. The figure shows that measuring correlations along a Lagrangian path for a given time shift is difficult (it requires moving probes or full 3D unsteady data). Because of this difficulty, most experiments are based on an Eulerian frame with a single probe at a fixed point or consider aspects of a moving-Eulerian reference frame using two or more probes aligned along the same mean flow path.

Based on (6.54) and (6.57), one may define the temporal correlation coefficients associated with the Eulerian, Lagrangian, and moving-Eulerian reference frames in terms of the time shift as follows:

$$\Upsilon_{\mathcal{E}}(\tau) \equiv \Upsilon_{\mathrm{i}}(0, \tau) \tag{6.58a}$$

$$\Upsilon_{\mathcal{L}}(\tau) \equiv \Upsilon_{\mathrm{i}}(x_{\mathcal{L}}, \tau) \qquad \textit{for HIST}. \tag{6.58b}$$

$$\Upsilon_{m\mathcal{E}}(\tau) \equiv \Upsilon_{\mathrm{i}}(x_{m\mathcal{E}}, \tau) \tag{6.58c}$$

The velocity direction index is removed on the LHS for these temporal correlations since they are independent of the velocity direction for HIST. The correlation coefficients follow the same limits as noted in (6.55a,b):

$$\Upsilon_{\mathcal{E}} = \Upsilon_{\mathcal{L}} = \Upsilon_{m\mathcal{E}} = 1 \quad \textit{for } \tau = 0. \tag{6.59a}$$

$$\Upsilon_{\mathcal{E}} = \Upsilon_{\mathcal{L}} = \Upsilon_{m\mathcal{E}} = 0 \quad \textit{for } \tau = \infty. \tag{6.59b}$$

Similarly, the temporal correlation at intermediate time shifts will follow the trend of Figure 6.14. An example for the Lagrangian reference frame is presented in Figure 6.17, where the time scale is normalized by a turbulent time scale ($\tau_{\Lambda\mathcal{L}}$) to be defined in the next section.

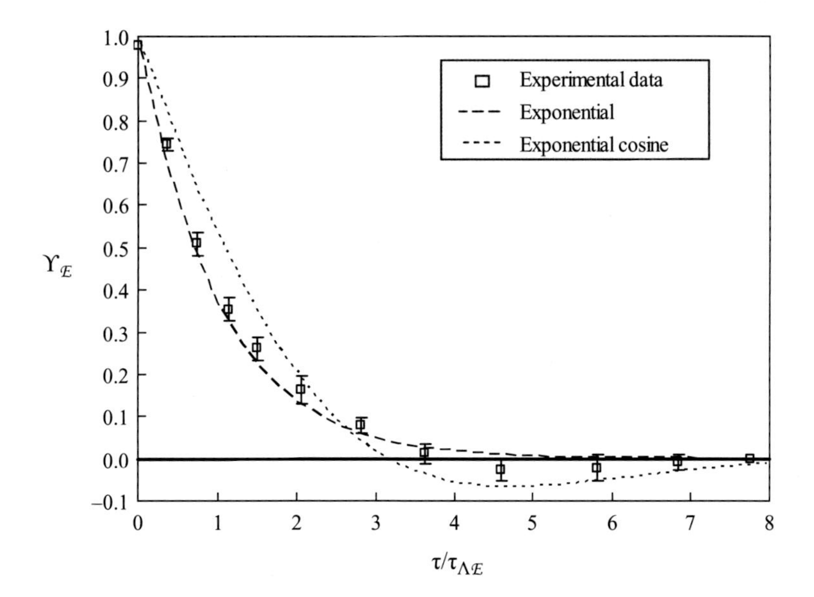

Figure 6.17 Eulerian correlation coefficient as a function of time shift based on measurements in grid-generated turbulence with nearly HIST conditions (Snyder and Lumley, 1971) compared with empirical models of the exponential function (which better captures the initial decay) and the exponential-cosine function (which qualitatively captures the negative loop).

6.4.3 Integral Time Scales

For stationary turbulence, the previous correlation coefficient can be used to define a key turbulent time scale (τ_{Λ}) by integrating the curve of Figure 6.14 over all time shifts (τ) as follows:

$$\tau_{\Lambda,i}(\mathbf{X},\mathcal{X}) \equiv \int_0^{\infty} \Upsilon_i(\mathbf{X},\mathcal{X},\tau)d\tau \quad \textit{where } i = x, y, \text{or } z. \tag{6.60}$$

Stemming from this integral definition, τ_{Λ} is termed the integral time scale for velocity fluctuations in a given direction. For NAST flows, the integral time scale can depend on location ($\mathbf{X}$) and on the direction of the velocity fluctuations (u_i).

As with the correlation coefficient, this time scale can be considered for a given reference frame. For example, the Eulerian, Lagrangian, and moving-Eulerian integral time scales for y-velocity fluctuations are defined as follows:

$$\tau_{\Lambda\mathcal{E},y}(\mathbf{X}) \equiv \int_0^{\infty} \Upsilon_{\mathcal{E},y}(\mathbf{X},\tau)\, d\tau = \int_0^{\infty} \frac{\overline{u'_y(\mathbf{X},t)u'_y(\mathbf{X},t+\tau)}}{u'_{y,rms}u'_{y,rms}} d\tau. \tag{6.61a}$$

$$\tau_{\Lambda\mathcal{L},y}(\mathbf{X}) \equiv \int_0^{\infty} \Upsilon_{\mathcal{L},y}(\mathbf{X},\tau)\, d\tau = \int_0^{\infty} \frac{\overline{u'_y(\mathbf{X},t)u'_y(\mathbf{X}+\mathcal{X}_{\mathcal{L}},t+\tau)}}{u'_{y,rms}u'_{y,rms}} d\tau. \tag{6.61b}$$

$$\tau_{\Lambda m\mathcal{E},y}(\mathbf{X}) \equiv \int_0^{\infty} \Upsilon_{m\mathcal{E},y}(\mathbf{X},\tau)\, d\tau = \int_0^{\infty} \frac{\overline{u'_y(\mathbf{X},t)u'_y(\mathbf{X}+\mathcal{X}_{m\mathcal{E}},t+\tau)}}{u'_{y,rms}u'_{y,rms}} d\tau. \tag{6.61c}$$

Because the integral time scale reflects how long a velocity fluctuation will last, it is often referred to as the "eddy lifetime"; for example, the eddy lifetime at a fixed Eulerian position is $\tau_{\Lambda\mathcal{E}}$. For HIST, the temporal correlations are independent of time and location (6.58) and are also independent of the velocity direction, so these time scales are isotropic. For example, the time scale in the Lagrangian reference frame is as follows:

$$\tau_{\Lambda\mathcal{L}} = \tau_{\Lambda\mathcal{L},x} = \tau_{\Lambda\mathcal{L},y} = \tau_{\Lambda\mathcal{L},z} \qquad \textit{for HIST}. \tag{6.62}$$

Since the Eulerian, Lagrangian, and moving-Eulerian temporal correlations are independent of time, location and direction, they are each a unique function of τ/τ_{Λ}. The functional dependence for the Eulerian integral scale is shown in Figure 6.17 based on measurements in a nearly HIST flow. This dependence is highly typical of nearly HIST flows, whereby the correlation coefficient decays nearly exponentially in time (as is the case for for many random processes). By assuming an exponential decay for the three reference frames of (6.61), the correlation coefficients for the reference frame shifts defined in (6.58) become

$$\Upsilon_{\mathcal{E}}(\tau) \approx \exp(-\tau/\tau_{\Lambda\mathcal{E}}). \tag{6.63a}$$

$$\Upsilon_{\mathcal{L}}(\tau) \approx \exp(-\tau/\tau_{\Lambda\mathcal{L}}). \tag{6.63b}$$

$$\Upsilon_{m\mathcal{E}}(\tau) \approx \exp(-\tau/\tau_{\Lambda m\mathcal{E}}). \tag{6.63c}$$

This is the most common analytical form for the correlation coefficients.

However, the exponential decay is only an approximation. In particular, Eulerian correlation measurements often reveal a slight negative correlation at longer times, such as for time shifts of $4 < \tau/\tau_{\Lambda E} < 7$ in Figure 6.17. This negative loop is a result of structural coherency and convection. For example, a counterclockwise vortex produces an upwash on the right side and a downwash on the left side. As the vortex moves past a fixed Eulerian point, the upwash (positive velocity fluctuation) is followed by the downwash (negative velocity fluctuation) during the time of the vortex passage. This sign reversal can result an anticorrelation ($\Upsilon < 0$) over this time shift. If the velocity fluctuations were purely periodic (as in a series of identical vortices passing the point), the correlation coefficient would also be periodic. For example, a velocity field that varies with $\sin(\pi t/\tau_q)$ at a point would yield a correlation coefficient of $\Upsilon = \cos(\pi\tau/\tau_q)$, which yields a negative loop followed by positive loop, and so on. In this case, the field would oscillate for all time shifts (instead of decaying to zero for $\tau \gg \tau_\Lambda$). However, turbulence is highly random, with many sizes, strengths, and shapes of vortices that evolve and distort as they move. As such, the randomness of turbulence is stronger than any periodicity. To combine the effects of randomness and periodicity, one may combine the exponential decay and the cosine function to create a hybrid correlation coefficient, as follows:

$$\Upsilon(\tau) = \exp(-\tau/\tau_{\Lambda ce})\cos(\pi\tau/\tau_{\Lambda ce}) \qquad \textit{for exponential-cosine form.} \tag{6.64}$$

As shown in Figure 6.17, this function captures the negative loop at long times ($t \gg \tau_\Lambda$). However, the simpler exponential function of (6.63) is more accurate for $t < \tau_\Lambda$ and thus is much more commonly used and generally preferred.

An important property of the correlation coefficients of (6.63) is the quantitative decay, whereby $\Upsilon = e^{-1}$ at $\tau = \tau_\Lambda$. And after two such lifetimes, only 13.5% of the original fluctuation information is retained. This loss of influence can be qualitatively seen in Figure 6.18, where most positive or negative velocity fluctuations are gone within one lifetime, and even the longest events only last a few eddy lifetimes.

Integral Time Scales in Typical NAST Flows

Interestingly, flows that are nonhomogeneous and anisotropic (but stationary) can still have integral time scales that are nearly isotropic and homogeneous. For example, the turbulent boundary-layer velocity fluctuations shown in Figure 6.5b have strong variations in location and direction for $y/\delta < 1$. However, Figure 6.19 shows that $\tau_{\Lambda L}$ based on velocity fluctuations in the y-direction is similar to the average for all three directions indicating nearly isotropic time scales. And while the integral time scale reduces closer to the wall for $y/\delta < 0.5$, these spatial variations are not as extreme as seen for the velocity fluctuations in Figure 6.5b. Furthermore, the integral time scale is found to be nearly homogeneous for $y/\delta > 0.5$ in Figure 6.19.

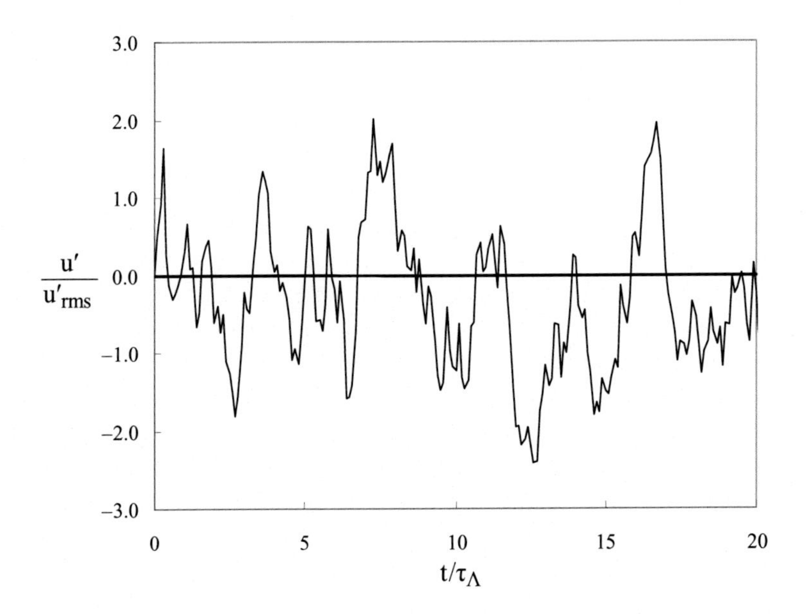

Figure 6.18 Sample turbulent flow time history of velocity (scaled with the rms value) as a function of time (scaled with the integral time scale) for data discretized at $\Delta t = 1/8\ \tau_\Lambda$ (higher-frequency fluctuations are filtered out). The distances between significant peaks tend to be on the order of τ_Λ.

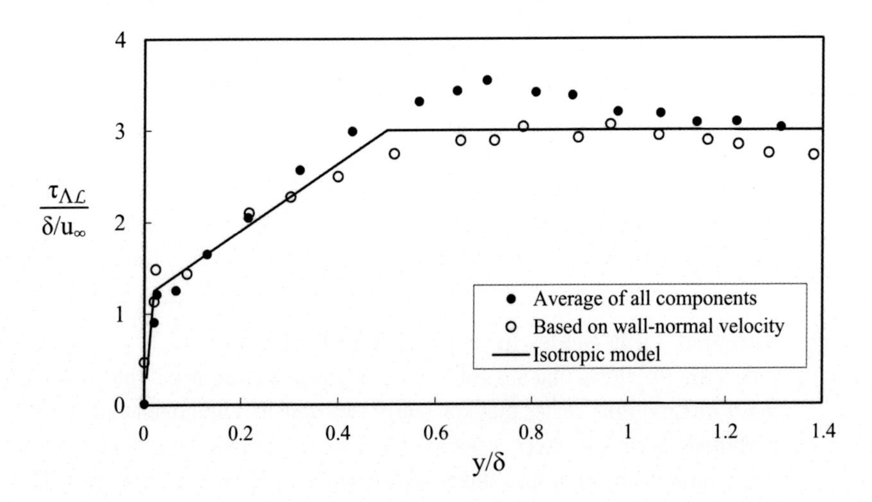

Figure 6.19 Lagrangian integral time scales normalized with the ratio of boundary-layer thickness (δ) and freestream velocity (u_∞) for a flat plate turbulent boundary layer at $Re_\delta = 4{,}500$ (Bocksell and Loth, 2006), where hollow circles are time scales based on wall-normal velocity fluctuations, the solid circles are on the average of all three integral time scales for the velocities in all three directions, and the solid line is a three-part empirical fit.

A second typical characteristic is that the integral time scale is proportional to the length and velocity scales of the overall flow. For example, Figure 6.19 shows that $\tau_{\Lambda L} \approx 3\delta/u_\infty$ for $y/\delta > 0.5$. The Eulerian integral time for this same flow also depends on these length and velocity scales with $\tau_{\Lambda E} \approx 0.3\delta/u_\infty$ for $y/\delta > 0.5$ (Bocksell and Loth, 2006). A similar proportionality of $\tau_{\Lambda E} \approx 0.25\delta/u_\infty$ was found at much higher Reynolds numbers ($Re_\delta > 10^8$) in the outer part of an atmospheric boundary layer (Hanna, 1981). As another example, measurements in a jet flow yielded $\tau_{\Lambda E} \approx 0.23\delta/u_c$, where u_c is the local mean centerline velocity and δ is the local jet thickness (Wang et al., 2008).

A third feature of integral time scales in typical turbulent flows is that the relative magnitude for the different reference frames is given as $\tau_{\Lambda m\mathcal{E}} > \tau_{\Lambda\mathcal{L}} \gg \tau_{\Lambda\mathcal{E}}$ for weak turbulence (e.g., u_Λ less than 10% of the mean velocity). In particular, this relationship can be approximately quantified (Corrsin, 1963; Wang et al., 2008) as follows:

$$\tau_{\Lambda\mathcal{L}}/\tau_{\Lambda m\mathcal{E}} \approx 0.6 \qquad \textit{for } u_\Lambda \ll \bar{u}. \tag{6.65a}$$

$$\tau_{\Lambda\mathcal{E}}/\tau_{\Lambda m\mathcal{E}} \approx 0.4(u_\Lambda/\bar{u}) \tag{6.65b}$$

The close relationship between the Lagrangian and moving-Eulerian integral time scales of (6.65a) occurs because their spatial shifts (per Figure 6.15) will be similar for low turbulence. However, $\tau_{\Lambda\mathcal{L}} < \tau_{\Lambda m\mathcal{E}}$ since the Lagrangian fluid path meanders and interacts with other structures, as shown in Figure 6.16. Notably, the ratio of 0.6 in (6.65a) is only a rough estimate and the exact ratio can vary significantly for different types of flow. For example, ratios ranging from 0.3–1.0 have been reported (e.g., Elghobashi and Truesdell, 1984; Squires and Eaton, 1990; Loth and Stedl, 1999; Coppen et al., 2001; Carlier and Stanislas, 2005; Wang and Manhart, 2012). This high variability for an empirical constant of proportionality between turbulent properties is unfortunately common.

The relationship of (6.65b) indicates that Eulerian time scale is the shortest time scale. This is because turbulent structures convect past the Eulerian reference frame, and so the coherency at this fixed point is more quickly reduced. For example, consider a counterclockwise vortex that retains its coherency while moving past a fixed position. When the upwash portion of the eddy is at this point, there will induce an upward velocity fluctuation, but once the eddy has moved downstream, this point will be in the downwash portion so a downward velocity fluctuation will be induced. Therefore, the Eulerian fluctuations decorrelate based on the convection speed and the eddy size (and whether the eddy is upstream or downstream). In contrast, the Lagrangian and moving-Eulerian paths approximately move with the vortex, so there is little decay due to convection and instead most of the decorrelation is due to deformation and evolution of the structure. Again, the ratio of 0.4 in (6.65b) is only a rough estimate and ratios ranging from 0.2–0.8 have been reported for various flows (e.g., Pasquil, 1974; Anfossi et al., 2006; Di Bernardino et al., 2017).

In the opposite extreme of neglible mean flow (no mean convection) and strong turbulence so that the velocity fluctuations dominate, the moving-Eulerian path will be approximately the same as the Eulerian path (per 6.57a,c). Accordingly, the integral time scale for this case is as follows:

$$\tau_{\Lambda\mathcal{L}}/\tau_{\Lambda m\mathcal{E}} \approx 0.6 \qquad \text{for } u_\Lambda \gg \bar{u}. \tag{6.66a}$$

$$\tau_{\Lambda\mathcal{E}} \approx \tau_{\Lambda m\mathcal{E}} \tag{6.66b}$$

Again, the 0.6 ratio of (6.66a) is only approximate. For intermediate turbulence levels, the Eulerian integral time scales are intermediate to the values given by (6.65b) and (6.66b).

In the following section, we turn our attention to spatial shifts and integral length scales. In doing so, the subscripts for the velocity correlations will indicate whether spatial or time shifts are being considered, as follows:

$$\Upsilon_{\mathcal{E}}(\tau), \Upsilon_{\mathcal{L}}(\tau), \Upsilon_{m\mathcal{E}}(\tau) \quad \textit{for time shifts.} \tag{6.67a}$$

$$\Upsilon_{\parallel}(\mathcal{X}_{\parallel}), \Upsilon_{\perp}(\mathcal{X}_{\parallel}) \quad \textit{for spatial shifts.} \tag{6.67b}$$

For HIST, the time shift correlations of (6.67a) are independent of velocity direction (6.58). However, the spatial shift correlations of (6.67b) will be dependent on velocity direction, even for HIST, as described in the next section.

6.4.4 Integral Length Scales

Similar to that for the integral time scale, the integral length scale (Λ_i) is defined as the integral of the spatial correlation coefficient over all streamwise spatial shifts for velocity fluctuations in the i-direction with zero temporal shift ($\tau = 0$), as follows:

$$\Lambda_i(\mathbf{X}) \equiv \int_0^\infty \Upsilon_i(\mathbf{X}, \mathcal{X}_{\parallel})d\mathcal{X}_{\parallel} \equiv \int_0^\infty \frac{\overline{u'_i(\mathbf{X}, t)u'_i(\mathbf{X} + \mathcal{X}_{\parallel}, t)}}{u'_{i,rms}u'_{i,rms}} d\mathcal{X}_{\parallel}. \tag{6.68}$$

As indicated by the subscript index on the LHS, an integral length scale is based on location and based on the direction of velocity fluctuations (e.g., in the streamwise direction or one of the lateral directions). However, as shown in Figure 6.20, this length scale always uses a streamwise spatial shift that is parallel to the mean flow.

For homogeneous turbulence, the integral length scales are homogeneous, that is, independent of location. However, unlike integral time scales, the integral length scales are anisotropic even for isotropic turbulence. This anisotropy occurs because

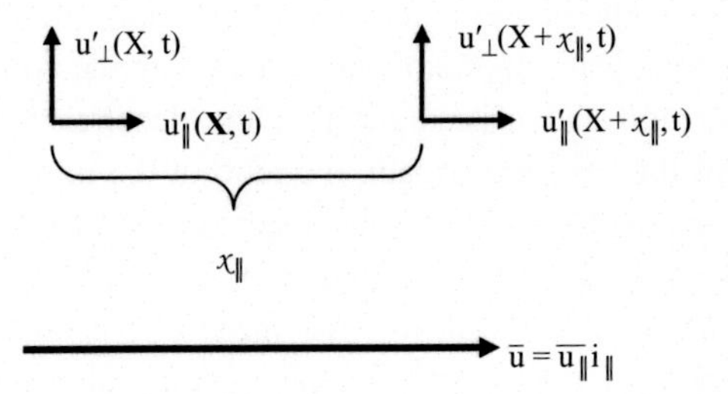

Figure 6.20 Schematic of streamwise spatial shift with streamwise velocity fluctuations that are parallel to the direction of the mean flow, while lateral velocity fluctuations are perpendicular.

the direction of the streamwise spatial shift is the same for streamwise velocity fluctuations but is different for lateral velocity fluctuations as follows:

$$\Upsilon_{\parallel}(x_{\parallel}) \equiv \frac{\overline{u'_{\parallel}(\mathbf{X},t)u'_{\parallel}(\mathbf{X}+x_{\parallel},t)}}{u'_{\parallel,rms}u'_{\parallel,rms}} \tag{6.69a}$$

for HIST.

$$\Upsilon_{\perp,2}(x_{\parallel}) = \Upsilon_{\perp,1}(x_{\parallel}) \equiv \frac{\overline{u'_{\perp,1}(\mathbf{X},t)u'_{\perp,1}(\mathbf{X}+x_{\parallel},t)}}{u'_{\perp,1,rms}u'_{\perp,1,rms}} \tag{6.69b}$$

The two lateral correlations of (6.69b) are equal since they are both perpendicular to the streamwise shift.

For the streamwise correlation, the dependency on spatial shift can be approximated with an exponential decay (as in 6.63), while that for the lateral correlation has more complex spatial dependency:

$$\Upsilon_{\parallel}(x_{\parallel}) \approx \exp(-x_{\parallel}/\Lambda_{\parallel}). \tag{6.70a}$$

$$\Upsilon_{\perp,1}(x_{\parallel}) = \Upsilon_{\perp,2}(x_{\parallel}) \approx [1 - x_{\parallel}/(2\Lambda_{\parallel})] \exp(-x_{\parallel}/\Lambda_{\parallel}). \tag{6.70b}$$

The differences between these two correlations are illustrated in Figure 6.21, where the lateral correlation includes a significant negative loop and a faster decay. The

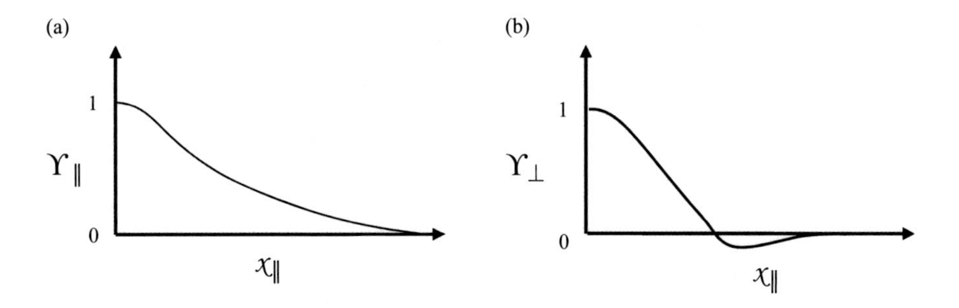

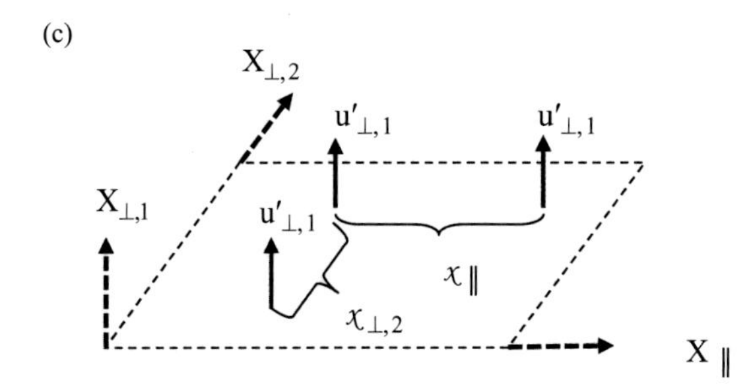

Figure 6.21 Schematics of (a) the streamwise correlation function (based on the streamwise shift); (b) the lateral correlation function (based on the streamwise shift), which includes a negative loop; and (c) lateral velocity perturbations on a plane parallel to the mean flow. For (c), upward velocity fluctuation at one part of this plane must be balanced by downward fluctuation at another part of this plane so that the net area-averaged flux is zero.

associated length scales based on the correlation integrals will therefore be different as well:

$$\Lambda_{\parallel} \equiv \int_0^{\infty} \Upsilon_{\parallel}(x_{\parallel})dx_{\parallel} \equiv \int_0^{\infty} \frac{\overline{u'_{\parallel}(\mathbf{X},t)u'_{\parallel}(\mathbf{X}+x_{\parallel},t)}}{u'_{\parallel,rms}u'_{\parallel,rms}} dx_{\parallel}. \tag{6.71a}$$

$$\Lambda_{\perp} \equiv \int_0^{\infty} \Upsilon_{\perp}(x_{\parallel})dx_{\parallel} \equiv \int_0^{\infty} \frac{\overline{u'_{\perp}(\mathbf{X},t)u'_{\perp}(\mathbf{X}+x_{\parallel},t)}}{u'_{\perp,rms}u'_{\perp,rms}} dx_{\parallel}. \tag{6.71b}$$

Based on the differences in (6.70), one may expect that the integral scale based on streamwise fluctuations will be longer than that based on lateral fluctuations.

To demonstrate why the negative loop occurs for the lateral correlation, consider incompressible flow where the instantaneous velocity divergence is zero (2.8). Using the velocity decomposition of (6.5) and taking a time average of the divergence, the mean velocity field must also be divergence-free. Subtracting these results, the divergence of the instantaneous velocity fluctuations is likewise zero:

$$\nabla \mathbf{u}' = \frac{\partial u'_{\parallel}}{\partial x_{\parallel}} + \frac{\partial u'_{\perp,1}}{\partial x_{\perp,1}} + \frac{\partial u'_{\perp,2}}{\partial x_{\perp,2}} = 0. \tag{6.72}$$

Next we consider velocity fluctuations for one of the lateral directions integrated over a surface plane defined by the other two directions (per Figure 6.21c) and hypothesize that this integral is zero:

$$\int_{-\infty}^{\infty}\int_{-\infty}^{\infty} u'_{\perp,1}(\mathbf{X},t)\, dx_{\parallel}dx_{\perp,2} = 0. \tag{6.73}$$

Since this plane is parallel to the streamwise direction and the mean flow, the time-averaged mass flux that is normal to this plane must be zero. By extension, area-averaged mass flux normal to this plane is also zero (for infinite integration), so the integral must also be zero, which supports the RHS hypothesis of (6.73). By further extension, the integral of the spatial correlation of two such velocity fluctuations in the same direction and shifted on the same plane is also zero, as follows:

$$\int_{-\infty}^{\infty}\int_{-\infty}^{\infty} \overline{u'_{\perp,1}(\mathbf{X},t)u'_{\perp,1}(\mathbf{X}+x_{\parallel},t)}\, dx_{\parallel}dx_{\perp,2} = 0. \tag{6.74}$$

A similar integral can be written for fluctuations in the other lateral direction. Combining those two integrals with (6.69) and (6.72) and assuming weak turbulence results in a relationship between the streamwise and lateral spatial correlations (Csanady, 1963), as follows:

$$\Upsilon_{\perp}(x_{\parallel}) = \Upsilon_{\parallel}(x_{\parallel}) + \frac{x_{\parallel}}{2}\frac{\partial \Upsilon_{\parallel}(x_{\parallel})}{\partial x_{\parallel}} \qquad \textit{for } u_{\Lambda} \ll \bar{u}. \tag{6.75}$$

Combining this relationship with (6.70a) yields (6.70b). In addition, the combination of (6.71) and (6.75) yields the following:

$$\Lambda_{\perp,1} = \Lambda_{\perp,2} = \tfrac{1}{2}\Lambda_{\parallel} \qquad \textit{for}\ u_{\Lambda} \ll \bar{u}. \tag{6.76}$$

As such, the integral length scales are anisotropic even when the turbulent velocity fluctuations are isotropic. In particular, the integral length scales for lateral velocity fluctuations are half that for streamwise velocity fluctuations (as can be seen by the faster decay in Figure 6.21b). For example, if the mean fluid flow is in the x-direction (such that y and z are the lateral directions), then

$$\Lambda_y = \Lambda_x = \tfrac{1}{2}\Lambda_x \qquad \textit{for}\ u_{\Lambda} \ll \bar{u}. \tag{6.77}$$

Since this difference stems from mass conservation and incompressible flow, it is often called the *continuity effect*.

While (6.76) is for HIST flow, this ½ ratio between the two length scales is also often realized for most NAST flows with weak turbulence, such as shown in Figure 6.22 for a turbulent boundary layer. Furthermore, the integral length scales are generally proportional to lengths of the overall flow. For example, Figure 6.22 shows $\Lambda_{\parallel} \approx 0.5\delta$ for a boundary layer, while $\Lambda_{\parallel} \approx 0.3\delta$ has been found for mixing layers and jets of local thickness δ (Wygnanski and Fiedler, 1969. 1970; Vanderwel and Tavoularis, 2014).

However, if one considers strong turbulence and weak mean flow with HIST, the integral scales will be similar (Hoque et al., 2016):

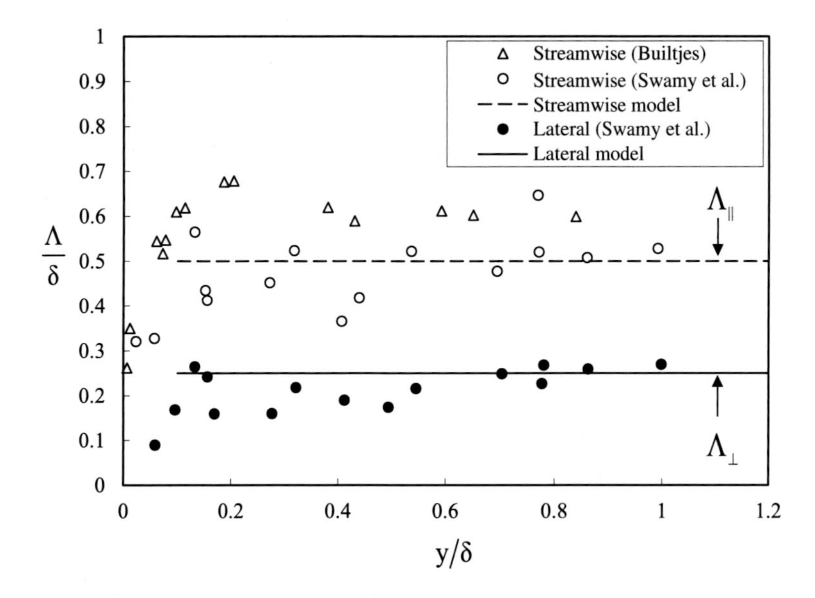

Figure 6.22 Measurements of integral length-scales normalized by turbulent boundary-layer thickness at $Re_\delta \approx 27{,}000$ (Builtjes, 1975) and $Re_\delta \approx 7{,}000$–$17{,}000$ (Swamy et al., 1979) showing the 2:1 ratio between the length scale based on streamwise velocity fluctuations versus that based on lateral velocity fluctuations is reasonable for $y/\delta > 0.1$.

$$\Lambda_{\perp} \approx \Lambda_{\parallel} \qquad \textit{for } u_{\Lambda} \gg \bar{u}. \tag{6.78}$$

For intermediate turbulence intensities, one may expect a transition from (6.76) to (6.78).

6.4.5 Integral-Scale Relationships and Frozen Turbulence

Integral-Scale Relationships

Given they are both based on correlation coefficients, one may expect integral length and time scales to be related (as was found for domain scales, where $D = u_D\tau_D$). In particular, dimensional analysis and physical reasoning suggest an integral-scale relationship, as follows:

$$\Lambda \sim u_{\Lambda}\tau_{\Lambda\mathcal{L}}. \tag{6.79}$$

The *turbulence structure parameter* (c_{Λ}) can be defined to quantify this relationship, as follows:

$$c_{\Lambda} \equiv \frac{\Lambda_{\perp}}{u_{\Lambda}\tau_{\Lambda\mathcal{L}}} = \frac{\Lambda_{\perp}}{\tau_{\Lambda\mathcal{L}}}\sqrt{\frac{3}{2k}}. \tag{6.80}$$

The RHS employs (6.18) to obtain the influence of the TKE. The value of the structure parameter (on the LHS) varies with flow type and location (Hinze, 1975). For example, the outer region of the turbulent boundary layer yields $c_{\Lambda} \approx 2$ based on the combination of $u_{\Lambda} \approx 0.04u_{\infty}$ (Figure 6.5), $\tau_{\Lambda} \approx 3\delta/u_{\infty}$ (Figure 6.19), and $\Lambda_{\perp} \approx 0.25\delta$ (Figure 6.22). However, the decaying wake flow of Snyder and Lumley (1971) yields $c_{\Lambda} \approx 0.5$. As such, there is no single value of the structure parameter for all flows, and instead we can only expect $c_{\Lambda} \sim 1$ for most flows, and that the exact value is a function of the local correlation coefficients.

If one again applies dimensional analysis and physical reasoning, the turbulent dissipation (6.33) can be expected to scale with the ratio of TKE and the integral time scale and that the dissipation can have a characteristic length scale related to TKE. These relationships can then be used to define the *dissipation parameter* (c_{ε}) and the *dissipation length scale* (Λ_{ε}) as follows:

$$c_{\varepsilon} \equiv \frac{\tau_{\Lambda\mathcal{L}}\varepsilon}{k} = \frac{\sqrt{3/2}\Lambda_{\perp}\varepsilon}{c_{\Lambda}k^{3/2}}. \tag{6.81a}$$

$$\Lambda_{\varepsilon} \equiv \frac{k^{3/2}}{\varepsilon} = \frac{\sqrt{3/2}}{c_{\varepsilon}c_{\Lambda}}\Lambda_{\perp}. \tag{6.81b}$$

Note that the RHS of (6.81a) and of (6.81b) employs (6.80) to express c_{ε} and Λ_{ε} in terms of the lateral integral length scale. As with the structure parameter, the dissipation parameter can also be expected to be of order unity for nearly HIST flows ($c_{\varepsilon} \sim 1$), but the quantitative value depends on the flow type and Reynolds number. For example, $c_{\varepsilon} \approx 0.2$–$0.3$ for grid-generated wake flows (Snyder and Lumley, 1971) while $c_\varepsilon \approx 0.3$–$0.5$ for pipe flows (Oesterle and Zaichik, 2004). Thus, c_ε and c_Λ characterize the relationships between integral scales but are not universal constants.

Frozen Turbulence Integral Scales

When the turbulence is very weak, the change in the turbulent flow is dominated by convection of the structures instead of evolution of the structures, so the fluctuations along the fluid path stay approximately constant. The limit for which only mean convection occurs and there are no evolution effects (i.e., the fluctuations do not change) is called *frozen turbulence*, given as follows:

$$\mathbf{u}'_{\mathrm{ft}}(\mathbf{X} + \bar{\mathbf{u}}\tau, t + \tau) \approx \mathbf{u}'(\mathbf{X}, t) \qquad \textit{for } u_\Lambda \ll \bar{u}. \tag{6.82}$$

For frozen turbulence (6.82), the velocity fluctuation field moves at the mean flow speed but does not evolve, as shown in Figure 6.23. As such, the turbulent structures are fixed in the moving-Eulerian reference frame (6.57c), so the associated correlation coefficient becomes

$$\Upsilon_{\parallel,mE}(\mathbf{X}, \tau) = \frac{\overline{u'_\parallel(\mathbf{X}, t) u'_\parallel(\mathbf{X} + \bar{\mathbf{u}}\tau, t + \tau)}}{u'_{\parallel,\mathrm{rms}} u'_{\parallel,\mathrm{rms}}} = 1 \qquad \textit{for frozen turbulence.} \tag{6.83}$$

In this limit, the turbulence stays fully correlated as it convects. This frozen-turbulence assumption is also known as Taylor's hypothesis (Hinze, 1975).

If we apply (6.57), (6.61a), and (6.71a) to relate the spatial shift to the temporal shift and assume HIST so there is temporal symmetry of the correlation coefficient,

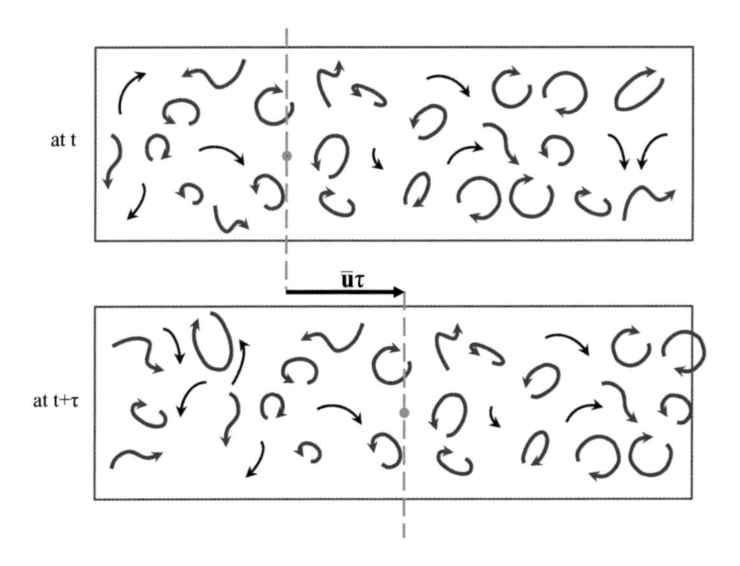

Figure 6.23 Frozen turbulence concept showing turbulent features with mean velocity subtracted (i.e., $\mathbf{u}'$) but where mean convection is much faster, so structure evolution can be ignored; the blue circles show the locations for a moving-Eulerian field where velocity correlation is always unity.

$\Upsilon(\mathbf{0}, \tau) = \Upsilon(\mathbf{0}, -\tau)$, the streamwise integral length scale for frozen turbulence can be expressed as follows:

$$\Lambda_{\|,\mathrm{ft}} = \bar{u} \int_0^\infty \Upsilon_{\|}(x_{\|}, 0) d\tau = \bar{u}\tau_{\Lambda\mathcal{E}} \qquad \textit{for frozen turbulence.} \qquad (6.84a)$$

$$\Lambda_{\|} \leq \Lambda_{\|,\mathrm{ft}} \qquad \textit{for HIST.} \qquad (6.84b)$$

The second expression states that the generalized integral scale is smaller than that for the frozen-turbulence limit. This is due to increased decorrelation from structure evolution.

6.4.6 Free-Shear Microscales

As noted previously, turbulence has a broad range of length scales and time scales (§2.6). The largest of these scales are the domain scales (5.12), while the most energetic scales are the integral scales (6.79). The smallest of the turbulent scales are the microscales, which become relatively smaller as the flow Reynolds number increases. To show this, consider a mixing layer whose Reynolds number is characterized by the velocity difference (ΔU) and the local shear-layer thickness (δ), as follows:

$$\mathrm{Re}_\delta = \frac{\rho_f \delta \Delta U}{\mu_f} \qquad \textit{for a mixing layer.} \qquad (6.85)$$

As shown in Figure 6.24, the shear-layer thickness, and thus Re_δ, grows as the flow convects downstream. In these flows, it can also be seen that the smallest structures become increasingly complex and smaller relative to δ as Re_δ increases from Figure 6.24a to Figure 6.24c. In addition, when Reynolds numbers are high enough to be fully developed (e.g., $\mathrm{Re}_\delta > 40{,}000$), the microstructure fine features do not have a preferred direction nor pattern, that is, they are locally homogeneous and isotropic. In contrast, the largest flow scales for all three shear layers have a preferred pattern, have a size comparable to δ (with eddy-and-braid structures aligned with the mean flow) and are not significantly influenced by Re_δ.

Similar results can be observed in other free-shear flows. For example, an axisymmetric jet, a plume, or a wake flow can be characterized by the difference between the local centerline velocity and the surrounding freestream velocity (ΔU) and the local jet, plume, or wake diameter (δ). In these flows, the largest scales are relatively independent of Re_δ and have consistent flow structure shapes aligned with the mean flow direction, while the smallest-scale features are more random in orientation and shape and become relatively smaller as the Re_δ number increases. Because the microstructures are driven by viscous dissipation, their relative size reduction is a consequence of the relative viscosity reduction as Re_δ increases (6.85).

In turbulent free-shear flows at high Re_δ, the smallest scales are called the *Kolmogorov microscales* (η). Since these scales are controlled by viscous dissipation (where the eddy energy is lost to heat), physical reasoning suggests that these scales

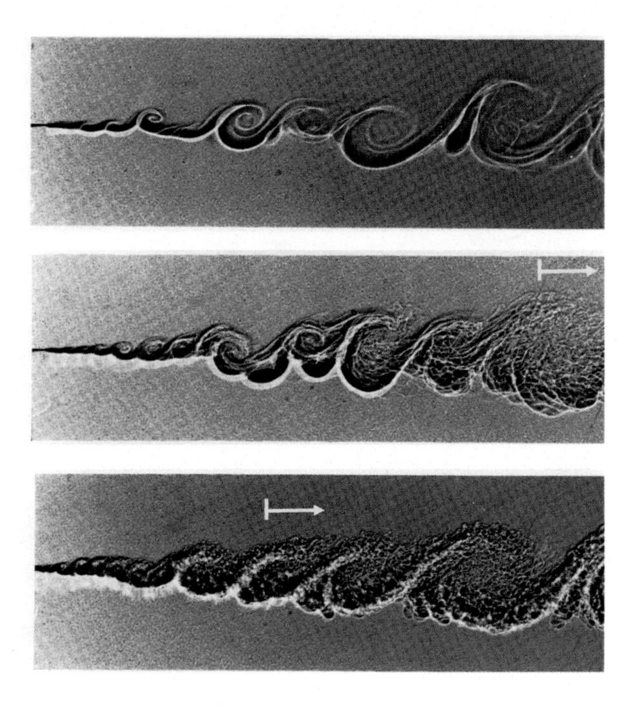

Figure 6.24 Free-shear layers with increasing local Reynolds numbers where fully developed turbulence is qualitatively observed for $Re_\delta > 40{,}000$ as indicated by yellow arrows (Brown and Roshko, 1974).

should be proportional to the turbulent dissipation (ε) and kinematic viscosity (ν). Based on dimensional analysis, this scaling can be used to define the Kolmogorov length scale as follows:

$$\eta \equiv \left(\nu_f^3/\varepsilon\right)^{1/4}. \tag{6.86}$$

Based on similar reasoning and dimensional analysis, the corresponding Kolmogorov velocity scale (u_η) can be similarly defined as follows:

$$u_\eta \equiv \left(\nu_f \varepsilon\right)^{1/4}. \tag{6.87}$$

Based on the combination of (6.86) and (6.87), the Kolmogorov Reynolds number is as follows:

$$Re_\eta \equiv \eta u_\eta/\nu_f = 1. \tag{6.88}$$

The RHS result ($Re_\eta = 1$) demonstrates that these microscale eddies have inertial forces that are comparable to the viscous forces. As such, structures smaller than η are dominated by viscous (versus inertial) forces so that they are quickly diffused and lost.

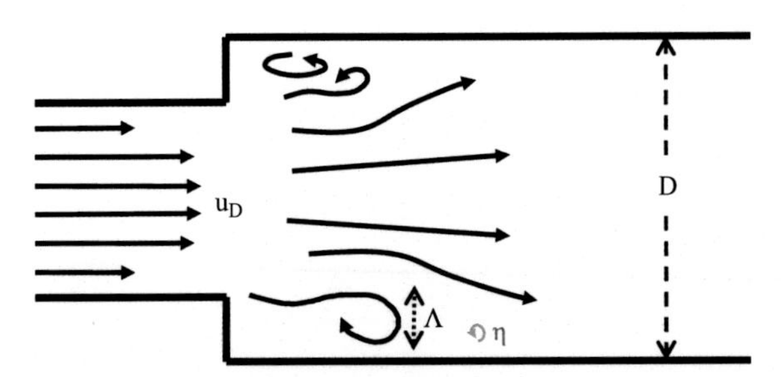

Figure 6.25 Flow in an expanding duct with a centerline velocity (u_D) notionally showing various length scales, including the domain scale based on a duct diameter (D), the integral length scale (Λ) based on most energetic structures, and the Kolmogorov length scale (η) based on the smallest turbulent structures.

Furthermore, the Kolmogorov length and velocity scales can also be used to define the Kolmogorov time scale, as follows:

$$\tau_\eta \equiv \eta/u_\eta = (\nu_f/\varepsilon)^{1/2}. \tag{6.89}$$

This is the time scale for the inertia of the smallest eddies to dissipate, that is, τ_η is the Kolmogorov eddy lifetime. Thus, η and τ_η act as spatial and temporal filters in the turbulence, whereby eddy structures of smaller size or smaller times (i.e., higher frequencies) cannot be sustained.

In general, we can expect that the velocity microscale (u_η) to be much smaller than both the domain and integral velocity scales (u_D and u_Λ). Similarly, we can expect that the length microscale (η) to be much smaller than both the domain and integral length scales (D and Λ), as qualitatively shown in Figure 6.25. To demonstrate the differences in the length scales, one may define an integral-scale Reynolds number and then express it as a length-scale ratio using (6.80), (6.81), and (6.86) as follows:

$$Re_\Lambda \equiv u_\Lambda \Lambda_\parallel/\nu_f \sim (\Lambda_\parallel/\eta)^{4/3}. \tag{6.90}$$

As such, Re_Λ » 1 for a fully developed turbulent flow indicates that Λ » η (and higher Reynolds numbers lead to higher ratios). For example, the yellow arrow location of the shear layers in Figure 6.24 yields $Re_\Lambda \approx 4{,}000$ and thus η ~ Λ/500 (too small to see in the photographs).

6.4.7 Turbulent Kinetic Energy Cascade

To characterize the rest of the turbulent scales in the flow, we define ℓ as a general turbulent length scale that is bounded by the smallest and largest turbulent scales:

$$\eta \le \ell \le D \qquad \textit{a general turbulent length scale.} \tag{6.91}$$

Next we define a *wavenumber* (*n*) based on the inverse of the turbulent wavelength, as follows:

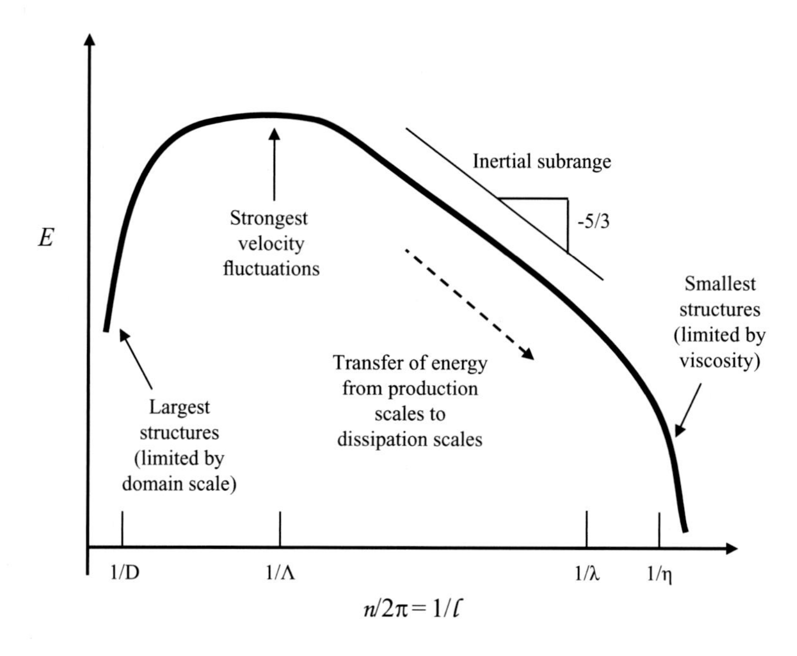

Figure 6.26 Schematic of the turbulence energy spectrum plotted on a *log–log scale* of specific turbulent kinetic energy as a function of the inverse of the wavelength. The turbulence is generally produced by instabilities at the domain scales (D) and become strongest at the integral scales (Λ). The energy then cascades down into smaller and smaller scales until it is dissipated by viscosity at the microscales (η). For a high enough Reynolds number, a well-developed inertial subrange (with a –5/3 slope on a log–log plot) will form between the integral scales and the Taylor scale (λ).

$$n \equiv 2\pi/\ell. \tag{6.92}$$

Thus, the smallest structures of the flow ($\ell = \eta$) will be associated with the highest wavenumbers ($n = 2\pi/\eta$), where n has units of 1/length. The *specific turbulent energy* (E) is next defined as the kinetic energy per wavenumber such that the integral of E over all wavenumbers equals the total kinetic energy, as follows:

$$\mathrm{k} = \int_0^\infty E(n)\,\mathrm{d}n. \tag{6.93}$$

The conceptual relationship between E and $n/2\pi$ is the turbulent kinetic energy spectrum and is notionally shown in Figure 6.26, where the area under this curve is $\mathrm{k}/2\pi$ based on (6.93).

The schematic of Figure 6.26 shows that the largest scales (smallest wavenumbers) are bounded by the macroscopic length scale (D) since all turbulent structures must fit inside the flow domain (e.g., the largest eddy in a pipe flow will generally be on the order of the pipe diameter). The smallest scales (highest wavenumbers) are limited by the Kolmogorov length scale (η), beyond which viscous dissipation effectively filters all the turbulence. The peak in the k distribution near $1/\Lambda$ indicates that the integral scales contain the highest portion of turbulent kinetic energy, that is, they are the most energetic. As such, the energy density decreases for wavenumbers that are either smaller or larger. In the lower range between 1/D and $1/\Lambda$, TKE is generated from fluid instabilities due to interaction with solid surfaces and/or by flow velocity gradients.

The upper range from $1/\Lambda$ to $1/\eta$ is often called the *cascade* of turbulent energy, which represents the transfer of turbulent kinetic energy generated at larger scales to that at smaller and smaller scales. For example, an eddy structure is often created at the largest scales associated with a mean flow gradient and becomes strongest at the integral scale. This eddy will have a finite lifetime before it changes (e.g., eddies of length Λ will have an eddy lifetime of τ_Λ), and generally the breakup will lead to smaller scales. The smaller eddies will also break up into even smaller scales, and so forth. Finally, the smallest eddies (Kolmogorov scale) do not break up but are instead dissipated (destroyed) by viscosity. For idealized HIST with a balance of production and dissipation of TKE per (6.50), the cascade reflects a continual flow of energy from the large scales (where production occurs) to the small scales (where dissipation occurs).

The turbulent cascade was qualitatively rendered by Leonardo DaVinci in Figure 6.27, where there is a continual process of large eddies forming, giving rise to smaller eddies and yet smaller eddies. The TKE energy cascade was also poetically and concisely summarized by L. F. Richardson in 1922: "Big whirls have little whirls that feed on their velocity. Little whirls have lesser whirls, and so on to viscosity" (Richardson, 1922) This refers to the nonlinear interactions that cause fluctuations at the larger scales (D and Λ) to generally break down into successively smaller scales, which are finally dissipated at the smallest (Kolmogorov) scales.

Figure 6.27 Water spilling into a pool from a drawing of Leonardo DaVinci, which indicates a cascade of turbulence with energy starting at large scales and moving into eddies at smaller and smaller scales.

The turbulent cascade also has specific characteristics if the integral-scale Reynolds number of (6.90) is high enough. In particular, the cascade of energy can give rise to an *inertial subrange* if the turbulent length scales satisfy $\Lambda \gg \ell \gg \eta$. Within this subrange, Kolmogorov hypothesized that the wavelengths (ℓ) depend on the specific energy (E) and on the turbulence dissipation (ε) but are independent of the viscosity since $\ell \gg \eta$. Using dimensional analysis with this hypothesis, the subrange wavelengths and wavenumbers are related as follows:

$$E \sim \varepsilon^{2/3} n^{-5/3} \sim \varepsilon^{2/3} \ell^{5/3} \qquad \textit{for } \Lambda \gg \ell \gg \eta. \tag{6.94}$$

This result indicates that the specific turbulent energy in the inertial subrange decreases as the wavenumber increases, yielding a –5/3 slope when considered on a log–log plot (Figure 6.26). In fact, the appearance of the –5/3 slope in the turbulent energy spectrum is often used as a critical test to determine whether a free-shear flow includes fully developed turbulence.

The inertial subrange of Figure 6.26 continues until the wavenumber is high enough that viscosity becomes important, leading to a faster decay of TKE. The Taylor length scale, λ, marks the end of the inertial subrange and the beginning of the dissipation range, where this negative slope increases. As such, λ is influenced by the turbulent energy, the dissipation, and the fluid viscosity and can be obtained as follows (Hinze, 1975):

$$\lambda = \mathrm{u}_\Lambda \sqrt{15 \nu_\mathrm{f}/\varepsilon} = \sqrt{15}\left(\mathrm{u}_\Lambda \tau_\eta\right). \tag{6.95}$$

The associated Taylor Reynolds number can then be defined as follows:

$$\mathrm{Re}_\lambda \equiv \mathrm{u}_\Lambda \lambda / \nu_\mathrm{f}. \tag{6.96}$$

Turbulent spectrums for several flows including free-shear flows and wall-bounded flows are shown in Figure 6.28 for a wide range of Taylor Reynolds numbers. Universally, an inertial subrange with a –5/3 slope is observed for $\mathrm{Re}_\lambda > 70$, which roughly corresponds to $\mathrm{Re}_\Lambda > 1{,}000$. In addition, higher wavenumbers for all these spectrums show a steeper negative slope, as in Figure 6.26, up to the Kolmogorov limit, where the spectrum stops.

While the preceding spectrum is based on wavenumber, one may also consider the spectrum in frequency space. In particular, the directional specific energy (F) is defined based on the velocity fluctuations in a single direction at a fixed point using energy per frequency so that the integral recovers the mean square of those velocity fluctuations. For HIST, the velocity fluctuations are isotropic so that directional specific energy is also isotropic, as follows:

$$\overline{\mathrm{u}'^2_\parallel} = \overline{\mathrm{u}'^2_{\perp,1}} = \overline{\mathrm{u}'^2_{\perp,2}} = \int_0^\infty F(f)\, df \qquad \textit{for HIST}. \tag{6.97}$$

Hinze (1975) proposed a model for the directional-specific energy frequency spectrum in terms of the Eulerian integral time scale as follows:

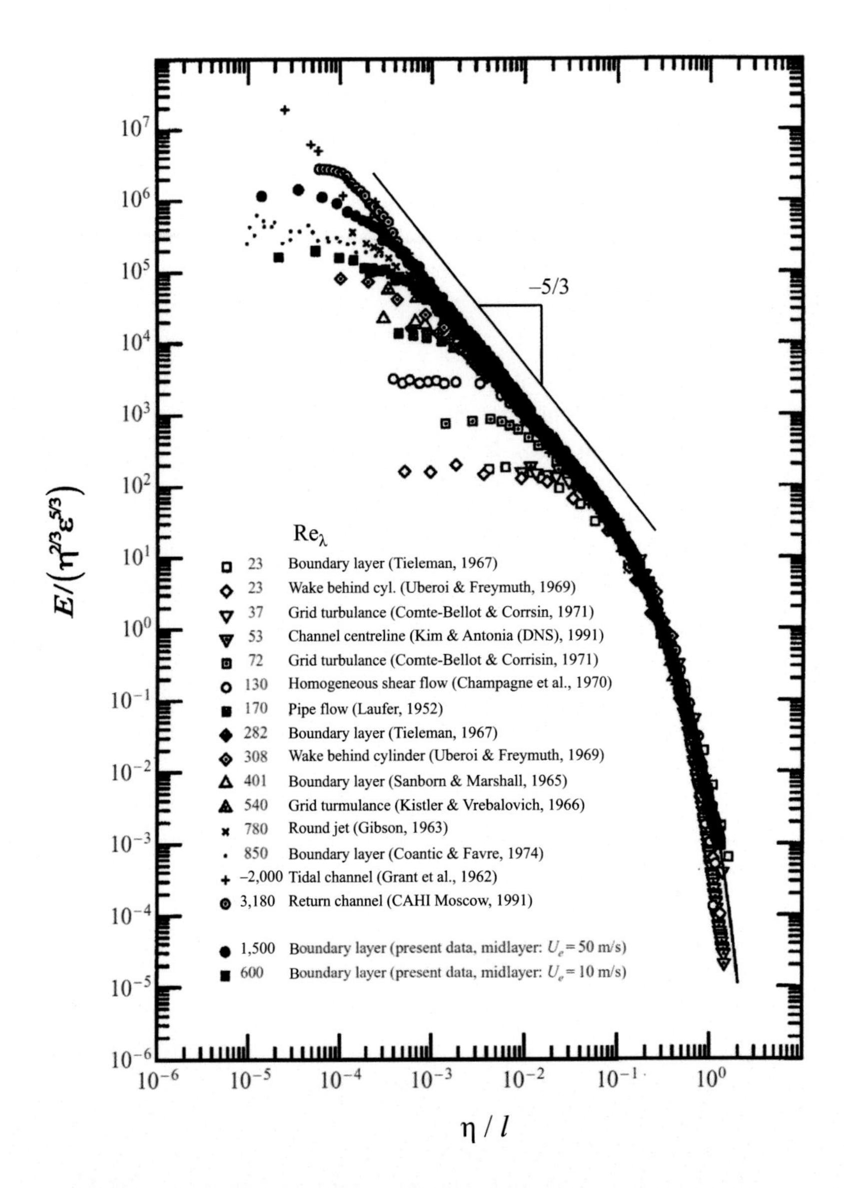

Figure 6.28 Experimental turbulence spectra in wavenumber space for various types of flows at a range of Taylor microscale Reynolds numbers (Pope, 2000).

$$F_{\mathcal{E}}(f) = \frac{4u_\Lambda^2 \tau_{\Lambda\mathcal{E}}}{1 + (2\pi f \tau_{\Lambda\mathcal{E}})^2}. \tag{6.98}$$

The spectra for the Lagrangian and the moving-Eulerian frame are the same except they use $\tau_{\Lambda\mathcal{L}}$ or $\tau_{\Lambda m\mathcal{E}}$ for the time scale. As shown in Figure 6.29, this simple model provides a reasonable representation of the overall spectrum shape, though the decay rate is –2 at high frequencies (greater than the –5/3 slope for the Kolmogorov spectrum seen in the experiments), and there is no low cutoff at the domain frequency nor a high cutoff at the Kolmogorov frequency.

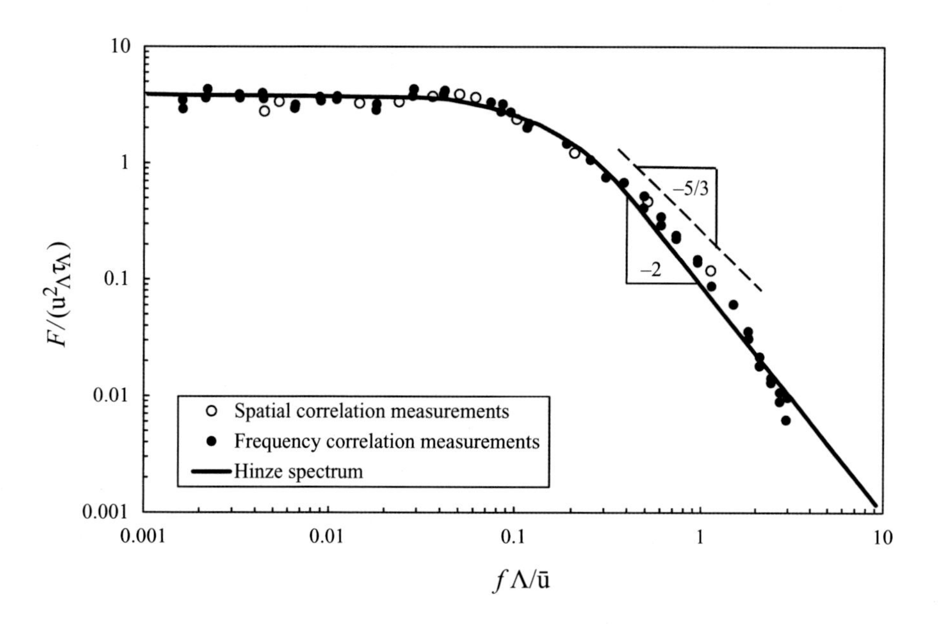

Figure 6.29 Turbulence directional-specific energy in frequency space comparing experimental data to a simplified spectrum model (Hinze, 1975).

The frequency of (6.97) can also be related to the wavenumber of (6.93) for weak turbulence by relating the time and length scales by mean convection per (6.84a) and the wavenumbers to the wavelengths by (6.92), as follows:

$$f \approx \frac{\bar{u}}{l} = \frac{\bar{u}n}{2\pi}. \tag{6.99}$$

For HIST, the respective frequency and wavenumber integrals of (6.97) and (6.93) are then related by (6.18), as follows:

$$\int_0^\infty E(n)\,dn = \frac{3}{2}\int_0^\infty F(f)\,df = \frac{3}{2}u_\Lambda^2 \quad \textit{for HIST and } u_\Lambda \ll \bar{u}. \tag{6.100}$$

Thus, $E(n)$ will have the same general shape as $F(f)$ for a turbulent flow: peaking at the integral scale, decaying at higher frequencies/wavenumbers, and terminating at the microscale. If an inertial subrange exists, the decay portion in the wavenumber spectrum ($E \sim n^{-5/3}$ as in Figure 6.28) will similarly be present in the frequency spectrum ($F \sim f^{-5/3}$ as shown by the experimental data in Figure 6.29). One may also convert the simplified Hinze frequency spectrum of (6.98) to a wavenumber spectrum in terms of an average integral scale, as follows:

$$E(n) = \frac{6u_\Lambda^2 \Lambda}{1 + (2\pi n \Lambda)^2} \tag{6.101a}$$

$$\textit{for HIST}.$$

$$\Lambda \approx \frac{1}{3}(\Lambda_\parallel + 2\Lambda_\perp) \tag{6.101b}$$

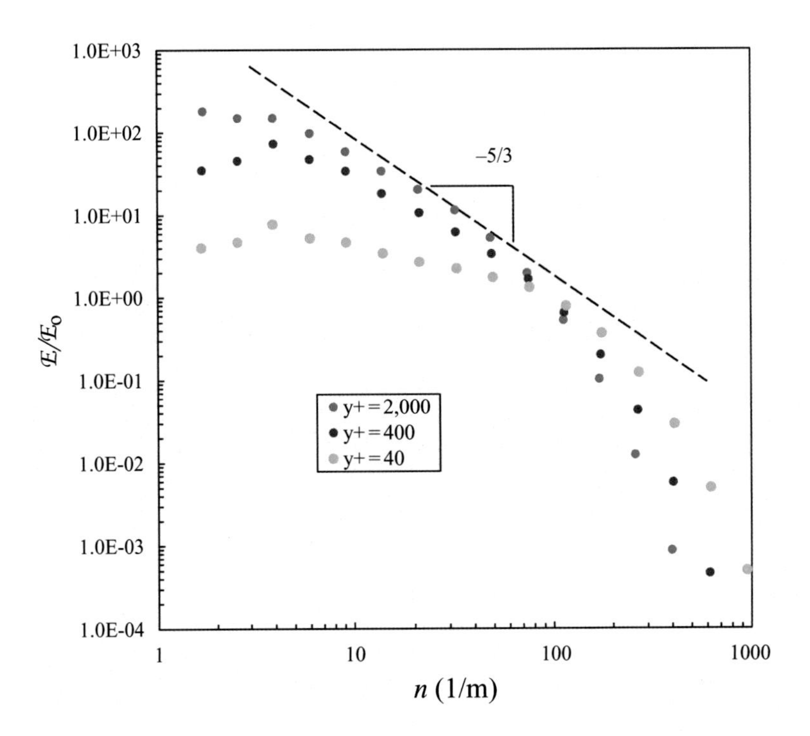

Figure 6.30 Turbulence spectra showing specific kinetic energy (normalized such that $\mathcal{E}/\mathcal{E}_o = 1$ at $n = 100$/m) as a function of wavenumber in a boundary layer at various heights above the wall surface for $Re_\delta = 31{,}000$ (Brown et al., 2017).

Again, this simplified Hinze spectrum is only approximate since it yields $E \sim n^{-2}$ at high frequencies, as shown in Figure 6.29, and since it does not appropriately stop at the Kolmogorov scale.

For wall-bounded flows, the turbulent spectrum will be modified when close to the wall. For example, the kinetic energy wavenumber spectrum for a turbulent boundary layer is shown in Figure 6.30 for different y^+ locations. Far from the wall (at $y^+ = 2{,}000$), the spectrum has an inertial subrange and a shape similar to that for free-shear flows (Figure 6.28). As shown in Figure 6.30, close to the wall (at $y^+ = 40$), the inertial subrange becomes lost, and the turbulent spectrum has less energy at lower frequencies (due to damping of turbulence by the wall). In addition, there is more energy at the higher frequencies due to a decrease in size of the structures, since the local turbulence tends more to the inner scales than the outer scales.

6.4.8 Computational Requirements of RANS, LES, and DNS

The turbulent kinetic energy cascade discussed in the preceding section can be used to estimate the computational resources for the turbulence numerical solutions of §6.3, which are illustrated in terms of the turbulent spectrum in Figure 6.31. The RANS approach is represented as a Dirac delta function since it effectively represents all the

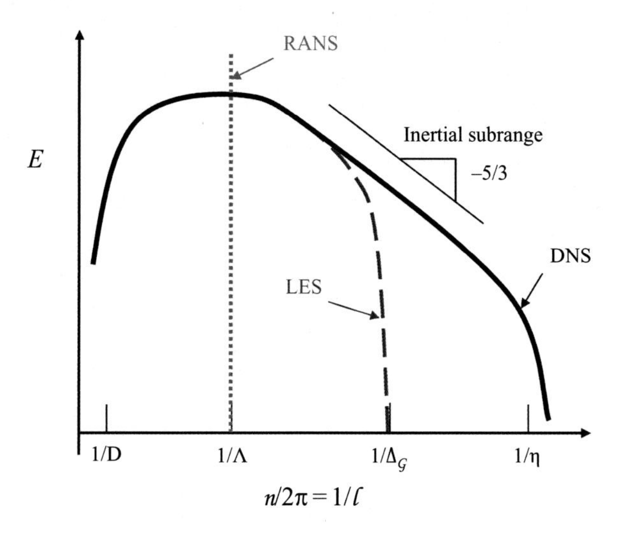

Figure 6.31 Schematic of the resolved portion of turbulence energy spectrum plotted on a log–log scale: RANS, which only captures a single length-scale (the mixing length); LES, which resolves most of the energy (from domain length scale to a subgrid length scale that is ideally in the inertial subrange); and DNS, which resolves all the energy (from domain scale down to the Kolmogorov scale).

kinetic energy in a single turbulent length scale (and thus a single frequency). DNS resolves all the length scales and frequencies in the turbulent spectrum. Intermediate approaches such as LES only resolve the largest scales of the turbulence (which contain most of the kinetic energy).

The ranges of scales resolved in the turbulent cascade can be used to estimate the minimum grid length scale (Δx_{min}) for the domain, as shown in Figure 6.32. This figure also shows that Δx_{min} is generally set for creeping flow and inviscid flow based on the minimum flow feature length scale (ℓ_{min}) in the domain. This length scale can be based on the minimum flow feature (e.g., shear-layer or boundary-layer thickness) and/or minimum geometric feature of a surface (e.g., radius of curvature or gap height). If one assumes that roughly 10 points are needed to resolve that feature, this yields the following:

$$\Delta x_{min} \sim \ell_{min}/10 \quad \textit{for laminar flow or inviscid flow.} \tag{6.102}$$

For turbulent free-shear flow with DNS where the Kolmogorov scale is much smaller than any geometric feature ($\eta \ll \ell_{min}$), the minimum grid scale is based on the Kolmogorov scale, as follows:

$$\Delta x_{min} \sim \eta \qquad \textit{for turbulent flow with DNS.} \tag{6.103}$$

This same criterion can also be used for transitional flow as shown in Figure 6.32. For LES, the filter width of (6.43) is typically smaller than the integral length scale and determines the minimum grid scale:

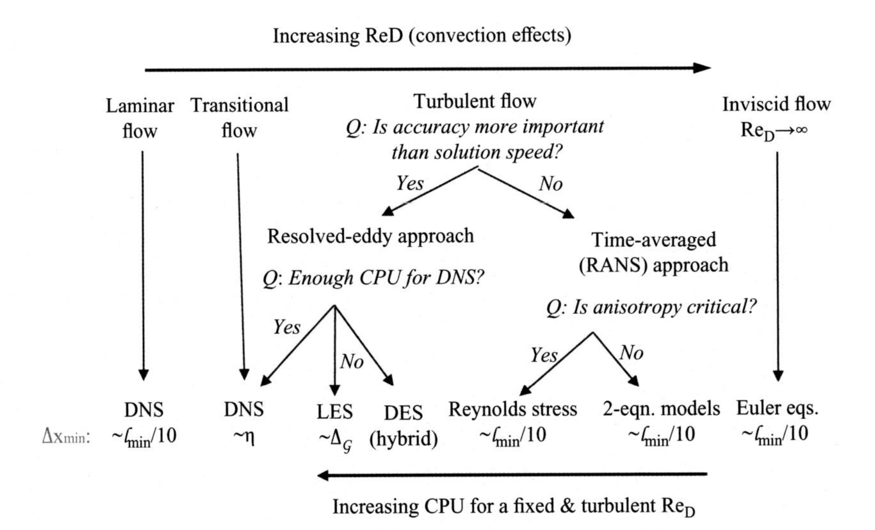

Figure 6.32 Schematic for selecting computational approaches for continuous-phase flow and the associated grid resolution for the free-shear regions.

$$\Delta x_{min} \sim \Delta_{\mathcal{G}} < \Lambda \quad \textit{for turbulent flow with LES.} \tag{6.104}$$

For the RANS approach, none of the turbulent structures are resolved, so

$$\Delta x_{min} \sim \ell_{min}/10 \quad \textit{for turbulent flow with RANS.} \tag{6.105}$$

As shown in Figure 6.32, to predict anisotropic aspects of turbulence (for increased accuracy), Reynolds stress turbulence models can be used. These do not increase the grid resolution requirements, but do increase the number of PDEs and thus are more computationally intensive.

The previous minimum mesh sizes can be combined with the flow Reynolds numbers to estimate the computational requirements for the RANS, LES, and DNS approaches. The most important factor for computational memory resources is typically the number of fluid nodes (N_f). This is because the flow variables at each grid point must generally be stored in the central processing unit (CPU) memory, and the required memory will scale linearly as a function of the number of grid points/nodes. In addition, the computational time requirements are generally based on a fixed number of floating-point operations per grid node per time step. Thus, overall CPU time is proportional to the product of the number of nodes (N_f) and the number of time steps (N_t). However, N_t can be difficult to predict since it will also depend on temporal accuracy and integration stability (Chung, 2002).

It should be noted that turbulent boundary layers require that the first set of grid points above a solid surface should be within $y^+ = 1$ (regardless of whether DNS, LES, or RANS is used). This y^+ requirement can influence the total number of nodes in the domain along with the criteria given in (6.103–6.105). A simpler approach is to note that the number of fluid nodes increase with Reynolds number.

The following estimates of N_f are based on the domain Reynolds number (which is generally proportional to the ratio of domain size to minimum grid scale). It should be noted that these are only very rough guides, since the actual computational resources can vary widely due to variations in grid type (structured versus unstructured), dimensionality (2D versus 3D), flow-field complexity (geometry and unsteady aspects), and the degree of grid stretching. Furthermore, the cost of computing depends markedly on the type of time stepping (explicit versus implicit, single-step versus multistep, etc.), as well as the computer processing speed and the computer architecture. In addition, the approximations that follow do not take into account substantial differences that may occur due to variations in boundary conditions (e.g., a complex domain shape often requires more computational cells than a simple brick-shaped domain) or due to spatial discretization accuracy (e.g., for a fixed number of nodes, solution accuracy can depend on the order of the spatial discretization scheme).

In the context of the preceding qualifying statments, the domain Reynolds number (Re_D) has the largest influence on CPU requirements. For 2D RANS, the data in Figure 6.33 for internal flows indicate that the number of grid nodes increases slowly with Re_D. Based on such results, the number of nodes for internal flows (inside passages) can be roughly approximated as follows:

$$N_f \sim 10^3\, Re_D^{1/5} \quad \textit{for 2D RANS}. \tag{6.106a}$$

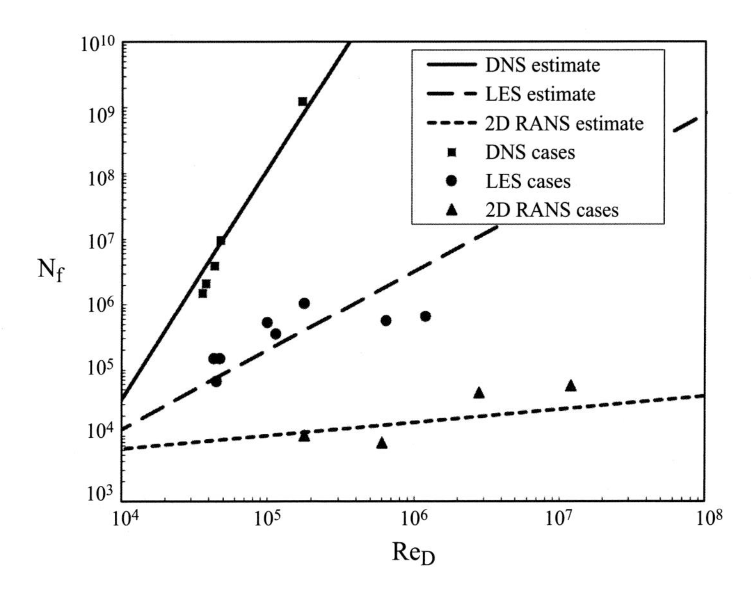

Figure 6.33 Typical number of computational fluid nodes for simple-geometry internal flows as a function of the domain Reynolds number for a streamwise length of domain (D) based on various sources including and Wu et al. (2012), Feldman and Avila (2018), and Peng et al. (2018), where actual number of nodes can be larger for complex geometries or for higher accuracy simulations.

$$N_f \sim 10^5 \, Re_D^{1/5} \qquad \textit{for 3D RANS.} \tag{6.106b}$$

For external flows (which have a far-field open boundary condition), the number of nodes will generally be larger, since the spanwise dimension is often larger.

For free-shear turbulence simulated with DNS, one must capture wavelengths over a ratio of Λ/η (from the integral scales to the Kolmogorov scales), which scales with $Re_\Lambda^{3/4}$ based on (6.90). As such, the number of nodes in all three dimensions tends to scale with $Re_\Lambda^{9/4}$ for the cascade region (a much more rapid scaling than for RANS). However, one must also capture wavelengths over a ratio of D/Λ (from the domain scale to the integral scales) and treat any near-wall regions. Because of these additional requirements, the number of nodes in all three dimensions for DNS tends to scale as $Re_D^{7/2}$ (Piomelli, 1997). This scaling is consistent with the DNS data in Figure 6.33 for internal flows, which can be estimated as follows:

$$N_f \sim (Re_D/500)^{7/2} \qquad \textit{for DNS.} \tag{6.107}$$

This scaling leads to an impractically high number of nodes (and CPU resources) for high Reynolds numbers (e.g., several billion nodes are needed for $Re_D > 10^6$). Thus, DNS is generally limited to flows with modest domain-scale Reynolds numbers.

For LES, the number of nodes will be related to D/Λ again, but the cascade region only needs to resolve a wavelength ratio of $\Lambda/\Delta_{\mathcal{G}}$ per (6.104). Piomelli (1997) estimated the nodal resolution for typical LES flows as $Re_D^{6/5}$. This scaling is consistent with the LES data in Figure 6.33, and the nodal requirements can be estimated as follows:

$$N_f \sim 0.2 \, Re_D^{1.2} \qquad \textit{LES wall-bounded requirements.} \tag{6.108}$$

From Figure 6.33, LES is much less expensive compared to DNS, but can still be quite expensive compared to RANS. The number of nodes for a hybrid RANS/LES approach can vary widely, but will be generally bounded by (6.106b) and (6.108).

In summary, as a computational approach resolves more of the turbulent structures (from RANS, to LES, to DNS), the computational resources for a given flow Reynolds number quickly increases. As such, the choice of a turbulent flow is a balance between available computational resources with the desired level of solution accuracy and robustness. This trade-off between fluid physics and computational cost depends on the user's priorities: Is it better to get an approximate solution quickly (e.g., in a day or so), or is it better to get a more accurate solution later (e.g., in a few weeks or so)?

6.5 Theoretical Turbulent Diffusivity of a Species

In this final section on turbulence, we connect the Lagrangian integral time scale (τ_{AL} of §6.4.3) to turbulent mass diffusivity (Θ_{turb} in §6.2.4) based on HIST theory. This connection will help serve as a basis for the first section in Chapter 7 and also provides a theoretical foundation for the Fickian diffusion model. To do so, we consider turbulent flow in the limit of zero molecular diffusion ($\Theta_{\mathcal{M}} = 0$), so (6.42b) becomes

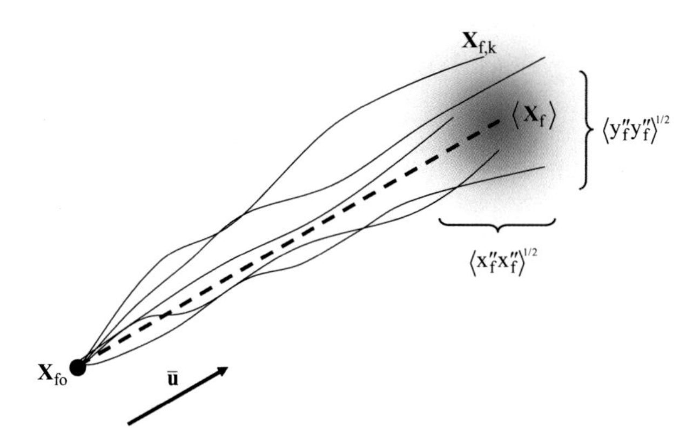

Figure 6.34 Schematic of fluid tracer trajectories (solid lines) injected at different times from a fixed-point source, where the ensemble-average trajectory (the dashed line) is based on the mean velocity and where all the tracer locations for a fixed time shift form a cloud (the shaded region) distributed about this ensemble average, with a cloud size based on the rms of the positional deviations (shown by brackets).

$$\frac{\mathcal{D}\mathcal{M}}{\mathcal{D}t} = \frac{\partial \mathcal{M}}{\partial t} + \nabla \cdot (\mathcal{M}\mathbf{u}) = 0 \qquad \textit{for } \Theta_{\mathcal{M}} = 0. \tag{6.109}$$

Since the species moves at the local instantaneous fluid speed ($\mathbf{u}$) with no molecular diffusion, it can be considered as a fluid tracer, that is, it will have a constant value along the time-dependent path of a fluid element. As such, species turbulent diffusion can be quantified by considering the turbulent diffusion of the fluid tracer paths.

Consider the fluid tracers shown in Figure 6.34, which emanate at a fixed point in space ($\mathbf{X}_{fo}$) and are released at various times (t_o) but have a consistent time shift (τ) since release (i.e., they travel from t_o to $t_o + \tau$). Since the flow is turbulent with a constantly changing velocity field, each release time will yield a different path. Along each path, we may consider the change of an arbitrary property q of the fluid element, where q can represent position, velocity, temperature, etc. If we apply the ensemble-averaging concept of (5.26), the Lagrangian average can be obtained using a statistically large number of fluid tracer trajectories ($N_f \gg 1$), as follows:

$$\langle q(\tau) \rangle = \frac{1}{N_f} \sum_{k=1}^{N_f} q_k(\tau) \qquad \textit{ensemble average for a fluid tracer.} \tag{6.110}$$

As such, the instantaneous q for a fluid tracer of index number k can be decomposed into an ensemble average and a deviation from this average as follows:

$$q_k(\tau) = \langle q(\tau) \rangle + q_k''(\tau) \qquad \textit{decomposition into mean and deviation.} \tag{6.111}$$

To determine the fluid tracer spread, consider q as the position of an individual fluid tracer ($\mathbf{X}_f$). If we set the release point as the origin ($\mathbf{X}_{fo} = 0$), then $\mathbf{X}_{f,k}$ is the movement relative to this point for tracer k and is controlled by the local velocity along its path:

$$\mathbf{X}_{f,k}(\tau) = \int_{t_o}^{t_o+\tau} \mathbf{u}(\mathbf{X}_{f,k}, t)\, \mathcal{D}t = \chi_{\mathcal{L},k}(t_o, \tau). \tag{6.112}$$

Per the RHS, the tracer position change is equal to the Lagrangian spatial shift of (6.57b) and is notionally shown by the solid lines in Figure 6.34. For a uniform mean velocity field, the ensemble-averaged tracer position based on (6.110) and (6.112) can then be obtained as follows:

$$\langle \mathbf{X}_f(\tau)\rangle \equiv \frac{1}{N_f}\sum_{k=1}^{N_f} \mathbf{X}_{f,k}(t_{o,k}, \tau) = \tau\bar{\mathbf{u}} = \chi_{mE}(\tau) \quad \textit{for uniform mean flow}. \tag{6.113}$$

Per the RHS, the ensemble-averaged tracer trajectory is equal to the moving-Eulerian spatial shift of (6.57c) and is notionally shown by the dashed line along the man fluid velocity in Figure 6.34. Thus, this mean path is the centroid of all the individual paths, and the instantaneous positions for a given time shift form a cloud of points around this ensemble-averaged path.

To quantify the size of this cloud, we define the deviation from this average for each tracer based on (6.111) and break this deviation into the three Cartesian components:

$$\mathbf{X}''_{f,k} \equiv \mathbf{X}_{f,k} - \langle \mathbf{X}_f(\tau)\rangle \equiv x''_{f,k}\mathbf{i}_x + y''_{f,k}\mathbf{i}_y + z''_{f,k}\mathbf{i}_z. \tag{6.114}$$

The average of each deviation is zero (by definition), but the rms of each deviation will be nonzero and will be equal in all directions for HIST:

$$\langle x''_f x''_f\rangle^{1/2} = \langle y''_f y''_f\rangle^{1/2} = \langle z''_f z''_f\rangle^{1/2}. \tag{6.115}$$

These cloud widths are shown notionally in Figure 6.34. Similar to the Brownian diffusivity of (5.32), the turbulent diffusivity (with units of length^2/time) is the growth rate of the mean square of the tracer location deviations and will be the same in all directions for HIST:

$$\Theta_{turb}(\tau) \equiv \frac{1}{2}\frac{\mathcal{D}\langle x''_f x''_f\rangle}{\mathcal{D}\tau} = \frac{1}{2}\frac{\mathcal{D}\langle y''_f y''_f\rangle}{\mathcal{D}\tau} = \frac{1}{2}\frac{\mathcal{D}\langle z''_f z''_f\rangle^{1/2}}{\mathcal{D}\tau} \quad \textit{for HIST}. \tag{6.116}$$

Next we relate this turbulent diffusivity to the kinetic energy and the Lagrangian integral time scale. To do this, we employ Taylor's theory of turbulent diffusion (Reeks, 1977), which converts the diffusivity based on particle spread (6.116) to a diffusivity based on the time-shifted integral of the ensemble-averaged velocity correlation, as follows:

$$\Theta_{turb}(\tau) = \int_{t_o}^{t_o+\tau} \langle u'_y(t_o, 0)u'_y(t, \mathbf{X}_f)\rangle \mathcal{D}t. \tag{6.117}$$

Following Reeks (1977), the RHS integral can then be expressed in terms of the turbulent strength (6.18) and the Lagrangian correlation coefficient (6.58b) as follows:

$$\Theta_{\text{turb}}(\tau) = u_\Lambda^2 \int_{t_0}^{t_0+\tau} \Upsilon_{\mathcal{L}}(t)\, \mathcal{D}t. \tag{6.118}$$

Assuming an exponential form of the correlation coefficient (6.63b) and employing the Lagrangian integral time scale of (6.61b), the turbulent diffusivity is then

$$\Theta_{\text{turb}}(\tau) = \tau_{\Lambda\mathcal{L}} u_\Lambda^2 \left(1 - e^{-\tau/\tau_{\Lambda\mathcal{L}}}\right). \tag{6.119}$$

This is the Hinze theory for species turbulent diffusion (Hinze, 1975), and the time evolution is shown in Figure 6.35a, where the spread rate initially increases with time but then plateaus to a constant. Also shown in this figure is the diffusivity for "short times" via $\tau \ll \tau_{\Lambda\mathcal{L}}$ and for "long times" via $\tau \gg \tau_{\Lambda\mathcal{L}}$. For short times, the correlation is approximately unity (Figure 6.17), yielding the following:

$$\Theta_{\text{turb,o}}(\tau) \approx u_\Lambda^2 \tau \qquad \textit{for } \tau \ll \tau_{\Lambda\mathcal{L}}. \tag{6.120}$$

The "o" subscript on the LHS identifies the short time limit and the RHS shows the diffusivity increases linearly with the time shift. For long times, the diffusivity is given as follows:

$$\Theta_{\text{turb},\infty} \approx u_\Lambda^2 \tau_{\Lambda\mathcal{L}} \quad \textit{for } \tau \gg \tau_{\Lambda\mathcal{L}}. \tag{6.121}$$

The "∞" subscript on the LHS identifies the long-time limit and the RHS shows that this diffusivity reaches a constant that is proportional to the Lagrangian integral time scale. To investigate the accuracy of (6.119), the mean square of fluid-tracer position deviations can be obtained by integrating in time (τ) the diffusivity of (6.116) to yield

$$\langle x_f'' x_f'' \rangle = \langle y_f'' y_f'' \rangle = \langle z_f'' z_f'' \rangle = 2u_\Lambda^2 \tau_{\Lambda\mathcal{L}}^2 \left[\tau/\tau_{\Lambda\mathcal{L}} + e^{-\tau/\tau_{\Lambda\mathcal{L}}} - 1\right]. \tag{6.122}$$

This theoretical result is shown in Figure 6.35b and compares quite favorably with measurements for a nearly HIST flow.

Finally, the long-time theoretical diffusivity of (6.121) can be expressed in terms of kinetic energy and dissipation using (6.18) and (6.81a), as follows:

$$\Theta_{\text{turb},\infty} \approx \frac{2c_\varepsilon}{3}\frac{k^2}{\varepsilon} \qquad \textit{for } \tau \gg \tau_{\Lambda\mathcal{L}}. \tag{6.123}$$

For $c_\varepsilon = 0.2$, this is consistent with experimental values of Figure 6.35b, and with the Fickian diffusion model of (6.41) showing its theoretical foundation. However, as noted in §6.2.4, such models may not be accurate for complex flows.

6.6 Chapter 6 Problems

Show all steps and list any needed assumptions:

(P6.1) Consider a flat plate boundary layer of water at NTP with a constant freestream velocity of 2 m/s at downstream distance (x) of 1 m for fully

(a)

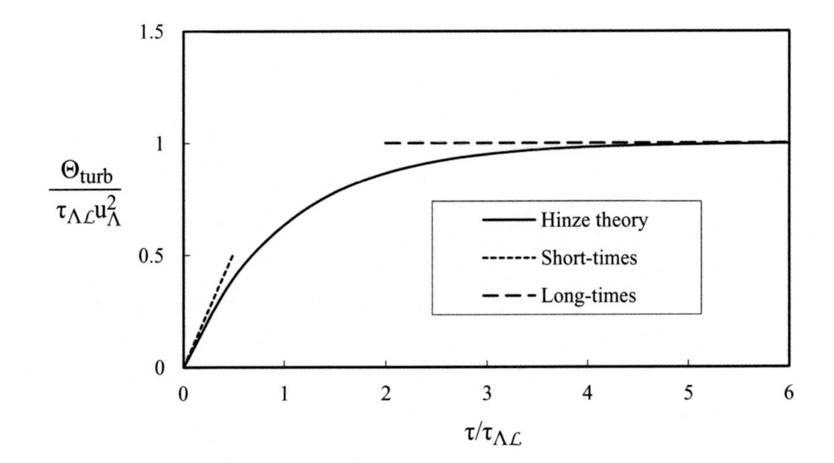

(b)

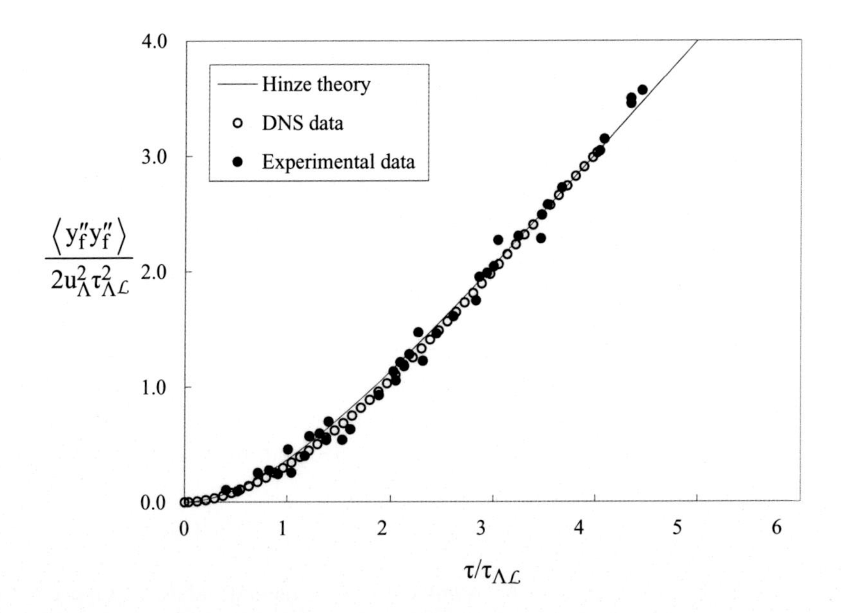

Figure 6.35 Hinze theory for turbulent spread of fluid tracers in nearly homogeneous isotropic turbulence in terms of release time for (a) turbulent diffusivity (spread rate), and (b) diffusion (spread) in the y-direction given a mean flow in the x-direction (Hinze, 1975; Dorgan and Loth, 2004).

laminar conditions and for fully turbulent conditions. (a) Determine whether the turbulent flow satisfies $u_\infty \gg u_{fr}$, $\delta \gg y_{fr}$, and $\tau_\delta \gg \tau_f$. (b) On a linear–linear chart, plot the boundary–layer velocity profiles using a plot range of 0 to 2 m/s on the horizontal axis (for u) and 0 to 25 mm on the vertical axis (for y) with a black line for laminar flow and various colored dashed lines for turbulent flow for the power-law, log-law, and laminar sublayer regions with appropriate y ranges. (c) Repeat (a) but on a log–log plot with y ranging from 10 μm to 10 cm, and (d) use the plots to discuss the qualitative differences between the laminar and turbulent flows in terms of the fluid physics.

(P6.2) For a turbulent flow in a pipe with a diameter (D) of 10 cm and an air velocity (u_D) of 10 m/s at NTP, (a) determine the pressure loss per meter based on the wall shear stress; (b) quantify $D^+ = D/y_{fr}$, $u_D{}^+ = u_D/u_{fr}$, and $\tau_D{}^+ = \tau_D/\tau_{fr}$, based on $\tau_D = D/u_D$ to compare ratio of outer scales to inner scales; and (c) express D^+ as an empirical function of Re_D and $u_D{}^+$ for turbulent pipe flows.

(P6.3) Express $\overline{\mathbf{u} \cdot \mathbf{u}}$ as a function of $\bar{u}$ and k.

(P6.4) Obtain (6.22a) and (6.22b) based on (6.21).

(P6.5) Starting with the instantaneous species fraction transport of (2.33b), obtain the time-averaged PDEs of (6.25) and (6.39), then comment on the validity of this final transport equation.

(P6.6) Starting with the instantaneous momentum transport of (2.73b), obtain the time-averaged PDEs of (6.24) and (6.28).

(P6.7) Starting from the instantaneous momentum equations of (2.84a) and (2.84b) obtain the time-averaged version with the Boussineq approximation.

(P6.8) Consider a planar turbulent jet of nitrogen into an otherwise still air whose mean velocity point is in the y-direction and varies in the x-direction as shown in the following schematic. In terms of u_x and u_y fluctuations at point **A**, use a schematic as in Fig. 6.6 to identify the key correlation and gradient; then apply and discuss the three principles of momentum mixing.

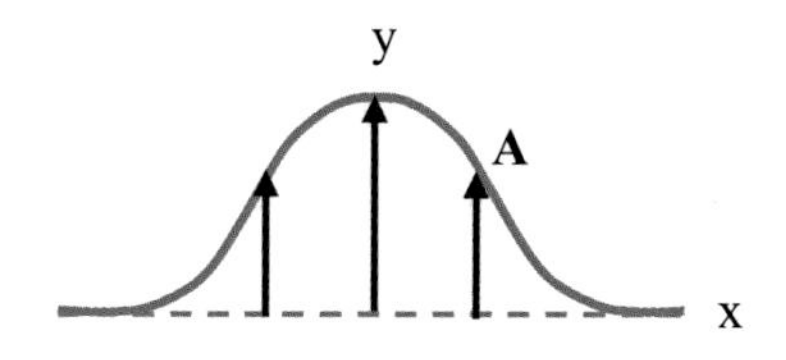

(P6.9) Consider a flat plate turbulent boundary layer of air at NTP with a freestream velocity of 75 m/s and a thickness of 6 mm. (a) Obtain Λ_{mix}/y_{fr} and Λ_{mix}/δ, at $y^+ = 5$ and $y^+ = 5000$, then comment on these results. (b) Obtain the equations for the velocity gradients and plot μ_{turb}/μ_f versus y/δ on a log–log plot for the power-law, log-law, and laminar sublayer regions and discuss the trends.

(P6.10) Consider the planar turbulent jet of (P6.8). (a) In terms of the mass fraction of nitrogen at point **A**, use a schematic like Fig. 6.8 to identify the key correlation and gradient; then apply and discuss the three principles of species mixing. (b) Discuss the different roles for molecular and turbulent mixing of a species in this flow.

(P6.11) Derive the transport PDE for the kinetic energy (6.31) by first taking the dot product of the instantaneous velocity with (2.73b).

(P6.12) List and discuss the primary pros and cons for solving the fluid momentum equations for a turbulent flow using (a) RANS, (b) DNS, and (c) LES.

(P6.13) Consider a constant channel flow with a diameter (D) of 10 cm that passes through a wire mesh that has a mesh spacing (ℓ_{mesh}) of 2.54 cm. Away from the thin boundary layers along the wall, the mean velocity in the x-direction is 6.55 m/s and the turbulent wake between $30 < x/\ell_{mesh} < 170$ can be approximated as follows:

$$\overline{u'_x u'_x} = \bar{u}_x^2[42.4(x/\ell_{mesh} - 16)]^{-1}.$$
$$\overline{u'_y u'_y} = \overline{u'_z u'_z} = \bar{u}_x^2[39.4(x/\ell_{mesh} - 12)]^{-1}.$$

(a) Obtain $\varepsilon(x)$ based on streamwise TKE decay. (b) By plotting the LHS values of (6.51a) and (6.51b) for $30 < x/\ell_{mesh} < 170$, determine the range of x/ℓ_{mesh} that is nearly homogeneous and the range that is nearly isotropic.

(P6.14) (a) Show that the exponential form of (6.63c) is consistent with the integral definition of $\tau_{\Lambda m\mathcal{E}}$, and (b) find the constant A if the correlation coefficient has the following form:

$$\Upsilon(\tau) = \cos(A\tau/\tau_{\Lambda m\mathcal{E}})\exp(-A\tau/\tau_{\Lambda m\mathcal{E}})$$

(P6.15) (a) Starting with (6.61a) and (6.82), obtain the result of (6.84a) and (b) determine the value of $\tau_{\Lambda m\mathcal{E}}$ in the limit of frozen turbulence.

(P6.16) Show that (6.70a) and (6.75) yields (6.70b) and (6.76).

(P6.17) Use both physical reasoning and dimensional analysis with E, n, and ε to explain why the specific turbulent energy in the inertial subrange must have a –5/3 slope on a log–log plot.

(P6.18) Consider the flow of (P6.13) with $c_\Lambda = 1$ and the following empirical description of the integral length scale for streamwise velocity fluctuations:

$$\Lambda_{\|} = (\ell_{mesh}/183)(x/\ell_{mesh} + 161).$$

(a) Obtain $\tau_{\Lambda\mathcal{L}}$ and c_ε at $x/\ell_{mesh} = 100$ and (b) determine what range of wavelengths, if any, would satisfy Kolmogorov's inertial subrange criterion of $\Lambda \gg \ell \gg \eta$ at $x/\ell_{mesh} = 100$ if "»" can be approximated as at least ten times larger.

(P6.19) Consider a turbulent air flow in a nearly spherical volume of 20 cm in diameter with continuous agitation so it is nearly HIST and ergodically stationary with a turbulent strength of 1.6 m/s and an integral scale of 3 cm. Assuming the turbulence structure parameter and dissipation parameter are both unity, (a) quantitatively plot the specific kinetic energy ($\mathcal{E}$) as a function of angular wavenumber ($n/2\pi$) with correct dimensional units on a log–log chart for the simplified Hinze spectrum (6.101); (b) qualitatively adjust this plot to consider the inertial sub-range as well as the minimum and maximum wavenumbers of the flow; and (c) quantitatively plot the Eulerian directional specific energy as a function of frequency (f) with correct dimensional units on a log–log chart for the simplified Hinze spectrum.

(P6.20) Consider a turbulent water flow in a river at NTP where k is 0.1 m^2/s^2, $\tau_{\Lambda\mathcal{L}}$ is 0.05 s, and $c_\varepsilon = c_\Lambda = 1$, determine (a) the Kolmogorov time, length, and velocity scales; (b) the integral-scale Reynolds number; and (c) the Taylor length scale and Taylor Reynolds number.

(P6.21) For a circular pipe flow with a domain Reynolds number of 10^6, (a) estimate the number of fluid nodes needed for 2D RANS, 3D RANS, LES, and DNS; and (b) discuss the accuracy of the crude LES empirical model for fluid nodes using a recently published computational study of a turbulent pipe flow, and do the same for DNS.

(P6.22) (a) Starting from (6.118), obtain the short-time turbulent diffusivity of (6.120) and the long-time turbulent diffusivity of (6.121), and show this is equivalent to (6.123). (b) Using (6.116), obtain the intermediate-time tracer spread of (6.122).

(P6.23) For the flow of (16.20), plot the turbulent diffusivity from 0 to 0.5 s. If the river speed is 1.5 m/s, plot the tracer cloud width from 0–2 meters downstream to show how a chemical in the river will spread.

7 Multiphase Turbulent Flow

Turbulence is an effective way to spread particles in a fluid. For example, internal combustion engines often induce high turbulence levels so that injected fuel droplets will be widely distributed to allow more uniform combustion. Combining the intricacies of turbulence with the complexities of particle motion is a field of research that continues to grow, as there are many unresolved issues. Fortunately, most of the basic aspects are reasonably understood and can be related to the turbulent Stokes numbers and the drift parameter, which are defined in §7.1. Using these parameters, the next sections focus on one-way coupling, including turbulent particle diffusion (§7.2), particle velocity fluctuations (§7.3), turbulent particle velocity bias (§7.4), and turbulence-induced deformation and breakup (§7.5). The last three sections characterize two-way coupling (§7.6), three-way coupling (§7.7), and four-way coupling (§7.8) in turbulent flows.

7.1 Turbulent Stokes Numbers and Particle Dispersion

Just as particle inertia determines a particle's reaction to domain-scale flow variations (§5.2), particle inertia also determines the reaction to turbulent flow fluctuations. This inertia influence can be characterized by a set of turbulent Stokes numbers based on the various turbulent time scales. For small inertia particles that tend to follow the flow, the turbulent time scale most likely to be "seen" along its path is the fluid Lagrangian integral time scale (the time scale seen by larger particles is more difficult to predict). As such, the Lagrangian integral time scale ($\tau_{\Lambda\mathcal{L}}$ of 6.61b) and the particle response time (τ_p of 5.11) are commonly used to define the integral-scale turbulent Stokes number as follows:

$$\mathrm{St}_\Lambda \equiv \frac{\tau_p}{\tau_{\Lambda\mathcal{L}}} = \frac{\left(\rho_p + c_\forall \rho_f\right) d^2}{(18\mu_f f)\tau_{\Lambda\mathcal{L}}} = \frac{w_{term}}{g|1-R|\tau_{\Lambda\mathcal{L}}}. \tag{7.1}$$

The RHS employs (3.93) to describe this parameter using the particle terminal velocity.

This Stokes number is the drag-based response time for the particle velocity compared to the time duration of a significant fluid velocity fluctuation seen by the

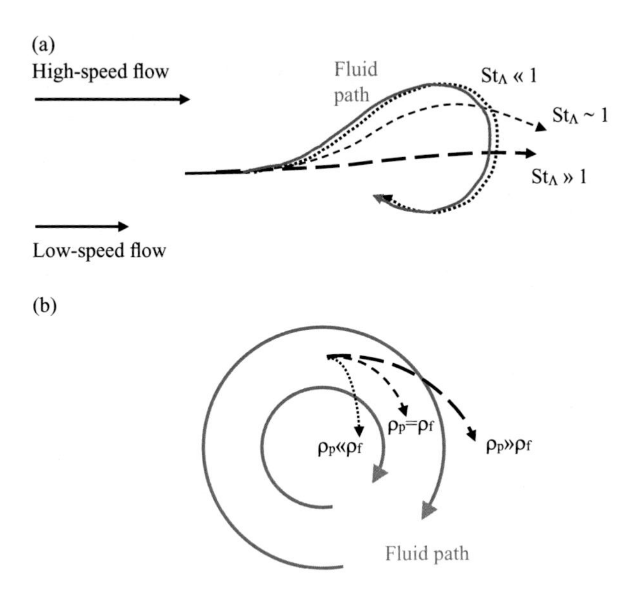

Figure 7.1 Schematic of particle trajectories (dashed lines) relative to the fluid path (the blue line) for (a) various Stokes numbers for a high-density particle ($\rho_p \gg \rho_f$) in an eddy, and (b) various density ratios for an intermediate inertia particle ($St_\Lambda \sim 1$) in a circular vortex, where high-density particles are centrifuged out, while low-density particles (such as bubbles) are pulled inward to the lower pressure.

particle. As in (5.14), the influence of the integral-scale Stokes number on particle dynamics can be summarized as follows:

$$St_\Lambda \ll 1 \textit{ particle closely follows integral scale structures.} \tag{7.2a}$$

$$St_\Lambda \sim 1 \textit{ particle motion partly follows integral scale structures.} \tag{7.2b}$$

$$St_\Lambda \gg 1 \textit{ particle motion is weakly affected by the integral scale structures.} \tag{7.2c}$$

These regimes are illustrated in Figure 7.1a for a high-density particle in the eddy of a turbulent mixing layer, where increasing St_Λ leads to a decreasing influence of the continuous-phase eddy structure on the particle motion. As such, a small inertia particle nearly follows the eddy, while a large inertia particle tends to follow its initial path. For intermediate inertia, the particle significantly responds to the eddy but still has a significant deviation from the fluid path.

While the Stokes number determines the degree of trajectory deviation, the particle density ratio determines the direction of this deviations, as shown in Figure 7.1b. In particular, high-density particles ($\rho_p \gg \rho_f$) tend to move to the outside edges of vortices due to centrifuging effects. This is because high-density particles tend to be directed in a straight line by their inertia and only curve partially inward due to the drag force. On the other hand, low-density particles ($\rho_p \ll \rho_f$) tend to move to the center of vortices, where the flow pressure is lower. This is because these low-density particles are much

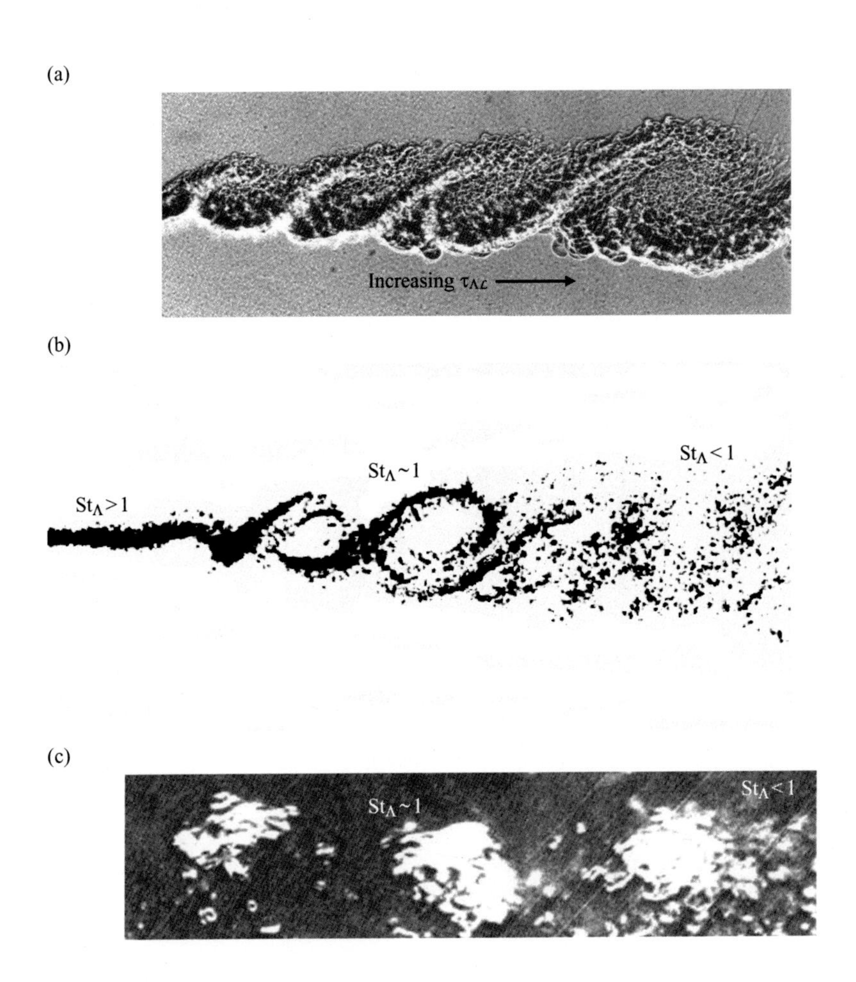

Figure 7.2 Photographs of turbulent shear layers with the high-speed flow on the top and with growing structures in the downstream direction (leading to increasing $\tau_{\Lambda\mathcal{L}}$) for (a) unladen fluid (Roshko and Brown, 1974); (b) 40 μm diameter solid particles in air (Wen et al., 1992); and (c) 1–5 mm bubbles in a water jet (Sene et al., 1994).

more likely to be affected by the fluid-stress force (3.64), which will pull them in the direction of the negative pressure gradient, that is, radially inward. Intermediate density particles ($\rho_p \sim \rho_f$) will have paths similar to that of the surrounding fluid since the centrifugal and fluid-stress effects will tend to balance out.

To visualize these effects, consider the turbulent shear layer of Figure 7.2a whereby the eddy and braid structures grow as they move downstream (as do the associated integral length and time scales). The addition of high-density particles ($\rho_p \gg \rho_f$) to this shear layer is shown in Figure 7.2b. Notably, τ_p is constant but the turbulent structure growths yields longer time scales so St_Λ is decreasing in the streamwise direction (per 7.1). The particles are initially at $St_\Lambda > 1$ and so tend to move with the mean flow. As the particles move downstream, they become characterized by $St_\Lambda \sim 1$ and are centrifuged toward the outside edges of eddies. Finally, the fluid turbulent integral-scale time is large enough that the particles are characterized by $St_\Lambda < 1$ and become dispersed throughout, consistent with that of the fluid itself. For the bubbles in water ($\rho_p \ll \rho_f$) of Figure 7.2c, the bubbles tend to be pulled into the low-pressure eddy cores at $St_\Lambda \sim 1$ and then become more dispersed downstream when $St_\Lambda < 1$. As such, the particle patterns in Figures 7.2b and 7.2c are consistent with Stokes effects shown in Figure 7.1a and the density ratio effect of Figure 7.1b.

While the preceding discusses the impact of the integral scales of turbulence, particles can also be influenced by the microscales of the turbulence. For a free-shear layer, the smallest temporal microscale is the Kolmogorov time scale (τ_η of 6.89), which can be used to define a microscale Stokes number as follows:

$$St_\eta \equiv \frac{\tau_p}{\tau_\eta} = \tau_p\sqrt{\frac{\varepsilon}{\nu_f}}. \tag{7.3}$$

This Kolmogorov Stokes number indicates how closely a particle will follow the smallest scales in the turbulence and can be used to establish response regimes (similar to 5.14), as follows:

$St_\eta \ll 1$ *particle motion closely follows all turbulent structures.* (7.4a)

$St_\eta \sim 1$ *particle motion partly follows the microstructures.* (7.4b)

$St_\eta \gg 1$ *particle motion is weakly affected by the microstructures.* (7.4c)

For a boundary layer, the smallest temporal microscale is the inner time scale (τ_{fr} of 6.9) so the corresponding microscale Stokes numbers can be defined as follows:

$$St^+ \equiv \frac{\tau_p}{\tau_{fr}} = \frac{\tau_p u_{fr}^2}{\nu_f}. \tag{7.5}$$

As in (7.4), a similar set of response regimes can be identified using St^+ for these flows.

The influences of the integral-scale and microscale Stokes numbers for a near-HIST flow are illustrated in Figure 7.3 for experiments of solid particles in a gas ($\rho_p \gg \rho_f$) in microgravity conditions (where the relative velocities are only driven by particle inertia and turbulent velocity fluctuations). In Figure 7.3a, particles with $St_\Lambda \sim 1$ are distributed in large-scale concentration patterns at scales consistent with the integral-scale turbulence. However, the small particles of Figure 7.3b with $St_\eta \sim 1$ yield fine-scale patterns with focused regions of high particle concentration. This phenomenon of fine-scale particle clustering in thin bands is known as *preferential concentration* (Eaton and Fessler, 1994). A similar result is shown in Figure 7.4 for simulations of high-density particles in a near-HIST flow with $St_\eta \sim 1$. The thin bands of preferential concentration coincide with regions of low vorticity due to the centrifuge effect, and the band thicknesses range from η to 4η (as was also the case with Figure 7.3b). In summary, $St_\Lambda \sim 1$ leads to large-scale particle concentration patterns consistent with the integral scales while $St_\eta \sim 1$ leads to preferential concentration of particles in highly focused bands with thicknesses consistent with the Kolmogorov length scale.

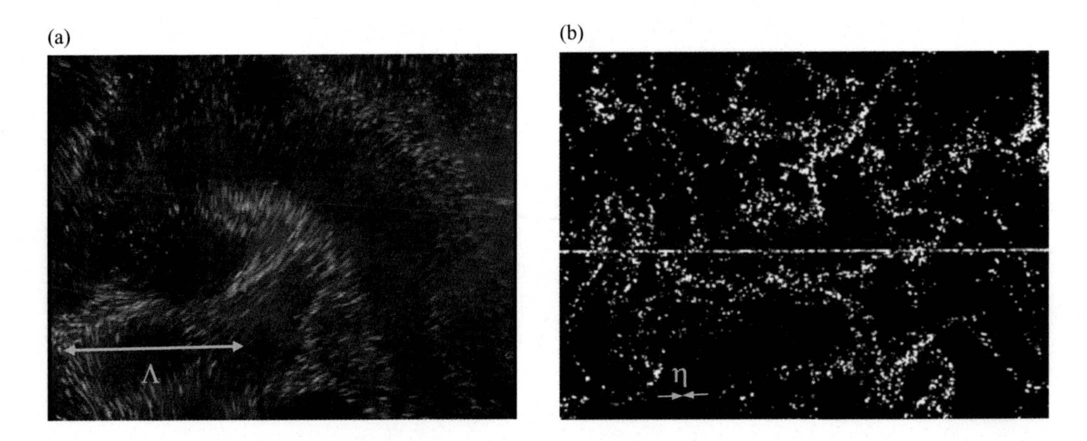

Figure 7.3 Illuminated particle fields whose concentration is correlated to the underlying turbulent structures for approximately homogeneous isotropic turbulence of particles in a gas ($\rho_p \gg \rho_f$) in microgravity conditions (Groszmann, 1999) for (a) $St_\Lambda \approx 1$ and $St_\eta \approx 10$, where particles concentration patterns are associated with the integral-scale structures; and (b) $St_\eta \approx 1$, where the particles cluster in tight bands associated with the microscales. Notional lengths of the integral scale and Kolmogorov scales are shown by orange arrows, but the actual scales have a greater range ($\Lambda = 22$ mm and $\eta = 0.6$ mm).

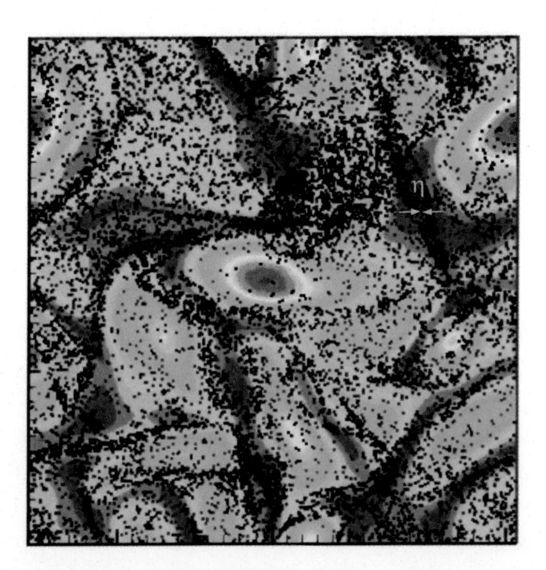

Figure 7.4 Two-dimensional slice in homogeneous isotropic decaying turbulence for $St_\eta = 1$, $g = 0$, and $\rho_p \gg \rho_f$ from a three-dimensional unsteady DNS prediction showing particle locations (black dots) and vorticity contours (red is maximum and blue is minimum), where particles are centrifuged out of vortices into bands of low vorticity (L. Y. M. Gicquel, 2005, personal communication, Centre Européen de Recherche et de Formation Avancée en Calcul Scientifique, http://cerfacs.fr/~lgicquel/gallery_images.html).

The concentration patterns seen in instantaneous realizations of Figures 7.2–7.4 all reflect *structural turbulent dispersion* of the particles based on the underlying instantaneous turbulent features. However, another useful viewpoint is to consider the concentration of the particles in a time-averaged sense. In this case, the turbulent structures are no longer individually resolved, and the average particle concentration spreads smoothly by *turbulent diffusion*. Figure 7.5 highlights the differences between the unsteady structural dispersion of particles (Figure 7.5a) and mean turbulent diffusion of particles (Figure 7.5b). While structural turbulent dispersion uncovers the fluid physics of the particle motion, turbulent diffusion is often more important to quantify system-level performance of a multiphase system. In the following, the mean spread (turbulent diffusion) of particles is analyzed in terms of the particle inertia and particle terminal velocity.

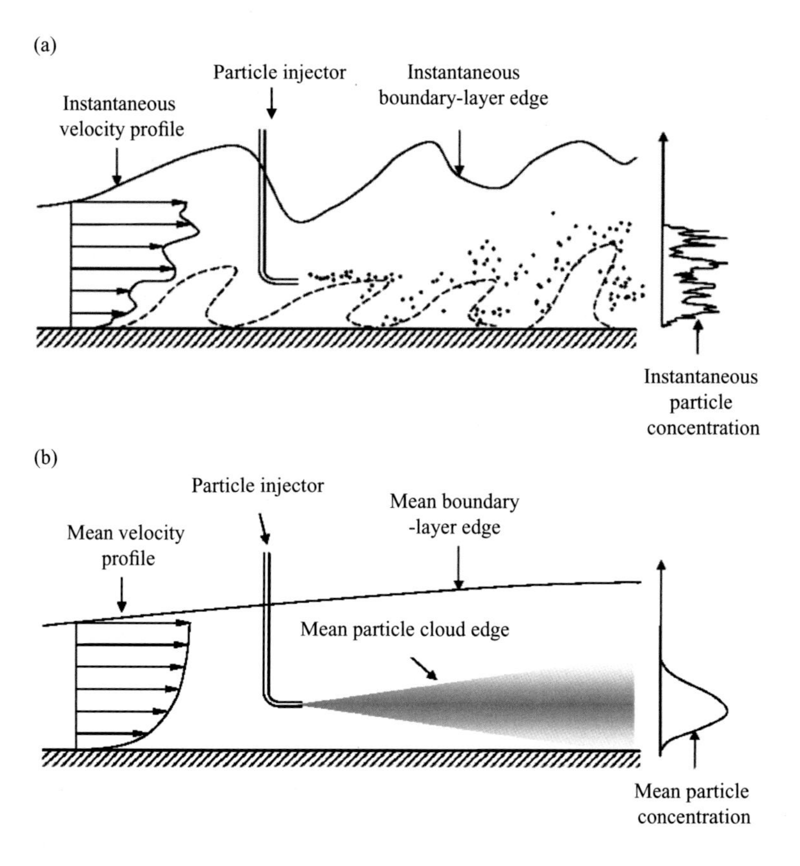

Figure 7.5 Schematic of particle dispersion in a turbulent boundary layer showing (a) instantaneous turbulent flow features and particle distribution demonstrating structural dispersion; and (b) time-averaged continuous-fluid features and particle concentration demonstrating mean diffusion.

7.2 Turbulent Diffusion of Particles

7.2.1 Particle Diffusivity Characteristics and Limits

To quantify turbulent diffusion for particles, consider a group of particles originating from a fixed single point in space ($\mathbf{X}_{po}$) but released at different times (t_o). Owing to different turbulent structures appearing at various times, each particle will follow a different trajectory, as shown in Figure 7.6 (similar to Figure 6.34). The Cartesian particle coordinates (x_p,y_p, and z_p) define the particle position vector, while the particle time shift (τ) along the trajectory is defined relative to the release time, as follows:

$$\mathbf{X}_p = x_p\mathbf{i}_x + y_p\mathbf{i}_y + z_p\mathbf{i}_z. \tag{7.6a}$$

$$\tau = t - t_o. \tag{7.6b}$$

These can be combined to specify the particle path spatial shift for a given time shift based on the integral of the particle velocity over its path:

$$\mathcal{X}_{@p}(\tau) \equiv \mathbf{X}_p(t_o + \tau) - \mathbf{X}_{p,o}(t_o) = \int_{t_o}^{t_o+\tau} \mathbf{v}(t)dt. \tag{7.7}$$

Following (6.110) and (6.111), a statistically large number of particle trajectories (N_p) can be used to define the time-varying ensemble-averaged trajectory and the associated trajectory decomposition, as follows:

$$\langle\mathbf{X}_p(\tau)\rangle \equiv \frac{1}{N_p}\sum_{k=1}^{N_p}\mathbf{X}_{p,k}(t_{o,k} + \tau) \qquad \textit{mean particle trajectory}. \tag{7.8a}$$

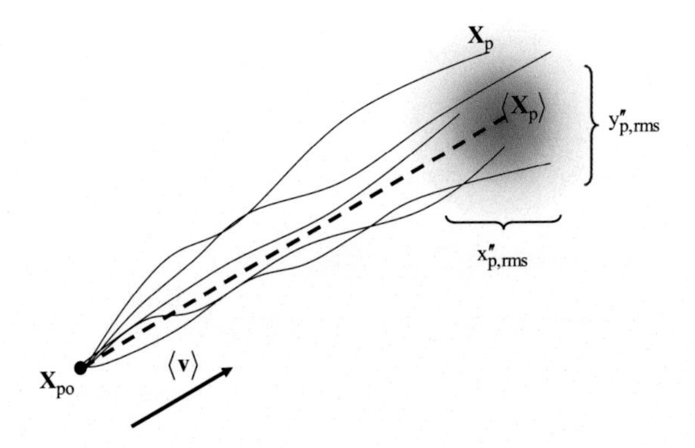

Figure 7.6 Schematic of particle trajectories (solid lines) injected at different times (t_o) from a point source and traveling until $t_o + \tau$, for a total travel time of τ. The ensemble average of these trajectories is the mean trajectory (the dashed line), and the spread of the particle concentration (the shaded region) can be quantified by the root mean square of the particle position deviation at a given time (e.g., in the y-direction, as shown by the vertical bracket).

$$\mathbf{X}_{p,k}(t_{o,k}+\tau) = \langle \mathbf{X}_{p,k}(\tau)\rangle + \mathbf{X}''_{p,k}(\tau) \qquad \textit{trajectory decomposition.} \tag{7.8b}$$

Note that $\mathbf{X}''_{p,k}$ is the trajectory deviation for a single particle with index k for a given time shift.

As an analogy, consider cars in a parking lot as particles. The cars leave the lot at different times (t_o), and we consider the car positions based on a time shift since leaving (τ). The average position of the cars after this shift is the ensemble average (7.8a). For example, the average car position 10 minutes after leaving the parking lot could be 5 km west. Any trajectory deviation is relative to this average. Therefore, if one of the cars had a postion 3 km east ten minutes after leaving the parking lot, its trajectory deviation is 8 km east (per 7.8b).

If we consider the y-component of the particle position and drop the index notation, it can be decomposed into a mean and a deviation, from which ensemble-averaged mean particle spread and the turbulent diffusivity are defined as follows:

$$y_p(\tau) = \left\langle y_p(\tau)\right\rangle + y''_p(\tau) \tag{7.9a}$$

$$y''_{p,rms}(\tau) = \sqrt{\left\langle y''_p(\tau) y''_p(\tau)\right\rangle} \tag{7.9b}$$

$$\Theta_{p,y}(\tau) \equiv \frac{1}{2}\frac{d\left\langle y''_p(\tau) y''_p(\tau)\right\rangle}{dt}. \tag{7.9c}$$

Similar expressions can be obtained for the decomposition, the spread and the diffusivity in the x- and z-directions. As shown in Figure 7.6, the mean spread of (7.9b) characterizes the size of particle cloud size at a given time shift. This spread can be used to quantity the diffusivity per (7.9c).

To obtain the particle velocity fluctuations, we use a decomposition as in (7.8b) to yield the following:

$$\mathbf{v}(\mathbf{X}_p, t_o+\tau) = \langle \mathbf{v}(\tau)\rangle + \mathbf{v}''(\mathbf{X}_p, \tau) \qquad \textit{for releases at } \mathbf{X}_{po}. \tag{7.10}$$

The velocity deviation can used for particle diffusivity via Taylor's theory of turbulent diffusion (6.117). For example, the y-component of velocity with (7.9c) can be used to obtain diffusivity in the y-direction as follows:

$$\Theta_{p,y}(\tau) = \int_{t_o}^{t_o+\tau} \left\langle v''_y(t_o) v''_y(t_o+\tau)\right\rangle dt. \tag{7.11}$$

The particle velocity fluctuations will be influenced by the fluid velocity fluctuations seen along the particle paths.

For HIST, the fluid velocity correlations are not a function of space, so they are only a function of the time and spatial shifts. As such, the correlation of the u_y fluctuations along the particle path will be equal to the time-averaged values, as follows:

$$\left\langle u''_y(\mathbf{X}_{po}, t_o) u''_y(\mathbf{X}_{po}+\mathcal{X}_p, t_o+\tau)\right\rangle = \overline{u'_y(\mathbf{X}_{po}, t_o) u'_y(\mathbf{X}_{po}+\mathcal{X}_p, t_o+\tau)}. \tag{7.12}$$

Therefore, the u_y correlation coefficient and the integral time-scale averaged along all particle paths can be defined as in (6.61b), except that the spatial shift of $\chi_\mathcal{L}$ (6.57b) is replaced with χ_p (7.7), yielding

$$\Upsilon_{@p,y}(\tau) \equiv \frac{\overline{u'_y(\mathbf{X},t)u'_y(\mathbf{X}+\chi_p,t+\tau)}}{u'_{y,rms}u'_{y,rms}}. \tag{7.13a}$$

$$\tau_{@p,y} \equiv \int_0^\infty \Upsilon_{@p,y}(\tau)\,d\tau. \tag{7.13b}$$

The "@p" subscript denotes the unhindered fluid velocity interpolated to the particle centroid. The integral time scale for the fluid turbulence along the particle path can be defined more generally as follows:

$$\tau_{@p,i} \equiv \int_0^\infty \Upsilon_{@p,i}(\tau)\,d\tau = \int_0^\infty \frac{\overline{u'_i(\mathbf{X},t)u'_i(\mathbf{X}+\chi_p,t+\tau)}}{u'_{i,rms}u'_{i,rms}}d\tau \qquad \textit{where } i = x, y, \text{or } z. \tag{7.14}$$

As such, $\tau_{@p,i}$ is the average time it takes for fluid velocity fluctuations in the i-direction to decay along the particle path.

In the limit of a zero-inertia particle ($St_\Lambda = 0$), the fluid and particle paths are equal ($\mathbf{x}_p = \mathbf{x}_f$ and $\mathbf{v} = \mathbf{u}$), so the particle turbulent diffusivity equals the fluid turbulent diffusivity and the integral scale seen along the particle path equals the fluid Lagrangian integral scale:

$$\Theta_{p,i} \rightarrow \Theta_{turb,i} \qquad \textit{for } St_\Lambda \rightarrow 0. \tag{7.15a}$$

$$\tau_{@p,i} \rightarrow \tau_{\Lambda\mathcal{L},i} \tag{7.15b}$$

Furthermore, $\tau_{\Lambda L}$ and Θ_{turb} are isotropic for HIST (per 6.62 and 6.116), so

$$\begin{aligned} \Theta_{turb} &= \Theta_{p,x} = \Theta_{p,y} = \Theta_{p,z} \\ \tau_{\Lambda\mathcal{L}} &= \tau_{@p,x} = \tau_{@p,y} = \tau_{@p,z} \end{aligned} \qquad \textit{for } St_\Lambda \rightarrow 0 \textit{ and HIST}. \tag{7.16}$$

As such, the tracer particle cloud will grow as a sphere and at the same rate as a fluid tracer. However, the diffusivity will differ for particles with significant inertia and/or significant mean relative velocity. These effects can be quantified by defining the ratio of particle to tracer turbulent diffusivity as follows:

$$\Theta^* \equiv \frac{\Theta_p}{\Theta_{turb}}. \tag{7.17}$$

The mean relative velocity is often driven by the terminal velocity. In this case, the effect is characterized via the *drift parameter*, defined as follows:

$$\mathcal{g} \equiv \frac{w_{term}}{u_\Lambda} = \frac{\tau_p|1-R|g}{u_\Lambda}. \tag{7.18}$$

The drift parameter and the initial particle relative velocity can then be used to define two limits of turbulent diffusion: *weak drift* and *strong drift*. The weak-drift limit occurs when both the gravitational and initial relative velocity effects are negligible, such that the ensemble-averaged particle relative velocity is small compared to the turbulent velocity strength. The strong-drift limit occurs when the particle terminal velocity or initial velocity is high, so the ensemble-averaged particle relative velocity is much larger than the turbulent velocity strength. These two extremes are given as follows:

$$\Theta_{\text{wd}} : \textit{weak drift limit} : g \ll 1 \ \textit{and} \ \mathrm{w_o} \ll \mathrm{u}_\Lambda \ \ \textit{so} \ \langle \mathbf{w} \rangle \ll \mathrm{u}_\Lambda. \tag{7.19a}$$

$$\Theta_{\text{sd}} : \textit{strong drift limit} : g \gg 1 \ \textit{or} \ \mathrm{w_o} \gg \mathrm{u}_\Lambda \ \ \textit{so} \ \langle \mathbf{w} \rangle \gg \mathrm{u}_\Lambda. \tag{7.19b}$$

These weak-drift and strong-drift limits will be characterized in the next two sections in terms of the Stokes number and the drift parameter, and then a following section will combine these results for the generalized case of intermediate drift.

7.2.2 Weak-Drift Limit and Inertial Effects

In the following, we overview the theoretical turbulent diffusivity of particles by Hinze (1975) for the weak-drift limit. In this limit, the mean particle relative velocity is small relative to the turbulence (7.19a), so the ensemble-averaged particle velocity and average spatial shift along the particle path can be approximated with the fluid velocity:

$$\langle \mathbf{v} \rangle_{\text{wd}} \approx \langle \mathbf{u}_{@\text{p}} \rangle \qquad \textit{for weak drift.} \tag{7.20a}$$

$$\langle x \rangle_{\text{wd}} \approx \langle \mathbf{u}_{@\text{p}} \tau \rangle \tag{7.20b}$$

Therefore, the fluid turbulence correlation coefficient and integral time scale along the particle paths tend to those for the fluid Lagrangian paths for small-inertia particles and to the moving-Eulerian paths for large-inertia particles, where the latter can be expressed as

$$\Upsilon_{@\text{p}}(\tau) \approx \Upsilon_{mE}(\tau) \qquad \textit{for weak drift and } \mathrm{St}_\Lambda \rightarrow \infty. \tag{7.21a}$$

$$\tau_{@\text{p}} \approx \tau_{\Lambda mE} \tag{7.21b}$$

The Hinze theory (1975) ignores this correlation dependence on inertia, but this effect is empirically incorporated in the model of Wang and Stock (1993). For small-inertia particles, the turbulent frequency spectrum along an average particle path ($\mathcal{F}_{@\text{p}}$) will be the same as that along an average fluid path ($\mathcal{F}_{\mathcal{L}}$). For small-inertia particles, both spectra can be approximated using (6.98) with the integral scale along the fluid path ($\tau_{\Lambda\mathcal{L}}$) as follows:

$$F_{@\text{p}}(f) = F_{\mathcal{L}}(f) = \frac{4\mathrm{u}_\Lambda^2 \tau_{\Lambda\mathcal{L}}}{1 + (2\pi f \tau_{\Lambda\mathcal{L}})^2}. \tag{7.22}$$

To obtain the finite-inertia particle response to this spectrum, one may use the particle Equation of Motion (EOM) of (3.88c) but without gravitational effects (based on

7.19a). This EOM can be expressed using the instantaneous or the fluctuating particle velocities (using 7.10) as follows:

$$\frac{d\mathbf{v}}{dt} = -\frac{\mathbf{v}-\mathbf{u}_{@p}}{\tau_p} + R\frac{D\mathbf{u}_{@p}}{Dt} \qquad (7.23a)$$

$$\frac{d\mathbf{v}''}{dt} + \frac{\mathbf{v}''}{\tau_p} = \frac{\mathbf{u}''}{\tau_p} + R\frac{d\mathbf{u}''}{dt} \qquad (7.23b)$$

for weak drift.

The latter equation is obtained by subtracting the instantaneous EOM from the ensemble-averaged EOM and assumes linear drag (so that τ_p is constant along the particle path).

Next we assume the fluid and particle velocity rms fluctuations are isotropic for weak drift and that their instantaneous magnitudes in a given direction can be represented by Fourier series:

$$u_i'' = 2\pi \int_0^\infty [\ell_1 \cos(2\pi f t) + \ell_2 \sin(2\pi f t)]_i \, df. \qquad (7.24a)$$

$$v_i'' = 2\pi \int_0^\infty [\ell_3 \cos(2\pi f t) + \ell_4 \sin(2\pi f t)]_i \, df. \qquad (7.24b)$$

These expressions include four length scales where ℓ_1 and ℓ_2 depend only on f to satisfy the fluid spectrum, while ℓ_3 and ℓ_4 also depend on the particle characteristics. In order to obtain the length scales, one may substitute the fluctuations of (7.24) with (7.22) into the fluctuating particle velocity EOM of (7.23b). Carrying out the differentiations and integrals, then setting all the coefficients of the sine terms for a given frequency to be equal and doing the same for the cosine terms yields the following:

$$\ell_3 = \left[1 + \frac{4\pi^2 f^2 \tau_p^2 (R-1)}{1 + 4\pi^2 f^2 \tau_p^2}\right]\ell_1 + \left[\frac{2\pi f \tau_p (R-1)}{1 + 4\pi^2 f^2 \tau_p^2}\right]\ell_2. \qquad (7.25a)$$

$$\ell_4 = -\left[\frac{2\pi f \tau_p (R-1)}{1 + 4\pi^2 f^2 \tau_p^2}\right]\ell_1 + \left[1 + \frac{4\pi^2 f^2 \tau_p^2 (R-1)}{1 + 4\pi^2 f^2 \tau_p^2}\right]\ell_2. \qquad (7.25b)$$

Considering (7.24) and (7.25), the length scales are a function of frequency and particle characteristics and can be used to relate the particle velocity frequency spectrum (F_p) to the fluid velocity frequency spectrum seen by the particles ($F_{@p}$), as follows:

$$\frac{\ell_3^2 + \ell_4^2}{\ell_1^2 + \ell_2^2} = 1 + \frac{4\pi^2 f^2 \tau_p^2 (R-1)}{1 + 4\pi^2 f^2 \tau_p^2} + \frac{4\pi^2 f^2 \tau_p^2 (R-1)^2}{\left(1 + 4\pi^2 f^2 \tau_p^2\right)^2} + \frac{16\pi^4 f^4 \tau_p^4 (R-1)^2}{\left(1 + 4\pi^2 f^2 \tau_p^2\right)^2}. \qquad (7.26a)$$

$$F_p(f) = F_{@p}(f)\left[\frac{\ell_3^2 + \ell_4^2}{\ell_1^2 + \ell_2^2}\right]. \qquad (7.26b)$$

This result can be combined with (7.22) to obtain the particle velocity fluctuations, which can then be substituted into (7.11) for the weak-drift particle diffusivity. Following Hinze's assumption that the integral time scale seen by the particles is based on the fluid Lagrangian times scale (true for small-inertia particles), the result is as follows:

$$\Theta_{wd}(\tau) = \tau_{\Lambda\mathcal{L}} u_\Lambda^2 \left[1 - \frac{\left(\tau_{\Lambda\mathcal{L}}^2 - R^2\tau_p^2\right)e^{-\tau/\tau_{\Lambda\mathcal{L}}} - \tau_p^2\left(1 - R^2\right)e^{-\tau/\tau_p}}{\tau_{\Lambda\mathcal{L}}^2 - \tau_p^2} \right]. \tag{7.27}$$

Since the integral time scales and velocity fluctuations are isotropic for weak drift, the particle turbulent diffusivity is also isotropic, as follows:

$$\Theta_{wd} = \frac{1}{2}\frac{d\left\langle x_p'' x_p'' \right\rangle}{dt} = \frac{1}{2}\frac{d\left\langle y_p'' y_p'' \right\rangle}{dt} = \frac{1}{2}\frac{d\left\langle z_p'' z_p'' \right\rangle}{dt}. \tag{7.28}$$

Thus, the particles will diffuse uniformly in all directions, that is, the positional spread about the mean particle trajectory can be represented as a spherical particle cloud (Figure 7.6), and Θ_p characterizes the rate at which the cloud size increases.

Using the preceding with (6.119) and (7.17), the following is the normalized weak-drift particle diffusivity:

$$\Theta_{wd}^* \equiv \frac{\Theta_{wd}}{\Theta_{turb}} = 1 - \frac{\left(1 - R^2\right)St_\Lambda^2\left[e^{-\tau/\tau_{\Lambda\mathcal{L}}} - e^{-\tau/(\tau_{\Lambda\mathcal{L}}St_\Lambda)}\right]}{\left(1 - St_\Lambda^2\right)\left[1 - e^{-\tau/\tau_{\Lambda\mathcal{L}}}\right]}. \tag{7.29}$$

This diffusivity ratio depends on the acceleration parameter (R), the integral-scale Stokes number (St_Λ), and the ratio of the particle path time to the integral time scale ($\tau/\tau_{\Lambda\mathcal{L}}$). The influence of the latter two parameters is presented in Figure 7.7 for high-density particles ($R \rightarrow 0$). For very short times ($\tau \ll \tau_{\Lambda\mathcal{L}}$), increasing St_Λ reduces the

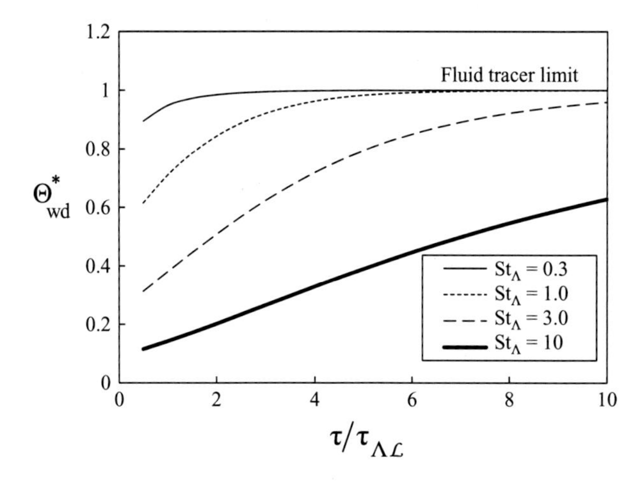

Figure 7.7 Normalized particle diffusivity as a function of time with various Stokes numbers for high-density particles ($\rho_p \gg \rho_f$) in weak-drift conditions.

diffusivity ratio, that is, increasing particle inertia reduces the particle turbulent spread rate compared to that for the fluid particles. For very long times (compared to both the fluid integral scale and the particle response time), this plot and (7.29) shows that the normalized diffusivity for weak-drift particles approaches unity:

$$\Theta^*_{wd} = 1 \qquad \textit{for } \tau \gg \tau_{\Lambda\mathcal{L}} \textit{ and } \tau \gg \tau_p. \tag{7.30}$$

As such, the long-time particle turbulent spread rate tends to that of a fluid, regardless of particle inertia. This (perhaps nonintuitive) result occurs because the long-time limit is long enough for the particles to respond to all scales of the turbulence. However, the long-time limit is not easily met for large inertia particles; for example, it is not yet met for $St_\Lambda = 10$ (where $\tau = \tau_p$) in Figure 7.7. In addition, high inertia particles will see an integral time scale based on the fluid moving-Eulerian integral time scale and so can have even larger diffusivity (Reeks, 1977).

The weak-drift particle positional spread in the y-direction can be obtained by substituting the particle spread of (7.27) into the relationship of (7.9c) and integrating to yield the following:

$$\frac{\left\langle y''_p y''_p \right\rangle_{wd}}{2u^2_\Lambda \tau^2_{\Lambda\mathcal{L}}} = \frac{\tau}{\tau_{\Lambda\mathcal{L}}} + \left[\frac{1 - St^2_\Lambda R^2}{1 - St^2_\Lambda}\right]\left[e^{-\tau/\tau_{\Lambda\mathcal{L}}} - 1\right] - St^3_\Lambda \left[\frac{1 - R^2}{1 - St^2_\Lambda}\right]\left[e^{-\tau/(\tau_{\Lambda\mathcal{L}} St_\Lambda)} - 1\right]. \tag{7.31a}$$

$$\left\langle x''_p x''_p \right\rangle_{wd} = \left\langle y''_p y''_p \right\rangle_{wd} = \left\langle z''_p z''_p \right\rangle_{wd}. \tag{7.31b}$$

The spread is isotropic in all directions consistent with the isotropy of (7.28).

The Hinze theory spread of (7.31a) is shown in Figure 7.8a for particles in a gas ($\rho_p \gg \rho_f$) at various Stokes numbers. For small Stokes numbers, the spread tends to the fluid-tracer limit, while larger Stokes numbers (higher inertia particles) yield lower spreads, especially at earlier times. This initial lag prevents the finite-inertia particles from ever catching up to that of fluid tracers (as in $St_\Lambda = 1$ for $\tau/\tau_{\Lambda\mathcal{L}} > 4$), even though the spread rates (slopes of the curves) are equal by this time. Therefore, this turbulent spread of high-density particles will always be less than that of fluid tracers.

It is important to note that the weak-drift results of Figures 7.7 and 7.8a assume particle terminal velocities can be neglected relative to the fluid turbulence intensity. However, in many turbulent flows, the weak-drift limit is *not* appropriate due to significant terminal velocities. For example, the experimental particle diffusion of Figure 7.8b is overpredicted by the weak-drift theory, especially at the higher Stokes numbers. This is due to significant gravitational drift, which can be incorporated in particle diffusion theory as follows.

7.2.3 Strong-Drift Limit and Eddy-Crossing Effects

As noted in (7.19b), the strong-drift limit occurs when either the terminal velocity (w_{term}) and/or initial velocity (w_o) is much greater than the turbulent velocity

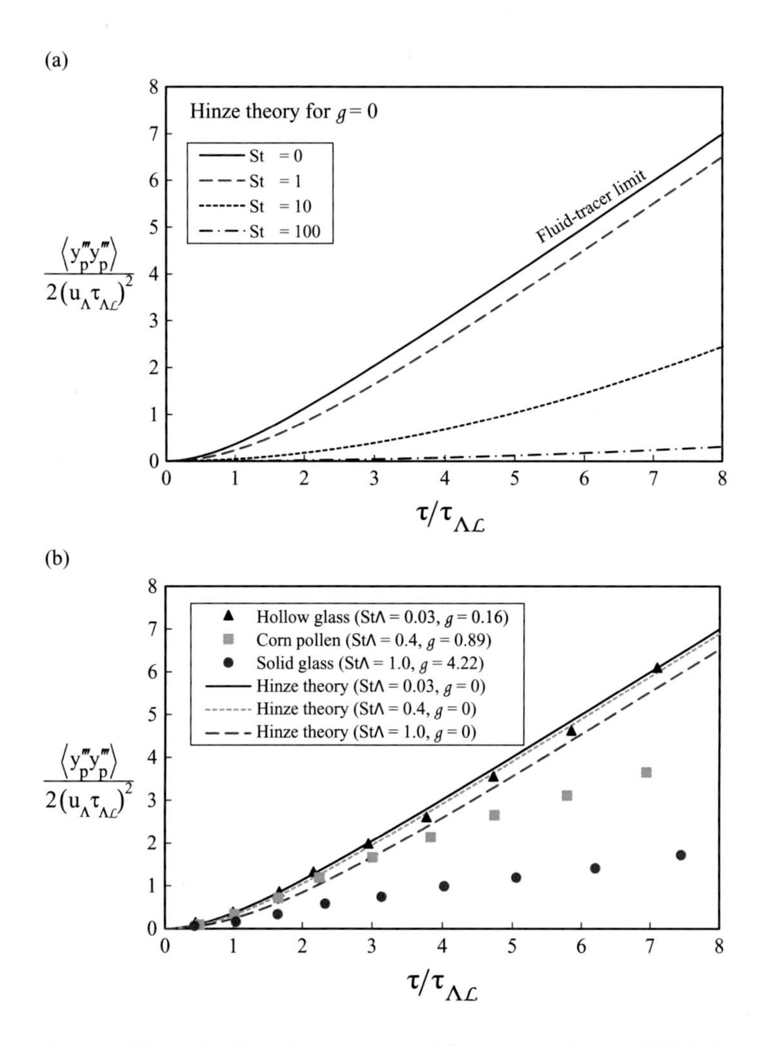

Figure 7.8 Theoretical lateral particle spread for $\rho_p \gg \rho_f$ and weak-drift limit ($g = 0$) as a function of time showing (a) the influence of integral-scale Stokes number; and (b) a comparison to experimental data at finite relative velocities ($g > 0$).

fluctuations (u_Λ). As such, the mean particle path differs significantly from the mean fluid path. The impact on turbulent diffusion is a reduction of the turbulent integral time scales seen by the particle, especially for lateral particle velocity fluctuations. To understand these reductions, recall that the fluid velocity fluctuations can be decomposed into streamwise and lateral components per (6.56) as follows:

$$\mathbf{u} = \bar{\mathbf{u}} + \mathbf{u}' = \left(\bar{u}_{\parallel} + u'_{\parallel}\right)\mathbf{i}_{\parallel} + u'_{\perp,1}\mathbf{i}_{\perp,1} + u'_{\perp,2}\mathbf{i}_{\perp,2}. \tag{7.32}$$

Note that the streamwise component includes the mean fluid velocity. Using this notation, the following are the correlation coefficients along the particle path for the fluid streamwise and lateral velocity fluctuations:

$$\Upsilon_{@p,\parallel}(\tau) = \frac{\overline{u'_{\parallel}(\mathbf{X}_{po}, t_o)u'_{\parallel}(\mathbf{X}_{po} + \mathcal{X}_p, t_o + \tau)}}{u'_{\parallel,rms}u'_{\parallel,rms}}. \tag{7.33a}$$

$$\Upsilon_{@p,\perp}(\tau) = \frac{\overline{u'_{\perp}(\mathbf{X}_{po}, t_o)u'_{\perp}(\mathbf{X}_{po} + \mathcal{X}_p, t_o + \tau)}}{u'_{\perp,rms}u'_{\perp,rms}}. \tag{7.33b}$$

As a consequence of the high relative velocity in the strong-drift limit, a particle will move across turbulent structures before they have time to change. As shown in Figure 7.9, the fluid velocity fluctuations are approximately "frozen" from the perspective of the particles. As such, they do not evolve and are only are shifted in time by the mean velocity per (6.82) so that

$$\mathbf{u}'(\mathbf{X}_{po}, t_o) = \mathbf{u}'_{ft}(\mathbf{X}_{po} + \overline{\mathbf{u}}\tau, t_o + \tau) \quad \textit{for frozen turbulence}. \tag{7.34}$$

This equality is conceptually shown as the top blue dashed line in Figure 7.10. If a spatial shift due to relative particle velocity is added to both sides of (7.34) as indicated by the vertical solid black lines of Figure 7.10, this becomes

$$+\mathbf{w}\tau \quad = \quad +(\mathbf{v} - \overline{\mathbf{u}})\tau. \tag{7.35a}$$

$$\mathbf{u}'(\mathbf{X}_{po} + \mathbf{w}\tau, t_o) = \mathbf{u}'_{@p,ft}(\mathbf{X}_{po} + \mathbf{v}\tau, t_o + \tau). \tag{7.35b}$$

This equivalence is represented by the lower blue dashed line in Figure 7.10.

For homogeneous turbulence, the spatial correlation with a spatial shift of $\mathbf{w}\tau$ at t_o will be the same as at $t_o + \tau$, as follows:

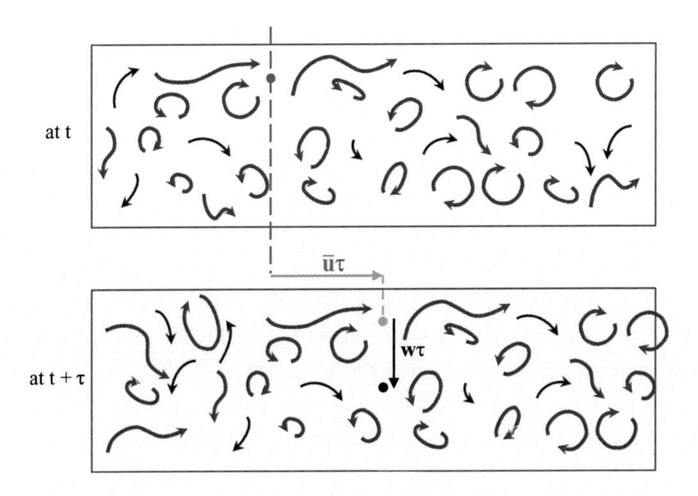

Figure 7.9 Particle in frozen turbulence where the fluid structures move at the mean velocity and do not evolve and where the light blue circle shows a later position for a fluid in the moving Eulerian reference frame (for which the fluid velocity fluctuations are preserved) while the black circle shows the later position of the particle which also includes relative velocity.

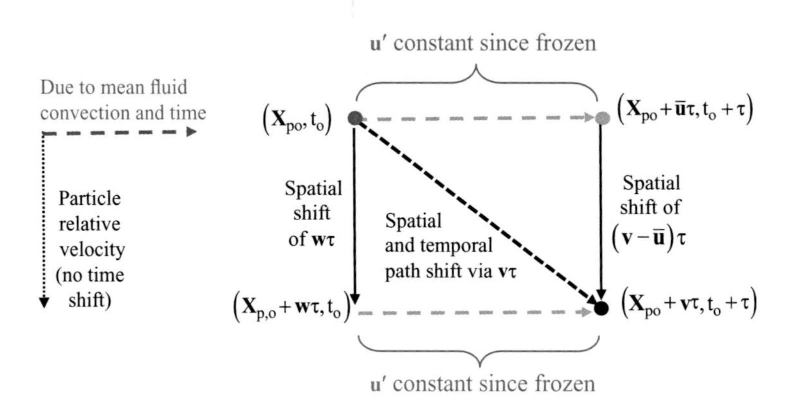

Figure 7.10 Schematic of a particle falling down while being convected by a mean flow to the right, where blue lines are the mean fluid paths for which velocity fluctuations are constant for frozen turbulence, the solid lines are for the particle relative velocity (e.g., terminal velocity), and the dashed black line is the resultant particle path for which the change in fluid turbulence is the same as that for the solid lines.

$$\overline{\mathbf{u}'(\mathbf{X}_{po}, t_o)\mathbf{u}'(\mathbf{X}_{po}+\mathbf{w}\tau, t_o)} = \overline{\mathbf{u}'(\mathbf{X}_{po}, t_o+\tau)\mathbf{u}'(\mathbf{X}_{po}+\mathbf{w}\tau, t_o+\tau)}. \quad (7.36)$$

The LHS correlation is represented by the left vertical black line of Figure 7.10. For the correlation along the particle path, we are interested in the correlation along the diagonal black dashed line. Fortunately, the lower blue line allows this correlation to be the same as the left vertical line (spatial shift only). Therefore, the parallel and perpendicular correlation coefficients can be expressed as follows:

$$\Upsilon_{@p,\parallel}(\tau) = \frac{\overline{u'_\parallel(\mathbf{X}_{po}, t_o)u'_\parallel(\mathbf{X}_{po}+\mathbf{w}\tau, t_o)}}{u'_{\parallel,rms}u'_{\parallel,rms}} \equiv \Upsilon_\parallel(\mathbf{w}\tau, 0) \quad (7.37a)$$

for strong drift.

$$\Upsilon_{@p,\perp}(\tau) = \frac{\overline{u'_\perp(\mathbf{X}_{po}, t_o)u'_\perp(\mathbf{X}_{po}+\mathbf{w}\tau, t_o)}}{u'_{\perp,rms}u'_{\perp,rms}} \equiv \Upsilon_\perp(\mathbf{w}\tau, 0) \quad (7.37b)$$

The LHS Υ's are based on the average particle path, while the RHS Υ's are based only on spatial shifts and are equal to those of (6.69). Integrating the LHS Υ's over all the time shifts yields the fluid integral time scale seen by strong-drift particles, while integrating the RHS Υ's over all the time shifts can be equated to an integral over spatial shifts:

$$\tau_{sd,\parallel} \equiv \int_0^\infty \Upsilon_{@p,\parallel}(\tau)\,d\tau = \int_0^\infty \Upsilon_\parallel(\mathbf{w}\tau, 0)\,d\tau = \int_0^\infty \Upsilon_\parallel(\mathbf{w}\tau, 0)\frac{d\chi}{w}. \quad (7.38a)$$

$$\tau_{sd,\perp} \equiv \int_0^\infty \Upsilon_{@p,\perp}(\tau)\,d\tau = \int_0^\infty \Upsilon_\perp(\mathbf{w}\tau, 0)\,d\tau = \int_0^\infty \Upsilon_\perp(\mathbf{w}\tau, 0)\frac{d\chi}{w}. \quad (7.38b)$$

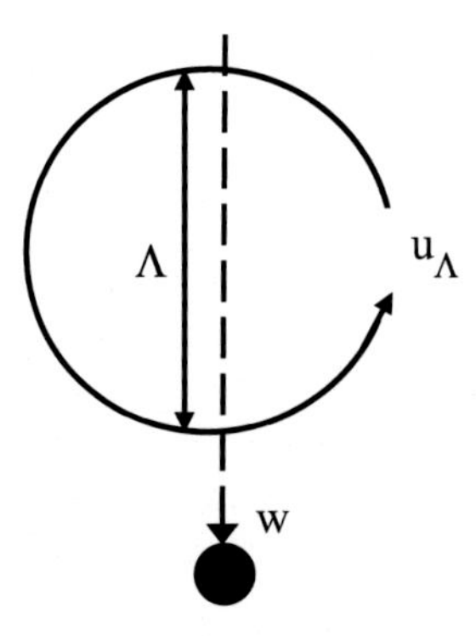

Figure 7.11 Schematic of eddy-crossing effect, where Λ is the integral length scale and w is the relative speed traversing this structure. If the turbulence is frozen, the effective integral time scale is $\tau_{sd} = \Lambda/w$.

If one assumes that **w** is approximately constant and equal to the ensemble-averaged value and that the spatial correlations are independent of position (homogeneous turbulence), the RHS integrals can be related to the integral length scales using (6.71), as follows:

$$\tau_{sd,\parallel} = \frac{\Lambda_{\parallel}}{\langle w \rangle} \tag{7.39a}$$

$$\tau_{sd,\perp} = \frac{\Lambda_{\perp}}{\langle w \rangle} \qquad \textit{for strong drift.} \tag{7.39b}$$

These relationships can be qualitatively inferred from Figure 7.11, where τ_{sd} is notionally the time it takes for a particle to cross a turbulent structure of size Λ while moving at a speed w. Based on this description, τ_{sd} is often termed the *eddy-crossing time*. Notably, this time depends on whether streamwise or lateral fluid velocity fluctuations are considered.

For HIST, it was noted that both lateral length scales are half of the streamwise length scale of continuity effects (6.76). Combining this result with (7.39), the integral time scales for the strong-drift limit also differ by the same factor:

$$\tau_{sd,\perp,1} = \tau_{sd,\perp,2} \approx \frac{1}{2}\tau_{sd,\parallel} \qquad \textit{for HIST}. \tag{7.40}$$

For example, suppose that the mean fluid flow is in the x-direction (so y and z are the lateral directions):

$$\tau_{sd,y} = \tau_{sd,z} \approx \frac{1}{2}\tau_{sd,x} \qquad \textit{for HIST}. \tag{7.41}$$

The LHS can be compared to the fluid Lagrangian time scale using (6.80), as follows:

$$\frac{\tau_{sd,\perp}}{\tau_{\Lambda\mathcal{L}}} = \frac{\Lambda_{\perp}/\langle w \rangle}{\Lambda_{\perp}/(u_{\Lambda}c_{\Lambda})} \approx \frac{u_{\Lambda}}{\langle w \rangle}. \tag{7.42}$$

The RHS assumes that the turbulence structure parameter of (6.80) is approximately unity. For strong drift, the RHS is much less than unity (per 7.19b), so the LHS is also much less than unity ($\tau_{sd} \ll \tau_{\Lambda\mathcal{L}}$). Considering also (7.40), both of the eddy-crossing times are much less than the fluid Lagrangian time scale for a strong-drift particle.

The turbulent frequency spectrum of (7.22) can be modified by replacing the fluid integral scale along a fluid path ($\tau_{\Lambda\mathcal{L}}$) with that seen along the path of strong-drift particles (τ_{sd}), as follows:

$$F_{@p,sd}(f) = \frac{4u_\Lambda^2 \tau_{sd}}{1 + (2\pi f \tau_{sd})^2} \qquad \textit{for strong drift.} \tag{7.43}$$

As a result of this eddy-crossing effect, strong-drift particles see fluid fluctuations with higher frequencies (since $\tau_{sd} \ll \tau_{\Lambda\mathcal{L}}$) but with the same turbulent kinetic energy (due to HIST). For example, consider a particle that is held in place while mean flow and turbulence are convected past it. This stationary particle would see more rapid temporal changes in the fluid fluctuations than a particle that moves with the mean flow. Similarly, a pedestrian standing by a highway will see cars passing at a higher frequency than a passenger in a car moving with the mean speed of the traffic.

If we adopt the strong-drift integral time scale with the approach of §7.2.2, the strong-drift particle diffusivity becomes

$$\Theta_{sd}(\tau) = \tau_{sd} u_\Lambda^2 \left[1 - \frac{\left(\tau_{sd}^2 - R^2 \tau_p^2\right) e^{-\tau/\tau_{sd}} - \tau_p^2 \left(1 - R^2\right) e^{-\tau/\tau_p}}{\tau_{sd}^2 - \tau_p^2} \right]. \tag{7.44}$$

Similarly, the long-time strong-drift limit becomes

$$\Theta_{sd}(\tau) = \tau_{sd} u_\Lambda^2 \qquad \textit{for } \tau \gg \tau_{\Lambda\mathcal{L}} \textit{ and } \tau \gg \tau_p. \tag{7.45}$$

Since τ_{sd} is anisotropic (7.40), particle diffusivity in the streamwise and lateral direction are also different and can be expressed as follows:

$$\Theta_{sd,\parallel} = \tau_{sd,\parallel} u_\Lambda^2 = \frac{\Lambda_\parallel}{\langle w \rangle} u_\Lambda^2 \quad \textit{for long-time streamwise diffusion.} \tag{7.46a}$$

$$\Theta_{sd,\perp} = \tau_{sd,\perp} u_\Lambda^2 = \frac{\Lambda_\perp}{\langle w \rangle} u_\Lambda^2 \quad \textit{for long-time lateral diffusion.} \tag{7.46b}$$

Thus, the lateral particle spread rate is half of the streamwise spread rate for strong-drift particles in HIST. This is due to a combination of the eddy-crossing and the integral length scale isotropy. As such, an ensemble-averaged strong-drift particle cloud at a given time shift will have a longer dimension in the streamwise direction than that in the two lateral directions.

7.2.4 Particle Diffusivity for Intermediate Drift

Next we consider the case of intermediate drift (between the limits of weak and strong drift), where the particle diffusivity will also be in between these limits. The only modification needed to consider intermediate drift is to generalize the integral scales seen by the particles ($\tau_{@p,i}$). For HIST, the general case can be approximated using a combination of the weak and strong limits (Csanady, 1963), as follows:

$$\frac{1}{\tau^2_{@p,i}} \approx \frac{1}{\tau^2_{wd}} + \frac{1}{\tau^2_{sd,i}} = \frac{1}{\tau^2_{\Lambda\mathcal{L}}} + \frac{\langle w\rangle^2}{\Lambda_i^2} = \frac{1}{\tau^2_{\Lambda\mathcal{L}}}\left(1 + \frac{\langle w\rangle^2\tau^2_{\Lambda\mathcal{L}}}{\Lambda_i^2}\right) \quad \textit{where } i = \parallel \text{ or } \perp. \tag{7.47}$$

This model yields the isotropic weak-drift limit (7.21b) when $w \ll \Lambda/\tau_{\Lambda\mathcal{L}}$ and yields the anisotropic strong-drift limit (7.42) when $w \gg \Lambda/\tau_{\Lambda\mathcal{L}}$. This result employs the same fluid Lagrangian integral scale as that seen by the particles, which is only reasonable for small-inertia particles (Reeks, 1977). Using this model, the integral time scale representing parallel or perpendicular fluid velocity fluctuations (relative to mean motion of the particles) can be expressed as follows:

$$\tau_{@p,\parallel} \approx \tau_{\Lambda\mathcal{L}}\Big/\sqrt{1 + \langle w\rangle^2\tau^2_{\Lambda\mathcal{L}}/\Lambda^2_\parallel}. \tag{7.48a}$$

$$\tau_{@p,\perp} \approx \tau_{\Lambda\mathcal{L}}\Big/\sqrt{1 + \langle w\rangle^2\tau^2_{\Lambda\mathcal{L}}/\Lambda^2_\perp}. \tag{7.48b}$$

These time scales, which incorporate eddy-crossing effects, can then be used to describe the fluid turbulent frequency spectrum seen by the particles, as follows:

$$F_{@p,\parallel}(f) = \frac{4u_\Lambda^2\tau_{@p,\parallel}}{1 + (2\pi f\tau_{@p,\parallel})^2} \tag{7.49a}$$

for intermediate drift.

$$F_{@p,\perp}(f) = \frac{4u_\Lambda^2\tau_{@p,\perp}}{1 + (2\pi f\tau_{@p,\perp})^2} \tag{7.49b}$$

Again, "@p" represents the fluid velocity fluctuations seen along the particle path.

Similarly, particle turbulent diffusivity can be obtained following the weak-drift result of (7.27), which includes inertia and density effects, by replacing $\tau_{\Lambda\mathcal{L}}$ with $\tau_{@p,i}$ as follows:

$$\frac{\Theta_{p,i}(\tau)}{u_\Lambda^2} = \tau_{@p,i} - \frac{\left(\tau^2_{@p,i} - R^2\tau_p^2\right)e^{-\tau/\tau_{@p,i}} - \tau_p^2(1-R^2)e^{-\tau/\tau_p}}{\tau_{@p,i} - \tau_p^2/\tau_{@p,i}} \quad \textit{where } i = \parallel \text{ or } \perp. \tag{7.50}$$

For HIST, these particle diffusivities are anisotropic because of the eddy-crossing effect caused by the anisotropy of the integral length scales (even though the fluid/tracer diffusivities are isotropic).

The long-time normalized diffusivity for the particles is also anisotropic and simplifies to the following:

$$\Theta_i^* \equiv \frac{\Theta_{p,i}}{\Theta_{turb}} = \frac{\tau_{@p,i}}{\tau_{\Lambda\mathcal{L}}} \approx \left[1 + \left(\frac{\langle w\rangle\tau_{\Lambda\mathcal{L}}}{\Lambda_i}\right)^2\right]^{-1/2} \quad \textit{for long-time diffusion.} \tag{7.51}$$

If the mean relative velocity is dominated by the terminal velocity, the term in parentheses on the RHS of (7.51) can be expressed in terms of the drift parameter (g of 7.18) as follows:

$$\frac{\tau_{\Lambda\mathcal{L}}\langle w\rangle}{\Lambda_{\perp}} \approx \frac{w_{term}}{u_{\Lambda}c_{\Lambda}} = \frac{g}{c_{\Lambda}}. \tag{7.52a}$$

$$\frac{\tau_{\Lambda\mathcal{L}}\langle w\rangle}{\Lambda_{\parallel}} \approx \frac{g}{2c_{\Lambda}}. \tag{7.52b}$$

The first expression makes use of the turbulence structure parameter of (6.80), and the second expression further uses (6.76). Substituting these into (7.51), the long-time longitudinal and lateral diffusivity for particles in HIST can be written as follows:

$$\Theta_{\perp}^{*} \approx \left[1 + (g/c_{\Lambda})^{2}\right]^{-1/2} \tag{7.53a}$$

$$\Theta_{\parallel}^{*} \approx \left[1 + \left(\frac{1}{2}g/c_{\Lambda}\right)^{2}\right]^{-1/2} \quad \textit{for long-time diffusion with } \langle w\rangle \approx w_{term}. \tag{7.53b}$$

These diffusivity ratios are plotted in Figure 7.12 for $c_{\Lambda} = 1$ (a typical value). For small drift parameters, the diffusivity ratios approach unity, consistent with the weak-drift limit (7.30). As the drift parameter increases (due to higher w_{term}), there is a decrease in the longitudinal diffusivity (due to the eddy-crossing effect) and an even larger reduction in the lateral diffusivity (due to the combined eddy-crossing effect and the continuity effect).

Figure 7.13 shows lateral diffusivity from nearly HIST experiments and simulations for a variety of particle density ratios. In general, the results are consistent with

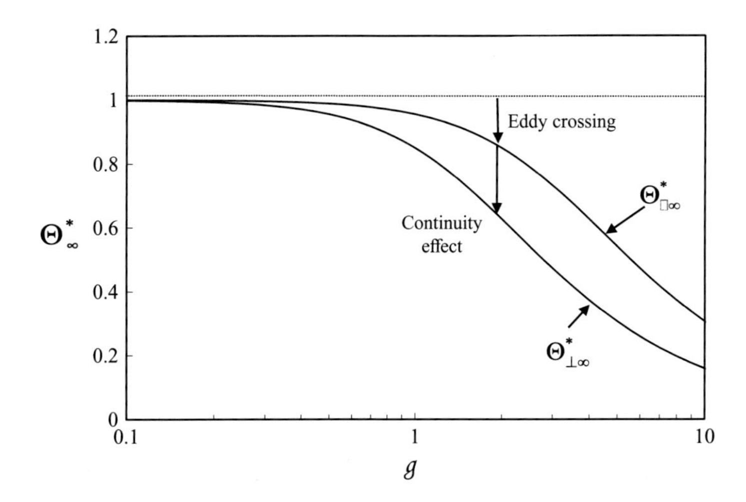

Figure 7.12 Lateral and longitudinal particle diffusivity for $c_{\Lambda\tau} = 1.0$ based on the Csanady model and Hinze theory, where eddy crossing reduces the diffusion in the streamwise direction, but the continuity effect reduces it even more for the perpendicular direction.

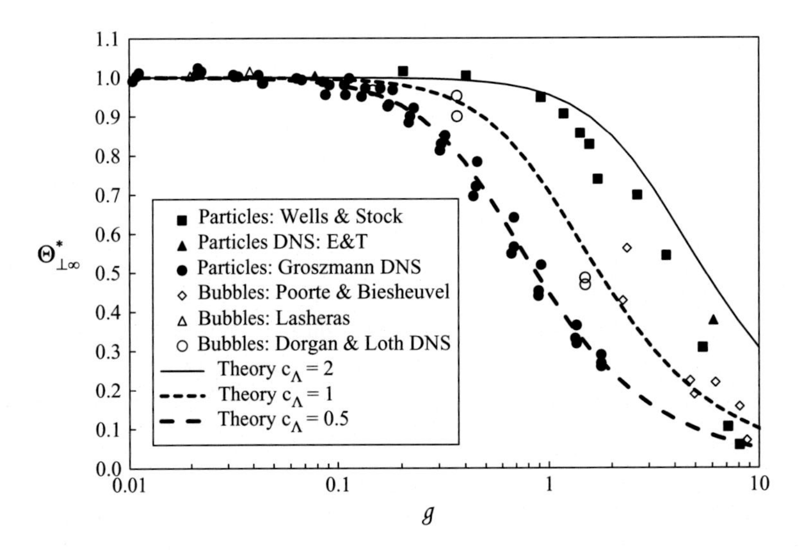

Figure 7.13 Particle lateral diffusivity based on experimental data of Wells and Stock (1983), Lasheras (1998), and Poorte and Biesheuvel (2002), as well as DNS data of E&T (Elghobashi and Truesdell, 1984), Groszmann et al. (1999), and Dorgan and Loth (2009) compared against the Hinze theory with the Csanady time-scale model for various c_Λ values.

the Hinze theory combined with the Csanady time-scale model if c_Λ ranges from 0.5–2.0. The large spread in the data indicates that a single value of c_Λ cannot be used for all turbulent flows (Wang and Stock, 1993). In addition, secondary effects that can cause disagreement include (but are not limited to) nonlinear drag, history force, lift force, as well as turbulence anisotropy and/or nonhomogeneity.

If one neglects these secondary effects, the lateral spread for high-density particles can be obtained by substituting $\tau_{@p,i}$ for $\tau_{\Lambda\mathcal{L}}$ in (7.31a) (so as to account for eddy-crossing and continuity effects). For a flow with mean velocity in the x-direction, the lateral spread will be in the y- and z-directions. In this case, the spread in the (lateral) y-direction based on integrating (7.31a) and (7.48b) is given as follows:

$$\frac{\left\langle y_p'' y_p'' \right\rangle}{2u_\Lambda^2 \tau_{\Lambda\mathcal{L}}^2} \approx \frac{\tau_{@p,y}\tau}{\tau_{\Lambda\mathcal{L}}^2} + \frac{\tau_{@p,y}^4\left(e^{-\tau/\tau_{@p,y}} - 1\right)}{\tau_{\Lambda\mathcal{L}}^2\tau_{@p,y}^2 - \tau_{\Lambda\mathcal{L}}^2\tau_p^2} - \frac{\tau_p^3\left(e^{-\tau/\tau_p} - 1\right)}{\tau_{\Lambda\mathcal{L}}^2\tau_{@p,y}} \quad \textit{for } \rho_p \gg \rho_f. \tag{7.54}$$

This prediction is shown in Figure 7.14 and compares well with a nearly HIST experimental data set for $c_\Lambda = 1.5$. Comparing this result to Figure 7.8b shows that incorporating the eddy-crossing effect is critical to model the particle turbulent diffusion when g is significant. Again, it should be noted that other experiments may be better predicted with another c_Λ, as no single c_Λ is expected to robustly predict spread for all turbulent flows.

In summary, there are four key results for particle turbulent diffusivity in HIST that incorporate effects of inertia, long-time limits, relative velocity, and anisotropy. as follows:

(a) *Inertia effect*, which reduces Θ^* for increasing St_Λ at short and intermediate times.

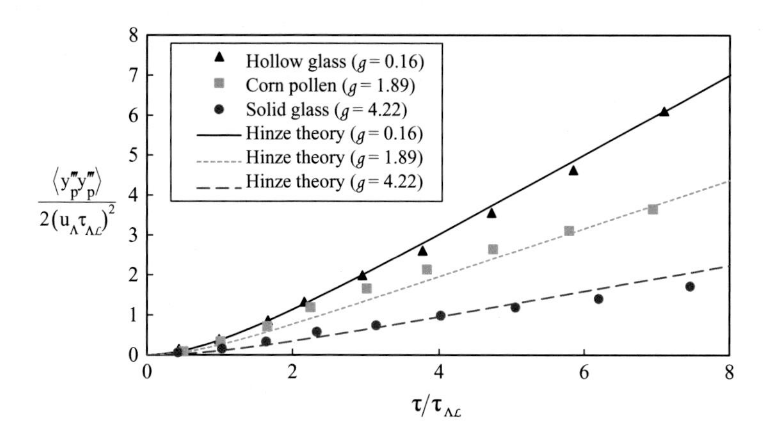

Figure 7.14 Lateral particle spread Snyder and Lumley (1971) measurements for particles in air ($\rho_p \gg \rho_f$) for which $\tau_{\Lambda L} \approx 0.25k/\varepsilon$ compared to the Hinze theory and Csanady model for $c_\Lambda = 1.5$.

(b) *Long-time limit,* which yields $\Theta^* = 1$ for long times at weak drift ($g \ll 1$).
(c) *Eddy-crossing effect*, which reduces Θ^* for increasing g.
(d) *Continuity effect*, which yields $\Theta_\perp^* \approx \frac{1}{2}\Theta_\parallel^*$ for increasing St_Λ and g.

In the following, the preceding theoretical approach is leveraged for other particle characteristics.

7.3 Turbulence Impact on Particle Velocity Fluctuations

7.3.1 Particle Turbulent Kinetic Energy

Just as the mean square of the fluid velocity fluctuations defines the fluid turbulent kinetic energy (6.17), the mean square of the particle velocity fluctuations defines the particle turbulent kinetic energy. In particular, this energy can be defined using ensemble averaging:

$$\mathrm{k_p} \equiv \frac{1}{2}\langle \mathbf{v}'' \cdot \mathbf{v}'' \rangle = \frac{1}{2}\left(\langle \mathrm{v}_x''\mathrm{v}_x''\rangle + \left\langle \mathrm{v}_y''\mathrm{v}_y''\right\rangle + \langle \mathrm{v}_z''\mathrm{v}_z''\rangle\right). \tag{7.55}$$

For HIST, the lateral components are equal but may be different from the streamwise components due to eddy-crossing effects, so

$$\mathrm{k_p} = \frac{1}{2}\left(\langle \mathrm{v}_\parallel''\mathrm{v}_\parallel''\rangle + 2\langle \mathrm{v}_\perp''\mathrm{v}_\perp''\rangle\right) \qquad \textit{for HIST}. \tag{7.56}$$

To obtain these velocity fluctuations, we can combine the particle velocity frequency spectra (7.26) with the fluid spectra seen for particles with generalized drift (7.49) as follows:

$$\langle v''_{\parallel} v''_{\parallel} \rangle = \int_0^\infty F_{p,\parallel}(f)\, df = \int_0^\infty \left(\frac{\ell_3^2 + \ell_4^2}{\ell_1^2 + \ell_2^2} \right) \left(\frac{4u_\Lambda^2 \tau_{@p,\parallel}}{1 + 4\pi^2 \tau_{@p,\parallel}^2 f^2} \right) df. \tag{7.57a}$$

$$\langle v''_{\perp} v''_{\perp} \rangle = \int_0^\infty F_{p,\perp}(f)\, df = \int_0^\infty \left(\frac{\ell_3^2 + \ell_4^2}{\ell_1^2 + \ell_2^2} \right) \left(\frac{4u_\Lambda^2 \tau_{@p,\perp}}{1 + 4\pi^2 \tau_{@p,\perp}^2 f^2} \right) df. \tag{7.57b}$$

Employing the dependence of the length scales on τ_p and R from (7.25) and integrating, the long-time streamwise and lateral particle velocity fluctuations (Hinze, 1975) are as follows:

$$\frac{\langle v''_{\parallel} v''_{\parallel} \rangle}{u_\Lambda^2} = \frac{1 + R^2 (\tau_p/\tau_{@p,\parallel})}{1 + (\tau_p/\tau_{@p,\parallel})}. \tag{7.58a}$$

$$\frac{\langle v''_{\perp} v''_{\perp} \rangle}{u_\Lambda^2} = \frac{1 + R^2 (\tau_p/\tau_{@p,\perp})}{1 + (\tau_p/\tau_{@p,\perp})}. \tag{7.58b}$$

By using St_Λ of (7.1) and defining $St_\parallel$ and $St_\perp$ as the particle-path Stokes numbers for parallel and perpendicular fluid velocities, the long-time velocity fluctuations become

$$\frac{\langle v''_{\parallel} v''_{\parallel} \rangle}{u_\Lambda^2} = \frac{1 + R^2 St_\parallel}{1 + St_\parallel} \quad \textit{where} \quad St_\parallel \approx St_\Lambda \sqrt{1 + (\tau_{\Lambda\mathcal{L}} \langle w \rangle / \Lambda_\parallel)^2}. \tag{7.59a}$$

$$\frac{\langle v''_{\perp} v''_{\perp} \rangle}{u_\Lambda^2} = \frac{1 + R^2 St_\perp}{1 + St_\perp} \quad \textit{where} \quad St_\perp \approx St_\Lambda \sqrt{1 + (\tau_{\Lambda\mathcal{L}} \langle w \rangle / \Lambda_\perp)^2}. \tag{7.59b}$$

If the relative velocity is approximated by the terminal velocity (per Eq. 7.52), these particle-path Stokes numbers can be expressed using c_Λ of (6.80) as follows:

$$St_\parallel \approx St_\Lambda \sqrt{1 + \frac{1}{4} (g/c_\Lambda)^2}. \tag{7.60a}$$

$$St_\perp \approx St_\Lambda \sqrt{1 + (g/c_\Lambda)^2}. \tag{7.60b}$$

As such, the Stokes numbers and the particle velocity fluctuations are isotropic for the weak-drift limit but are anisotropic in the strong-drift limit.

The ratio of the long-time particle to fluid turbulent kinetic energy (TKE) by combining (7.56) and (7.59) becomes

$$\frac{k_p}{k} = \frac{\langle v''_{\parallel} v''_{\parallel} \rangle}{3u_\Lambda^2} + \frac{2\langle v''_{\perp} v''_{\perp} \rangle}{3u_\Lambda^2} = \frac{1 + R^2 St_\parallel}{3 + 3St_\parallel} + \frac{2 + 2R^2 St_\perp}{3 + 3St_\perp} \quad \textit{for HIST}. \tag{7.61}$$

For high-density particles (R→0) with weak drift, the particle fluctuations are isotropic, yielding the following:

$$\frac{k_p}{k} = \frac{\langle v''_{\parallel} v''_{\parallel} \rangle}{u_\Lambda^2} = \frac{\langle v''_{\perp} v''_{\perp} \rangle}{u_\Lambda^2} \approx \frac{1}{1 + St_\Lambda} \quad \textit{for } g = 0 \textit{ and } \rho_p \gg \rho_f. \tag{7.62}$$

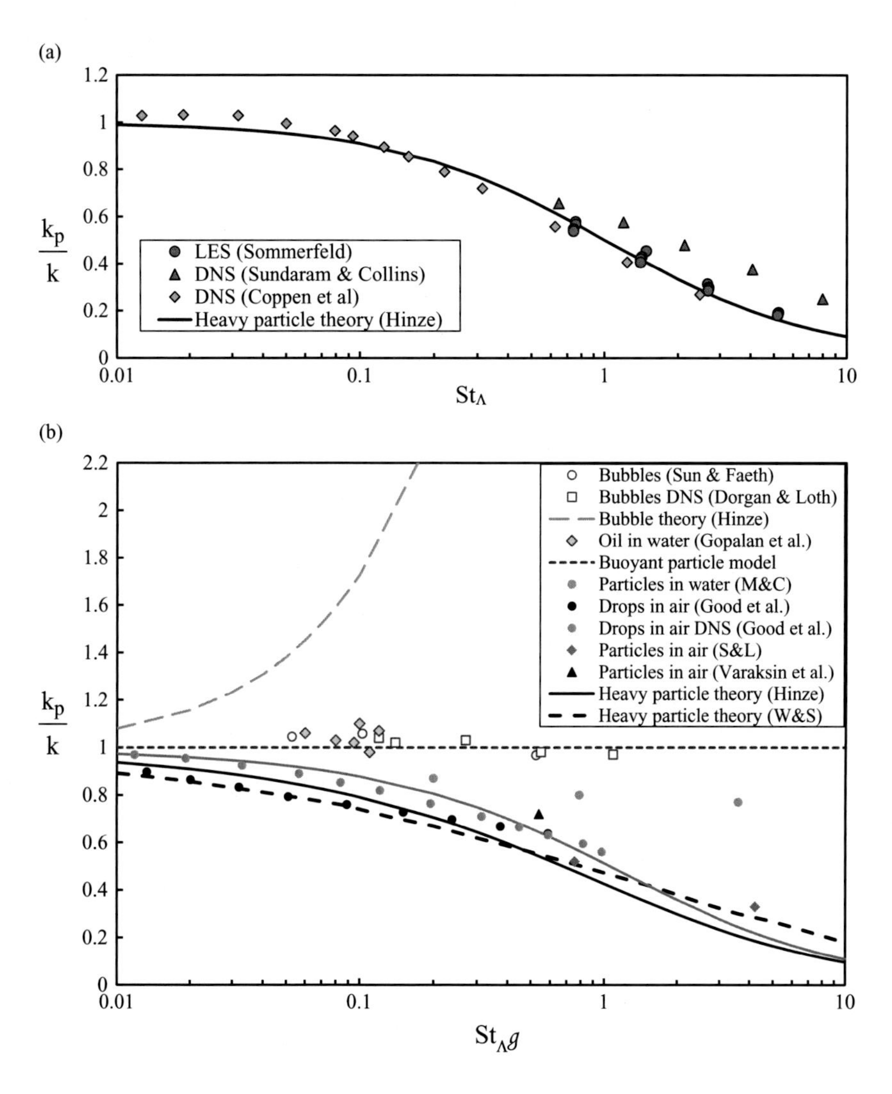

Figure 7.15 Particle kinetic energy: (a) high-density particles with no drift ($g = 0$) via simulations (Sundaram and Collins, 1997; Coppen et al., 2001; Sommerfeld, 2001) and Hinze theory; and (b) various density particles with finite drift ($g > 0$) via measurements (Snyder and Lumley, 1971; Sun and Faeth, 1986; Varaksin et al., 2000; Gopalan et al., 2008; Good et al., 2014; Mena and Curtis, 2020); DNS (Dorgan and Loth, 2009; Good et al., 2014); Hinze (1975) theory (with $c_\Lambda = 0.5$, & = 2); Wang and Stock (1993) theory and the empirical buoyant particle model ($k_p = k$).

This limit was obtained by Kulick et al. (1994) and reasonably predicts the TKE of high-density weak-drift particles over a wide range of Stokes numbers, as shown in Figure 7.15a. It has also been found to be reasonable for nonlinear drag (Sommerfeld, 2001; Squires, 2007). Interestingly, this turbulent result is qualitatively similar to the laminar flow result of (5.23).

It is also interesting that increasing St_Λ (particle inertia) reduces the long-time particle velocity fluctuations (7.62) but does not impact the long-time turbulent particle diffusion (7.30). To understand this, we recall that the turbulent diffusion is a product of the integral time scale and the velocity fluctuations (e.g., 7.45), so the reduction in the particle TKE must therefore be accompanied by an increase in the integral time scale of the *particle* velocity fluctuations. This increase is attributed to the particle inertia that filters out smaller frequencies from the fluid turbulence leading to longer correlation times for the particle velocity.

To consider more general drift conditions for high-density particles ($R \to 0$), one may apply the terminal velocity assumptions of (7.60) with (7.56) and (7.59) to yield the following:

$$\frac{k_p}{k} = \frac{1}{3 + 3\mathrm{St}_\Lambda\sqrt{1 + \frac{1}{4}(g/c_\Lambda)^2}} + \frac{2}{3 + 3\mathrm{St}_\Lambda\sqrt{1 + (g/c_\Lambda)^2}} \quad \textit{for } \rho_p \gg \rho_f. \tag{7.63}$$

By combining the strong-drift and weak-drift limits, the long-time TKE ratio can be approximated as follows:

$$\frac{k_p}{k} \approx \frac{1}{1 + \mathrm{St}_\Lambda + 3\mathrm{St}_\Lambda g/(4c_\Lambda)} \quad \textit{for } \rho_p \gg \rho_f. \tag{7.64}$$

A semi-empirical model was also developed by Wang and Stock (1993) using the strong-drift theory as well as $c_\Lambda = 1/2$ and an empirical model for $\mathrm{St}_\perp$ as follows:

$$\frac{k_p}{k} \approx \frac{1}{3 + 3\mathrm{St}_\Lambda g} + \frac{2 - 1.29(1 + \mathrm{St}_\Lambda)^{0.4(1+0.1)\mathrm{St}_\Lambda}}{3 + 6\mathrm{St}_\Lambda g} \quad \textit{for } g \gg 1 \textit{ and } \rho_p \gg \rho_f. \tag{7.65}$$

The first term on the RHS is the contribution from streamwise particle velocity fluctuations, while the second term is from the lateral fluctuations.

The results of (7.63)–(7.65) show that the particle TKE tends to the fluid TKE when both the Stokes numbers and the drift parameter tend to zero, but otherwise particle TKE is reduced as the Stokes number and/or drift parameter increases (Poelma et al., 2007). The combined trend is shown in Figure 7.15b for nearly HIST flows where the Hinze theory predictions for c_Λ values of 1/2 (the black solid line) and 2 (the blue solid line) generally correlate with experimental data for high-density particles, as does the strong-drift predictions of Wang and Stock. Notably, these models do not account for effects associated with anisotropic and/or nonhomogenous turbulence. Furthermore, (7.63)–(7.65) assume a relative velocity based on terminal velocity via (7.60), which will not be valid if the relative velocity is primarily dictated by wall reflections (Kussin and Sommerfeld, 2001).

For the low density of bubbles ($\rho_p \ll \rho_f$) where $R \to 3$, (7.61) becomes

$$\frac{k_p}{k} = \frac{1 + 9\mathrm{St}_\parallel}{3 + 3\mathrm{St}_\parallel} + \frac{2 + 18\mathrm{St}_\perp}{3 + 3\mathrm{St}_\perp} \quad \textit{for } \rho_p \ll \rho_f. \tag{7.66}$$

At large Stokes numbers, this yields $k_p = 9k$! However, $k_p > k$ has not been generally observed experimentally for bubbles. Instead, bubble data from experiments and DNS as shown in Figure 7.15b indicate that bubble and fluid TKE are approximately equal, at least for $St_{\Lambda g} < 1$. A similar result was found for the TKE of slightly buoyant oil drops in water. As such, a simple empirical model for buoyant particle kinetic energy can be given as follows:

$$k_p = k \qquad \textit{for } \rho_p < \rho_f \textit{ and } St_{\Lambda g} < 1. \tag{7.67}$$

The difference between this observation-based model and the Hinze theory of (7.66) can be attributed to several factors. First, lift force is neglected by the Hinze theory, but it can be significant for bubbles. To remedy this, Spelt and Biesheuvel (1997) considered lift forces for small-inertia bubbles in an updated theory, but this again yielded $k_p > k$ for significant Stokes numbers, which is inconsistent with observations. Another factor is the history force, which reduces the bubble velocity fluctuations (as noted in Figure 5.5 for laminar oscillatory flow), though this still only explains part of the discrepancy. A third factor that the Hinze theory neglects is preferential concentration, which tends to drive bubbles toward eddy cores (as in Figures 7.1b and 7.2c), where the fluid velocity fluctuations are often lower and so can reduce bubble TKE. A fourth factor is the deformation of the bubbles by the turbulence, which can lead to trajectory oscillations (§4.4.2) that tend to dominate the bubble velocity fluctuations (Mathia et al., 2018). Finally, nonlinear drag effects can impact the particle velocity fluctuations (Good et al., 2014). More research is needed to understand how these factors, all likely important, contribute to the overall differences between experiments and theory.

7.3.2 Particle Relative Velocity Fluctuations

The relative velocity dynamics of particles are also important, as they control many turbulent interphase coupling aspects, such as momentum, heat, and mass transfer to the particle as well as breakup for drops and bubbles. Fortunately, the previous analysis can be extended to determine these relative fluctuations. To start, the relative velocity of a single particle in one direction is represented as a Fourier integral per (7.24), as

$$w_i'' \equiv v_i'' - u_i'' = 2\pi \int_0^\infty [(\ell_3 - \ell_1)\cos(2\pi f t) + (\ell_4 - \ell_2)\sin(2\pi f t)]_i df. \tag{7.68}$$

Using the results from (7.25), these length scale differences can be expressed as follows:

$$\ell_3 - \ell_1 = \frac{4\pi^2 f^2 \tau_p^2 (R-1)}{1 + 4\pi^2 \tau_p^2 f^2}\ell_1 + \frac{2\pi f \tau_p (R-1)}{1 + 4\pi^2 \tau_p^2 f^2}\ell_2. \tag{7.69a}$$

$$\ell_4 - \ell_2 = -\frac{2\pi f \tau_p (R-1)}{1 + 4\pi^2 \tau_p^2 f^2}\ell_1 + \frac{4\pi^2 f^2 \tau_p^2 (R-1)}{1 + 4\pi^2 \tau_p^2 f^2}\ell_2. \tag{7.69b}$$

Using (7.24), (7.26b), and (7.57a), the ensemble-average of the relative velocity fluctuations in the streamwise direction for HIST can then be obtained as

$$\left\langle w''_{\parallel} w''_{\parallel} \right\rangle = \int_0^{\infty} \left(\frac{(\ell_3 - \ell_1)^2 + (\ell_4 - \ell_2)^2}{\ell_1^2 + \ell_2^2} \right) \left(\frac{4u_{\Lambda}^2 \tau_{@p,\parallel}}{1 + 4\pi^2 f^2 \tau_{@p,\parallel}^2} \right) df. \tag{7.70}$$

The RHS can then be integrated using (7.69) to yield the fluctuations in terms of the Stokes number of (7.59a) as follows:

$$\frac{\left\langle w''_{\parallel} w''_{\parallel} \right\rangle}{u_{\Lambda}^2} = \frac{St_{\parallel}(R-1)^2}{1 + St_{\parallel}}. \tag{7.71}$$

A similar result can be written for the perpendicular velocity fluctuations using (7.57b) and the Stokes number of (7.59b). Next, we consider the high-density and the high-buoyancy particle limits.

For high-density particles (R→0, e.g., drops in a gas), the following are the relative velocity fluctuations in the parallel and perpendicular directions:

$$\frac{\left\langle w''_{\parallel} w''_{\parallel} \right\rangle}{u_{\Lambda}^2} \approx \frac{St_{\parallel}}{1 + St_{\parallel}} \tag{7.72a}$$

$$\frac{\left\langle w''_{\perp} w''_{\perp} \right\rangle}{u_{\Lambda}^2} \approx \frac{St_{\perp}}{1 + St_{\perp}} \quad \textit{for } \rho_p \gg \rho_f. \tag{7.72b}$$

For St→0, the particles have negligible inertia so that their relative velocities are negligible compared to the fluid velocity fluctuations. For St→∞, the particles will not respond to the turbulence, so their relative velocities approach that of the fluid velocity fluctuations. To illustrate the latter limit, consider a particle released in a flow with no mean velocity with such high inertia particle that it remains nearly stationary ($\mathbf{v} \approx 0$). In this case, the relative velocity fluctuations are nearly equal to the surrounding fluid fluctuations.

The result of (7.72) indicates that intermediate Stokes numbers will have relative velocity fluctuations between these two limits. These theoretical trends are shown in Figure 7.16 and are generally consistent with available data for particles and bubbles, whereby relative velocity fluctuations increase as the inertia gets larger.

To obtain the strength of the relative particle velocity fluctuations, the sum in all three directions can be written in vector form for HIST as

$$\frac{\langle \mathbf{w}'' \cdot \mathbf{w}'' \rangle}{\langle \mathbf{u}'' \cdot \mathbf{u}'' \rangle} = \frac{\left\langle w''_{\parallel} w''_{\parallel} \right\rangle + 2\left\langle w''_{\perp} w''_{\perp} \right\rangle}{3u_{\Lambda}^2} \approx \left(\frac{St_{\parallel}}{3 + 3St_{\parallel}} + \frac{2St_{\perp}}{3 + 3St_{\perp}} \right). \tag{7.73}$$

Applying the terminal velocity assumptions of (7.60) for the Stokes numbers yields

$$\frac{\langle \mathbf{w}'' \cdot \mathbf{w}'' \rangle}{\langle \mathbf{u}'' \cdot \mathbf{u}'' \rangle} \approx \left[\frac{St_{\Lambda}\sqrt{1 + \frac{1}{4}g^2}}{3 + 3St_{\Lambda}\sqrt{1 + \frac{1}{4}g^2}} + \frac{2St_{\Lambda}\sqrt{1 + g^2}}{3 + 3St_{\Lambda}\sqrt{1 + g^2}} \right] \quad \textit{for } \rho_p \gg \rho_f. \tag{7.74}$$

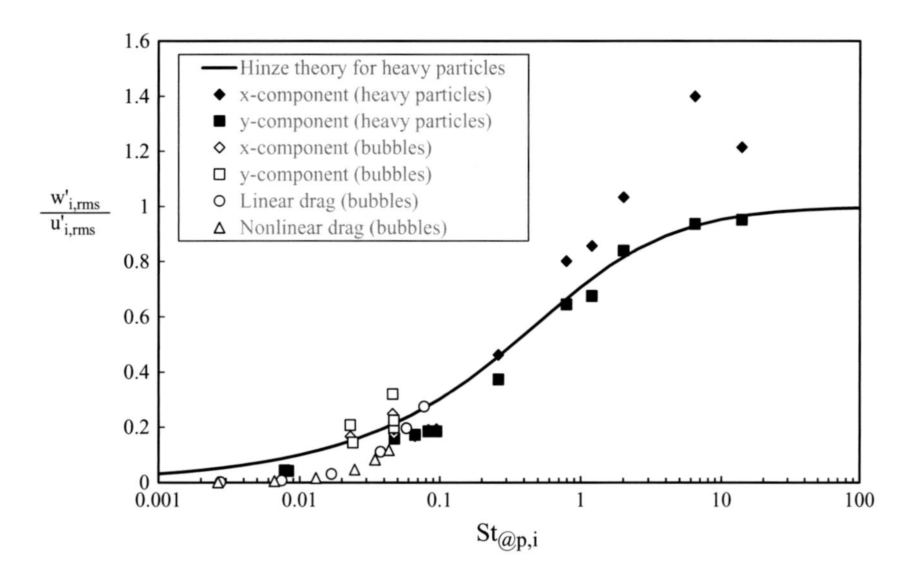

Figure 7.16 Relative particle velocity fluctuation amplitudes (transverse and streamwise) along the particle trajectory normalized by that of the fluid Eulerian fluctuations (transverse and streamwise) comparing theoretical prediction for heavy particles ($\rho_p \gg \rho_f$) with DNS results for high-density particles (Dorgan and Loth, 2004; Dorgan et al., 2005) and for bubbles (Dorgan and Loth, 2009) in a boundary layer for x-component and y-component results, as well as DNS from Zhang (2019) in isotropic turbulence for linear and nonlinear drag bubbles.

This result will be used later to estimate the mean particle Reynolds number in turbulent flow.

For the high-buoyancy limit of bubbles in a liquid ($R \rightarrow 3$), (7.71) indicates that the relative velocity fluctuations will substantially exceed that of the fluid. However, the data in Figure 7.16 suggest bubbles (surprisingly) tend to behave similarly to that of high-density particles per (7.72) for small and moderate Stokes numbers. This suggests that the factors that caused differences between theory and observation for bubble TKE are also important for the bubble relative velocities.

Mean Reynolds Number

As discussed earlier, the particle relative velocity can have strong fluctuations that depend significantly on the particle inertia. Since the particle Reynolds number depends on the relative velocity, turbulence can similarly lead to oscillations in Re_p. As discussed in the following, these oscillations can cause the mean relative velocity to deviate from the terminal velocity value. To show this, consider the ensemble-averaged Reynolds number compared to that for terminal conditions:

$$\langle Re_p \rangle \equiv \rho_f d\langle w \rangle / \mu_f = Re_{p,term} \langle w \rangle / w_{term}. \tag{7.75}$$

The relative velocity ensemble average and its fluctuation per (7.10) together comprise the instantaneous relative velocity, as follows:

$$\mathbf{w}(\tau) \equiv \langle \mathbf{w}(\tau) \rangle + \mathbf{w}''(\tau). \tag{7.76}$$

If one assumes a HIST flow, a linear particle drag (small Re_p), and neglects the effects of initial conditions, the average of the drag and gravitational forces will be equal, so the *vectors* of average relative velocity and terminal velocity are also approximately equal:

$$\langle \mathbf{w} \rangle \approx \mathbf{w}_{\text{term}}. \tag{7.77}$$

However, the *magnitudes* of the average relative velocity and terminal velocity will be different since the relative velocity includes fluctuations. This can be shown by assuming the terminal value is in the y-direction, so the instantaneous and averaged relative velocity magnitude can be related to the fluctuations as follows:

$$\text{w} = \sqrt{\mathbf{w} \cdot \mathbf{w}} = \sqrt{\left(\text{w}''_\text{x}\right)^2 + \left(\langle \text{w} \rangle + \text{w}''_\text{y}\right)^2 + \left(\text{w}''_\text{z}\right)^2}. \tag{7.78a}$$

$$\langle \text{w} \rangle = \left\langle \sqrt{\left(\text{w}''_\text{x}\right)^2 + \left(\text{w}_{\text{term}} + \text{w}''_\text{y}\right)^2 + \left(\text{w}''_\text{z}\right)^2} \right\rangle \approx \text{w}_{\text{term}} + \left\langle \sqrt{\mathbf{w}'' \cdot \mathbf{w}''} \right\rangle. \tag{7.78b}$$

The RHS approximation of (7.78b) is the sum of the limits when the terminal velocity dominates and that when the turbulent fluctuations dominate. Comparing the LHS and RHS of (7.78b), the relative velocity *magnitude* will be greater than the terminal velocity due to the relative particle velocity fluctuations. As such, the average Reynolds number would be greater than the terminal velocity Reynolds number per (7.75):

$$\left\langle \text{Re}_\text{p} \right\rangle = \text{Re}_{\text{p,term}} \left(1 + \frac{\left\langle \sqrt{\mathbf{w}'' \cdot \mathbf{w}''} \right\rangle}{\text{w}_{\text{term}}} \right) \approx \text{Re}_{\text{p,term}} \left(1 + \frac{\sqrt{\left\langle \mathbf{w}'' \cdot \mathbf{w}'' \right\rangle}}{\text{w}_{\text{term}}} \right). \tag{7.79}$$

The RHS employs a rough approximation based on weak velocity fluctuations and allows incorporation of the model of (7.74) to yield the following:

$$\frac{\left\langle \text{Re}_\text{p} \right\rangle}{\text{Re}_{\text{p,term}}} = \frac{\langle \text{w} \rangle}{\text{w}_{\text{term}}} \approx 1 + \frac{1}{g} \sqrt{\frac{\text{St}_\Lambda \sqrt{1 + \frac{1}{4} g^2}}{1 + \text{St}_\Lambda \sqrt{1 + \frac{1}{4} g^2}} + \frac{2\text{St}_\Lambda \sqrt{1 + g^2}}{1 + \text{St}_\Lambda \sqrt{1 + g^2}}}. \tag{7.80}$$

As shown in Figure 7.17a, the Reynolds number ratio based on this model increases as the particle Stokes number increases and as the drift parameter decreases. This increase can be important because many multiphase processes are proportional to the average Reynolds number, such as the average particle heat and mass transfer rates discussed in §3.5.

In the limit of zero drift parameter ($g \to 0$), the mean relative velocity (defined in 7.77) will be negligible, but the relative velocity *magnitude* (defined in 7.78b) can be significant due to particle inertia, and will depend on turbulence intensity and the Stokes number as follows:

$$\frac{\langle \text{w} \rangle}{\text{u}_\Lambda} \approx \sqrt{\frac{3\text{St}_\Lambda}{1 + \text{St}_\Lambda}} \quad \textit{for } \text{w}_{\text{term}} \ll \text{u}_\Lambda. \tag{7.81}$$

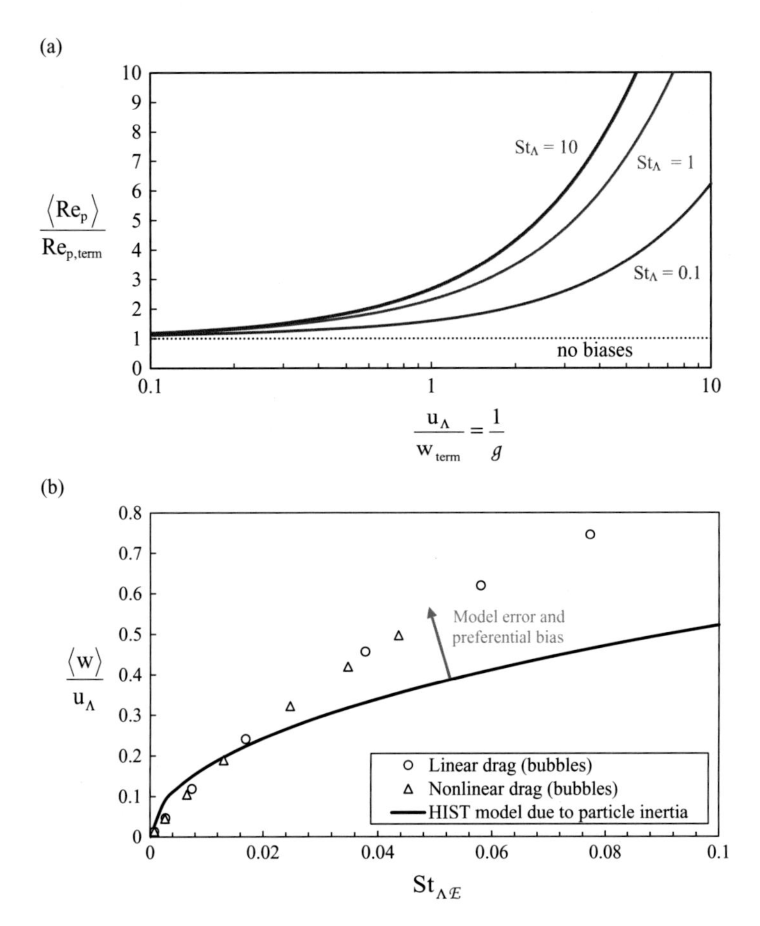

Figure 7.17 Results of the path-averaged relative values enhanced by turbulent fluctuations: (a) Reynolds number relative to terminal values as a function of the drift parameter, and (b) velocity relative to the fluid turbulence intensity for increasing Stokes numbers for zero drift ($g = 0$) compared to DNS results for bubbles (Zhang, 2019).

As shown in Figure 7.17b, the average relative velocity of weak drift bubbles ($g \to 0$) in isotropic turbulence is reasonably predicted by (7.81) for small Stokes numbers ($St_\Lambda < 0.03$), for both linear and nonlinear drag. However, higher Stokes numbers can yield even higher values. This difference can be attributed to errors in the approximations used in (7.78) and (7.79) but also can include preferential bias effect, which is one of the turbulent biases discussed next.

7.4 Turbulent Biases on Mean Particle Velocity

The particle settling velocity is defined as the ensemble-averaged particle velocity, and can be decomposed into ensemble averages of the relative velocity and the fluid velocity seen by the particles, as follows:

$$\mathbf{v}_{\text{settle}} \equiv \langle \mathbf{v} \rangle = \langle \mathbf{w} \rangle + \langle \overline{\mathbf{u}}_{@p} \rangle. \tag{7.82}$$

If there are no turbulent biases, the ensemble-averaged relative velocity will be governed by (7.77), and the fluid velocity will be based on the time-averaged value so the average particle velocity becomes:

$$\langle \mathbf{v} \rangle = \mathbf{w}_{\text{term}} + \overline{\mathbf{u}}_{@p} \qquad \textit{if no turbulent biases}. \tag{7.83}$$

This will be reasonable in conditions of linear drag, one-way coupling, homogenous particle concentration, and homogenous turbulence. However, there are at least five turbulent bias mechanisms that can modify the ensemble-averaged particle velocity, including the following:

(1) *Nonlinear drag bias* ($\Delta\mathbf{w}_{\text{Re}}$), which decreases the mean relative velocity for intermediate and high Re_p.
(2) *Preferential bias* ($\Delta\mathbf{v}_{\text{pref}}$), which augments the particle velocity in the direction of gravity.
(3) *Clustering bias* ($\Delta\mathbf{v}_{\text{clus}}$), which increases the mean relative velocity at high concentration.
(4) *Diffusiophoresis* ($\Delta\mathbf{v}_{\alpha}$), which biases particle velocity toward low concentrations.
(5) *Turbophoresis* ($\Delta\mathbf{v}_{k}$), which biases particle velocity to low TKE regions.

The nonlinear drag bias and turbophoresis impact the ensemble-averaged relative velocity, and the other biases primarily impact the ensemble-averaged fluid velocity:

$$\langle \mathbf{w} \rangle = \mathbf{w}_{\text{term}} + \Delta\mathbf{w}_{\text{Re}} + \Delta\mathbf{v}_{k}. \tag{7.84a}$$

$$\langle \mathbf{u}_{@p} \rangle = \overline{\mathbf{u}}_{@p} + \Delta\mathbf{v}_{\text{pref}} + \Delta\mathbf{v}_{\text{clus}} + \Delta\mathbf{v}_{\alpha}. \tag{7.84b}$$

Note that the cluster bias impacts the local fluid velocity per (7.84b), but any three-way coupling clustering effects (to be discussed) would also impact (7.84a). The above biases can be linearly combined using (7.82)–(7.84) as:

$$\langle \mathbf{v} \rangle = \mathbf{w}_{\text{term}} + \overline{\mathbf{u}}_{@p} + \Delta\mathbf{w}_{\text{Re}} + \Delta\mathbf{v}_{\text{pref}} + \Delta\mathbf{v}_{\text{clus}} + \Delta\mathbf{v}_{\alpha} + \Delta\mathbf{v}_{k}. \tag{7.85}$$

This myriad of biases can fortunately be neglected in certain conditions, such as the following:

- Nonlinear drag bias can be neglected for creeping-flow conditions ($\text{Re}_p \ll 1$).
- Preferential bias can be neglected for small or large inertias ($\text{St}_\eta \ll 1$ or $\text{St}_\eta \gg 1$).
- Clustering bias can be neglected for small particle concentrations ($\alpha \ll 1$ and $m \ll 1$).
- Diffusiophoresis can be neglected for weak concentration gradients ($\overline{\alpha} \approx$ const.).
- Turbophoresis can be neglected for weak TKE gradients ($k \approx$ const.).

To more easily quantify these biases with experiments and simulations, a stirred box flow condition is often employed, where the mean flow is zero but turbulence is finite. In that case, the settling velocity equals the terminal velocity plus any biases:

$$\mathbf{v}_{\text{settle}} = \mathbf{w}_{\text{term}} + \Delta\mathbf{w}_{\text{Re}} + \Delta\mathbf{v}_{\text{pref}} + \Delta\mathbf{v}_{\text{clus}} + \Delta\mathbf{v}_{\alpha} + \Delta\mathbf{v}_{k} \qquad \textit{for } \overline{\mathbf{u}} = 0. \tag{7.86}$$

The following sections discuss the theory and models as well as experimental and numerical results for these different biases.

7.4.1 Nonlinear Drag Bias

For a particle with nonlinear drag, the mean particle settling velocity can be reduced due to turbulence via the nonlinear drag bias. To demonstrate this, consider many particles averaged over long times settling in a HIST flow with no mean fluid velocity, so the effective gravitational forces balance out the ensemble-averaged drag:

$$\mathbf{F}_{\mathrm{g,eff}} = \forall_{\mathrm{p}}\left(\rho_{\mathrm{p}} - \rho_{\mathrm{f}}\right)\mathbf{g} = \langle \mathbf{F}_{\mathrm{D}} \rangle \qquad \mathit{for}\ \overline{\mathbf{u}} = 0. \tag{7.87}$$

The LHS can be expressed with the terminal velocity using (3.94b), while the RHS can be expressed as an average of using the instantaneous relative velocity via (3.55a), yielding the following:

$$\frac{1}{8}\pi d^2 \rho_{\mathrm{f}} C_{\mathrm{D,term}} \mathbf{w}_{\mathrm{term}} w_{\mathrm{term}} = \frac{1}{8}\pi d^2 \rho_{\mathrm{f}} \langle C_{\mathrm{D}} \mathbf{w} w \rangle. \tag{7.88}$$

Further assuming a Newton-based drag, for which the drag coefficient is constant (as in 3.57) and taking gravity in the y-direction yields the following:

$$w_{\mathrm{term}}^2 = \langle w_y w \rangle \quad \mathit{for}\ C_{\mathrm{D}} = \mathrm{const.} \tag{7.89}$$

The product on the RHS can then be expanded to include the ensemble mean and fluctuations, as follows:

$$w_{\mathrm{term}}^2 = \left\langle \left(\langle w_y \rangle + w_y''\right)\sqrt{\left(w_x''\right)^2 + \left(\langle w_y \rangle + w_y''\right)^2 + \left(w_z''\right)^2} \right\rangle. \tag{7.90}$$

The RHS can then be approximated as the sum of the limits for very weak and very strong relative velocity isotropic fluctuations. Further assuming isotropic fluctuations, the RHS yields the following:

$$w_{\mathrm{term}}^2 \approx \langle w_y \rangle^2 + \langle w_y \rangle \left\langle \sqrt{\left(w_x''\right)^2 + \left(w_y''\right)^2 + \left(w_z''\right)^2} \right\rangle \approx \langle w_y \rangle^2 + \langle w_y \rangle \left\langle \sqrt{\mathbf{w}'' \cdot \mathbf{w}''} \right\rangle. \tag{7.91}$$

The first term on the RHS is the square of the settling velocity magnitude. This equation can be rearranged to solve for this velocity and then approximated for weak fluctuations as follows:

$$\langle w_y \rangle \approx \sqrt{w_{\mathrm{term}}^2\left(1 - \sqrt{\mathbf{w}'' \cdot \mathbf{w}''}\right)} \approx w_{\mathrm{term}}\left(1 - \frac{1}{2}\sqrt{\mathbf{w}'' \cdot \mathbf{w}''}\right). \tag{7.92}$$

Thus, the settling velocity (LHS) for a particle with Newton-based drag and no other biases will be less than the terminal velocity due to turbulence fluctuations. Based on (7.84a) and HIST, this reduction is the *nonlinear drag bias* and denoted as $\Delta \mathbf{w}_{\mathrm{Re}}$. If one employs the approximation of (7.73) and considers first-order terms, this bias can be expressed as follows:

$$\Delta \mathbf{w}_{\mathrm{Re}} \equiv \langle \mathbf{w} \rangle - \mathbf{w}_{\mathrm{term}} \approx -\frac{\mathbf{w}_{\mathrm{term}}}{2g}\sqrt{\frac{\mathrm{St}_{\parallel}}{1+\mathrm{St}_{\parallel}}+\frac{2\mathrm{St}_{\perp}}{1+\mathrm{St}_{\perp}}} \quad \textit{for } \mathrm{C_D} = \mathrm{const.} \tag{7.93}$$

High-density particles with Newton-based drag therefore will fall slower in turbulence than in steady-flow conditions. Similarly, buoyant particles with Newton-based drag will rise slower in turbulence than in steady-flow conditions. This result can also be written in terms of the drift parameter using the approximation in (7.74).

To generalize this bias for intermediate particle Reynolds numbers, an implicit model was developed by Mei (1994). Or one may use an explicit model with (4.20), (7.52), (7.59) with $c_{\Lambda} \approx 1$ and (7.88) as

$$\frac{\Delta \mathrm{w}_{\mathrm{Re}}}{\mathrm{w}_{\mathrm{term}}} = \frac{|\langle \mathbf{w} \rangle - \mathbf{w}_{\mathrm{term}}|}{\mathrm{w}_{\mathrm{term}}} \approx -\frac{(\mathrm{Y}_{\mathrm{term}}-1)}{2g}\sqrt{\frac{\mathrm{St}_{\parallel}}{1+\mathrm{St}_{\parallel}}+\frac{2\mathrm{St}_{\perp}}{1+\mathrm{St}_{\perp}}}. \tag{7.94a}$$

$$\frac{\langle \mathrm{C_D} \rangle_{\mathrm{Re}}}{\mathrm{C_{D,term}}} \equiv \frac{\mathrm{w}_{\mathrm{term}}^2}{\langle \mathrm{w_y} \rangle^2} \approx 1 + \sqrt{\mathbf{w}'' \cdot \mathbf{w}''} = 1 + \frac{(\mathrm{Y}_{\mathrm{term}}-1)}{g}\sqrt{\frac{\mathrm{St}_{\parallel}}{1+\mathrm{St}_{\parallel}}+\frac{2\mathrm{St}_{\perp}}{1+\mathrm{St}_{\perp}}}. \tag{7.94b}$$

The second equation shows that the ensemble-averaged drag coefficient increases due to turbulence (consistent with the decrease in the relative velocity). As expected, this model yields no bias for the linear drag regime ($\mathrm{Y}_{\mathrm{term}} = 1$) and yields (7.93) for the Newton-based regime ($\mathrm{Y}_{\mathrm{term}} = 2$). Results from experiments, simulations, and the (7.94) model for intermediate Y values are shown in Figure 7.18 for spherical particles, where the settling velocity reduces more as Reynolds number and turbulence intensity increase. Similar trends were found for nonspherical particles by Byron et al. (2019). Some of the differences between the model and the experiments in Figure 7.18 can be related to effects neglected by (7.94), including any anisotropy of the turbulence, the history, and added-mass forces, and interactions with container boundaries. However, some of the differences can also be attributed to preferential concentration bias, discussed in the next section.

7.4.2 Preferential Bias and Clustering

As shown in Figures 7.1–7.4, small particles tend to concentrate in certain regions of the flow yielding nonuniform particle concentration. This tendency is defined as preferential concentration and is especially strong for Kolmogorov Stokes numbers near unity ($\mathrm{St}_{\eta} \sim 1$). An important consequence of this phenomena is that the average fluid velocity seen by the particles in these concentrated regions can be different from that averaged over all regions. If the mean particle concentration is small and uniform, (7.84b) can be used for the velocity difference as follows:

$$\Delta \mathbf{v}_{\mathrm{pref}} = \langle \mathbf{u}_{@\mathrm{p}} \rangle - \overline{\mathbf{u}}_{@\mathrm{p}} \quad \textit{for preferential concentration.} \tag{7.95}$$

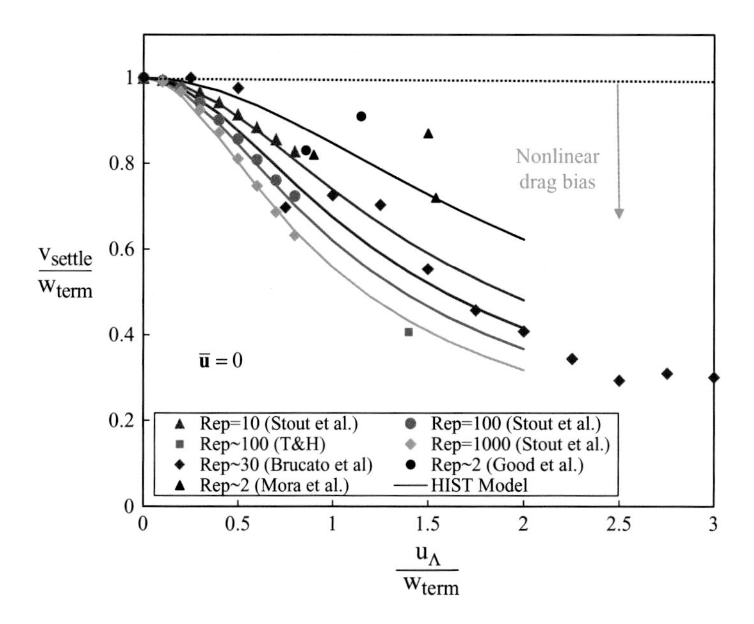

Figure 7.18 Impact of the nonlinear drag effect on normalized settling velocity (for zero mean fluid velocity) as a function of turbulence intensity based on experiments of Tunstall and Houghton (1968), Brucato et al. (1998), Kawanisi and Shiozaki (2008), Good et al. (2014), and Mora et al. (2020), as well as stochastic simulations of Stout et al. (1995) for $St_{\Lambda@p} > 1$ and $St_\eta \gg 1$.

The LHS is termed *preferential bias* and can change the particle settling velocity independently of the nonlinear drag bias per (7.86).

For high-density particles ($\rho_p \gg \rho_f$), preferential bias generally arises because particle positions are more likely to "see" downward-moving fluid, as shown in Figure 7.19. The mechanism for this bias is shown schematically in Figure 7.20a where high-density particles are centrifuged to the outside of eddies (Figure 7.1b), which causes them to be preferentially located in downdrafts. Thus, the fluid velocity along the particle path is biased downward, in the direction of gravity. This *preferential bias* therefore increases the particle-settling velocity compared to the case where there is no turbulence. On the other hand, bubbles are attracted to the low-pressure eddy cores (Figure 7.1c) due to fluid-stress forces. As shown in Figure 7.20b, this causes the trajectories to be held up, so their rise speed is inhibited by the preferential concentration.

To model effects of preferential bias in a dilute distribution of particles, consider HIST flow with a uniform mean velocity and linear particle drag (so nonlinear drag bias can be neglected). To obtain a relationship between the velocities of (7.95), we define the Eulerian ensemble average for an arbitrary quantity q based on the average value seen by all particles that pass through a location $\mathbf{X}$ (at different times), as follows:

$$\langle q(\mathbf{X}) \rangle \equiv \frac{\sum_{k=1}^{N_{pm}} q(\mathbf{X})}{N_{pm}} \qquad \textit{ensemble average for given location.} \qquad (7.96)$$

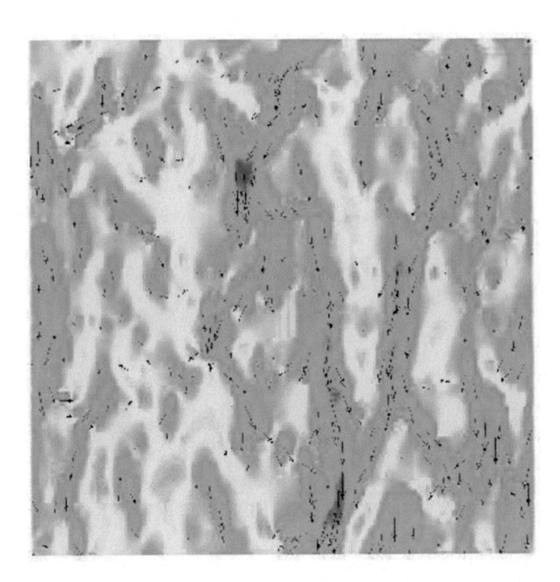

Figure 7.19 High-density particles (shown as dots) and falling in homogeneous isotropic turbulence with zero mean velocity with vertical velocity fluctuations shown by color contours, where preferential concentration of particles tend to occur in the blue-colored downdrafts as opposed to the red-colored updrafts (Couder, 2005).

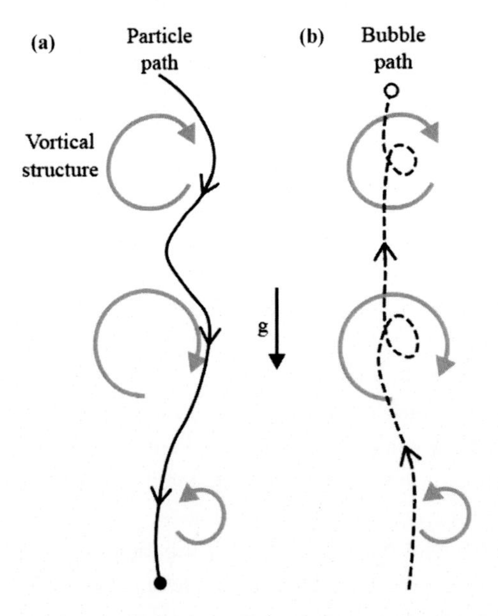

Figure 7.20 Preferential paths of particles in a turbulent flow with zero mean velocity, showing (a) a falling particle with high relative density catches eddy downdrafts, which increases the fall velocity; and (b) a bubble with low relative density migrates to eddy cores, which decreases the rise velocity.

This average thus weights a quantity (q) by the number of particles. If the particles all have the same volume, this becomes a particle volumetric concentration weighting (based on 5.1 and 5.2) as follows:

$$\langle q(\mathbf{X})\rangle = \frac{\sum_{k=1}^{N_{pm}} \forall_p q(\mathbf{X})}{\forall_p N_{pm}} = \frac{\overline{\alpha q}(\mathbf{X})}{\overline{\alpha}(\mathbf{X})} \quad \textit{for uniform particle volume.} \tag{7.97}$$

If we consider q as the fluid velocity seen by the particle and drop the **X** location notation, then

$$\langle \mathbf{u}_{@p}\rangle = \frac{\overline{\alpha \mathbf{u}_{@p}}}{\overline{\alpha}} = \frac{\overline{\alpha}\,\overline{\mathbf{u}}_{@p} + \overline{\alpha' \mathbf{u}'_{@p}}}{\overline{\alpha}} = \overline{\mathbf{u}}_{@p} + \frac{\overline{\alpha' \mathbf{u}'_{@p}}}{\overline{\alpha}}. \tag{7.98}$$

The RHS decomposes the fluid velocity into mean and fluctuating parts. Combining with (7.95), the preferential bias for the particle velocity becomes

$$\Delta \mathbf{v}_{pref} = \overline{\alpha' \mathbf{u}'_{@p}}/\overline{\alpha}. \tag{7.99}$$

The RHS correlation will be nonzero if the particles are preferentially located. In this case, the fluid velocity along the particle path is biased in the direction of gravity for both high-density and high-buoyancy particles (per Figure 7.20), so the LHS satisfies the inequality:

$$\Delta \mathbf{v}_{pref} \cdot \mathbf{g} > 0. \tag{7.100}$$

In the following, results (from theory, models, experiments, and numerical simulations) for this bias are considered first for one-way coupling ($\alpha \rightarrow 0$) and then for clustering conditions where both two-way and three-way coupling effects become important.

Preferential Bias for One-Way Coupling of High-Density Particles

For sparse conditions with one-way coupling, the preferential bias effect can be obtained theoretically for high-density particles with linear drag and small Kolmogorov-scale Stokes number in a HIST flow with weak turbulence relative to terminal velocity (Maxey, 1987). For these conditions, the bias is a function of a preferential turbulent length scale (ℓ_{pref}), as follows:

$$\Delta \mathbf{v}_{pref} \approx \frac{3\pi}{8} \frac{u_\Lambda^2 \tau_p}{\ell_{pref}} \frac{\mathbf{g}}{|\mathbf{g}|} \quad \textit{for } \rho_p \gg \rho_f, St_\eta \ll 1, \,\&\, \mathcal{g} \gg 1. \tag{7.101}$$

Based on Poorte and Biesheuvel (2002), this length scale can be approximated by the Taylor length scale ($\ell_{pref} \approx \lambda$). Employing (6.95), the Maxey theoretical preferential bias then becomes

$$\frac{\Delta \mathbf{v}_{pref}}{u_\Lambda} \approx \frac{3\pi St_\eta}{8\sqrt{15}} \frac{\mathbf{g}}{|\mathbf{g}|} \approx 0.3 St_\eta \frac{\mathbf{g}}{|\mathbf{g}|} \quad \textit{small-particle preferential bias.} \tag{7.102}$$

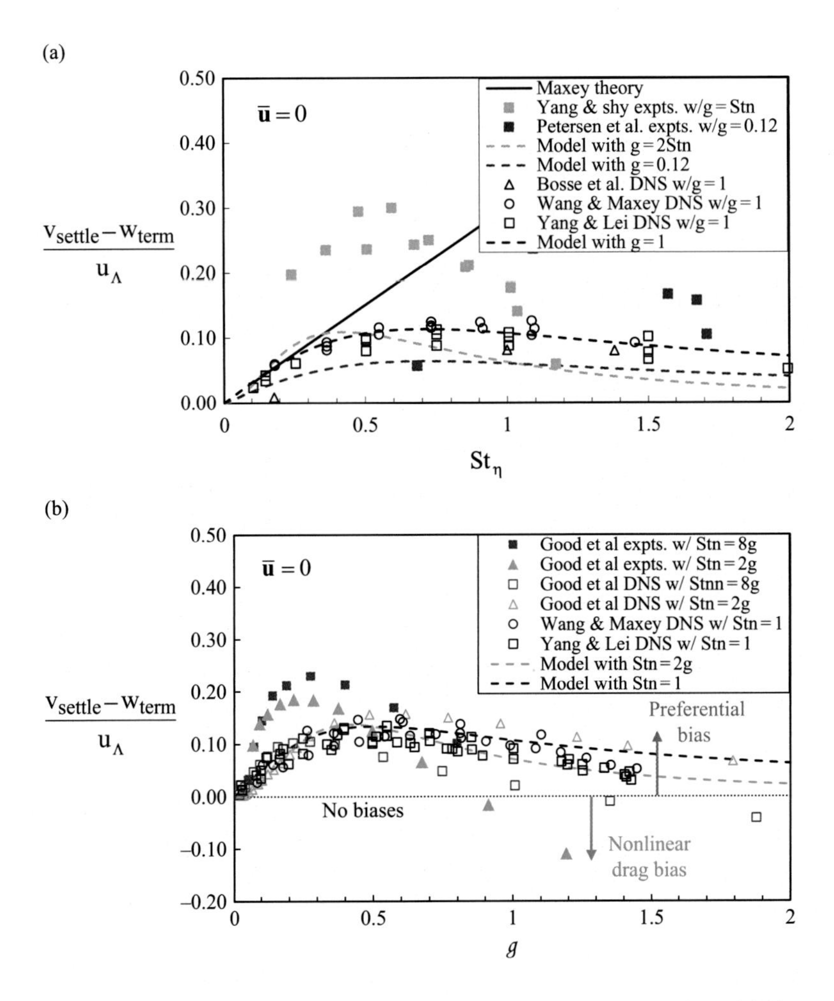

Figure 7.21 Change in settling velocity for heavy particles (where preferential bias model and DNS all assume one-way coupling): (a) as a function of the Kolmogorov Stokes number for Maxey theory (Maxey, 1987), experiments (Yang and Shy, 2003; Petersen et al., 2019), one-way coupling DNS (Wang and Maxey, 1993; Yang and Lei, 1998; Bosse et al., 2006), and preferential bias model; and (b) as a function of the drift parameter for experiments and DNS with nonlinear drag (Good et al., 2014), for linear drag DNS (Wang and Maxey, 1993; Yang and Lei, 1998), and for the preferential bias model.

If there are no other biases, this bias equals the change in the setting velocity:

$$\Delta \mathbf{v}_{\text{pref}} = \mathbf{v}_{\text{settle}} - \mathbf{w}_{\text{term}} \qquad \textit{for}\ \bar{\mathbf{u}} = 0. \tag{7.103}$$

This velocity difference is compared to experiments and simulations in Figure 7.21. As shown in Figure 7.21a, the Maxey theoretical bias (solid line) qualitatively explains the settling velocity increase as St_η initially grows, but it does not explain the peak at around 0.7 and the bias decrease for $\text{St}_\eta > 1$. In addition, the theory does not include an influence of the drift parameter, though this effect is significant as seen in Figure 7.21b, where the data show a settling velocity peak at around $g = 0.5$. As such, both St_η and g are important to preferential bias (whereas the small-particle theory only considers the effect of St_η). Wang and Maxey (1993) noted that these peaks occur because particles with lower St_η and g tend to closely follow the fluid, resulting in reduced opportunities for preferential concentration, while particles with higher St_η and g have increased eddy-crossing relative to the microscales, which also reduces this effect.

The linear drag DNS results of Figure 7.21 tend to have less scatter than the experimental data sets and can be used for an empirical one-way coupling model for high-density linear drag particles in HIST as follows:

$$\frac{\Delta \mathbf{v}_{pref}}{u_\Lambda} \approx 1.6\left(\frac{St_\eta}{1+2St_\eta^2}\right)\left[\frac{g}{1+4g^2}\right]\frac{\mathbf{g}}{|\mathbf{g}|} \quad \textit{for } \rho_p \gg \rho_f. \tag{7.104}$$

This model predicts a peak bias at $St_\eta = 0.7$ and $g = 0.5$. The combined effects of the Stokes number and drift parameter for one set of DNS results are also shown in Figure 7.22, where the peak bias similarly occurs near $St_\eta = 1$ and $g = 0.5$.

In general, this model and the one-way coupling DNS predictions have the same trends of the measured data, but the experimental biases in Figure 7.21a tend to be larger. Good et al. (2014) suggested that the increased experimental bias may be due to turbulent flow gradients and flow complexity around the particle that modifies their drag forces. In addition, clustering bias (which augments the preferential bias as discussed later) may have a significant role. On the other hand, the experimental biases in Figure 7.21b for $g > 1$ are lower than predicted by DNS or the model. This can be partly explained by the presence of nonlinear drag bias, which reduces fall velocity (and counteracts the preferential bias increases). The competing trends for these two effects can be combined based on (7.86) as follows:

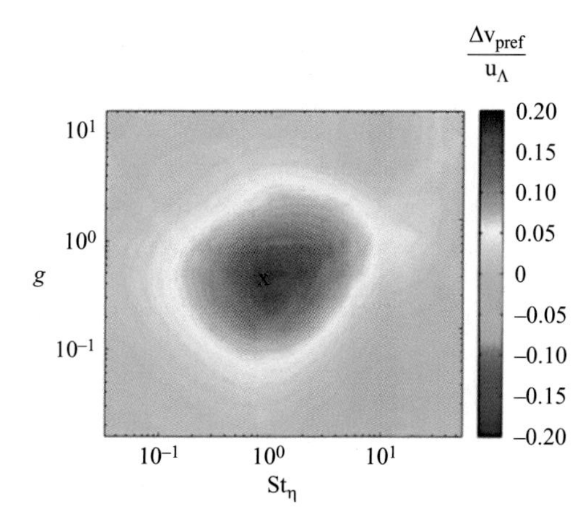

Figure 7.22 Contours of preferential bias normalized by fluid turbulent strength as a function of the particle drift parameter (g) and Kolmogorov Stokes number (St_η) based on DNS of Good et al. (2014) for heavy particles with linear drag, where "x" indicates the peak value for the empirical model for heavy linear drag particles.

$$\mathbf{v}_{\text{settle}} = \mathbf{w}_{\text{term}} + \Delta\mathbf{w}_{\text{Re}} + \Delta\mathbf{v}_{\text{pref}} \qquad \textit{for } \overline{\mathbf{u}} = 0. \tag{7.105}$$

However, combining (7.94) and (7.104) only gives qualitive predictions when compared to experiments. Moreover, simulations (Wang and Maxey, 1993) and experiments (Rosa et al., 2016) indicated an influence of Re_λ, especially for $\text{Re}_\lambda < 100$ (Mora et al., 2020). Because of such complications, a theory or model that can robustly predict preferential bias is still needed. In the following, we discuss the qualitative effect of clustering bias and then summarize all bias effects for high-density particles.

Clustering Bias (Two-Way and Three-Way Coupling) of High-Density Particles

The preferential bias becomes even more complicated when both two-way and three-way coupling effects occur. This overall coupling mechanism is called the *clustering bias* and can magnify the preferential bias effect for high-density particles. For example, Figure 7.23a shows much higher fall velocities than predicted by the preferential bias model of (7.102) for particle mass loadings as small as 1% (volume fractions as low as 10^{-5}). The reason for this increase can be traced to both two-way and three-way coupling. For two-way coupling, a cluster of high-density particles at local high concentration acts as a single mixed-fluid entity that has an effective density higher than that of the surrounding mixture (Aliseda et al., 2002). As a result, the group of particles induces a local downward velocity of the local fluid (as discussed in §5.4), so the local particles fall faster as well. In addition, the particles are closer to each other and therefore can have reduced relative drag via drafting effects (as discussed in §5.5), which can further increase the average relative fall velocity beyond that of the one-way coupling settling velocity. While more work is needed to develop a theory or model to robustly predict cluster bias, DNS with two-way coupling has been found to reasonably represent the trends associated with increased mass loading (Monchaux and Dejoan, 2017), as shown in Figure 7.23b.

An overview of the effects of the turbulent biases discussed earlier (nonlinear drag bias, preferential bias, and clustering bias) for high-density particles is shown in Figure 7.24. In general, the biases grow stronger as the turbulence level increases (as the drift parameter decreases) but can act in opposite directions, to either increase or decrease the particle relative velocity.

Preferential Bias for Buoyant Particles

For bubbles, preferential concentration reduces the rise velocity but is further complicated by additional forces, such as fluid-stress, added-mass, history, and lift forces. A theoretical model for preferential bias was developed by Spelt and Biesheuvel (1997) for a linear drag law drag with small St_η, and the model can be generalized to include the added-mass and fluid-stress forces (to account for continuous-phase velocity acceleration along the particle path). The result is a function of the acceleration parameter and increases with St_η, as follows:

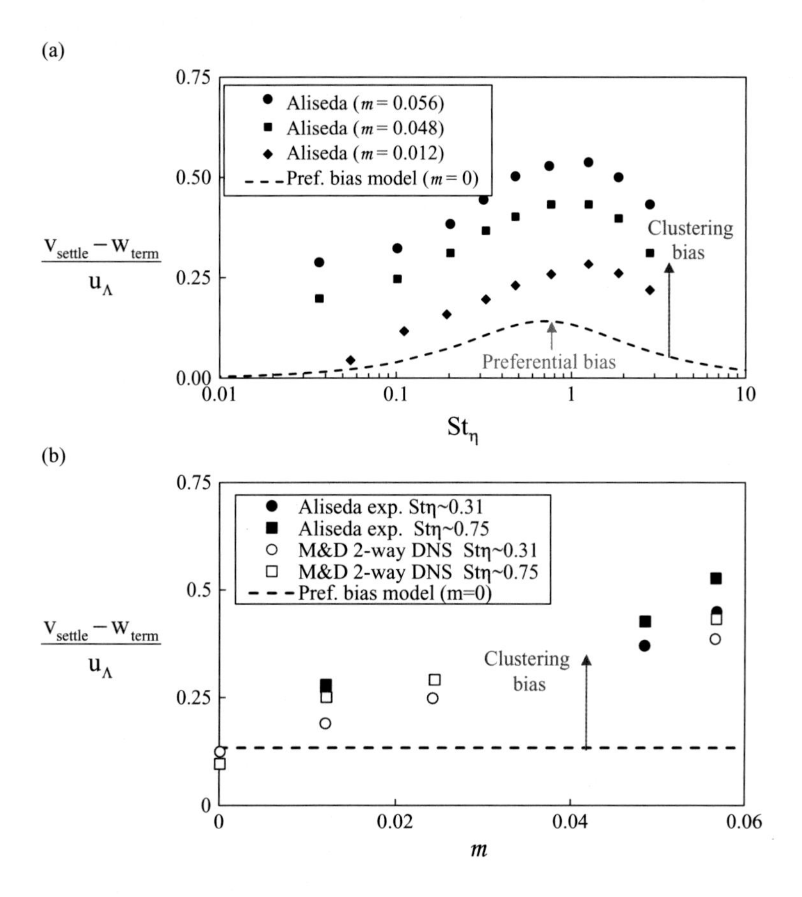

Figure 7.23 Heavy particle settling velocity changes compared to the one-way coupled preferential bias model: (a) as a function of the Kolmogorov Stokes number for droplets in air with $g = 0.02$–0.8 (Aliseda et al., 2002), for solid particles in air with $g = 0.2$–0.7 (Yang and Shy, 2005), and for a bias model with $g = 0.5$; and (b) as a function of mass loading for Aliseda et al. experiments, for two-way coupling DNS of Monchaux and Dejoan (2017), and for a bias model with $g = 0.5$ and $\mathrm{St}_\eta = 0.5$.

$$\frac{\Delta \mathbf{v}_{\mathrm{pref}}}{\mathrm{u}_\Lambda} \approx \frac{3\pi|1-\mathrm{R}|\mathrm{St}_\eta}{8\sqrt{15}}\frac{\mathbf{g}}{|\mathbf{g}|} \approx 0.3|1-\mathrm{R}|\mathrm{St}_\eta \frac{\mathbf{g}}{|\mathbf{g}|} \qquad \textit{for } \mathrm{St}_\eta \ll 1 \tag{7.106}$$

For high-density particles (R→0), this theoretical result reverts to (7.102), predicting an increase in settling velocity, as discussed earlier. For buoyant particles ($\mathrm{R} > 1$), this theory predicts a downward bias (reduction of rise velocity), as shown in Figure 7.25a for bubbles ($\mathrm{R} = 3$). This trend is consistent with experiments and simulations for bubbles, but the normalized bias tends to a limiting value of about –0.5 for large St_η values. To describe this trend, Zhang (2019) proposed an empirical preferential bias model for bubbles, as follows:

$$\frac{\Delta \mathbf{v}_{\mathrm{pref}}}{\mathrm{w}_{\mathrm{term}}} \approx \frac{1}{2}\left[\mathrm{e}^{-1.2\mathrm{St}_\eta} - 1\right] \qquad \textit{for } \rho_\mathrm{p} \ll \rho_\mathrm{f}. \tag{7.107}$$

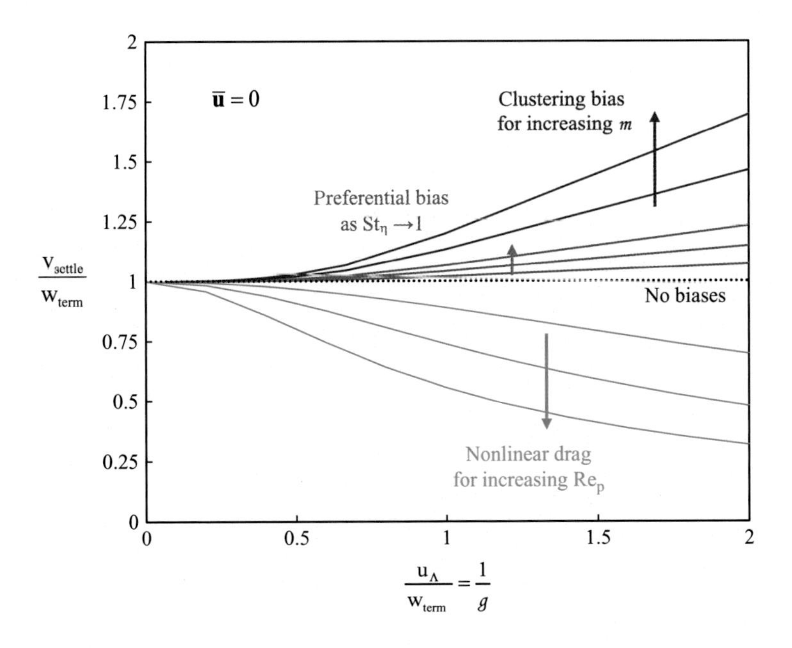

Figure 7.24 Various biases for the path-averaged settling velocity of heavy particles in HIST with zero mean fluid velocity as flow turbulence increases, including nonlinear drag bias (which increases with particle Reynolds number), preferential bias (which is maximized when St_η tends to unity), and clustering bias (which increases as mass loading increases).

As shown in Figure 7.25a, this model is roughly reasonable when compared to the experimental bubble and DNS data, but there is a large variance in the data indicating that additional effects, such as history and lift forces as well as clustering, may be important.

For slightly buoyant particles, the theory of (7.106) predicts negligible bias magnitudes (since $R \sim 1$). However, experimental results of oil drops in water as shown in Figure 7.25b indicate very large biases, which vary widely. In general, the biases for these buoyant particles are mostly positive for $St_\eta g > 1$ (peaking near $0.75w_{term}$), suggesting strong clustering effects that increase rise but are mostly negative for $St_\eta g < 1$ (as low as $-0.85w_{term}$), suggesting strong preferential bias effects. However, there is no relevant theory that suitably predicts these effects, and the data are too scattered to justify an empirical fit. As such, additional research is needed to understand preferential bias for slightly buoyant particles.

7.4.3 Diffusiophoresis and Turbophoresis

There are also two gradient-based turbulent drift mechanisms that can lead to a bias in particle velocity: *diffusiophoresis*, caused by a mean gradient in particle concentration, and *turbophoresis*, caused by a mean gradient in turbulent kinetic energy. Both mechanisms cause a net drift of particles in the direction opposite to the gradient.

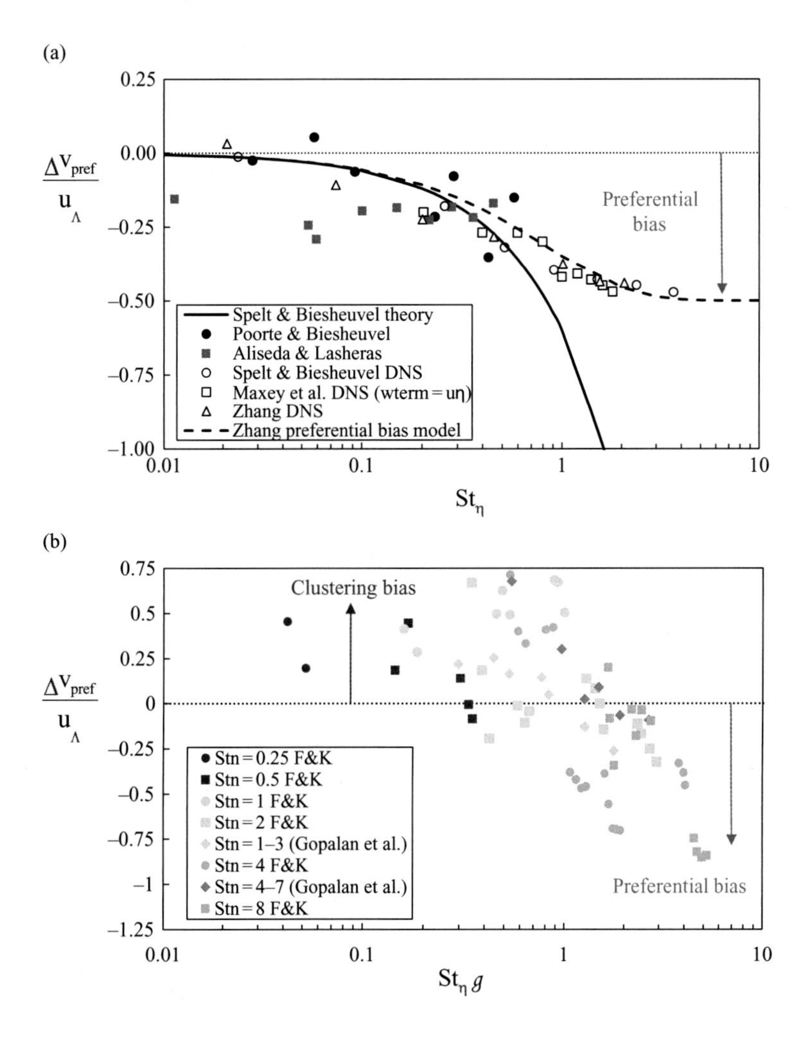

Figure 7.25 Preferential bias of buoyant particles: (a) bubbles based on experiments of Poorte and Biesheuvel (2002) and Aliseda and Lasheras (2011), DNS data of Maxey et al. (1997), Spelt and Biesheuvel (1997), and Zhang (2019), as well as theory and empirical models for bias; and (b) slightly buoyant oil drops in water based on experiments of F&K (Friedman and Katz, 2002) and Gopalan et al. (2008).

As such, these biases (unlike those in §7.4.1 and §7.4.2) are not related to gravitational effects. Rather, diffusiophoresis and turbophoresis effects are related to the turbulent diffusion of particles discussed in §7.2.

Concentration Gradient Bias (Diffusiophoresis)

As was discussed in §6.2.4 and illustrated in Figure 6.9, a species concentration gradient in HIST will, over time, reduce as the concentration distribution is equilibrated due to turbulent mixing. The same equilibrating effect is true for particles, where the net particle transport toward regions of low particles is termed *diffusiophoresis*. This is

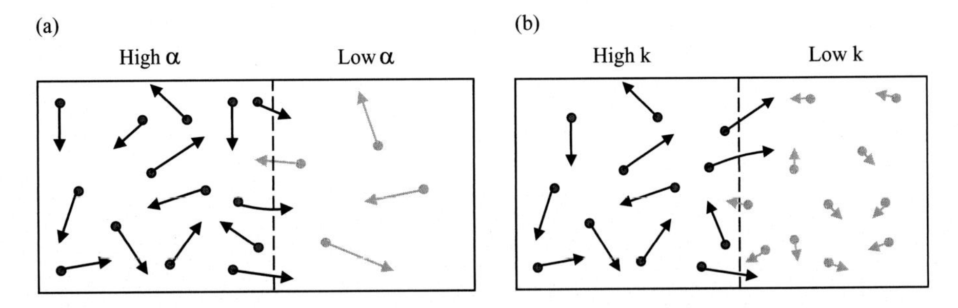

Figure 7.26 Illustration of particle concentration flux in a stirred box flow (turbulence with no mean fluid velocity) where a rightward drift of particles will be caused: (a) a gradient in particle concentration (diffusiophoresis); and (b) a gradient in turbulence intensity (turbophoresis). In both case, black particles are more likely to move across the dashed flux boundary than gray particles, which leads to a rightward mean flow of particles.

illustrated in Figure 7.26a for a stirred box flow. The concentration gradient (more particles on the left than on the right) coupled with particle velocity fluctuations caused by turbulence will yield a net particle flux to the right (since there are more trajectory opportunities for this to occur due to more particles in the right).

To model diffusiophoresis, consider an HIST flow with linear drag ($\Delta \mathbf{w}_{\text{Re}} = 0$) with no preferential concentration ($\Delta \mathbf{v}_{\text{pref}} = 0$) but a gradient in particle concentration. In this case, the ensemble average of the particle velocity can be expressed using same approach as (7.98), yielding

$$\langle \mathbf{v} \rangle = \overline{\alpha \mathbf{v}}/\overline{\alpha} = \overline{(\overline{\alpha} + \alpha')(\overline{\mathbf{v}} + \mathbf{v}')}/\overline{\alpha} = \overline{\mathbf{v}} + \overline{\alpha' \mathbf{v}'}/\overline{\alpha}. \tag{7.108}$$

The last term on the RHS defines the concentration-gradient bias and can be related to the settling velocity per (7.85) if there are no other biases:

$$\Delta \mathbf{v}_{\alpha} \equiv \overline{\alpha' \mathbf{v}'}/\overline{\alpha} = \langle \mathbf{v} \rangle - \left(\overline{\mathbf{u}}_{@p} + \mathbf{w}_{\text{term}} \right). \tag{7.109}$$

The concentration-gradient bias can be shown to be proportional to the particle concentration gradient as similarly found for species turbulent diffusion via Figure 6.8, by simply replacing $\mathcal{M}$ with α, and u with v. As such, the three principles for particle turbulent diffusion due to a particle concentration gradient can be obtained as follows:

Principle 1: $\overline{\alpha' v_i'}$ is proportional to the particle concentration gradient in the i-direction.
Principle 2: $\overline{\alpha' v_i'}$ transport is in the opposite direction of this gradient.
Principle 3: $\overline{\alpha' v_i'}$ is proportional to the strength of the particle velocity fluctuations.

As such, the RHS correlation can be modeled using a Fickian turbulent diffusion hypothesis (per 6.38 and 6.41) based on the mean particle concentration gradient (instead of species gradient) and the particle turbulent kinetic energy (instead of the fluid kinetic energy), as follows:

$$\Delta \mathbf{v}_{\alpha,i} \approx -\frac{1}{\overline{\alpha}} \frac{c_\mu k_p^2}{Sc_{turb}\varepsilon} \frac{\partial \overline{\alpha}}{\partial X_i}. \tag{7.110}$$

The particle TKE can be related to the surrounding fluid TKE by (7.61). Further details on this bias are discussed by Elghobashi and Abou-Arab (1983).

Turbulence Gradient Bias (Turbophoresis)

If we now consider a flow with a uniform particle concentration but a gradient in the fluid TKE, then turbophoresis can be important. As shown in Figure 7.26b, if there is higher turbulence on the left-hand side, particles on this side will have more random motion and are more likely to be flung to the right-hand side (compared to particles in the low-turbulence region being flung to the left-hand side). As such, nonhomogeneous turbulence will drive particles from high-turbulence regions into low-turbulence regions via turbophoresis.

A simple version of the turbophoretic velocity can be modeled by combining (3.89a) and (3.93) for the particle Equation of Motion and then taking a time average of all terms:

$$\overline{\frac{d\mathbf{v}}{dt}} = -\frac{\overline{\mathbf{v}} - \overline{\mathbf{u}}_{@p} - \mathbf{w}_{term}}{\tau_p} \qquad \textit{for } \rho_p \gg \rho_f. \tag{7.111}$$

For small response times ($St_\Lambda \ll 1$) in a flow with steady and uniform mean flow, the LHS mean particle acceleration at a fixed location can be approximated by the mean acceleration of the continuous phase along the fluid path. This average Lagrangian particle acceleration can then be decomposed into an Eulerian time derivative (which is zero for steady mean flow) and a spatial gradient term per (1.19a), which is only nonzero for the velocity fluctuations if the mean flow is uniform:

$$\overline{\frac{d\mathbf{v}}{dt}} \approx \overline{\frac{\mathcal{D}\mathbf{u}}{\mathcal{D}t}} = \overline{\frac{\partial \mathbf{u}}{\partial t}} + \overline{(\mathbf{u}\cdot\nabla)\mathbf{u}} = \overline{(\mathbf{u}'\cdot\nabla)\mathbf{u}'} \qquad \textit{for uniform and steady } \overline{\mathbf{u}}. \tag{7.112}$$

If one further assumes that the turbulence is isotropic (6.49), the RHS can be written as 2/3 of the TKE gradient. This result can be combined with (7.111) to yield a model for the turbophoretic velocity bias, as follows:

$$\Delta \mathbf{v}_k \equiv \overline{\mathbf{v}} - \left(\mathbf{w}_{term} + \overline{\mathbf{u}}_{@p}\right) \approx -\frac{2}{3}\tau_p \nabla k \qquad \textit{for } St_\Lambda \ll 1 \;\&\; \rho_p \gg \rho_f. \tag{7.113}$$

This bias model is qualitatively consistent with Figure 7.26b, where turbulence moves particles from regions of high-turbulence intensity to regions of low-turbulence intensity and has been found to be quantitatively consistent with particles with $St^+ < 2$ and $\mathcal{M} < 10^{-3}$, while larger-inertia particles and higher-volume concentrations generally have a reduced bias (Johnson et al., 2020). To incorporate larger particle inertias, the fluid TKE gradient can be replaced by the particle TKE for isotopic fluctuations or, more accurately, by the lateral velocity fluctuations for lateral turbophoresis (Reeks, 2014), as follows:

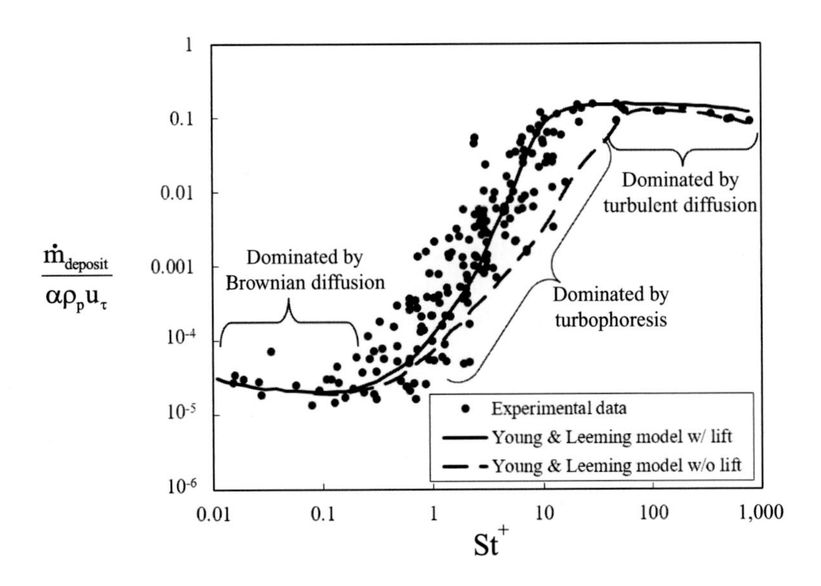

Figure 7.27 Particle deposition mass flux per surface area for turbulent channel and pipe flows (Young and Leeming, 1997).

$$\Delta \mathbf{v}_k \approx -\frac{2}{3}\tau_p \nabla k_p \qquad \textit{for } \rho_p \gg \rho_f \textit{ and isotropic fluctuations.} \tag{7.114a}$$

$$\Delta v_{k,\perp} \approx -\tau_p \nabla \langle v''_\perp v''_\perp \rangle \qquad \textit{for } \rho_p \gg \rho_f \textit{ and lateral fluctuations.} \tag{7.114b}$$

Because the RHS gradient terms reduce with St_Λ and g (per 7.64 and 7.59b), the net turbophoresis tends to peak at about unity and decrease for larger values.

The effects of turbophoresis, turbulent diffusion, and Brownian diffusion are shown for a turbulent boundary layer in Figure 7.27. Since this flow has high wall-normal gradients of TKE just above the wall (Figure 6.5), turbophoresis will drive particles away from this region, both toward the freestream and toward the wall. The latter causes particle deposition, as shown in Figure 7.27, where turbophoresis is the primary driver for $1 < St^+ < 20$. However, deposition at $St^+ > 20$ is driven by lateral turbulent diffusion whereby particle relative velocity fluctuations become independent of the Stokes number and scale with the turbulent velocity (7.72b), which scales with u_τ for a turbulent boundary layer. At the other extreme, wall deposition at $St^+ < 1$ is driven by Brownian diffusion since the other two diffusion effects are negligible for very small particles. The lift force is also important for intermediate Stokes numbers, and this force is discussed further in §10.1.

7.5 Turbulent Deformation and Breakup of Drops and Bubbles

Turbulent kinetic energy can serve as a mechanism for breakup of fluid dynamic particles (drops and bubbles). Since turbulence is associated with a wide range of

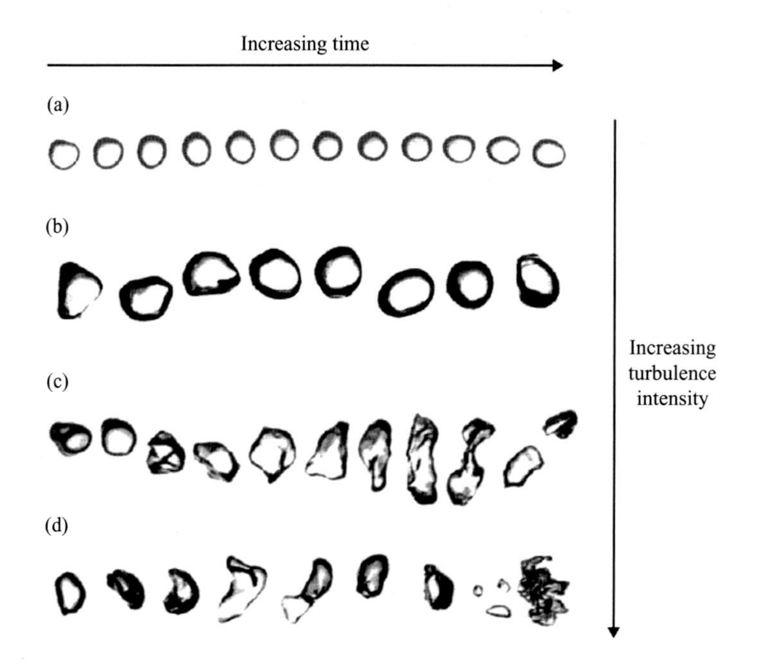

Figure 7.28 Microgravity turbulent conditions showing four bubbles in a time sequence, where each are undergoing shape oscillation, shear deformation, and sometimes breakup (Risso and Fabre, 1998) for which the relative turbulent intensity increases from (a) to (d). The bubble breaks up into two parts in (c) and into many parts in (d).

length and time scales and flow nonlinearities (§6.4), the resulting deformation and breakup in turbulent flows can be driven by a complex mixture of mechanisms. However, the key drivers tend to be frequency-induced shape harmonics, shear-induced elongation, and instability-induced deformation.

For example, bubbles subjected to nearly homogeneous isotropic turbulence are shown in Figure 7.28 in the order of increasing turbulence intensity. At modest turbulence intensity (Figure 7.28a), deformations approximately yield a $\mathcal{S} = 2$ shape mode, while higher turbulence (Figure 7.28b) also induces a combination of the $\mathcal{S} = 2$ and $\mathcal{S} = 3$ shape harmonics owing to stronger turbulent fluctuations at these frequencies. At still higher turbulence (Figure 7.28c), bubble elongation due to shearing occurs. These elongated shapes are then receptive to instabilities at the locations where curvature (and surface tension) is weak. The indentation along the middle of the bubble suggests a Kelvin–Helmholtz instability, which is amplified and eventually leads to breakup. At the highest turbulence (Figure 7.28d), the shapes are even more complex, and the breakup leads to several smaller bubbles.

For drops in a liquid, the deformations due to turbulence can lead to similar complexity but with even higher deformations before breakup (Figure 7.29). This is due to the damping influence associated with the droplet viscosity that can delay breakup until the deformations are very large. However, the probability of breakup still increases with the level of turbulence intensity and as surface tension effects decrease.

Figure 7.29 Time sequence of a droplet deforming in a turbulent liquid flow ($\mu_p{}^* \approx 9$), where the initial ligament is approximately 1 mm long (Clift et al., 1978).

To consider when breakup occurs, it is helpful to define a turbulent Weber number based on the ratio of the local fluid velocity fluctuations and the surface tension energy. Hinze hypothesized this interaction using the local wavelength of the turbulence (ℓ). In particular, turbulent structures much larger than the drop diameter ($\ell \gg d$) are assumed to present a nearly linear gradient of flow in the vicinity of the drop and are thus too large to significantly deform the particle. On the other hand, turbulent structures much smaller than the drop ($\ell \ll d$) are assumed to be too weak and small to induce substantial deformation. Therefore, the turbulent wavelength that corresponds to the drop diameter ($\ell \approx d$) is expected to have the most influence on deformation and breakup. This wavelength can be used to identify the relevant portion of the kinetic energy ($k_{\ell=d}$) and define a relevant turbulent Weber number based on the form of (4.41), as follows:

$$We_{\ell=d} \sim \rho_f k_{\ell=d} d/\sigma. \tag{7.115}$$

When this Weber number is high, deformation due to turbulence is expected to cause breakup, since the restoring surface tension effects will be weak in comparison.

To employ (7.115), the kinetic energy at a given wavelength is proportional to the ratio of specific turbulent energy to wavelength per (6.92) and (6.93) and is described for the inertial subrange per (6.94) as follows:

$$k_\ell \sim E/\ell \sim (\varepsilon \ell)^{2/3}. \tag{7.116}$$

This result can be combined with (7.115) to define Hinze's dissipation-based Weber number (We_ε), as follows:

$$We_\varepsilon \equiv 2\rho_f \varepsilon^{2/3} d^{5/3}/\sigma. \tag{7.117}$$

Based on experimental studies, the critical Weber number was found empirically as $We_{\varepsilon,crit} \approx 1.2$ so that the critical diameter (d_{crit}) above which breakup will occur is

$$d_{crit} \approx \left(\frac{1.2\sigma}{2\rho_f \varepsilon^{2/3}}\right)^{3/5}. \tag{7.118}$$

In fully developed turbulence, this formulation has been found to be quite robust for both drops (Clift et al., 1978) and bubbles (Martinez-Bazan et al., 1999b; Millies and Mewes, 1999).

However, turbulent dissipation is often not as easily measured nor predicted as compared to turbulent kinetic energy. As such, other breakup models have been proposed using a Weber number based on TKE. For example, Risso and Fabre (1998) proposed a model that includes the integral length scale based on streamwise velocity fluctuations (6.71a), as follows:

$$\mathrm{We_k} \equiv \frac{0.7\rho_f k d^{5/3}}{\sigma \Lambda_{\|}^{2/3}}. \tag{7.119}$$

Based on experiments for breakup, they empirically obtained $\mathrm{We_{k.crit}} \approx 3$, and this threshold can then be used to determine d_{crit} (similar to 7.118). Other models have been proposed to empirically incorporate viscous damping effects (via an Ohnesorge number correction) or turbulence spectra distribution (Martinez-Bazan et al., 1999b; Qiand, 2003). However, the results from these models are generally similar to those given by (7.117) or (7.119).

7.6 Two-Way Turbulence Coupling by Particles

If the flow is turbulent, two-way coupling associated with the continuous-phase TKE is often referred to as turbulence *modulation*. When particles are introduced, there are now two sets of production and destruction terms, as shown in Figure 7.30: those associated with the continuous-phase fluid interactions, per (6.50), and those associated with the particle interactions. As such, TKE production can be associated with continuous-phase flow gradients but also by the particle wakes (especially for large particles). In addition, viscous dissipation occurs at the Kolmogorov scales via ε but also on the particle surfaces (especially for small particles). The overall change in TKE generation with the four mechanisms identified in Figure 7.30 can be formalized as follows:

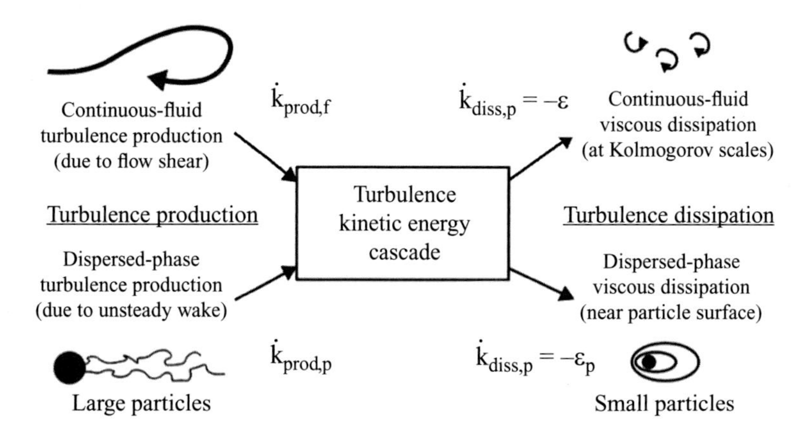

Figure 7.30 Influence of production and dissipation on the continuous-phase turbulence in a multiphase flow (adapted from Crowe et al., 1998) where production terms are positive and associated with energy being added *to* the turbulence while dissipation terms are negative and are associated with work done *by* the turbulence.

$$\dot{k} = \dot{k}_{prod,f} + \dot{k}_{prod,p} + \dot{k}_{diss,f} + \dot{k}_{diss,p} = \dot{k}_{prod,f} + \dot{k}_{prod,p} - \varepsilon - \varepsilon_p. \quad (7.120)$$

The two production terms reflect energy gains, whereas the two dissipation terms reflect TKE losses (ε and ε_p are defined as positive for losses).

The primary mechanisms for turbulence modulation by particles (enhancement or reduction) in Figure 7.30 can be conceptualized by considering the hypothetical example of flow past stationary particles (fixed in space). If the incoming flow has turbulent structures with mean velocity gradients, the presence of the small, fixed particles can serve to reduce those gradients by increasing dissipation thereby reducing TKE. In fact, this is the principle of flow conditioning screens used to reduce turbulence levels and velocity gradients in wind tunnels. However, if the incoming flow has low turbulence, large stationary particles will produce turbulent wakes, which will increase the downstream turbulence.

Unfortunately, the actual production and dissipation mechanisms are much more complicated because the particles move and react to the turbulence. In addition, most flows have regions of high mean velocity gradients, anisotropic turbulence and TKE production. Therefore, all four terms on the RHS of (7.120) can change due to particles, making it difficult to assess individual effects of the particles on the turbulence. As such, it is often difficult to predict whether particles will qualitatively increase or decrease a flow's TKE, let alone predict the magnitude of the change (Eaton, 1999). Despite this, some trends can be identified.

For example, gas turbulence in a vertical pipe for the addition of particle of different sizes and mass loadings is shown in Figure 7.31. For small particles (Figure 7.31a), the turbulence is reduced due to work by the fluid on the particle motion. For large particles (Figure 7.31b), the turbulence is enhanced due to particle wakes. Intermediate-size particles can lead to turbulence enhancement for some portions of the flow and turbulence damping in other regions (Tsuji et al., 1984). In all these cases, the changes in turbulent kinetic energy (via production by large particles or dissipation by small particles) are intensified as the particle mass loading increases.

To understand the physics of these changes, it is helpful to consider the spectra of the turbulent kinetic energy. An example of the influence of particles on a measured streamwise specific energy spectrum is shown in Figure 7.32. The single-phase flow spectrum is consistent with the Hinze model of Figure 6.29 with a nearly -5/3 slope in the inertial subrange. However, the presence of particles yields two major changes: a reduction at the low frequencies (attributed to particle dissipation) and an increase at high frequencies (attributed to the passage of particle wakes). This combination of energy decreasing at low frequencies (where the particles are small relative to the integral scales) and increasing at high frequencies (where the particles approach the size of the turbulent scales) has been observed in several other turbulent flows (Tsuji et al., 1984; Poelma et al., 2007). Thus, turbulence modulation does not manifest in a simple across-the-board increase or decrease in TKE, but instead the turbulent spectrum changes locally depending on the relative particle scales.

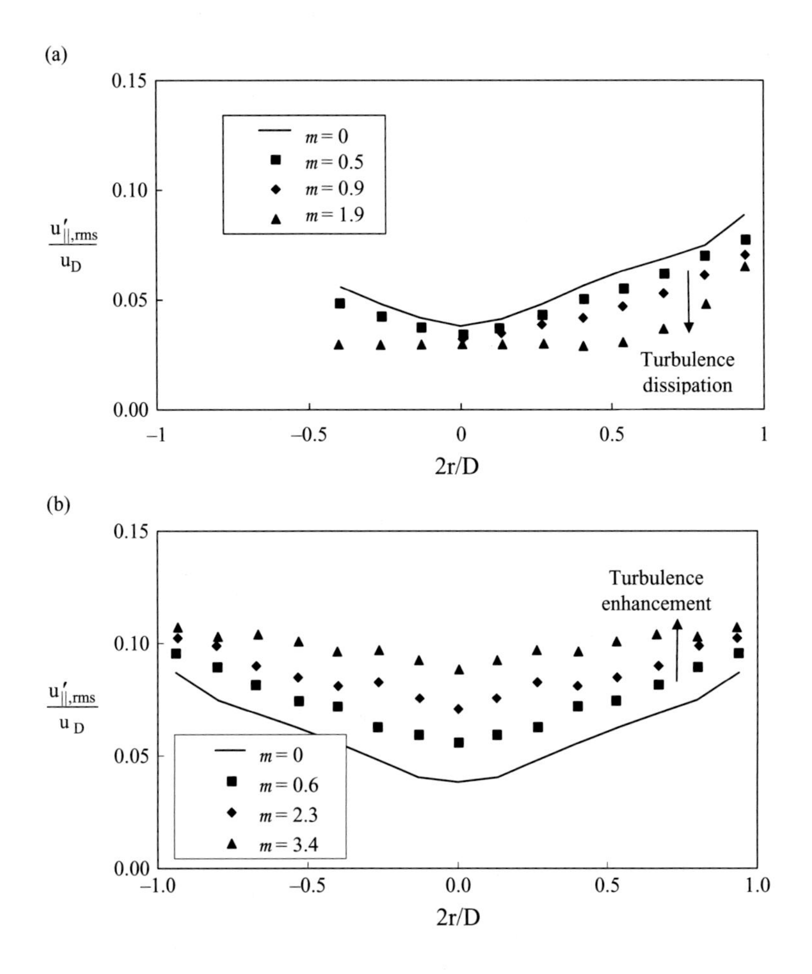

Figure 7.31 Streamwise turbulence in a circular pipe with upward gas flow and falling polystyrene particles (Tsuji et al., 1984) with diameters of (a) 200 μm (small particles that reduce the turbulence) and (b) 2,780 μm (large particles that increase the turbulence).

Theoretical Treatments

While a unified robust theoretical treatment of turbulence modulation has proved elusive, some basic mechanisms can be analyzed. We first consider turbulent dissipation, followed by turbulence production.

If the particles have no significant relative velocity with the fluid, then the change in dissipation is due to the mixed fluid viscosity via particle volume fraction (5.8a). However, this effect is generally weak for high-density particles since the volume fraction is generally very small. Therefore, dissipation enhancement at low volume fractions is generally attributed to relative velocity effects (Tanaka and Eaton, 2010). To describe this turbulent dissipation by particle relative velocity (ε_p), one may employ an energy/work principle whereby the viscous losses at particle surfaces equal the rate of fluid work applied by the moving particles. This work for a single particle can be based on the product of the interphase force and the fluid velocity. Since the

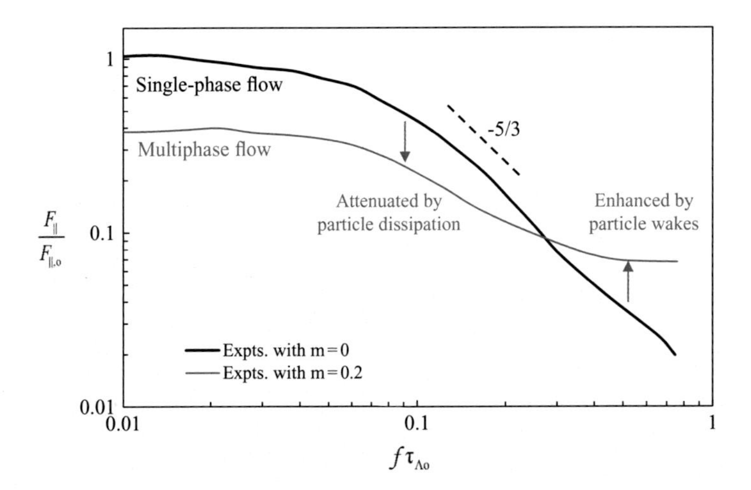

Figure 7.32 Specific kinetic energy spectrum for streamwise velocity fluctuations in a boundary layer in the presence of copper particles (70 μm in diameter) at a mass loading of $m = 0.2$ compared to that for the unladen flow at $m = 0$ (Rogers and Eaton, 1991).

drag force is generally the primary interphase fluid dynamic force and can be related to the particle response time (3.88a), the work rate per particle is

$$\mathbf{F}_\mathrm{D} \cdot \mathbf{u} = \left(\mathrm{m}_\mathrm{eff}\mathbf{w}/\tau_\mathrm{p}\right) \cdot \mathbf{u}. \tag{7.121}$$

If one assumes that τ_p is constant based on a liner drag law, as is often reasonable for small particles, which tend to drive dissipation, the ensemble average of the work rate by the particles per unit mass of fluid becomes

$$\frac{\mathrm{n}_\mathrm{p}\langle \mathbf{F}_\mathrm{D} \cdot \mathbf{u}\rangle}{\rho_\mathrm{f}} = \frac{\mathrm{n}_\mathrm{p}\mathrm{m}_\mathrm{eff}}{\rho_\mathrm{f}\tau_\mathrm{p}}\langle \mathbf{w} \cdot \mathbf{u}\rangle. \tag{7.122}$$

This expression includes the work done by the mean and fluctuating particle motions on the mean and fluctuation fluid velocity. If we consider only the particle and fluid fluctuations, the resulting particle-induced kinetic energy dissipation can be expressed in terms of a velocity correlation as follows:

$$\varepsilon_\mathrm{p} \equiv -\dot{\mathrm{k}}_\mathrm{diss,p} = -\frac{m_\mathrm{eff}}{\tau_\mathrm{p}}\langle \mathbf{w}'' \cdot \mathbf{u}''\rangle. \tag{7.123}$$

The LHS term is defined as positive to be consistent with (6.33), while the RHS employs an effective mass loading (m_eff) that extends the definition of (5.6) to include added mass:

$$m_\mathrm{eff} \equiv \frac{\alpha\rho_\mathrm{p} + \alpha c_{\forall}\rho_\mathrm{f}}{(1-\alpha)\rho_\mathrm{f}} \approx \frac{\mathrm{n}_\mathrm{p}\mathrm{m}_\mathrm{eff}}{\rho_\mathrm{f}}. \tag{7.124}$$

The RHS approximation assumes small volumetric particle fractions (as in 5.7a) to allow proportionality with the effective particle mass.

If we consider the high-density particle limit (e.g., particles in air such that $m_{\text{eff}} = m$) and apply the relative velocity definition, the particle-induced kinetic energy dissipation is as follows:

$$\varepsilon_{\text{p}} = \frac{m}{\tau_{\text{p}}}\left(\langle \mathbf{u}'' \cdot \mathbf{u}'' \rangle - \langle \mathbf{v}'' \cdot \mathbf{u}'' \rangle\right) \qquad \textit{for}\ \rho_{\text{p}} \gg \rho_{\text{f}}. \tag{7.125}$$

This result was obtained by Mando et al. (2009), where the RHS expansion is based on (7.68). Note that the RHS term in parentheses is generally positive, since particle velocity fluctuations are less than fluid velocity fluctuations in the high-density particle limit (Figure 7.15). Therefore, this is a mechanism for increased dissipation, which can translate into reduced TKE for small particles (qualitatively consistent with Figure 7.31a). Notably, (7.125) can be generalized to include correlations of particle concentration fluctuations (Fessler and Eaton, 1999), which can be important if preferential concentration is significant.

To estimate the overall impact of particles on dissipation, one may compare this enhanced dissipation at the particle surfaces (ε_{p}) with the turbulent dissipation at the Kolmogorov scales for an unladen HIST flow (ε_{o}). The ratio of these dissipations can be expressed using the approach of §7.3.1 to model the RHS velocity correlations of (7.125) to yield the following:

$$\frac{\varepsilon_{\text{p}}}{\varepsilon_{\text{o}}} = \frac{2m}{c_{\varepsilon,\text{o}}\text{St}_{\Lambda,\text{o}}}\left[1 - \frac{1}{\sqrt{1 + \text{St}_{\Lambda,\text{o}} + 3\text{St}_{\Lambda,\text{o}}g_{\text{o}}/(4c_{\Lambda,\text{o}})}}\right] \qquad \text{for}\ \rho_{\text{p}} \gg \rho_{\text{f}}. \tag{7.126}$$

The RHS includes $c_{\varepsilon,\text{o}}$ as the unladen dissipation parameter from (6.81a) as well as the integral Stokes number and drift parameter based on the unladen turbulence ($\text{St}_{\Lambda,\text{o}}$ and g_{o}). The result indicates that smaller Stokes numbers and larger mass loadings will yield larger dissipation, and both trends are qualitatively consistent with more TKE damping for small particles. However, careful experimental measurements in a HIST flow by Tanaka and Eaton (2010) have shown that overall turbulent dissipation actually reduces very little for such conditions. This suggests that increases in ε_{p} as m increases are counteracted with decreases in ε, so the turbulent spectrum is asymmetrically modified (consistent with Figure 7.32). As such, changes in overall dissipation rates (sum of these two) are often weak and are not the primary factor for TKE attenuation (Polema et al. 2007; Tanaka and Eaton, 2010).

Instead, the changes in TKE can be related to changes in the overall spectrum shape via the size of the turbulent integral scales. In particular, the ratio of the fluid TKE with particles (k) and compared to that for no particles (k_{o}) can be related the ratio of the laden and unladen dissipation rates and dissipation length scales per (6.81b) as follows:

$$\frac{\text{k}}{\text{k}_{\text{o}}} \approx \left(\frac{\varepsilon}{\varepsilon_{\text{o}}}\frac{\Lambda_{\varepsilon}}{\Lambda_{\varepsilon,\text{o}}}\right)^{2/3}. \tag{7.127}$$

Thus, weak changes in the dissipation indicate that significant changes in the kinetic energy must be related to changes in the turbulent length scales. This hypothesis is consistent with measurements that showed a reduction in the integral length scales as the TKE decreased (Tanaka and Eaton, 2010). Similarly, larger particles were found to increase the turbulent length scale and were associated with increased TKE (Hoque et al., 2016).

Turning now to turbulence production via $k_{prod,p}$ of (7.120), it is helpful to consider the mechanical work added to the fluid based on the interphase force and relative velocity of a particle. If the drag force dominates the interphase fluid force, the work may be estimated as follows:

$$\mathbf{F}_D \cdot \mathbf{w} = \left(m_{eff}\mathbf{w}/\tau_p\right) \cdot \mathbf{w}. \tag{7.128}$$

This work can be used to change the mean momentum of the fluid, as shown, for example, in Figure 5.8. These changes can, in turn, change the mean velocity gradients, which will therefore change the shear-induced turbulence production ($k_{prod,f}$). However, some of work of (7.128) can be translated into turbulence addition if there is unsteadiness in the particle wake (which especially occurs for larger particles at higher Reynolds numbers). If we define the fraction of this work converted to wake unsteadiness as c_{prod}, the associated turbulence production (obtained via the same approach used to obtain (7.123) becomes

$$\dot{k}_{prod,p} = c_{prod}\, m_{eff} \left\langle \frac{\mathbf{w} \cdot \mathbf{w}}{\tau_p} \right\rangle. \tag{7.129}$$

The RHS indicates that increasing particle mass loading will increase turbulence production, which is qualitatively consistent with Figure 7.31b (for large particles). The RHS velocity correlation can be simplified by assuming linear drag (so τ_p is constant) and assuming $\mathbf{w}$ is dominated by the steady terminal velocity ($\mathbf{w}_{term}$) so that only c_{prod} (the fraction of energy resulting in wake unsteadiness) is needed to apply (7.129).

If we consider the solid sphere particles in a uniform steady flow as described in §3.2.4, the wakes are steady for $Re_p < 130$ such that one may expect $c_{prod} = 0$ for these conditions. Such a hypothesis is consistent with experiments (Wu and Faeth, 1994) and resolved-surface simulations (Bagchi and Balachandar, 2004) which indicated that significant particle wake turbulence is only induced for $Re_p > 210$. However, Hoque et al. (2016) observed particle-induced turbulence increases in grid-generated turbulence for Re_p values in the range of 10–100. Similarly, Parthasarathy and Faeth (1990) observed increases with a fully developed cascade (including a –5/3 slope for the inertial subrange) for Re_p as low as 38. This suggests that the surrounding turbulence can be enhanced even when the particle wakes are steady. As such, the mechanism of turbulence augmentation by unsteady particle wakes is not well understood, and this has hindered a theoretical framework.

Turbulence Modulation Trends and Models

To characterize the influence of particle size and turbulent length scale on turbulence modulation, Crowe (2000) broadly reviewed pipe flows with particles in air and found that the ratio of particle diameter to an integral length scale generally dictates whether

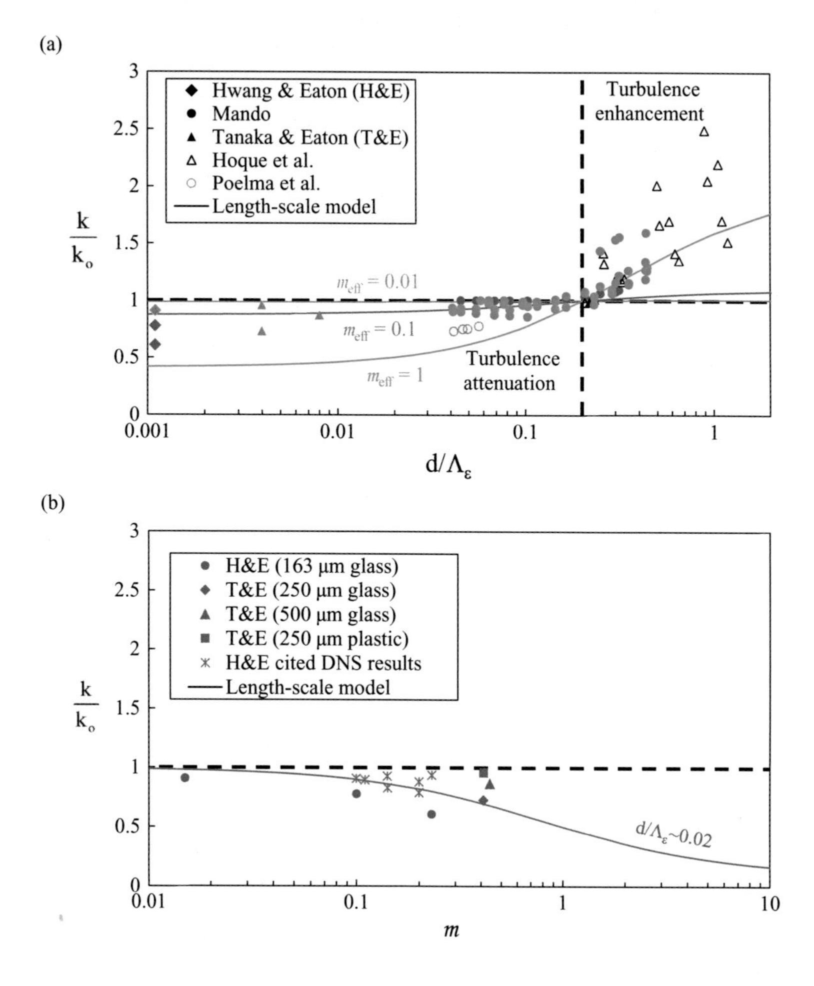

Figure 7.33 Modulation for particles in air (solid symbols) or water (hollow symbols) relative to the length-scale modulation model as a function of (a) particle size for various mass loadings from measurements in nearly HIST flows (Poelma et al. 2007; Tanaka and Eaton, 2010; Hoque et al. 2016) and in a jet flow (Mando, 2009); and (b) mass loading for small particles ($d < 0.1\Lambda_\varepsilon$) in nearly HIST flows based on experiments of Hwang and Eaton (2006) and Tanaka and Eaton (2010) and their cited DNS results.

TKE is reduced or augmented. However, the pipe flow has large gradients in the mean flow and kinetic energy, as well as significant turbulence anisotropy, which makes it difficult to develop a quantitative robust model from such data. If one considers flows that are nearly HIST, a more fundamental picture of modulation can emerge. Modulation in nearly HIST flows are shown in Figure 7.33a in terms of change in kinetic energy (k/k_o) as a function of the particle diameter relative to the local dissipation length scale (d/Λ_ε, defined with 6.81b) and the effective mass loading (defined in 7.124). The surveyed experimental data can all be characterized by the following criteria:

$$d/\Lambda_{\varepsilon,o} < 0.2 \quad \textit{turbulence attenuation due to particles.} \tag{7.130a}$$

$$d/\Lambda_{\varepsilon,o} > 0.2 \quad \textit{turbulence augmentation due to particles.} \tag{7.130b}$$

As expected from (7.123) and (7.129), the attenuation and augmentation effects are enhanced with larger effective mass loadings, such as shown in Figure 7.33b.

A simple empirical model that employs the dependence on effective mass loading of (7.129) and the influence of turbulent length scale of (7.127) while following the trends of (7.130) and Figure 7.33 is as follows:

$$\frac{k}{k_o} \approx \frac{1 + 1.4(1 + m_{eff})(5d/\Lambda_{\varepsilon,o})}{1 + 1.4(m_{eff} + 5d/\Lambda_{\varepsilon,o})}. \tag{7.131}$$

While this *length-scale modulation model* (which uses an empirical factor of 1.4) captures the general trends in TKE change of Figure 7.33 for high-density particles, there is significant experiment scatter, indicating that this model is only qualitative in terms of both the impact of mass loading and of size ratio.

One may also investigate this model for pipe and channel flows, keeping in mind that these flows have anisotropic and nonhomogeneous turbulence. Given these complexities, the TKE and turbulent length scale are best considered using an area average or at the pipe half-radius to avoid focusing on regions dominated by dissipation (at the pipe centerline) or by TKE production (at the pipe wall). Using this approach, available pipe and channel modulation data for particles in air (where $m = m_{eff}$) are shown in Figure 7.34. All these experimental data are well characterized by the criteria of (7.130), and again the model of (7.131) is qualitatively consistent.

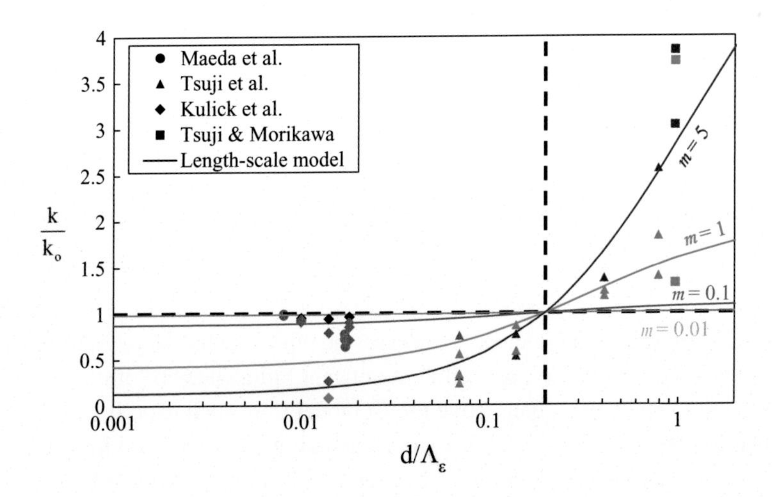

Figure 7.34 Turbulence modulation for pipe flows of air ($m = m_{eff}$) measured at or near the half-radius (r/R = 0.5) in terms of particle size relative to the dissipation length scale (Maeda et al., 1980; Tsuji and Morikawa, 1982; Tsuji et al., 1984; Kulick et al., 1994) compared to the turbulence modulation prediction based on the length-scale modulation model.

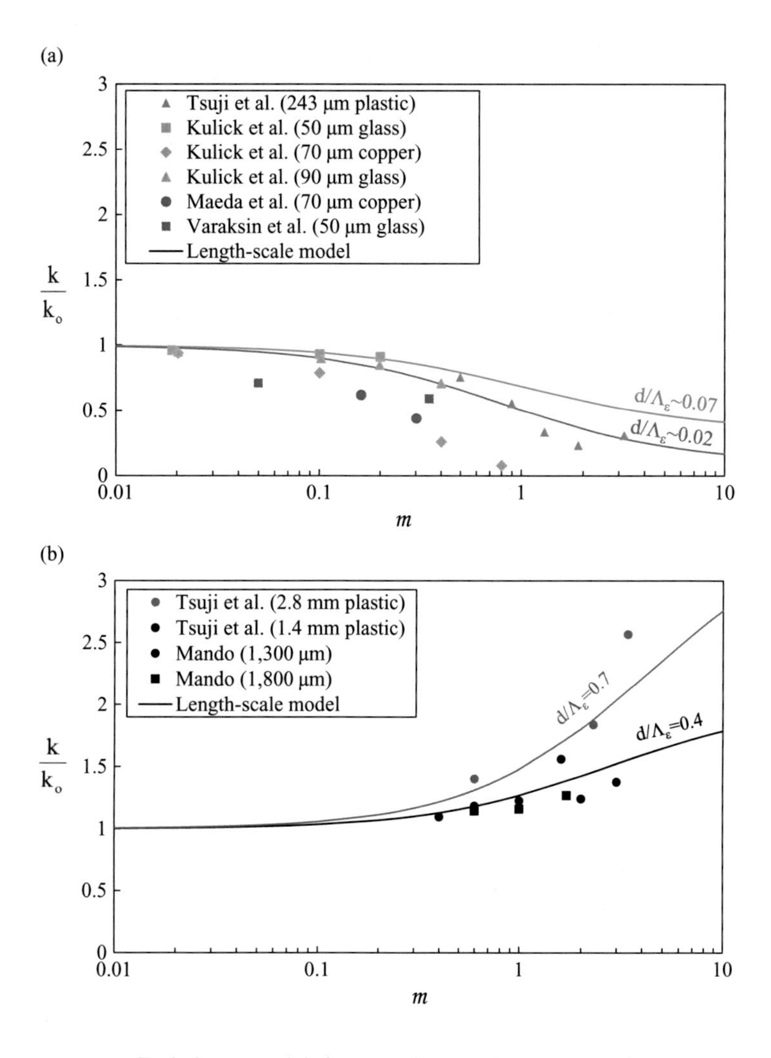

Figure 7.35 Turbulence modulation at various particle mass loadings in air based on measurements for pipe flows at or near the half-radius and a jet flow (Maeda et al., 1980; Mando, 2009; Tsuji and Morikawa, 1982; Tsuji et al., 1984; Kulick et al., 1994; Varaksin et al., 2000) and the length-scale modulation model for (a) small particles ($d/\Lambda_\varepsilon < 0.2$) and (b) large particles ($d/\Lambda_\varepsilon > 0.2$).

A more complex empirical model developed by Crowe (2000) provides similar trends. but is not as robust.

In Figure 7.35, the pipe flow data are divided into small and large particles conditions as a function of mass loading. These results and the model of (7.131) can be used to set an approximate criterion for turbulence modulation, as follows:

$$m_{\text{eff}} \gtrsim 0.01 \qquad \textit{significant turbulence modulation for } \rho_p \gg \rho_f. \tag{7.132}$$

This criterion is similar to that for macroscopic density changes (5.37), and similarly should only be used as a qualitative guide.

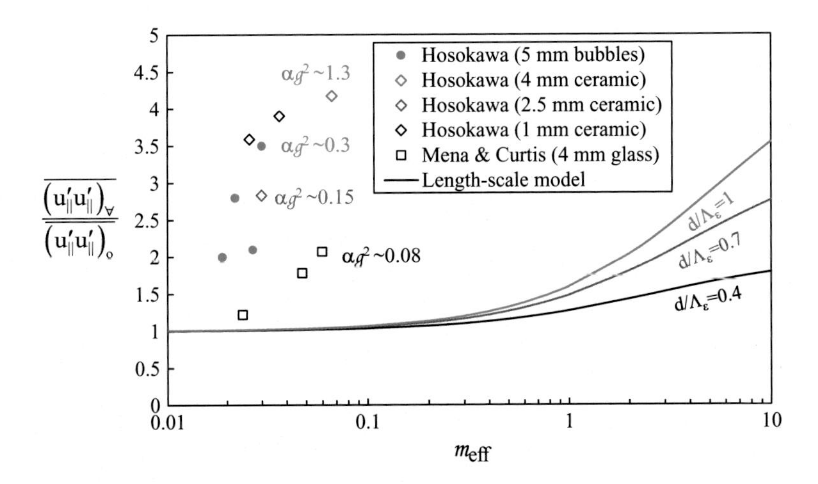

Figure 7.36 Measured turbulence modulation due to large particles ($d/\Lambda_\varepsilon > 0.2$) in water based on measured fluid vertical velocity fluctuations (Hosokawa et al., 2004; Mena and Curtis, 2020) compared to the length-scale modulation model.

The situation is even more complicated when the density of the particles is no longer high relative to the surrounding fluid. For example, turbulence modulation caused by large particles and bubbles in water, as shown Figure 7.36, results in large increases in the fluid velocity fluctuations even for effective mass loadings of only a few percent. This suggests that (7.132) is not valid for such conditions. Instead, the increases in velocity fluctuations for Figure 7.36 tend to be correlated to significant levels of αg^2. Fortunately, this augmentation has a theoretical foundation, as discussed in the following subsection.

Pseudoturbulence

For particles with densities similar to or less than that of the continuous phase, a unique phenomenon that can increase surrounding fluid velocity fluctuations is *pseudoturbulence*. As illustrated in the time sequence of Figure 7.37 for a bubble rising in an otherwise still bath, these fluctuations are caused by the fluid associated with the added mass, where such fluid tends to move at the speed of the bubble. As these added-mass regions pass by an Eulerian (fixed point in space), that point will temporarily see an upward velocity fluctuation. As such, these fluctuations induced by added mass can occur in laminar flow, where they do not reflect real turbulence (hence the name "pseudoturbulence"). By analogy, a single-phase laminar flow can also have fluctuations due to unsteady events that are not related to turbulence, such as pulsatile flow. And because this effect, shown in Figure 7.37, is especially pronounced for bubbles (where added mass dominates particle mass), it is also sometimes called *burbulence*. Importantly, these fluctuations caused by the random arrival and departure of added-mass regions at a measuring volume do not have the spectrum of

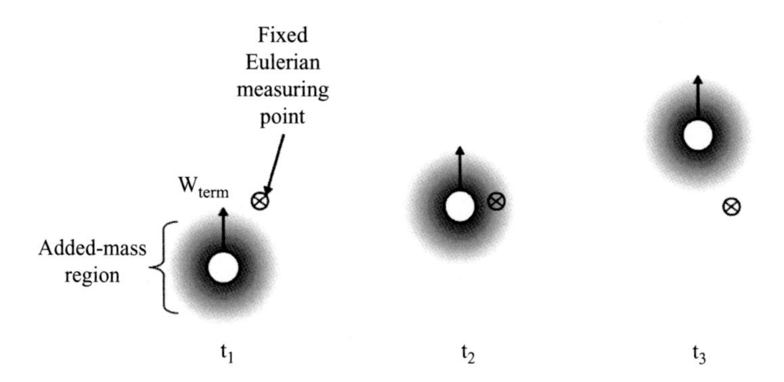

Figure 7.37 Schematic of time sequence of a rising bubble and a nearby Eulerian measuring point for an otherwise still bath where the added mass tends to move at the bubble speed and temporarily causes an upward velocity perturbation at the measuring point, demonstrating the pseudoturbulence effect.

structures and wavelengths of *true* turbulence. As such, pseudoturbulence is not associated with a TKE production rate or any turbulent dissipation.

Pseudoturbulence may be theoretically modeled for a bubble rising at a steady velocity by noting that the added mass must also travel at the same speed. For a sphere in an otherwise still fluid, the fluid kinetic energy associated with the added mass based on (3.71) and (3.76) is as follows:

$$\mathrm{KE}_{\forall}/\mathrm{sphere} \equiv \frac{1}{2} c_{\forall} \rho_f v^2 \qquad \textit{potential flow in otherwise still fluid} \tag{7.133}$$

This kinetic energy induced by added-mass passage at an Eulerian measuring point can be linearly superimposed for all passing particles by assuming a sparse concentration (Biesheuvel and van Wijngaarden, 1984). If one further assumes the particle moves at the terminal velocity and considers vertical fluctuation contributions to the added-mass kinetic energy (van Wijngaarden, 1998), the result is as follows:

$$\overline{\left(u'_{\parallel} u'_{\parallel}\right)_{\forall}} = c_{\forall,\parallel} \alpha w_{term}^2 \qquad \textit{for potential flow and sparse concentration.} \tag{7.134}$$

The RHS generalizes the added-mass coefficient to be the value associated with vertical movement. For a sphere, this coefficient simply reverts to ½. However, a deformed bubble will often be spheroidal and will rise in a broadside orientation, causing an increase in the vertical added-mass coefficient. As will be discussed in §10.2, this added-mass coefficient is typically in the range 1–2 for ellipsoidal bubbles in water, so these values are used with (7.134) to compare with experimental results in Figure 7.38. The higher added-mass coefficients are more appropriate for higher drift parameters, which correspond to larger and more deformed bubbles, as expected.

If one considers the pseudoturbulence vertical velocity fluctuations of (7.134) relative to that for single-phase flow turbulence, the result in terms of the drift parameter becomes

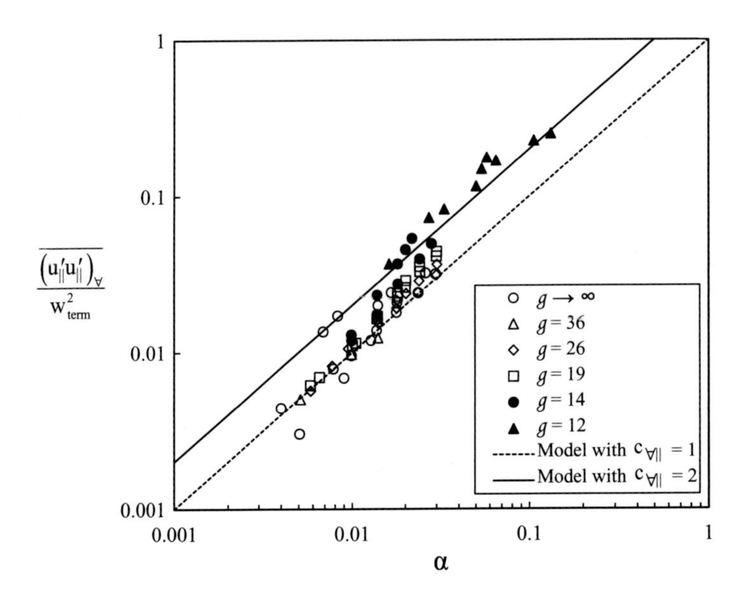

Figure 7.38 Excess continuous-fluid streamwise velocity fluctuations in turbulent flows as measured by Theofanous and Sullivan (1982) and Lance and Bataille (1991) for ellipsoidal bubbles at various terminal velocity ratios with $300 < \mathrm{Re_p} > 1{,}000$, where $g \to \infty$ corresponds to experiments in laminar flow.

$$\frac{\overline{(\mathrm{u}'_{\|}\mathrm{u}'_{\|})_{\forall}}}{\overline{(\mathrm{u}'_{\|}\mathrm{u}'_{\|})_{\mathrm{o}}}} = c_{\forall\|}\alpha g^2 \qquad \textit{for} \text{ turbulent flow } \textit{and sparse concentration}. \tag{7.135}$$

Vertical velocity fluctuations for bubbly flows in turbulence are shown in Figure 7.39, where pseudoturbulence is seen to dominate at high drift parameters, while the remaining TKE increase can be attributed to cluster-based modulation. To explain this, we consider the measurements by Martínez-Mercado et al. (2007) for small bubbles with $\mathrm{Re_p} < 80$ in still water, which yielded velocity fluctuations severalfold higher than that predicted by (7.134). This is attributed to these small bubbles tending to toward each other such that the high-concentration cluster induces an upward flow region, resulting in a higher effective added mass than that for a simple linear sum. Such clustering can be especially evident in turbulent flows, due to preferential concentration creating a cluster-based modulation effect that enhances pseudoturbulence, as noted in Figure 7.39.

It should be noted that the KE predicted by (7.133) is only associated with vertical velocity, but in general lateral fluctuations can also occur due to the displacements caused by the bubble width as they rise. Pseudoturbulence measurements by Hosokawa and Tomiyama (2013) indicated that the lateral velocity fluctuating energy is approximately half that for the vertical velocity. This can be used to estimate the pseudoturbulence kinetic energy as follows:

$$\overline{(\mathrm{u}'_{\perp}\mathrm{u}'_{\perp})_{\forall}} \approx \frac{1}{2}\overline{(\mathrm{u}'_{\|}\mathrm{u}'_{\|})_{\forall}}. \tag{7.136a}$$

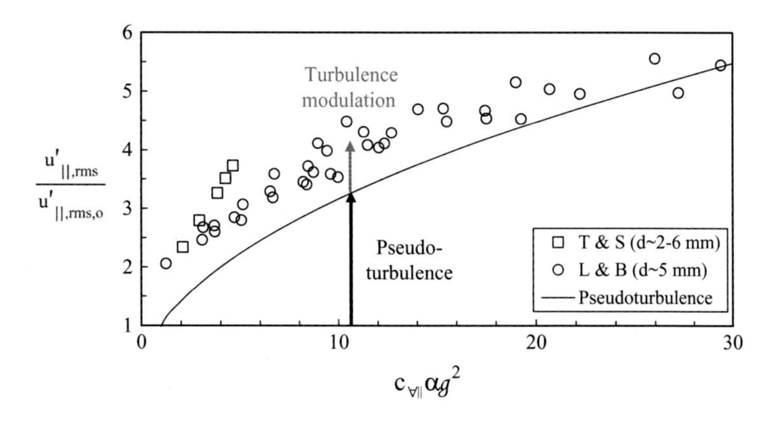

Figure 7.39 Impact of volume fraction and drift parameter on vertical velocity fluctuations for bubbly flows based on an added-mass coefficient of unity comparing increases associated with pseudoturbulence and with true turbulence modulation (enhancement due to bubble wakes).

$$k_\forall = \frac{1}{2}\left[\overline{(u'_\parallel u'_\parallel)_\forall} + 2\overline{(u'_\perp u'_\perp)_\forall}\right] \approx \overline{(u'_\parallel u'_\parallel)_\forall} = c_{\forall,\parallel}\alpha w^2_{\text{term}}. \tag{7.136b}$$

The effective total kinetic energy can then be modeled as the sum of the true turbulence and the pseudoturbulence (Mudde and Saito, 2001):

$$k_{\text{eff}} = k + k_\forall \approx \frac{3}{2}u^2_\Lambda + c_{\forall,\parallel}\alpha w^2_{\text{term}}. \tag{7.137}$$

Notably, the true turbulence (k) can be affected by the presence of the bubbles through the mechanisms outlined in Figure 7.30.

7.7 Three-Way and Four-Way Turbulence Coupling

In the following, three-way coupling and four-way coupling (previously discussed for the laminar flow in §5.5 and §5.6 are considered in the presence of turbulence. In particular, three-way turbulence coupling occurs due to turbulent fluid dynamic interactions between particles (without direct particle contact), while four-way coupling considers collisions between particles due to turbulence (direct particle contact).

Three-Way Coupling in Turbulence

For three-way coupling, one particle's local flow field impacts that of another particle. In steady flows, this can be manifested via drafting, as in Figure 5.11a, and the same principles can occur in turbulence. As such, the three-way coupling criteria of (5.46) can be reasonably applied for turbulence by considering the mean volume fraction:

$$\bar{\alpha} \gtrsim 0.01 \quad \textit{for significant mean three-way coupling.} \tag{7.138a}$$

$$\bar{\alpha} \gtrsim 0.1 \quad \textit{for dense flow via three-way coupling.} \tag{7.138b}$$

Notably, drafting effects can be reduced by turbulence since it can increase the mixing and diffusion of a particle's wake before this wake intersects with another particle (Bagchi and Balachandar, 2004). However, turbulence can also cause regions of increased particle concentration via clustering, which can reduce local gaps between particles and enhance localized drafting effects. Clustering can be particularly pronounced via preferential concentration for small particles, as shown in Figures 7.3 and 7.4. This clustering can occur for effective mass loadings as low as 1%, as shown in Figures 7.23 and 7.36. As such, a three-way coupling clustering criterion for turbulence can be as follows:

$$m_{\text{eff}} \tilde{>} 0.01 \quad \textit{for significant three-way coupling at } \text{St}_\eta \sim 1. \tag{7.139}$$

Note that the criterion of (7.139) for particles in a gas is approximately equivalent to $\alpha > 10^{-5}$ based on (7.124). This is far more restrictive than the three-way coupling criteria of (7.138a). As such, turbulent conditions can be more sensitive to three-way coupling effects, especially for particles that exhibit strong clustering.

Collisions Due to Turbulence

Particle collisions can also be significantly enhanced by the presence of turbulence due to increased particle velocity fluctuations. The associated collisions can be characterized per (5.48) and (5.54) by the overall collision frequency ($\mathcal{N}$), the collision frequency per particle (f_{coll}), and the collision Stokes number, given as follows:

$$\mathcal{N} \equiv n_p f_{\text{coll}}. \tag{7.140a}$$

$$\text{St}_{\text{coll}} \equiv \tau_p f_{\text{coll}}. \tag{7.140b}$$

For turbulent flow, there are two important limits for f_{coll}: (1) the *kinetic-based limit* for large-inertia particles due to integral-scale turbulence; and (2) the *diffusion-based limit* for small-inertia particles due to microscale turbulence. The large-inertia limit and the small-inertia limit are first considered, followed by an approximation for intermediate-inertia particles.

For large-inertia kinetic-based collisions, particle trajectories are assumed to intersect based on random fluctuations of the particles in turbulent flow. This occurs when the particles emanate from different eddies and have significant inertias (e.g., $\text{St}_\Lambda > 1$), as shown conceptually by Figure 5.13d. To analyze this condition, Abrahamson (1975) assumed the two particle trajectories are uncorrelated (since they are associated with different turbulent structures) and the particle velocity fluctuations have a Gaussian distribution. In this case, the collision velocity of (5.50) is proportional to the rms of the particle velocity fluctuations. For uniform particle sizes, the collision velocity due to turbulence (v_{coll}) and the associated collision frequency per particle ($f_{\text{coll,k}}$) in the kinetic-based limit can be expressed as follows:

$$v_{\text{coll}} = \frac{4}{\sqrt{\pi}} v'_{\text{rms}} \tag{7.141a}$$

$$f_{\text{coll,k}} = 4\sqrt{\pi} d^2 n_p v'_{\text{rms}} \quad \textit{for kinetic-based collisions.} \tag{7.141b}$$

As expected, this frequency scales with particle number density.

For high-density particles with weak eddy-crossing effects ($g \ll 1$), (7.62) indicates that these particle velocity fluctuations are isotropic and are related to the fluid TKE and particle Stokes number. The fluid TKE can be related to the integral time and length scales using (6.80) to yield a kinetic-based frequency and collisional Stokes number (using 7.140b), as follows:

$$f_{\text{coll,k}} \approx \frac{24\alpha}{\sqrt{\pi}c_\Lambda\sqrt{1+\text{St}_\Lambda}}\frac{\Lambda_\perp}{d\tau_\Lambda} \approx \frac{13\alpha}{\sqrt{1+\text{St}_\Lambda}}\frac{\Lambda_\perp}{\tau_\Lambda d} \qquad (7.142a)$$

$$\text{St}_{\text{coll,k}} \equiv \tau_p f_{\text{coll,k}} \approx \frac{13\alpha\,\text{St}_\Lambda}{\sqrt{1+\text{St}_\Lambda}}\frac{\Lambda_\perp}{d} \qquad (7.142b)$$

for $g \ll 1$.

Following (5.49), $\text{St}_{\text{coll,k}} > 1$ indicates collision-dominated trajectories.

In the other limit of very small particle inertias, Saffman and Turner (1956) investigated a diffusion-based collision mechanism by considering two particles in a straining region associated with the same Kolmogorov-scale eddy. The particles are considered to have small sizes and response times relative to these eddies, that is, $d \ll \eta$ and $\text{St}_\eta \ll 1$. Integrating over the probability distribution function of relative particle velocities for this flow, the resulting diffusion-based collision frequency per particle (f_ε) was obtained theoretically as follows:

$$f_{\text{coll},\varepsilon} = \sqrt{\frac{2\pi}{15}}\frac{d^3 n_p}{\tau_\eta} \qquad \textit{for diffusion-based collisions.} \qquad (7.143)$$

This diffusion-based collision frequency increases rapidly with particle diameter but varies inversely with the Kolmogorov time scale (defined in 6.89). For monodisperse particles, the corresponding diffusion-based collisional Stokes number based on (5.2), (5.48), and (7.3) is as follows:

$$\text{St}_{\text{coll},\varepsilon} \equiv \tau_p f_{\text{coll},\varepsilon} \approx 1.24\alpha\text{St}_\eta. \qquad (7.144)$$

Thus, the relative importance of collisions increases with both particle concentration and inertia. Comparing (7.144) to (7.142), the diffusion-based Stokes number is generally smaller than the kinetic-based Stokes number. This difference is consistent with particles being less likely to collide if their trajectories are controlled by the same turbulent eddy.

For intermediate particle sizes, one may expect that the collision frequency will have an upper bound given by the large-inertia kinetic-based theory and a lower bound given by the small-inertia diffusion-based theory. Such a trend is shown in Figure 7.40, where the predictions of collision rates indeed tend toward the diffusion theory (7.144) for small Stokes numbers and to the kinetic theory (7.142) for larger Stokes numbers. The intermediate regime can be complex, as it includes effects of particle inertia, preferential concentration, and integral-scale turbulence (Ferry and Balachandar, 2002) and thus does not readily lend itself to theoretical analysis. Therefore, empirical models typically use a combination of the two extreme limits (Williams and Crane, 1983; Sundaram and Collins, 1997; Kruis and Kusters, 1997). One such combination empirical model for the turbulent collisional frequency is given as follows:

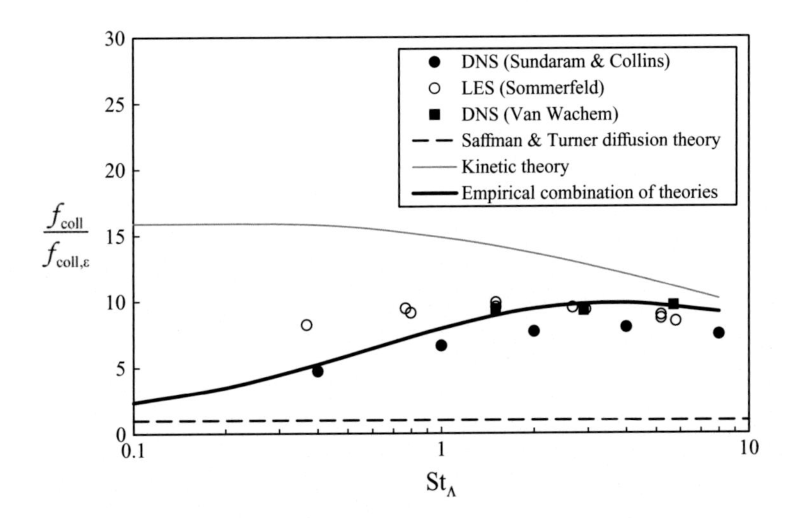

Figure 7.40 Collision frequency relative to that for the diffusion limit as a function of various particle inertias for limiting theories, DNS results (Sundaram and Collins, 1997; van Wachem, 2020), LES results (Sommerfeld, 2001), and an empirical fit that combines the theories.

$$f_{coll} \approx \frac{f_{coll,\varepsilon} + St_\Lambda f_{coll,k}}{1 + St_\Lambda} \quad \textit{for collisions in turbulence.} \tag{7.145}$$

This model yields the diffusion limit when St_Λ is very small and the kinetic limit when St_Λ is very large. As shown in Figure 7.40, this simple combination provides qualitative agreement with simulations. Unfortunately, no direct measurements are available for near-HIST flows (given the diagnostic difficulties) for experimental validation. Substituting (7.142) and (7.143), the model of (7.145) can be expressed in terms of the turbulent collisional Stokes numbers as follows:

$$St_{coll} \equiv \tau_p f_{coll} \approx \frac{St_{coll,\varepsilon} + St_\Lambda St_{coll,k}}{1 + St_\Lambda} \approx \frac{\alpha}{1 + St_\Lambda}\left(1.2 St_\eta + \frac{13 St_\Lambda^2}{\sqrt{1 + St_\Lambda}} \frac{\Lambda_\perp}{d}\right). \tag{7.146}$$

Based on (5.49), particle collisions will be significant for $St_{coll} > 0.1$ and will dominate for $St_{coll} > 1.0$. However, preferential concentration effects can cause increased local concentrations, such that the models for St_{coll} may underestimate collisions for $St_\eta \sim 1$.

Notably, collisional Stokes numbers in turbulence are much larger than that for other nonturbulent collision mechanisms. As such, four-way coupling effects (particle collisions) are more likely to occur once turbulence is introduced to a flow.

7.8 Multiphase Coupling Summary

The various coupling regimes for turbulent flow are summarized in the following for one-way coupling (turbulence affects the particle motion, but not vice versa), two-way coupling (particles affect the turbulence), three-way coupling (the particles interact fluid dynamically in the turbulence), and four-way coupling (turbulence causes particle collisions). For the one-way coupling, the turbulent Stokes numbers are the primary indicator of the particle's ability to follow the flow (§7.1–7.2) and this coupling can be summarized in order of increasing particle size/inertia as follows:

$$\begin{aligned}
&\mathrm{St}_\eta \ll 1 \ \textit{particles closely follow all turbulent flow structures.}\\
&\mathrm{St}_\eta \sim 1 \ \textit{particles can lead to preferential concentration.}\\
&\mathrm{St}_\Lambda \ll 1 \ \textit{particles closely follow integral-scale structures.}\\
&\mathrm{St}_\Lambda \sim 1 \ \textit{particles party follow integral-scale structures.}\\
&\mathrm{St}_\Lambda \gg 1 \ \textit{particles are only weakly affected by turbulent structures.}
\end{aligned} \tag{7.147}$$

The two-way, three-way, and four-way turbulent coupling criteria based on parameters identified in (7.132), (7.135), (7.138), (7.139), and (7.146) can also be summarized as follows:

$$\begin{aligned}
&\mathrm{m_{eff}} > 0.05 \quad \textit{turbulence modulation is significant.}\\
&\alpha g^2 > 0.05 \quad \textit{pseudoturbulence coupling is significant.}\\
&\bar{\alpha} > 0.01 \quad \textit{mean three-way coupling is significant.}\\
&m_{\mathrm{eff}} > 0.01 \,\&\, \mathrm{St}_\eta \sim 1 \quad \textit{local clustering bias is significant.}\\
&\alpha \gtrsim 0.1 \quad \textit{three-way coupling dominates.}\\
&\mathrm{St_{coll}} > 0.1 \quad \textit{four-way coupling becomes significant.}\\
&\mathrm{St_{coll}} > 10 \quad \textit{four-way coupling dominates.}
\end{aligned} \tag{7.148}$$

Importantly, these coupling criteria do not account for detailed aspects such as domain geometry, nonlinear drag, eddy-crossing effects, and anisotropic and/or nonhomogeneous turbulence. These criteria are therefore simply qualitative guides.

Based on the qualitative criteria of (7.147) and (7.148), notional regime maps as a function of particle size and concentration can provide approximate boundaries for the various coupling regimes. The two regime maps consider water drops in air ($\rho_p/\rho_f \sim 800$) for Figure 7.41, and air bubbles in water ($\rho_p/\rho_f \sim 0.001$) for Figure 7.42. Both maps assume terminal velocity (based on the drag coefficients of §3.2) and monodisperse particles. The particles diameters for the two maps range from 10 μm (close to the minimum size for a continuum flow around the particle) to 10 mm (close to the maximum size studied in laboratories). The domain sizes are set as 10 cm and 5 cm, which are typical scales for many laboratory air and water flows and ensure a large D/d ratio. The air and water flow speeds are set as 5 m/s and 1 m/s, yielding similar domain Reynolds numbers. The flow turbulence is considered to be homogeneous, isotropic, and ergodogically stationary, with $u_\Lambda \approx 0.05 u_D$ and $\tau_\eta \approx 0.2\tau_\Lambda$ (typical for fully developed turbulence).

The regime map for the case of drops in air (Figure 7.41) employs mass loading (m) as the particle concentration variable. At mass loading on the order of 1%, one-way coupling takes place. The smallest droplet diameters (on the order of 10 μm) are consistent with $\mathrm{St}_\eta \sim 1$, where preferential concentration effects can be significant.

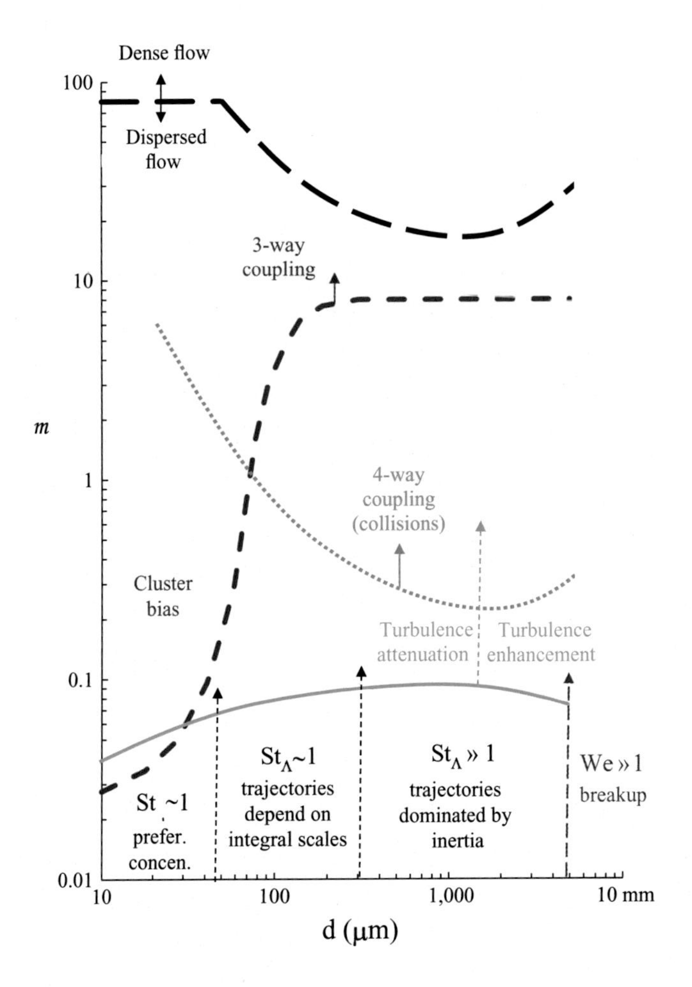

Figure 7.41 Notional regime map for water drops in a turbulent air flow with $u_D = 5$ m/s, $u_\Lambda = 0.25$ m/s, $D = 0.1$ m, $\Lambda_\varepsilon = 0.01$ m, $\tau_D = \tau_\Lambda = 0.02$ s, and $\tau_\eta = 0.004$ s.

At larger diameters (on the order of 100 μm), particle response times will be similar to integral-scale times so that overall turbulent diffusivities and kinetic energy for drops will be less than that for the fluid. Yet larger drops with diameters on the order of 1,000 μm will have large integral Stokes numbers and drift parameters much greater than unity, so they tend to cut through the flow field at high terminal velocities and are only weakly influenced by the turbulence. Once the drop sizes reach 5 mm or so, the Weber number based on terminal velocity will tend to cause breakup. If the flow speeds were higher (e.g., 50 m/s or more), the breakup criterion would be instead driven by a Weber number based on turbulence.

As mass loading increases for Figure 7.41, turbulence modulation first occurs whereby smaller drops or solid particles with the same density ratio cause a reduction in TKE while larger particles increase TKE. However, small particles can also be affected by three-way coupling due to preferential concentration and cluster bias

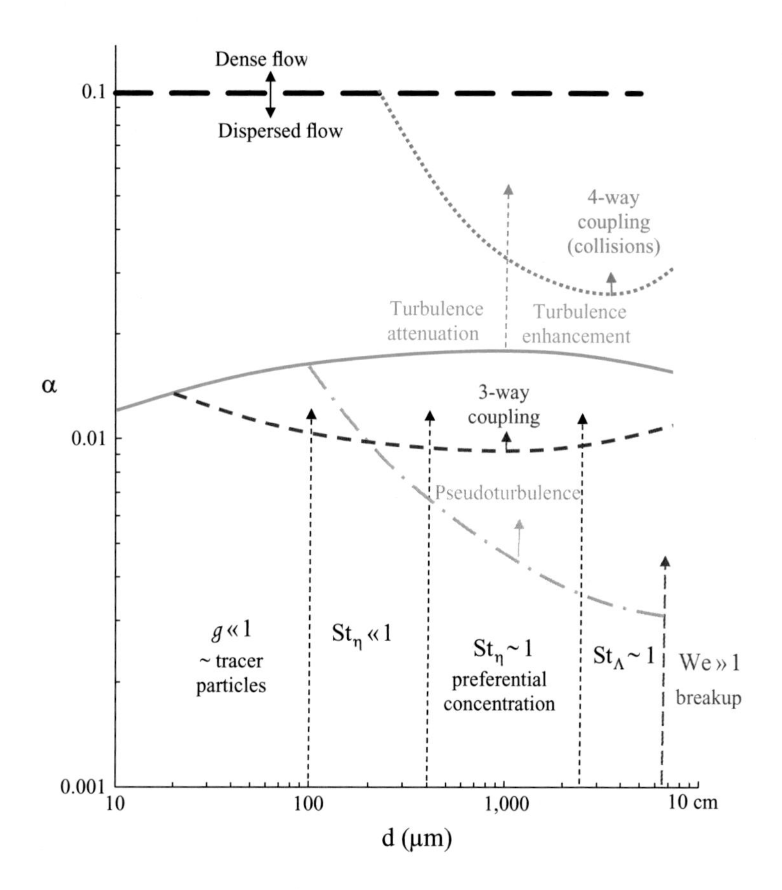

Figure 7.42 Notional regime map for monodisperse air bubbles in a turbulent water flow with $u_D = 1$ m/s, $u_\Lambda = 0.05$ m/s, $D = 0.05$ m, $\Lambda_\varepsilon = 0.005$ m, $\tau_D = \tau_\Lambda = 0.05$ s, and $\tau_\eta = 0.01$ s.

effects. At still higher mass loadings, particle–particle collisions due to turbulence become important, particularly for intermediate-sized particles due to a combination of significant inertia and particle number densities. Notably, preferential concentration can increase local collisions, making this boundary difficult to predict for small particles. In addition, a polydisperse size distribution would further increase collision frequencies so that the four-way coupling boundary would occur at lower mass loadings. At higher mass loadings, the collisions become so prevalent that they dominate particle motion resulting in dense flow (especially for smaller particles) and/or four-way coupling (especially for larger particles).

The regime map for air bubbles for this water flow (Figure 7.42) employs volume fraction for the particle concentration variable. In comparison with Figure 7.41, there are significant differences. For example, bubbles in water have lower terminal velocities (compared to drop in a gas for the same diameter). In addition, time scales in water tend to be larger since the flow speeds are slower. This combination generally results in lower Stokes numbers, so that bubbles in water are more likely to follow the flow as compared to drops in air of the same size. For example, bubbles less than 100 μm are nearly tracer particles, and bubbles on the order of 1 mm for this flow are still characterized by $St_\eta \sim 1$. Bubbles with diameters of 2–7 mm yield $St_\Lambda \sim 1$ while even larger bubbles tend to break up based on turbulent Weber numbers.

As the bubble volume fraction increases, pseudo-turbulence first becomes important for the large bubbles since these have the highest drift parameters. Next, three-way coupling becomes important at volume fractions of 1% or more, followed by two-way turbulence coupling (modulation) at volume fractions of about 2%. For bubbles sizes below 100 μm, fluid velocity fluctuations will generally reduce due to enhanced dissipation. For bubbles sizes between 100 μm and 1 mm, true turbulence will still tend to be reduced, but pseudoturbulence will be important, making the net change in fluid velocity fluctuations difficult to predict. For bubbles larger than 1 mm, both true and pseudoturbulence will be active so fluid TKE will significantly increase. At yet higher volume fractions, particle collisions become important. Finally, volume fractions of about 10% yield dense flow (where three-way and/or four-way coupling dominates). The regime boundaries in this figure are also approximately appropriate for solid particles in water with a specific gravity of two (the same density difference), except that there would not be a breakup boundary.

Again, these sample regime maps indicate the qualitative capacity of a multiphase flow to incur coupling, but do not dictate the exact size and concentration boundaries or the magnitude of the modifications. However, they are helpful to understand when specific coupling mechanisms may be important to consider.

7.9 Chapter 7 Problems

Show all steps and list any needed assumptions:

(P7.1) Consider a chemical vat of a liquid with a density of 900 kg/m^3 and a kinematic viscosity of 10^{-4} m^2/s stirred with $u_\Lambda = 0.3$ m/s, $\tau_{\Lambda\mathcal{L}} = 0.05$ s, dissipation and turbulence structure parameters of unity, and no mean flow. How long will it take for 1 mm neutrally buoyant particles introduced in the center to spread an average of 30 cm.

(P7.2) (a) Starting with the normalized weak-drift particle diffusivity of (7.27), obtain the analytical solution in the limit $St_\Lambda \rightarrow 1$; as a hint, use L'Hopital's rule. (b) On a single chart, plot the diffusivity as a function of $t/\tau_{\Lambda\mathcal{L}}$ up to 10 for very dense particles ($R \rightarrow 0$), for neutrally buoyant particles ($R \rightarrow 1$), and for bubbles ($R \rightarrow 3$). Then discuss the trends.

(P7.3) Consider NTP turbulent air flowing at a mean velocity of 2 m/s in the x-direction with $\tau_{\Lambda\mathcal{L}}$ of 0.1 s, $k = 0.06$ m^2/s^2, $c_\varepsilon = 0.2$, and $c_\Lambda = 1$. For 200 μm diameter water droplets released into this flow at the mean velocity, (a) compute St_Λ and St_η and comment on the qualitative features of the expected particle concentration patterns; (b) obtain the equations for the streamwise and lateral diffusivities (with units of m^2/s) and then plot both on a single chart as a function of time since release from 0–1 s; and (c) plot the rms of the streamwise and lateral mean spread lengths as a function of mean streamwise distance (x) since release from 0–2 m.

(P7.4) Consider a river with a mean velocity of 1 m/s and $\tau_{\Lambda\mathcal{L}}$ of 0.5 s, $k = 0.015\ m^2/s^2$, $c_\varepsilon = c_\Lambda = 1$. For centrally-released sand particles with a density of 2.3 kg/m^3 and an average volumetric diameter of 300 μm, determine the the lateral diffusivity rate at a downstream distance of 1 m, and then compare this to the diffusivity of (zero-inertia) tracer particles and discuss the qualitative differnece.

(P7.5) For the flow conditions of P7.3, (a) compute the particle turbulent kinetic energy; (b) qualitatively plot the specific turbulent kinetic energy as in Fig. 6.26 spectra as a function of frequency for the fluid and for the particles (with appropriate high frequency cutoffs in both cases), then discuss the difference.

(P7.6) For the flow conditions of (P7.3), compute the rms of the relative velocity fluctuations and the average particle Reynolds numbers.

(P7.7) For flow conditions of (P7.4), (a) determine which of the five turbulent velocity bias terms can be neglected assuming HIST with small particle concentration levels, and give your reasoning; and (b) estimate the ensemble-averaged settling velocity in the middle of particle cloud.

(P7.8) Starting from (7.111), derive the turbophoresis result of (7.113) assuming a nearly uniform mean velocity, incompressible flow, and isotropic turbulence.

(P7.9) Consider air (instead of particles) injected into the chemical vat of (P7.1) and estimate the maximum bubble diameter due to turbulent breakup where the air-liquid surface tension is half that for water.

(P7.10) Consider ethanol drops injected at 20 m/s into a chamber of air at NTP with a mean fluid velocity of zero. (a) Determine the maximum drop diameter due to Kelvin–Helmholtz instability. (b) If the chamber is filled with turbulent air that has a Kolmogorov length scale of 20 μm, determine the maximum drop diameter due to turbulent breakup, and then determine whether the maximum size from (a) or (b) will determine the actual drop size.

(P7.11) Consider the conditions of (7.1), then use the length-scale modulation model to determine n_p for which the fluid TKE would change by 10%, and whether it will increase or decrease.

(P7.12) Same as (P7.11) but for the flow and particle conditions of (P7.3).

(P7.13) Consider water flowing upward in a pipe at a vertical velocity of 0.5 m/s where the centerline flow region is nearly HIST with $u_\Lambda = 0.05$ m/s, $\tau_{\Lambda\mathcal{L}} = 0.1$ s, $c_\varepsilon = 2.0$, and $c_\Lambda = 1.5$. If 1.2 mm diameter contminated bubbles are added in this region with a volume fraction of 1%, determine the effective total kinetic energy of the liquid velocity fluctuations.

(P7.14) For (P7.1) flow with $m = 0.1$, determine turbulence-based St_{coll} and f_{coll}.

8 Multiphase Flow Numerical Approaches

This chapter provides a broad survey of numerical approaches for multiphase flow based on particle reference frame (§8.1), relative velocity magnitude (§8.2), and particle size relative to the grid size (§8.3). These approaches are then considered for particle diffusion due to Brownian motion and turbulent flow (§8.4). Throughout this chapter, the numerical approaches are discussed based on their ability to capture physics and their computational requirements for a given flow (to help select the best tools for the job). For more details of multiphase flow computational methods, the reader is referred to Elghobashi (1994), Shirolkar et al. (1996), Shyy et al. (1997), Tomiyama (1998), Prosperetti and Tryggvason (2007), and Tryggvason et al. (2011).

8.1 Lagrangian versus Eulerian Approaches for Point-Force Particles

Numerically, the particles (be they solid, liquid, or gas) are often denoted as the dispersed phase while the surrounding fluid is denoted as the continuous phase. While the continuous-flow formulations are typically solved in an Eulerian reference using continuum-based PDEs, the dispersed-phase characteristics (velocity, concentration, diameter, etc.) are generally treated with either Lagrangian discrete-based ODEs or Eulerian continuum-based PDEs. The Lagrangian versus Eulerian particle reference frame approaches are illustrated in Figure 8.1 (relative to an Eulerian two-dimensional grid for the fluid phase). The choice between these two reference frame options is often the most important for a multiphase numerical approach.

The Lagrangian representation for particles (Figure 8.1a) is a natural approach since the particle positions are declared and updated along the particle paths. These paths are defined by the center of mass of the particle (or center of mass for a group of particles, as will be discussed) and governed by the Lagrangian equations of motion. In particular, a particle's position, velocity, and other characteristics can be described with ordinary differential equations (ODEs) along the particle trajectory. Computationally, this is also referred to as a *discrete path* approach since each trajectory is resolved. When an Eulerian representation of the continuous phase is coupled with a point-force Lagrangian approach for the particles, this combined treatment is termed the *Eulerian-Lagrangian* approach.

In contrast, the Eulerian approach for the particles (Figure 8.1b) uses computational cells that represent cell-contained particles with cell-averaged characteristics, such as

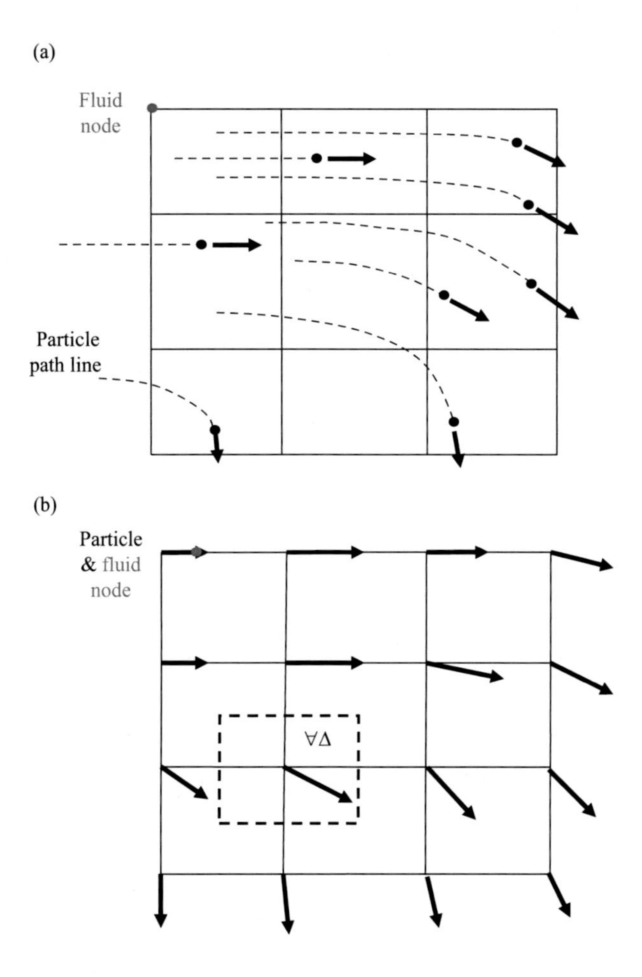

Figure 8.1 Particle reference frames in the context of an Eulerian continuous-phase grid: (a) Lagrangian velocity vectors at particle positions (as black dots) compared with the fluid nodes (e.g., the blue dot at the intersection of fluid grid cells); (b) Eulerian particle velocity vectors representing an average for the discrete control volume (the dashed line) centered at a discrete fluid grid nodes.

particle concentration and velocity. These cell-averaged characteristics are typically obtained on the same continuous-phase grid nodes as used for the surrounding fluid. The particle cell-averaged characteristics are then solved using continuous field partial differential equations (PDEs). Numerically, this is considered a *continuum* approach since the particle field is assumed to be represented at all points in space. When an Eulerian point-force approach for the particles is coupled with an Eulerian representation of the continuous phase, this combined treatment is termed the *Eulerian-Eulerian* approach.

The two different particle approaches of Lagrangian versus Eulerian lead to significant differences in benefits and limitations. The Lagrangian approach allows straightforward inclusion of particle collisions with walls and with other particles.

However, the Lagrangian approach requires that the Eulerian-based continuous-phase properties of the surrounding fluid be interpolated to the particle centroids (and the reverse if particle concentration and two-way and/or three-way coupling effects are to be predicted). In contrast, the Eulerian approach for the particles allows direct prediction of particle concentration and fluid coupling at the same cells as the surrounding fluid but does not resolve particle trajectories explicitly.

While uncommon, there are also hybrid approaches (mixing Lagrangian and Eulerian) that treat the particle phase with both continuum-based and discrete-based approaches (Hirche et al., 2019). However, most numerical treatments for particles fall into the two standard categories of Lagrangian (discrete) or Eulerian (continuum) approaches, and this chapter will focus only on those two options, first discussing details of the Lagrangian approach (§8.1.1) and then the Eulerian approach (§8.1.2). Subsequently, the two approaches are compared to each other in terms of their ability to handle particle size distribution (§8.1.3), two-way or three-way coupling (§8.1.4), and four-way coupling (§8.1.5).

8.1.1 Lagrangian (Discrete) Particle Approaches

Lagrangian Particle Equation of Motion

In the Lagrangian methodology, ODEs for the particle properties are based on particle-path derivatives. Generally, the effects of particle or wall collision are treated as discontinuous events (the ODE stops at the wall and then restarts based on restitution values). As such, forces due to collisions are not integrated directly in the ODEs. If particle spin is important, a torque ODE is included for angular momentum (see §10.1.3).

The Lagrangian ODEs for the particle trajectory can be summarized in vector form for position (3.1), linear momentum (3.2), angular momentum (10.26), mass transfer (3.97), and energy transfer (3.107), as follows:

$$\frac{d\mathbf{X}_\mathrm{p}}{dt} \equiv \mathbf{v}. \tag{8.1a}$$

$$\frac{d(\mathrm{m}_\mathrm{p}\mathbf{v})}{dt} = \mathbf{F}_\mathrm{body} + \mathbf{F}_\mathrm{surf}. \tag{8.1b}$$

$$\frac{d(I_\mathrm{p}\boldsymbol{\Omega})}{dt} = \mathcal{T}_\mathrm{surf}. \tag{8.1c}$$

$$\frac{d\mathrm{m}_\mathrm{p}}{dt} \equiv \dot{\mathrm{m}}_\mathrm{p}. \tag{8.1d}$$

$$\mathrm{m}_\mathrm{p}\frac{d\mathrm{e}_\mathrm{p}}{dt} = \dot{Q}_\mathrm{p} + \dot{\mathrm{m}}_\mathrm{p}h_\mathrm{phase}. \tag{8.1e}$$

The LHS terms include Lagrangian time derivatives for particle position ($\mathbf{X}_\mathrm{p}$), linear velocity ($\mathbf{v}$), angular velocity ($\boldsymbol{\Omega}_\mathrm{p}$), mass ($\mathrm{m}_\mathrm{p}$), and specific energy ($\mathrm{e}_\mathrm{p}$). The latter is linearly proportional to the particle temperature if the particle specific heat is constant ($\mathrm{e}_\mathrm{p} = c_\mathrm{p,p}\mathrm{T}_\mathrm{p}$). The RHS terms include the body force ($\mathbf{F}_\mathrm{body}$), fluid dynamic surface force and torque ($\mathbf{F}_{surf}$ and $\mathcal{T}_{surf}$), the mass transfer and heat transfer rates ($\dot{m}_p$, $\dot{Q}_p$), as well the enthalpy for phase change ($\hbar_{phase}$).

If the particle concentration is low enough that three-way and four-way coupling effects are negligible (e.g., based on 7.148), the RHS of (8.1b) can be decomposed per (3.6) to apply a point-force approach:

$$\mathbf{F}_{body} = m_p\mathbf{g} = \rho_p\forall_p\mathbf{g}. \tag{8.2a}$$

$$\mathbf{F}_{surf} = \mathbf{F}_D + \mathbf{F}_L + \mathbf{F}_\forall + \mathbf{F}_H + \mathbf{F}_S + \mathbf{F}_{Br} + \mathbf{F}_{\nabla T}. \tag{8.2b}$$

The surface forces of (8.2b) can be described by the models in Chapters 3 and 9–11. For example, the following are surface forces for incompressible flow past a solid sphere based on (3.84):

$$\mathbf{F}_{surf} = -3\pi d\mu_f(\mathbf{v} - \mathbf{u}_{@p})f + \rho_f\forall_p\left[(1 + c_\forall)\frac{\mathcal{D}\mathbf{u}_{@p}}{\mathcal{D}t} - c_\forall\frac{d\mathbf{v}}{dt} - \mathbf{g}\right]. \tag{8.3}$$

The drag correction (f) can be described using a drag model, such as the models for a smooth sphere in incompressible flow from §3.2.4, or one of the models in Chapter 9 for particular flow and particle conditions. These drag models are generally a function of the particle Reynolds number of (3.8):

$$Re_p = \rho_f d|\mathbf{v} - \mathbf{u}_{@p}|/\mu_f. \tag{8.4}$$

The Reynolds number and (8.3) both employ $\mathbf{u}_{@p}$, the unhindered surrounding fluid velocity properties, as discussed in §1.4.

Similarly, the RHS terms of (8.1d) and (8.1e) can be modeled based on (3.103), (3.106), (3.109), and (3.112) as follows:

$$\dot{m}_p = -\pi d\rho_f B\Theta_{\mathcal{M}}Sh \tag{8.5a}$$

$$Sh = \left(2 + 0.6 \cdot Re_p^{1/2} \cdot Sc_f^{1/3}\right)\frac{\ln(1 + B)}{B}. \tag{8.5b}$$

$$\dot{Q}_p = \pi d k_f\left(T_{f@p} - T_p\right)Nu. \tag{8.5c}$$

$$Nu = \left(2 + 0.6 \cdot Re_p^{1/2} \cdot Pr_f^{1/3}\right)\frac{\ln(1 + B)}{B}. \tag{8.5d}$$

These equations for the particle dynamics employ unhindered surrounding fluid properties. In particular, the Sherwood number (Sh) and Nusselt number (Nu) employ Re_p (and thus $\mathbf{u}_{@p}$), and (8.5c) uses $T_{f@p}$ while the Spalding number (B) is a function of $\mathcal{M}_{@p}$ (via 3.100).

Numerically, the unhindered values can be obtained by linearly interpolating the fluid properties at the surrounding nodes to the particle centroid ($\mathbf{X}_p$), as shown in Figure 8.2a. To properly employ the unhindered fluid properties (which neglect the presence of the particle), the Eulerian computational volumes of the surrounding fluid must be well posed, that is, they must be at least large enough to fully contain a single particle, per (5.9):

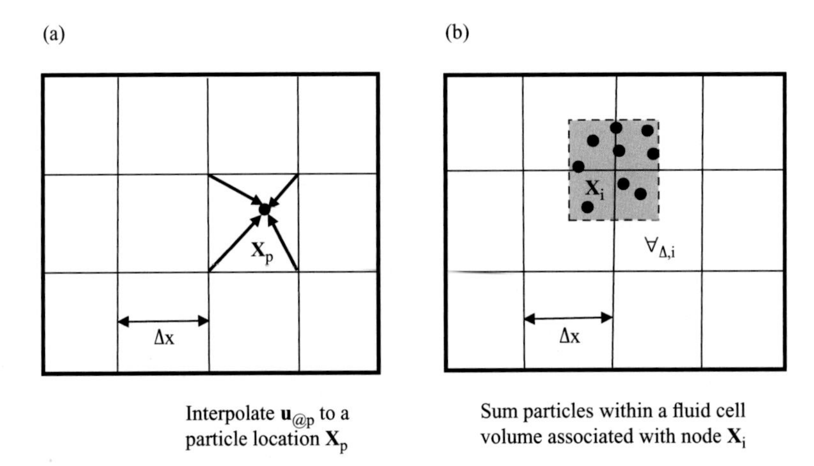

Figure 8.2 Schematic of Lagrangian point-force particles in a two-dimensional Eulerian continuous-phase grid with a typical grid length Δx, showing (a) interpolation of the fluid properties of the surrounding nodes to the particle position at $\mathbf{X}_p$; and (b) particle centroids within the computational fluid volume associated with a node $\mathbf{X}_i$.

$$\forall_\Delta > \forall_p \quad \textit{for a well-posed discrete particle concentration.} \tag{8.6a}$$

$$\Delta x > d \tag{8.6b}$$

The second expression assumes an isotropic grid ($\Delta x = \Delta y = \Delta z$). The constraint of (8.6) effectively limits particle size for a given discretization of the continuous phase for a point-force approximation. For larger particles, one must consider other approaches, as discussed in §8.3.

If two-way coupling effects are to be included numerically (see §8.1.4), then the surrounding fluid properties will themselves depend on the particle mass concentration. This concentration can be numerically obtained by summing the particles masses for a computational cell, as shown in Figure 8.2b. If the three-way coupling effects (particles influencing each other through the fluid dynamic interactions) are important, the local particle volumetric concentration (α) can be used to modify the particle surface forces (per §11.2) as follows:

$$\mathbf{F}_{surf,\alpha} = (1-\alpha)^{-3.5}\mathbf{F}_D + (1-\alpha)(\mathbf{F}_L + \mathbf{F}_\forall + \mathbf{F}_H) + \mathbf{F}_S. \tag{8.7}$$

The expressions for mass and heat transfer (8.5a and 8.5c) can also be corrected for three-way coupling as discussed by Michaelides (2006). Another three-way coupling effect is the modification of particle drag and lift force due to the proximity of wall, as discussed in §11.1. Four-way coupling (particle collisions with walls or other particles) can be incorporated via discontinuous changes in the particle velocity and spin as discussed in §11.3–§11.5.

Lagrangian Treatment of Particle Concentration

As noted previously, it is often important to predict particle concentration at the nodes associated with the surrounding fluid Eulerian mesh for two-way and three-way coupling. Even for one-way coupling conditions, the discrete particle concentration (based on the fluid computational cells) can be an important; for example, it determines particle flux rates to surfaces. This concentration can be obtained when tracking individual particles and when tracking groups of particles.

First, we consider obtaining the concentration when tracking individual particles. To obtain discrete particle concentrations at the fluid Eulerian mesh points, all the particles whose centroids are within a fluid cell volume (Figure 8.2b) are summed. For example, the discrete particle number concentration ($n_{p,\Delta}$) is equal to the number of particles in a cell ($N_{p,\Delta}$) relative to the fluid cell volume (per 5.1), as follows:

$$n_p = \frac{N_{p,\Delta}}{\forall_\Delta}. \tag{8.8}$$

To obtain the discrete cell-averaged particle volume fraction, the sum of the particle volumes in the cell (of index i) is used (per 5.2a) as follows:

$$\alpha = \frac{1}{\forall_\Delta}\sum_{j=1}^{N_{p,\Delta}} \forall_{p,j}. \tag{8.9}$$

Similar cell-averages can be obtained for the particle mass fraction and the mass loading using (5.4) and (5.6) and also extended to identify the discrete mixture density, mixture viscosity, and mixture conductivity with (5.5) and (5.8).

Computing particle concentration using the preceding method requires tracking all the particles trajectories in the domain. However, in many multiphase flows, the number of physical particles in the domain is very high. For example, the number of droplets expelled by a human cough can correspond to millions of aerosol particles per cubic centimeter (Zayas et al., 2012). In such cases, it may be computationally impossible or impractical to track all of the particles with individual Lagrangian ODEs. An alternative approach is to computationally track representative groups, often called *parcels*, *computational particles*, or *super particles*. These groups are essentially a small cloud of particles that are assumed to have the same Lagrangian trajectory. If we denote $N_{P,\Delta}$ as the number of parcels in a cell and $N_{p/P}$ as the number of particles per parcel, the particle concentration per volume and particle volume fraction can be evaluated as follows:

$$N_{P,\Delta} \equiv \frac{\text{number of parcels}}{\text{computational cell}}. \tag{8.10a}$$

$$N_{p/P} \equiv \frac{\text{number of particles}}{\text{parcel}}. \tag{8.10b}$$

$$n_p = \frac{1}{\forall_\Delta}\sum_{j=1}^{N_{P,\Delta}} N_{p/P,j}. \tag{8.10c}$$

$$\alpha = \frac{1}{\forall_\Delta}\sum_{j=1}^{N_{P,\Delta}} N_{p/P,j}\forall_{p,j}. \tag{8.10d}$$

Generally, a single parcel is composed of particles with the same properties (i.e., diameter, velocity, temperature, mass, etc.). For the corresponding ODEs for the trajectory and momentum (for constant particle mass per 3.117), the parcel velocity is denoted as $\mathbf{v}_P$, so

$$d\mathbf{X}_P/dt = \mathbf{v}_P. \tag{8.11a}$$

$$m_p d\mathbf{v}_P/dt = \mathbf{F}_{body} + \mathbf{F}_{surf}. \tag{8.11b}$$

In this case, the trajectory and momentum equations for a single parcel are the same as those for a single particle (8.1a and 8.1b are equivalent to 8.11a and 8.11b). If there is no mass transfer, breakup, or coalescence, the number of particles per parcel will be constant along its trajectory.

The key advantage for the parcel approach is that it can dramatically reduce the number of computational trajectories, thereby reducing computational cost. For example, $N_{p/P} = 100$ allows the number of computed trajectories to be reduced 100-fold. However, there is an upper limit to $N_{p/P}$ for the size of the parcel cloud in order to use the fluid properties from a single cell. In particular, a well-posed parcel cloud should have a size less than the cell computational volume:

$$\forall_P \leq \forall_\Delta \quad \textit{for a well-posed discrete particle concentration.} \tag{8.12}$$

The LHS is the parcel cloud volume and represents the mixed-fluid volume occupied by the particles and their surrounding fluid.

To initialize parcels in a computational domain so they are well-posed, a common approach is to set the parcel volume at the initial locations equal to the local fluid computational volume. This assumes that the grid is well refined in this region so (8.13) will be satisfied at later positions and this initialization is given by:

$$\forall_{P,o} \equiv \text{initial parcel volume} = \forall_{\Delta,o}. \tag{8.13}$$

In many cases, it is reasonable to assume that the parcel cloud stays approximately constant. However, the parcel volume can change in size due to gradients in the particle velocity. To track this change, one may obtain an ODE for the parcel volume as in (2.5) but using the particle path derivative (and assuming no volume diffusion nor generation), as follows:

$$\frac{d\forall_P}{dt} = \forall_P(\nabla \cdot \mathbf{v}). \tag{8.14}$$

This equation can be integrated along with the other particle Lagrangian equations and considered at each time step in terms of the volume constraint of (8.12). When this constraint is significantly violated, that particular parcel can be numerically subdivided into a number of smaller parcels that are then distributed among the region of original parcel volume.

To release particles or parcels over a given time period, the injection surface can be discretized into a set of computational cells with face area A_Δ and surface normal $\mathbf{n}_\Delta$ for which the initial particle concentration (α_{inj}) and injection velocity ($\mathbf{v}_{inj}$) are to be specified. Using (5.3), the prescribed injection mass flux (mass of particles per unit area and time) can be related to the mass of particles released per time step (Δt) relative to a discrete face area, as follows:

$$\frac{\text{mass of particles released per time step}}{A_{\Delta}\Delta t_{inj}} = \alpha_{inj}\rho_p(\mathbf{v}\cdot\mathbf{n})_{inj}. \quad (8.15)$$

Using this result and (1.4), the number of particles (N) or the number of particles per parcel ($N_{p/P}$) that are injected at each time step is given as follows:

$$\begin{matrix} N_{inj} \\ N_{p/P,inj} \end{matrix} = \alpha_{inj}(\mathbf{v}\cdot\mathbf{n})_{inj}A_{\Delta}\Delta t_{inj}/\forall_p \quad \begin{matrix} \textit{for discrete particles injected at cell face.} & (8.16a) \\ \textit{for a single parcel injected at cell face.} & (8.16b) \end{matrix}$$

Particles or parcels can be released consistently over the injection period using this approach. Notably, the injection time step need not be the same as the fluid time step. For parcels, the release time step can be based on the size of the parcel and the fluid cell via (8.13), whereby each parcel is released at $\Delta t_{inj} = |\mathbf{v}\cdot\mathbf{n}|/\Delta x$.

One may also compute the average particle volumetric concentration at a discrete flux surface over a flux averaging time (τ_{flux}) by summing the particle volume flux based on the total number of fluxed particles ($N_{p,flux}$) or the total number of fluxed parcels ($N_{P,flux}$), as follows:

$$\alpha = \frac{\text{volume of fluxed particles}}{A_{\Delta}\tau_{flux}} = \begin{cases} \dfrac{1}{A_{\Delta}\tau_{flux}}\displaystyle\sum_{k=1}^{N_{p,flux}}\frac{\forall_{p,k}}{(\mathbf{v}\cdot\mathbf{n})_k} & \textit{for particles.} \quad (8.17a) \\ \dfrac{1}{A_{\Delta}\tau_{flux}}\displaystyle\sum_{k=1}^{N_{P,flux}}\frac{\forall_{p,k}N_{p/P,k}}{(\mathbf{v}\cdot\mathbf{n})_k} & \textit{for a parcel.} \quad (8.17b) \end{cases}$$

Other average particle concentrations (e.g., mass loading) can be obtained with similar flux averages using the concentration parameters defined in §5.1.

8.1.2 Eulerian (Continuum) Particle Approaches

In the Eulerian approach for the particles, the particle concentration and momentum are described by transport equations in a manner similar to the density and momentum for the continuous phase. To obtain these transport equations, we assume that the particle characteristics (concentration, velocity, and energy) can be described as a continuum in the domain.

Based on this continuum assumption, the conservation equations for the Eulerian dispersed-phase mass, momentum, and energy can be derived using the Reynolds transport theorem of §2.1. If Brownian motion and particle collision effects are neglected, the PDEs for particle mass, momentum, and energy can be obtained in the same manner as (2.6a), (2.13a), and (2.27b) but with mass and heat transfer effects included, as follows:

$$\frac{\partial(m_p n_p)}{\partial t} + \nabla \cdot (m_p n_p \mathbf{v}) = n_p \dot{m}_p. \tag{8.18a}$$

$$\frac{\partial(m_p n_p \mathbf{v})}{\partial t} + \nabla \cdot [m_p n_p \mathbf{v} \otimes \mathbf{v}] = n_p \mathbf{F}_{body} + n_p (\mathbf{F}_{surf} + \dot{m}_p \mathbf{v}). \tag{8.18b}$$

$$\frac{\partial(m_p n_p e_p)}{\partial t} + \nabla \cdot (m_p n_p e_p \mathbf{v}) = n_p \left[\dot{Q}_p + \dot{m}_p \hbar_{phase} + \dot{m}_p e_p\right]. \tag{8.18c}$$

For example, (8.18a) equates the rate of change of particle mass in a control volume to the mass flux through its boundaries and to any mass source term within the control volume. The LHS time derivatives for (8.18) describe Eulerian changes of particle mass, momentum, and energy per unit of mixed-volume fluid. Furthermore, the convective term for the momentum equation uses the outer product notation per (2.14), as follows:

$$\nabla \cdot (m_p n_p \mathbf{v} \otimes \mathbf{v}) = m_p n_p (\mathbf{v} \cdot \nabla) \mathbf{v} = \mathbf{v} \nabla \cdot (m_p n_p \mathbf{v}). \tag{8.19}$$

Notably, the particle concentrations obtained with such Eulerian PDEs are statistical and are not quantized by individual particles; for example, a cell may contain 45.27456 particles. Further, a cell with 0.13 particles is equivalent to a 13% probability that one particle is in the cell at a given time.

These Eulerian PDEs can also be written in terms of particle volume fraction using (5.2), as follows:

$$\frac{\partial(\alpha \rho_p)}{\partial t} + \nabla \cdot (\alpha \rho_p \mathbf{v}) = \frac{\alpha}{\forall_p} \dot{m}_p. \tag{8.20a}$$

$$\frac{\partial(\alpha \rho_p \mathbf{v})}{\partial t} + \nabla \cdot [\alpha \rho_p \mathbf{v} \otimes \mathbf{v}] = \frac{\alpha}{\forall_p} \mathbf{F}_{body} + \frac{\alpha}{\forall_p} (\mathbf{F}_{surf} + \dot{m}_p \mathbf{v}). \tag{8.20b}$$

$$\frac{\partial(\alpha \rho_p e_p)}{\partial t} + \nabla \cdot (\alpha \rho_p e_p \mathbf{v}) = \frac{\alpha}{\forall_p} \left[\dot{Q}_p + \dot{m}_p (\hbar_{phase} + e_p)\right]. \tag{8.20c}$$

The RHS values are given by (8.2)–(8.5). Detailed derivations of the particle Eulerian PDEs for various conditions and forms are given by Drew (1983), Zhang and Prosperetti (1994), Prosperetti (2007), and Crowe et al. (2011).

These Eulerian conservation equations such as (8.18) or (8.20) can then be discretized such that each node in the computational domain includes cell-averaged particle-phase variables (e.g., α and $\mathbf{v}$), along with the cell-averaged surrounding fluid variables (e.g., p and $\mathbf{u}$). This allows the spatial discretization and time marching schemes of the dispersed-phase PDEs to be consistent with those used for the continuous-phase PDEs. Since both fluid and particle phases have similar PDEs and numerical approaches and use the same grid, this Eulerian-Eulerian multiphase treatment is often termed the *two-fluid* approach. For two-way coupling, the numerical consistency and the co-location of fluid and particle variables at the same nodes

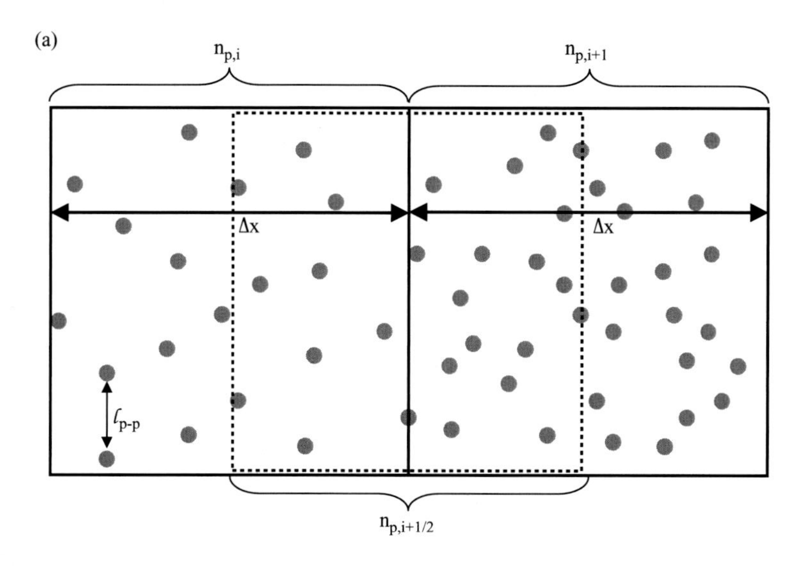

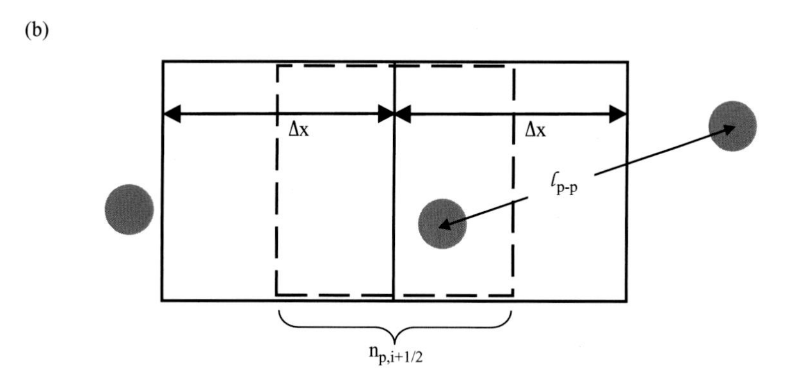

Figure 8.3 Two-dimensional Eulerian grid that contains discrete particles in adjoining computational control volumes: (a) $N_{p,\Delta} \gg 1$, allowing a "continuum" approximation; and (b) $N_{p,\Delta} \sim 1$ so that a "continuum" approximation is *not* appropriate.

improves the coupling accuracy and efficiency (see §8.1.4), which is an advantage over Lagrangian methods.

However, the Eulerian approach for particle concentration has a downside (compared to Lagrangian methods), since numerical diffusion occurs due to truncation terms from spatial discretization of the PDE gradient terms. To demonstrate this, consider two neighboring control volumes with respective particle concentrations $n_{p,i}$ and $n_{p,i+1}$ along with an intermediate concentration of $n_{p,i+1/2}$, as shown in Figure 8.3a. Assuming continuum conditions, a forward Taylor series expansions with a shift of $+\Delta x/2$ can be applied for $n_{p,i+1}$ as follows:

$$n_{p,i+1} = n_{p,i+1/2} + \frac{\Delta x}{2}\left(\frac{\partial n_p}{\partial x}\right)_{i+1/2} + \frac{1}{2!}\left(\frac{\Delta x}{2}\right)^2\left(\frac{\partial^2 n_p}{\partial x^2}\right)_{i+1/2} + O(\Delta x)^3 \ldots \quad (8.21)$$

The last term on the RHS corresponds to all terms of order Δx^3 or higher. A similar expansion for $n_{p,i}$ and $n_{p,i+1/2}$ can be obtained using a backward shift ($-\Delta x/2$). Combining the two expansions yields an expression for the particle concentration gradient between the nodes, as follows:

$$\left(\frac{\partial n_p}{\partial x}\right)_{i+1/2} = \frac{n_{p,i+1} - n_{p,i}}{\Delta x} + O(\Delta x)^2. \tag{8.22}$$

By dropping the terms of order Δx^2 or higher, we obtain a one-dimensional second-order spatial discretization for the particle concentration gradient. Such expressions can be used to discretize the spatial gradients, as needed for the LHS of (8.18). However, these expressions will introduce errors if there are nonlinear gradients or a particle concentration discontinuity. For example, consider an upward flow with an inflow particle concentration discontinuity as shown in Figure 8.4a. If the velocity is uniform and there is no diffusion due to turbulent or Brownian motion, the discontinuity will propagate downstream for the "exact" solution, as shown in Figure 8.4a. This is straightforward to capture with a Lagrangian approach for the particles. However, Eulerian difference schemes will have truncation term errors. In general, first-order schemes are numerically diffusive and will cause will n_p to spread laterally, as shown in Figure 8.4b and Figure 8.4c. Second-order schemes such as (8.22) have less diffusion but can lead to artificial undershoots and overshoots (numerical ringing) about the discontinuity, as shown by short-dashed line in Figure 8.4c. Nonlinear schemes (which often combine first-order and second-order schemes with limiters) can avoid the ringing and reduce but not eliminate the diffusion, as shown by medium-dashed line in Figure 8.4c. As such, numerical diffusion is an inherent consequence of an Eulerian approach for the particle concentration.

8.1.3 Particle Size Distribution Aspects

In the preceding formulations, the particle size was assumed to be uniform. However, most multiphase flows include a variety of particle sizes. If the variation in particle diameters is narrow (per 4.22 or 4.23), one may employ a single effective particle diameter (see §4.1.2). However, a broad particle size variation generally requires explicit characterization of the size distribution. This numerical prescription differs, depending on whether the particles are treated in an Eulerian or Lagrangian reference frame.

For a Lagrangian treatment of the dispersed phase where all particles paths are followed, including a prescribed size distribution can be achieved by initializing or releasing the particles with different sizes using random sampling from a prescribed number-based size PDF ($\mathcal{P}_N$ as defined in 4.1). For a Lagrangian parcel approach, the particle size distribution can be broken into a set of discrete bins, where N_b is the number of bins. A set of parcels can then be assigned to represent each bin, where the number of particles per parcel is based on the discrete $\mathcal{P}_N$ value. In this case, each initial parcel cloud volume can still be set equal to the computational cell per (8.13),

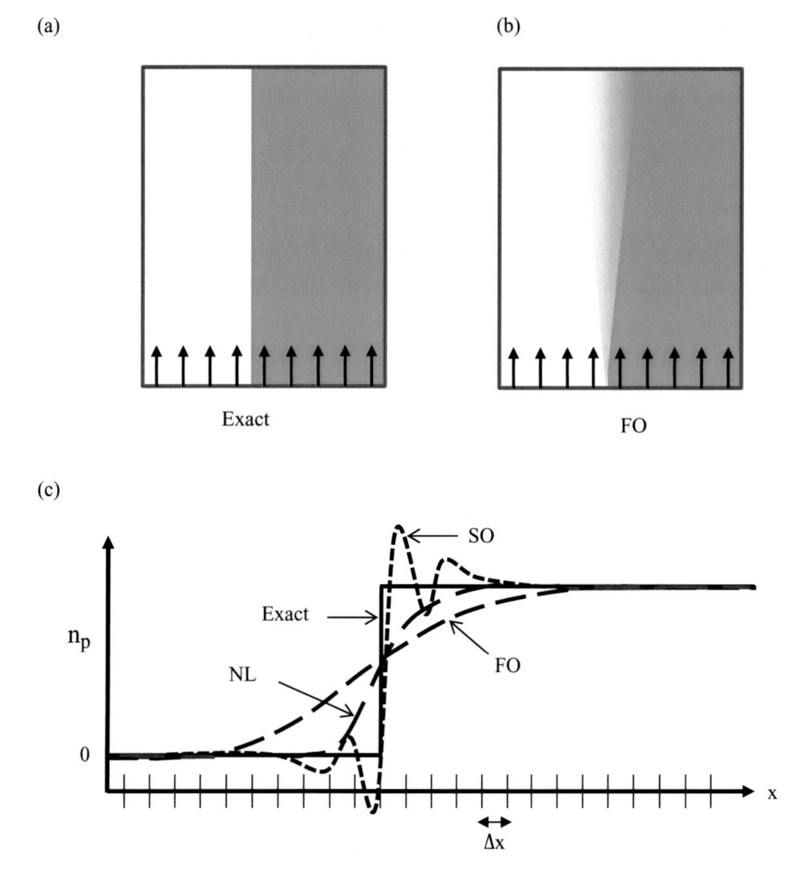

Figure 8.4 Treatment of a particle number concentration (n_p) discontinuity for an upward flow with uniform velocity and no molecular or turbulent diffusion: (a) exact solution where discontinuity is preserved downstream; (b) a first-order (FO) numerical solution where discontinuity is artificially diffused (this nonphysical diffusion is a numerical error), and (c) lateral profiles at a downstream plane for exact solution as well as numerical solutions that are FO, second-order (SO), and nonlinear (NL).

where each cell now initially contain N_b parcels. For example, if the size PDF is broken into five bins, then five different parcels would initialized in a cell, where each parcel has a different particle diameter. If the parcels are instead to be released continuously over time, N_b parcels can be simultaneously released per (8.16), where the number of particles per parcel ($N_{p/P}$) for a given bin is a again based on the discrete $\mathcal{P}_N$ value. This bin-based approach is computationally efficient for polydisperse distributions since it only requires one trajectory per bin per release cell. As such, the number of bins used should be large enough to reasonably represent the size distribution while being limited to help reduce computational requirements.

The Eulerian treatment of polydisperse particles is often not as efficient. One option is to use N_b number of Eulerian particle field PDEs, where each field represents particles of a given size range with a mean diameter of d_k, where $N_b = 1,k$. Each field would have a volume fraction (α_k) and number density ($n_{p,k}$) where the sum of all the volume fractions represents the total:

$$\alpha = \sum_{k=1}^{N_b} \alpha_k = \frac{\pi}{6} \sum_{k=1}^{N_b} n_{p,k} d_k^3. \tag{8.23}$$

Each particle field will require PDE solution for their velocity ($\mathbf{v}_k$), thus requiring N_b transport equations for each Eulerian particle group, as follows:

$$\frac{\partial \alpha_k}{\partial t} + \nabla \cdot (\alpha \mathbf{v})_k = 0. \tag{8.24a}$$

$$\frac{\partial (\alpha \mathbf{v})_k}{\partial t} + \nabla \cdot [\alpha \mathbf{v} \otimes \mathbf{v}]_k = \left[\frac{\alpha}{m_p} \left(\mathbf{F}_{body} + \mathbf{F}_{surf} \right) \right]_k. \tag{8.24b}$$

The overall concept is called the *multigroup* or *multibin* Eulerian approach for particles. Unfortunately, it can be costly to solve multiple sets of PDEs for all the individual Eulerian fields when N_b is large (to accommodate a broad size distribution), so this approach is not often used. Other Eulerian approaches have been developed to more efficiently incorporate polydispersivity using the method of moments (Marchisio and Fox, 2013). The most common version is the Direct Quadrature Method of Moments (DQMOM), developed by Yuan and Fox (2011), which employs a PDE for a set of moments of the size distrbution, which allows fewer PDEs compared to the multigroup method for the same level of accuracy. The DQMOM is especially well suited to dense flows with many particle collisions, but the complexity of the method has limited its usage for dilute conditions.

As such, Lagrangian methods for the particles are generally preferred (over Eulerian methods) for broad particle distributions with one-way coupling since they are more efficient and do so without numerical diffusion. However, Eulerian approaches for the particles can have significant advantages for two-way coupling, as discussed in the next section.

8.1.4 Two-Way and Three-Way Coupling Aspects

For two-way coupling, the surrounding fluid (continuous-phase) velocity, pressure and/or temperature fields are influenced by the presence of the particles. This coupling can be integrated into the surrounding fluid PDEs by including the interphase effects and volume fraction effects. For the latter aspect, the particles are part of the mixed-fluid volume (5.2) so that the surrounding fluid mass per mixed-fluid volume is $(1-\alpha)\rho_f$. Similarly, the surrounding fluid momentum and energy per mixed-fluid volume is $(1-\alpha)\rho_f\mathbf{u}$ and is $(1-\alpha)\rho_f e_f$. Application of the Reynolds transport theorem for these intensive properties combined with the addition of the interphase forces, heat, and mass transfer yields the two-way coupled PDEs for the surrounding fluid (using n_p for the interphase effects):

$$\frac{\partial[(1-\alpha)\rho_f]}{\partial t} + \nabla \cdot [(1-\alpha)\rho_f \mathbf{u}] = -n_p \dot{m}_p. \tag{8.25a}$$

$$\frac{\partial[(1-\alpha)\rho_f \mathbf{u}]}{\partial t} + \nabla \cdot [(1-\alpha)\rho_f \mathbf{u} \otimes \mathbf{u}] = (1-\alpha)\rho_f \mathbf{g} + (-\nabla p + \mathbf{G}) - n_p\left(\mathbf{F}_{surf} + \dot{m}_p \mathbf{v}\right). \tag{8.25b}$$

$$\frac{\partial[(1-\alpha)\rho_f e_f]}{\partial t} + \nabla \cdot [(1-\alpha)\rho_f e_f \mathbf{u}] = (1-\alpha)\mathbf{G} \cdot \mathbf{u} - \nabla \cdot [(1-\alpha)p\mathbf{u}] + \nabla \cdot [k_m \nabla T] - n_p\left[\dot{Q}_p + \dot{m}_p\left(h_{phase} + e_p\right)\right]. \tag{8.25c}$$

For the continuous-phase fluid mass transport PDE of (8.25a), the RHS is the rate of surrounding fluid mass transferred to the particles per mixed-fluid volume. This term is equal and opposite to the RHS term for the particle mass transport PDE (8.18a).

For the RHS of the momentum PDE (8.25b), the first term is the gravitational force acting on the fluid mass per mixed-fluid volume, and the second term is the fluid stress per mixed-fluid volume that would occur in the absence of particles. The third term is the fluid stress acting on the fluid per unit volume. It does not include a $(1-\alpha)$ factor since the fluid stress does not depend on whether the volume is occupied by particles or just by fluid (Prosperetti, 2007). The RHS of the energy PDE (8.25c) includes fluid work and energy transfer.

To employ the preceding Eulerian transport PDEs for the continuous-phase fluid, the particle concentrations are needed at the continuous-phase nodes. If the particle fields are treated with an Eulerian approach that uses the same grid as the surrounding fluid, the particle concentrations are already available at such nodes. As such, the Eulerian-Eulerian formulation is numerically convenient and accurate since the data transfer between the phases are communicated directly at the coincident node locations (there is no interpolation error).

If Lagrangian ODEs are instead used for the particles, the coupling to a continuous-phase node can be computed based on a summation, as shown in Figure 8.2b. As such, the particle concentration needed for (8.25) can be obtained with (8.8) for a Lagrangian particle approach and with (8.10) for a Lagrangian parcel approach. This summation approach can be applied to other interphase terms, such as for the surface force for particle and parcel approaches, as follows:

$$\left(n_p \mathbf{F}_{surf}\right)_\Delta = \begin{cases} \dfrac{1}{\forall_\Delta}\displaystyle\sum_{j=1}^{N_{p,\Delta}} \mathbf{F}_{surf,j} & \textit{for particles in the cell volume.} \quad (8.26a) \\ \dfrac{1}{\forall_\Delta}\displaystyle\sum_{j=1}^{N_{P,\Delta}} N_{p/P,j}\mathbf{F}_{surf,j} & \textit{for parcels in the cell volume.} \quad (8.26b) \end{cases}$$

However, a key question with either of these Lagrangian approaches is whether the particle concentration and interphase forces employ a statistically significant number of particles or parcels in a cell volume to ensure that the interphase coupling terms (with n_p) and gradient terms (with α) are accurate. This accuracy will be satisfied if the *particle-phase continuum* (Drew and Passman, 1999) is achieved by using a large number of particles or parcels in a host cell:

$$N_{p,\Delta} \gg 1 \quad \textit{for a continuum particle concentration field.} \tag{8.27a}$$

$$N_{p,\Delta} \gg 1 \quad \textit{for a continuum parcel concentration field.} \tag{8.27b}$$

To demonstrate the continuum principle in terms of the accuracy of particle concentration gradients, consider again the control volumes shown in Figure 8.3a. Since the computational cells contain many particles in this case, the number density will vary smoothly between cells, that is, n_p will be continuously differentiable. However, a much different result arises if $N_{p,\Delta}$ is of order unity, as in Figure 8.3b. In this case, the grid resolution is on the order of the local interparticle spacing ($\Delta x \sim \ell_{p\text{-}p}$), which leads to discontinuities in particle concentration depending on the position of the grid cell relative to the particle. For example, $n_{pi+1} = 1$ while $n_{p,i} = 0$, as shown in Figure 8.3b. Clearly, a spatial discretization such as (8.22) will not provide a concentration gradient that is continuous and effectively independent of the cell position for such conditions. Hence, (8.27) is important for accurate two-way coupling with a Lagrangian particle approach.

If parcels are instead used, the particle-phase continuum issue can be alleviated if the parcel contributions are distributed to multiple nodes, which is reasonable when the parcel volume is similar in size to the computational cell volume. This reduces the requirements for statistical significance. In general, four or more parcels per cell are typically used for two-way coupling, but 10 or so are often needed to be consistent with the accuracy of an Eulerian-Eulerian approach (Sivier et al., 1994). As such, the Eulerian-Lagrangian approach is much less efficient for two-way coupling as compared to the Eulerian-Eulerian approach (Druzhinin and Elghobashi, 1998, 1999). In addition, three-way coupling effects as in (8.7) also require accurate representations of the particle concentration, and so are also more computationally accurate with Eulerian-Eulerian approaches.

8.1.5 Four-Way Coupling Aspects

Four-way coupling generally refers to particles colliding with each other, but herein we generalize this to include particles impacting a surface as well. Numerical implementation for many particles in the domain generally uses an event-based approach, where the particle velocity changes between precollision and postcollision are obtained with using jump conditions based on restitution coefficients. In the following, the jump conditions are considered for linear momentum of spherical particles, where more detailed models of restitution coefficients (which include nonspherical and angular momentum aspects as well as particle spin, adhesion, and fluid viscosity) are discussed in §11.3–§11.5.

(a)

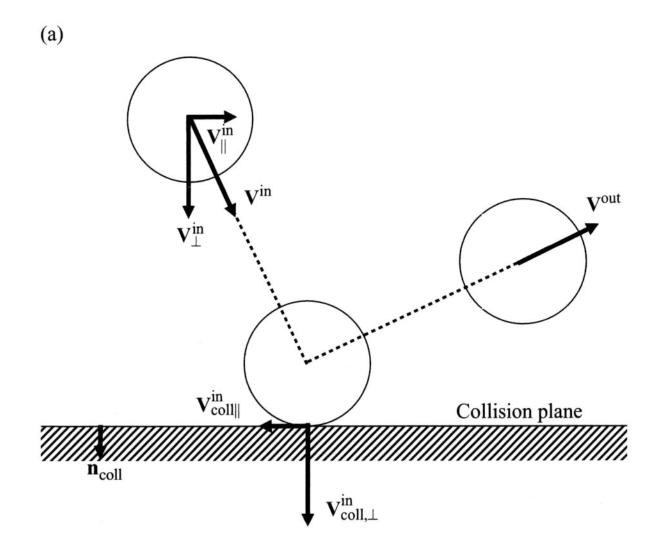

(b)

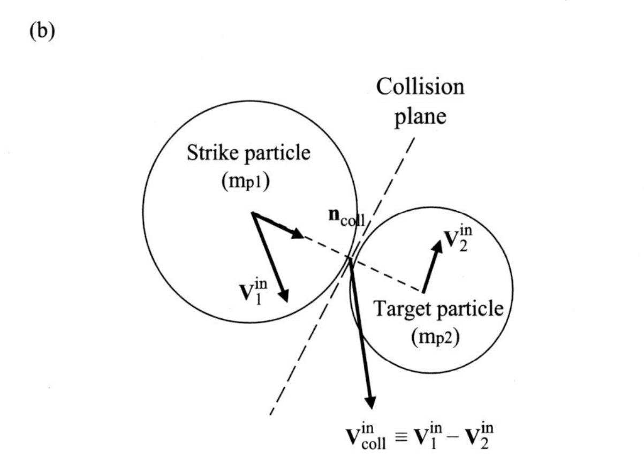

Figure 8.5 Schematics of collisions of spherical particles without spin identifying the incoming and outgoing velocities as well as the collision normal and velocity vectors: (a) particle–wall interaction; and (b) two particles with centroid locations $\mathbf{x}_{p1}$ and $\mathbf{x}_{p2}$.

For the case of a particle impacting a wall as shown in Figure 8.5a, the incoming and outgoing velocities are $\mathbf{v}^{in}$ and $\mathbf{v}^{out}$, so the jump change between these two states is as follows:

$$\Delta\mathbf{v} \equiv \mathbf{v}^{out} - \mathbf{v}^{in}. \tag{8.28}$$

For two particles colliding (as shown in Figure 8.5b), we define a strike particle of velocity $\mathbf{v}_1$ with mass m_{p1} and a target particle of velocity $\mathbf{v}_2$ with mass m_{p2}, so there are two jump conditions and a collision vector based on the difference of the incoming velocities:

$$\Delta \mathbf{v}_1 \equiv \mathbf{v}_1^{out} - \mathbf{v}_1^{in}. \tag{8.29a}$$

$$\Delta \mathbf{v}_2 \equiv \mathbf{v}_2^{out} - \mathbf{v}_2^{in}. \tag{8.29b}$$

$$\mathbf{v}_{coll}^{in} \equiv \mathbf{v}_1^{in} - \mathbf{v}_2^{in}. \tag{8.29c}$$

The particle–wall collision can be considered in the limit where the target particle has infinite mass and zero velocity, so the collision vector is simply the incoming particle velocity to the wall.

Based on the preceding definition, we define the collision normal vector ($\mathbf{n}_{coll}$) as perpendicular to the collision plane. For a sphere impacting a wall, the collision vector is opposite to the wall outward normal, as shown in Figure 8.5a. For two colliding spherical particles, as shown in Figure 8.5b, the unit collision vector points from the strike particle centroid ($\mathbf{X}_{p1}$) to the target particle centroid ($\mathbf{X}_{p2}$):

$$\mathbf{n}_{coll} \equiv \frac{\mathbf{X}_{p2} - \mathbf{X}_{p1}}{\left|\mathbf{X}_{p2} - \mathbf{X}_{p1}\right|}. \tag{8.30}$$

The collision normal vector allows us to decompose the particle velocity into normal and tangential components as follows:

$$\mathbf{v}_{\perp} \equiv (\mathbf{v} \cdot \mathbf{n}_{coll})\mathbf{n}_{coll}. \tag{8.31a}$$

$$\mathbf{v}_{\parallel} \equiv \mathbf{v} - \mathbf{v}_{\perp}. \tag{8.31b}$$

These components are shown in Figure 8.5 for both particle–wall and particle–particle impacts.

For a binary collision of two particles of the same material, a normal coefficient of restitution can be combined with the impulse equations to determine the normal particle velocity changes:

$$m_{p2}\Delta \mathbf{v}_{\perp,2} = -m_{p1}\Delta \mathbf{v}_{\perp,1} = (1 + e_{\perp})m_{coll}\mathbf{v}_{coll,\perp}^{in}. \tag{8.32a}$$

$$m_{coll} \equiv \frac{m_{p1}m_{p2}}{m_{p1} + m_{p2}}. \tag{8.32b}$$

Recall that a particle impacting a wall is given by $\mathbf{v}_2 = 0$ and $m_{coll} = m_{p1}$. The tangential particle momentum can be similarly related to a tangential coefficient of restitution as follows:

$$m_{p2}\Delta \mathbf{v}_{\parallel,2} = -m_{p1}\Delta \mathbf{v}_{\parallel,1} = \frac{2}{7}(1 + e_{\parallel})m_{coll}\mathbf{v}_{coll,\parallel}^{in}. \tag{8.33}$$

As discussed in §11.4, normal restitution coefficients for hard surfaces in a gas can be as high as 0.9. As discussed in §11.5, parallel restitution coefficients for hard, smooth surfaces in a gas can be as high as 0.35. Using these values on the RHS of (8.32a) and (8.33) allows the outgoing velocities for both particles to be obtained. However, restitution coefficients are generally lower for softer materials, plastic deformation, and irregular surfaces (§11.4). Effects of surrounding fluid viscosity and adhesion can also reduce the restitution, leading to a slide outcome or immediate deposition for solid particles (§11.4–§11.5) while high surface wettability can also lead to deposition outcomes for drops (§11.6).

Lagrangian Approach for Collisions

To determine when a particle collision occurs for a Lagrangian reference, each particle path can be considered in terms of the neighboring particle positions for each time step to determine potential collision partners (as shown in Figure 5.15). Once a collision is identified, the collision dynamics can be obtained based on the collision vector and restitution coefficients. Three basic outcomes for a particle hitting a wall are illustrated in Figure 8.6 and include deposition (zero restitution), accommodation (zero normal restitution but nonzero restitution), and reflection (nonzero normal and tangential restitutions). The particle flux rate at a discrete area for the deposition condition can be computed based on a simple summation, as in (8.17).

If the number of particles in the system is large (e.g., 10^5 or more) such that parcels are employed, wall-interaction collisions can be treated in the same fashion as individual particles. However, particle–particle collisions are more complicated for a parcel approach. Interparticle collisions (between two sets of particles) can be treated by considering the time period when two parcel clouds will overlap and then using models for collision rates (e.g., given in §5.6 and §7.7) to determine the probability of collisions.

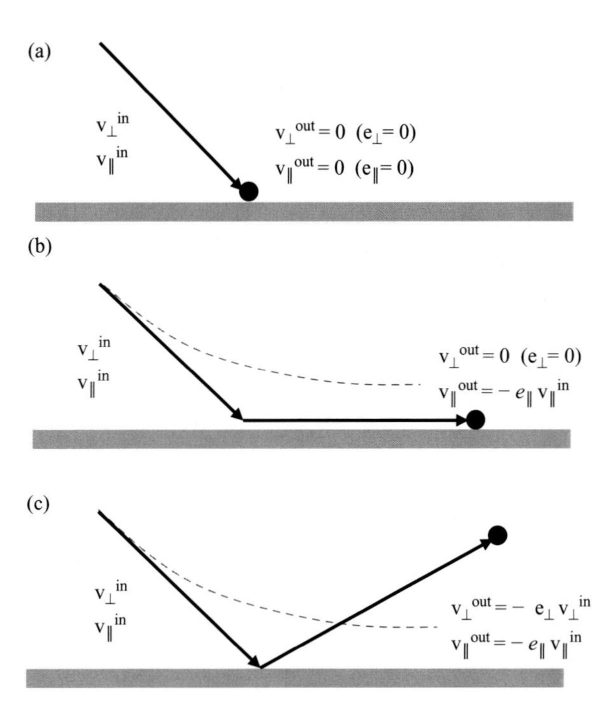

Figure 8.6 Particle–wall interaction shown as Lagrangian outcomes (solid lines) and Eulerian outcomes (dashed lines) for (a) absorption (stick), (b) accommodation (rolling), and (c) reflection (bouncing).

For example, if the overlap period is consistent with a 10% probability of collision, then 10% of the particles employ collision outcomes (two new sets of parcels are created that will change directions), while 90% of the intersections will have no collisions. Thus, the original parcels, but with only 90% of the original number of particles, will pass by each other unaffected. For the collision outcomes, the velocity of the two original parcels can be used to obtain $\mathbf{v}_{coll}$ of (8.29) and m_{coll} of (8.32b) in order to determine the two new parcels. However, turbulent flow conditions generally require additional consideration of intraparcel collisions (collisions among particles within a parcel). Recent parcel-based techniques have been developed to handle these cases (Johnson, 2020), but very high collision rates may be best handled with Eulerian methods in terms of computational efficiency.

Eulerian Approach for Collisions

An Eulerian-Eulerian approach can use an averaged approach for each time step whereby the particles outcomes for the percent of particles that are expected to collide are averaged with the outcomes for the rest of the particles in a given computational cell. As such, the cell outcome represents a number-based group average outcome. When the flow approaches a dense condition in terms of particle–particle collisions ($St_{coll} \gg 1$), the ensemble-averaged particle collisions in an Eulerian-Eulerian framework can be modeled as effective stresses on the particle concentration. These stresses can be incorporated into the particle transport equations using a *granular temperature*. Additional details for dense flow treatment are given by Gidaspow (1994), Crowe et al. (2011), and Marchisio and Fox (2013).

For particle–wall collisions in an Eulerian-Eulerian framework, the deposition conditions (Figure 8.6a) can be implemented numerically by simply letting $\mathbf{v}$ be "free" (i.e., unconstrained) so that there is a finite velocity at the wall. The particle flux rate for a given cell face on the wall can then be computed from the wall-normal velocity (based on a dot product) and the particle concentration via (5.3a), as follows:

$$\frac{\text{mass flux of particles}}{\text{discrete mixed fluid area}} = \left[(\mathbf{v} \cdot \mathbf{n})\alpha\rho_p\right]_\Delta. \tag{8.34}$$

However, it is difficult to accurately incorporate the accommodation or reflecting wall boundary conditions of Figures 8.6b and 8.6c in an Eulerian particle approach. This is because the Eulerian representation with a single PDE for particle momentum cannot represent particles moving both toward and away/along from the wall (i.e., it cannot incorporate a model for nonzero coefficient of restitution) within a single cell. One approach is to use a multifield where a new set of PDEs is generated for each set of collisions, but this can be complex and computationally expensive. To avoid this, an efficient but approximate remedy for the Eulerian dispersed-phase approach is to apply a zero-flux condition to the wall to ensure mass conservation:

$$\mathbf{v}_\perp = 0 \qquad \textit{Eulerian zero-flux boundary condition.} \tag{8.35}$$

The result of such a boundary condition is shown qualitatively by the dashed lines in Figures 8.6b and 8.5c, where it can be seen to be more reasonable for the slide outcome than for the reflection case. While reflective wall interactions are not easily captured with the Eulerian dispersed-phase approach, such interactions are often not important if the particles have small inertia, in which case further simplifications may be possible, as discussed in the next section.

8.2 Mixed-Fluid and Drift-Flux Eulerian Approaches

The preceding Eulerian and Lagrangian numerical treatments considered the particle acceleration based on a particle momentum equation, where the surface forces employed a relative particle velocity. When this relative velocity becomes very small due to small particle inertia, the particle dynamics and acceleration relative to the discrete fluid can be ignored. To determine when this occurs, we define a cell-based Stokes number for the particle (St_Δ) based on the particle response time (τ_p of 3.88a) and the fluid computational time scale (τ_Δ), defined as follows:

$$St_\Delta \equiv \tau_p/\tau_\Delta. \tag{8.36b}$$

$$\tau_\Delta = \min(\Delta t, \Delta x/u). \tag{8.36c}$$

The fluid computational time scale (τ_Δ) will generally be driven by the fluid time step for unsteady flows and by the fluid cell size/velocity for steady flow. This time scale represents the time over which a significant change in $\mathbf{u}_{@p}$ can occur. Based on the resulting cell-based Stokes number, two approaches may be reasonable for the particle relative velocity:

$$St_\Delta < 1 \textit{ and } w < u \qquad \textit{drift-flux approach.} \tag{8.37a}$$

$$St_\Delta \ll 1 \textit{ and } w \ll u \qquad \textit{mixed-fluid approach.} \tag{8.37b}$$

The *drift-flux approach* assumes that the particle relative velocity is steady (due to lack of dynamics) and is generally set as the terminal velocity (3.93). For the *mixed-fluid approach*, the particle relative velocity can be neglected altogether, such that both phases are assumed to move in unison. Notably, the mixed-fluid and drift-flux approaches both eliminate the need for an Eulerian PDE for particle momentum. As such, they are more numerically efficient than approaches with point-force dynamics, and should be employed when appropriate.

The differences between the mixed-fluid and separated-fluid approaches are conceptually illustrated in Figure 8.7, and details for these two approaches are given in the following subsections.

Mixed-Fluid Approach

The mixed-fluid approximation assumes that the dispersed phase and the continuous phase have effectively the same velocity, that is, the differences between the two phases are negligible in comparison to variations of the overall flow field. The mixed-fluid approach has several other names, including *locally homogeneous flow*,

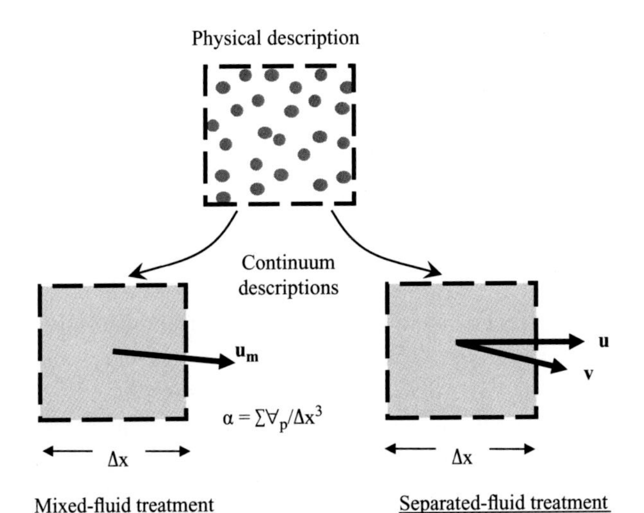

Mixed-fluid treatment:

- Uses $\mathbf{u}_m$ throughout (w«u)
- Does not employ particle diameter, shape, relative velocity, etc.
- One set of PDEs for mixed-fluid
- For very small particles with negligible inertia ($St_\Delta \ll 1$)

Separated-fluid treatment:

- Allows $\mathbf{v} \neq \mathbf{u}$ and uses both throughout
- Includes relative velocity effects such as drag, lift, St influence, etc.
- PDEs needed for each phase field
- Particles only limited by grid size ($d < \Delta x$)

Figure 8.7 Comparison of Eulerian mixed-fluid and separated-fluid treatments for a computational cell relative to the actual physical description.

single-fluid scalar transport, *modified-density*, and *phase equilibrium*. For particles in air, it is also called the *dusty gas approach* (Balachandar and Eaton, 2010). All these approaches employ a single velocity field for both the particles and the surrounding fluid. If there are also thermal variations for the surrounding fluid, this equilibrium is generally extended to the temperatures of the two phases since thermal response times and momentum response times are generally similar for particles (3.115). Using these assumptions, the mixed-fluid values are defined as equal to the respective continuous-phase and particle values at the centroid locations, as follows:

$$\mathbf{u}_m(\mathbf{X},t) \equiv \left[\mathbf{u}_{@p}(\mathbf{X},t) = \mathbf{v}(\mathbf{X},t)\right] \qquad (8.38a)$$

$$T_m(\mathbf{X},t) \equiv \left[T_{f@p}(\mathbf{X},t) = T_p(\mathbf{X},t)\right] \qquad \textit{mixed-fluid treatment.} \qquad (8.38b)$$

Such a mixed-fluid approach is equivalent to neglecting particle momentum and thermal inertias relative to the time scales of the surrounding fluid. In terms of momentum, this is consistent with the criteria of (8.37b) for very small particles. Special considerations can be included to incorporate effects of Brownian or turbulent diffusion, as discussed in §8.4.

Since both phases move at the same velocity, the mixture employs the mixed-fluid density of (5.5), which represents the combined mass of all phases per unit volume:

$$\rho_m \equiv \rho_p\alpha + \rho_f(1-\alpha). \qquad (8.39)$$

Since both phases have the same temperature, one may similarly define a mixed-fluid energy per unit mass (e_m) using the respective volumetric fractions and specific heats of each phase, as follows:

$$e_m \equiv \left[\alpha c_{\forall.p} + (1-\alpha) c_{\forall.f}\right] T_m. \tag{8.40}$$

As such, the particles and surrounding fluid are considered as an *immiscible* mixture distributed throughout a domain.

Using these mixed-fluid properties and the assumption of (8.38) results in a single set of transport Eulerian PDEs (as opposed to one set for the continuous phase and one set of the dispersed phase). In particular, the transport of mass, momentum, and energy can be obtained by combining (8.20) with (8.25), yielding the following:

$$\frac{\partial \rho_m}{\partial t} + \nabla \cdot (\rho_m \mathbf{u}_m) = 0. \tag{8.41a}$$

$$\frac{\partial (\rho_m \mathbf{u}_m)}{\partial t} + \nabla \cdot (\rho_m \mathbf{u}_m \otimes \mathbf{u}_m) = \rho_m \mathbf{g} - \nabla p + \mathbf{G}_m. \tag{8.41b}$$

$$\frac{\partial (\rho_m e_m)}{\partial t} + \nabla \cdot (\rho_m e_m \mathbf{u}_m) = \mathbf{G}_m \cdot \mathbf{u}_m - \nabla \cdot (p \mathbf{u}_m) + \nabla \cdot (k_m \nabla T). \tag{8.41c}$$

The momentum equation uses the vector operation of (2.14), while both the momentum and energy equations employ a mixed-fluid viscous stress gradient, which can be written in tensor form using the mixed-fluid velocity as follows:

$$G_{m,i} \equiv \frac{\partial K_{m,ij}}{\partial x_j} = \frac{\partial}{\partial x_j}\left[\mu_m \left(\frac{\partial u_{m,i}}{\partial x_j} + \frac{\partial u_{m,j}}{\partial x_i} - \frac{2}{3}\delta_{ij} \nabla \cdot \mathbf{u}_m\right)\right] \quad \textit{for}\; j = 1, 2, \textit{and}\; 3. \tag{8.42}$$

This expression uses the mixed-fluid viscosity (μ_m) of (5.8a), while the RHS of (8.41c) uses the mixed-fluid thermal conductivity (k_m) of (5.8b).

The mixed-fluid treatment thus includes the presence of the particles through the volume fraction and through the mixture density, viscosity, and conductivity (α, ρ_m, μ_m, and k_m). The particle volume fraction (α) is obtained by a transport equation, using (8.20a) and (8.38), as follows:

$$\frac{\partial \left(\alpha \rho_p\right)}{\partial t} + \nabla \cdot \left(\alpha \rho_p \mathbf{u}_m\right) = \alpha \dot{m}_p / \forall_p. \tag{8.43}$$

As such, the mixed-fluid approach allows a single set of PDEs that is similar to that for single-phase flow, except that a particle concentration PDE is added. This avoids the need for interphase momentum and heat transfer (e.g., no models are needed for drag force nor for collisions) and features the numerical simplicity described on the LHS of Figure 8.7.

Drift-Flux Approach

In between the separated-flow and mixed-fluid flow approaches is the drift-flux approach. This approach is also called the *partially mixed*, *terminal velocity*, *weakly separated*, or *Eulerian equilibrium* approach (Balachandar and Eaton, 2010). The simplest and most common version of the drift-flux method assumes that the relative velocity is based on the balance of drag force relative to gravitational or centrifugal forces (Ishii, 1975 and Johansen et al., 1990). In the case of the gravitational forces, the relative velocity is simply approximated by the terminal velocity (via a point-force drag model) as follows:

$$\mathbf{v}(\mathbf{X},t) = \mathbf{u}_{@p}(\mathbf{X},t) + \mathbf{w}_{term}. \tag{8.44}$$

This approach effectively serves as the particle momentum equation so the dispersed phase (for no mass transfer) only requires an Eulerian particle concentration PDE:

$$\frac{\partial(\alpha\rho_p)}{\partial t} + \nabla \cdot \left[\alpha\rho_p\left(\mathbf{u}_{@p} + \mathbf{w}_{term}\right)\right] = 0. \tag{8.45}$$

Additional terms can be added to the particle drift velocity to incorporate weak particle acceleration (Rani an Balachandar, 2003; Zhao et al. 2009) as well as effects of turbulence (discussed further in §8.4.2). This drift-flux method is nearly as efficient as the mixed-fluid method (which also requires a particle concentration transport equation), but it also captures mean relative velocity effects. However, the drift-flux approach cannot capture any effects associated with particle dynamics due to fluid variations or particle collisions.

8.3 Point-Force, Distributed-Force, and Resolved-Surface Approaches

The previously discussed methods use a surface force on the particle ($\mathbf{F}_{surf}$) based on a relative velocity as defined by the point-force velocity fields of (1.15). Numerically, this requires that the particles be smaller than the grid scale, per Figure 8.8a and the criterion of (8.6b):

$$d < \Delta x \qquad \textit{to allow point-force treatment}. \tag{8.46}$$

In addition, incorporation of two-way and three-way coupling requires that the particle concentration be well posed with several particles per cell (8.27). However, sometimes these criteria cannot be met (e.g., local flow features require high spatial resolution due to domain geometry or turbulent scales), causing particles to be of similar size or larger than the local mesh size.

In such cases, the multiphase flow can be treated with either a *distributed-force approach* when $d \sim \Delta x$ or a *resolved-surface approach* when $d \gg \Delta x$. These approaches increase the computational requirements per particle but have the advantage of providing more accurate descriptions of $\mathbf{F}_{surf}$, especially for two-way and three-way coupling effects. Because of this advantage, these approaches are sometimes employed as means to improve solution fidelity even when (8.46) can be employed. For example, predicting the drag force on a nonspherical particle close to a wall (for which there are no robust empirical point-force models) is most accurately

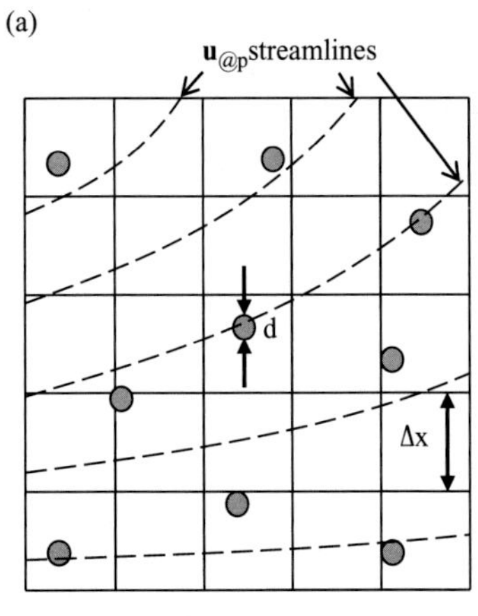

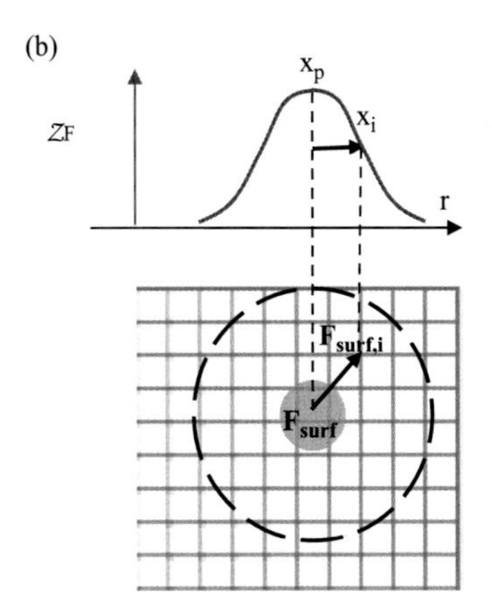

Point-force treatment

- Requires $d < \Delta \mathbf{x}$ for one-way coupling
- Requires $N_{p,\Delta} >> 1$ for two-way and three-way coupling
- Assumes definition of $\mathbf{u}_{@p}$ to obtain relative velocity (does not resolve individual particle disturbances)
- Requires models for drag, lift, etc.
- Ideal for many small particles

Distributed-force treatment

- Allows $d \sim \Delta \mathbf{x}$
- Influence of nonlinear gradients in the surrounding fluid represented with surface averages (and volume averages)
- Distributes force of particle on fluid to a distributed region for two-way coupling
- Requires models for drag, lift, etc.
- Moderate overhead for many particles

Figure 8.8 Comparison of point-force and distributed-force representations for a particle.

handled by reducing the fluid mesh size around the particle and employing a resolved-surface approach. The distributed-force and resolved-surface methods are discussed in the following two sections.

8.3.1 Lagrangian Distributed-Force Approaches

The distributed-force technique is employed when the particle diameter is on the order of the spatial resolution of the continuous phase, as follows:

$$d \sim \Delta x \qquad \textit{for distributed-force treatment.} \tag{8.47}$$

This treatment can describe the particle response to complex nonlinear flow features (whereas the point-force methods generally assume linear variations of the fluid pressure or velocity). To implement the distributed-force technique, the fluid dynamic interaction between the fluid and the particle is extended over several grid cells, as shown in Figure 8.8b.

For one-way coupling where only the fluid affects the particle motion, a distributed interaction captures the effects of flow-field variations in the vicinity of the particle. For creeping flow, Faxen (1922) showed that the quasisteady drag, history, and lift forces in a nonlinear flow are consistent with a surface average of the unhindered fluid velocity while the fluid-stress and added mass are consistent with a volume average of the unhindered fluid acceleration. Using this theoretical result, the discrete version can replace the centroid values (used for the point-force approach) with averages based on surface and volume integrals (Loth and Dorgan, 2009), as follows:

$$\mathbf{w}(t) = \mathbf{v}(t) - \mathbf{u}_{@surf}(t) \tag{8.48a}$$

$$\mathbf{u}_{@surf}(t) \equiv \frac{1}{A_p}\iint[\mathbf{u}(\mathbf{X},t)]dA_p \quad \textit{for the distributed-force approach.} \tag{8.48b}$$

$$\frac{\mathcal{D}\mathbf{u}_{@vol}}{\mathcal{D}t}(t) \equiv \frac{1}{\forall_p}\iint\left[\frac{\mathcal{D}\mathbf{u}}{\mathcal{D}t}(\mathbf{X},t)\right]d\forall_p \tag{8.48c}$$

These integrals can be discretized over the particle surface and in the volume that would otherwise be occupied by the particle to describe $\mathbf{F}_{surf}$ for one-way coupling, and can be reasonably used for particle sizes of $\Delta x < d < 5\Delta x$. While designed for creeping flow, this surface-average approach works well for Re_p at least up to 10 (Climent and Maxey, 2003).

For two-way coupling, the interphase particle force on the fluid is included in the fluid momentum PDE of (8.25b) with a distributed version of $\mathbf{F}_{surf}$. In particular, the distributed-force approach divides $\mathbf{F}_{surf}$ for a single particle into discrete portions distributed over several fluid cells in the vicinity of the particle, as shown in Figure 8.8b. Generally, a force transfer function ($\mathcal{Z}_F$) is used to make this distribution at fluid nodes ($\mathbf{X}_i$) for a particle centroid (at $\mathbf{X}_p$). This force transfer can be achieved with a clipped Gaussian function (Maxey and Patel, 2001), as follows:

$$\mathcal{Z}_F(\mathbf{X}_i,\mathbf{X}_p) = \exp\left[-\frac{1}{2}\left|\mathbf{X}_i - \mathbf{X}_p\right|^2/r^2\right] \quad \textit{for } r < 3r_p. \tag{8.49a}$$

$$\mathbf{F}_{surf,i}(\mathbf{X}_i) = \frac{\sum_{i=1}^{N_F}\mathcal{Z}_F(\mathbf{X}_i,\mathbf{X}_p)\mathbf{F}_{surf}(\mathbf{X}_p)}{\sum_{i=1}^{N_F}\mathcal{Z}_F(\mathbf{X}_i,\mathbf{X}_p)} \quad \textit{for } r < 3r_p. \tag{8.49b}$$

In this expression, r is the distance from the particle centroid. As such, the limit of $r < 3r_p$ indicates that $\mathbf{F}_{surf}$ is distributed up the three particle radii from the particle centroid. By the numerical cutoff of $r = 3r_p$, the transfer function of (8.49a) has reduced to 1.1%. In comparison, the theoretical pressure influence of the particle on the fluid reduces to 11% for creeping flow per (3.20) but reduces to 0.13% for inviscid flow per (3.47). Thus, (8.49a) is intermediate to these reductions, and the distribution form is based on numerical convenience for two-way coupling. The transfer function can also be modified to incorporate three-way coupling effects (Maxey and Patel, 2001; Reichardt et al., 2017). Similar spatial distributions can be applied for mass and heat transfer rates for a particle to the fluid mesh. However, it is difficult to capture effects of Re_p, particle shape, and particle–wall interactions and any particle–particle collision with the distributed-force approaches. Therefore, a more accurate technique to handle these effects is to use a resolved-surface approach, as discussed in the next section.

8.3.2 Resolved-Surface Approaches

For large or complex particle interactions, the resolved-surface approach can be used. In this case, the particle surface boundary (e.g., nonslip for a solid particle) is explicitly specified in the computational domain, and the detailed local flow over the particle surface is numerically discretized and computed. Therefore, the concepts of unhindered fluid velocity field and particle relative velocity are no longer employed, and instead the resolved fields of (1.12) are used:

$$\mathbf{U} \equiv \text{continuous-phase velocity external to the particle surface.} \tag{8.50a}$$

$$\mathbf{V} \equiv \text{dispersed-phase velocity within the particle surface.} \tag{8.50b}$$

The fields can be solved via the PDEs as discussed in Chapter 2. The associated discrete shear stresses and pressure stresses on the particle surface can then be computationally integrated them via (3.4) to predict $\mathbf{F}_{surf}$. This method, also called the *full DNS approach*, requires sufficient local computational resolution to resolve the flow details over the particle surface. As such, the discrete cells must be much smaller than the particle, as follows:

$$\Delta x \ll d \textit{ for resolved-surface treatment.} \tag{8.51}$$

Applying this criterion in all three directions requires many local fluid nodes (hundreds or thousands) around each individual particle depending on the particle geometry and the particle Reynolds number. As such, this technique is only reasonable when the number of particles in the computational domain is modest.

The primary benefit if this approach is that it removes all empiricism in computing the drag, lift, and added-mass forces since all these effects are directly incorporated by the discrete surface integration. This allows accurate predictions of $\mathbf{F}_{surf}$, regardless of particle shape, particle Reynolds number, particle or flow acceleration, flow gradients, and so on (with no need to determine $\mathbf{u}_{@p}$ or $\mathbf{u}_{@surf}$). In addition, internal fluid dynamics for a drop or bubble can be directly simulated using internal discretization to better describe the surface boundary conditions via any internal recirculation. Furthermore, two-way, three-way, and four-way coupling are automatically included in the resolved-surface approach since the interstitial fluid and contact dynamics can be fully resolved as well. As such, the resolved-surface technique is the most accurate and physically realistic surface force method but is also the most computationally intensive.

As shown in Figure 8.9, the two primary treatments for the resolved-surface method are the *Gridded Interface Method* (GIM) and the *Immersed Interface Method* (IIM). The GIM approach uses a computational grid that is fitted to the surface on the particles, as shown in Figure 8.9b, while IIM allows a single structured Eulerian grid for the entire computational domain, as shown in Figure 8.9c. Details of these two approaches are discussed in the following.

The Gridded Interface Method approach is typically used when the particle shape is simple (e.g., a sphere or an ellipsoid) so that the grid creation around this surface is

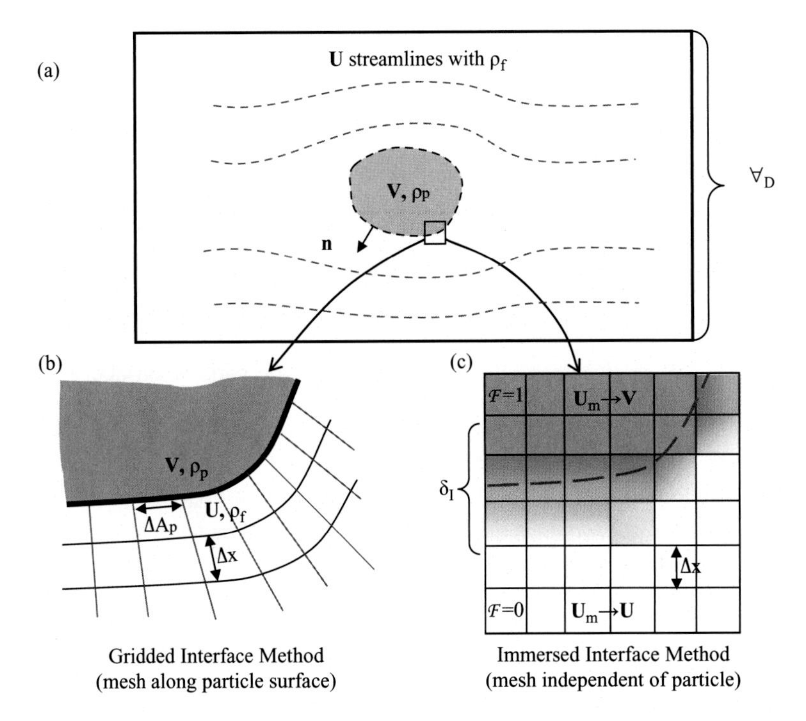

Resolved-surface treatments

- Requires d » **Δx** (high CPU/particle)
- Particle surface force automatically captured by flow around particle
- Does not require empirical models for surface forces
- Ideal for complex and/or deforming particle shapes and complex flow

Figure 8.9 Resolved-surface approaches showing (a) a schematic of a particle in a computational domain, along with (b) near-surface close-ups of a GIM mesh and (c) of a IIM mesh superimposed on the marker function distribution.

straightforward. For a solid particle in a continuum fluid, the PDEs for **U** and P are then solved with a no-slip boundary condition at the particle radius. An example GIM solution is shown in Figure 8.10a for a spherical particle, where the wake details are well described. For a fluid particle with significant internal recirculation, the particle interior can also be discretized and solved along with the PDEs for **V** and P_p, using boundary conditions that match the interfacial stresses on the particle surface with **U** and P, as in (3.32) for steady flow. Once the flow field around the particle surface is known, the surface force can be determined by numerically integrating the pressure and the continuous-phase shear stress over the discretized surface elements:

$$\mathbf{F}_{surf} = \sum_{k=1}^{N_k} \left(-P\mathbf{n} + K_{ij}n_j\right)_k \Delta A_{p,k}. \tag{8.52}$$

In this equation, ΔA_p is a discrete particle surface area element. This can be combined with the Lagrangian Equation of Motions (8.1b) to determine the particle acceleration

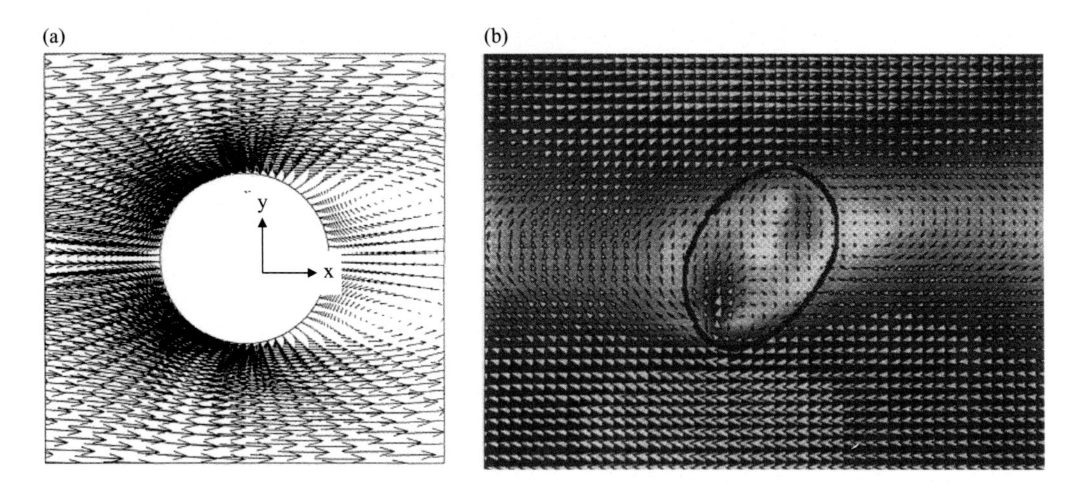

Figure 8.10 Examples of resolved-surface velocity fields relative to particle centroid velocity: (a) flow past a solid spherical particle using GIM for **U** vectors (Kurose and Komori, 1999); and (b) a deforming bubble near an eddy center using IIM for $\mathbf{U}_\mathrm{m}$ vectors (Loth et al., 1997).

and new particle velocity so that (8.1a) can be used to move the particle to the next position (and the grid adapted to this new position).

The Immersed Interface Method has the advantage that the grid does not need to adapt to the particle surface, since the particle surface boundary conditions are enforced within a cell. This enables the flow PDEs to be solved efficiently on simple Cartesian grids and handle complex deformations. An example IIM solution is shown in Figure 8.10b, where a complex bubble shape interacts with a vortex. To embed the surface within the fluid cells, a common approach is to use a *marker function* ($\mathcal{F}$), defined as the fraction of particle matter within a cell. This function is defined throughout the domain and is designed to vary smoothly across a numerically thickened interface. As such, $\mathcal{F} = 0$ outside of the particle, $\mathcal{F} = 1$ inside the particle, and the mixed-fluid density and viscosity for any cell are given as follows:

$$\rho_\mathrm{m} = (1 - \mathcal{F})\rho_\mathrm{f} + \mathcal{F}\rho_\mathrm{p}. \tag{8.53a}$$

$$\mu_\mathrm{m} = (1 - \mathcal{F})\mu_\mathrm{f} + \mathcal{F}\mu_\mathrm{p}. \tag{8.53b}$$

The mixed-fluid properties can then be used to establish the PDEs for a mixed-fluid velocity ($\mathbf{U}_\mathrm{m}$), which reverts to the surrounding fluid or particle velocities outside of the interface:

$$\mathcal{F} = 0 \quad \rightarrow \quad \mathbf{U}_\mathrm{m} = \mathbf{U} \ \ (\textit{outside of the particle}). \tag{8.54a}$$

$$\mathcal{F} = 1 \quad \rightarrow \quad \mathbf{U}_\mathrm{m} = \mathbf{V} \ \ (\text{inside of the particle}). \tag{8.54b}$$

As discussed by Prosperetti and Tryggvason (2007), the numerical interface thickness is prescribed to be as thin as possible while still allowing a stable continuum variation of velocity, density, and viscosity across the grid. This typically results in a prescribed thickness of two to four computational cells, where larger thicknesses are generally needed when there is a large variation in density (e.g., between a gas and a liquid). For deformable fluid particles, the surface tension is then added as an embedded force within the interface thickness that depends on the local curvature. For solid particles, sharp interface methods can be used to reduce the interface thickness to a single computational cell since the interior velocity field is no longer needed (Gibou et al., 2019).

8.3.3 Computational Node Requirements

In the following, we consider the computational requirements of particle numerical approaches based on the particle characteristics (size, speed, and number). In a broad sense, there are three primary numerical approaches for particles as shown in Figure 8.11: Eulerian treatment (using cell-averaged particle concentration), Lagrangian point-force treatment (using trajectories based on a force model and a relative velocity), and resolved-surface approaches (using discrete elements around the particle surface to determine the fluid dynamic surface forces). As shown in this figure and discussed earlier, the Eulerian treatment is more appropriate for a large number of small particles, while the resolved-surface treatment is more appropriate for a few larger particles. To provide a more detailed guidance for selection, it is helpful to estimate the required computational resources in terms of the number of nodes for the dispersed phase (N_d) by comparing the number of particles (N_p) relative to the number of fluid nodes (N_f).

When the number of particles is high relative to the fluid nodes ($N_p \gg N_f$), an Eulerian approach for particle concentration is often preferred, so the dispersed-phase characteristics can be computed on the same computational mesh (and same spatial resolution) as that of the continuous phase, and thus both phases will have the same number of nodes:

$$N_d = N_f \qquad \textit{Eulerian point-force approach} \left(\textit{for}\, N_p \gg N_f\right). \tag{8.55}$$

Notably, N_d is independent of the total number of particles (N_p) in the domain, allowing high computational efficiency per particle. For example, a combustion spray may include 10^{12} drops (N_p) at a given time, while discretization of the combustion chamber may only require 10^7 fluid cells (N_f). In this case, there are about 10^5 particles per cell, so the Eulerian-Eulerian point-force approach is highly efficient. The computational resources can be reduced even further for small particle inertias and relative velocities (per 8.37) by using the mixed-fluid or drift-flux approaches so that only the particle concentration PDEs must be solved.

However, if one wishes to capture particle–wall effects (§8.1.5) and/or more accurately capture diffusion (to be discussed in §8.4) with many particles ($N_p \gg N_f$), a Lagrangian parcel approach can be used, where the parcel node count (N_P) typically scales with the fluid nodes, as follows:

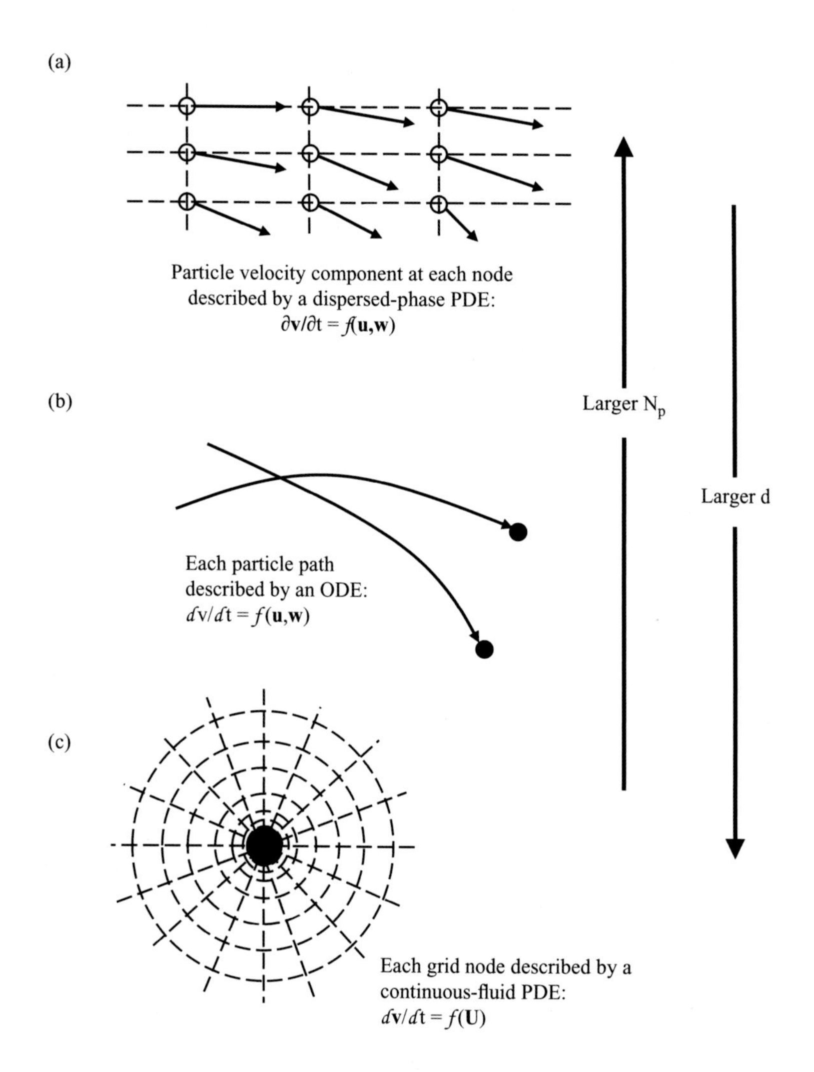

Figure 8.11 Comparison of particle treatments for (a) an Eulerian approach defined on Eulerian computational nodes; (b) a Lagrangian approach defined on particle centroids; and (c) a Lagrangian resolved-surface approach with a surface-fitted grid

$$N_d = N_P \sim N_f \qquad \textit{Lagrangian parcel point-force approach } \left(N_p \gg N_f\right). \qquad (8.56)$$

As such, the computational efficiency per particle is similar to the Eulerian approach of (8.55), but the Lagrangian approach avoids unphysical numerical diffusion of particle concentration (§8.1.2) and can more efficiently describe a broad particle size distribution. However, accurately incorporating two-way coupling with the Lagrangian approach generally requires several parcels per computational cell (8.27b), for example:

$$N_d = N_P \sim 8N_f \qquad \textit{Lagrangian parcel point-force for two-way coupling}. \qquad (8.57)$$

The RHS value of 8 is rough approximation and depends on the multiphase flow details (the appropriate value for a given flow can be obtained with a parcel resolution study). This result indicates that Eulerian approaches are often more efficient when two-way coupling dominates.

If the number of particles is similar to or less than the number of fluids nodes ($N_p \sim N_f$ or $N_p < N_f$) but is still large ($N_p \gg 1$), then it is generally more accurate and efficient to track each particle trajectory using a point-force approach with models for $\mathbf{F}_{surf}$, as follows:

$$N_d = N_p \qquad \textit{Lagrangian particle point-force approach}\ \left(N_p \gg 1\right). \tag{8.58}$$

This allows direct capturing of collision with other particles and walls and can more readily incorporate two-way and three-way coupling effects when there are many small particles.

For all the previous point-force approaches, the computational resources for the particles are assumed to scale with the number of dispersed nodes. However, a resolved-surface approach does not employ discrete nodal values for the particles, but instead changes N_f due to the presence of the particles.

When there are a few large particles, a resolved-surface approach can be used where each particle trajectory is based on discrete integration of its surface forces. In this case, several continuous-phase nodes are required on the particle surface and in the fluid dynamic interaction region. Relative to the number of fluid nodes required in absence of the particles (N_{fo}), the actual number of fluid nodes will be greater when the particles are added. The extra number of nodes required per particle depends strongly on the Reynolds number and shape of the particles. As a rough estimate, a spherical particle at moderate Reynolds numbers may require about 10^4 fluid nodes, while a similar number of nodes may be needed for complex shaped particles at lower Re_p. Based on this rough guide, the total number of fluid nodes can be estimated as follows:

$$N_f = N_{f,o} + 10^4 N_p \qquad \textit{Resolved-surface particle approach.} \tag{8.59}$$

Because the total number of fluid nodes can grow quickly with the number of particles, this approach is generally only reasonable for a small number of particles (e.g., $N_p < 10^3$).

The guides to select an approach are summarized in Figure 8.12. Particle number, size, and speed are the primary determinants, and the required computational resources tend to increase as the number and/or size of the particles increase.

8.4 Simulating Diffusion of Particles

In general, particles can be diffused by Brownian motion and by turbulence, and this leads to additional numerical considerations. As discussed in the following two subsections, these two mechanisms employ similar numerical techniques for mean diffusion.

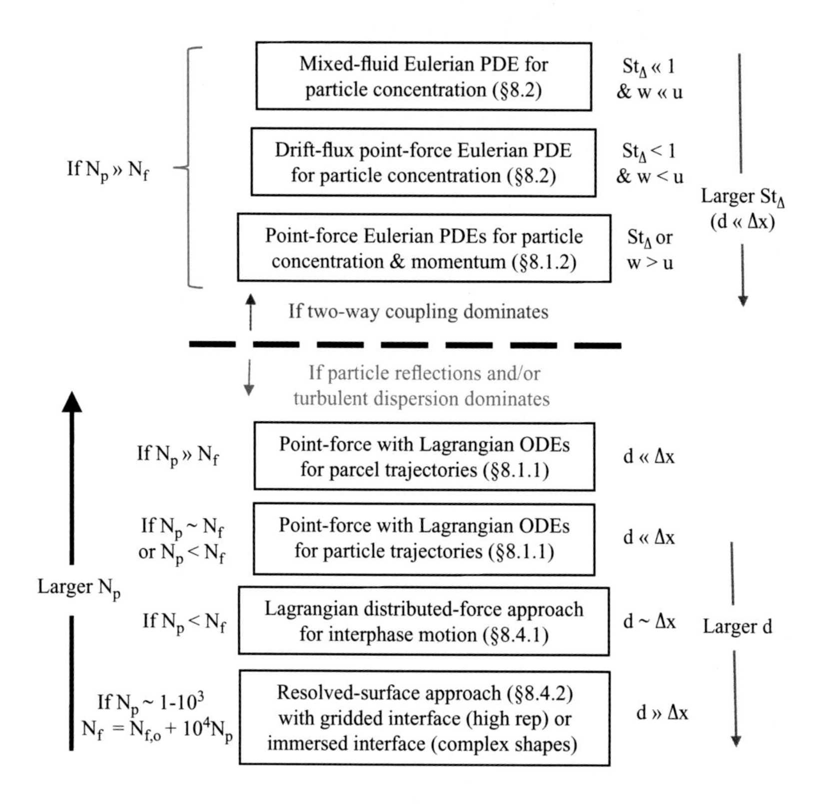

Figure 8.12 Computational particle approaches as a function of the number of particles in the domain (N_p), the number of fluid nodes in the domain (N_f), grid-based Stokes number (St_Δ), the particle diameter (d), and continuous-phase grid resolution (Δx).

8.4.1 Point-Force Brownian Diffusion of Particles

To capture Brownian motion (which can be significant for small particles), the mean diffusion associated with the random molecular fluid interactions can be numerically predicted based on the theoretical models of §5.3 for both Eulerian and Lagrangian particle reference frames. For an Eulerian particle representation, Brownian diffusion can be applied to the particle mass transport PDE of (8.20a) in a manner similar to that for species diffusion (2.33a). The resulting particle transport PDEs based on long-time diffusion (5.32) becomes

$$\frac{\partial\left(\alpha\rho_p\right)}{\partial t} + \nabla\cdot\left(\alpha\rho_p\mathbf{v}\right) = \frac{\alpha}{\forall_p}\dot{m}_p + \nabla\cdot\left[\Theta_{Br}\nabla\left(\alpha\rho_p\right)\right]. \tag{8.60a}$$

$$\Theta_{Br} = \frac{\kappa T_f}{3\pi d\mu_f}. \tag{8.60b}$$

This Eulerian approach captures the Brownian diffusion directly and can use the same grid and numerical schemes used for the surrounding fluid. However, this Eulerian approach can be subject to numerical diffusion errors (§8.1.2), which can be avoided with a Lagrangian approach for the particle Brownian diffusion.

The most common Lagrangian approach for Brownian diffusion employs a stochastic force with appropriate statistical averaging. In particular, the particle Equation of Motion of (3.88c) can be extended to include the random Brownian force ($\mathbf{F}_{Br}$) based on (5.24), as follows:

$$\frac{d\mathbf{v}}{dt} = -\frac{\mathbf{v} - \mathbf{u}_{@p}}{\tau_p} + (1 - R)\mathbf{g} + R\frac{d\mathbf{u}_{@p}}{dt} + \frac{\mathbf{F}_{Br}}{m_p}. \tag{8.61}$$

In this expression, the fluid acceleration is taken along the particle path (reasonable per §3.3.2) for numerical convenience. After the continuous phase is marched forward from t^n to t^{n+1} based on a time step Δt, a first-order implicit scheme can be applied to (8.61) per Ounis et al. (1991) to march the particles as follows:

$$\frac{\mathbf{v}^{n+1} - \mathbf{v}^n}{\Delta t} = \frac{\mathbf{u}^{n+1}_{@p} - \mathbf{v}^{n+1}}{\tau_p} + (1 - R)\mathbf{g} + \frac{\mathbf{u}^{n+1}_{@p} - \mathbf{u}^{n}_{@p}}{\Delta t} + \frac{\mathbf{F}^{n+1}_{Br}}{m_p} + O(\Delta t). \tag{8.62a}$$

$$\frac{\mathbf{v}^{n+1} - \mathbf{v}^n}{\Delta t} = \frac{\mathbf{u}^{n+1}_{@p} - \mathbf{v}^{n+1}}{\tau_p} + \mathbf{g} + \frac{\mathbf{F}^{n+1}_{Br}}{m_p} + O(\Delta t) \qquad \textit{for } \rho_p \gg \rho_f. \tag{8.62b}$$

Note that (8.62a) is for a generalized particle density and (8.62a) is simplified for high-density particles (e.g., when the surrounding fluid is a gas). These equations can be rearranged to explicitly solve for the new particle velocity ($\mathbf{v}^{n+1}$), and the isotropic Brownian force can be modeled with a random Wiener process that satisfies the diffusion of (8.60). The resulting discretization written in component (instead of vector) form for the high-density particle case is as follows:

$$v_i^{n+1} = \frac{v_i^n \tau_p + u^{n+1}_{@p,i}\Delta t + g_i(\tau_p \Delta t) + \zeta_i\sqrt{2\Theta_{p,Br}\Delta t}}{\tau_p + \Delta t} \qquad \textit{for } \rho_p \gg \rho_f. \tag{8.63}$$

The RHS includes a random number (ζ), which is defined to have zero mean and unity variance:

$$\overline{\zeta_i} = 0. \tag{8.64a}$$

$$\zeta_{i,rms} = 1. \tag{8.64b}$$

A numerical random number generator is then used to sample ζ values for each time step and for each component of velocity. Predictions with this stochastic approach generally require many particles and many time steps to ensure statistical convergence, since the time step should be small fraction of the particle response time in order resolve the particle inertia effect ($\Delta t \ll \tau_p$). This can lead to a large number of time steps for very small particles, which may be a reason to instead use the Eulerian approach of (8.60). However, the Lagrangian approach of (8.63) has the benefit of efficiently handling broad particle size distributions (§8.1.3) and can explicitly capture particle collisions with other particles as well as reflections from walls (§8.1.5).

8.4.2 Point-Force RANS-Based Turbulent Particle Diffusion

Diffusiophoresis and Turbophoresis

As discussed in §7.4.3, diffusiophoresis causes particles to diffuse from regions of high particle concentration to regions of low particle concentration, while turbophoresis causes particles to move from regions with high fluid TKE to regions with low TKE. As with Brownian diffusion, numerical approaches for these two turbulent diffusion mechanisms for particles with RANS can employ (a) Eulerian approaches to add a diffusion term the particle concentration PDE, or (b) a Lagrangian approach to add a random term to the particle momentum ODE.

If the particles (dispersed phase) are treated with an Eulerian approach with a RANS flow, we employ a time average on (8.20a). For constant particle density, this yields the following:

$$\frac{\partial \overline{\alpha}}{\partial t} + \nabla \cdot (\overline{\alpha}\,\overline{\mathbf{v}}) = \overline{\alpha \dot{m}_p}/m_p - \nabla \cdot \overline{\alpha' \mathbf{v}'}. \tag{8.65}$$

The first term on the RHS represents the time-averaged mass transfer. The mass transfer rate can then be modeled with (8.5), where the Sherwood number is based on the time-averaged Reynolds number (which can be estimated from 7.80). The second term on the RHS represents the turbulent diffusion of the particles and can be treated with a Fickian diffusion model as with the species per (6.38) but using the particle diffusivity (Θ_p), as follows:

$$\overline{\alpha' \mathbf{v}'} = -\Theta_p \nabla \overline{\alpha}. \tag{8.66}$$

This diffusivity can be modeled based on long-time particle diffusion relative to species diffusion (7.51). Substituting this into (8.65) and using the species diffusivity of (6.123) and the relationship of (6.81a) yields the following:

$$\frac{\partial \overline{\alpha}}{\partial t} + \nabla \cdot (\overline{\alpha}\,\overline{\mathbf{v}}) = \nabla \cdot \left[\Theta_p \nabla \overline{\alpha}\right] + \frac{2}{3}\overline{\alpha}\tau_p \nabla \cdot \left[\nabla k_p\right]. \tag{8.67a}$$

$$\Theta_{p,i} = \frac{2c_\varepsilon k^2}{3\varepsilon}\left[1 + \left(\frac{\langle w \rangle c_\varepsilon k}{\varepsilon \Lambda_i}\right)^2\right]^{-1/2} \quad \textit{where}\ i = \parallel \text{ or } \perp. \tag{8.67b}$$

The two terms on RHS of (8.67a) effectively capture diffusiophoresis (using 8.67b) and turbophoresis (using 7.114a) with an Eulerian approach. These same diffusion terms can be added to the Eulerian approach for the drift-flux model (Zhao et al., 2009).

For a Lagrangian dispersed-phase approach, one may employ the instantaneous ODEs of (8.1) combined with a stochastic approach to mimic the velocity fluctuations seen by the particles. This is similar in principle to the Brownian Lagrangian approach (8.60)–(8.64), except that the diffusion is not generally isotropic (per 8.67b). If we decompose the fluid velocity into mean and fluctuating components, the high-density particle Equation of Motion becomes

$$\frac{d\mathbf{v}}{dt} = -\frac{\mathbf{v}-\overline{\mathbf{u}}}{\tau_p} + \mathbf{g} + \frac{\mathbf{u}'}{\tau_p} \quad \textit{for } \rho_p \gg \rho_f. \tag{8.68}$$

Applying the first-order scheme of (8.68) yields the following:

$$\frac{\mathbf{v}^{n+1}-\mathbf{v}^{n}}{\Delta t} = \frac{\overline{\mathbf{u}}^{n+1}-\mathbf{v}^{n+1}}{\tau_p} + \mathbf{g} + \frac{\mathbf{u}'^{n+1}}{\tau_p} + O(\Delta t). \tag{8.69}$$

Rearranging to solve for the new particle velocity ($\mathbf{v}^{n+1}$) and again applying a random Wiener process that satisfies (8.67b) as used in (8.63), the high-density particle result is as follows:

$$v_i^{n+1} = \frac{v_i^n \tau_p^n + \overline{u}_i^{n+1}\Delta t + g_i\left(\tau_p^n \Delta t\right) + \zeta_i\sqrt{2\Theta_{p,i}\Delta t}}{\tau_p^n + \Delta t} \quad \textit{for } \rho_p \gg \rho_f. \tag{8.70}$$

Note that this expression allows the particle response time to be updated at each time step to account for nonlinear drag. For a large number of time steps, this result also ensures the long-time anisotropic particle diffusion for HIST is realized. Again, the time step should be a small fraction of the particle response time, but the Lagrangian approach of (8.70) avoids numerical diffusion (§8.1.2), efficiently captures polydispersity (§8.1.3) and can explicitly incorporate collisions with other particles as well as reflections from walls (§8.1.5).

To improve the Lagrangian approach performance in regions with high gradients of turbulent kinetic energy and to allow for anisotropic turbulence, several higher-order stochastic forms for the turbulent diffusion have been developed. A popular model is the continuous random walk (CRW), where the fluid fluctuating velocity of (8.69) is modeled using a Markov chain (Legg and Raupach, 1982). The CRW approach again uses a random number (ζ) to create synthetic turbulence but also employs the previous fluid velocity fluctuation, the rms of the velocity fluctuations in each direction, and a correlation coefficient based on the integral time scale seen by the fluid ($\tau_{@p,i}$, defined in 7.48):

$$u_i^{n+1} = \overline{u}_i^{n+1} + u_i'^{n}\Upsilon_i + \left(u'_{i,rms}\zeta_i\right)\sqrt{1-\Upsilon_i^2} \quad \textit{for Lagrangian CRW approach.} \tag{8.71a}$$

$$\Upsilon_i = \exp\left(-\Delta t/\tau_{@p,i}\right) \tag{8.71b}$$

There are two key advantages of this approach (relative to the random Wiener process). First, the rms of the fluid velocity fluctuations can be anisotropic, which improves predictive performance of particle diffusion in NAsT flows (Bocksell and Loth, 2006). Second, the equation allows more generalized particle densities and drag models. In particular, the vector forms of the drag force (3.55a) and the relative velocity (1.15d) can be expressed as follows:

$$\mathbf{F}_D^{n+1} = -\frac{1}{8}\pi d^2 \rho_f \mathbf{w}^{n+1} w^{n+1} C_D^{n+1}. \tag{8.72a}$$

$$\mathbf{w}^{n+1} = \mathbf{v}^{n+1} - \left(\overline{\mathbf{u}}^{n+1} + \mathbf{u}'^{n+1}\right). \tag{8.72b}$$

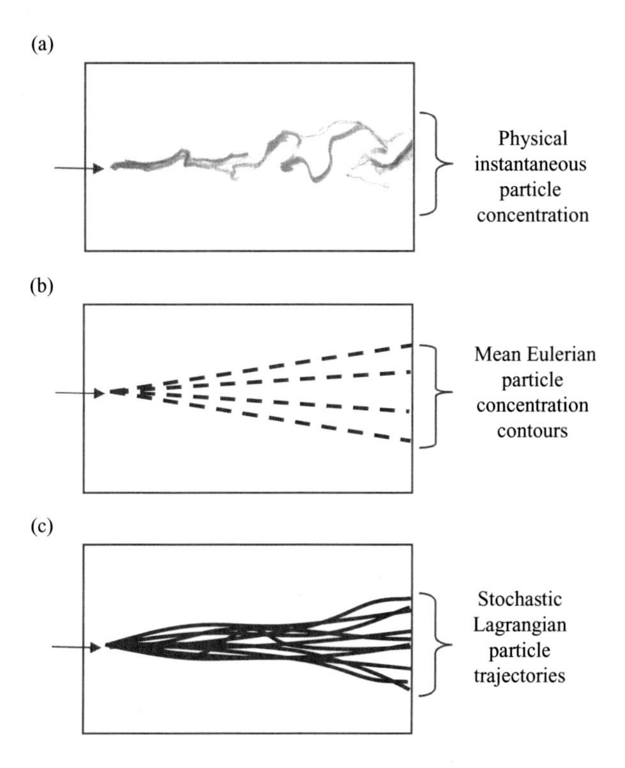

Figure 8.13 Particle point-force treatments in turbulent flow for particles released at the arrow location: (a) actual instantaneous particle distribution; (b) Eulerian mean diffusion based on θ_p and a steady RANS solution; and (c) Lagrangian stochastic diffusion based on a steady RANS solution with random numbers to represent turbulent fluctuations.

At each time step, the drag coefficient of (8.72b) can be computed using a drag model (e.g., 3.60) based on the instantaneous relative velocity of (8.72b). Furthermore, the continuity of the fluid velocity fluctuations between time steps provided by (8.71a) allows discrete fluid velocity accelerations so the CRW can be applied to low-density particles, such as bubbles, which include acceleration-based forces (Yang et al., 2021).

A comparison of the preceding Eulerian and Lagrangian turbulent diffusion approaches for a RANS flow is given schematically in Figure 8.13. The physical particle dispersion is due to a complex set of turbulent eddies with a broad range of lengths and time scales (Figure 8.13a). The Eulerian diffusion approach only predicts the time-averaged spread (Figure 8.13b), while the Lagrangian approach uses synthetic velocity fluctuations (no turbulent structures) with a statistically large number of particle paths (Figure 8.13c). Notably, both Eulerian and Lagrangian RANS approaches depend on RANS predictions of the continuous-phase mean velocity and turbulent kinetic energy. In terms of accuracy, this presents at least two problems. First, continuous-phase RANS predictions for mean velocities, turbulent kinetic energy, dissipation, etc., have been found to be problematic for complex geometries and/or NAsT flows (§6.2.3). Secondly, turbulent bias effects beyond diffusiophoresis and turbophoresis, as listed in (7.85), must be modeled and added to be included, as discussed in the following subsection.

Nonlinear Drag and Preferential Biases

As discussed in §7.4.1, nonlinear drag bias reduces the average particle relative velocity as turbulence increases. When treating the particles with an Eulerian numerical approach (including the drift-flux approach), this bias can be incorporated via an effective drag coefficient based on (7.94b), as follows:

$$\mathrm{F_D} = -\frac{1}{8}\pi d^2 \rho_f \langle \mathbf{w} \rangle \langle \mathrm{w} \rangle \langle \mathrm{C_D} \rangle \qquad (8.73a)$$

$$\langle \mathbf{w} \rangle = \mathbf{v} - \overline{\mathbf{u}}_{@p} \qquad \textit{for an Eulerian RANS approach.} \qquad (8.73b)$$

$$\frac{\langle \mathrm{C_D} \rangle}{\mathrm{C_D}} = 1 + \frac{u_\Lambda (Y-1)}{\langle \mathrm{w} \rangle}\sqrt{\frac{\mathrm{St}_\parallel}{1+\mathrm{St}_\parallel} + \frac{2\mathrm{St}_\perp}{1+\mathrm{St}_\perp}} \qquad (8.73c)$$

The drag force of (8.73a) employs an ensemble-averaged relative velocity (8.73b) and an effective drag coefficient (8.73c) that increases with turbulence (u_Λ) for nonlinear drag conditions, which occurs for $Y > 1$, where Y is given by (4.20). The unbiased C_D in the LHS denominator of (8.73c) employs a conventional drag model (e.g., 3.60) with an Re_p based on the relative velocity of (8.73b). As such, (8.73) models the nonlinear drag bias for an Eulerian approach. Fortunately, particles treated with an Lagrangian approach do not need to model this bias since fluid velocity fluctuations are included via synthetic turbulence as in (8.71).

As discussed in §7.4.2, preferential bias occurs for small particles when the Kolmogorov Stokes numbers are of order unity. This bias causes a net increase in particle velocity in the gravity direction and is not explicitly incorporated into either the Eulerian or the Lagrangian approaches for RANS, since such an approach does not incorporate Kolmogorov eddy structures. The preferential bias can instead be modeled within a RANS framework as follows:

$$\langle \mathbf{w} \rangle^{n+1} = \mathbf{v}^{n+1} - \overline{\mathbf{u}}_{@p} + \Delta \mathbf{v}_{pref} \qquad \textit{for Eulerian particles.} \qquad (8.74a)$$

$$\mathbf{w}^{n+1} = \mathbf{v}^{n+1} - \left(\overline{\mathbf{u}}^{n+1} + \mathbf{u}'^{n+1}\right) + \Delta \mathbf{v}_{pref} \qquad \textit{for Lagrangian particles.} \qquad (8.74c)$$

The RHS can then use an appropriate bias model, such as (7.104) for high-density particles. Note that cluster bias can also be included in this form, but robust models are not currently available.

While these approaches address turbulent biases for a RANS approach, the most accurate technique to incorporate these effects is to employ DNS, since this provides all the scales of the turbulence with high fidelity. Unfortunately, DNS requires very high computational resources at high Reynolds numbers (see Figure 6.33). Fortunately, an LES approach is not as computationally intensive as DNS and resolves the integral-scale turbulence (unlike RANS). However, LES may still require some diffusion and bias modeling for subgrid aspects, as discussed in the next section.

8.4.3 Point-Force LES-Based Turbulent Particle Diffusion

As discussed in §6.3.2, Large Eddy Simulation (LES) is achieved by spatially filtering the fluid PDEs with a filter length scale, $\Delta_{\mathcal{G}}$. Turbulent structures with sizes above this filter length are computed directly, while smaller turbulent structures are described using a subgrid turbulence model. As such, LES resolves turbulence wavelengths from D down to $\Delta_{\mathcal{G}}$ and models turbulence wavelengths in the range $\Delta_{\mathcal{G}}$ down to η. The subgrid length scale and kinetic energy can be used to define a subgrid turbulence time scale ($\tau_{\mathcal{G}}$) as follows:

$$\tau_{\mathcal{G}} \equiv \Delta_{\mathcal{G}}/k_{\mathcal{G}}^{1/2}. \tag{8.75}$$

The RHS (based on the same type of scaling using in 6.80) employs the subgrid TKE of (6.46c) and allows characterization of the LES in terms of the turbulent time scales. In particular, LES resolves time scales from τ_D down to $\tau_{\mathcal{G}}$ and models those from $\tau_{\mathcal{G}}$ down to τ_η. In general, most of the turbulent kinetic energy is captured directly with LES, so $\Lambda > \Delta_{\mathcal{G}}$ and $\tau_\Lambda > \tau_{\mathcal{G}}$.

For a surrounding fluid described by LES, the particles may be treated with either an Eulerian PDE based on particle concentration or Lagrangian ODEs based on particle path. For a particle with a drag response time of τ_p, the subgrid turbulence Stokes number ($St_{\mathcal{G}}$) can be defined as follows:

$$St_{\mathcal{G}} \equiv \tau_p/\tau_{\mathcal{G}}. \tag{8.76}$$

As such, particles with $\tau_p > \tau_{\mathcal{G}}$ have high enough inertia that they can only respond to the resolved scales so their dispersion can be captured explicitly. However, particles with $\tau_p < \tau_{\mathcal{G}}$ also can respond to the higher-frequency subgrid turbulence and thus can have subgrid particle diffusion in addition to the resolved-scale dispersion. Thus, the subgrid turbulence Stokes number influence can be summarized as follows:

$$St_{\mathcal{G}} > 1 \quad \textit{particle motion all predicted by the resolved turbulence.} \tag{8.77a}$$

$$St_{\mathcal{G}} < 1 \quad \textit{particle motion also affected by subgrid turbulence.} \tag{8.77b}$$

The particle subgrid diffusion associated with (8.77b) can be incorporated with Eulerian or Lagrangian approaches.

For an Eulerian treatment of the particles, an unsteady particle concentration PDE can include a subgrid particle diffusion model based on an isotropic version of (8.67b) and the dissipation relationship of (6.46e) for negligible inertia as follows:

$$\frac{\partial\widehat{\alpha}}{\partial t} + \nabla\cdot\left(\widehat{\alpha}\,\widehat{\mathbf{v}}\right) = \widehat{\alpha}\;\dot{m}_p/m_p + \Theta_{p,\mathcal{G}}\nabla\widehat{\alpha}. \tag{8.78a}$$

$$\Theta_{p,\mathcal{G}} = \frac{2}{3}c_\varepsilon k_{\mathcal{G}}^2/\varepsilon = \frac{2}{3}c_\varepsilon k_{\mathcal{G}}^{1/2}\Delta_{\mathcal{G}}. \tag{8.78b}$$

The Fickian diffusivity model of (8.78b) assumes isotropic diffusion (reasonable at small scales) and employs the relationship of (8.75).

For a Lagrangian treatment of the particles, one may employ a Weiner approach based on (8.70) to incorporate subgrid synthetic turbulence and inertia as follows:

$$v_i^{n+1} = \frac{v_i^n \tau_p + \widehat{u}_i^{n+1}\Delta t + g_i(\tau_p \Delta t) + \zeta_i \sqrt{2\Theta_{p,\mathcal{G}}\Delta t}}{\tau_p + \Delta t} \qquad \textit{for}\, R \to 0. \tag{8.79}$$

This LES subgrid model samples a random number at each time step for ζ.

The turbulent biases discussed in §8.4.2 are automatically captured with LES if $St_{\mathcal{G}} > 1$. For $St_{\mathcal{G}} < 1$, the Lagrangian and Eulerian approaches for particles in LES can use bias approaches similar to those used for RANS. For example, nonlinear drag bias for particles treated with an Eulerian numerical approach can be incorporated as in (8.73) by employing the resolved relative velocity:

$$\langle \mathbf{w} \rangle = \mathbf{v} - \widehat{\mathbf{u}}_{@p} \qquad \textit{for Eulerian LES approach.} \tag{8.80}$$

Again, particles treated with a Lagrangian approach and synthetic (random) turbulence as in (8.79) automatically account for the nonlinear drag bias. Preferential bias and turbophoresis for an LES approach can also be modeled for Eulerian and Lagrangian approaches as follows:

$$\langle \mathbf{w} \rangle^{n+1} = \mathbf{v}^{n+1} - \widehat{\mathbf{u}}_{@p} + \Delta \mathbf{v}_{pref} + \Delta \mathbf{v}_k \quad \textit{for Eulerian particles.} \tag{8.81a}$$

$$\mathbf{w}^{n+1} = \mathbf{v}^{n+1} - \left(\widehat{\mathbf{u}}^{n+1} + \underset{\smile}{\mathbf{u}}^{n+1}\right) + \Delta \mathbf{v}_{pref} \quad \textit{for Lagrangian particles.} \tag{8.81b}$$

As with RANS, turbophoresis effect is implicitly included with a Lagrangian approach, so no such term appears in (8.81b). Finally, subgrid diffusiophoresis bias for LES can be modeled as a diffusion term, as in (8.78a) or (8.79). Fortunately, most of the particle turbulent diffusion and biases will be captured by the resolved scale (regardless of St_Δ), so the bias models described here have only a secondary impact, and their accuracy is not critical for overall particle motion (and can often be neglected altogether).

8.4.4 Summary of Numerical Approaches for Particles in Turbulence

Finally, we summarize the numerical approaches for particles in turbulence based on the ability to capture various multiphase physics.

The most accurate numerical approach is to predict the particle dynamics with a resolved-surface technique (§8.3.2) and predict the turbulence with DNS (§6.3.1). The resolved-surface approach avoids empirical models or theoretical restrictions associated with the particle surface force, while DNS avoids all empiricism associated with turbulent diffusion and dispersion. Moreover, DNS exactly incorporates two-way, three-way, and four-way coupling effects. However, the resolved-surface method with DNS is an extremely computationally intensive combination. For example, a high

Reynolds number flow described with DNS combined with a large number of particles treated with a resolved-surface approach will require a very large number of computational grid points (perhaps billions) and will generally be impractical on most computing platforms.

The least computationally intensive and most common approach is to use point-force models for the particle surface forces (§8.1–§8.2) and RANS for the turbulence (§6.2). Point-force models generally rely on models for drag forces (which are generally empirical at high Reynolds numbers), while RANS does not resolve any of the turbulence (so the entire spectrum is treated with empirical coefficients and models for the integral properties). An LES approach is intermediate to the DNS and RANS approaches in terms of accuracy.

The choice of the turbulence approach strongly influences the approach for the particles and accuracy of particle turbulent diffusion (Balachandar and Eaton, 2010). For example, using a DNS approach for the fluid allows the full turbulent cascade to be resolved down to the Kolmogorov length scale (η of 6.86). The influence of these smallest flow scales on the partices is through the Kolmogorov Stokes number (St_η of 7.3), which can be related to the ratio of particle size to the Kolmogorov length scale (d/η) by assuming Stokesian drag ($f = 1$) and using (5.11) as follows:

$$\frac{d}{\eta} = \sqrt{18\frac{St_\eta}{\rho_p/\rho_f + c_\forall}} \sim \frac{d}{\Delta x} \quad \textit{for DNS treatment of the turbulence.} \tag{8.82}$$

The RHS is a result of the computational resolution needed for DNS (6.103) and also indicates the probable particle treatment per (8.37), (8.46), (8.47), and (8.51), which are repeated here:

$$\begin{aligned} &d > 10\Delta x && \textit{for resolved-surface treatment.} \\ &d \sim 3\Delta x && \textit{for distributed-force treatment.} \\ &d < \Delta x && \textit{for the point-force treatment.} \\ &St_\Delta < 1 \textit{ and } w < u && \textit{for the drift-flux approach.} \\ &St_\Delta \ll 1 \textit{ and } w \ll u && \textit{for the mixed-fluid approach.} \end{aligned} \tag{8.83}$$

By combining (8.82) and (8.83), a criterion based on relative particle size can be expressed in terms of the Stokes number for a given density ratio. For example, DNS with $\rho_p/\rho_f \approx 800$ (e.g., water drops in air) leads to the following particle approach criteria:

$$\begin{aligned} &d > 10\eta && St_\eta > 5{,}000 && \textit{resolved-surface approach with DNS.} \\ &d \sim 3\eta && St_\eta \sim 500 && \textit{distributed-force approach with DNS.} \\ &d \sim \eta && St_\eta \sim 50 && \textit{point-force approach with DNS.} \\ &d \sim 0.1\eta && St_\eta \sim 0.5 && \textit{point-force drift-flux approach with DNS.} \\ &d < 0.03\eta && St_\eta < 0.05 && \textit{point-force mixed-fluid approach with DNS.} \end{aligned} \tag{8.84}$$

These criteria are consistent with Bagchi and Balachandar (2003), who found that a resolved-surface technique with DNS was needed for $d > 10\eta$ but a point-force

Table 8.1 Different numerical approaches for treating particles in turbulence listed in order of decreasing computational resources and general decreasing accuracy, where the Lagrangian point-force and RANS is the most common.

	Interactions between particles & turbulence?			
Particle approach & fluid approach	**Nonlinear drag bias?**	**$\nabla\alpha$ & ∇k diffusion?**	**Prefer. bias?**	**2-/3-way coupling?**
Resolved-surface & DNS $d\mathbf{v}/dt = f(\mathbf{U}) + \ldots$ ($d \gg \eta$)	Yes	Yes	Yes	Yes
Point-force and DNS $d\mathbf{v}/dt = f(\mathbf{u}_{@p}) + \ldots$ ($d \sim \eta$)	Yes	Yes	Yes	Model
Mixed-fluid & DNS $\mathbf{v} = \mathbf{u}_m$ ($St_\eta \ll 1$)	Yes	Yes	Yes	Model
Point-force & RANS $d\tilde{\mathbf{v}}/dt = f(\bar{\mathbf{u}}) + \ldots$	Yes	Model	Model	Model
Drift-flux & RANS $\tilde{\mathbf{v}} = \tilde{\mathbf{u}} + \mathbf{w}_{term} + \ldots$ ($w/u \ll 1$)	Model	Model	Model	Model
Mixed-fluid & RANS $\mathbf{v} = \mathbf{u}_m$ ($w/u \approx 0$)	No	No	No	Yes

approach was reasonable for $d < \eta$. If one instead applies the guidelines for the density ratio of bubbles in water (with $c_\forall = 1/2$), the restrictions are even tighter:

$$\begin{array}{lll} d > 10\eta & St_\eta > 3 & \textit{resolved-surface approach with DNS.} \\ d \sim 3\eta & St_\eta \sim 0.3 & \textit{distributed-force approach with DNS.} \\ d < \eta & St_\eta < 0.03 & \textit{point-force mixed-fluid approach with DNS.} \end{array} \tag{8.85}$$

Therefore, caution should be used when applying a point-force technique to bubbles and other low-density particles when using DNS. For wall-bounded turbulent flows, the preceding criteria for high-density particles and for bubbles can be similarly obtained for DNS by combining (5.11) and (6.10) as follows:

$$d^+ \equiv \frac{d}{y_{fr}} = \sqrt{18\frac{St^+}{\rho_p/\rho_f + c_\forall}}. \tag{8.86}$$

As such, a resolved-surface approach that is needed for $d^+ > 10$ for DNS, is consistent with $St^+ > 3$ for bubbles in water.

When using an LES approach for the fluid, the computational resolution is instead based on the subgrid length scale and Stokes number (8.76), as follows:

$$\frac{d}{\Delta_\mathcal{G}} = \sqrt{18\frac{St_\Delta}{\rho_p/\rho_f + c_\forall}} \quad \textit{for LES treatment of the turbulence.} \tag{8.87}$$

Generally, particles in LES are small enough so that $d < \Delta_\mathcal{G}$ indicates that a point-force approach is appropriate, and the question of whether to simplify to a drift-flux or mixed-fluid approach can be based on (8.37). For RANS, none of the flow scales are resolved, so point-force approaches are generally appropriate.

Finally, Table 8.1 summarizes the choice of the approaches in terms of the turbulent biases, using the preceding guidelines relative to particle size, speed, and reference frame from §8.1–§8.3. It can be seen that going from the resolved-surface DNS approach to the mixed-fluid RANS approach (which decreases computational resources for a given flow Reynolds number) tends to result in decreased fidelity for the turbulent-particle interactions. As such, selecting numerical approaches often requires a careful balance between prediction fidelity and computational efficiency. In particular, a computational strategy should consider both the practical issues of computational resources with turnaround constraints as well as the physics-based issues of accuracy for the particular multiphase flow to be simulated. Fortunately, the continuing increase of computing power is allowing improved representation of the fluid physics to be more readily available.

8.5 Chapter 8 Problems

Show all steps and list any needed assumptions:

(P8.1) Discuss the pros and cons when comparing the following numerical approaches versus the other options:
(a) Lagrangian particles versus Lagrangian parcels.
(b) Lagrangian parcels versus Eulerian particle PDEs.
(c) Mixed-fluid versus drift-flux versus point-force approaches.
(d) Point-force versus distributed-force versus resolved-surface approaches.

(P8.2) Derive the two-way coupled continuous-phase PDEs of (8.25a) and (8.25b) starting from the Reynolds transport theorem of (2.3).

(P8.3) Show that the Eulerian mass, momentum, and energy transport equations for the dispersed and the continuous phases given by (8.20) and (8.25) combine to form the mixture transport equations (8.41) and explain any assumptions needed to obtain this result.

(P8.4) Consider soot particles (density of 2 g/cm^3) carried by downward laminar flow of air at NTP in a vertical cylindrical pipe with a length of 20 cm. At the inflow location, the pipe diameter is 20 mm, the air flow velocity has a Poiseuille parabolic profile with a centerline value of 0.2 m/s, the soot particles are moving at the same speed as the air flow, and they are uniformly distributed across the pipe cross-sectional area with a mass loading of 0.1%. At a downstream axial distance of 50% of the pipe length, the pipe cross-section undergoes a step contraction to a diameter of 10 mm and any particles that hit a pipe surface are assumed to stick. Using available multiphase flow software, select an appropriate Eulerian numerical approach for both continuous and dispersed phases (based on the available software options) and explain your reasoning. Use this approach to predict the surface mass flux (mass of particles impacting per unit time) and demonstrate that the results are numerically converged for particle diameter of 30 μm neglecting Brownian diffusion. For particles of 10, 30, 100, 300 and 1000 microns, plot the fluxes as a function of the particle Stokes numbers and qualitatively discuss the results.

(P8.5) Same as (P8.4) but with a Lagrangian approach for individual soot particles and also plot and discuss the particle trajectories.

(P8.6) Same as (P8.4) or (P8.5) but include Brownian diffusion and use particles diameters of 0.1, 0.3, 1, 3 and 10 microns.

(P8.7) Same as (P8.4), but using the drift-flux method (compare and discuss the results with those from P8.4, if available).

(P8.8) Consider still NTP air with a 5 mm diameter hollow sphere falling at 0.3 m/s. Use a resolved-surface approach with a Gridded Interface Method by creating an axisymmetric computational grid around the particle with a reference frame fixed to the particle. Using available fluid dynamics software, predict the steady flow around the particle and the drag (and demonstrate that the results are numerically converged). Then compare the drag to that given by an appropriate drag coefficient model and discuss your findings.

(P8.9) Consider downward flow of air at NTP in a vertical cylindrical pipe with a constant pipe diameter of 15 mm with a 5 mm diameter sphere falling along the centerline at 0.5 m/s and where the airflow centerline speed is 0.2 m/s far away from the particle. Use a resolved-surface approach with a Gridded Interface Method by creating an axisymmetric computational grid around the particle and inside the pipe with a reference frame fixed to the particle. Using available fluid dynamics software, predict the steady flow around the particle and the drag (and demonstrate that the results are numerically converged), then compare with (P8.8) to characterize the differences in the flow around the particle and the particle drag.

(P8.10) Consider a jet of air that issues upward into still air (both at NTP) with an initial jet diameter of 5 mm, initial uniform velocity of 20 m/s, and no initial turbulence. The jet is initially laden with uniformly dispersed 70 μm diameter water drops also moving at 20 m/s with an initial water mass flux rate of 3 g/s. After the jet diameter has increased to 10 mm, determine whether the flow conditions suggest one-way, two-way, three-way, and/or four-way coupling. Using available multiphase flow software, select an appropriate Eulerian RANS turbulence model and numerical approach and explain your reasoning. Use this approach to predict the evolution of the jet and the droplets to a distance of 20 cm and demonstrate that the results are numerically converged. At an axial location of 10 cm downstream, obtain the radial profiles of the mean axial velocity and TKE of the airflow, as well as those of the droplets. Discuss the differences in the context of integral Stokes numbers and the drift parameter. Also discuss any expected turbulent bias effects for the droplet velocity.

(P8.11) Same as P8.10 but with a Lagrangian parcel approach for the droplets.

(P8.12) Same as (P8.11) but with a large eddy simulation for the air turbulence (compare and discuss the results with those from P8.11, if available).

(P8.13) Consider a vertical pipe of 4 cm in diameter and a length of 1 m with a turbulent flow of NTP tap water moving upwards with an initial (at the bottom of the pipe) uniform velocity of 0.6 m/s. The flow also includes 2 mm diameter bubbles initially at the same velocity with an initially uniform 3% volume fraction. Determine whether the downstream flow conditions suggest one-way, two-way, three-way, and/or four-way coupling. Using available multiphase flow software, select an appropriate Eulerian RANS turbulence model and numerical approach for both phases (based on the available software options) and explain your reasoning. Use this approach to predict the evolution of the water and bubble velocities as well as the bubble volume fraction along the pipe length and demonstrate that your results are numerically converged. At an axial location of 50% of the pipe length, plot the radial distribution of the mean and rms of the water and bubble velocities as well as the bubble volume fraction distribution. Discuss the expected effects of pseudoturbulence as a function of radius at this location.

(P8.14) Same as P8.13 but with a Lagrangian particle approach for the bubbles (compare and discuss the results with those from P8.13, if available).

(P8.15) Starting from (8.20a), obtain the transport PDE of (8.67).

(P8.16) Write a code that solves the Lagrangian particle Equation of Motion using synthetic turbulence based on (8.70). Consider NTP HIST air flowing at a mean velocity of 2 m/s in the horizontal direction with $\tau_{\Lambda\mathcal{L}}$ of 0.1 s, TKE of $0.06\,\mathrm{m^2/s^2}$, $c_\varepsilon = 0.2$, and $c_\Lambda = 1$. Using the predicted particle trajectories, obtain the lateral spread of water droplets of sizes 30, 100, and 300 μm at an axial distance of 1 m from release, and demonstrate that the results are numerically converged for one of these drop sizes.

(P8.17) Choose one of the multiphase flows illustrated in Chapter 1 and choose an appropriate numerical method to solve the flow for a relevant set of conditions and then discuss the results.

(P8.18) Develop a guide like (8.84) for DNS of particles with $\rho_p = \rho_f$.

9 Drag Force on an Isolated Particle

In previous chapters, the description of particle drag was considered for three limited conditions that were all associated with incompressible uniform steady flow:

- A solid sphere as a function of Reynolds numbers (§3.2.4).
- A fluid sphere with internal recirculation at $Re_p \ll 1$ and a clean interface (§3.2.2).
- A deformed fluid particle in an oblate shape at high $Re_p > 100$ (§4.3.1 and §4.4.1).

In this chapter, many more conditions are considered for the drag force of an isolated particle, including the influence of

- Velocity gradients in the surrounding fluid for a solid sphere (§9.1.2).
- Particle Mach number and Knudsen number for a solid sphere (§9.2.2).
- Temperature gradients in the surrounding fluid for a solid sphere (§9.2.3).
- Particle spin and fluid vorticity for a solid particle (§9.2.4).
- Flow turbulence and particle roughness for a solid particle (§9.2.4).
- Nonspherical shape for a solid particle (§9.2.5 and §9.2.6).
- Contamination or internal recirculation for a spherical fluid particle (§9.3.1).
- Deformation of a fluid particle at various Reynolds numbers (§9.3.2).

When explicit theoretical descriptions are not available, experiments and resolved-surface simulations are used for empirical point-force models. However, such empirical models are subject to the uncertainties and test conditions of the experimental or numerical results. As such, there are complex flow regimes where no models can be recommended with confidence, such as the instantaneous drag of deformed bubbles undergoing shape oscillations or Mach number effects on nonspherical solid particles. In addition, drag forces due to chemical reaction, ionization, magnetic fields, electric fields, photonic fields, van der Waals forces, etc., are not considered, so the reader is referred to Clift et al. (1978), Soo (1990), and Marshall and Li (2014) for such discussion.

9.1 Decomposition of Point Forces

9.1.1 General Surface Point-Force Expression

In the point-force assumption, the fluid dynamic surface force ($\mathbf{F}_{surf}$) for a single particle is represented as a linear sum of several individual and independent forces. As noted in (3.6), this generally includes drag ($\mathbf{F}_D$), lift ($\mathbf{F}_L$), virtual mass force due to relative acceleration ($\mathbf{F}_\forall$), far-field fluid stresses acting on the particle surface ($\mathbf{F}_S$), history force ($\mathbf{F}_H$), Brownian force ($\mathbf{F}_{Br}$), and the surface force due to thermal gradients ($\mathbf{F}_{\nabla T}$). Neglecting the random Brownian force, the remaining point-force expression is deterministic and given by the following:

$$\mathbf{F}_{surf} = \mathbf{F}_D + \mathbf{F}_L + \mathbf{F}_S + \mathbf{F}_\forall + \mathbf{F}_H + \mathbf{F}_{\nabla T}. \tag{9.1}$$

In general, the force summation introduced by the RHS of (9.1) has three important caveats:

(1) *Linear composition is not guaranteed* since there can be nonlinear interactions between the various forces. Such interactions are not well understood but fortunately are often small enough to be neglected for many conditions (as in the theoretical treatments in §9.1.2).
(2) *Most point-force expressions at finite Reynolds numbers or for complex shapes are empirical* because theoretical treatments are often limited to particles with small Reynolds numbers and simple shapes. When the conditions do not allow a theoretical solution (as is often the case), surface force expressions are generally based on fits to experimental and/or numerical data, which are only applicable for validated test conditions and are often subject to some degree of uncertainty.
(3) *The point-force treatment assumes unhindered continuous-fluid properties are equivalent to far-field conditions*. This unhindered approach employs local undisturbed flow properties *hypothetically* extrapolated to the particle centroid to describe the flow seen by the particle. This assumes that the flow in the vicinity of the particle has weak spatial gradients relative to the particle diameter, as well as weak temporal gradients relative to the particle response time.

All three caveats apply for the point-force models of Chapters 8 and 9, where the first two caveats can be especially problematic for complex particle shapes or flow conditions, while the last caveat can be especially problematic for large particles. However, all three caveats may be removed by employing the resolved-surface simulation techniques discussed in Chapter 8. In particular, the surface forces can be computed by numerically resolving some or all of the flow around the particle itself, which can accurately account for highly complex conditions and/or shapes. When there are a large number of particles, such resolved-surface simulations can be computationally expensive and often impractical. Fortunately, the point forces described in Chapters 9–11 are often reasonable for many multiphase flow fields with a large number of particles.

9.1.2 Theoretical Point-Force Momentum for an Isolated Particle

There are many analytical expressions for the fluid dynamic surface force on a particle, starting from the seminal work of Stokes in 1851 represented in (3.30). In general, theoretical results are confined to simple geometries such as spheroidal shapes and limiting flow conditions such as creeping flow ($Re_p \to 0$) or inviscid flow ($\mu_f = 0$). The most commonly used theoretical surface force descriptors are for an isolated solid sphere surrounded by an incompressible continuum fluid. Of these, perhaps the two most important are the Maxey–Riley equation for creeping irrotational flow and the Auton–Hunt–Prud'homme equation for inviscid rotational flow. These two equations often serve as a baseline to which additional effects are incorporated. The two equations are discussed, in turn, in the following.

The Maxey–Riley derivation is more complex than the Stokes drag derivation of §3.2.1 as it includes flow unsteadiness and nonuniformity. In particular, this derivation showed that nonuniformity leads to an influence on drag, added-mass force, and history force via second-order gradients (represented by $\nabla^2 \mathbf{u}_{@p}$). These second-order gradients effects are referred to as Faxen effects, since Faxen (1922) was first to derive such contributions.

Regarding its assumptions, the Maxey–Riley equation (1983) was derived for a nonrotating rigid particle in an unsteady, incompressible flow. The particle Reynolds number is assumed to be very small, so that nonlinear convective terms can be neglected as in (2.80). The unhindered flow is also assumed to have weak spatial gradients compared to the particle scales. These assumptions can be summarized as follows:

$$Re_p \ll 1. \tag{9.2a}$$

$$\left|\nabla u_{@p}\right| \ll \frac{\nu_f}{d^2}. \tag{9.2b}$$

$$\left|\nabla^2 u_{@p}\right| \ll \frac{u_{@p}}{d^2}. \tag{9.2c}$$

Under these assumptions and using the decomposition of (9.1), the Maxey–Riley equation can be expressed as follows:

$$\mathbf{F}_{surf} = \mathbf{F}_D + \mathbf{F}_S + \mathbf{F}_\forall + \mathbf{F}_H. \tag{9.3a}$$

$$\mathbf{F}_D = -3\pi d \mu_f \mathbf{W}. \tag{9.3b}$$

$$\mathbf{F}_S = \rho_f \forall_p \left[\frac{\mathcal{D}\mathbf{u}_{@p}}{\mathcal{D}t} - \mathbf{g}\right]. \tag{9.3c}$$

$$\mathbf{F}_\forall = -\frac{\rho_f \forall_p}{2} \frac{d\left(\mathbf{w} - \frac{d^2}{40}\nabla^2 \mathbf{u}_{@p}\right)}{dt}. \tag{9.3d}$$

$$\mathbf{F}_H = -\frac{3}{2} d^2 \sqrt{\pi \rho_f \mu_f} \left[\int_0^t \left(\frac{d\mathbf{W}/d\tau}{\sqrt{t-\tau}}\right) d\tau + \frac{\mathbf{W}(0)}{\sqrt{t}}\right]. \tag{9.3e}$$

$$\mathbf{W} \equiv \mathbf{w} - \frac{d^2}{24}\nabla^2 \mathbf{u}_{@p}. \tag{9.3f}$$

This quasisteady drag of (9.3b) includes effects of flow gradients (which are neglected in 3.30) by defining a Faxen-corrected relative velocity ($\mathbf{W}$ of 9.3f), instead of using the conventional relative velocity ($\mathbf{w}$ of 1.15d). The fluid-stress term ($\mathbf{F}_S$) is the same as that given by (3.64), since this stems directly from (3.62). However, the added-mass force ($\mathbf{F}_\forall$) and the history force ($\mathbf{F}_H$) differ from (3.77) and (3.82) due to the Faxen corrections. Notably, the inclusion of an initial relative velocity via $\mathbf{W}(t = 0)$ in (9.3e) is based on later work by Maxey (1993) and was not in the original Maxey–Riley derivation (which assumed zero initial velocity). For uniform flow, the Maxey–Riley equation of (9.3) effectively reverts to the equation given in (3.83).

An important point about the Maxey–Riley equation is that the preceding combination of forces was obtained within a single theoretical framework. Thus, the decomposition described is exact and complete within the bounds of the previously stated assumptions.

Auton (1987) and Auton et al. (1988) also considered the forces on a moving sphere within a single theoretical framework but for inviscid and rotational fluid. In particular, these two studies assumed a far-field linear shear flow (Figure 2.9a) with slip conditions on the particle surface (Figure 2.10b). In the first study, Auton assumed steady flow and that the shear velocity gradient was small (i.e., weak) compared to $|w/d|$. Auton et al. then extended this to include unsteady and straining flows by additionally assuming that the frequency of the Eulerian unsteadiness is small (i.e., slow) compared to $|w/d|$. These two assumptions for weak spatial and temporal gradients can be expressed as follows:

$$|\boldsymbol{\omega}_{\text{shear}}| \ll \frac{w}{d}. \tag{9.4a}$$

$$\left|\frac{\partial \mathbf{w}}{\partial t}\right| \ll \frac{w^2}{d}. \tag{9.4b}$$

As was discussed in §3.2.3, inviscid flow leads to zero net drag on the body:

$$\mathbf{F}_D = 0 \qquad \textit{for inviscid flow.} \tag{9.5}$$

Similarly, the history force will be zero, so the resulting surface fluid force on the particle found by Auton et al. only includes fluid-stress, added-mass, and lift forces due to the fluid vorticity:

$$\mathbf{F}_{\text{surf}} = \mathbf{F}_S + \mathbf{F}_\forall + \mathbf{F}_L. \tag{9.6a}$$

$$\mathbf{F}_S = \rho_f \forall_p \left[\frac{\mathcal{D}\mathbf{u}_{@p}}{\mathcal{D}t} - \mathbf{g}\right]. \tag{9.6b}$$

$$\mathbf{F}_\forall = \rho_f \forall_p c_\forall \left[\frac{\mathcal{D}\mathbf{u}_{@p}}{\mathcal{D}t} - \frac{d\mathbf{v}}{dt}\right]. \tag{9.6c}$$

$$\mathbf{F}_L = \frac{1}{2}\rho_f \forall_p (\boldsymbol{\omega} \times \mathbf{w}). \tag{9.6d}$$

This result is named the AHP particle dynamic equation, referring to the authors: Auton, Hunt, and Prud'homme. As with the Maxey–Riley equation, it is an exact theoretical result for the stated assumptions that allows linear composition.

It is interesting to compare the AHP equation to the Maxey–Riley (or the BBO) equation. The biggest difference is that Maxey–Riley includes drag and history forces and Faxen effects, whereas AHP includes instead a lift force. However, the other terms are quite similar if one neglects Faxen effects. In particular, we note that the fluid stress is the same in both cases (since this is a direct consequence of using the incompressible Navier–Stokes equations). The added-mass terms are slightly different between (9.3d) and (9.6c), though Maxey (1993) noted that the differences are negligible in the creeping-flow limit if $\nabla^2\mathbf{u}_{@p}$ is negligible (i.e., the gradients are weak).

Since the drag force is typically the most important surface force, it will be the primary focus for the remainder of this chapter and considered for the case of an isolated particle in quasisteady conditions. The isolated history force (an unsteady-drag force related to high flow and/or particle accelerations), along with lift, added-mass, and fluid-stress forces, will be discussed in Chapter 10, while the influence of walls and neighboring particles on drag is discussed in Chapter 11. In the following, drag is considered first for a solid particle under various conditions (§9.2), then for a fluid particle, such as a drop or bubble (§9.3).

9.2 Drag Force for an Isolated Solid Particle

9.2.1 Influence of Reynolds Number for a Sphere

The most common condition to consider for drag is that of a smooth, isolated solid sphere with no rotation surrounded by a continuum incompressible fluid.

In terms of the fluid dynamics, the Reynolds number regimes discussed in §3.2.4 are listed in Table 9.1. For increasing Re_p, the flow about the particle in laminar flow transitions from fully attached and steady, to separated but still steady, to an unsteady separated wake with an upstream laminar boundary layer, and then to a condition where the upstream boundary layer transitions to turbulent before separation occurs. The evolution from steady attached flow to steady separated flow is shown in Figure 9.1 (for $9 < Re_p < 120$), where it can be seen that the separation angle (defined in Figure 9.1g) moves upstream as Re_p increases. At $Re_p > 130$, perturbations form in the wake due to instabilities of the shear layers of the separation bubble causing

Table 9.1 Viscous flow regimes for incompressible continuum flow past a solid sphere.

Reynolds range	Flow physics
$Re_p \leq 0.01$	Creeping (Stokes) flow without inertial effects
$0.01 \leq Re_p \leq 22$	Laminar flow with steady fully attached flow
$22 \leq Re_p \leq 130$	Laminar boundary layer and steady *separated* wake
$130 \leq Re_p < 1{,}000$	Laminar boundary layer and *unsteady* separated wake
$130 \leq Re_p < Re_{p,crit}$	Laminar boundary layer and *turbulent* separated wake
$Re_p > Re_{p,crit}$	*Turbulent* boundary layer and turbulent separated wake

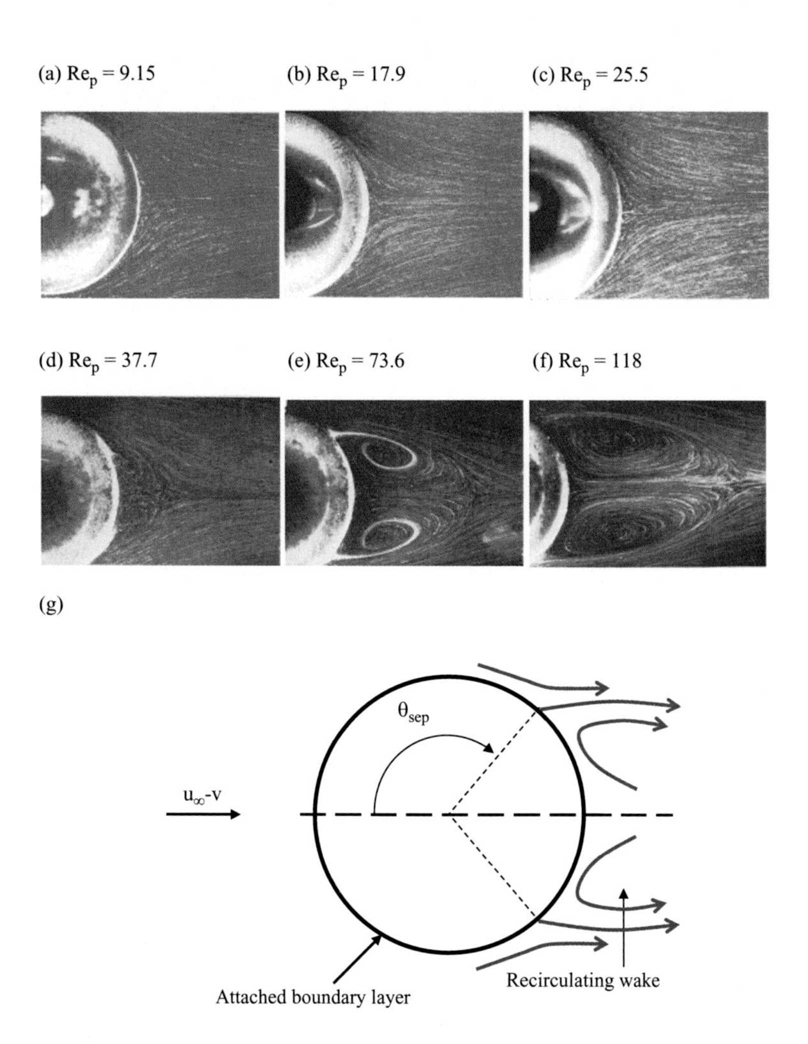

Figure 9.1 (a)-(f) Photographs of streamlines for a solid sphere at steady laminar flow conditions for various Reynolds numbers (Batchelor, 1967). The separation point (θ_{sep}) defined in (g) is about 135° at $Re_p = 73.6$, as shown in (e).

unsteadiness. As Re_p increases further, the unsteady vortex formation in the wake moves upstream and sheds with a frequency proportional to w/d (Clift et al., 1978). These vortices reach the sphere surface at about Re_p of 6,000, for which the shedding frequency (normalized by w/d) is maximized. For $Re_p > 6{,}000$, the most upstream separation point rotates around the sphere surface at the shedding frequency, yielding coherent structures with a highly turbulent wake. When the sphere reaches $Re_p > Re_{p,\,crit}$, the surface boundary layer becomes turbulent upstream, causing separation to be delayed (it moves rearward), resulting in a smaller wake with less fluctuations in the separation points on the sphere surface.

In terms of the drag coefficient, there is a creeping-flow solution but empirical models are needed to adjust the drag at finite Reynolds numbers, whereby the drag force can be written in terms of a Stokesian drag correction (f) or a drag coefficient (C_D) per (3.55), as follows:

$$\mathbf{F}_D = -3\pi d\mu_f f\mathbf{w} = -(\pi/8)d^2\rho_f C_D\mathbf{w}|\mathbf{w}|. \tag{9.7}$$

The drag coefficient is related to the drag correction, as follows:

$$C_D = \frac{24}{Re_p}f. \tag{9.8}$$

When Re_p is large, C_D is more commonly used since it tends to a constant. When Re_p is small, f is more commonly used since it tends to a constant. For a solid sphere, f_{Re} represents the ratio of drag at finite Reynolds number to that for creeping flow as follows:

$$f_{Re} \equiv \frac{F_D(Re_p)}{F_D(Re_p \to 0)} = \frac{F_D}{3\pi d\mu_f w}. \tag{9.9}$$

For small but finite particle Reynolds numbers, Oseen (1910) included linearized convection per (2.81) to solve the flow and obtained a surface pressure coefficient defined by (3.51) as follows:

$$C_p = \frac{6\cos\theta}{Re_p} - \frac{1}{2} + \frac{3\cos^2\theta}{2} \qquad \textit{for } Re_p \ll 1. \tag{9.10}$$

This can be compared to the Stokes surface pressure coefficient based on (3.23) as:

$$C_p = \frac{6\cos\theta}{Re_p} \qquad \textit{for } Re_p \to 0. \tag{9.11}$$

The differences between these two expressions do not contribute to a change in drag since they are symmetric fore-to-aft (so the form drag is the same). However, the Oseen solution also includes viscous surface stresses due to convective terms that are not symmetric yielding an extra drag correction term:

$$f_{Re} = 1 + \frac{3}{16}Re_p \qquad \textit{for } Re_p \ll 1. \tag{9.12}$$

A higher-order theoretical solution with an additional correction term was obtained by Proudman (1969). However, the effective differences between the Stokes, Oseen, and Proudman expressions are small in their range of applicability, as shown in Figure 9.2 for $Re_p < 1$. Based on experimental measurements, it is reasonable to apply either the Stokes or the Oseen result for $Re_p < 0.5$.

Since analytical solutions are not appropriate at high Re_p values (due to nonlinear convection effects), several empirical drag models have been proposed for this regime (Clift et al., 1978). A common and accurate model for particles with laminar and transitional wakes ($Re_p < 1{,}000$) is given by the Schiller–Naumann drag model of (3.58) as follows:

$$f_{Re} \equiv 1 + 0.15\,Re_p^{0.687} \qquad \textit{for } Re_p < 1{,}000. \tag{9.13}$$

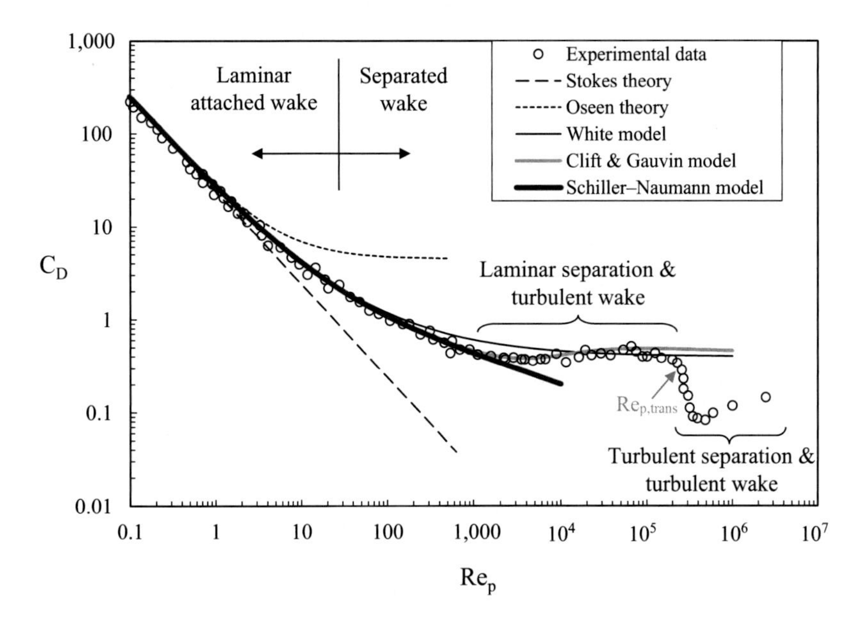

Figure 9.2 Drag coefficient for a smooth solid sphere at various Reynolds numbers for incompressible flow regimes with experimental data reported in White (2016) compared to various theories and empirical models.

As shown in Figure 9.2, this model gives an accurate description in this $\mathrm{Re_p}$ range but does not capture the higher subcritical regime, where the solid sphere drag in incompressible flow instead tends to the following:

$$C_D = C_{D,crit} \approx 0.42 \qquad \textit{for } \mathrm{Re_{p,sub}} < \mathrm{Re_p} < \mathrm{Re_{p,crit}} = 2 \times 10^5. \tag{9.14}$$

This regime and those spanning down to the creeping-flow regime are captured by the White drag model of (3.60b), and the even simpler (though less accurate) drag model by Dallavalle (1948) as follows:

$$f_{Re} = \left[1 + \sqrt{\frac{0.4\,\mathrm{Re_p}}{24}}\right]^2 \qquad \textit{for } \mathrm{Re_p} < 2 \times 10^5. \tag{9.15}$$

A more accurate (and complex) model that is valid from creeping flow up to $\mathrm{Re_{p,crit}}$ is that of Clift and Gauvin (1970):

$$f_{Re} = 1 + 0.15\,\mathrm{Re_p}^{0.687} + \left(\frac{\mathrm{Re_p}}{24}\right)\frac{0.42}{1 + \frac{42,500}{\mathrm{Re_p}^{1.16}}} \qquad \textit{for } \mathrm{Re_p} < 2 \times 10^5. \tag{9.16}$$

The first two terms are the same as the Schiller–Naumann drag correction of (9.13), while the remaining terms incorporate the slight dip near $\mathrm{Re_p}$ of 1,000, as shown in Figure 9.2, which the White model does not capture. The Clift and Gauvin model is

Table 9.2 Flow regimes based on Schaaf and Chambre (1958) with boundary conditions and creeping-flow drag correction for a stationary spherical particle assuming $c_\theta \approx 1.22$.

Knudsen range	Flow physics	Boundary conditions	Stokes correction
$Kn_p \leq 0.01$	Continuum flow	No-slip, $U_\theta = 0$	$f \approx 1$
$0.01 \leq Kn_p \leq 0.1$	Slip flow	Slip, $U_\theta \sim O(Kn_p)$	$f \approx 1 - 2.44Kn_p$
$0.1 \leq Kn_p \leq 10$	Transition flow	Boltzmann distribution	$f \sim \dfrac{1}{1 + 2.514Kn_p}$
$Kn_p \geq 10$	Free-molecular flow	Molecular reflection	$f \approx 0.301/Kn_p$

within 6% error of experimental data and is perhaps the most accurate single-equation drag model that captures the regimes spanning from creeping to critical flow.

For $Re_p > Re_{p,crit}$, the reduced separation of the wake causes the drag coefficient to rapidly decrease due to the drag crisis. The Reynolds number when C_D first drops to 0.3 is defined as $Re_{p,trans}$ and is about 350,000 (Figure 9.2).

While the preceding discussion was based on incompressible continuum conditions, the next section addresses effects of compressibility and rarefaction.

9.2.2 Influence of Flow Compressibility and Rarefaction

Both flow compressibility and noncontinuum will change the drag of a spherical particle. Generally, flow compressibility becomes more important as the Mach number increases, while the rarefaction becomes more important as the Knudsen number increases. For an ideal gas, the particle Mach number is based on the relative particle velocity and the gas sound speed (2.44 and 2.45), while the particle Knudsen number is the ratio of the mean-free-path length of the surrounding molecules to the particle diameter as noted in (1.20), which uses (1.25):

$$M_p \equiv \frac{w}{a_g} = \frac{w}{\sqrt{\gamma \mathcal{R}_g T_g}}. \tag{9.17a}$$

$$Kn_p \equiv \frac{\ell_{m,coll}}{d} = \frac{\mu_g \sqrt{\pi}}{\rho_g \sqrt{2\mathcal{R}_g T_g} d}. \tag{9.17b}$$

For flow over spheres, compressibility becomes significant when the relative Mach number is no longer small, such as $M_p > 0.3$. This corresponds to $w > 100$ m/s for air at NTP and causes an increase of C_D. Such conditions can occur for particles interacting with high-speed flows. Examples include metal particles in a plasma spray, combustion particles in a rocket engine, surface ablation in a hypersonic boundary layer, and particles subjected to a shock wave. In these cases, the duration of high relative Mach numbers scales with the particle response time, so compressibility effects tend to be limited to $t < \tau_p$ (Sivier et al., 1994).

For flow over spheres, the Kn_p leads to the conditions noted in Table 9.2 where small Kn_p values ($<10^{-2}$) are consistent with a no-slip approximation (to within 2%

accuracy for drag), but larger values can cause noncontinuum effects. The noncontinuum condition corresponds to $d < 7\,\mu m$ for air at NTP, as discussed in §1.5. For these conditions, the collision rates of molecules on the surface are no longer very high, so that there can be a difference between the mean molecular velocity and the mean particle surface velocity, which can cause a reduction in drag. For *slip flow* ($0.01 \leq Kn_p \leq 0.1$), the effect is weak and can be considered a small departure from the no-slip condition (allowing theoretical analysis). However, this effect is large for *free-molecular flow* ($Kn_p > 10$), where the molecules primarily impact and reflect from the particle but not each other (allowing a different theoretical analysis). The *transition* regime ($0.1 \leq Kn_p \leq 10$) is the most complex to analyze (it does not allow theoretical analysis).

To consider the combination of compressibility and rarefaction effects, it is helpful to relate the Knudsen, Mach, and Reynolds numbers for an ideal gas to each other as follows:

$$\frac{Kn_p\, Re_p}{M_p} = \sqrt{\tfrac{1}{2}\pi\gamma}. \tag{9.18}$$

For $Re_p \gg 1$, $Kn_p/M_p \ll 1$, so compressibility is more likely to dominate than rarefaction. On the other hand, $Re_p \ll 1$ leads to $Kn_p/M_p \gg 1$, so rarefaction is more likely to dominate. The effects of these two regimes on the drag coefficient are illustrated in Figure 9.3, which shows that rarefaction reduces drag while compressibility tends to increase drag (relative to the incompressible continuum conditions with $M_p \to 0$ and $Kn_p \to 0$). This figure also indicates a seminexus of the compressibility and rarefaction effects at a single Reynolds number, as follows:

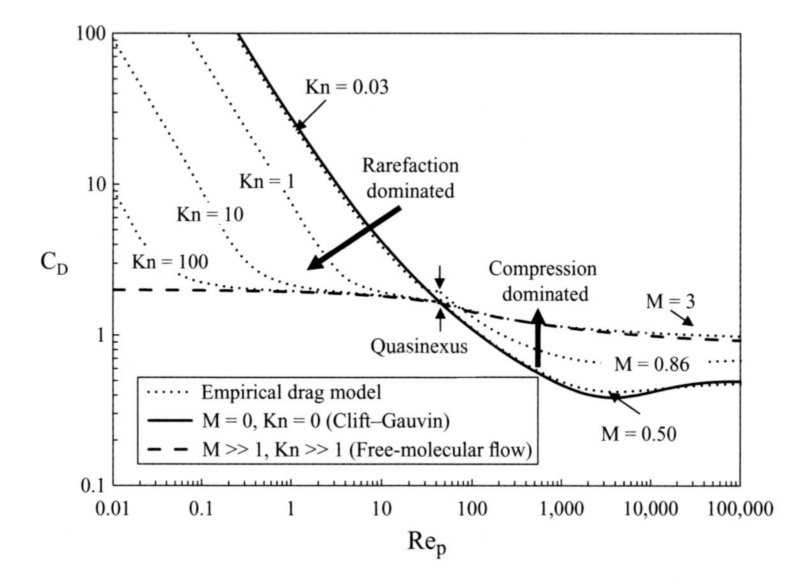

Figure 9.3 Rarefaction and compression effects on drag coefficient of a spherical particle at isothermal conditions ($T_p = T_{g,\infty}$).

Table 9.3 Compressibility flow regimes for a sphere.

Mach range	Flow physics
$M_p \leq 0.1$	Incompressible flow
$0.1 \leq M_p \leq 0.65$	Compressible subsonic flow
$0.65 \leq M_p \leq 1.2$	Transonic flow
$1.2 \leq M_p < 5$	Supersonic flow
$M_p \geq 5$	Hypersonic flow

$$C_D \sim 1.63 \quad at \quad Re_p = 45 \quad for\ all\ M_p\ and\ Kn_p. \tag{9.19}$$

This seminexus condition can be used to help model the drag with separate equations for $Re_p < 45$ (governed by rarefaction effects) and for $Re_p > 45$ (governed by compressibility effects). In the following, the compression-dominated regime is first considered followed by the rarefaction-dominated regime.

Compression-Dominated Regime

The Mach number is used to define the five flow regimes shown in Table 9.3, and example flow fields for a subsonic and two supersonic cases are shown in Figure 9.4. The associated drag changes for subcritical Reynolds numbers ($Re_{p,sub} < Re_p < Re_{p,crit}$) can be quantified by defining a compressibility drag ratio as follows:

$$C_M \equiv \frac{C_{D,crit}}{C_{D,crit,M=0}} = \frac{C_{D,crit}}{0.42}. \tag{9.20}$$

The RHS employs the incompressible critical drag coefficient for a sphere.

For Mach numbers less than 0.1, the flow over a sphere is considered incompressible, so $C_M = 1$. For $0.1 < M_p < 0.65$, the flow around a sphere stays everywhere subsonic, but increased compressibility leads to earlier boundary-layer separation, which moderately increases drag (Schlichting and Gersten, 2017), as shown in Figure 9.5. Once $M_p > 0.65$, the high-speed region (just above the boundary layer at $\theta = 90°$, per 3.47c) generally reaches a sonic speed (Hoerner, 1965). As the Mach number increases further, gas dynamic waves appear in this region, as shown in Figure 9.4a, initially with a lambda-shock pattern. This shock interaction with the boundary layer further promotes flow separation, resulting in a rapid drag coefficient increase (Figure 9.5). As the particle Mach number becomes supersonic ($M_p > 1$), a bow shock forms in front of the sphere (Figure 9.4b), which eliminates the shock boundary-layer interaction, causing the wake separation point to move backward, especially at higher Mach numbers (Figure 9.4c). The bow shock effect increases the subcritical drag coefficient, while the rearward movement of the separation causes a decrease, which is more pronounced at higher Mach numbers. As a result of this combination, the compressibility drag ratio reaches a maximum of 2.25 at about $M_p \sim 1.45$. It then decreases until it reaches the hypersonic regime, where the drag coefficient ratio is nearly constant and equal to 2 from Mach 5 to Mach 10, as

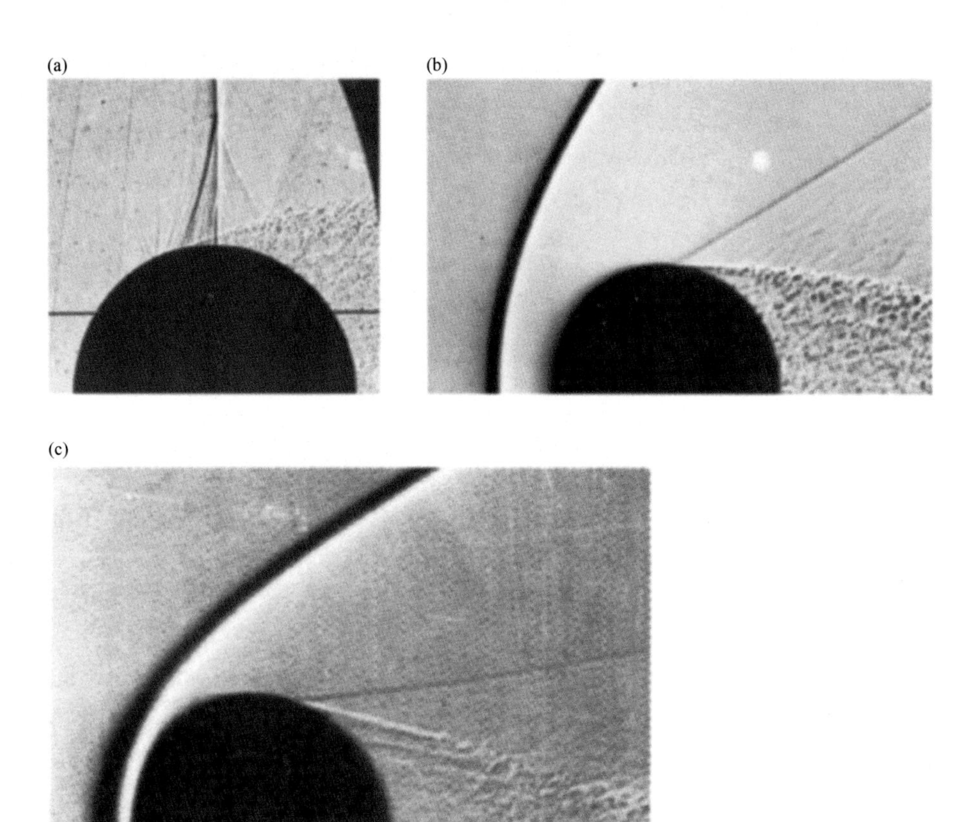

Figure 9.4 Shadowgraphs of spheres from van Dyke (1982) with very high particle Reynolds numbers (approximately 900,000) at (a) $M_p = 0.86$, (b) $M_p = 1.53$, and (c) $M_p = 3$.

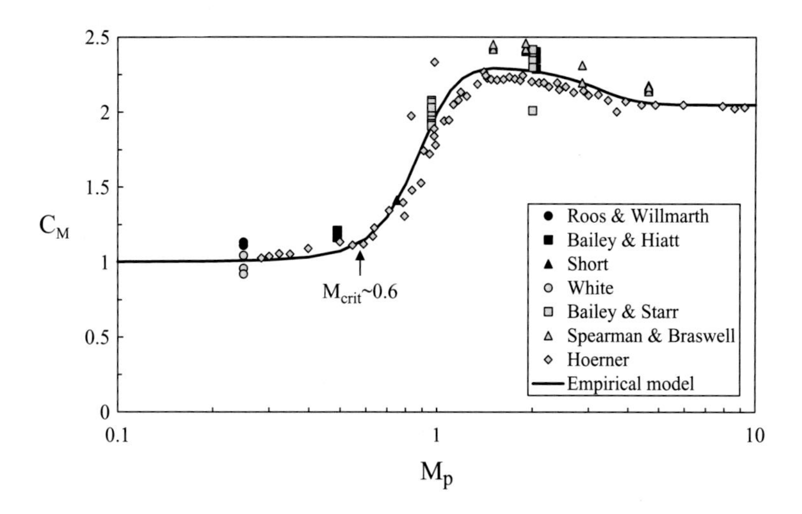

Figure 9.5 Drag ratio as a function of particle relative Mach numbers in the Newton drag regime ($10^4 < Re_p < 10^6$) for a variety of experimental data sets compared to an empirical model (Loth et al., 2021).

shown in Figure 9.5. Also shown is an empirical model for the compressibility drag ratio at subcritical Reynolds numbers (Loth et al., 2021), as follows:

$$C_M = 1.65 + 0.65\tanh\left(4M_p - 3.4\right) \qquad \textit{for } M_p \le 1.5. \tag{9.21a}$$

$$C_M = 2.18 - 0.13\tanh\left(0.9M_p - 2.7\right) \qquad \textit{for } M_p > 1.5. \tag{9.21b}$$

This two-part fit for $Re_p > 100{,}000$ is similar to the high Re_p model of Crowe et al. (1972).

For lower Reynolds numbers down to the nexus point ($45 < Re_p < Re_{p,crit}$), the influence of Mach number can be incorporated with a modified Clift–Gauvin drag expression, as follows:

$$C_D = \frac{24}{Re_p}\left[1 + 0.15\,Re_p^{0.687}\right]H_M + \frac{0.42C_M}{1 + 42,500/Re_p^{1.16C_M} + G_M/Re_p^{0.5}}. \tag{9.22}$$

This function is constructed to recover the incompressible Clift–Gauvin limit as M_p approaches zero and to the result given by (9.20) and (9.21) at subcritical Reynolds numbers. The form of (9.22) includes two more parameters, H_M and G_M, which are modeled as follows:

$$G_M = 20 - 10.9M_p + 3.29M_p^2 + 166M_p^3 \qquad \textit{for } M_p < 0.8. \tag{9.23a}$$

$$G_M = 5 + 40M_p^{-3} \qquad \textit{for } M_p > 0.8. \tag{9.23b}$$

$$H_M = 1 - 0.074M_p + 0.212M_p^2 + 0.0239M_p^3 \qquad \textit{for } M_p < 1.0. \tag{9.23c}$$

$$H_M = 0.93 + 1/\left(3.5 + M_p^5\right) \qquad \textit{for } M_p > 1.0. \tag{9.23d}$$

Comparisons with experimental data show that this model is quite robust (Loth et al., 2021), and examples are shown in Figure 9.6 for two M_p values, where the Re_p influence becomes weaker as the Mach number increases.

Rarefaction-Dominated Regime

For particle Reynolds numbers to the left of the nexus ($Re_p < 45$), rarefaction effects are more likely to dominate, which cause a finite slip condition on the particle surface, resulting in lower shear stress. The impact of this slip on the drag can be expressed in terms of a Knudsen-based Stokes correction as follows:

$$f_{Kn} \equiv \frac{F_D\left(Kn_p, Re_p \to 0\right)}{3\pi\mu_f d}. \tag{9.24}$$

This ratio is often referred to as the Cunningham (1910) correction factor. In the following, we consider conditions for $Kn_p \ll 1$, for $Kn_p \gg 1$, and then the bridge for this gap. The resulting boundary conditions and Stokes corrections for these regimes are summarized in Table 9.2.

For small but finite Knudsen numbers ($Kn_p \ll 1$) with creeping incompressible flow ($Re_p \to 0$ and $M_p \to 0$), slip will occur due to a thin layer over the surface, known as the

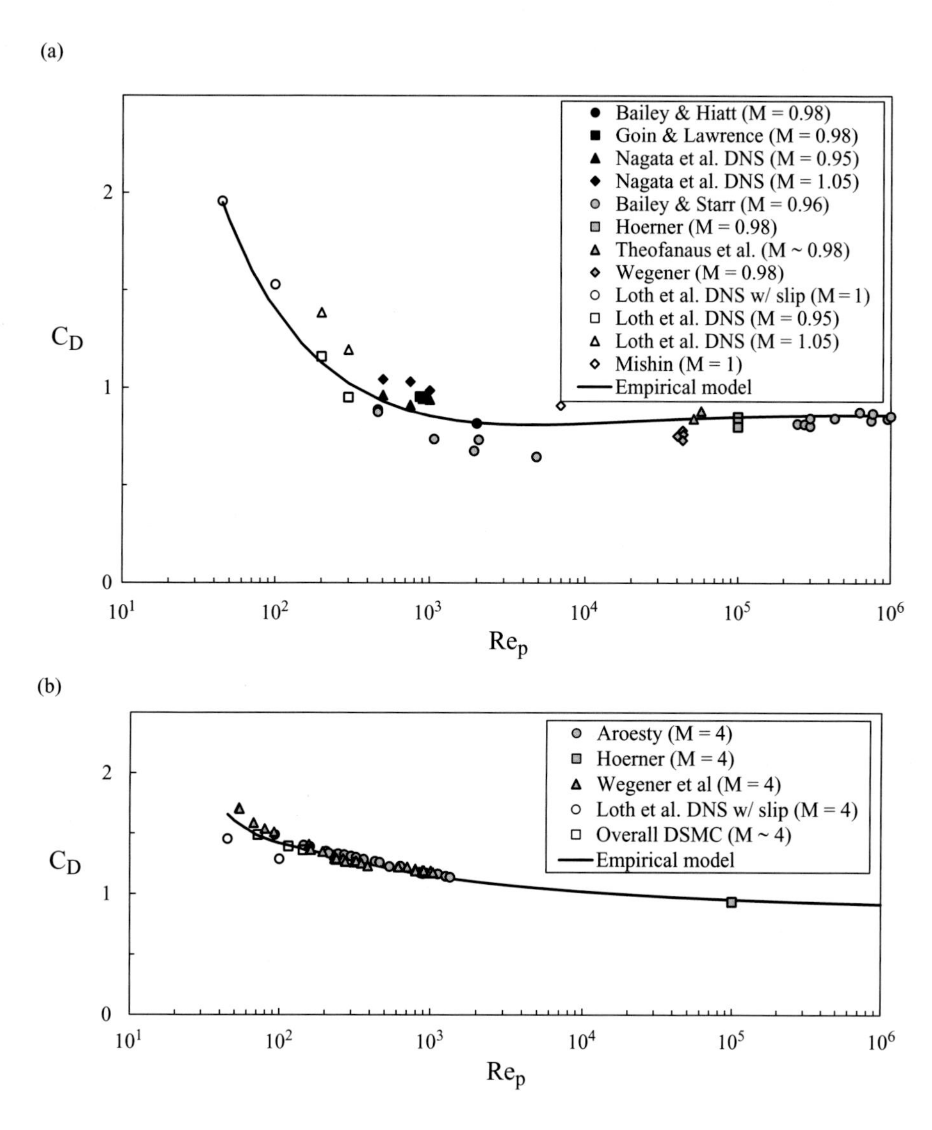

Figure 9.6 Drag coefficient as a function of Reynolds number (Aroesty, 1962; Hoerner, 1965; Bailey and Hiatt, 1972; Bailey and Starr, 1976; Mishin 1997; Wegener and Ashkenas, 2006; Theofanous et al., 2018; Nagata et al., 2020) compared to empirical model (Loth et al., 2021): (a) for transonic conditions (M_p of about 1), and (b) for high supersonic conditions (M_p of about 4).

Knudsen layer, which has a thickness of about one free path length. The resulting tangential slip velocity is a function of the tangential momentum coefficient (c_θ) as well as the velocity and temperature gradients (Talbot et al., 1980), as follows:

$$U_\theta \approx c_\theta \mathrm{Kn}_p \left[\frac{\partial U_r}{\partial \theta} - r^2 \frac{\partial}{\partial r}\left(\frac{U_\theta}{r}\right) \right]_{r=r_p} + c_\theta \left(\frac{\nu_g}{\mathcal{R}_g T_g} \right) \left(\frac{\partial T_g}{\partial \theta} \right)_{r=r_p}. \tag{9.25}$$

The first term of the RHS is negligible for continuum flow, while the second term is negligible for uniform gas temperature near the particle surface (no heat transfer). Overall, the slip increases with c_θ, which is related to the accommodation coefficient (c_{accom}) as follows:

$$c_{accom} = \frac{2}{c_\theta + 1} = \frac{v_{g\parallel}^{in} - v_{g\parallel}^{out}}{v_{g\parallel}^{in}}. \tag{9.26}$$

The accommodation coefficient is defined as the fraction of surface colliding gas molecules that are diffusively reflected (whose tangential energy is adsorbed) while the remaining fraction of molecules have specular reflection (whose tangential energy is preserved). As noted by the RHS, this coefficient is based on the average difference between the incoming and outgoing gas tangential velocities using (8.31) nomenclature. For a molecularly smooth particle, the tangential energy will be fully accommodated and lost (adsorbed) so $c_{accom} = 1$. However, a particle with some molecular roughness will only have partial accommodation, so $c_{accom} < 1$. Often, the coefficient range is 0.8–1.0, so a common estimate is $c_{accom} \approx 0.9$, leading to $c_\theta \approx 1.22$ (Talbot et al., 1980).

Theory for the resulting Stokes drag correction is available for very small and very large Knudsen numbers. Assuming weak slip and uniform gas temperature, Basset (1888) solved the velocity field around the particle using (9.25) and obtained the drag correction as follows:

$$f_{Kn \ll 1} = 1 - 2c_\theta Kn_p + O\left(Kn_p^2\right) \quad \textit{for } Kn_p \ll 1. \tag{9.27}$$

At the other extreme of free molecular flow ($Kn_p \gg 1$), Epstein (1929) assumed a Maxwell distribution of the molecule velocities that will impact the particle surface with a single collision, yielding a drag correction as follows:

$$f_{Kn,fm} \rightarrow \frac{8 + 2\pi/(c_\theta + 1)}{36 Kn_p} \quad \textit{for } Kn_p \gg 1. \tag{9.28}$$

To bridge the gap from very low to very high Knudsen numbers, Phillips (1975) proposed an intermediate Knudsen number model, as follows:

$$f_{Kn} = \frac{15c_\theta - 6c_\theta^2 Kn_p + (16 + 16c_\theta + 4\pi)(c_\theta^2 + 2)Kn_p^2}{15c_\theta + 24c_\theta^2 Kn_p + 36c_\theta(c_\theta^2 + 1)Kn_p^2 + 72(c_\theta^2 + 2)(c_\theta + 1)Kn_p^3}. \tag{9.29}$$

A simpler empirical representation for a wide range of Knudsen numbers based on $c_\theta \approx 1.22$ is given (Clift et al., 1978) as follows:

$$f_{Kn} = \frac{1}{1 + Kn_p\left[2.514 + 0.8\exp(-0.55/Kn_p)\right]}. \tag{9.30}$$

These models yield a drag reduction as the noncontinuum effects (proportional to Kn_p) increase, owing to the increasing surface slip. As shown in Figure 9.7, the Clift et al. model compares favorably with the data, including the famous experiments of Millikan (1911, 1923) for which the charge of an electron was first measured.

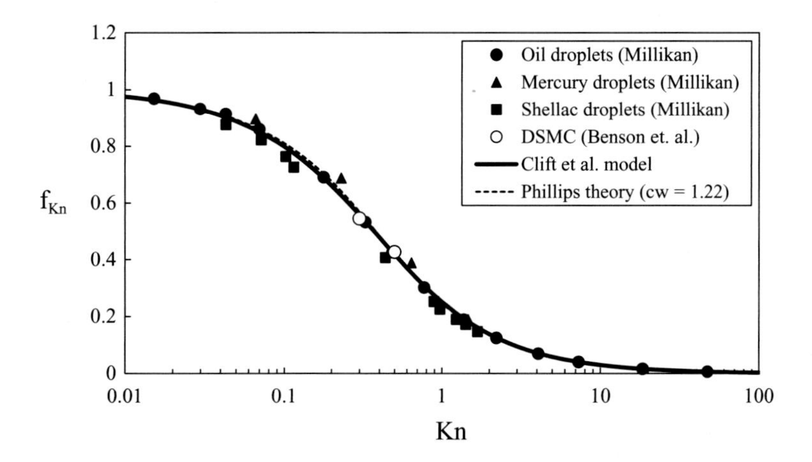

Figure 9.7 Knudsen number effects on the Stokesian drag correction for $Re_p \ll 1$ and $M_p \ll 1$ from Millikan (1911, 1923) and Benson et al (2004) compared to the theory of Phillips (1975) and empirical model of Clift et al. (1978).

It should be noted that the preceding results are appropriate for creeping flow ($Re_p \to 0$). To account for finite Reynolds numbers, the Knudsen number effect can be combined with a Schiller–Naumann correction (9.13) for a combined effective drag coefficient, as follows:

$$C_{D,Kn,Re} = f_{Kn}\frac{24}{Re_p}\left(1 + 0.15\,Re_p^{0.687}\right) \qquad \textit{for } Re_p < 45 \textit{ and } M_p \ll 1. \tag{9.31}$$

To include compressibility effects (finite Mach number) in the rarefaction-dominated regime, we define the molecular speed ratio as follows:

$$s \equiv M_p\sqrt{\gamma/2}. \tag{9.32}$$

Assuming equal tangential and normal accommodation coefficients in free-molecular flow, Stadler and Zurick (1951) and Schaaf and Chambre (1958) derived a drag coefficient for $s \geq 1$ as follows:

$$C_{D,fm} = \frac{(1+2s^2)\exp(-s^2)}{s^3\sqrt{\pi}} + \frac{(4s^4+4s^2-1)\mathrm{erf}(s)}{2s^4} + \frac{2}{3s}\sqrt{\frac{\pi T_p}{T_{f@p}}} \qquad \textit{for}\, Kn_p \gg 1. \tag{9.33}$$

The first two terms on the RHS are associated with diffuse reflection while the third term is associated with specular reflection based on the ratio of particle temperature (T_p) to the unhindered gas temperature ($T_{f@p}$). These temperatures are generally different if a particle undergoes rapid heating or cooling (e.g., due to injection conditions, shock passage). For a temperature ratio of unity (isothermal conditions) with $s \gg 1$ and $Re_p \to 0$, the drag coefficient approaches a minimum given by $C_{D,fm} \to 2$, as shown in Figure 9.8. The isothermal free molecular limit can then be empirically corrected for finite Reynolds numbers as follows:

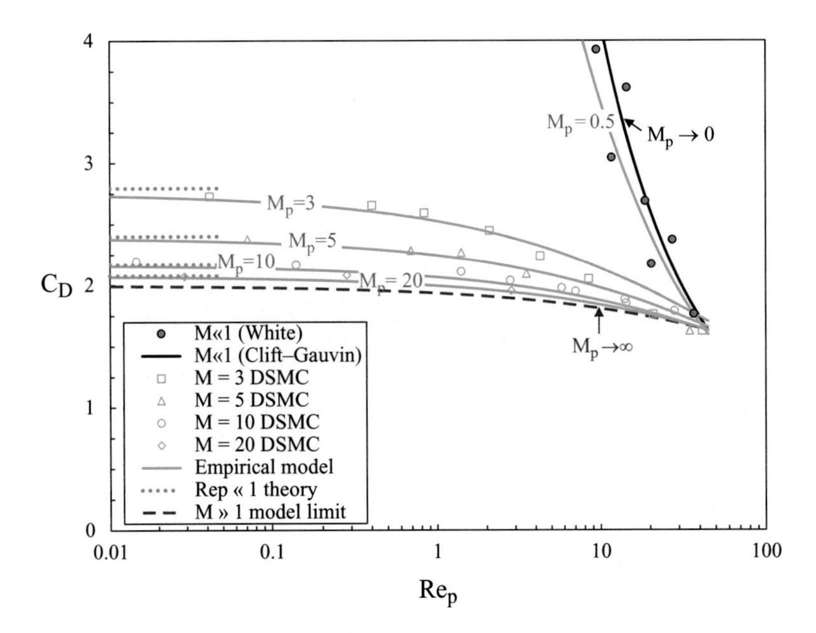

Figure 9.8 Mach number effects within the isothermal rarefaction regime ($Re_p < 45$) based on DSMC data of Macrossan (2004) as well as the empirical model of Loth et al. (2021).

$$C_{D,fm,Re} = C_{D,fm}\left[1 + \left(\frac{C_{D,fm}}{J_M} - 1\right)\sqrt{\frac{Re_p}{45}}\right]^{-1} \quad \textit{for } Kn_p \gg 1. \tag{9.34}$$

This expression from (Loth et al., 2021) uses a Mach number function (J_M) given as follows:

$$J_M = 2.26 - 0.1M_p^{-1} + 0.14M_p^{-3} \quad \textit{for } M_p < 1.0. \tag{9.35a}$$

$$J_M = 1.6 + 0.25M_p^{-1} + 0.11M_p^{-2} + 0.44M_p^{-3} \quad \textit{for } M_p > 1.0. \tag{9.35b}$$

This result can be combined with (9.31) to handle all Mach, Knudsen, and Reynolds number effects below the nexus, as follows:

$$C_{D,fm,M,Re} = \frac{C_{D,Kn,Re}}{1 + M_p^4} + \frac{M_p^4 C_{D,fm,Re}}{1 + M_p^4} \quad \text{for } Re_p \leq 45. \tag{9.36}$$

The isothermal result is shown in Figure 9.8, where increasing M_p (decreasing Kn_p) causes the drag coefficient to reduce and tend to the free molecular limit. The effect of temperature ratio is shown in Figure 9.9, where increased particle temperature leads to increased drag, especially at low Re_p.

9.2.3 Thermophoretic (Not Drag) Force

Next we temporarily divert our discussion on drag force to discuss the thermophoretic force, since this force is strongly related to Knudsen number effects. Note that the drag

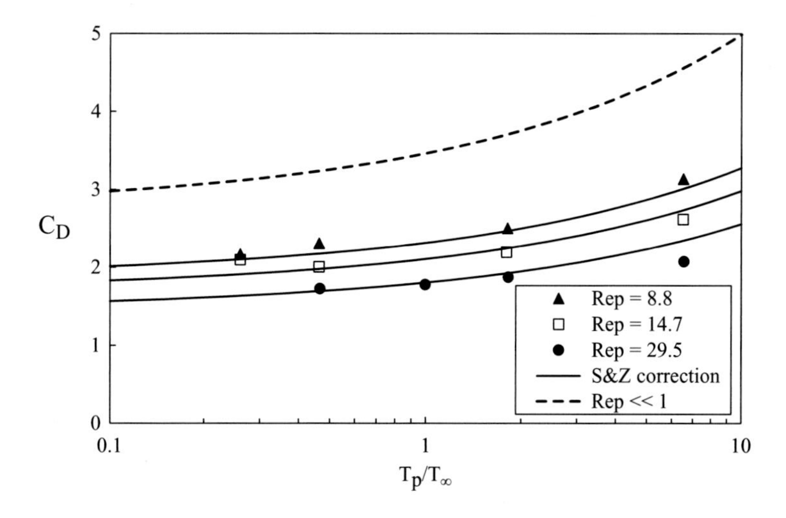

Figure 9.9 Influence of particle temperature ratio at $M_p = 2$ based on data of Bailey and Hiatt (1972) for various particle Reynolds numbers compared to the Stadler and Zurick (S&Z) correction.

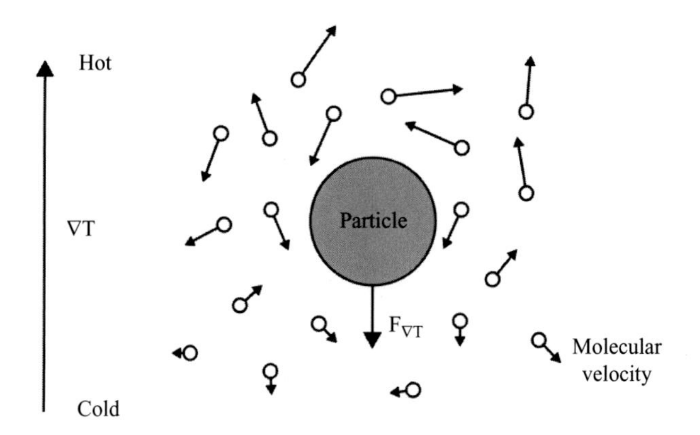

Figure 9.10 A particle surrounded by a gas with a temperature gradient creates a thermophoretic force on the particle.

force is caused by a relative velocity, while the thermophoretic force is caused by a fluid temperature gradient, which represents the gradient in the kinetic energy of the surrounding molecules (Figure 9.10). As such, the "hot side" of a particle experiences collisions with molecules that have higher velocities compared to that of the "cool side" per (1.25b). This yields a force in the direction opposite to the temperature gradient, that is, thermophoresis drives particles toward the colder regions. For example, kerosene lamps can yield soot on the colder glass when first lit, and a car heater can cause droplets to fog up a cold windshield when initially turned on. Also, thermal precipitators use cold walls to collect and filter nano-sized particles from gas flows, and microprocessor fabrication systems use heated surfaces to avoid particle deposition that can cause impurities.

As with the drag force, the thermophoretic force depends on the particle Knudsen number $\left(\mathrm{Kn_p}\right)$ and can be described by three primary regimes: near-continuum $\left(\mathrm{Kn_p} \ll 1\right)$, transitional $\left(\mathrm{Kn_p} \sim 1\right)$, and approximately free-molecular $\left(\mathrm{Kn_p} \gg 1\right)$. However, thermophoresis is also influenced by the ratio of gas to particle thermal conductivities, given as follows:

$$k^* \equiv \frac{k_g}{k_p} = \frac{15\mu_g R_g}{4k_p}. \tag{9.37}$$

If the gas thermal conductivity is large compared to that of the particle $(k^* \gg 1)$, then the largest temperature gradients will be inside the particle. However, most gas/particle combinations are associated with $k^* \ll 1$, so that the particle has an approximately uniform temperature and the largest temperature gradients are in the gas (Epstein, 1929).

For the near-continuum regime $\left(\mathrm{Kn_p} \ll 1\right)$, Brock (1962) used the surface boundary condition of (9.25) with a uniform temperature gradient in the far field to impose a heat flux on the sphere surface based on k^*, and solved the gas and particle temperature fields for creeping flow with a temperature accommodation coefficient (c_T) to yield a theoretical thermophoresis force:

$$\mathbf{F}_{\nabla T} = -\frac{6\pi\mu_g \nu_g d\left(k^* + 2\mathrm{Kn_p}c_T\right)(2-c_\theta)/c_\theta}{\left(1+2k^* + 4\mathrm{Kn_p}c_T\right)\left[1+6\mathrm{Kn_p}(2-c_\theta)/c_\theta\right]}\left(\frac{\nabla T_g}{T_g}\right)_{@p} \quad \textit{for } \mathrm{Kn_p} \ll 1. \tag{9.38}$$

The negative sign shows that this force is opposite to the unhindered gas temperature gradient, and typical surface temperature accommodations can be estimated with $c_T \approx 2.17$ (Loyalka, 1992).

For the free-molecular limit $\left(\mathrm{Kn_p} \gg 1\right)$ at the opposite extreme, the theoretical force for monatomic gases (Waldmann, 1961) is inversely proportional to the Knudsen number and given as follows:

$$\mathbf{F}_{\nabla T,\mathrm{fm}} = -\frac{\pi}{2}\mu_g \nu_g \frac{d}{\mathrm{Kn_p}}\left(\frac{\nabla T_g}{T_g}\right)_{@p} \quad \textit{for } \mathrm{Kn_p} \gg 1. \tag{9.39}$$

While Brock's theory (9.38) for small Knudsen number happens to equal the free-molecular limit (9.39) at a temperature accommodation coefficient of unity and $\mathrm{Kn_p} \gg 1$, there is no theory for the intermediate $\mathrm{Kn_p}$ conditions due to nonlinear effects. However, data from resolved-surface simulations based on the Boltzmann equation (Loyalka, 1992) are available for this intermediate regime. These results are shown in Figure 9.11 in terms of the dimensionless thermophoretic force defined as follows:

$$F^*_{\nabla T} \equiv \frac{F_{\nabla T}}{F_{\nabla T,\mathrm{fm}}}. \tag{9.40}$$

The results indicate that the Brock theory is appropriate for $\mathrm{Kn_p} < 0.01$, while the free molecular theory is appropriate for $\mathrm{Kn_p} > 20$. In between, the dimensionless force

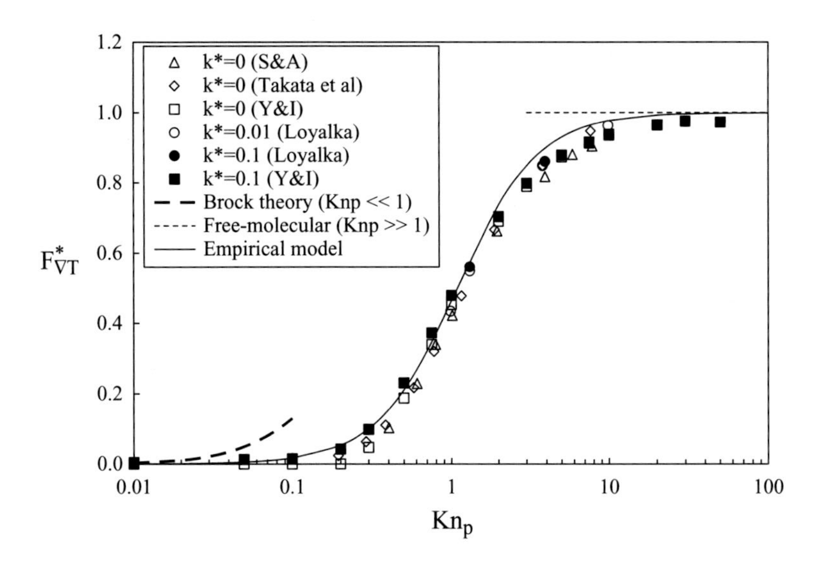

Figure 9.11 Thermophoretic force as a function of particle Knudsen number for resolved-surface simulations (Sone and Aoki, 1983; Yamamato and Ishihara, 1988; Loyalka, 1992; Takata et al., 1994) as well as various theories and models (where $k^* \leq 0.01$ are filled symbols, and $k^* \geq 0.1$ are closed symbols).

increases with Knudsen number (decreasing particle size) and is only a weak function of the conductivity ratio for small k^*. There are several empirical models for this intermediate regime (Zheng, 2002), but one model that is both simple and robust is given as

$$\mathbf{F}_{\nabla T} = -\frac{\pi\mu_g \nu_g d}{2}\left(\frac{\nabla T_g}{T_g}\right)_{@p}\left(\frac{Kn_p^{0.7}}{1.15 + Kn_p^{1.7}}\right) \qquad for\, k^* < 0.1. \tag{9.41}$$

As shown in Figure 9.11, this empirical model gives reasonable agreement with simulation data.

9.2.4 Influence of Spin, Vorticity, Turbulence, and Roughness

Influence of Particle Spin and Flow Vorticity

While vorticity and particle rotation primarily produce a lift force (perpendicular to the relative velocity) as will be discussed in Chapter 10, they generally have a weak effect on drag. In the following, the potential effects of fluid vorticity on drag are considered for linear shear (Figure 2.9a) and for a vortex (Figure 2.9b).

For a particle in a linear shear flow, creeping-flow theories have indicated no theoretical impact on drag (Saffman, 1965) and a similar lack of influence was found for Re_p up to 10 (Dandy and Dwyer, 1990; Bagchi and Balachandar, 2002a). For $100 < Re_p < 500$, a small increase in drag (a few percent) was predicted in very high shear rates (Kurose and Komori, 1999), but these changes have not been confirmed experimentally. Thus, it is generally reasonable to assume that linear shear does not have a significant influence on the drag of solid spherical particles.

For a particle in a rotational vortex flow in small but finite Reynolds numbers (linearized inertia terms), Heron et al. (1975) obtained the following theoretical drag increase:

$$\mathrm{f}_{\mathrm{vortex}} = 1 + \frac{5}{14}\sqrt{\frac{\omega_{\mathrm{vortex}}\mathrm{d}^2}{2\nu_{\mathrm{f}}}} \quad \textit{for}\ \mathrm{Re}_{\mathrm{p}} \ll 1. \tag{9.42}$$

However, experiments by Sridhar and Katz (1995) for particles in a vortex flow at $20 < \mathrm{Re}_{\mathrm{p}} < 80$ indicated little influence of rotational vorticity on the drag coefficient, suggesting that the effect on drag is confined to smaller Re_{p} and ω_{vortex} values, where its effect is generally weak anyway.

Since the difference between a shear flow and a vortex flow is the presence of strain, Bagchi and Balachandar (2002c) investigated the influence of strain alone on drag. They noted a strain-induced drag increase for $10 < \mathrm{Re}_{\mathrm{p}} < 300$ that was qualitatively consistent with a model based on potential flow theory for $\mathrm{Re}_{\mathrm{p}} < 20$ for which there was no flow separation. However, the net changes in drag were generally small (on the order of a few %), indicating that the neglect of strain on drag is generally reasonable.

Next, we consider the impact of particle spin on drag. For the case of a spherical particle rotating at small but finite Reynolds numbers, Rubinow and Keller (1961) derived an expression for drag and found no effect of spin, that is, they recovered the Oseen correction (9.12). At finite Re_{p} values, there have been several experimental and computational studies investigating particle spin, but the data include significant scatter with some conflicting trends and/or little influence for most conditions (Loth, 2008b). In summary, it is often reasonable to follow the guideline of Crowe et al. (2011) to simply ignore the effect of particle spin and flow vorticity on the drag force.

Influence of Turbulence and Particle Roughness

Freestream turbulence and particle roughness can both lead to a drag reduction at a high particle Reynolds number because they can introduce instabilities that cause the particle's boundary layer to transition sooner to turbulent flow, thereby reducing the degree of flow separation. For example, dimples on a golf ball are used to reduce drag, and this same reduction is found when there is upstream flow turbulence. These two parameters are considered in terms of the relative turbulent intensity (t) and the relative surface roughness (r), defined as follows:

$$t \equiv \mathrm{u}'_{\mathrm{rms}}/\mathrm{w}. \tag{9.43a}$$

$$r \equiv \left(\mathrm{r}_{\mathrm{I}} - \mathrm{r}_{\mathrm{p}}\right)_{\mathrm{rms}}/\mathrm{r}_{\mathrm{p}}. \tag{9.43b}$$

The former is the inverse of the drift parameter (7.18) for terminal velocity conditions, while the latter parameter employs the local interface radius (r_{I}), which is the distance from the particle centroid to a point on the surface. Their effect on drag can be

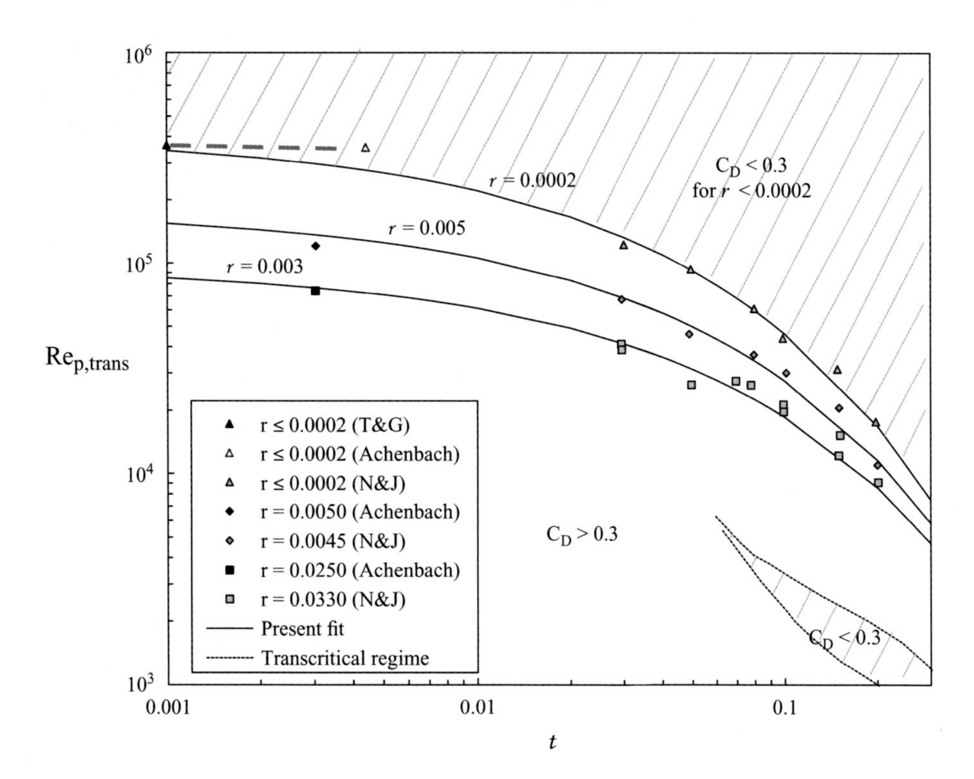

Figure 9.12 Combined dependence on roughness and turbulence levels on transcritical drag coefficient based on experimental data (Torobin and Gauvin, 1960; Clamen and Gauvin, 1969; Achenbach, 1974, Neve and Jaafar, 1982), where the blue dashed line is $\mathrm{Re}_{p,trans}$ for a smooth sphere in steady flow.

quantified via changes in the transcritical Reynolds number ($\mathrm{Re}_{p,trans}$), defined where C_D drops to 0.3 (as part of the drag crisis). As shown in Figure 9.2, the transcritical Reynolds number is about 3.5×10^5 for a smooth sphere with steady flow. The combined effect of nondimensional roughness and turbulence on $\mathrm{Re}_{p,trans}$ is shown in Figure 9.12, where it can be seen that $\mathrm{Re}_{p,trans}$ decreases for $r > 0.0002$ and for $t > 0.001$. The following is an empirical expression that correlates the combined effects:

$$\log_{10} \mathrm{Re}_{p,trans} = 4.5(r + 0.004)^{-0.03} - 2.0t^{0.55}(r + 0.004)^{-0.07}. \tag{9.44}$$

As shown in Figure 9.12, this gives reasonable predictions of $\mathrm{Re}_{p,trans}$ for moderate roughness and turbulence levels. However, high turbulence levels of 5% can introduce an additional "transcritical" regime for which the drag coefficient goes below 0.3 for a narrow range of low Reynolds numbers. For example, transcritical flow occurs at $2{,}200 < \mathrm{Re}_p < 3{,}300$ for $t = 0.1$. This is because high freestream turbulence can cause a surface laminar boundary layer to detach for a short distance, creating a laminar separation bubble (Clamen and Gauvin, 1969). This bubble delays the turbulent separation over the surface (increasing θ_{sep}), thereby reducing C_D. However, further increases in Re_p prevent the laminar separation bubble from reattaching, leading to a

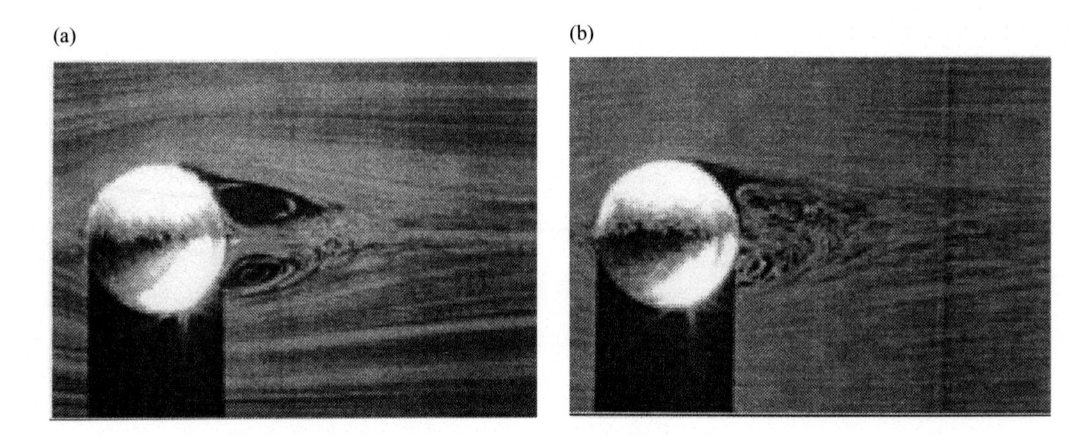

Figure 9.13 Sphere wakes at $Re_p = 80$ and turbulence levels of (a) 0.7% and (b) 7.0% (Elamworawutthikul and Gould, 1999).

massively separated condition (reducing θ_{sep}). This causes the C_D to rise again above 0.3, ending the transcritical Reynolds number regime (Loth, 2008c).

The reduction in $Re_{p,trans}$ is generally associated with same amount of reduction for $Re_{p,crit}$. However, for Reynolds numbers much less than $Re_{p,crit}$, the effect of freestream turbulence and surface roughness on the drag of particles is weak so long as $r < 1\%$ and $t < 1\%$. At higher t values, the available data for changes in drag for subcritical conditions contains large scatter and conflicting trends (Warnica et al., 1995; Crowe et al., 2011). At low Reynolds numbers, where the wake flow is normally steady and symmetric, turbulence can cause unsteadiness and asymmetry, as shown in Figure 9.13. However, the weak change in the mean separation typically leads to weak (and not well-defined) changes in drag. As such, one may not confidently propose a robust model for the quantitative impact of turbulence, beyond that given by (9.44). However, when the roughness becomes very large, the particle can be considered nonspherical, and such effects of particle shape are discussed in the following section.

9.2.5 Nonspherical Regularly Shaped Solid Particle Drag

The drag of nonspherical solid particles will depend on the type of shape and degree of nonsphericity. Nonspherical solid particles may be classified as having shapes (ellipsoids, cones, disks, etc., as shown in Figure 4.11 and 4.12) or asymmetric irregular shapes (nonsymmetric rough surfaces, as shown in Figure 4.10c). The drag associated with these particle shapes can be related to a Stokes correction that includes both shape and Reynolds number effects defined as follows:

$$f_{shape,Re} \equiv \frac{F_{D,shape}(Re_p, d)}{F_{D,sphere}(Re_p \to 0, d)} = \frac{F_{D,shape}(Re_p, d)}{3\pi d\mu_f w}. \tag{9.45}$$

Predictions of this correction are discussed in the following subsections, beginning with spheroids and ellipsoids.

Table 9.4 Stokes correction factors for various oblate and prolate spheroids for flow parallel and for flow perpendicular to the axis of symmetry.

Spheroid shape	$f_{E\parallel}$	$f_{E\perp}$
Oblate exact (E < 1)	$\dfrac{(4/3)\,E^{-1/3}\left(1-E^2\right)}{E+\dfrac{\left(1-2E^2\right)\cos^{-1}E}{\sqrt{1-E^2}}}$	$\dfrac{(8/3)\,E^{-1/3}\left(E^2-1\right)}{E-\dfrac{\left(3-2E^2\right)\cos^{-1}E}{\sqrt{1-E^2}}}$
Oblate approx. (0.25 < E < 1)	$\left(\dfrac{4}{5}+\dfrac{E}{5}\right)E^{-1/3}$	$\left(\dfrac{3}{5}+\dfrac{2E}{5}\right)E^{-1/3}$
Disk (E < 0.25)	$\dfrac{8}{3\pi}E^{-1/3}$	$\dfrac{16}{9\pi}E^{-1/3}$
Prolate exact (E > 1)	$\dfrac{(4/3)\,E^{-1/3}\left(1-E^2\right)}{E-\dfrac{\left(2E^2-1\right)\ln\left(E+\sqrt{E^2-1}\right)}{\sqrt{E^2-1}}}$	$\dfrac{(8/3)\,E^{-1/3}\left(E^2-1\right)}{E+\dfrac{\left(2E^2-3\right)\ln\left(E+\sqrt{E^2-1}\right)}{\sqrt{E^2-1}}}$
Prolate approx. (6 > E > 1)	$\left(\dfrac{4}{5}+\dfrac{E}{5}\right)E^{-1/3}$	$\left(\dfrac{3}{5}+\dfrac{2E}{5}\right)E^{-1/3}$
Needle (E > 6)	$\dfrac{(2/3)\,E^{2/3}}{\ln(2E)-1/2}$	$\dfrac{(4/3)\,E^{2/3}}{\ln(2E)-1/2}$

Spheroids and Ellipsoids in Creeping Flow

As discussed in §4.2.1, spheroids and ellipsoids are the most common regular particle shapes, where an ellipsoid has an ellipsoidal cross section for all three orthogonal axes, while a spheroid has a circular cross section along one axis (the axis of symmetry) with ellipsoidal cross sections along the other two orthogonal axes. As such, a spheroid is defined by a single aspect ratio (E) based on the ratio of the lengths along and about the axis of symmetry (Figure 4.11). Spheroids are categorized as oblate if $E < 1$ (with the limiting shape $E\rightarrow 0$ corresponding to a disk) and prolate if $E > 1$ (with the limiting shape $E\rightarrow\infty$ corresponding to a needle). A spheroid with $E = 1$ corresponds to a sphere.

Based on (9.45), the Stokes correction for a spheroid (f_E) is the ratio of creeping-drag force to that of a sphere (where both have the same volume equivalent diameter):

$$f_{shape} = f_E \equiv \frac{F_D\left(E, Re_p \rightarrow 0\right)}{3\pi d\mu_f w}. \tag{9.46}$$

Thus $f_E\rightarrow 1$ as $E\rightarrow 1$. Exact, limiting and approximate solutions for the drag on spheroids at creeping-low conditions ($Re_p\rightarrow 0$) were derived by Oberbeck (1876) for flow parallel to the axis of symmetry ($f_{E\parallel}$) and for flow perpendicular to the axis of symmetry ($f_{E\perp}$). These drag corrections are given in Table 9.4 along with approximations for small departures from the spherical shape, that is, $|E - 1| \ll 1$, as well as the limiting solutions for $E\rightarrow 0$ (disk) or $E\rightarrow\infty$ (needle).

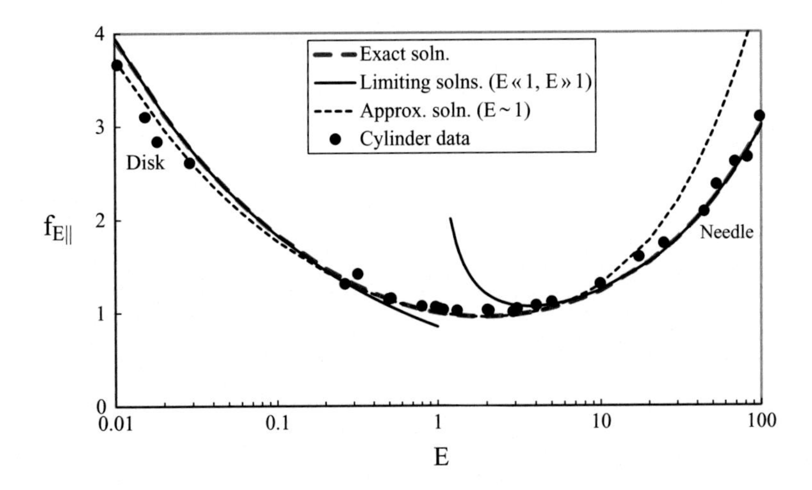

Figure 9.14 Influence of eccentricity on spheroid drag correction based on cylinder data from Dressel (1985) and Clift et al. (1978), where limiting solutions are theories for very low and very high aspect ratios, while the approximate solution is for the moderate aspect ratio case.

The creeping flow parallel drag corrections as a function of aspect ratio are shown in Figure 9.14, where the approximate and limiting solutions are reasonable over a wide range of aspect ratios (relative to the exact solutions). The figure also shows that large variations in aspect ratio yield only moderate changes in the drag correction (less than a fourfold change in f_E for $0.01 < E < 100$). This is because increases in friction drag for longer bodies are counteracted by decreases in pressure drag, while the reverse is true for shorter bodies. This low sensitivity allows the spheroid expression to also be reasonable for a cylinder in creeping flow, as shown in Figure 9.14.

Generally, spheroids tend move in a broadside orientation (parallel for oblate and perpindicular for prolate) as this is more stable (Figure 4.14). However, other orientations are possible due to initial conditions, trajectory instabilities, or nonuniform flow. In such cases, the effective drag force is a linear combination as:

$$\mathbf{F}_{\mathrm{D,eff}} = -3\pi\mu_f(\mathbf{w}_{\parallel} f_{E\parallel} + \mathbf{w}_{\perp} f_{E\perp}). \tag{9.47}$$

The RHS includes the relative velocity components that are parallel and perpendicular to the particle's axis of symmetry (and whose vector sum is $\mathbf{w}$). For an orientation that is neither exactly parallel nor perpindicular, $\mathbf{F}_{\mathrm{D,eff}}$ will *not* be parallel to $\mathbf{w}$ if the two correction factors are not equal. Because of this, a particle falling with an orientation angle to gravity will also have a glide trajectory angle (White, 2016). Since the drag force is formally defined to act in the opposite direction of $\mathbf{w}$, the additional smaller component that is perpendicular to $\mathbf{w}$ should be formally considered as a lift force (to be discussed in §10.1.4).

In Brownian motion, the particle orientation will be chaotic due to the random molecular interactions. All orientations can be assumed to be equally possible for many realizations (or long times) and the average correction factor can be obtained by integrating over all possible orientations as follows:

$$\langle f_E \rangle = \frac{E^{-1/3}\sqrt{1-E^2}}{\cos^{-1}E} \qquad \text{for } E < 1. \tag{9.48a}$$

$$\langle f_E \rangle = \frac{E^{-1/3}\sqrt{E^2-1}}{\ln\left(E+\sqrt{E^2-1}\right)} \qquad \text{for } E > 1. \tag{9.48b}$$

This average drag correction is also reasonable (to within a few percent) for the wide range of particle orientations associated with the Jeffrey orbits (Davis, 1991). Drag corrections for ellipsoids (in all three directions) and other orthotropic particles are given by Clift et al. (1978).

Other Symmetric Particles in Creeping Flow

Symmetric nonspheroidal particles with edges (such as cones, cubes, etc., of Figure 4.12) have a geometric surface that can be described analytically but often do not have analytical solution for the drag in the creeping-flow limit ($Re_p \rightarrow 0$). However, as a first approximation, the shapes and corresponding drag corrections of many particles may be approximated as spheroids by determining an effective aspect ratio (E). For example, a spheroidal approximation is reasonable for cylinders, as shown in Figure 9.14. However, edges tend to increase the drag. For example, a cylinder with E = 1 has a drag coefficient (based on volumetric diameter), which is about 14.4% larger than that for a sphere, primarily due to the increased surface area for the cylinder.

Such differences become more profound when the particles have additional edges and indentations (concave regions), as this causes extra surface area (A_{surf}) and/or projected area (A_{proj}) compared to that of a sphere of the same volume. As such, it is helpful to define the surface and projected areas normalized by that of a sphere with the same volume, as follows:

$$A^*_{surf} \equiv A_{surf}/\left(\pi d^2\right). \tag{9.49a}$$

$$A^*_{proj} \equiv A_{proj}/\left(\frac{1}{4}\pi d^2\right). \tag{9.49b}$$

These area ratios chracterize the degree of nonsphericity, where higher values will increase drag relative to that of a sphere (Ganser, 1993; Madhav and Chhabra, 1995; Xie and Zhang, 2001). By assuming that the 2:1 ratio of friction drag to form drag for creeping flow also extends to nonspherical particles, Leith (1987) developed an empirical drag correlation using the area ratios of (9.49), as follows:

$$f_{shape} = \frac{1}{3}\left(2\sqrt{A^*_{surf}} + \sqrt{A^*_{proj}}\right). \tag{9.50}$$

The surface and projected area ratios can be obtained analytically for most regular shapes. Table 9.5 lists several regularly shaped nonellipsoidal particle shapes along

Table 9.5 Projected and surface area ratios along with Stokesian drag corrections for symmetric particle shapes from data of Pettyjohn and Christiansen (1948) and Ahmadi (2005) with predicted corrections based on the technique of Leith (1987).

			f_{shape}	
Shape	A^*_{proj}	A^*_{surf}	**Predicted**	**Measured**
Sphere	1	1	1	1
Cube octahedron	~1	1.10	1.03	1.03
Octahedron	~1	1.17	1.05	1.07
Cube	~1	1.24	1.08	1.08
Tetrahedron	~1	1.49	1.15	1.19
2-sphere cluster	1.26 (broadside)	1.26	1.12	1.12
3-sphere cluster	1.44 (broadside)	1.44	1.20	1.27
4-sphere cluster	1.59 (broadside)	1.59	1.26	1.32
4-sphere cluster	1.59 (broadside)	1.59	1.26	1.17

with measured and predicted Stokes correction factors based on (9.50). The relations are generally reasonable but increased nonsphericity (especially with concave features) tends to result in decreased accuracy, often with an underprediction. In addition, highly flattened or elongated shapes ($E \ll 1$ or $E \gg 1$) tend to be overpredicted by this approximation (Ganser, 1993).

Nonspherical Particles in the Newton-Drag Regime

Next we consider nonspherical drag at high Reynolds numbers (before proceeding to intermediate values). As with spherical particles, nonspherical solid particles also have a Newton-based drag regime where the mean drag coefficient is approximately constant for a wide range of Reynolds numbers (e.g., 10^3–10^5). These drag coefficients can be normalized by that of a sphere (0.42 from the Clift–Gauvin expression) to define a shape-based Newton drag correction:

$$C_{shape} \equiv \frac{C_{D,shape,crit}}{C_{D,sphere,crit}} \approx \frac{\langle C_{D,shape} \rangle}{0.42}. \tag{9.51}$$

This correction has been called the "scruple" by Thompson and Clark (1991) and is generally related to the shape of the particle Cross Section (C/S) in the plane parallel to $\mathbf{w}$. Cylinders and prolate spheroids with $E > 1$ in free fall tend to have a steady broadside motion (perpindicular to the axis of symmetry), so this relative cross section will be circular. For such particles at high Reynolds numbers (Christiansen and Barker, 1965; Jayweera and Mason, 1965), the measured drag is approximately equal to the projected area (area that is perpindicular to $\mathbf{w}$) times the dynamic pressure. This is consistent with a sectional-based drag coefficient of unity. Therefore, the Newton particle drag correction for such shapes can be approximated as follows:

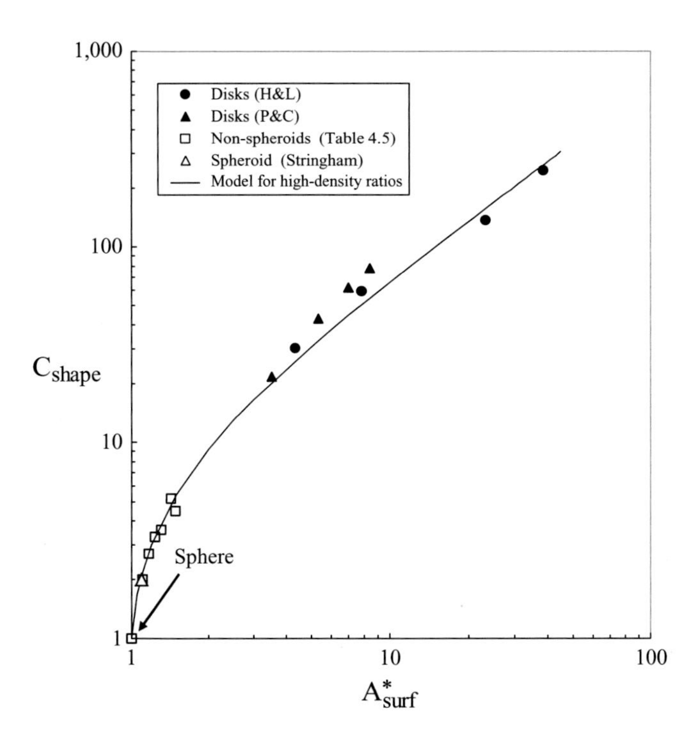

Figure 9.15 Newton drag correction for regularly shaped particles in liquids for $E \leq 1$ based on experimental data (Pettyjohn and Christiansen, 1948; Stringham et al., 1969; Haider and Levenspiel, 1989) and an empirical model (Loth, 2008c).

$$C_D \approx A^*_{proj} = 0.42 C_{shape} \qquad \textit{for circular } C/S \ (\textit{in the plane parallel to } \mathbf{w}). \tag{9.52}$$

Particles that are substantially oblate (e.g., disks or spheroids with $E < 0.5$) will tend to fall broadside (parallel to the axis of symmetry) or will undergo significant tumbling, so their relative C/S is a noncircular. As such, the surface area ratio is more relevant and correlates better with drag changes for oblate particles. For these particles in liquids, as shown in Figure 9.15, which includes nonspheroids such as cubes and cones, the shape-based Newton drag correction can be modeled as follows:

$$C_{shape} \approx 1 + 1.5\left(A^*_{surf} - 1\right)^{1/2} + 6.7\left(A^*_{surf} - 1\right) \quad \textit{for noncircular } C/S \textit{ in liquids.} \tag{9.53}$$

Comparisons with measured values for free-fall trajectories of different geometries are given in Table 9.6, where it can be seen that (9.53) is often reasonable, except for the highly angular shapes. Other models for drag corrections (as well as surface and projected area ratios) for specific regularly shaped geometries are given by Clift et al. (1978), Lasso and Weidman (1986), and Loth (2008c).

Notably, drag is more sensitive to shape at higher Reynolds numbers, e.g., C_{shape} values in Table 9.6 are much larger than the f_{shape} values of Table 9.5 for a given particle shape. This is because creeping conditions ensures attached flow over the particle regardless of shape, while high Re_p conditions yield significantly increased

Table 9.6 Predicted and measured Newton drag corrections for various regular particle shapes based on data from Pettyjohn and Christiansen (1948), Stringham et al. (1969), Clark et al. (1989), and Gogus et al. (2001), which are shown as open squares in previous figure.

			C_{shape}	
Particle shape	A^*_{surf}	$C_{D,crit}$	**Predicted**	**Measured**
Sphere	1.00	0.42	1.0	1.0
Spheroid (E = 0.5)	1.095	0.84	2.1	2.0
Cube octahedron	1.10	0.83	2.1	2.0
Octahedron	1.17	1.14	2.8	2.7
Cube	1.24	1.40	3.3	3.3
Equilateral prism	1.43	2.20	4.9	5.2
Tetrahedron	1.49	1.90	5.3	4.5

flow separation at higher nonsphericity. In the following, intermediate Re_p conditions are considered by blending these two limits.

Nonspherical Regularly Shaped Particles at Intermediate Re_p

To integrate the Stokes and Newton drag regimes for intermediate Re_p, Ganser (1993) defined transformed values of the drag coefficient and Reynolds number as follows:

$$C_D^* \equiv \frac{C_D}{C_{shape}}. \tag{9.54a}$$

$$Re_p^* \equiv \frac{C_{shape}\,Re_p}{f_{shape}}. \tag{9.54b}$$

As shown in Figure 9.16, particle data for intermediate Re_p conditions using these transformed parameters tend to collapse along two different curves: one for circular C/S, and one for noncircular C/S. The particles with a circular C/S have a Reynolds number dependence like that of spheres and so can use a modified form of (9.16):

$$C_D^* = \frac{24}{Re_p^*}\left[1+0.15\left(Re_p^*\right)^{0.687}\right] + \frac{0.42}{1+42{,}500\left(Re_p^*\right)^{-1.16}} \quad \textit{for circular}\ \mathrm{C/S}. \tag{9.55}$$

In contrast, particles with a noncircular C/S produce unsteady wake separation at much lower Re_p than that for a sphere, and the flow separation point does not vary strongly with Reynolds number. As such, these particles are more likely to reach a Newton-based drag regime at a smaller Re_p, as shown in Figure 9.16, and these intermediate Re_p conditions are better fitted by the following:

$$C_D^* = \frac{24}{Re_p^*} + 0.4 \qquad \textit{for noncircular}\ \mathrm{C/S}. \tag{9.56}$$

The transformed approach of (9.54) with (9.55) or (9.56) works well for a wide variety of other regularly shaped geometries, including cubes, cones, and clusters (Loth, 2008c).

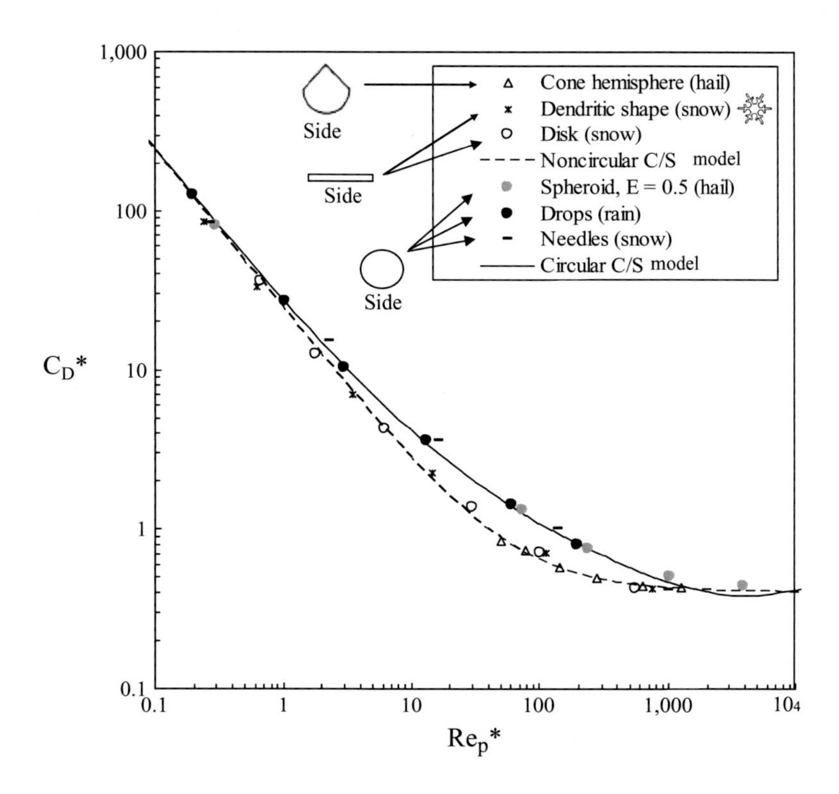

Figure 9.16 Normalized drag relationships for various forms of precipitation with symmetric shapes where the cross-sectional shape in the side view determines the intermediate Reynolds number variation (Jayaweera and Mason, 1965; Beard and Prupacher, 1969; Stringham et al., 1969; List and Schemenauer, 1971).

9.2.6 Irregularly Shaped Solid Particle Drag

Many naturally occurring particles have an irregular shape that cannot be defined by a symmetric geometric surface, such as the angular particle shapes in Figures 4.10a and 9.17. These angular irregular shapes often stem from crushing or random agglomeration. For such particles, computing the area ratios of (9.49) is difficult. A more convenient measure of nonsphericity for thesc particles is the Corey Shape Factor (CSF), which uses dimensions that can be measured for a particle lying flat on a surface, as shown in Figure 9.17. These include the shortest dimension (d_{min}), the medium dimension (d_{med}), and the longest dimension (d_{max}) combined as follows:

$$\mathrm{CSF} \equiv \frac{d_{min}}{\sqrt{d_{max} d_{med}}}. \tag{9.57}$$

As such, CSF reduces as the shape becomes less spherical. However, it should be noted that CSF = 1 for both a sphere and a cube, so CSF does not uniquely define the shape.

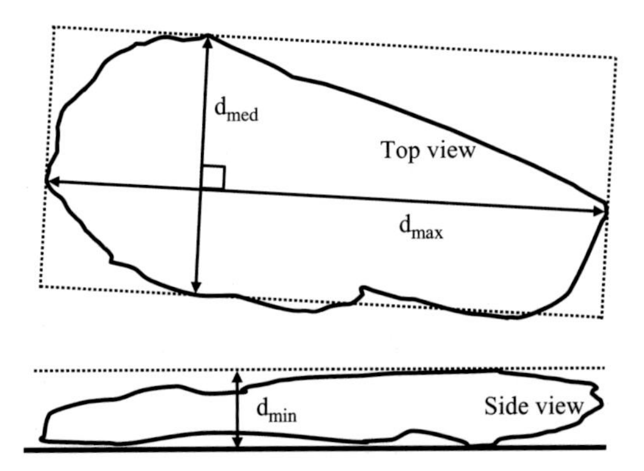

Figure 9.17 Schematic of an irregularly shaped particle lying flat on a surface to allow Corey Shape Factor (CSF) measurement, where the top view provides the longest dimension (d_{max}) and the perpendicular intermediate dimension (d_{med}), while the side view height is the shortest dimension (d_{min}).

For a set of particles with random shapes, a statistically large number of CSF values and time durations are needed to obtain an average characterization. This is because irregular and angular particles have shape asymmetries that cause torque imbalances resulting in complex tumbling and random orientations. The resulting complex fluid dynamics forces lead to drag force variations as a function of time. As such, only approximate and probabilistic drag models are possible for highly irregular particles. Notably, decreased sphericity and CSF will generally cause increase drag. For example, average CSF values are about 0.9 for smooth, well-rounded shapes, are about 0.7 for naturally rounded shapes with some edges, are about 0.55 for angular test dusts (no rounding), and are about 0.4 for crushed sediment with sharp edges (Jimenez and Madsen, 2003; Connolly et al., 2020). While several other shape characterization parameters besides CSF have been proposed (not-roundedness, anisometry, bulkiness ratio, circularity, shape entropy, polygonal harmonics, etc.), these are often more complex to characterize and do not generally improved drag predictions (Clift et al., 1978; Dressel, 1985; Smith and Cheung, 2003; Tran-Cong et al., 2004).

In the following, the influence of CSF on drag is considered for various Reynolds numbers. For creeping flow (at $Re_p \rightarrow 0$), experimental measurements for f_{shape} as a function of CSF for irregularly shaped particles in free fall are given in Figure 9.18. As expected, there is significant scatter, and decreasing CSF is correlated with increasing drag (Dressel, 1985). A model to represent the average of many orientations and particle samples (Connolly et al., 2020) follows:

$$\langle f_{shape} \rangle = 1/(CSF)^{0.18} \qquad \textit{for irregular shapes at } Re_p \rightarrow 0. \tag{9.58}$$

Measurements of the drag increase for the Newton regime in Figure 9.19 show higher scatter and a stronger sensitivity to shape changes than that for creeping flow (as was

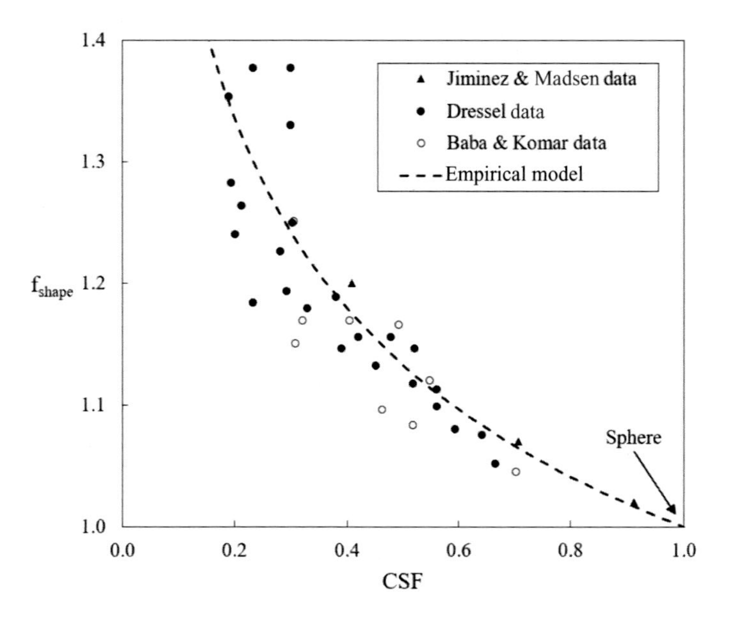

Figure 9.18 Stokesian drag correction for irregular particles in terms of the Corey Shape Factor with data from Dressel (1985), Baba and Komar (1981), and Jimenez and Madsen (2003) for $Re_p < 0.5$ and empirical model (Connolly et al., 2020).

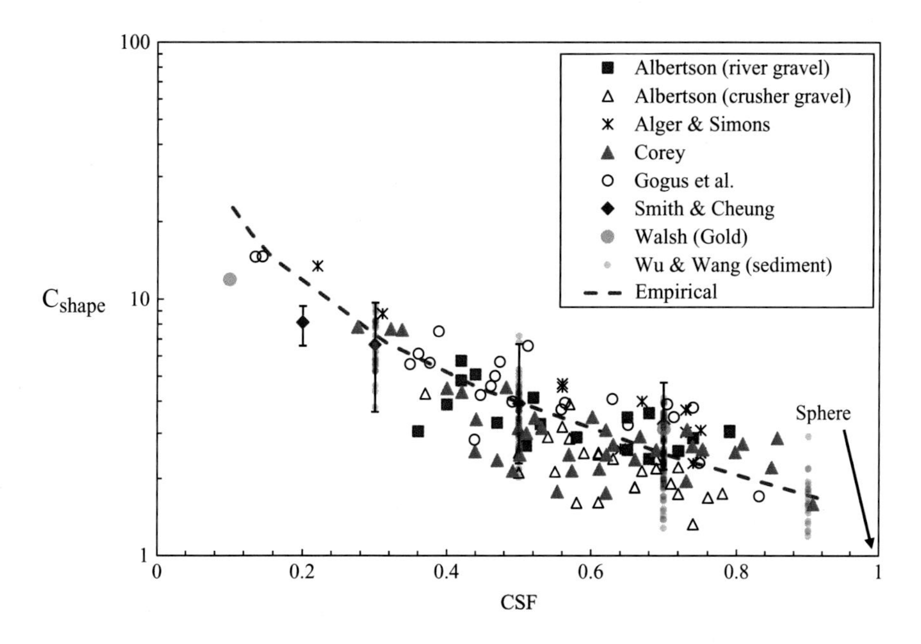

Figure 9.19 Newton drag correction for irregularly shaped particles in terms of the Corey Shape Factor based on data of Albertson (1952), Alger and Simons (1968), Walsh (1988), Gogus et al. (2001), Smith and Cheung (2003), and Wu and Wang (2006), as well as the empirical model of Connolly et al. (2020).

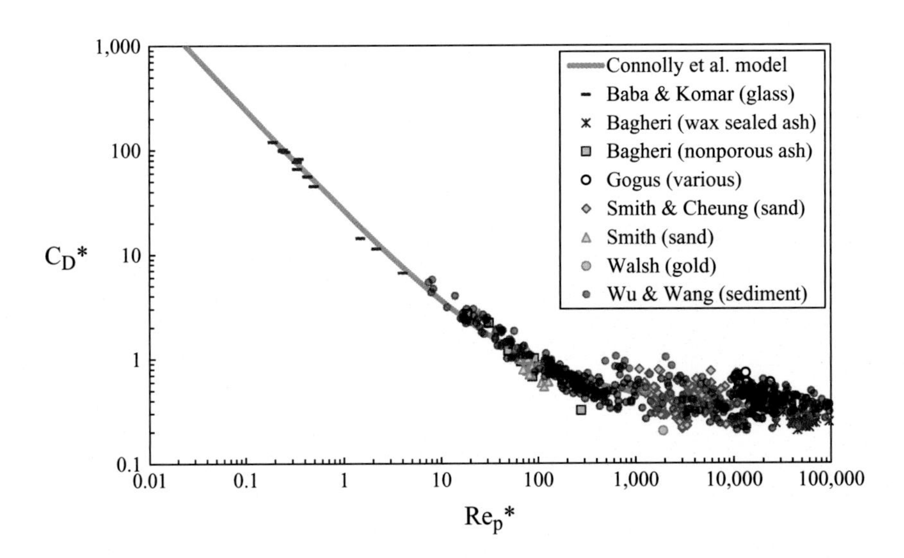

Figure 9.20 Drag correction for irregularly shaped particles for a range of normalized Reynolds numbers (Connolly et al., 2020).

also noted for regularly shaped particles). This Newton regime can be described with an empirical model, given as follows:

$$\langle C_{\text{shape}} \rangle = \sqrt{6}/\text{CSF} - 1 \qquad \textit{for irregular shapes for } \text{Re}_{\text{p,sub}} < \text{Re}_{\text{p}} < \text{Re}_{\text{p,crit}}. \tag{9.59}$$

To obtain the drag coefficient at intermediate Re_{p} values, the transformation of (9.54) can again be employed, but with the following transformed drag curve (as shown in Figure 9.20):

$$C_D^* = \frac{24}{\text{Re}_p^*}\left[1 + 0.1\left(\text{Re}_p^*\right)^{0.6}\right] + 0.3 \quad \textit{for irregular shapes for } \text{Re}_{\text{p}} < \text{Re}_{\text{p,crit}}. \tag{9.60}$$

As can be seen, this is only approximate as there is substantial scatter, especially at high Re_{p}.

9.3 Drag Force for an Isolated Fluid Particle

If there is weak internal recirculation and a spherical shape, the flow for fluid particles can be approximated with a no-slip condition so that drag expressions for a spherical solid particle of §9.2.1 are appropriate. This is reasonable if the Weber number is small such that shape deformation is negligible, and the particle has high viscosity, as noted in (3.42), or is fully contaminated such that internal recirculation is negligible, as noted in (3.44). An example of this is shown in Figure 9.21 for drops in air and for

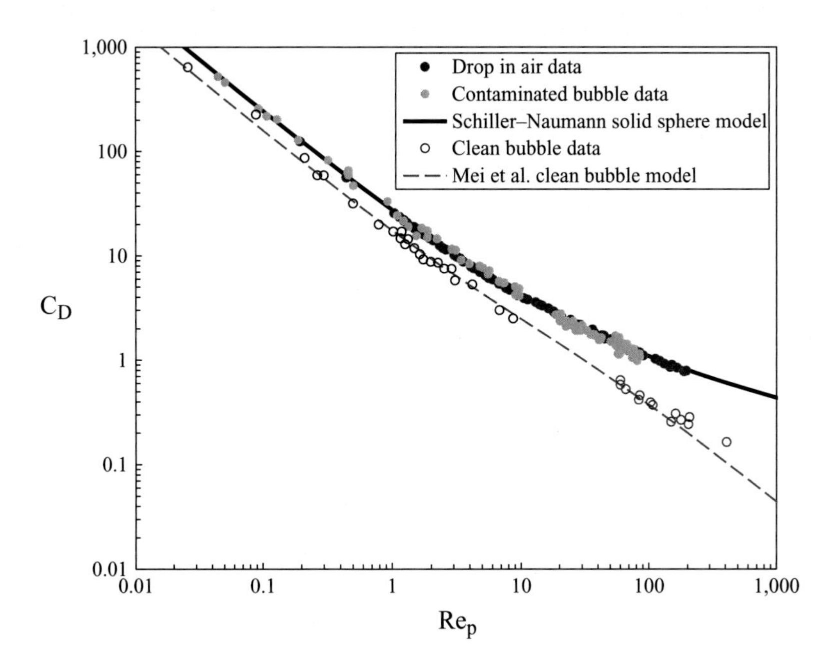

Figure 9.21 Experimental results of drag coefficient of spherical fluid particles, including drops in air as blue solid symbols (Beard and Prupacher, 1969), clean bubbles in various liquids as open symbols (Haberman and Morton, 1953; Miyahara and Takahashi, 1985), and contaminated bubbles as gray symbols (Garner and Hammerton, 1954; Sridhar and Katz, 1995), with comparison to a solid sphere model (Schiller and Naumann, 1933), and a clean bubble model (Mei et al., 1994).

contaminated bubbles in liquids. In the following, the effects of internal recirculation on fluid sphere drag are considered in (§9.3.1), followed by the effects of deformation in terms of aspect ratio (§9.3.2) and drag (§9.3.3).

9.3.1 Spherical Drops and Bubbles

For a clean or partially contaminated fluid particle, relative velocity will cause a finite tangential velocity along the particle interface that drives internal circulation. This tangential velocity reduces the surface shear stress and the viscous drag and can also reduce the form drag. These effects are considered in the following, first for low Reynolds numbers and then for higher Reynolds numbers.

Fluid Particles in the Creeping-Flow Limit

If the fluid sphere is clean with no contaminants, then an exact analytical solution can be obtained for the creeping-flow regime as discussed in §3.2.2. The creeping-flow Hadamard–Rybczynski drag force for a clean fluid particle yields the result of (3.40). Brenner and Cox (1963) included linearized inertial terms to extend this to include a first-order correction:

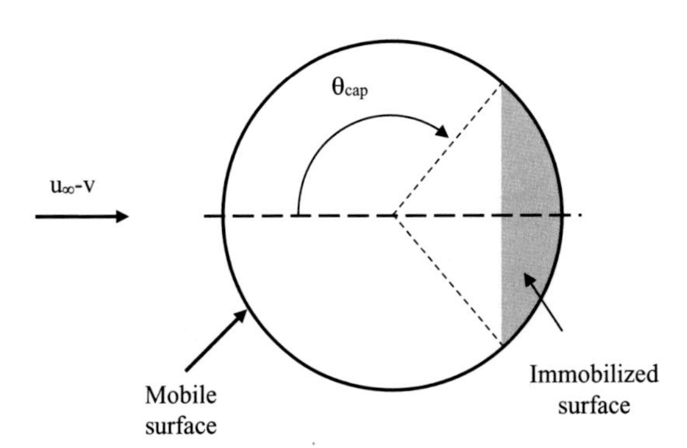

Figure 9.22 Schematic of a stagnant cap with θ_{cap} as the angle from the leading edge at which the interface becomes immobile.

$$f_{\mu *} = \frac{2+3\mu_p^*}{3+3\mu_p^*} + \frac{3}{16}\mathrm{Re}_p\left(\frac{2+3\mu_p^*}{3+3\mu_p^*}\right)^2 + o\left(\mathrm{Re}_p^2 \ln \mathrm{Re}_p\right) \quad \textit{for clean conditions.} \tag{9.61}$$

For very high particle viscosity (e.g., a drop in a gas), the drag correction approaches the Oseen expression of (9.12). The correction for very low particle viscosity (e.g., clean gas bubble in a liquid) approaches $2/3 + \mathrm{Re}_p/12$. Experimental results for clean bubbles are shown in Figure 9.21, where a 2/3 reduction is reasonable for $\mathrm{Re}_p < 1$.

In general, partially contaminated bubbles have drag coefficients intermediate to the clean and fully contaminated cases. A partially contaminated surface is illustrated in Figure 9.22, where the immobilized region is in the rear portion. This is because surfactants that impact in the front generally attach themselves to the interface and are swept by the finite tangential velocity to the aft region, where their concentration builds up (Bel Fdhila and Duineveld, 1996; Cuenot et al., 1997). This immobilized region can be modeled as a stagnant cap, which starts at the cap angle, θ_{cap}, while the upstream portion is assumed to be fully mobile (Griffith, 1962). As such, the cap angle for a partially contaminated condition is between the fully contaminated and clean limits, that is, $0° < \theta_{cap} < 180°$. Using this model, Sadhal and Johnson (1983) obtained an exact solution for partial contamination with a variable viscosity case at the creeping-flow limit, as follows:

$$f_{cap} = 1 - \frac{6\theta_{cap} + 3\sin\theta_{cap} + 3\sin 2\theta_{cap} - \sin 3\theta_{cap}}{18\pi + 18\mu_p^*}. \tag{9.62}$$

The RHS reverts to 1 for fully contaminated conditions ($\theta_{cap} = 0°$) and to 2/3 for clean conditions ($\theta_{cap} = 180°$). To use this expression, it would be ideal if θ_{cap} could be predicted as a theoretical function of the surfactant concentration and the surrounding liquid properties. Unfortunately, this angle is not easily predicted. Even determining

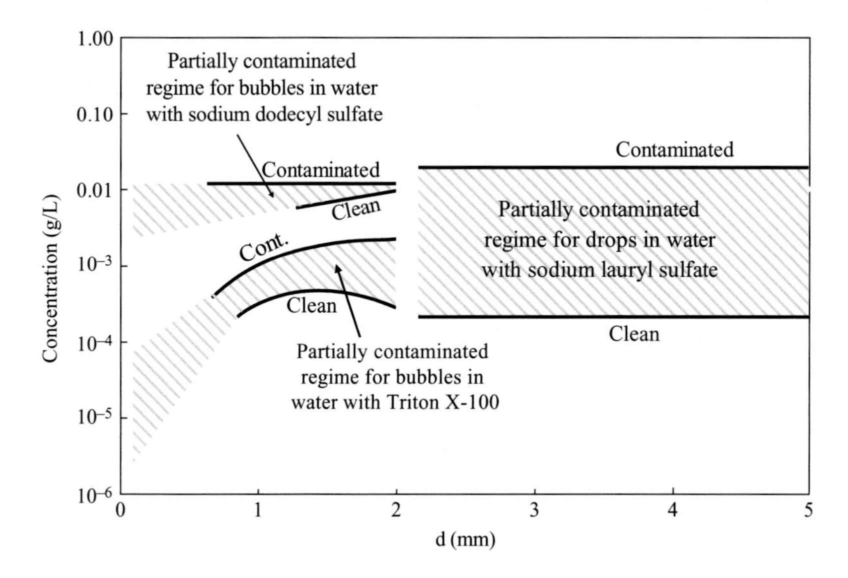

Figure 9.23 Summary of surfactant concentration regimes for various surfactants in water where the limits between clean and fully contaminated conditions are respectively indicated by the solid lines at the bottom and top of the hatched lines (Bel Fdhila and Duineveld, 1996). Boundaries for diameters below 0.7 mm are extrapolated estimates.

the surfactant concentration boundaries for fully contaminated or clean conditions is difficult, since they can be influenced by solubility, surface pressure, cation concentration, organic concentration, molecular weight, etc. As a result, different surfactants can have different levels of effectiveness, as demonstrated by Figure 9.23. For example, a 1 mm bubble with a 10^{-3} g/L surfactant concentration can be considered clean for sodium dodecyl sulfate but fully contaminated for Triton X-100. This is further complicated by the fact that θ_{cap} can be both time and shape dependent (Wu and Gharib, 1998).

Fluid Particles at Intermediate and Large Reynolds Numbers

When the Reynolds number becomes much greater than unity, the flow around a a fluid particle becomes an even a stronger function of the internal recirculation. For example, Figure 9.24 shows the dependence of the viscosity ratio on the surrounding flow streamlines for $Re_p = 100$. In particular, a very high viscosity ratio (as for a drop in a gas) of Figure 9.24a yields significant flow separation and a wake like that of a solid particle (Figure 9.1). This consistent with the surface pressure distributions of Figure 9.25, where that for a liquid sphere in a gas (drop) is nearly identical to that for a solid sphere, indicating recirculation effects are very weak. As such, the drag for spherical drops or contaminated bubbles is close (within 2%) to that for a solid sphere, as shown in Figure 9.21.

In contrast, Figure 9.24b for a very low viscosity ratio (as in a clean gas bubble in a liquid) has no flow separation due to the reduced surface shear stress resulting from

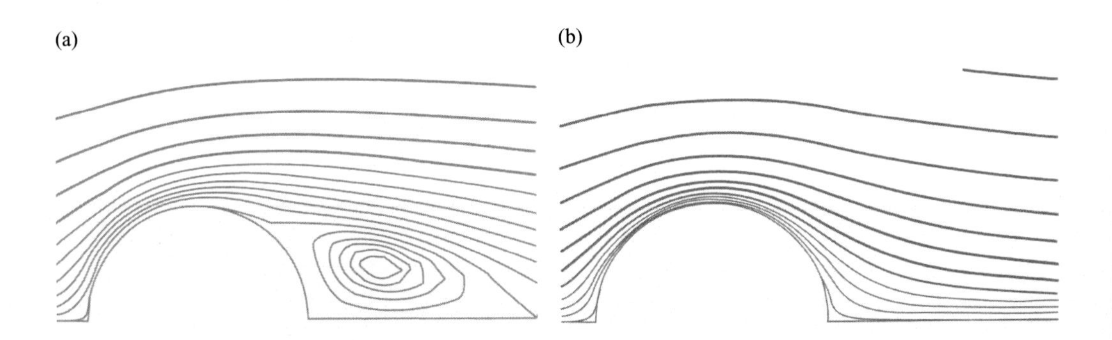

Figure 9.24 Streamlines for steady flow past a sphere at $Re_p = 100$ for (a) $\mu_p \gg \mu_f$ and (b) $\mu_p \ll \mu_f$ (Clift et al., 1978).

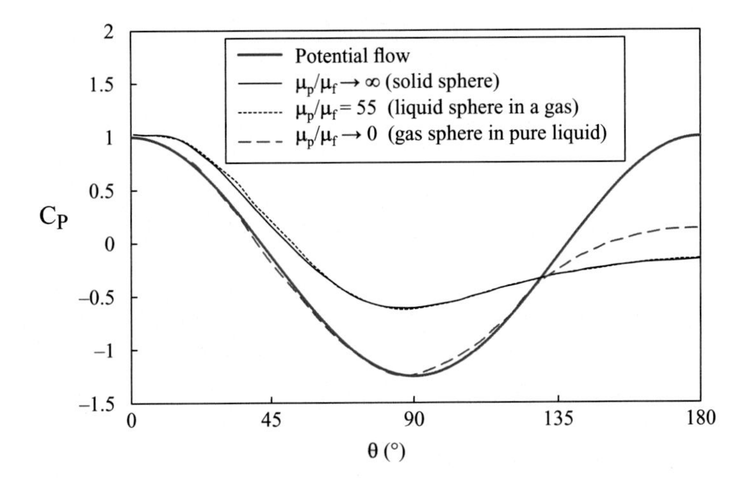

Figure 9.25 Pressure distribution over spheres for different viscosity ratios at $Re_p = 100$ with comparison to potential flow solution (Clift et al. 1978).

high internal recirculation. In fact, a clean spherical gas bubble will have no wake separation *for all Reynolds numbers*, so the flow outside of the boundary layer at high Re_p will be more like that given by potential flow theory. This is consistent with the pressure coefficient shown in Figure 9.25 for a clean gas sphere that follows the inviscid potential flow solution up until about 130°, after which the pressure recovery is not as high as the inviscid case owing to viscous losses.

The lack of flow separation for a clean bubble fortunately allows closed-form theoretical solutions. In particular, Levich (1949, 1962) assumed a thin boundary layer on the surface and in the wake and approximated the streamlines outside of these viscous regions with a potential flow solution. By assuming high Reynolds numbers and a steady fluid particle surface boundary condition (3.36), the first-order drag corrections (f and C_D) were obtained as follows:

$$f_{Levich} = 2 + o\left(Re_p^{-1/2}\right) \quad (9.63a)$$

for clean bubbles and $Re_p^{1/2} \gg 1$.

$$C_{D,Levich} = 48/Re_p + o\left(Re_p^{-3/2}\right) \quad (9.63b)$$

This linear drag result stems from the fully attached flow. In contrast, contaminated bubbles at high Reynolds numbers have flow separation that causes a nonlinear drag law consistent with the Schiller–Naumann fit (9.13).

To improve the accuracy of (9.63) for clean bubbles at intermediate Reynolds numbers, Moore (1963) extended the theory to include an additional higher-order term, yielding a drag correction as follows:

$$f_{Moore} = 2\left[1 - 2.21/\sqrt{Re_p}\right] + o\left(Re_p^{-11/6}\right) \quad \textit{for clean bubbles with } Re_p \gg 1. \quad (9.64)$$

The term in the square brackets represents a correction to the Levich drag, and Moore's expression is reasonable for clean bubbles with Re_p values as low as 100. To bridge the gap between the Moore and Hadamard–Rybczynski theoretical results of (3.41), Mei et al. (1994) proposed an empirical blend (which properly tends to each limit) as follows:

$$f_{Mei} = \frac{2}{3}\left\{1 + \left[\frac{8}{Re_p} + \frac{1}{2}\left(1 + \frac{3.315}{\sqrt{Re_p}}\right)\right]^{-1}\right\} \quad \textit{for spherical clean bubbles.} \quad (9.65)$$

As shown in Figure 9.21, this combination of theories (good for all Re_p) compares well with experimental data and resolved-surface simulations. This figure also shows that the drag difference between spherical clean and contaminated bubbles becomes exaggerated as Re_p increases. For partially contaminated bubbles, one may interpolate between the limits for a solid sphere (9.13) and a clean sphere (9.65) using (9.62) as follows:

$$f = \left(3f_{cap} - 2\right)f_{Re} + \left(3f_{cap} - 3\right)f_{Mei} \quad \textit{for partially contaminated bubbles.} \quad (9.66)$$

This interpolation is qualitatively consistent with numerical results (Magnaudet and Eames, 2000).

Notably, the clean data of Figure 9.21 are not for bubbles in water. This is because it is difficult to achieve clean bubble conditions in water, especially at small sizes (recall Table 3.1). As such, the terminal drag of bubbles in distilled water, as shown in Figure 9.26, is close to the solid particle limit for $Re_{p,term} < 50$ and then tends to the clean bubble drag limit of (9.65) for $Re_{p,term}$ approaching 430 (after which deformation effects cause a drag increase). Based on these data, the following is a rough approximation for spherical bubbles in distilled water:

$$f_{Mei,distilled} = (3/2)f_{Mei} \quad \textit{for spherical bubbles in distilled water.} \quad (9.67)$$

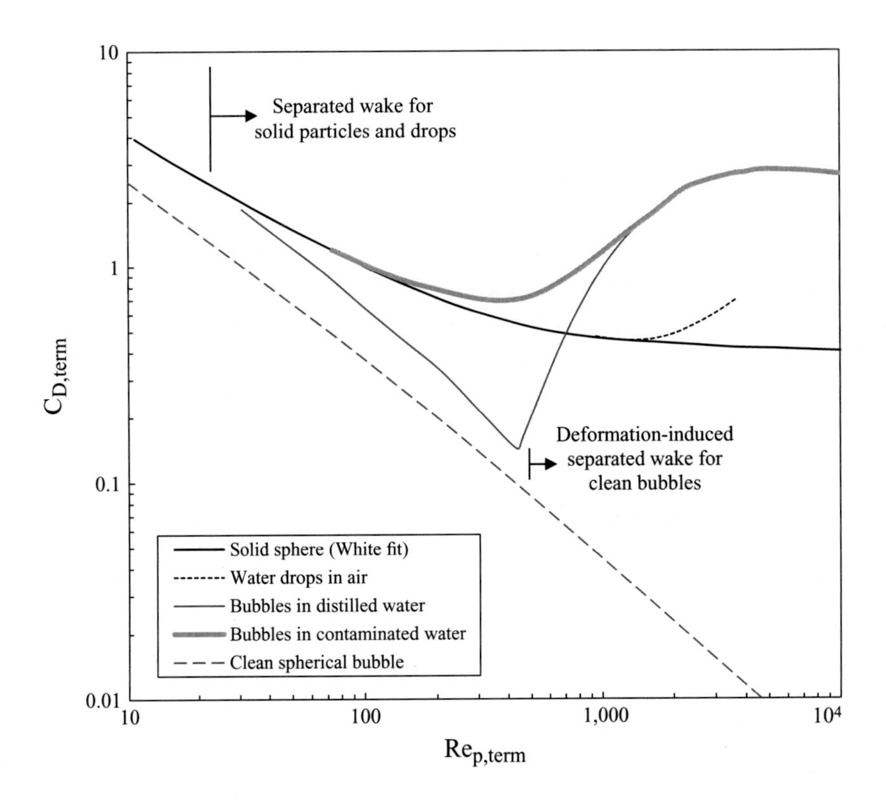

Figure 9.26 Terminal drag coefficients for solid and fluid particles.

For small $\mathrm{Re_p}$, this reverts to the fully contaminated drag ($f = 1$) consistent with small bubbles in distilled water (Table 3.1), while large $\mathrm{Re_p}$ yields a 50% increase relative to Moore's drag ($f = 3$) consistent with the trends of Figure 9.26.

For a spherical immiscible drop in a liquid ($\mu_p \sim \mu_f$), the drag correction will be bounded between that for clean bubbles ($\mu_p \gg \mu_f$) and for drops ($\mu_p \gg \mu_f$). Unfortunately, analytical solutions are only available for creeping flow (3.41), so guidance at larger Reynolds numbers must be taken from experiments and resolved-surface simulations (Figure 9.27). Using these results, drag of spherical immiscible drop in a liquid can be modeled (Loth, 2008a) using a combination of the solid sphere and clean bubble results and a viscosity ratio function (u):

$$\frac{f_{\mathrm{Re},u} - f_{\mathrm{Mei}}}{f_{\mathrm{Re}} - f_{\mathrm{Mei}}} = \frac{u + 0.01\,\mathrm{Re_p}(0.4u - 0.8u^2 + 1.4u^3)}{1 + 0.01\,\mathrm{Re_p}}. \tag{9.68a}$$

$$u \equiv \frac{\mu_p}{\mu_p + \mu_f} \tag{9.68b}$$

Notably, this correlation is limited to spherical fluid particles. Deformation becomes important as the drop or bubble size increases due to increases in Weber number (§4.3–§4.5). The aspect ratio and drag for deformed fluid particles are discussed in the following subsection.

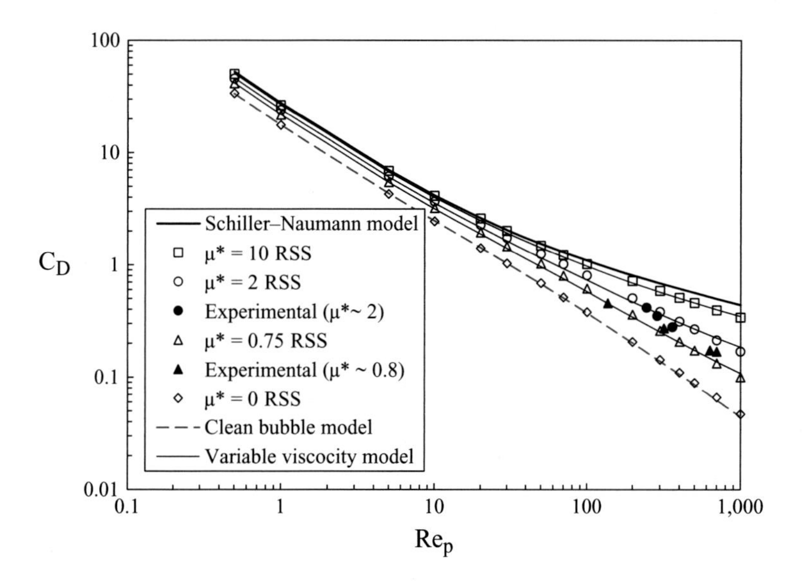

Figure 9.27 Drag coefficient from experiments of drops in gas (Beard and Prupacher, 1969) of clean bubbles (Haberman and Morton, 1953; Miyahara and Takahashi, 1985) and of intermediate viscosity drops in liquids (Winnikow and Chao, 1966). Also shown are resolved-surface simulations (RSS) of Feng and Michealidis (2001) as well as an empirical variable viscosity model relative to that for a solid particle and a clean bubble.

9.3.2 Aspect Ratio and Drag of Deformed Fluid Particles

For drops in air and bubbles in distilled water moving at terminal velocity (Figure 9.28), deformation onset occurs at diameters of roughly 1 mm. However, bubbles in tap water require somewhat larger diameters because the drag increase due to contamination leads lowers their terminal velocities and Weber numbers. For larger diameters bubbles, there will be a transition to an oblate spheroid and then to a spherical cap shape at $Re_p \sim$ 4,700, for which the level of contamination is no longer important because shape and drag are largely governed by bluff body separation. For larger diameter drops, there will also be a transition to oblate spheroids, beyond which instabilities lead to breakup (recall Figure 4.18). In both cases, the parameters that govern these shape transitions and their impact on drag are the Weber number (ratio of relative hydrodynamic stress to surface tension stress) and the Reynolds number (ratio of relative hydrodynamic stress to viscous stress), as discussed in §4.3.1.

The overall effect of Re_p and We on fluid particle aspect ratio is shown in Figure 9.29, where the solid lines indicate boundaries based on theory, and the dashed lines indicate boundaries based on experimental results. In general, the figure shows that particles at low Re_p and We will tend to be spherical while those at high Re_p and We take on spherical-cap shapes or breakup (for drops in a gas). Similar maps based on Morton number and Bond number (as shown in many texts) are much more complex and are only appropriate for terminal velocity conditions (§4.3.1). As such, the following subsection discusses the aspect ratio and drag for fluid particles for Re_p and We.

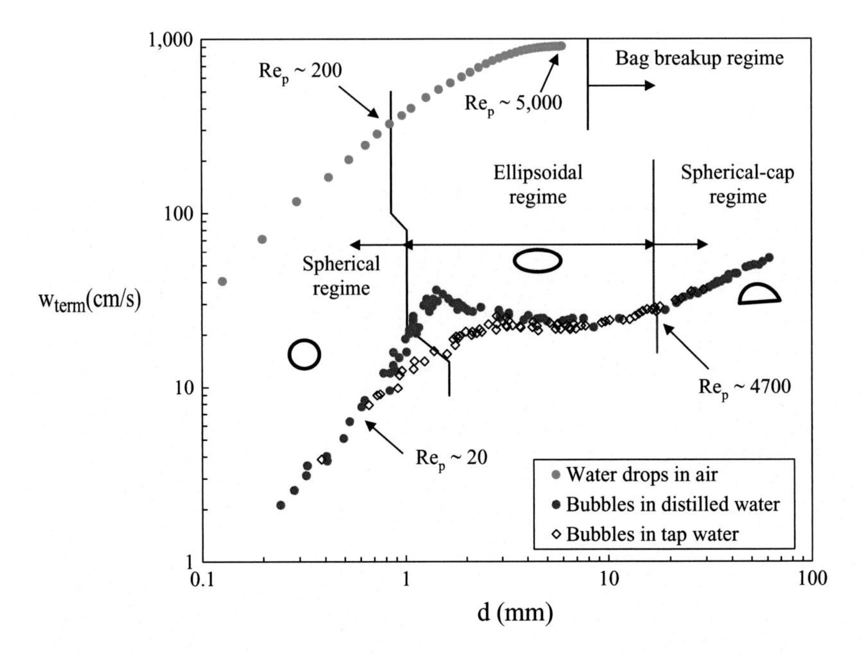

Figure 9.28 Terminal velocities of water drops in air from Beard (1976) and air bubbles in water from Haberman and Morton (1954) as well as Clift et al. (1978).

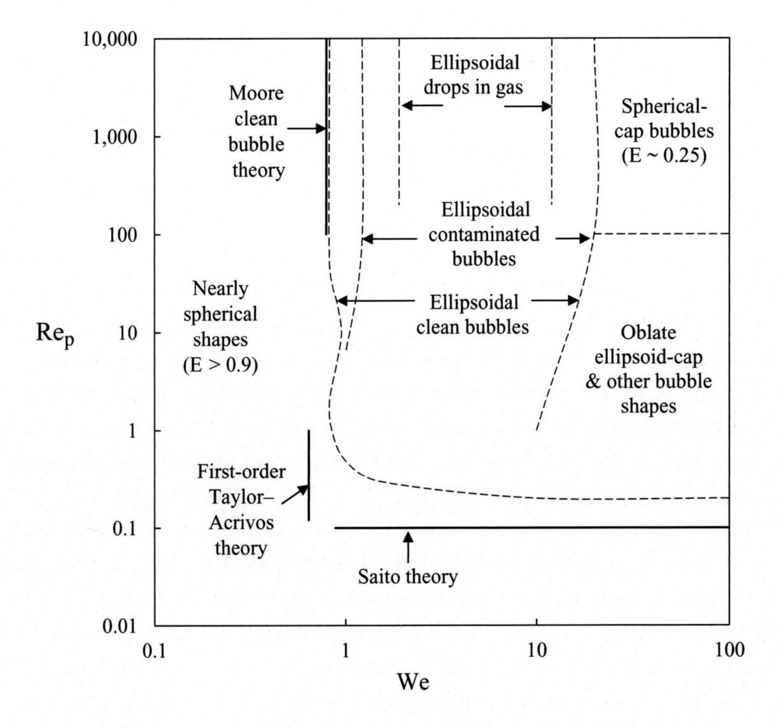

Figure 9.29 Qualitative boundaries for shapes of bubbles and drops where solid lines are theoretical bounds and dashed lines are empirical bounds based on Loth (2008a).

Theoretical Deformation and Drag for Low Reynolds Numbers

The shape of fluid particles has been investigated theoretically by many researchers and perhaps the most important low Reynolds number results are given by Saito (1913) and Taylor and Acrivos (1964). Saito examined showed that a fluid particle of any viscosity ratio and Weber number will remain spherical for the creeping-flow limit (where inertial terms are neglected). This simple criterion can be qualitatively expressed as follows:

$$\mathrm{Re_p} \to 0 \qquad \textit{particles are spherical.} \tag{9.69}$$

In Figure 9.29, this relationship is quantified as $\mathrm{Re_p} < 0.1$. However, the experimentally observed limit (designated by the dashed curve in Figure 9.29) suggests that spherical conditions persist to at least $\mathrm{Re_p} < 0.2$ for clean bubbles and can be even higher for contaminated bubbles and drops (Pan and Acrivos; 1968, Grace, 1973; Grace et al., 1976; Bhaga and Weber, 1981).

Taylor and Acrivos (1964) extended Saito's work to obtain a theoretical aspect ratio for weak convection and deformation ($\mathrm{Re_p} \ll 1$ and $\mathrm{We} \ll 1$) via the particle viscosity and density ratios:

$$E \approx 1 - 3c_{\mathrm{We}}\mathrm{We}\left[1 + \frac{2\left(11\mu_p^* + 10\right)}{7\left(10\mu_p^* + 10\right)}\frac{\mathrm{We}}{\mathrm{Re_p}}\right]. \tag{9.70a}$$

$$c_{\mathrm{We}} = \frac{1}{16\left(\mu_p^* + 1\right)^3}\left[\frac{81}{80}\mu_p^{*3} + \frac{57}{20}\mu_p^{*2} + \frac{103}{40}\mu_p^* + \frac{3}{4} - \frac{\left(\rho_p^* - 1\right)\left(\mu_p^* + 1\right)}{12}\right]. \tag{9.70b}$$

The first-order theory of (9.70a) (which ignores the term in the square brackets) indicates that deformation increases linearly with Weber number, while second-order theory (effects of terms in the square brackets) yields even stronger deformation. Comparison with numerical simulation in Figure 9.30 indicates the first-order theory is reasonable for We = 0.5 but overpredicts deformation for We = 1. As such, nonlinear effects at finite $\mathrm{Re_p}$ reduce deformation compared to that of creeping-flow theory of (9.70).

For clean bubbles, comparisons of the theories to experimental and numerical results for clean bubbles are shown in Figure 9.31 for low Reynolds numbers, where both theories again generally overpredict deformation. If one defines the transition from spherical to oblate shapes to occur at $E < 0.9$, theoretical transition for a bubble occurs at $\mathrm{We} > 0.64$ as indicated by the vertical solid-line boundary in Figure 9.29. If one instead bases $E < 0.9$ on low $\mathrm{Re_p}$ experiments and simulations (as in Figures 9.30 and 9.31), then the transition occurs at a higher Weber number, as indicated by the dashed curve of in Figure 9.29.

For drops in a gas, the c_{We} of (9.70b) is about 20% larger compared to that for bubbles in a liquid, so that drops are theoretically expected to be more deformed than

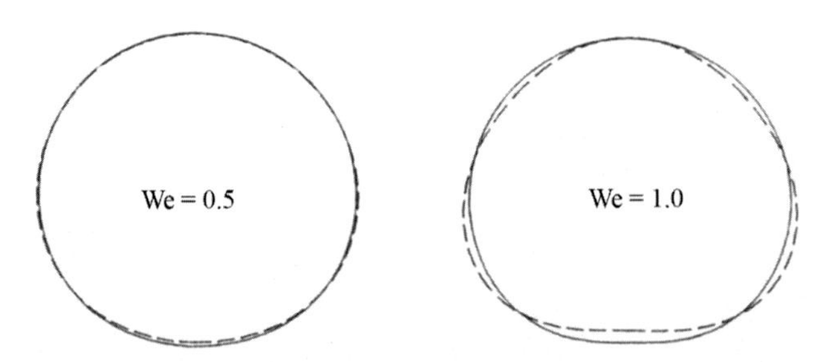

Figure 9.30 Comparison of clean bubble shapes at $\mathrm{Re_p} = 0.5$ based on Taylor and Acrivos (1964) first-order theory, as shown by dashed line, and simulation by Ryskin and Leal (1984), as shown by solid line.

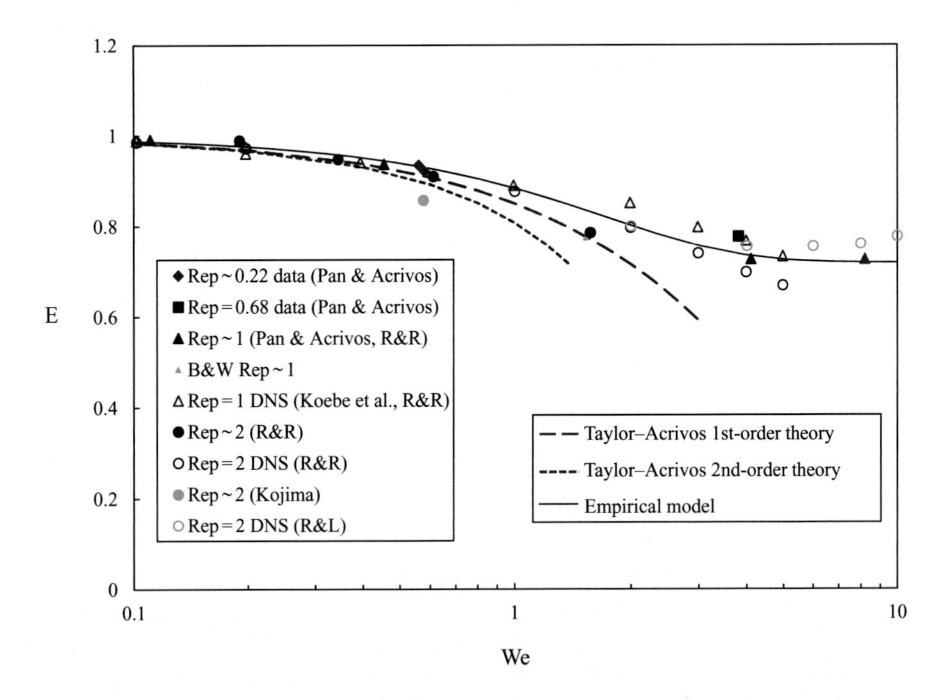

Figure 9.31 Aspect ratios of clean bubbles at low Reynolds numbers ($\mathrm{Re_p} \leq 2$), where solid symbols are experimental data (Kojima et al., 1968; Pan and Acrivos, 1968; Bhaga and Weber, 1981; Raymond and Rosant, 2000) and hollow symbols are resolved-surface simulations (Ryskin and Leal, 1984; Raymond and Rosant, 2000; Koebe et al., 2003) compared to Tylor-Acrivos theories and the empirical fit (Loth, 2008a).

bubbles at the same We and $\mathrm{Re_p}$. However, there are few, if any, experimental results for droplet deformation in a gas at low Reynolds numbers (since deformation tends to occur for $\mathrm{Re_p} > 200$).

For the drag along the axis of symmetry (broadside motion for an oblate shape), Taylor and Acrivos (1964) used the aspect ratios predicted by (9.70) to extend the Brenner and Cox theory of (9.61) to include Weber number effects, as follows:

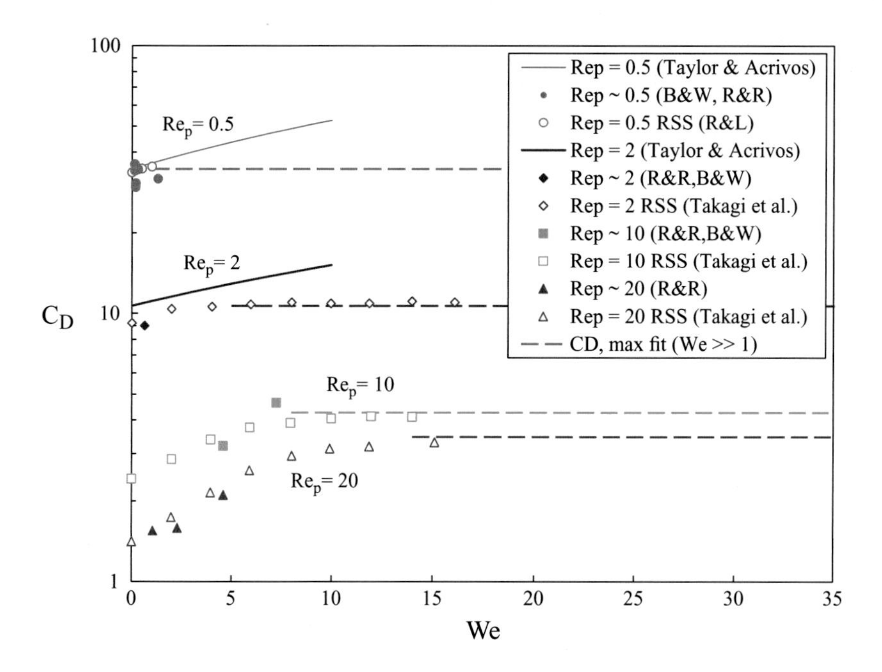

Figure 9.32 Drag coefficients of clean bubbles at low to moderate Reynolds numbers with theory, experimental data, Resolved Surface Simulations (Loth, 2008a).

$$f_{\|} = \frac{2+3\mu_p^*}{3+3\mu_p^*} + \frac{3}{16}\left(\frac{2+3\mu_p^*}{3+3\mu_p^*}\right)^2 \mathrm{Re_p} + c_{\mathrm{We}}\frac{6}{5}\frac{\left(8+3\mu_p^{*2}-\mu_p^*\right)}{\left(3+3\mu_p^*\right)^2}\mathrm{We}. \tag{9.71}$$

The last term represents the drag increase due to deformation. Comparisons of this theoretical drag with experimental values and resolved-surface simulations are shown in Figure 9.32. As expected, the theory is reasonable for small $\mathrm{Re_p}$ and small We, but overpredicts the drag coefficient for $\mathrm{Re_p} > 1$ and/or $\mathrm{We} > 1$.

For drops in a gas, the effects of deformation on drag from (9.71) are theoretically larger than that for bubbles in a liquid, but there is a lack of experiments and resolved-surface simulations to characterize these differences at intermediate Reynolds and Weber numbers. In the following, we consider models and theories at large Reynolds numbers and Weber numbers for clean bubbles, for contaminated bubbles and drops in a liquid, and finally for drops in a gas.

Clean Bubbles at Larger Reynolds Numbers

For clean bubbles at high $\mathrm{Re_p}$, a low-order theoretical result was obtained by Moore (1965) for aspect ratio based on effects of the viscous boundary layer on the shape but assuming a potential flow pressure based on a sphere and is given as follows:

$$\frac{1}{E} = 1 + \frac{9}{64}\mathrm{We} + o\left(\mathrm{We}^2\right). \tag{9.72}$$

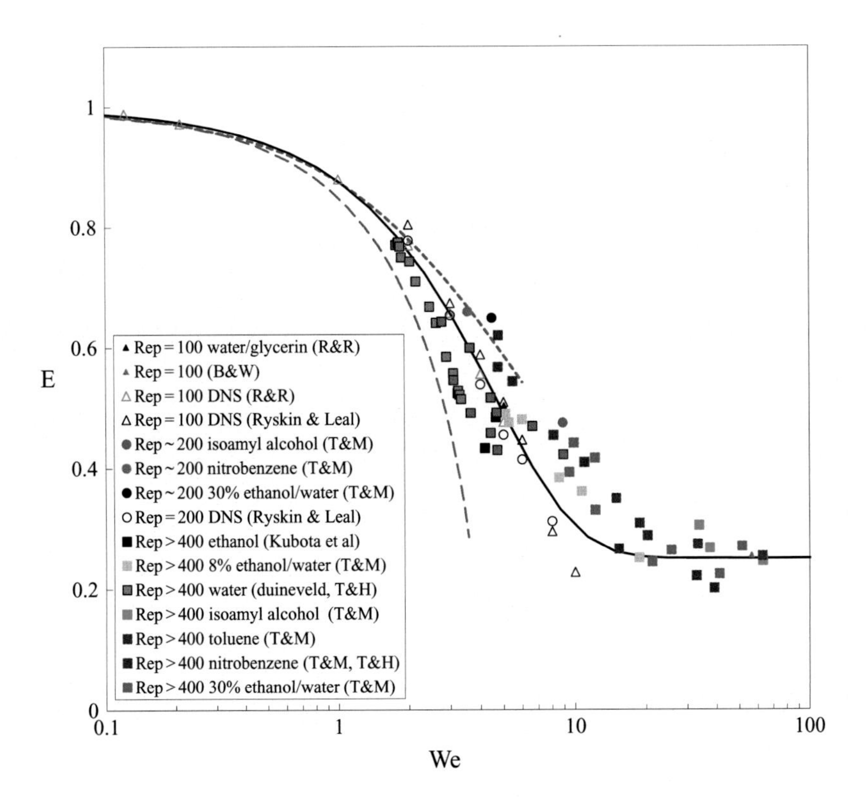

Figure 9.33 Aspect ratios at high particle Reynolds numbers (> 100) for clean bubbles, where the long-dashed line is a high-order theory, the short-dashed line is a low-order theory (Moore, 1965), and the solid line is an empirical model (Loth, 2008a).

To improve accuracy, Moore's high-order theory for aspect ratio instead used a potential flow pressure distribution based on an oblate spheroid (Lamb, 1945). The result was used to determine deformation and provide an implicit function of aspect ratio with the Weber number. This high-order theory can be approximated by an explicit relationship (to within 1%) for moderate deformations (Loth, 2008a) follows:

$$\frac{1}{\mathrm{E}} \approx 1 + \frac{9}{64}\mathrm{We} - 0.01\mathrm{We}^2 + 0.03\mathrm{We}^3. \tag{9.73}$$

In general, the data in Figure 9.33 for $\mathrm{Re_p} > 100$ are reasonably bounded by the low-order and high-order theories for $\mathrm{We} < 3$ (i.e., $\mathrm{E} > 0.6$). At higher Weber numbers and deformations, Moore's approach breaks down since the bubbles cannot be considered as oblate spheroids (the fore-aft symmetry is broken). As such, this low-order theory tends to overpredict E (underpredict deformation) at higher Weber numbers. On the other hand, the high-order theory assumes attached flow and tends to underpredict E (overpredict deformation) due to not accounting for the flow separation that occurs at large deformation.

To capture the experimentally observed influence of Weber number at both low and high Reynolds numbers (seen in both Figures 9.31 and 9.33), the aspect ratio can be modeled empirically (Loth, 2008a) in terms of a minimum aspect ratio (E_{min}) and an aspect ratio coefficient (c_E) as follows:

$$E = 1 - (1 - E_{min})\tanh(c_E \mathrm{We}) \qquad \textit{for clean bubbles.} \tag{9.74}$$

The modeled minimum aspect ratio and the aspect ratio coefficient depend on the Reynolds number, as follows:

$$E_{min} = 0.25 + 0.55\exp(-0.09\,\mathrm{Re_p}) \tag{9.75a}$$
$$c_E = 0.165 + 0.55\exp(-0.3\,\mathrm{Re_p}) \qquad \textit{for clean bubbles.} \tag{9.75b}$$

This model reasonably represents the experimental data of Figure 9.31 for $0.5 < \mathrm{Re_p} < 20$ and that of Figure 9.33 for $100 < \mathrm{Re_p} < 4{,}000$, and reverts to $E_{min} \approx 0.25$ of (4.56) for $\mathrm{Re_p} > 100$. However, the skirted bubble (Figure 4.26) does not follow the E_{min} limit (though this condition is rare, as it requires $\mathrm{We} > 100$ and very still environments).

Theoretical drag at high Reynolds numbers was also obtained by Moore (1965) assuming small deformations and inviscid pressure fields with a viscous thin attached boundary layer, yielding the following:

$$C_{D,Moore} = \frac{48}{\mathrm{Re_p}} G_E \left(1 - \frac{2.21\,H_E}{\mathrm{Re_p^{1/2}}}\right). \tag{9.76}$$

Note that the ½ exponent in the denominator of the last term is accidentally missing in Loth (2008) and that the parameters G_E and H_E can be given explicitly for $E > 0.5$ as follows:

$$G_E \approx 0.1287 + 0.4256E^{-1} + 0.4466E^{-2}. \tag{9.77a}$$

$$H_E \approx 0.8886 + 0.5693E^{-1} - 0.4563E^{-2}. \tag{9.77b}$$

As with the aspect ratio results, the Moore theory for drag (assuming $\mathrm{We} \ll 1$) reverts to the spherical bubble drag of (9.64) for $\mathrm{We} = 0$ (since $G_E \rightarrow 1$ and $H_E \rightarrow 1$ as $E \rightarrow 1$) and provides reasonable predictions for clean bubbles with $\mathrm{We} < 1$ for $\mathrm{Re_p} > 100$ (Loth, 2008a).

A theoretical maximum drag coefficient is also available for a bubble in the limit of very large Weber and Reynolds number ($\mathrm{We} \gg 1$ and $\mathrm{Re_p} \gg 1$). In particular, Joseph (2006) assumed a spherical-cap bubble shape with $E \approx 0.25$ and negligible surface tension, which assumed the pressure inside the bubble was constant. The inviscid liquid pressure distribution over top surface then results from a balance of hydrostatic and hydrodynamic effects, while the aft surface liquid pressure can be approximated as constant. The resulting theoretical drag for high $\mathrm{Re_p}$ is $C_{D,max} = 8/3$ and compares reasonably with experiments for $\mathrm{We} \gg 1$ and $\mathrm{Re_p} \gg 1$, as shown in Figure 9.34.

Joseph's theory can also be combined with the clean spherical bubble drag for creeping ($C_D = 16/\mathrm{Re_p}$) to reasonably describe intermediate $\mathrm{Re_p}$ conditions (Darton and Harrison, 1974; Clift et al., 1978) as shown by the green dashed line in

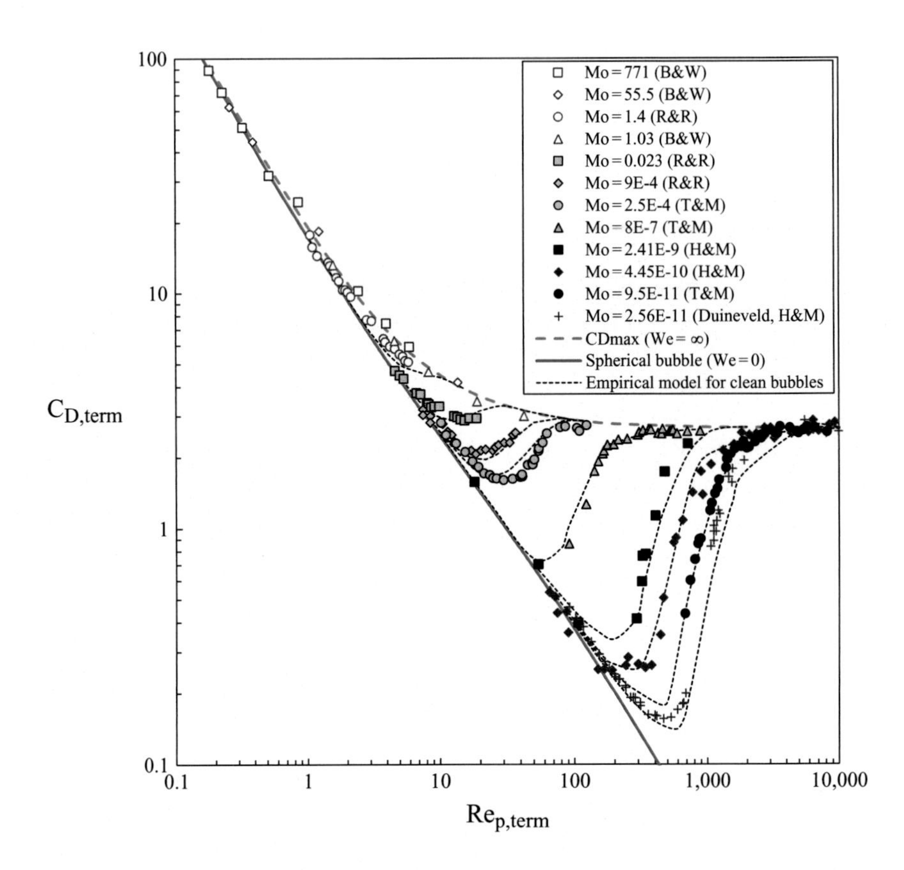

Figure 9.34 Experimental terminal drag coefficients for various Morton numbers of clean bubbles (Loth, 2008a) compared to an empirical fit based on combining multiple theoretical limits.

Figure 9.34. This blended $C_{D,max}$ of (9.78a) can be further combined with the spherical drag model of Mei (9.65) to represent drag of deforming clean bubbles at a wide range of Reynolds and Weber numbers:

$$C_{D,\max} = \frac{8}{3} + \frac{16}{Re_p}. \tag{9.78a}$$

$$C_{D,Mei} = \frac{16}{Re_p}\left\{1 + \left[\frac{8}{Re_p} + \frac{1}{2}\left(1 + \frac{3.315}{\sqrt{Re_p}}\right)\right]^{-1}\right\} \quad \textit{for clean bubbles}. \tag{9.78b}$$

$$C_D = C_{D,Mei}\left(1 - \Delta C_D^*\right) + C_{D,\max}\,\Delta C_D^*. \tag{9.78c}$$

$$\Delta C_D^* \approx \tanh\left[0.0008\left(WeRe_p^{0.2}\right)^{1.45}\right]. \tag{9.78d}$$

While (9.78a) is an empirical blend of two theories (as is the case for 9.78b), it is very similar to the theoretical result given in (4.58a). Predictions of the combined model of (9.78) for clean bubbles compared to the experimental results are shown in

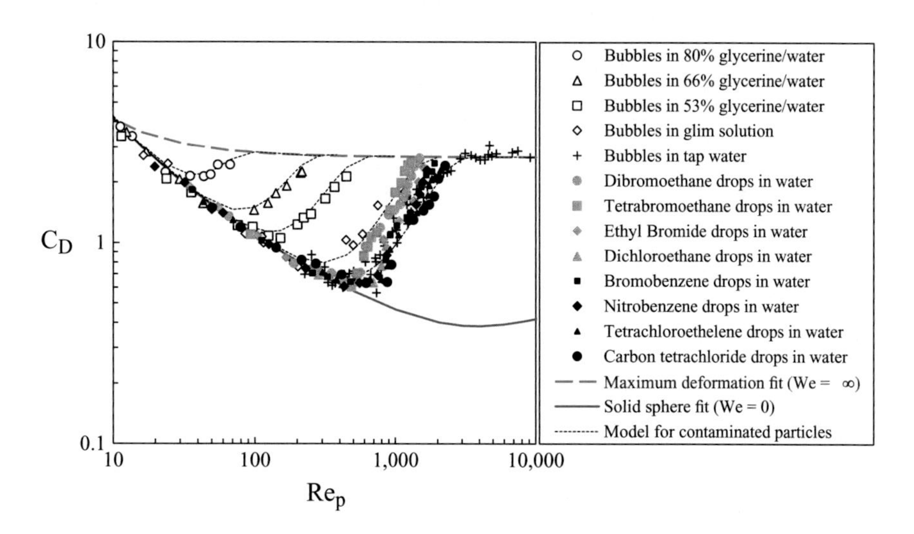

Figure 9.35 Drag measurements and predictions for contaminated particles in liquids (open symbols are bubbles and closed symbols are drops) where each dashed line corresponds to a fixed Morton number, and data for the curve are associated with increasing deformation as the Weber number increases (based on Loth, 2008a).

Figure 9.34 for a wide range of Morton numbers (i.e., liquid viscosities). The agreement is generally good, indicating robustness, and shows that the maximum increase in drag due to deformation strongly increases with Re_p.

Drops and Contaminated Bubbles in Liquids

For contaminated bubbles and drops in a liquid, qualitative measurements of aspect ratio due to deformation are generally only available for $Re_p > 80$ (Loth, 2008a). For such conditions, increasing Weber numbers first leads to the oblate spheroids and finally a spherical-cap shape, where the aspect ratio decreases until it reaches a minimum value, with $E_{min} \approx 0.25$ per (4.56) as shown by the contaminated drop and bubble data in Figure 4.20. This figure also shows that this aspect ratio data (for $Re_p > 100$) is well predicted with the empirical model of (4.57), which uses (4.56). At the other extreme of $Re_p < 1$ and $We < 1$, the Taylor and Acrivos theory was found to be reasonable. To bridge these two conditions for intermediate Reynolds numbers, it may be reasonable to use E_{min} based on (9.75) with (4.57). However, experiments and/or resolved-surface simulations are needed to support such a model.

The drag of contaminated bubbles and drops in a liquid at high Reynolds numbers ($Re_p > 100$) is related to a combination of Weber and Reynolds number per (4.58) and as shown in Figure 4.21. This model turns out to be reasonable for a wide range or Reynolds and Weber numbers, as shown in Figure 9.35. However, there is significant experimental scatter and differences relative to the predictions for bubbles in water. As such, the accuracy of (4.58d) is only approximate, with typical uncertainties of 10% (and sometimes exceeding 20%).

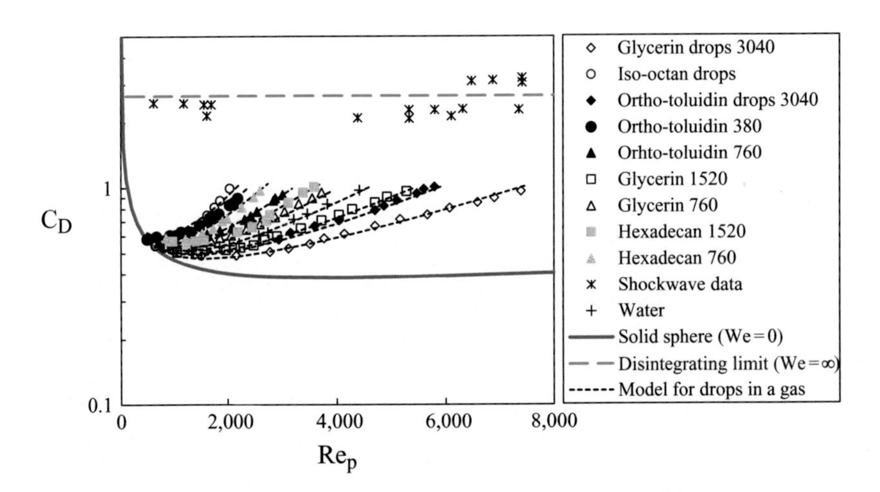

Figure 9.36 Drag coefficient as a function of the Reynolds number for terminal conditions as well as for disintegrating drops for various liquids with pressures varying from 0.5–4 atm (based on Loth, 2008a).

Drops in a Gas at Larger Reynolds Numbers

For drops in a gas, experimental deformation results are generally only available for $Re_p > 200$. For such conditions, increasing the Weber number first leads to the oblate spheroids with decreasing aspect ratio, and the flow around the drop is generally separated, so no theoretical drag predictions are available. As such, aspect ratio and drag predictions generally employ empirical models based on surveys of experimental and resolved-surface simulation data. Fortunately, the aspect ratios correlate well with the Weber number, as shown in Figure 4.20, and are well described by (4.47) for $We < 12$. At higher Weber numbers, a variety of unsteady shapes are possible, such as a bag shape (Figure 4.18) or lenticular shape (Figure 4.20). However, these occur during breakup and are highly transitory (and thus difficult to model).

For drag of deformed drops in gas, experimental results are again limited to high Reynolds numbers, but the drag increment correlates well with a combination of Weber and Reynolds numbers, as shown in Figure 4.21, and is well described by (4.48) for $We < 12$. Predictions with this empirical model for drop drag coefficient are compared to experiments in Figure 9.36 for a wide range of liquids, where the good agreement indicates high robustness. The trends also show that steady-state results are generally limited to $C_D < 1$, except during breakup, where the drag coefficients tend to the theoretical maximum of 8/3 at the disintegrating limit, consistent with (9.78).

10 Lift, Added-Mass, and History Forces on a Particle

This chapter discusses lift, added-mass, and history forces for an isolated particle to complement the Chapter 9 discussion of drag force. Lift force is first discussed in the context for vorticy-induced lift (§10.1.1) and spin-induced lift (§10.1.2) for a spherical particle. These two types of lift are combined for fixed-torque and free-spin particles (§10.1.3), and then lift for nonspherical particles is considered (§10.1.4). The added-mass force is next considered for solid, fluid, and nonspherical particles (§10.2) followed by that for history force (§10.3).

The impact of walls and other particles (via three-way and four-way coupling) is discussed in Chapter 11. For both Chapters 10 and 11, the particles are assumed to be surrounded by an incompressible continuum fluid.

10.1 Lift Force and Torque

While drag is generally the largest surface force acting on a particle, lift can also be substantial in certain cases (Leal, 1980; Drew, 1983). In fact, there are a number of multiphase flows for which lift is vital to the particle trajectories: microcentrifuges (Heron et al., 1975), particle deposition in boundary layers (Young and Leeming, 1997), radial distribution of bubbles in pipes (Kariyasaki, 1987), and rotating bioreactors (Ramirez et al., 2003). In this section, the point-force lift is described in terms of the local flow field and particle conditions, based on the particle's relative velocity, vorticity, and spin.

For a particle in a surrounding fluid, the two primary drivers for the lift force are (1) vorticity-induced lift due to continuous fluid vorticity ($\boldsymbol{\omega}_f$) and (2) spin-induced lift due to rotation of the particle ($\boldsymbol{\Omega}_p$). As shown in Figure 10.1, both conditions create positive lift by causing a higher speed on the top of the particle compared to the bottom. This higher speed creates a low-pressure region on the top of the particle, creating an upward force coupled with a downward deflection of the wake flow. As such, lift is taken as positive when the force is in the direction of $\boldsymbol{\omega}_f \times \mathbf{w}$ for vorticity-induced lift or in the direction of $\boldsymbol{\Omega}_p \times \mathbf{w}$ for spin-induced lift

An example of vorticity-induced lift is the stable levitation of a ping-pong ball with a jet (e.g., from a hair dryer). The ball remains in the jet center since deviation to either side will expose it to a shear flow that will provide a restoring lift force back to the to the high-speed region, that is, the center of the jet. An example of spin-induced lift is

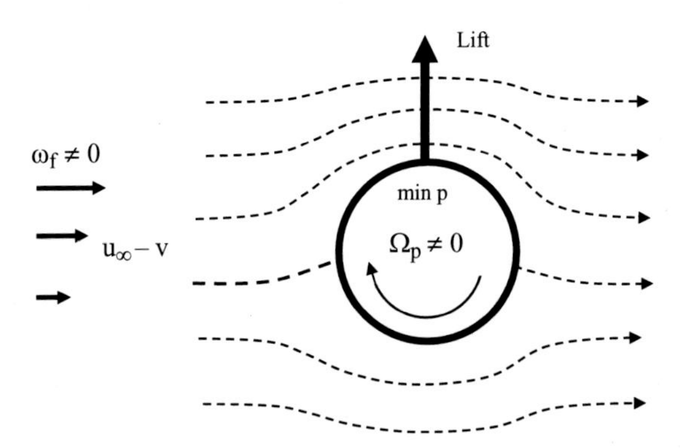

Figure 10.1 Particle lift due to continuous-phase shear (ω_f) and/or particle rotation (Ω_p). Note that for flow moving left to right and a positive lift in the upward direction, the fluid dynamic convention is to set clockwise rotation and clockwise vorticity both as positive values.

that of a tennis ball hit with topspin. This spin causes faster fluid motion below the ball that creates a downward force so the trajectory can arc down over the net, even when at a high speed. A similar trajectory result occurs for a baseball pitch with topspin, where the lift will cause it to curve down. In both cases, roughness of the ball surface can accentuate the lift force, as will be discussed.

The dimensionless lift coefficient uses a normalization like that for the drag coefficient (3.54), and the shear-induced and spin-induced components are generally assumed to be linearly additive (e.g., Saffman, 1968), allowing the following general form:

$$C_L \equiv F_L / \left(\frac{\pi}{8} \rho_f w^2 d^2 \right). \tag{10.1a}$$

$$\mathbf{F}_L = \mathbf{F}_{L,\omega} + \mathbf{F}_{L,\Omega} = \frac{\pi}{8} \rho_f w^2 d^2 \left[C_{L,\omega} \frac{\boldsymbol{\omega}_f \times \mathbf{w}}{|\boldsymbol{\omega}_f \times \mathbf{w}|} + C_{L,\Omega} \frac{\boldsymbol{\Omega}_p \times \mathbf{w}}{|\boldsymbol{\Omega}_p \times \mathbf{w}|} \right]. \tag{10.1b}$$

Since the continuous-phase angular velocity ($\boldsymbol{\Omega}_f$) is half the fluid angular rotation rate ($\boldsymbol{\omega}_f$) via (2.54), the unhindered fluid rotation at the particle centroid can be expressed as follows:

$$\boldsymbol{\Omega}_{f@p} \equiv \frac{1}{2} \boldsymbol{\omega}_{f@p}. \tag{10.2}$$

When both vorticity-induced and spin-induced lift are acting on a particle, this result can be used to obtain the relative particle spin ($\boldsymbol{\Omega}_{p,rel}$) as follows:

$$\boldsymbol{\Omega}_{p,rel} \equiv \boldsymbol{\Omega}_p - \boldsymbol{\Omega}_{f@p} = \boldsymbol{\Omega}_p - \frac{1}{2} \boldsymbol{\omega}_{f@p}. \tag{10.3}$$

As discussed in §1.4, the unhindered surrounding flow properties neglect the presence of the particle itself, that is, they neglect the local flow disturbances caused by the individual particle at $\mathbf{x}_p$.

Since the fluid vorticity and the particle rotation have units of 1/time, they can be made dimensionless by a characteristic time scale. Often, the most appropriate time scale for this normalization is the relative convection time past the particle (d/w). In this case, the dimensionless vorticity and spin are written as follows:

$$\omega^* \equiv \omega_f(d/w). \tag{10.4a}$$

$$\Omega^* \equiv \Omega_p(d/w). \tag{10.4b}$$

In the following point-force theories and models, shear-induced lift ($C_L = C_{L\omega}$) will be proportional ω^*, while spin-induced lift ($C_L = C_{L\Omega}$) will be proportional Ω^*. However, these lift coefficients can also be influenced by Reynolds number, nonsphericity, deformation, internal circulation, flow complexity, etc.

It should be kept in mind that the lift force of (10.1b) is subject to the point-force caveats described in §9.1.1. To avoid these caveats, one may instead numerically resolve the flow around the particle, as discussed in Chapter 8. This resolved-surface approach can significantly improve accuracy when the particle and/or flow conditions are complex, such as for particles at a high Reynolds number, with surface roughness or complex shapes, or in turbulence with strong gradients at wavelengths similar to the particle diameter. When such complex conditions do not exist, the point forces described in this chapter are, fortunately, generally accurate.

The following sections deal with a spherical particle subjected to vorticity-induced lift (§10.1.1), to spin-induced lift (§10.1.2), and then to lift induced by combined vorticity and spin (§10.1.3). Finally, lift for nonspherical particles is discussed (§10.1.4).

10.1.1 Vorticity-Induced Lift

For vorticity-induced lift in viscous conditions, there are two fundamental types of flows that have been analyzed: (1) linear shear flow with $\boldsymbol{\omega}_f = \boldsymbol{\omega}_{shear}$ (Figure 2.9a) for which $C_{L\omega} = C_{L,shear}$; and (2) rigid vortex flow with $\boldsymbol{\omega}_f = \boldsymbol{\omega}_{vortex}$ (Figure 2.9b) for which $C_{L\omega} = C_{L,vortex}$. Since the fundamental theory for linear shear lift was obtained by Saffman (1965), this lift is often termed as *Saffman lift*. Since the fundamental theory for vortex lift was obtained by Heron et al. (1975), this lift is often termed as *Heron lift*. The conditions for linear shear flow and for rigid vortex flow are considered separately in the following subsections.

Linear Shear (Saffman) Lift for Solid Spheres

For linear shear flow, the vorticity ($\boldsymbol{\omega}_{shear}$) stems from a uniform gradient of the velocity in one direction, as in (2.53). In this flow, one may define a particle shear Reynolds number as follows:

$$Re_\omega \equiv \rho_f \omega_f d^2/\mu_f = Re_p \omega^*_{shear}. \tag{10.5}$$

The RHS employs the dimensionless shear of (10.4a). For $Re_p \ll Re_\omega^{1/2} \ll 1$, Saffman (1965) used a matched expansion of an inner and outer solution to solve for the particle lift force:

$$\mathbf{F}_{L,\omega} = \mathbf{F}_{L,Saff} = 1.615\rho_f d^2\sqrt{\nu_f/\omega_{shear}}\left(1 - 0.334\sqrt{Re_\omega}\right)\boldsymbol{\omega}_{shear} \times \mathbf{w}. \tag{10.6}$$

For weak shear (small Re_ω), a first-order result can be obtained by neglecting the term in the parentheses. Using the normalizations of (10.1a) and (10.4a), the associated first-order shear-induced Saffman lift coefficient is given as follows:

$$C_{L,Saff} \equiv \frac{12.92}{\pi}\sqrt{\frac{\omega_{shear}\nu_f}{w^2}} = \frac{12.92}{\pi}\sqrt{\frac{\omega^*_{shear}}{Re_p}} \quad \textit{for } Re_p \ll Re_\omega^{1/2} \ll 1. \tag{10.7}$$

Since the lift-to-drag force ratio can be expressed as C_L/C_D based on (3.54) and (10.1a), this force ratio can be obtained by comparing (10.7) with (9.8) for $f = 1$ to yield the following:

$$\frac{C_{L,Saff}}{C_D} = 0.1714\sqrt{Re_p\omega^*_{shear}}. \tag{10.8}$$

As such, the Saffman lift to drag ratio becomes larger as the Reynolds number and nondimensional shear increase. However, to consider Reynolds numbers without the limit noted by (10.7) requires a modified theory.

In particular, McLaughlin (1991) extended Saffman's theory using Fourier analysis to eliminate the restriction of $Re_p \ll Re_\omega^{1/2}$ while still employing $Re_p \ll 1$. McLaughlin's solution yields an integral that was numerically evaluated for the lift force ratio, defined as J^*, which was then empirically represented by Mei and Klausner (1994) with the following fit:

$$J^* \equiv C_{L,McL}/C_{L,Saff}. \tag{10.9a}$$

$$J^* \approx 0.3\left\{1 + \tanh\left[\frac{5}{2}\left(\log_{10}\sqrt{\frac{\omega^*_{shear}}{Re_p}} + 0.191\right)\right]\right\}\left\{\frac{2}{3} + \tanh\left[6\sqrt{\frac{\omega^*_{shear}}{Re_p}} - 1.92\right]\right\}. \tag{10.9b}$$

Results for J^* are shown in Figure 10.2, where it can be seen that (10.9b) is a good fit to the theoretical result, and correlates well with the experimental data, which are for $Re_p < 1$. It can also be seen that J^* tends to unity (to Saffman's solution) at high ω^*/Re_p where lift effects can be significant. For example, $\omega^* = 1$ and $Re_p = 0.5$ will produce a lift-to-drag ratio of 12% based on (10.8). At the other extreme, J^* tends to zero at small dimensionless vorticity due to convection effects. The lift ratio (J^*) even becomes slightly negative at very low vorticity values. This is not captured by the fit of (10.9b), but shear lift is generally negligible in these conditions (Mei and Klausner, 1994).

For $Re_p > 1$, experimental data are scarce, so much of our understanding comes from resolved-surface simulations (RSS). As shown in Figure 10.3a, such data indicate that the McLaughlin lift (derived for $Re_p \ll 1$) is reasonable for Re_p as high as 50, as follows:

$$C_{L,shear} \approx C_{L,McL} = J^* C_{L,Saff} \qquad \textit{for } Re_p \leq 50. \tag{10.10}$$

However, Kurose and Komori (1999) found that $Re_p > 50$ can result in small negative lift coefficients, as shown in Figure 10.3b, which are attributed to asymmetric pressure

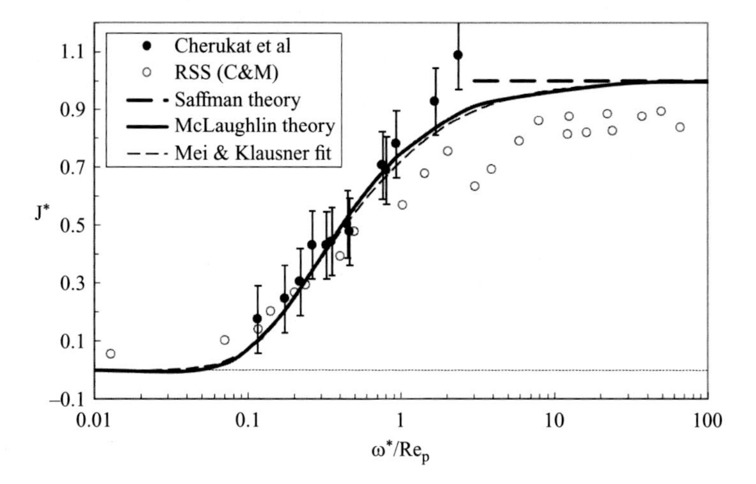

Figure 10.2 Shear-induced lift theories for solid spheres assuming $Re_p \ll 1$ compared to data at $0.1 < Re_p < 1$, where open symbols are RSS (Cherukat et al., 1999) filled symbols are experimental data (Cherukat et al., 1994), and the dashed line is the fit to the McLaughlin theory (Mei and Klausner, 1994).

and viscous contributions stemming from flow separation. A fit to their results (Loth, 2008b) is given as follows:

$$C_{L,shear} \approx -\left(\omega^*_{shear}\right)^{1/3}\left\{0.0525 + 0.0575\tanh\left[5\log_{10}\left(\frac{Re_p}{120}\right)\right]\right\} \quad \textit{for } Re_p > 50. \tag{10.11}$$

A similar fit was proposed by Shi and Rzehak (2019) based on further simulation results, but both fits should be used with some caution since neither are supported by quantitative experimental data.

Linear Shear Lift for Clean Drops and Bubbles

The preceding shear-based results are based on a no-slip condition, which is appropriate for solid spheres, spherical drops in a gas, and contaminated drops and bubbles in a liquid. For clean drops or bubbles in a liquid, the lift force will generally be reduced due to internal flow circulation (recall Figure 3.2). For clean fluid particles at small Re_p, Legendre and Magnaudet (1997) showed that the McLaughlin lift is modified based on the viscosity ratio, as follows:

$$C_{L,shear} = C_{L,McL}\left(\frac{2+3\mu_p^*}{3+3\mu_p^*}\right)^2 = C_{L,Saff}J^*\left(\frac{2+3\mu_p^*}{3+3\mu_p^*}\right)^2 \qquad \textit{for } Re_p \ll 1. \tag{10.12}$$

This result reverts to the Saffman and McLaughlin lifts for contaminated or solid particles ($\mu_p \gg \mu_f$) but yields a 4/9 reduction for a bubble ($\mu_f \gg \mu_p$) at this creeping-flow condition.

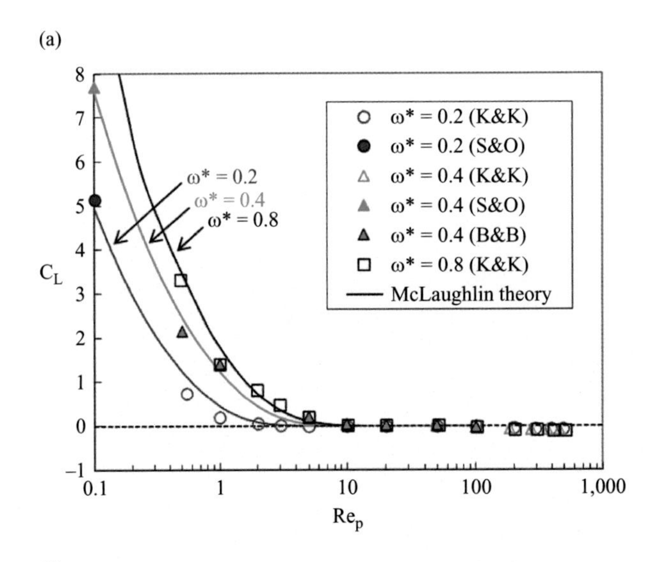

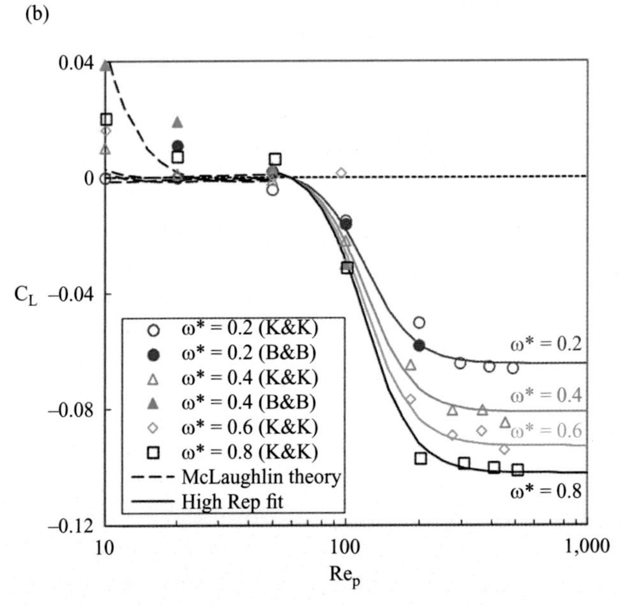

Figure 10.3 Shear-induced lift for solid spheres as a function of Reynolds number for RSS compared to McLaughlin theory and to Kurose–Kumori high Re_p fit compared to data (Salem and Osterlé, 1998; Kurose and Komori, 1999; Bagchi and Balchandar, 2002a).

For the other extreme of inviscid flow, Auton derived a lift and found it to be linearly proportional to the product of vorticity and the relative velocity (9.6d) with the following lift coefficient:

$$C_{L,Auton} = \frac{2}{3}\omega^* \qquad \textit{for inviscid}. \tag{10.13}$$

This inviscid Auton lift does not depend on whether the fluid vorticity arises from a linear shear flow or from a pure vortex flow (unlike the Saffman lift). It also has a linear dependence on vorticity (unlike Saffman lift), which suggests the definition of another lift coefficient that is shear-normalized as follows:

$$C^*_{L,\omega} \equiv \frac{F_L}{\forall_p \rho_f \omega_f w} = \frac{F_L}{\left(\frac{\pi}{6} d^3\right) \rho_f \omega_f w} = \frac{3C_L}{4\omega^*}. \tag{10.14a}$$

$$C^*_{L,\omega,\mathrm{Auton}} = 1/2. \tag{10.14b}$$

This different normalization, compared to (10.1a), yields a constant value in the Auton limit (10.14b). To determine the transition between the creeping-flow limit of McLaughlin and the inviscid limit of Auton, a study by Legendre and Magnaudet (1998) investigated clean bubbles for Re_p of 100–500 and found that flow over the surface remains attached and leads to a thin wake that deflects downward. As shown in Figure 10.4, the resolved-surface lift forces tended to Auton's inviscid result at high Re_p and to McLaughlin's at low Re_p, and they proposed an empirical combination for the intermediate conditions, as follows:

$$C^*_{L,\omega,L\&M} = \left\{ \left[\frac{1}{2} \frac{(1 + 16/Re_p)}{(1 + 29/Re_p)} \right]^2 + \left[\frac{4}{9} \frac{C^*_{L,\omega,McL}}{\left(1 + 0.2\,Re_p/\omega^*_{shear}\right)^{3/2}} \right]^2 \right\}^{1/2}. \tag{10.15}$$

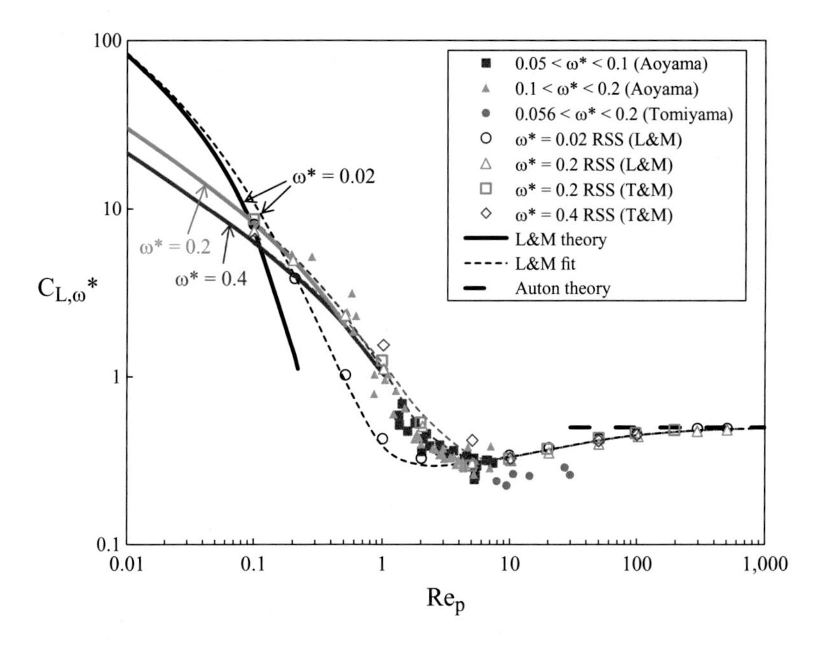

Figure 10.4 Lift coefficient for clean bubbles in a shear flow where open symbols are RSS flow (Legendre and Magnaudet, 1998; Takagi and Matsumoto, 1999) and filled symbols are experimental data (Tomiyama et al., 2002b; Aoyama et al., 2017), and the empirical fit provides a reasonable transition from creeping flow to inviscid flow (from Loth, 2008b).

As shown in Figure 10.4, experimental results generally follow this blended lift coefficient model, but the data for $1 < Re_p < 30$ yield measured lift coefficients lower than predicted (10.15). This tended to occur for bubbles in highly viscous liquids ($Mo > 10^{-4}$), and potential reasons for these differences were discussed by Aoyama et al. (2017).

Vortex (Heron) Lift for Solid Spheres

The lift of a particle in a vortex with solid body rotation as in (2.54) and Figure 2.9b for the limit of small but finite Re_p was obtained theoretically by Heron et al. (1975), as follows:

$$C_{L,\omega} = C_{L,vortex} = 5.091\sqrt{\omega^*_{vortex}/Re_p} \qquad \textit{for } Re_p \ll 1. \tag{10.16}$$

This circular-vortex lift coefficient has the same functional dependence as the linear-shear version of (10.7) but is 24% larger. The vortex-based lift coefficient does not drop with the Reynolds number (as seen for linear-shear version), and instead the theory gives reasonable results for Re_p as high as 100 for contaminated bubbles (which tend to behave like solid particles), as shown in Figure 10.5. Comparing Figures 10.3 and 10.5 for $25 < Re_p < 100$ shows that the vortex-based lift coefficient is generally *two orders of magnitude* greater than shear-induced lift for comparable values of vorticity at these high Reynolds numbers. However, conditions in a turbulent boundary layer tend to yield higher values of shear ($\omega_{shear} \gg \omega_{vortex}$) so that both shear-based lift and vortex-based lift may be important depending on the Re_p and the flow conditions (Bagchi and Balachandar, 2002b).

Figure 10.5 also indicates a lack of experimental and RSS data for $Re_p > 100$. This makes it unclear whether the vortex-based lift for solid particles at high Re_p will tend to the inviscid Auton limit of (10.14b) (as for clean bubbles in linear shear in Figure 10.4) or will perhaps tend to negative lift coefficient (as for solid particles in linear shear in Figure 10.3). The lift behavior of clean fluid particles in a pure vortex is even less understood. As such, further experiments and studies are thus needed for vortex-based lift for a variety of conditions.

10.1.2 Spin-Induced Lift

The lift due to particle rotation is often termed *Robins–Magnus lift* based on experimental observations for spheres by Robins in 1742 and for cylinders by Magnus in 1853, though the concept of lift due to spin was first postulated by Newton in 1672 (Barkla and Auchterlonie, 1971).

In general, particle spin will cause increased velocity on the top portion of the particle, which leads to a low pressure at this location and an upward lift (Figure 10.1). The flow field can change dramatically at high relative spin rates. This is qualitatively illustrated in Figure 10.6 using streamlines based on the potential flow solution for a cylinder with a prescribed circulation. At low spin rates (Figure 10.6a), the stagnation points tend to move downward, and the streamlines above are more closely spaced,

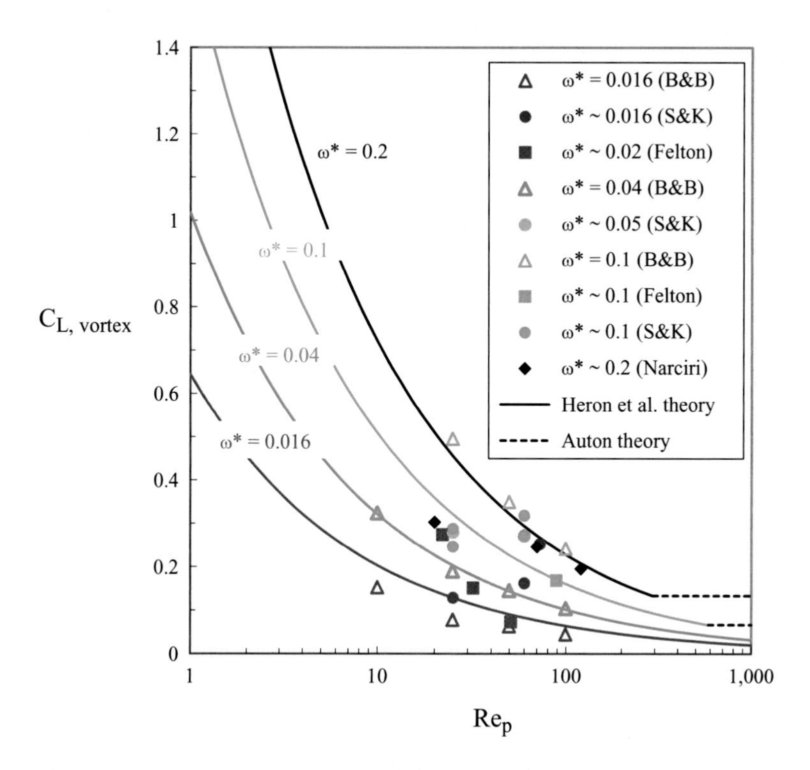

Figure 10.5 Theoretical lift coefficient for a solid-body rotation vortex of Heron et al. (1975) at $Re_p \ll 1$ and for Auton inviscid lift, where these two theories are compared to experimental contaminated bubble data as filled symbols (Narciri, 1992; Sridhar and Katz, 1995; Felton and Loth, 2001) as well as RSS of solid spheres as open symbols (Bagchi and Balachandar, 2002b).

indicating the higher velocity and lower pressure that results in upward lift. As the normalized spin rates tend toward unity (Figure 10.6b), the stagnation points move to the bottom, leading to even greater top-to-bottom differences in flow velocity and pressure, which produce higher lift. However, much higher rotation rates (Figure 10.6c) yield a closed streamline region around the particle for which lift may no longer increase linearly with spin rate (as will be shown).

For small spin rates (Figure 10.6a) and small Re_p, Rubinow and Keller (1961) obtained a theoretical solution for lift due to particle spin with a no-slip condition:

$$\mathbf{F}_{L,R\&K} = \frac{\pi}{8} d^3 \rho_f \left(\boldsymbol{\Omega}_p \times \mathbf{w} \right). \tag{10.17}$$

Based on (10.1b) and (10.4b), the associated lift coefficient becomes

$$C_{L,R\&K} = \Omega^* \qquad \textit{for } \Omega^* \ll 1 \textit{ and } Re_p \ll 1. \tag{10.18}$$

This result shows that creeping-flow spin-based lift for a solid particle (or contaminated fluid particle) is independent of the surrounding fluid viscosity (unlike creeping-flow shear-based lift). In addition, spin lift is linearly proportional to the normalized

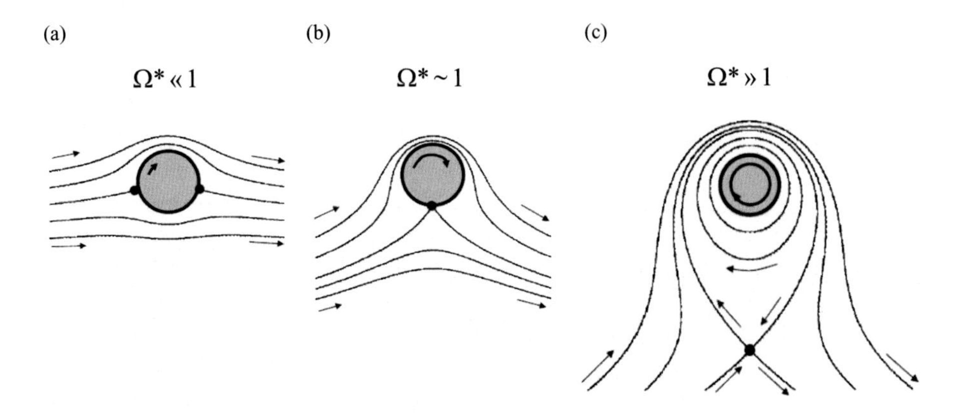

Figure 10.6 Schematic of continuous-fluid streamlines for various spin rates based on inviscid flow circulation of a cylinder, where all cases result in positive (upward) lift.

spin rate. As such, it is convenient to define a spin-normalized lift coefficient by the theoretical result of (10.18) as follows:

$$C^*_{L,\Omega} \equiv \frac{C_{L,\Omega}}{\Omega^*}. \tag{10.19}$$

For $Re_p \ll 1$, this spin-normalized lift coefficient is simply unity, as shown in Figure 10.7. For $Re_p > 1$, no theoretical solution exists for spin-based solid particle lift, so models are generally developed using experiments and resolved-surface simulations, such as in the data shown in Figure 10.7. This figure shows that the spin-normalized lift coefficient generally decreases as Re_p increases from 1 to 1,000 and reduces as $\Omega^* > 1$ (consistent with the flow physics of Figure 10.6), especially at higher Re_p. To represent these trends, several empirical models have been put forth (Tanaka et al., 1990; Tri et al., 1990; Bagchi and Balachandar, 2002a), including a model by Loth (2008b), as follows:

$$C^*_{L,\Omega} = 1 - \{0.675 + 0.15(1 + \tanh[0.28(\Omega^* - 2)])\}\tanh\left[0.18\,Re_p^{1/2}\right]. \tag{10.20}$$

This model tends to (10.18) for $Re_p \ll 1$ and then transitions to the Tanaka model for Re_p of 1,000 and generally fits the experimental data. Shi and Rzehak (2019) surveyed recent lift predictions from resolved-surface simulations and provided a similar fit with improved correlation for simulations at $Re_p > 100$, though (10.20) gives better predictions for experiments. Some of this difference between experiments and simulations may be explained by wake unsteadiness, which can occur for $Re_p > 100$ at high shear and leads to large relative lift fluctuations (stronger than those for drag). Differences can also arise due to fixed trajectories (simulations) versus free trajectories (experiments).

At $Re_p > 10{,}000$ with high particle spin, the lift force can be even more unsteady and is generally associated with complex wakes, which can even cause a reversal in the direction of time-averaged lift. This is illustrated by the time-averaged streamlines

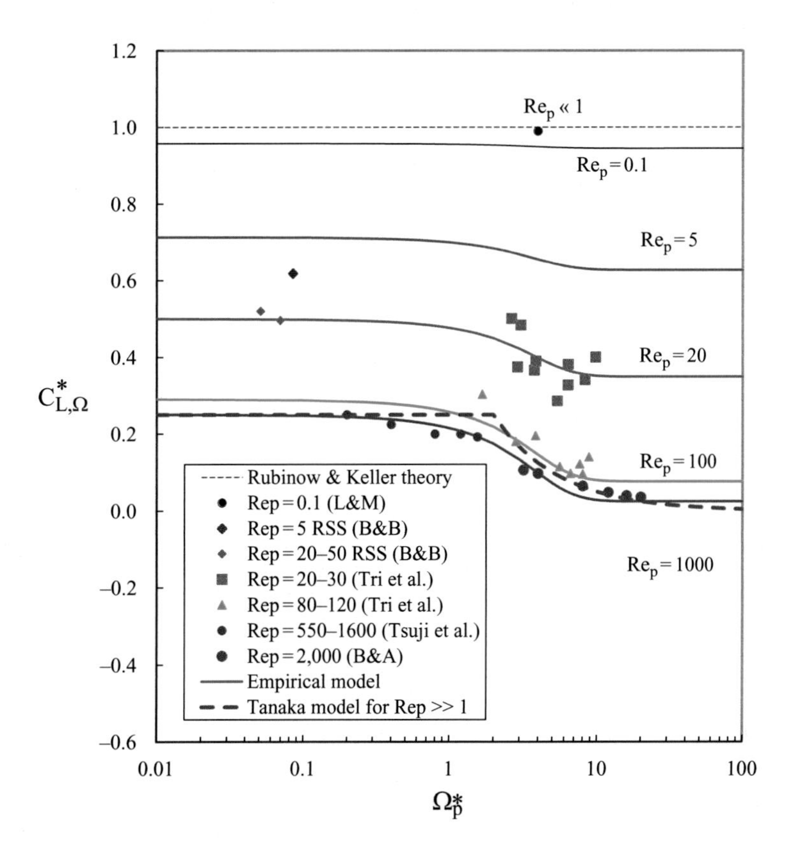

Figure 10.7 Spin lift force coefficient predictions for $Re_p < 2{,}000$ compared to data (Barkla and Auchterlonie, 1971; Tri et al., 1990; Tsuji et al., 1985; Legendre and Magnaudet, 1998; Bagchi and Balachandar 2002a) and the model by Loth (2008b), where open symbols are RSS and filled symbols are experimental data.

in Figure 10.8 for $Re_p \approx 100{,}000$. For no-spin ($\Omega^* = 0$), this flow corresponds to subcritical conditions (the boundary layer is laminar until separation) with symmetric streamlines and a symmetric separation region (Figure 10.8a). At a spin rate of $\Omega^* = 0.68$ (Figure 10.8b), the separation region and overall wake shift downward. This leads to a higher-speed region above the particle with an upward lift force (and positive lift coefficient) as expected. However, increasing the spin further to $\Omega^* = 1.06$ (Figure 10.8c) results in a smaller separation region due to turbulent transition and reattachment on the bottom surface. As a result, the downstream wake actually deflects slightly upward, which causes a downward lift that Kim et al. (2014) termed the "inverse Magnus effect," which corresponds to a negative lift coefficient. At further increases in the sphere spin (Figure 10.8d), the upper surface separation point also moves downstream, resulting an even smaller separation, but the wake now deflects downward again, yielding a positive lift coefficient. Therefore, particle spin near the critical Re_p can lead to complex separation and changes in the lift force direction.

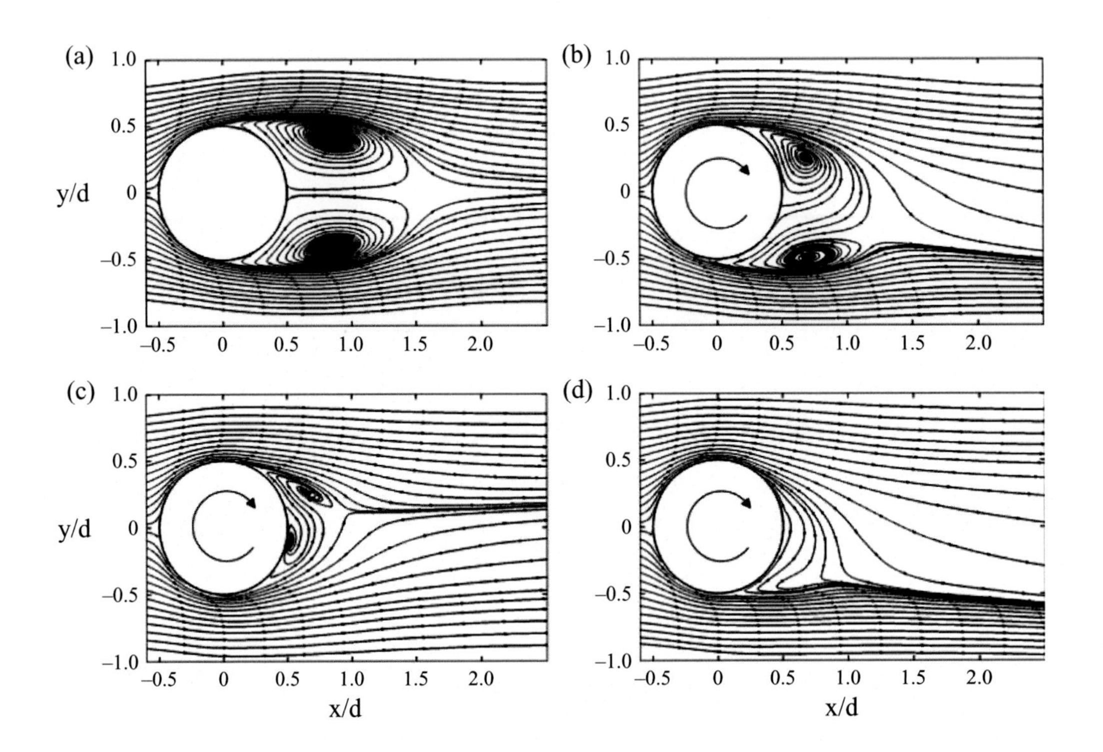

Figure 10.8 Measurements of time-averaged streamlines for a spinning sphere with flow moving left to right at $Re_p \approx 100{,}000$ (Kim et al., 2014) for (a) $\Omega^* = 0.0$, where lift is zero; (b) $\Omega^* = 0.68$, where lift is upward (positive); (c) $\Omega^* = 1.06$, where lift is downward (negative); (d) $\Omega^* = 1.60$, where lift is again upward (positive).

This complexity is reflected by the variety of experimental spin-normalized lift coefficients of solid spheres at $50{,}000 < Re_p < 500{,}000$ shown in Figure 10.9a. The results can be generally grouped into two categories: subcritical (where the surface is smooth and where laminar boundary-layer separation occurs) and supercritical (where the surface is rough, tripped, or at a high enough Re_p to ensure turbulent transition of the boundary layer before separation). The supercritical tripped flow data (black symbols) yield positive lift coefficients of order unity, with magnitudes much higher than indicated by the Tanaka fit of Figure 10.7. This data can be roughly represented by a fit proposed by Sawicki et al. (2003) for baseballs given as follows:

$$C^*_{L,\Omega} = \min[0.75, 0.3 + 0.09/\Omega^*] \qquad \textit{for}\ Re_p > 10^5. \tag{10.21}$$

Due to substantial scatter in the data, this supercritical model is rather qualitative. For nominally subcritical conditions (red symbols), the lift coefficient has even more scatter, is generally smaller, and can be negative for $\Omega^* < 1$ (the inverse Magnus effect). These results can be approximated with the subcritical fit by Loth (2008b) as follows:

$$C^*_{L,\Omega} = \min\left[0.25\tanh(\Omega^* - 0.5) - 0.1, \frac{1}{2\Omega^*}\right] \ \textit{for}\ 5\times10^4 < Re_p < 10^5. \tag{10.22}$$

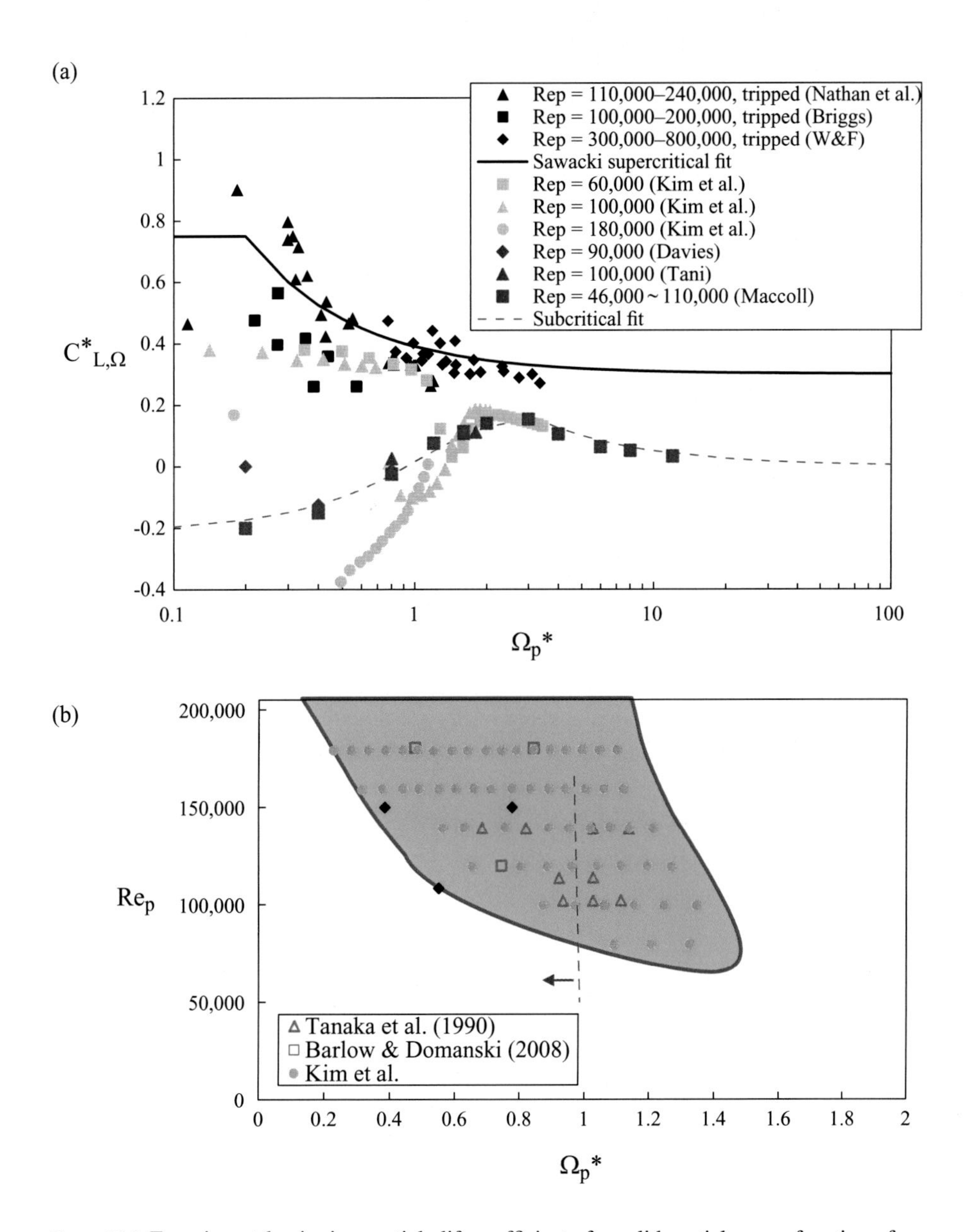

Figure 10.9 Experimental spinning particle lift coefficients for solid particles as a function of nondimensional rotation rate for (a) supercritical conditions (Tani, 1950; Briggs, 1959; Watts and Ferrer, 1987; Nathan et al., 2006), subcritical conditions (Maccoll, 1928; Davies, 1949), and transcritical conditions (Kim et al., 2014); and (b) reported negative lift coefficients as a green-shaded region (Tanaka et al. 1990; Barlow and Domanski, 2008; Kim et al., 2014) and a subcritical model for negative lift as the red dashed line.

However, data taken by Kim et al. (2014) with very low incoming turbulence ($< 0.4\%$) as shown by gray symbols indicates much different lift coefficients for $\Omega^* < 1$. As with Figure 10.8, this is attributed to asymmetric boundary layer transition, giving rise to "transcritical" conditions.

Based on the preceding, the lift of a spinning sphere near the critical Reynolds numbers can be highly complex and may only be qualitatively predicted by (10.21) or (10.22). To characterize the conditions where the lift coefficient *may* go negative, a regime map is given in Figure 10.9b. The map shows that the inverse Magnus effect can occur for $80{,}000 < \mathrm{Re_p} < 180{,}000$, but is only qualitatively described by the criterion of (10.22), shown as a vertical red dashed line. Clearly, more research is needed to develop accurate spin-induced lift models for solid particles at $\mathrm{Re_p} > 1{,}000$, which should also account for quantitative levels of freestream turbulence and/or surface roughness. In addition, spin-based lift for clean fluid particles is not well understood.

10.1.3 Lift and Torque for Combined Particle Spin and Fluid Vorticity

In this section, the combination of particle spin and fluid shear is considered for particle lift using two torque conditions: *prescribed spin,* based on a given particle rotation rate (Figure 10.10a); and *free spin*, where the particle attains an equilibrium rotation relative to the local fluid vorticity such that there is no net torque on the particle (Figure 10.10b). Prescribed spin is often set in experiments and simulations, whereas free (equilibrium) spin is more consistent with particles moving in a fluid with small rotational inertias relative to the time scales of the flow seen by the particle. First, we consider the prescribed spin combination.

Combined Fluid Vorticity and Prescribed Spin

Saffman (1965, 1968) considered a spinning particle in shear flow at low Reynolds numbers assuming $\mathrm{Re}_\omega \ll 1$, $\mathrm{Re_p} \ll \mathrm{Re}_\omega^{1/2}$, $\Omega^* \ll 1$, and $\mathrm{Re_p} \ll 1$ whereby the particle rotation and fluid vorticity are in the same direction and perpendicular to the relative velocity. The resulting first-order lift coefficient is given by the following:

$$C_L = \frac{12.92}{\pi}\sqrt{\frac{\omega^*_{\text{shear}}}{\mathrm{Re_p}}} + \Omega^* \qquad \textit{for } \mathrm{Re_p} \ll 1. \tag{10.23}$$

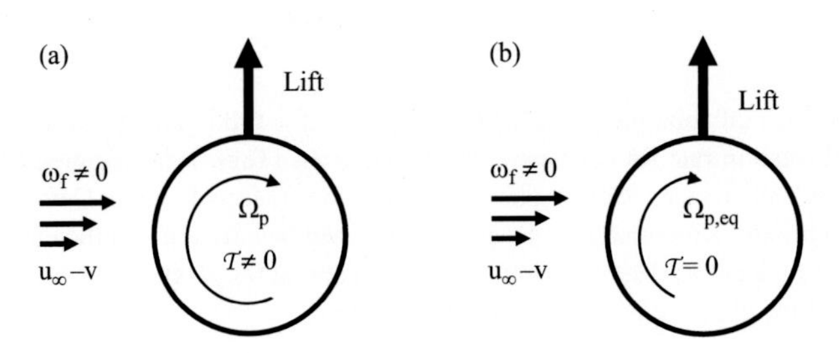

Figure 10.10 Schematic of particle lift in continuous-phase shear with (a) prescribed spin and (b) free spin (no net torque).

This result indicates that the spin lift of Rubinow and Keller (10.18) can be linearly combined with the shear-induced lift of Saffman (10.7) for these conditions. This is consistent with the linear addition of these two lifts per (10.1b). The result of (10.23) also indicates that shear-induced lift will tend to dominate spin-induced lift if $Re_p \ll 1$.

Salem and Osterlé (1998) found that the shear-lift portion result can be empirically extended to $Re_p < 50$ by incorporating the shear correction of (10.9). Furthermore, the spin-lift portion can be corrected using model of (10.20) so the net lift coefficient, normalized based on (10.19), is as follows:

$$C_L = \frac{12.92}{\pi} J^* \sqrt{\frac{\omega^*_{shear}}{Re_p}} + \Omega^* C^*_{L,\Omega} \qquad \textit{for } Re_p < 50 \tag{10.24}$$

Similarly, Bagchi and Balachandar (2002a) showed that a linear combination of shear lift and spin lift applies for normalized shear and spin rates as high as 0.4 and for Re_p values as high as 100, even when the shear and spin are not in the same direction. Therefore, the general form of lift can be expressed as follows:

$$\mathbf{F}_L(\omega,\Omega) \approx \mathbf{F}_{L\omega}\left(\omega_{shear} \neq 0, \Omega_p = 0\right) + \mathbf{F}_{L\Omega}\left(\omega_{shear} = 0, \Omega_p \neq 0\right). \tag{10.25}$$

This again supports the lift combination used in (10.1b).

Torque and Angular Momentum

To use the prescribed spin lift of (10.25), the particle rotation must be known along the particle path. This can be obtained from the particle equation of angular motion, which equates the rate of change of angular momentum to the fluid dynamic torque ($\mathcal{T}_{surf}$) acting on the particle:

$$\frac{d\left(I_p \mathbf{\Omega}_p\right)}{dt} = \mathcal{T}_{surf}. \tag{10.26}$$

Here, I_p is the angular moment of inertia about the particle centroid and the torque neglects any effects of collision with other particles or walls (to be discussed in §11.3–11.6). For a sphere, the moment of inertia is based on the particle mass and diameter, as follows:

$$I_p = \frac{m_p d^2}{10} = \frac{\pi \rho_p d^5}{60}. \tag{10.27}$$

The fluid dynamic torque for a sphere is based on the viscous stresses of (2.18b):

$$\mathcal{T}_{surf} = r_p \int \mathbf{i}_r \times (K_{r\theta} \mathbf{i}_\theta) dA_p. \tag{10.28}$$

Note that there is no pressure contribution since pressure stress only acts radially on the surface and therefore does not contribute to torque for a sphere.

As with fluid dynamic point-force approximation of (9.1), one may also obtain a point-torque expression for the angular momentum equations. Theoretical work in this area has focused on particles spinning with no relative velocity ($w = 0$). In this case, the particle spin Reynolds number (Re_Ω) best characterizes the ratio of convection to viscous effects and is defined as follows:

$$Re_\Omega \equiv \frac{\rho_f d^2 |\mathbf{\Omega}_p|}{\mu_f}. \tag{10.29}$$

For a sphere with weak spin ($Re_\Omega \ll 1$) and no shear ($\omega = 0$), Basset (1888) derived the unsteady fluid dynamic torque in quiescent flow, as follows:

$$\begin{aligned} \mathcal{T}_{surf} = & -\pi\mu_f d^3 \mathbf{\Omega}_p - \frac{1}{6} d^4 \sqrt{\pi\mu_f\rho_f} \int_0^t \left(\frac{d\mathbf{\Omega}_p/d\tau}{\sqrt{t-\tau}} \right) d\tau \\ & + \frac{4\pi\mu_f d}{3} \int_0^t \left(\frac{d\mathbf{\Omega}_p}{d\tau} \exp\left[\frac{4\nu_f(t-\tau)}{d^2} \right] \operatorname{erf} \sqrt{\frac{4\nu_f(t-\tau)}{d^2}} \right) d\tau. \end{aligned} \tag{10.30}$$

This relationship between torque and spin is similar to that between drag and relative velocity (3.83), and the two integral terms represent torque history forces. Happel and Brenner (1973) extended this to a flow with small but finite fluid vorticity and obtained the torque for particle spin and surrounding vorticity (without the history forces), as follows:

$$\mathcal{T}_{surf} = -\pi\mu_f d^3 \left(\mathbf{\Omega}_p - \frac{1}{2} \boldsymbol{\omega}_{f@p} \right) = -\pi\mu_f d^3 \mathbf{\Omega}_{p,rel} \qquad \textit{for } Re_\Omega \ll 1 \textit{ and } Re_\omega \ll 1. \tag{10.31}$$

Thus, the torque is proportional to the relative spin of the particle to the fluid (10.3).

To extend Happel and Brenner's result to finite Reynolds numbers, one may introduce a torque-based spin correction (f_Ω) and a torque-based vorticity correction (f_ω) as follows:

$$f_\Omega \equiv \left(\frac{\mathcal{T}_{surf}}{\pi\mu_f d^3 \Omega_p} \right)_{\omega_f = 0}. \tag{10.32a}$$

$$f_\omega \equiv \left(\frac{\mathcal{T}_{surf}}{\frac{1}{2}\pi\mu_f d^3 \omega_f} \right)_{\Omega_p = 0}. \tag{10.32b}$$

Using these corrections with (10.31), the fluid dynamic torque at finite Re_p is as follows:

$$\mathcal{T}_{surf} = -\pi\mu_f d^3 \left(f_\Omega \mathbf{\Omega}_p - \frac{1}{2} f_\omega \boldsymbol{\omega}_f \right). \tag{10.33}$$

To first determine f_Ω, one may define the nondimensional torque coefficient as follows:

$$C_\Omega \equiv \frac{\mathcal{T}_{surf}}{\frac{1}{2}\rho_f \Omega_p^2 r_p^5}. \tag{10.34}$$

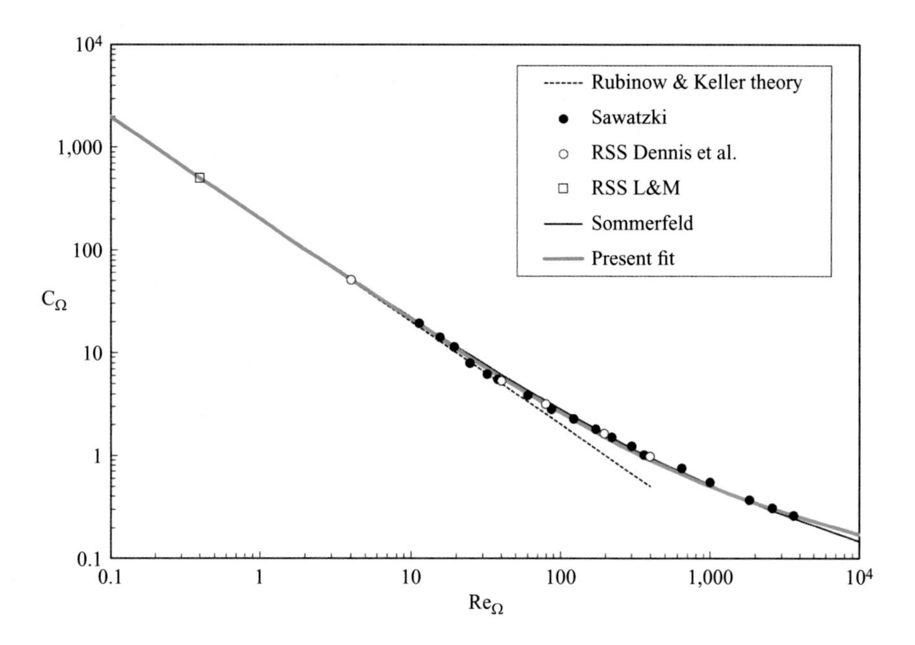

Figure 10.11 Particle torque coefficient as a function of the spin Reynolds number for $Re_p = 0$, based on theory (Rubinow and Keller, 1961), experiments (Sawatzki, 1970), simulations (Dennis et al., 1980; Legendre and Magnaudet, 1998), and a previous model (Sommerfeld, 2001) as well as present model.

This is related to the torque-based spin correction and spin Reynolds number as

$$C_\Omega = \frac{64\pi f_\Omega}{Re_\Omega} \qquad \textit{for } Re_\omega \ll 1. \tag{10.35}$$

The theoretical result for the steady creeping-flow value with no relative velocity (Rubinow and Keller, 1961) is given by $f_\Omega = 1$. The torque coefficient based on experiments and predictions for Re_Ω up to 2,000 is shown in Figure 10.11 where the theoretical value tends to underpredict the torque at high Re_Ω. The model by Sommerfeld (2001) works well for $Re_\Omega > 30$, and a more generalized model (blue line) is given by the following:

$$f_\Omega = 1 + 0.012\,Re_\Omega^{0.7} \qquad \textit{for } Re_\Omega < 2{,}000. \tag{10.36}$$

To investigate the impact of Re_p on f_ω, Salem and Osterlé (1998) employed simulations in shear flow for no particle spin and found $f_\omega \approx 1$ for ω^* as high as 0.6 and Re_p up to 40. As such, one may use $f_\omega \approx 1$ as a rough approximation when employing (10.33).

When applying the angular Equation of Motion, it is useful to consider the spin response time of the particle (τ_Ω). Based on (10.31) for creeping flow ($f_\Omega = f_\omega = 1$), this can be defined as follows:

$$\tau_\Omega \equiv \frac{I_p|\boldsymbol{\Omega}_{rel}|}{|\mathcal{T}_{surf}|} = \frac{\rho_p d^2}{60\mu_f}. \tag{10.37}$$

This spin response time can be compared to the drag response time of (3.88a) for creeping flow (f = 1) to yield the following:

$$\frac{\tau_\Omega}{\tau_p} = \frac{0.3\rho_p}{\rho_p + c_\forall \rho_f}. \tag{10.38}$$

As such, low-density particles ($\rho_p \ll \rho_f$), have very fast angular responses ($\tau_\Omega \ll \tau_p$) and thus are likely to rapidly adjust to equilibrium conditions so the net fluid torque is zero (Figure 10.10b). For high-density particles ($\rho_p \gg \rho_f$), the angular response is still relatively small, only 30% of the drag response, indicating that such particles may also tend to reach spin equilibrium. The torque-free spin-equilibrium condition is discussed in the following.

Free Spin Conditions

As noted previously, low-inertia particles tend to spin-equilibrium conditions ($\mathcal{T}_{surf} = 0$), especially in liquids. The equilibrium spin rate at zero torque can be obtained from (10.31) for small Reynolds numbers as follows:

$$\Omega_{p,eq} = \frac{1}{2}\omega_f \qquad \textit{for } \mathrm{Re}_\Omega \ll 1 \textit{ and } \mathrm{Re}_\omega \ll 1. \tag{10.39}$$

An Oseen-like first-order correction for this equilibrium particle spin rate for finite shear was obtained theoretically by Lin et al. (1970) as follows:

$$\Omega_{p,eq} = \frac{1}{2}\omega_{shear}\left(1 - 0.0385\,\mathrm{Re}_\omega^{3/2}\right) \qquad \textit{for } \mathrm{Re}_p = 0. \tag{10.40}$$

Shi and Rzehak (2019) found that resolved surface simulations are consistent with this theory up to $\mathrm{Re}_\omega < 0.1$ and $\mathrm{Re}_p < 0.1$. They also used resolved-surface simulation results (Homann et al., 2013) for conditions up to $\mathrm{Re}_\omega < 220$ and $\mathrm{Re}_p < 200$ to develop an empirical model for solid sphere equilibrium spin rate as follows:

$$\Omega_{p,eq} = \frac{1}{2}\omega_{shear}\{1 + 0.4[\exp(-0.0135\,\mathrm{Re}_\omega) - 1]\}\left(1 - 0.07026\,\mathrm{Re}_p^{0.455}\right). \tag{10.41}$$

Given such an equilibrium spin rate for an imposed shear, the combined spin-induced and shear-induced lift can be computed using (10.1b) and (10.24) (Shi and Rzehak, 2019). For such conditions, Bagchi and Balachandar (2002a) found that the lift-to-drag ratio for solid spheres at $\mathrm{Re}_p < 40$ can be as high as one fourth (L/D~1/4), indicating the potential significance of lift for solid and contaminated particles. For clean fluid particles, the spin-based lift is unfortunately not well understood, so a reasonable option may be to employ just the shear-based lift for such particles.

10.1.4 Lift for Nonspherical Particles

Lift and torque for nonspherical particles have received less attention than the spherical case. This is attributed to two reasons: (1) high-fidelity trajectories of such particles require time integration of the angular momentum equations of motion using the angles of orientation and torque in three dimensions; and (2) limited theoretical and experimental information is available for the lift and torque of nonspherical particles. In the following, lift is considered for three cases: spheroidal solid particles, deformable fluid particles in shear, and oscillating trajectories of ellipsoidal bubbles. Additional cases are discussed in the review by Leal (1980).

Lift for Solid Particles

Even without flow shear or particle rotation, nonspherical particles create lift if their geometry is not symmetric with respect to the relative velocity. For example, consider a spheroid in creeping flow with differences in the drag correction for the cases of parallel versus perpendicular orientation (such as $f_{E\parallel}$ versus $f_{E\perp}$ as listed in Table 9.4). If this spheroid's axis of symmetry is oriented at an angle of attack (a) to the relative velocity ($\mathbf{w}$), this velocity can be decomposed into portions that are parallel and perpendicular to the spheroid's axis of symmetry to obtain the net resistance force, $\mathbf{F}_{D,eff}$ of (9.47), which can be broken into the component that is parallel to the relative velocity (drag) and that which is perpendicular (lift), as follows:

$$\mathbf{w} = \mathbf{w}_{\parallel} + \mathbf{w}_{\perp} = w(\cos a)\mathbf{i}_{\parallel} + w(\sin a)\mathbf{i}_{\perp}. \tag{10.42a}$$

$$\mathbf{F}_{D,eff} = -3\pi\mu_f(\mathbf{w}_{\parallel} f_{E\parallel} + \mathbf{w}_{\perp} f_{E\parallel}) = \mathbf{F}_D + \mathbf{F}_L. \tag{10.42b}$$

$$\mathbf{F}_D = -3\pi\mu_f \mathbf{w}\left[f_{E\parallel} + (f_{E\perp} - f_{E\parallel})(\sin^2 a)\right]. \tag{10.42c}$$

$$\mathbf{F}_L = -3\pi\mu_f \mathbf{w}(f_{E\perp} - f_{E\parallel})(\sin a)(\cos a). \tag{10.42c}$$

This angle of attack effect was also investigated for spheroids for Re_p up to 2,000 and used to develop an empirical model that reverts to (10.42) for creeping flow (Sanjeevi et al., 2018). At even larger particle Reynolds numbers, List et al. (1973) measured lift and moment coefficients for spheroids at various orientations and aspect ratios at $40{,}000 < Re_p < 400{,}000$.

If shear flow is introduced, then a nonspherical particle can have lift even with no angle of attack. For ellipsoidal particles in shear flow at small but finite Re_p, a theoretical solution can be obtained for the three components of lift force and torque (Jeffrey, 1922) as summarized by Gavze and Shapiro (1997). If the fluid shear is constant in space and time, the torque equations can be explicitly integrated to yield the classical Jeffrey cyclic motion in linear shear flow (Gauthier et al., 1998).

Shear Lift for Deformable Fluid Particles

Fluid particles in shear can also lead to a lift if they deform due to significant Weber numbers. An interesting consequence of such deformation is that the lift direction can change direction for $We > 1$, as seen by comparing the trajectories of Figure 10.12a to Figure 10.12b. This is because fluid particles subjected to a relative velocity tend to deform such that their cross section resembles a hydrofoil (due to asymmetric pressure loads), as shown in Figure 10.12c. This shape, combined with the oncoming relative velocity, creates a force opposite to the direction expected from a Saffman-based lift for spheres. As a result, a low Weber number bubble rising in a shear moves to the right (consistent with a positive C_L per 10.6), while the deformed hydrofoil-shaped bubble moves to the left (consistent with a negative C_L relative to the shear direction).

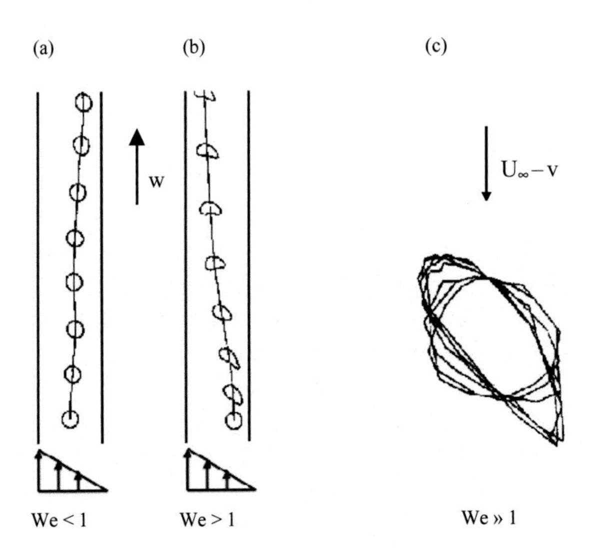

Figure 10.12 Simulation of deformable bubbles rising in a shear flow at different Weber numbers showing shape evolution (Tomiyama, 1998).

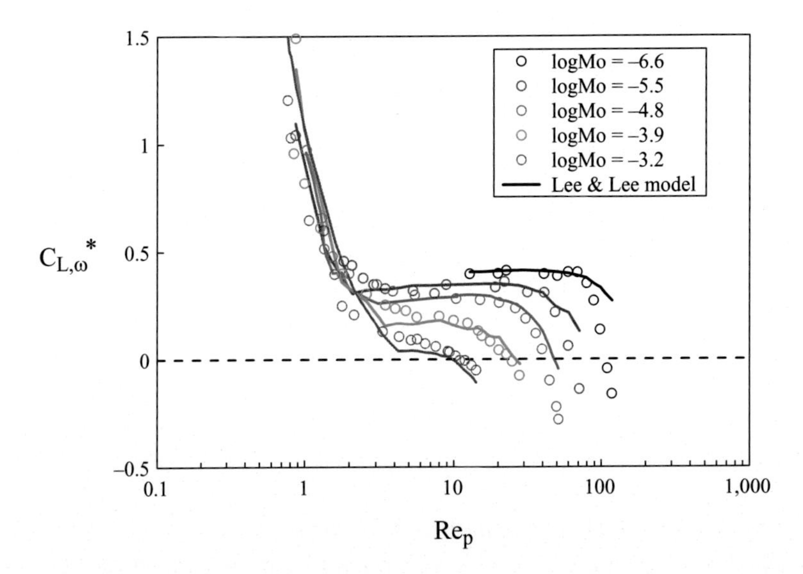

Figure 10.13 Experimental shear-induced lift coefficients for clean deformable air bubbles in high-viscosity liquids (Aoyama et al., 2017) compared with model of Lee and Lee (2020).

This lift force direction reversal due to deformation has been observed for drops and bubbles for a wide range of conditions, including $0.01 < Re_p < 2{,}000$ (Kariyasaki, 1987; Tomiyama et al., 2002b; Ford and Loth, 1998).

Detailed measurements of lift for deformable clean bubbles in highly viscous liquids ($Mo > 10^{-4}$) and moderate viscosity liquids ($10^{-4} < Mo < 10^{-6}$ are shown in Figure 10.13, which shows the lift changing from positive to negative at high

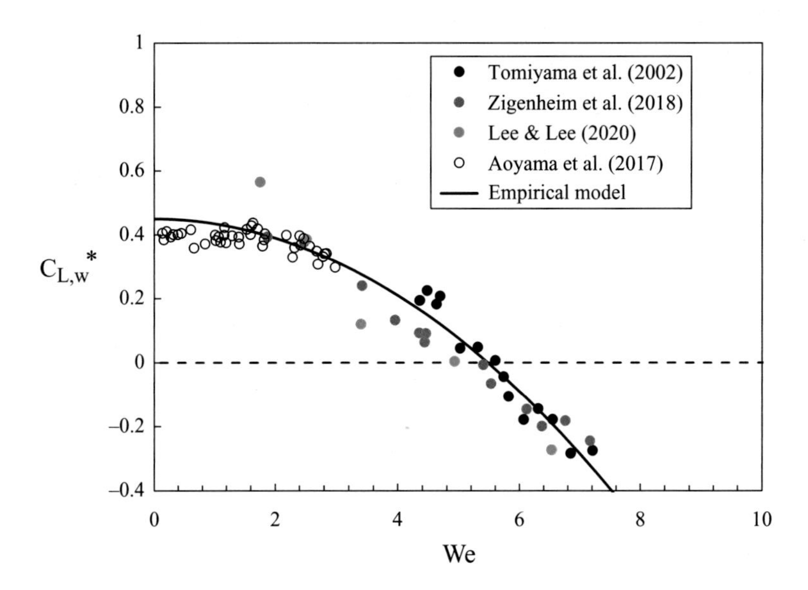

Figure 10.14 Experimental shear-induced lift coefficients for deformable air bubbles in water at $200 < \mathrm{Re_p} < 2{,}000$ for $\mathrm{We} < 20$ (Tomiyama, 2002b; Ziegenhein et al., 2018; Lee and Lee, 2020) and in low-viscosity fluids at $50 < \mathrm{Re_p} < 120$ for $\mathrm{We} < 3$ (Aoyama et al., 2017) compared to the Weber-based empirical model.

Weber numbers. These lift trends were empirically modeled (Lee and Lee, 2020) using the shear-normalized lift coefficient of (10.14a) combined with the fluid Ohnesorge number of (4.54b), as follows:

$$C^*_{L,\omega} = \mathrm{Max}\left(C^*_{L,L\&M}, 0.5 - 2.8E^{-2.2}\mathrm{Oh_f}\right) \qquad \textit{for}\ \mathrm{Re_p} \leq 4. \tag{10.43a}$$

$$C^*_{L,\omega} = 0.5 - 2.8E^{-2.2}\mathrm{Oh_f} \qquad \textit{for}\ 4 < \mathrm{Re_p} < 120. \tag{10.43b}$$

These expressions depend on the bubble aspect ratio, which can be obtained from (9.74) and (9.75), and compare well with the data of Figure 10.13.

For low-viscosity liquids with $\mathrm{Mo} < 10^{-5}$, such as water, the changes in lift coefficient due to deformation are controlled more by Weber number (since viscous effects become very small), as shown in Figure 10.14, where this dependence can be empirically modeled as follows:

$$C^*_{L,\omega} = 0.45 - 0.015\mathrm{We}^2 \qquad \textit{for}\ \mathrm{Mo} < 10^{-5}\ \textit{and}\ \mathrm{We} < 7. \tag{10.44}$$

Figure 10.14 also shows that this Weber-based model works well for $\mathrm{Re_p}$ as low as 50 if limited to $\mathrm{We} < 3$. Notably, the models of (10.43) and (10.44) are nearly the same (and both reasonable) for moderate viscosity liquids with ($10^{-4} < \mathrm{Mo} < 10^{-6}$).

Oscillating Lift for Ellipsoidal Bubbles

As discussed in §4.4.2, ellipsoidal bubbles at high $\mathrm{Re_p}$ undergo helicoidal and/or sinusoidal trajectory oscillations due the double-threaded wake instabilities. The effect is often neglected for a point-force approximation since the lift is not due to shear, spin, or a steady angle of attack. However, the oscillatory side force (effectively an oscillatory lift force) can be important to local bubble–bubble and bubble–wall interactions. To capture this effect with a point-force method, one may impose an oscillating side force of zero mean to the bubble dynamic equation. This side force can be stochastic based on sampling a Gaussian distribution of velocity fluctuations (Tomiyama, 1998). The tumble of disks and solid objects could also be modeled in this vein. The side force can also be deterministic by superposing a lift force or a trajectory oscillation using the oscillation amplitude ($\mathcal{A}_{osc}$) and frequency (f_{osc}) from (4.59) for sinusoidal motion (Loth, 2000). To consider a symmetric helicoidal (spiral) motion, a second orthogonal component can be added that is $\pi/2$ out of phase, but with the same amplitude (Ellingsen and Risso, 2001). Superposing the oscillating lift to the quasisteady shear-based lift has been found to be reasonable for bubbles in laminar flow (Lee and Lee, 2020) and in turbulent flow (Ford and Loth, 1998).

10.2 Added-Mass Force

The added mass (also called the *virtual mass* or *apparent mass*) represents the effective portion of the surrounding continuous-fluid mass that is accelerated along with the particle (Figure 3.9). It is especially important for particles in liquids; for example, the virtual mass of a spherical bubble causes an effective gravitational acceleration of $-2\mathbf{g}$, as noted in §3.4. The added-mass force can be obtained by employing a fluid work principle as discussed in §3.3.2, where the added mass for a sphere in inviscid flow is given by (3.75) and for creeping flow by (3.77). These two expressions both give an added-mass coefficient of ½ but have small differences in the acceleration vector, which are negligible when the fluid gradients are linear (Maxey, 1993). Several studies have investigated added mass at intermediate $\mathrm{Re_p}$ conditions and found that the inviscid form is reasonable for both solid and fluid spheres, even when some flow separation is present (Bataille et al., 1991; Mei et al., 1991; Mei and Klausner, 1992, 1994; Legendre and Magnaudet, 1998; Kim et al., 1998; Wakaba and Balachandar, 2005).

However, a nonspherical particle introduces additional complexity to the added mass since the added-mass coefficient must be replaced by an added-mass tensor ($c_{\forall,ij}$) to relate the added-mass force in the i-direction to the acceleration in the j-direction (Yih, 1969):

$$F_{\forall,i} = c_{\forall,ij}\rho_f\forall_p\left(\mathcal{D}u_{j,@p}/\mathcal{D}t - d u_j/d t\right) \qquad for\, j = 1, 2, and\ 3. \tag{10.45}$$

For a spheroid, there are two added-mass coefficients for acceleration parallel or perpendicular to the axis of symmetry so long as the acceleration is only in one of these two directions. The added-mass coefficient for parallel acceleration can be expressed using the aspect ratio (E) for prolate and oblate spheroids (Lai and Mockros, 1972) as follows:

$$c_{\forall\parallel} = \frac{\ln(2E)}{E^2} \qquad \textit{for } E > 1. \tag{10.46a}$$

$$c_{\forall\parallel} = \frac{E\cos^{-1}E - \sqrt{1-E^2}}{E^2\sqrt{1-E^2} - E\cos^{-1}E} \qquad \textit{for } E < 1. \tag{10.46b}$$

This coefficient can also be used to obtain the coefficient for perpendicular acceleration (Loewenberg, 1993), as follows:

$$c_{\forall\perp} = 1/(1 + 2c_{\forall\parallel}). \tag{10.47}$$

Although (10.46) and (10.45) are also for inviscid flow, these added-mass expressions are generally appropriate for particles in creeping flow (Clift et al., 1978; Benjamin, 1987) and are expected to be reasonable for bodies at a wide range of Reynolds numbers so long as flow separation is small.

However, objects with large separated wakes can result in complex added-mass coefficients that are difficult to predict (Brennen, 2005). Fortunately, one analytically tractable condition is the spherical-cap shape (which is typical of bubbles at We » 1 and Re_p » 1). This is treated by assuming a closed wake flow region based on a Hill's vortex, and by linearly combining the added masses for the spherical fore section and the recirculating closed-wake aft section (Kendoush, 2003). In terms of geometry, the spherical-cap aspect ratio (E) is the ratio of height to the maximum diameter, and the cap angle (θ_{cap}) can be defined as the angle from the axis of symmetry to that where the flat rear is located. Either of these parameters can be used to obtain a cap coefficient as follows:

$$c_{cap} = \cos(\theta_{cap}) = 1 - 2E \qquad \textit{for } E \geq 0.5. \tag{10.48a}$$

$$c_{cap} = \cos(\theta_{cap}) = (1 - 4E^2)/(1 + 4E^2) \qquad \textit{for } E \leq 0.5. \tag{10.48b}$$

For example, a hemisphere ($E = 0.5$) corresponds to $\theta_{cap} = 90^o$ and $c_{cap} = 0$. The theoretical added-mass coefficients for a spherical-cap shape are then given in terms of the cap coefficient as follows:

$$c_{\forall\parallel} = \frac{1144 + 940(c_{cap}) - 410(c_{cap})^3 - 135(c_{cap})^8 + 81(c_{cap})^{10}}{560 - 840(c_{cap}) + 280(c_{cap})^3}. \tag{10.49}$$

This theoretical result is compared to that for oblate spheroids and to experimental and resolved-surface data for deformed bubbles in Figure 10.15. In general, the data follow the added-mass theory for an oblate spheroid (10.46) for modest deformations, while the higher deformations (with larger 1/E values) tend to follow the spherical-cap added-mass theory of (10.49).

10.3 History Force

While the quasisteady drag force ($\mathbf{F}_D$) is the drag associated with relative velocity, the history force ($\mathbf{F}_H$) is the drag associated with relative acceleration, where both forces

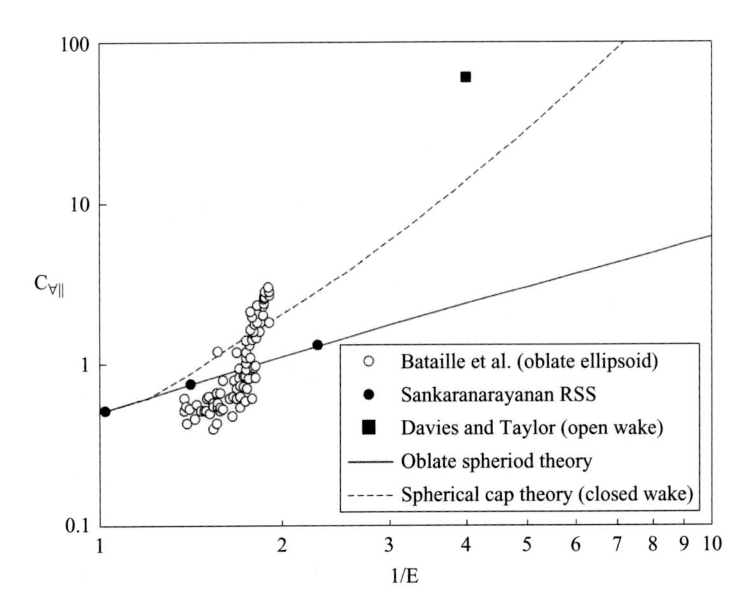

Figure 10.15 Added-mass coefficient as a function of bubble eccentricity (inverse of the aspect ratio) based on experimental data from Davies and Taylor (1950) and Bataille et al., (1991), as well as clean bubble Resolved-Surface Simulations from Sankaranarayanan et al. (2003).

act opposite to the relative velocity (**w**). The history force is often neglected but can be significant at high accelerations. In the following, the history force is discussed for solid spherical particles for finite Reynolds numbers. This is followed by the history force for fluid particles and then for nonspherical solid particles. Finally, the history effects on the lift force ($\mathbf{F}_L$) are discussed.

10.3.1 Drag History Force for Solid Spheres

To determine when the history force is important, it is helpful to define the time scale for diffusion (based on viscosity) as follows:

$$\tau_{diff} \equiv \rho_f d^2/\mu_f = d^2/\nu_f. \tag{10.50}$$

If one considers the creeping-flow regime ($Re_p \rightarrow 0$) for a particle accelerating from rest in uniform flow, the ratio of the history force (3.82) to the quasisteady drag force (3.30) can be expressed in terms of the diffusion time relative to the particle response time as follows:

$$\left|\frac{F_H}{F_D}\right| \sim \sqrt{\frac{\tau_{diff} t}{4\pi}} \left|\frac{dw/d\tau}{w}\right| \sim \sqrt{\frac{\tau_{diff}}{4\pi\tau_p}} \quad \textit{for creeping flow}. \tag{10.51}$$

The RHS approximation assumes the relative velocity changes at a time scale that is consistent with the drag-based particle response time. Based on the RHS and (10.50) and (5.11), a rough criterion for neglecting the history force in creeping flow becomes

$$18\rho_f \ll \left|\rho_p + c_{\forall}\rho_f\right| \qquad \textit{for negligible Basset history force.} \tag{10.52}$$

Thus, the history force can generally be neglected for high-density particles ($\rho_p \gg \rho_f$) but can be significant for moderate or low-density particles, such as bubbles in a liquid, for creeping flow.

To consider the history force at generalized Reynolds numbers, we define a history force kernel (H) for the time integral and note that this kernel reverts to the Basset kernel (H_{Basset}) for the creeping-flow limit (based on 3.82) as follows:

$$\mathbf{F}_H = -3\pi\mu_f d \int_{-\infty}^{t} H \frac{d\mathbf{w}}{d\tau} d\tau \qquad \textit{for general } \mathrm{Re}_p. \tag{10.53a}$$

$$H_{Basset} = \sqrt{\frac{d^2\rho_f}{4\pi(t-\tau)\mu_f}} = \sqrt{\frac{1}{4\pi}\left(\frac{\tau_{diff}}{t-\tau}\right)} \qquad \textit{for } \mathrm{Re}_p \to 0. \tag{10.53b}$$

In (10.53b), $(t - \tau)$ is the time since a particular acceleration has occurred, and this kernel is proportional to $(t - \tau)^{-1/2}$, indicating a slow decay of influence from previous times. However, high Re_p conditions will cause a faster decay at long times. In particular, the generalized kernel (H) of (10.53a) will be proportional to $(t - \tau)^{-2}$ based on convection-dominated conditions so long as the particle does not reverse direction (Sano, 1981).

The change in the decay rate with the Reynolds number is shown schematically in Figure 10.16 using a log–log plot of the history force kernel as a function of the time since an acceleration has occurred. If the acceleration is a constant, the history force is proportional to the area under the curve, per (10.53a). In this log–log plot, it can be

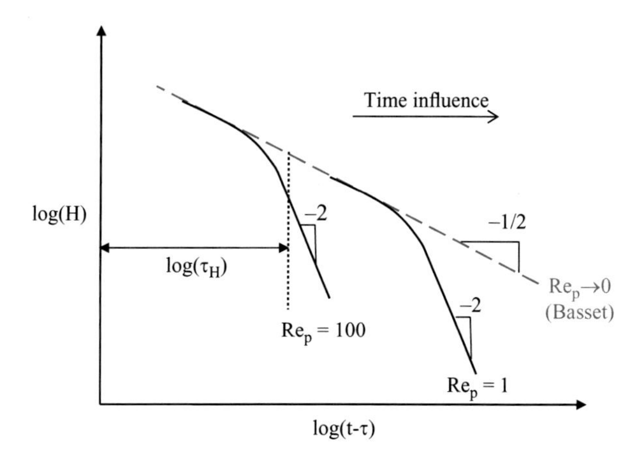

Figure 10.16 Schematic of a log–log plot of the history kernel as a function of time, where recent histories (high frequencies) can be approximated with the creeping-flow kernel of Basset, but very early events (low frequencies) have smaller kernels, that is, they result in faster decays. The history time scale (τ_H) indicates the transition between these kernel regimes and is also shown for $\mathrm{Re}_p = 100$ as a vertical dashed line.

seen that the creeping-flow case (Basset theory) has a –½ slope consistent with (10.53b). However, increasing the Reynolds number causes the kernel to drop off quicker (faster decay) and tends to a –2 slope (the convection limit). As a result, the area under the curve decreases as Re_p increases, indicating that the history force effect gets weaker as Re_p increases.

To model the transition between decay rates, Mei and Adrian (1992) proposed a generalized history force kernel that combines the creeping-flow and convection limits as follows:

$$H_{Re} = \left[(H_{Basset})^{-1/c_{H1}} + (H_{conv})^{-1/c_{H1}}\right]^{-c_{H1}}. \tag{10.54a}$$

$$H_{conv} = \frac{1}{\pi}\left(\frac{\tau_{diff}}{t-\tau}\right)^2\left[\frac{3}{4Re_p} + c_{H2}\right]^3. \tag{10.54b}$$

These expressions use two empirical coefficients that were proposed as $c_{H1} = 2$ and $c_{H2} = 0.105$ based on resolved-surface simulations for Re_p values up to 100 (Mei et al., 1991). Using this same form on (10.54), Dorgan and Loth (2007) investigated a wide variety of experimental data, including $9 < Re_p < 853$ and ρ_p/ρ_f in the range of 1.17–10.32, and proposed $c_{H1} = 2.5$ and $c_{H2} = 0.2$ for improved accuracy. An example of the differences for these constants is shown in Figure 10.17, where the Dorgan and Loth kernel is a slightly better fit to the data.

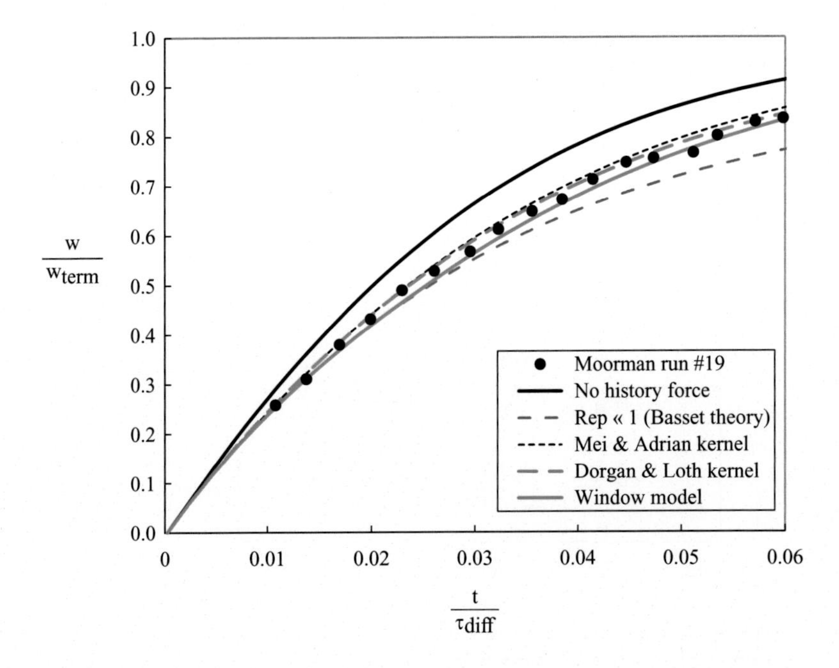

Figure 10.17 Velocity of a particle accelerating to terminal velocity at $Re_{p,term} = 166$ and a particle density ratio of 3.69 for which $\tau_{diff} \approx 0.5\tau_p$ based on experiments of Moorman (1955), as well predictions with various history force expressions.

Figure 10.17 also illustrates the importance of the history decay rate for a particle falling from rest in a quiescent liquid by comparing experimental results to predictions using only a quasisteady drag (no history forces) and to predictions with various forms of the history force. The history force must be included to properly model the change in relative velocity and is initially well described by the Basset theory (10.53b). However, as the particle $\mathrm{Re_p}$ increases with time, the history force effect is reduced consistent with the transition described by (10.54).

Employing the history force in numerical simulations generally requires small time steps (relative to the smaller of τ_p and τ_{diff}) as well as storage of all previous accelerations seen by the particle. If there are a large number of particles that are integrated over long times, this can result in high computational time and memory issues. To address this, a "window-based" approach can be used to limit the required time of integration. This is achieved by defining the history time scale, τ_H, as shown in Figure 10.17, such that the area under the curve using the Basset kernel (H_{Basset}) up to this time is equal to the area under the curve for the generalized kernel (H_{Re}) integrated over all times, as follows:

$$\int_{t-\tau_H}^{t} H_{Basset}\, d\tau \equiv \int_{-\infty}^{t} H_{Re}\, d\tau. \tag{10.55}$$

This definition is convenient since the LHS integral can be evaluated analytically using (10.53b), while the RHS can be numerically integrated using (10.54). For $c_{H1} = 2.5$ and $c_{H2} = 0.2$, the history time scale is approximately as follows (Dorgan and Loth, 2007):

$$\tau_H \approx \tau_{diff}\left(\frac{0.502}{\mathrm{Re_p}} + 0.123\right)^2. \tag{10.56}$$

This model yields $\tau_H \rightarrow \infty$ for creeping flow and $\tau_H \rightarrow 0.015\tau_{diff}$ for high $\mathrm{Re_p}$ conditions. As such, numerical integration and memory storage of past accelerations is only needed for $(t - \tau) < \tau_H$, as follows:

$$\mathbf{F}_H = -3\pi\mu_f d \int_{t-\tau_H}^{t} \sqrt{\frac{1}{4\pi}\left(\frac{\tau_{diff}}{t-\tau}\right)}\frac{d\mathbf{w}}{d\tau}\, d\tau \qquad \textit{for general } \mathrm{Re_p}. \tag{10.57}$$

This windows-based approach can reduce CPU requirements for history force by an order of magnitude compared to that using (10.54), while staying reasonably accurate, as shown in Figure 10.17. Parmar et al. (2018) developed a more accurate (though complex) approach for history force integration that also reduces storage requirements.

The case of a particle that reverses direction is more complex since the particle will ingest its own wake and can lead to decay rate of t^{-1} for some conditions (Lawrence and Mei, 1995; Lovalenti and Brady, 1995). Such wake ingestion effects abound for particles oscillating back and forth. However, the Mei and Adrian model of (10.54) has been found to be reasonable as long as $\tau_{osc} < \tau_H$, where τ_{osc} is the oscillation period. The model is not reasonable for slow oscillations ($\tau_{osc} \gg \tau_H$) due to wake ingestion effects, but the history force in this case is rather weak anyway. For example, the history force for $\tau_{osc} = 2\tau_H$ at $Re_p = 20$ represents less than 1% of the total drag (Mei et al. 1991).

10.3.2 Drag History Force for Fluid Spheres and Solid Spheroids

For fluid particles of variable viscosity, a closed-form solution of the history force is available for the clean bubble limit (Lovalenti and Brady, 1995) in creeping-flow conditions:

$$\mathbf{F}_H = -4\pi\mu_f d \int_{-\infty}^{t} \frac{d\mathbf{w}}{d\tau} \exp\left[36\nu_f(t-\tau)/d^2\right] \operatorname{erfc}\left[6\sqrt{\nu_f(t-\tau)}/d\right] d\tau. \tag{10.58}$$

This history force is smaller than that for a solid sphere due to the existence of finite slip on the bubble surface. For Reynolds numbers up to 300, Mei et al. (1994) proposed a detailed empirical correction that included short-time and long-time components based on resolved-surface simulations and found reasonable agreement for bubble trajectories in uncontaminated liquids (Park et al., 1995). However, the history force is generally weak for clean bubbles with $Re_p > 50$ so that it can often be reasonably neglected (Magnaudet and Eames, 2000).

For spheroids, Lai and Mockros (1972) obtained the history force for movement parallel to the axis of symmetry (creeping-flow) conditions and found it to be related to the correction for the quasisteady drag, f_E (Table 9.4) and the spherical Basset kernel (10.53b):

$$\mathbf{F}_{H,E\parallel} = -3\pi\mu_f d (f_{E\parallel})^2 \int_{-\infty}^{t} H_{Basset} \frac{d\mathbf{w}}{d\tau} d\tau. \tag{10.59}$$

As such, the history force is more likely to be important (compared to quasisteady drag) if a particle is nonspherical, since this drag correction can be as high as fourfold if the shape is highly oblate or highly prolate (per Figure 9.14). This same type of correction can be generalized to other nonspherical solid particles in creeping flow (Lawrence and Mei, 1995).

10.3.3 Lift History Force for Solid Spheres

As is the case with the drag force, the lift force can have an additional component when there are high accelerations that cause the viscous layer over the particle to evolve. Asmolov and McLaughlin (1999) derived an expression for an unsteady Saffman lift force in frequency space. Similarly, Coimbra and Kobayashi (2002) derived the history lift force integral for $Re_p \ll 1$ and $Re_\omega \ll 1$ to extend the steady lift of (10.6), as follows:

$$\mathbf{F}_{\mathrm{L}} = 1.615\mathrm{d}^2\sqrt{\frac{\mu_{\mathrm{f}}\rho_{\mathrm{f}}}{\omega}}(\boldsymbol{\omega} \times \mathbf{w}) + 1.615\frac{\rho_{\mathrm{f}}\mathrm{d}^3}{2\sqrt{\omega\pi}}\left[\int_0^t \left\{\frac{d(\boldsymbol{\omega} \times \mathbf{w})/d\tau}{\sqrt{t-\tau}}\right\} d\tau\right]. \quad (10.60)$$

To study the relative importance of the history lift force in creeping flow, Lim et al. (2005) simulated particle lift dynamics in rotating flows. They noted that the history effects are often negligible but can be significant for high frequencies and small values of ρ_p/ρ_f.

For finite Re_p values, there is no analytical solution for the history lift force, so one must rely primarily on experiments and simulations. For example, Wakaba and Balachandar (2005) examined the unsteady lift component for an accelerating solid particle with Re_p ranging from 5–125. They found that the unsteady lift kernel at finite Re_p was substantially smaller than the creeping-flow predictions, a result consistent with the unsteady drag behavior at finite Re_p.

11 Particle Interactions with Walls and Other Particles

In many multiphase flows, a particle can interact with other particles' fluid dynamically or by collisions. Similarly, multiphase pipe and boundary-layer flows can produce high concentrations of particles near a wall where they can interact with the wall via fluid dynamics or via contact mechanics. This chapter describes both fluid dynamic and collision interactions of particles with other particles or with walls. In the following, the wall-induced fluid dynamic modifications on drag, lift, and added mass on a particle are considered (§11.1), followed by the fluid dynamics of particles on each other via three-way coupling (§11.2). For collisions of particle with other particles or walls, the momentum change formulation is first established (§11.3), followed by restitution mechanics for normal collisions (§11.4) and oblique and/or spin collisions of solid particles (§11.5) as well as the outcome of collision of deformable fluid particles (§11.6).

11.1 Fluid Dynamic Influence of a Wall on a Particle

Fluid dynamic forces on a particle can be modified due to the proximity to the wall, especially at distances on the order of the particle diameter. For the fluid dynamic interactions discussed herein, the wall is approximated as a planar surface, as shown in Figure 11.1, where the closest distance of the particle centroid to the wall is ℓ_{wall} and that from the particle surface to the wall is ℓ_{gap}. Wall corrections can be particularly important for particles at low Reynolds numbers ($\text{Re}_p \rightarrow 0$) since these conditions have viscous interactions that can extend several diameters from the particle surface. Fortunately, these conditions often allow theoretical solutions. For higher Reynolds numbers (e.g., $\text{Re}_p > 1$), theoretical solutions are generally not available. In the following, we discuss the wall effects on the drag forces for solid particles and clean bubbles, the lift forces for solid particles and clean fluid particles, and then added-mass corrections. In these cases, generally the creeping-flow conditions are first considered followed by empirical extension to higher Reynolds numbers.

11.1.1 Drag Corrections for Solid Particles

For Stokesian flow with a nonrotating spherical particle, wall proximity causes additional fluid dynamic stresses on the surface of the particle (and on the wall) due

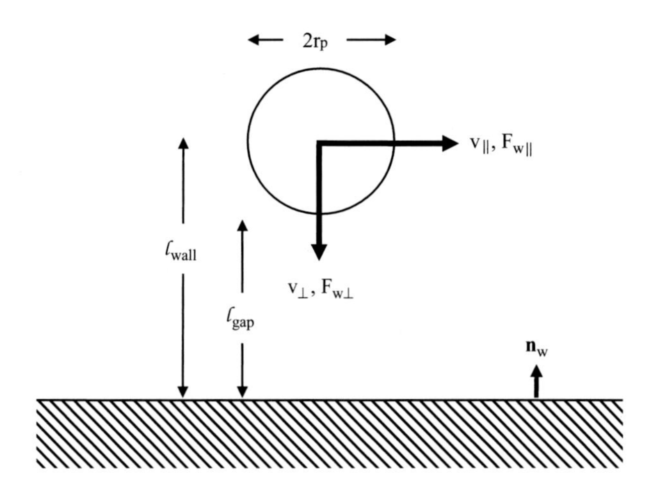

Figure 11.1 Schematic of particle distance, velocity, and force near a wall showing distance from particle centroid to the wall (ℓ_{wall}) and particle surface to the wall (ℓ_{gap}).

to confinement of the fluid in the gap region. These increased surface stresses increase the particle drag. The drag corrections for movement parallel and perpendicular to the wall for small Reynolds numbers and large wall distances ($\ell_{wall} \gg r_p$) can be obtained theoretically (Happel and Brenner,1973) as follows:

$$f_{\parallel} = 1 + \frac{9}{16}\frac{r_p}{\ell_{wall}} + \frac{81}{256}\frac{r_p^2}{\ell_{wall}^2} + \frac{217}{4096}\frac{r_p^3}{\ell_{wall}^2} \tag{11.1a}$$

$$f_{\perp} = 1 + \frac{9}{8}\frac{r_p}{\ell_{wall}} + \frac{81}{64}\frac{r_p^2}{\ell_{wall}^2} + \frac{473}{512}\frac{r_p^3}{\ell_{wall}^2} \tag{11.1b}$$

$$\textit{for}\ \mathrm{Re}_p \to 0\ \&\ \ell_{wall} \gg r_p.$$

The first-order correction (the second term on the RHS) represents most of the drag increase but is relatively small (< 6%) even when the particle is within $10r_p$ of the wall, as shown in Figure 11.2.

For the condition of very small gaps relative to the particle radius ($\ell_{gap} \ll r_p$) where the drag corrections are much larger, Goldman et al. (1967) obtained a theoretical creeping-flow correction for movement parallel to the wall based on lubrication theory, as follows:

$$f_{\parallel} = 0.9588 - \frac{8}{15}\ln\left(\frac{\ell_{gap}}{r_p}\right) \qquad \textit{for}\ \mathrm{Re}_p \to 0\ \&\ \ell_{gap} \ll r_p. \tag{11.2}$$

The first term on the RHS was included to allow an improved asymptotic behavior at larger gaps. Note that Goldman's theoretical model predicts infinite drag as the gap length goes to zero. This suggests that a small particle would never impact the wall since the viscous effects would prevent contact for creeping-flow conditions.

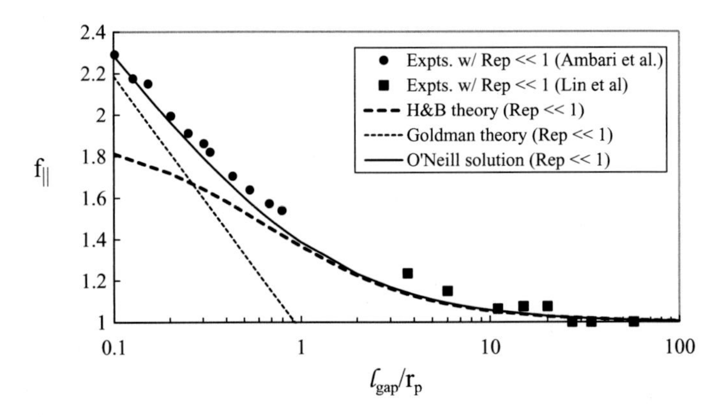

Figure 11.2 Wall correction for drag of a solid particle moving parallel to the wall with experimental data from Ambari et al. (1984) and Lin et al. (2000) compared to various theoretical predictions.

However, surface roughness combined with particle inertia can allow contact collisions to take place, as will be discussed in §11.4. To bridge the preceding two theories for very small and very large gaps, O'Neill (1964) obtained an exact solution in integral form for intermediate gaps, and solved this numerically. An empirical model that represents the exact intermediate-gap solution of O'Neill was given by Zeng et al. (2009), as follows:

$$f_{\parallel} \approx 1.028 - \frac{0.07r_p^2}{r_p^2 + \ell_{gap}^2} - \frac{8}{15}\ln\left(\frac{135\ell_{gap}}{135r_p + 128r_p}\right) \qquad \textit{for}\ \mathrm{Re}_p \to 0. \tag{11.3}$$

As shown in Figure 11.2, the O'Neill bridge model of (11.3) tends to the appropriate theoretical limits for very small gaps (11.2) and very large gaps (11.1), and matches experimental data for intermediate gaps at small Reynolds numbers.

To extend this result to large Reynolds numbers, the wall correction was obtained using matched asymptotic expansions for small gaps (Vasseur and Cox, 1977). Using these results and resolved-surface simulations, this correction was generalized to a broad range of Reynolds numbers and gaps (Re_p up to 200 and ℓ_{gap} down to $0.01r_p$) by Zeng et al. (2009), as follows:

$$f = f_{\parallel} f_{Re} = f_{\parallel}\left\{1 + 0.15\left[1 - \exp\left(-\sqrt{\ell_{gap}/2r_p}\right)\right]\mathrm{Re}_p^{\,0.687+0.313\exp\left(-\sqrt{2\ell_{gap}/r_p}\right)}\right\}. \tag{11.4}$$

The correction uses the product of the creeping-flow result (11.3) and a Reynolds number correction (f_{Re}) based on the form of the Schiller–Naumann correction (3.58b). The predicted changes are consistent with finite Re_p experiments

(Takemura et al., 2002) and yield a reduced wall influence as Re_p increases. For example, the wall-induced drag increase at $\ell_{gap} = r_p$ is ~40% for creeping flow but is ~5% for $Re_p = 200$.

For nonspherical shapes, the wall interactions can be complex, and they can give rise to a modified resistance tensor, which can give rise to rotational-translational coupling (Gavze and Shapiro, 1997). However, the average wall correction generally decreases as the particle's nonsphericity increases. Therefore, the preceding wall corrections for spherical particles can be considered as an upper bound to the corrections for nonspherical particles.

11.1.2 Drag Corrections for Clean Bubbles

For spherical clean bubbles (with internal circulation), the Stokesian drag correction for large wall distances ($\ell_{wall} \gg r_p$) was theoretically obtained (Magnaudet, 2003) as follows:

$$f_{\parallel} = \frac{2}{3} + \frac{1}{4}\left(\frac{r_p}{\ell_{wall}}\right) + \frac{3}{32}\left(\frac{r_p}{\ell_{wall}}\right)^2 + \frac{9}{256}\left(\frac{r_p}{\ell_{wall}}\right)^3 \quad (11.5a)$$

$$f_{\perp} = \frac{2}{3} + \frac{1}{2}\left(\frac{r_p}{\ell_{wall}}\right) + \frac{3}{8}\left(\frac{r_p}{\ell_{wall}}\right)^2 + \frac{9}{32}\left(\frac{r_p}{\ell_{wall}}\right)^3 \quad (11.5b)$$

for $Re_p \to 0$ & $\ell_{wall} \gg r_p$.

The first term on the RHS (=2/3) is the Hadamard–Rybczynski theory (3.41) such that the following terms are the wall corrections. For small but finite Reynolds number effects, an Oseen theory for a clean spherical bubble with wall corrections was obtained in integral form by Takemura et al. (2002).

At very high Re_p values, the flow over a clean bubble can be approximated with a thin attached boundary layer and elsewhere a potential flow (§9.3.1). Based on this approach, Kok (1993) extended the first-order clean bubble drag (9.63b) to obtain the drag coefficient in the limit of weak effects ($\ell_{wall} \gg r_p$) as follows:

$$C_{D\parallel} = \left(\frac{48}{Re_p}\right)\left[1 + \frac{1}{4}\left(\frac{r_p}{\ell_{wall}}\right)^3\right] \quad (11.6a)$$

$$C_{D\perp} = \left(\frac{48}{Re_p}\right)\left[1 + \frac{1}{8}\left(\frac{r_p}{\ell_{wall}}\right)^3\right] \quad (11.6b)$$

for $Re_p \gg 1$ & $\ell_{wall} \gg r_p$.

Measurements of dynamic bubble trajectories (Tsao and Koch, 1997; de Vries et al., 2001) indicate that this correction is reasonable for $Re_p > 200$. To generalize the clean spherical bubble drag for a wide range of Reynolds numbers and gaps ($0.1 < Re_p <$ 1,000 and $0.5 < \ell_{gap}/r_p < 8$), a seven-equation composite fit was developed by Shi et al. (2020).

For deformable bubbles, at high Reynolds numbers, the effects of the wall interactions can lead to modified shapes and high deformation result in a nearly constant drag coefficient of $C_D \approx 0.7$ (Barbosa et al., 2019).

11.1.3 Lift Corrections for Solid Particles and Fluid Particles

The presence of a wall can also induce a normal force (i.e., lift) away from the wall due to the asymmetry of the pressure distribution over the particle surface. This repulsive wall-induced lift can be linearly combined with the lift due to flow shear and/or particle spin (Shi et al., 2020). For a particle moving parallel to the wall surface with a small gap ($\ell_{gap} \ll r_p$), the wall-induced theoretical lift coefficient for solid spheres for creeping flow conditions (Magnaudet et al., 2003) is as follows:

$$C_L \approx \frac{1}{2}\left[1+\frac{1}{8}\left(\frac{r_p}{\ell_{wall}}\right)^3-\frac{33}{64}\left(\frac{r_p}{\ell_{wall}}\right)^2\right] \quad \textit{for } \mathrm{Re}_p \to 0 \ \& \ \ell_{gap} \ll r_p. \tag{11.7}$$

The theory was generalized for a variety of wall distances and viscosity ratios for solid and fluid particles by Takemura and Magnaudet (2003) using a normalized wall distance (ℓ^*) and then empirically described as follows:

$$C_L \approx \left(\frac{2+3\mu^*}{3+3\mu^*}\right)^2\left[\frac{9}{8}+5.78\mathrm{x}10^{-6}(\ell^*)^{4.58}\right]\exp(-0.292\ell^*) \quad \textit{for } \ell^* < 10. \tag{11.8a}$$

$$C_L \approx \left(\frac{2+3\mu^*}{3+3\mu^*}\right)^2 8.94(\ell^*)^{-2.09} \quad \textit{for } 10 < \ell^* < 300. \tag{11.8b}$$

$$\ell^* \equiv \ell_{wall}\, w/\nu_f = \ell_{wall}\, \mathrm{Re}_p/d. \tag{11.8c}$$

As such, the predicted clean bubble lift is 4/9 of that for a solid particle (or a contaminated bubble) due to the viscosity ratio effect. This difference is seen in Figure 11.3, where both cases tend to zero lift coefficient at large normalized wall distances. This figure also shows the creeping-flow lift of (11.8) compares surprising well with experiments for Re_p up to 90.

However, Figure 11.3 shows that clean bubbles far from the wall at high Reynolds numbers will yield small negative lift coefficients. This causes bubbles to be pulled to the surface and is consistent with inviscid effects whereby an increased velocity in the gap causes a lower pressure, which results in attraction forces. The inviscid wall-induced attractive lift force at large gaps distance was theoretically described (Takemura and Magnaudet, 2003) as follows:

$$C_L = -\frac{3}{8}\left(\frac{r_p}{\ell_{wall}}\right)^4\left[1+\frac{1}{8}\left(\frac{r_p}{\ell_{wall}}\right)^3\right] \quad \textit{for clean bubbles at } \mathrm{Re}_p \gg 1. \tag{11.9}$$

For intermediate Re_p values, Takemura and Magnaudet (2003) developed an empirical combination of the inviscid result (11.9) and the low Re_p solution (11.8) that bridges the attractive and repulsive regimes. However, the magnitude of the measured lift coefficient where this applies is generally very small (< 0.025).

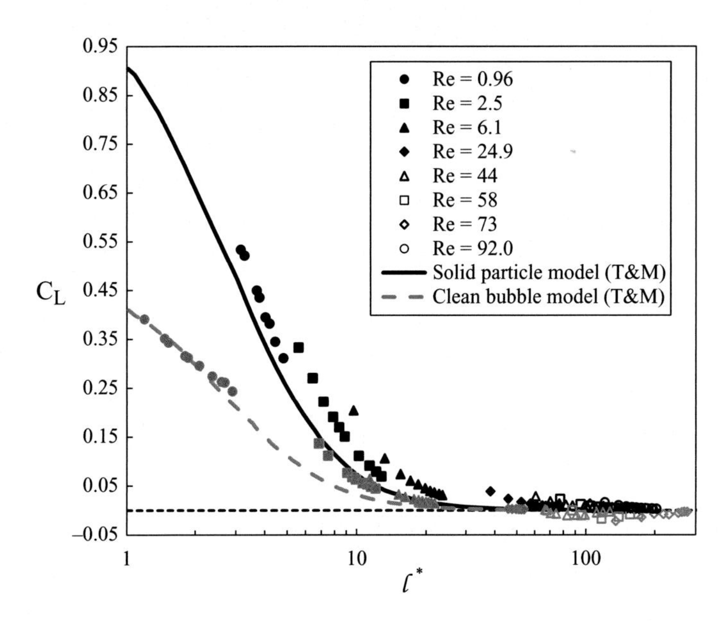

Figure 11.3 Wall-induced lift coefficients for a solid particle (and fully contaminated bubbles) in terms of normalized wall distance with experiments as black symbols, the creeping-flow model as a solid black line, clean bubble experiments as blue symbols (at similar Reynolds numbers) and the model as a dashed blue line (Takemura and Magnaudet, 2003).

11.1.4 Added-Mass Corrections for Spheres

As discussed in §3.3.2 and §10.2, the added-mass corrections obtained for inviscid flow and creeping flow can generally be used for a wide range of Reynolds numbers for a sphere. As such, the inviscid wall effects on the added-mass forces can be expected to be broadly appropriate and were obtained by Soo (1990) based on potential flow. In particular, a mirror image method was used by considering two spheres moving on opposite sides of an imaginary wall and examining the kinetic energy of the fluid caused by the motion of both spheres. The theoretical added-mass force components in directions parallel and perpendicular to the wall are given as follows:

$$F_{\forall,\parallel} = \frac{1}{2}\rho_f \forall_p \frac{dv_\parallel}{dt} + \rho_f \forall_p \left(\frac{r_p}{\ell_{wall}}\right)^3 \left(\frac{3}{32}\frac{dv_\parallel}{dt} - \frac{9}{32} v_\parallel \frac{v_\perp}{\ell_{wall}}\right). \tag{11.10a}$$

$$F_{\forall,\perp} = \frac{1}{2}\rho_f \forall_p \frac{dv_\perp}{dt} + \rho_f \forall_p \left(\frac{r_p}{\ell_{wall}}\right)^3 \left(\frac{3}{16}\frac{dv_\perp}{dt} - \frac{9}{32}\frac{v_\perp^2}{\ell_{wall}} + \frac{9}{64}\frac{v_\parallel^2}{\ell_{wall}}\right). \tag{11.10b}$$

In these RHS expressions, the first acceleration term is the conventional added-mass force for a sphere in an infinite fluid, while the remaining terms are corrections due to wall effects and include both acceleration and velocity terms.

11.2 Fluid Dynamic Influence of Other Particles

This section focuses on fluid dynamic interactions between groups of particles. The interstitial fluid dynamics are generally modeled in an ensemble fashion (for many particles) using the finite volume fraction (α). Following the decomposition (9.1), the fluid dynamic surface forces that include three-way coupling effects are each denoted with a subscript α, as follows:

$$\mathbf{F}_{surf,\alpha} = \mathbf{F}_{D,\alpha} + \mathbf{F}_{L,\alpha} + \mathbf{F}_{\forall,\alpha} + \mathbf{F}_{H,\alpha} + \mathbf{F}_{S,\alpha}. \tag{11.11}$$

The quasisteady drag force with three-way coupling ($\mathbf{F}_{D,\alpha}$) is the most sensitive to volume fraction and is discussed in the following subsection in the context of changes in terminal velocity.

11.2.1 Effect of Volume Fraction on Quasisteady Drag

As discussed in §5.5, the average drag of particles at finite volume fraction ($\mathbf{F}_{D,\alpha}$) is generally larger than that for an isolated particle ($\mathbf{F}_D$). The difference can be described by a three-way coupling drag coefficient ($C_{D,\alpha}$) or by a combination of a three-way coupling Stokes correction (f_α) and a Reynolds number correction (f_{Re}), where both are based on the relative velocity of the particle to the surrounding fluid ($\mathbf{w}_\alpha$), as follows:

$$\mathbf{F}_{D,\alpha} \equiv -\frac{\pi}{8}\rho_f d^2 \mathbf{w}_\alpha w_\alpha C_{D,\alpha} = -3\pi d\mu_f \mathbf{w}_\alpha f_\alpha f_{Re}. \tag{11.12a}$$

$$\mathbf{w}_\alpha \equiv (\mathbf{v} - \mathbf{u})_\alpha. \tag{11.12b}$$

Using (11.12a), the three-way coupling Stokes correction can be related (Loth, 2023) to the ratio of the drag forces for a fixed relative velocity and the ratio of the drag coefficients for a fixed Re_p as follows:

$$f_\alpha \equiv \left(\frac{F_{D,\alpha}}{F_D}\right)_{w=const.} = \left(\frac{C_{D,\alpha}}{C_D}\right)_{Rep=const.}. \tag{11.13}$$

This correction in drag due to neighboring particles can depend on the relative arrangement of the particles (Sirignano, 1993; Yuan and Prosperetti, 1994). For example, the drag will generally reduce for two particles moving along the same path (one is trailing in the other's wake) but will generally increase for particles that are spaced laterally (creating blockage and forcing higher fluid velocities between gaps). Averaged over all arrangements, the latter effect dominates for spherical particles, so the three-way coupling results in a net drag increase ($f_\alpha > 1$ and $C_{D,\alpha} > C_D$).

Expressions for f_α or $C_{D,\alpha}$ are often based on the effect of the terminal velocity ratio (χ), which is the ratio of the ensemble-averaged terminal velocity at finite particle volume fraction ($w_{term,\alpha}$) to the terminal velocity of an isolated particle (w_{term}), where both velocities have the same particle density and diameter and the same surrounding fluid density and viscosity:

$$\chi \equiv \frac{w_{term,\alpha}}{w_{term}}. \tag{11.14}$$

This velocity ratio is typically obtained with a large number of particles to ensure statistical convergence. The physics that control χ can include both hydrodynamic interactions (three-way coupling) and collision interactions (four-way coupling), where the latter becomes more important at higher volume fractions. To relate the velocity ratio to the drag correction for generalized Reynolds numbers, one may balance the steady drag force with the gravity and hydrostatic fluid-stress forces and assume a nearly constant drag-curve slope to obtain (Loth, 2023) the following:

$$f_\alpha = (1-\alpha)\chi^{-Y}. \tag{11.15}$$

The RHS uses the drag-power parameter of (4.18), where $Y = 1$ for linear drag ($C_D \approx$ const./Re_p) and $Y = 2$ for Newton-based drag regime ($C_D \approx$ const.).

11.2.2 Drag of Solid Spherical Particles

For creeping flow ($Re_p \rightarrow 0$ and $Y = 1$) and $\alpha \ll 1$, theoretical relations have been derived for the terminal velocity ratio by Burgers (1942) and Batchelor (1972), as follows:

$$\chi = 1/(1+6.875\alpha). \tag{11.16a}$$

$$\chi = (1-6.55\alpha). \tag{11.16b}$$

These expressions indicate that a volume fraction increase yields a velocity ratio decrease and thus a particle drag increase, $f_\alpha > 1$ for $Y = 1$ based on (11.15). This is shown in Figure 11.4a for small Re_p values, where these two theoretical expressions agree with experiments for small-volume fractions ($\alpha < 1\%$), but are generally not accurate at higher-volume fractions.

To account for higher-volume fractions, several empirical models have been generally developed. The most common of these is the model of Richardson and Zaki (1954a,b), which uses a power relationship for the velocity ratio as

$$\chi = (1-\alpha)^b. \tag{11.17}$$

In this form, b is the Richardson–Zaki exponent (Zuber, 1964), where $b_o \approx 4.5$ is defined as the value for spheres in creeping flow and has been found to be a good fit for experiments at small Reynolds numbers, as shown in Figure 11.4a.

For solid spheres at high Re_p, there are no theoretical solutions for finite volume fraction effects, so empirical approaches are employed. For such approaches, several investigators (e.g., Maude and Whitmore, 1958; Wen and Yu, 1966; Wallis, 1969) assumed that the drag force has separable dependencies on Reynolds number, volume fraction, surrounding fluid density, and viscosity. Based on this assumption, the Richardson–Zaki exponent depends inversely on the drag parameter, and the drag correction becomes conveniently independent of Re_p per (11.15):

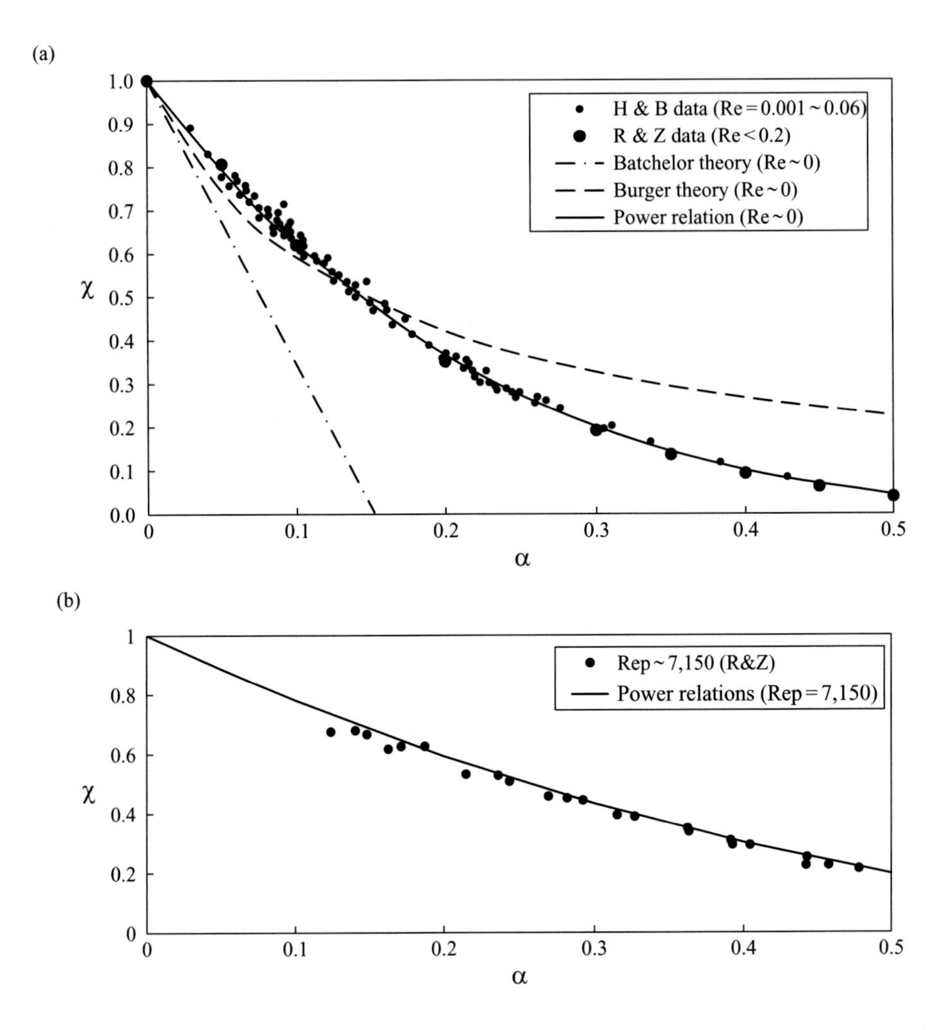

Figure 11.4 Terminal velocity ratios of solid spherical particles based on experimental data of Richardson and Zaki (1954a) and Happel and Brenner (1973) with the power-based model: (a) at a creeping-flow drag regime compared with theories and (b) in a Newton-based drag regime.

$$b = b_0/Y \approx 4.5/Y \qquad (11.18a)$$

$$f_\alpha = (1-\alpha)^{1-b_o} \approx (1-\alpha)^{-3.5} \quad \textit{for solid spheres at all } \mathrm{Re_p}. \qquad (11.18b)$$

The first relationship yields $b = 2.25$ in the Newton drag limit, which compares well with measured velocity ratios as shown in Figure 11.4b. Since fully contaminated spherical bubbles and drops have a drag equivalent to that of a solid sphere (per Figure 9.21), they are also well described a Richardson–Zaki coefficient, as shown in Figure 11.5, for a wide range of Reynolds numbers. Also shown is empirical model for Re_p influence on the Richardson–Zaki exponent (Rowe, 1987), as follows:

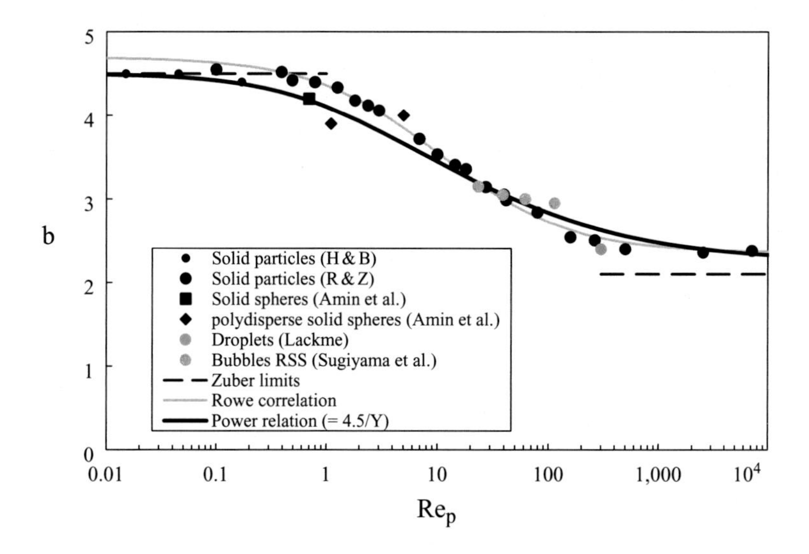

Figure 11.5 Variation of Richardson–Zaki exponent (used for the power relation) with Reynolds numbers for spherical particles, droplets, and bubbles with no recirculation (fully contaminated) based on spherical particle data (Richardson and Zaki 1954a; Happel and Brenner, 1973; Lackme, 1973; Syamlal et al., 1993; Sugiyama et al., 2001).

$$b = \frac{4.7 + 2.35\left(0.17\,\mathrm{Re}_p^{3/4}\right)}{1 + 0.17\,\mathrm{Re}_p^{3/4}} \qquad \textit{for solid spheres at all } \mathrm{Re}_p. \tag{11.19}$$

Similar fits have been proposed by Syamlal et al. (1993) and di Felice (1994) that smoothly vary from $b = 4.5$ for $\mathrm{Re}_p \ll 1$ and for $b = 2.25$ for $\mathrm{Re}_p \gg 1$. A primary benefit of (11.18) is that it is consistent with the particle drag curve and removes (and thus simplifies) the Re_p influence on f_α.

11.2.3 Drag of Solid Nonspherical Particles

During the three-way coupling of nonspherical solid particles, additional complexities arise due to changes in particle orientation as volume fraction increases. For example, disk-shaped particles (with $E \gg 1$) tend to group in horizontal layers (rather than being randomly located), as shown in Figure 11.6a, while long cylinders (with $E \ll 1$) lead to tangled networks, as shown in Figure 11.6b. As shown in Figure 11.7, these effects lead to higher Richardson–Zaki exponents (based on $5\% < \alpha < 15\%$) for $\mathrm{Re}_p \ll 1$. In particular, highly prolate particles yield $b \approx 9$ while disk-like particles yield $b \approx 6$. In addition, nonspherical regularly shaped particles with $E \sim 1$ (such as hexagonal prisms, cubes, plates, and irregularly shape particles) and irregularly shaped particles with $E \sim 1$ (sand particles and other crushed sediment) also lead to $b \approx 6$. Despite these differences at low Reynolds numbers, the exponents in the Newton drag regime (e.g., $\mathrm{Re}_p > 1{,}000$) for all the nonspherical particles approximately converge to the

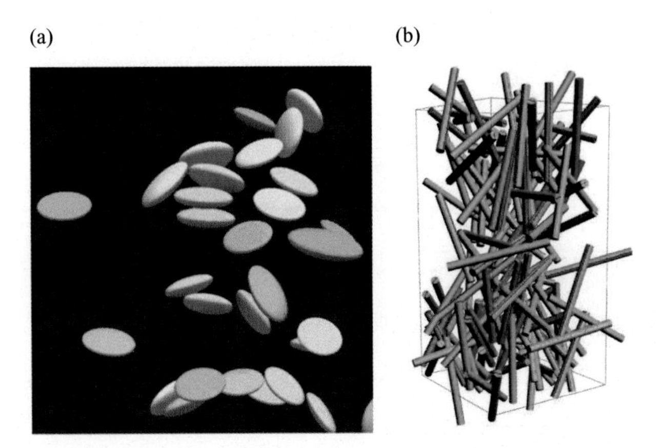

Figure 11.6 Instantaneous snapshots of nonspherical particles falling under gravity yielding a wide variety of orientations: (a) disk-like particles at $Re_p \sim 40$ for a global concentration of $\alpha = 0.5\%$ with a zoom-in of a local group of about 30 particles, which exhibits high clustering such that local concentration is $\alpha \sim 2.5\%$ (Fornari et al., 2018); and (b) cylinders at $Re_p \sim 50$ for a global concentration of $\alpha = 10\%$ (Derksen, 2020).

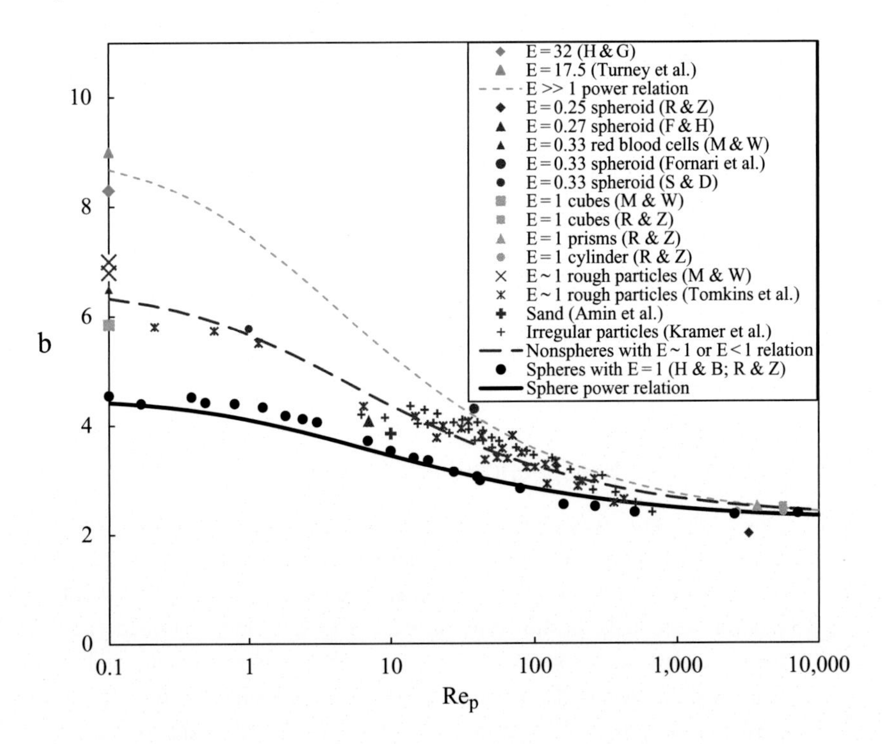

Figure 11.7 Variation of Richardson–Zaki exponent as a function of Re_p for various particle shapes: cylinders with E » 1 (black solid symbols), disks with E < 1 (open blue symbols), regular nonspheres with E ~ 1 (blue solid symbols), irregularly shaped particles with E ~ 1 (blue stars and "x" symbols), and spherical data of Figure 11.5, where data shown at $Re_p = 0.1$ includes data at $Re_p < 0.1$ (Richardson and Zaki, 1954a; Maude and Whitmore, 1958; Turney et al., 1995; Herzhaft and Guazzelli, 1999; Fonseca and Herrmann, 2004; Tomkins et al., 2005; Shardt and Derksen, 2012; Fornari et al. 2018; Kramer et al., 2019; Amin et al., 2021).

Newton spherical particle limit of $b \approx 2.25$. Based on such results, the Richardson–Zaki exponent for nonspherical particles can be modeled as follows:

$$b = 2b_o/Y^2 = 9/Y^2 \qquad \textit{for } E \gg 1 \textit{ (prolate)}. \tag{11.20a}$$

$$b = \sqrt{2}b_o/Y^{1.5} = 6.5/Y^{1.5} \qquad \textit{for } E \sim 1 \textit{ (nonspherical) or } E < 1 \textit{ (oblate)}. \tag{11.20b}$$

The exponents of (11.20) can be employed with the velocity ratio of (11.17) and the drag correction of (11.15) as follows:

$$f_\alpha = (1-\alpha)^{1-bY} \qquad \textit{for nonspherical (and spherical) particles}. \tag{11.21}$$

Taken with (11.20), this indicates that the drag increase due to particle volume fraction effects qualitatively increases as the particles become nonspherical, especially for prolate shapes. However, (11.20) and (11.21) should be considered very rough approximations since there is significant scatter in the data and a scarcity of data for prolate shapes at intermediate Reynolds numbers. In addition, the settling velocity often *increases* with volume fraction for nonspherical shapes at low concentrations, such as for $0.1\% < \alpha < 5\%$ (Herzhaft and Guazzelli, 1999; Fornari et al., 2018) due to clustering effects. As such, much more research is needed to understand the effects of particle shape on settling velocity, especially for high-aspect ratio prolate particles.

11.2.4 Drag of Clean and Deformed Bubbles

Unlike the case for solid particles, a clean spherical bubble at high Reynolds numbers has a thin attached wake that allows a theoretical solution for three-way coupling. In particular, the isolated drag coefficient of $C_D = 48/Re_p$ of (9.63b) can be extended for small volume fractions to obtain a theoretical first-order influence of bubble concentration (Sangani et al., 1991), as follows:

$$f_\alpha = 1 + 2.11\alpha \tag{11.22a}$$

$$\chi = (1-\alpha)/(1+2.11\alpha) \tag{11.22b}$$

for clean spherical bubbles with $Re_p \gg 1$.

The velocity ratio of (11.22b) combines (11.22a) and (11.15) with $Y = 1$ (since the isolated bubble drag varies as $1/Re_p$). As shown in Figure 11.8a, this theoretical result compares reasonably with measurements for nearly spherical bubbles at high Re_p for $\alpha < 15\%$. If one applies $Y = 1$ with (11.18a), the clean bubble Richardson–Zaki exponent would be constant at $b = 4.5$ for all Reynolds numbers (owing to an effectively linear drag law throughout). This result is shown by the long-dashed line in Figure 11.8a, which shows good agreement with experiments for $\alpha < 5\%$. However, higher-volume fractions are more consistent with the theory of (11.22). This suggests that the Richardzon-Zaki empirical model is only qualitative for clean bubble velocity ratios.

Once bubbles deform, the likelihood of flow separation is increased so that the transition to solid particle behavior occurs at even smaller α, as shown by Figure 11.8b

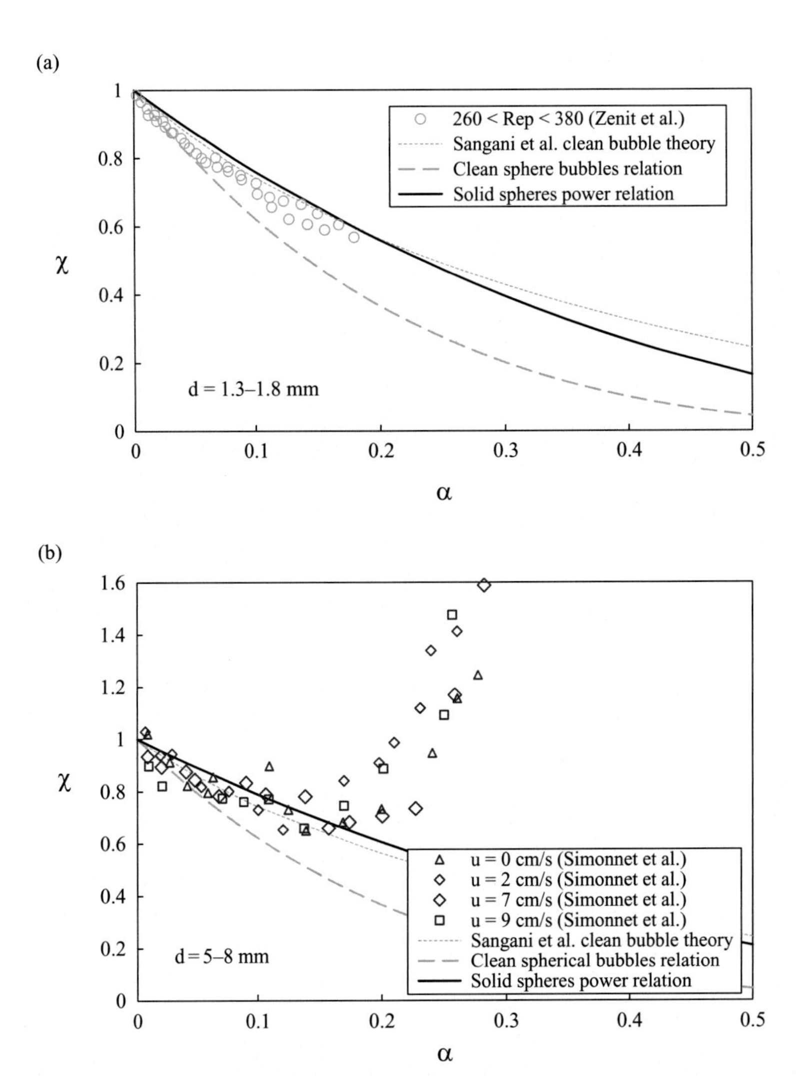

Figure 11.8 Experimental terminal velocity ratios of bubbles at large Reynolds numbers ($Re_p \gg 1$) as a function of volume fraction for measurements compared to theory of Sangani et al. (1991) and empirical power relations: (a) nearly clean bubbles from Zenit et al. (2001); and (b) bubbles with high deformation (Garnier et al., 2002; Simonnet et al., 2007).

(for bubbles with isolated Weber numbers given by $6 < We < 10$). In addition, it can be seen that the velocity ratio starts to increase for $\alpha > 10\%$, and even exceed unity for $\alpha > 20\%$. This is caused by changes in bubble shape as volume fraction increases. In particular, bubbles change from oblate to spherical as α increases, which decreases drag (Riboux et al., 2010; Loisy et al., 2017). In addition, deformable bubbles tend to line up vertically at higher-volume fractions, which leads to drafting effects that reduce drag and thus increase the velocity ratio (Sankaranarayanan et al., 2003).

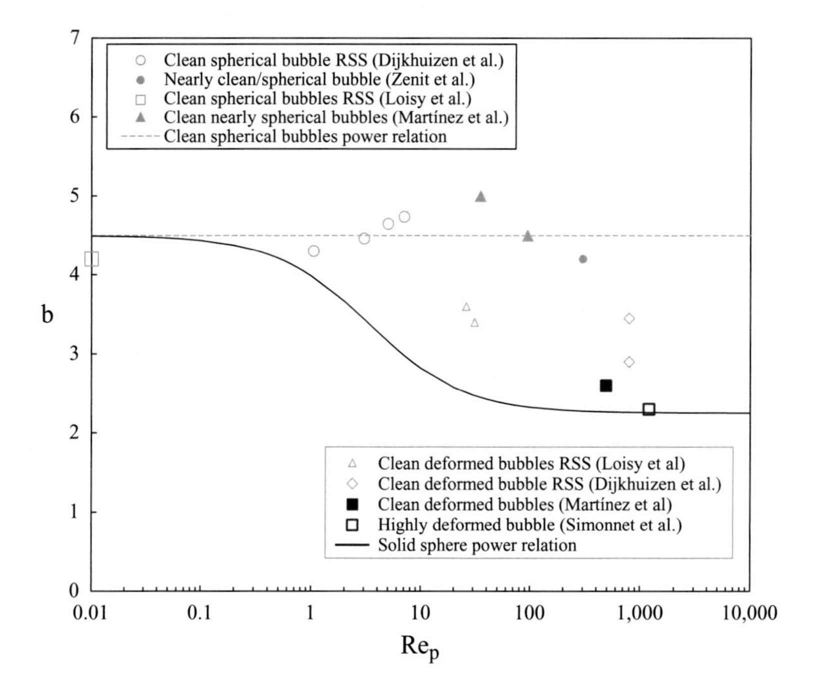

Figure 11.9 Richardson–Zaki exponent as a function of Reynolds numbers for deformed bubbles, including experiments (Zenit et al., 2001; Martínez-Mercado et al., 2007; Simonnet et al., 2007), resolved-surface simulations (Dijkhuizen et al., 2010; Loisy et al., 2017), and power relation exponents for clean spheres and solid spheres (where clean deformed bubbles generally lie between these two limits).

The complex conditions of high deformation and volume fraction are difficult to model.

If we restrict ourselves to We < 10 and $\alpha < 10\%$ (to avoid these complex conditions), the Richardson–Zaki exponents for clean bubbles are given in Figure 11.9, where spherical shapes are reflected qualitatively by $b = 4.5$ per (11.18a) with $Y = 1$, while deformed bubbles tend to the solid particle per (11.18a) with Y changing from 1 to 2 as Re_p increases. Fortunately, these two limits yield the same drag correction (11.18b), so

$$f_\alpha \approx (1-\alpha)^{-3.5} \qquad \textit{for bubbles } \mathrm{We} < 10 \text{ and } \alpha < 10\% \ \left(\textit{all } \mathrm{Re_p}\right). \tag{11.23}$$

As such, a single empirical expression can be used within these conditions regardless of the level of bubble contamination and deformation. Additional work is needed to consider higher Weber numbers and volume fractions beyond these conditions.

11.2.5 Effect of Volume Fraction on Other Fluid Dynamic Forces

The effect of volume fraction on the nondrag forces can also be significant, though not as strong as that for drag. For the added-mass force, theoretical results for creeping flow relative to the mixture velocity were obtained by Zuber (1964), van Wijndgaarden (1976), and Biesheuvel and Spoelstra (1989) with different results. However, Zhang and Prosperetti (1994) obtained an inviscid theoretical added-mass result for a spherical particle with relative acceleration to the surrounding fluid material, which is formally consistent with the change in the effective mixed fluid volume and is given as follows:

$$m_{\forall,\alpha} = \frac{1}{2}\rho_f(1-\alpha)\forall_p = c_{\forall,\alpha}\rho_f\forall_p. \tag{11.24a}$$

$$c_{\forall,\alpha} = c_\forall(1-\alpha). \tag{11.24b}$$

$$\mathbf{F}_{\forall,\alpha} = \mathbf{F}_\forall(1-\alpha). \tag{11.24c}$$

The inviscid lift can be shown to have this same volume fraction dependence (Prosperetti, 2007), as does the creeping-flow version of the history force (ten Cate and Sundaresan, 2006) so that

$$\mathbf{F}_{L,\alpha} = \mathbf{F}_L(1-\alpha). \tag{11.25a}$$

$$\mathbf{F}_{H,\alpha} = \mathbf{F}_H(1-\alpha). \tag{11.25b}$$

These relationships are simple and convenient but may only be expected to be reasonable for spheres at dilute concentrations (little is known about the appropriate forms at other conditions).

The final fluid dynamic force to consider for three-way coupling is the fluid-stress force. For an incompressible medium with weak flow gradients, the fluid-stress force for both inviscid flow (Drew, 1983) and viscous flow (Crowe et al., 2011) is based on the pressure and viscous stress gradients by (3.64), so

$$\mathbf{F}_{S,\alpha} = \mathbf{F}_S = \rho_f\forall_p\left(-\nabla p + \nabla \mathbf{K}_{ij}\right). \tag{11.26}$$

Therefore, the fluid-stress force on a single particle surrounded by other particles is the same as a single particle surrounded by no particles (Prosperetti, 2007). Linearly combining the preceding results in a manner similar to that of a single particle (9.1), the fluid dynamic force in dilute conditions for three-way coupling can be approximated as follows:

$$\mathbf{F}_{surf,\alpha} = f_\alpha\mathbf{F}_D + (1-\alpha)(\mathbf{F}_L + \mathbf{F}_\forall + \mathbf{F}_H) + \mathbf{F}_S. \tag{11.27}$$

However, surrounding particles can also lead to four-way coupling by the collision force, which is discussed in the next two sections, for both particle–wall interactions and particle–particle interactions.

11.3 Momentum Change Due to Particle Collisions

Particles colliding with walls or other particles are important in many multiphase flows, especially for systems which involve coating, deposition, filtering, and erosion, as discussed in §1.2.

If the flow can be considered dilute (per 5.49b), particle–particle collisions are generally are limited to two particles at a given time (so long as cohesive effects do not cause agglomerations of multiple particles). In addition, particle collisions with walls can be considered two-body collisions. These two-body collisions may be treated as individual events with a point-force model, as discussed later. However, high concentrations and strong cohesive forces can lead to particle agglomerations due to simultaneous collisions between three or more particles, as discussed by Endres et al. (2021). Furthermore, dense flow per (5.49c) can have many simultaneous collisions between three or more particles. In such cases, it is often best to employ the kinetic theory of granular fluids (rather than seeking to model each individual collision) as discussed by Jenkins and Savage (1983), Lun et al. (1984), and Ding and Gidaspow (1990).

For the case of two-body collisions between two particles or a particle and a wall, the interaction can be handled by either a *hard-sphere model* or a *soft-sphere model*. The soft-sphere model describes the collision as a continuous event with the unsteady particle deformation throughout the interaction, which can be solved numerically with discretized differential equations for the interior particle stress dynamics and small time steps (Crowe et al., 2011). This physics-based approach is generally more accurate since it allows for detailed mechanical response of the particle matter (and any interstitial fluid dynamics) based on material properties. However, it is numerically intensive for a single collision and is thus generally impractical to employ for a large number of collision (e.g., thousands or millions). The more computationally efficient hard-sphere model instead employs discontinuous "jump" relations to represent the net changes before and after the collision (particle deformation dynamics are not explicitly considered) using empirical models. As such, the difference between the hard-sphere model and the soft-sphere model is analogous to the difference between a point-force approximation and a resolved-surface approach. The hard-sphere model (more commonly used when there are many collisions) is discussed first.

For the hard-sphere approach, we adopt the nomenclature for the jump change in velocity (precollision versus postcollision) as described in §8.1.5, but generalize it to include particle spin. This generalized interaction is shown in Figure 11.10 for a particle hitting a wall, where the particle can have both a translational velocity and spin relative to the wall for both precollision and postcollision states. The incoming translational and angular velocities are defined as $\mathbf{v}^{\text{in}}$ and $\mathbf{\Omega}^{\text{in}}$, while the outgoing translational and angular velocities are defined as $\mathbf{v}^{\text{out}}$ and $\mathbf{\Omega}^{\text{out}}$. The jump change between these two states can be written as follows:

$$\Delta\mathbf{v} \equiv \mathbf{v}^{\text{out}} - \mathbf{v}^{\text{in}}. \tag{11.28a}$$

$$\Delta\mathbf{\Omega} \equiv \mathbf{\Omega}^{\text{out}} - \mathbf{\Omega}^{\text{in}}. \tag{11.28b}$$

For two particles colliding with each other, the jumps are similarly defined where the precollision state includes both a strike particle of velocities $\mathbf{v}_1$ and $\mathbf{\Omega}_1$ with a mass m_{p1} as well as a target particle of velocities $\mathbf{v}_2$ and $\mathbf{\Omega}_2$ with a mass m_{p2}. The net changes in the strike particle velocity translational and angular velocities ($\Delta\mathbf{v}_1$ and $\Delta\mathbf{\Omega}_1$) are defined as follows:

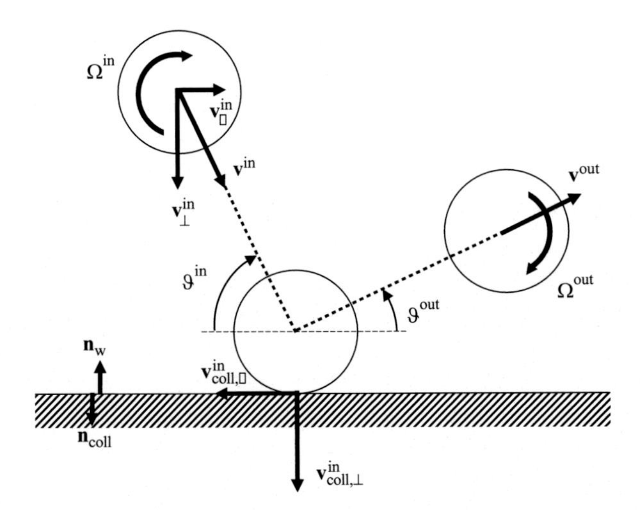

Figure 11.10 Schematics of a particle–wall interaction indicating incoming and outgoing characteristics, where clockwise rotation is considered positive.

$$\Delta \mathbf{v}_1 \equiv \mathbf{v}_1^{\text{out}} - \mathbf{v}_1^{\text{in}}. \tag{11.29a}$$

$$\Delta \boldsymbol{\Omega}_1 \equiv \boldsymbol{\Omega}_1^{\text{out}} - \boldsymbol{\Omega}_1^{\text{in}}. \tag{11.29b}$$

The net changes for the target particle are similarly defined as $\Delta \mathbf{v}_2$ and $\Delta \boldsymbol{\Omega}_2$. As in §8.1.5, the unit collision vector ($\mathbf{n}_{\text{coll}}$) is defined as the unit normal to the collision plane (8.30). For a wall collision, the collision vector is equal and opposite to the wall normal as shown in Figure 11.10.

If particle spin is neglected, the collision velocity ($\mathbf{v}_{\text{coll}}$) is simply based on the particle centroid velocities (as in Figure 8.5) and can be decomposed into perpendicular and parallel components based on (8.31). However, the inclusion of particle spin requires defining the collision velocity vector ($\mathbf{v}_{\text{coll}}$) as the *surface* velocity of the particle just before collision, as shown in Figure 11.10. Including both the translational (centroid-based) and angular (spin-based) components for a wall impact, the particle collision velocity (and its associated normal and tangential components) can be expressed as follows:

$$\mathbf{v}_{\text{coll}}^{\text{in}} \equiv \mathbf{v}_1^{\text{in}} + \mathrm{r_p}\boldsymbol{\Omega}_{\text{in}} \times \mathbf{n}_{\text{coll}} \tag{11.30a}$$

$$\mathbf{v}_{\text{coll},\perp}^{\text{in}} \equiv [\mathbf{v}_{\text{in}} \cdot \mathbf{n}_{\text{coll}}]\mathbf{n}_{\text{coll}} = \mathbf{v}_{\perp}^{\text{in}} \qquad \textit{for wall impact.} \tag{11.30b}$$

$$\mathbf{v}_{\text{coll},\parallel}^{\text{in}} \equiv \mathbf{v}_{\parallel}^{\text{in}} + \mathrm{r_p}\boldsymbol{\Omega}_{\text{in}} \times \mathbf{n}_{\text{coll}} = \mathbf{v}_{\text{coll}}^{\text{in}} - \mathbf{v}_{\text{coll},\perp}^{\text{in}} \tag{11.30c}$$

Notably, particle spin only modifies the tangential component of the collision velocity. A similar normal and tangential decomposition can be applied to the outgoing collision velocities.

For the collision of two particles with each other, one may similarly define the collision velocity vector ($\mathbf{v}_{\text{coll}}$) as the incoming *surface* velocity of the strike particle relative to that of the target particle (and its associated normal and tangential components) as follows:

$$\mathbf{v}_{coll}^{in} \equiv (\mathbf{v}_1^{in} - \mathbf{v}_2^{in}) + (\mathbf{r}_{p1}\boldsymbol{\Omega}_1^{in} + \mathbf{r}_{p2}\boldsymbol{\Omega}_2^{in}) \times \mathbf{n}_{coll} \quad (11.31a)$$

$$\mathbf{v}_{coll,\perp}^{in} \equiv [(\mathbf{v}_1^{in} - \mathbf{v}_2^{in}) \cdot \mathbf{n}_{coll}]\mathbf{n}_{coll} = \mathbf{v}_{\perp 1}^{in} - \mathbf{v}_{\perp 2}^{in} \quad \textit{for two particles.} \quad (11.31b)$$

$$\mathbf{v}_{coll,\parallel}^{in} \equiv \mathbf{v}_{\parallel 1}^{in} - \mathbf{v}_{\parallel 2}^{in} + (\mathbf{r}_{p1}\boldsymbol{\Omega}_1^{in} + \mathbf{r}_{p2}\boldsymbol{\Omega}_2^{in}) \times \mathbf{n}_{coll} = \mathbf{v}_{coll}^{in} - \mathbf{v}_{coll,\perp}^{in} \quad (11.31c)$$

Again, a similar decomposition can be applied to the outgoing collision velocities.

Given the preceding jump conditions, the outcome for particles with constant mass (no breakup, agglomeration, vaporization, etc.) is based on the collision impulse. The collision impulse ($\mathbf{I}_{coll}$) for two particles can be defined by the translational momentum change of the strike particle, which is equal and opposite to this change for the target particle. For spin, the cross product of the collision vector and the impulse determines the change in angular momentum for a given particle moment of inertia (I_p from 10.27) and radius as follows:

$$\mathbf{I}_{coll} \equiv \mathrm{m}_{p1}\Delta\mathbf{v}_1 = -\mathrm{m}_{p2}\Delta\mathbf{v}_2. \quad (11.32a)$$

$$\mathbf{n}_{coll} \times \mathbf{I}_{coll} \equiv \frac{-I_{p1}\Delta\boldsymbol{\Omega}_1}{\mathrm{r}_{p1}} = \frac{-I_{p2}\Delta\boldsymbol{\Omega}_2}{\mathrm{r}_{p2}}. \quad (11.32b)$$

These two equations can be solved for the outgoing translation and angular velocities of the strike and target particles based on the normal and tangential components of the impulse force. For a hard-sphere model, this impulse is generally described through normal and tangential restitution coefficients. In the following, we discuss these coefficients for (a) normal collisions of spheres with another sphere or a flat wall, (b) oblique and spin collisions of spheres with another sphere or a flat wall, and (c) collisions of nonspherical particles impacting a flat wall or particles impacting an irregular wall.

11.4 Normal Collisions of Solid Spheres

The basic normal wall collision is defined as a particle approaching a stationary wall with a perpendicular translational velocity and no spin: $\vartheta^{in} = 90°$ and $\Omega^{in} = 0$ (where ϑ^{in} and Ω^{in} are defined in Figure 11.10). In this case, the outgoing normal velocity direction is reversed and the particle speed is proportionally reduced by the normal coefficient of restitution ($e_\perp$), as follows:

$$\mathbf{v}_\perp^{out} = -e_\perp \mathbf{v}_\perp^{in}. \quad (11.33)$$

For a binary collision of two particles of the same material, the coefficient of restitution can be combined with the impulse equations to obtain the normal velocity changes in terms of the collision mass (m_{coll}) as noted in (8.32) and repeated here as follows:

$$\mathrm{m_{p2}}\Delta\mathbf{v}_{\perp,2} = -\mathrm{m_{p1}}\Delta\mathbf{v}_{\perp,1} = (1 + e_{\perp})\mathrm{m_{coll}}\mathbf{v}^{\mathrm{in}}_{\mathrm{coll},\perp}. \tag{11.34a}$$

$$\mathrm{m_{coll}} \equiv \frac{\mathrm{m_{p1}m_{p2}}}{\mathrm{m_{p1}} + \mathrm{m_{p2}}}. \tag{11.34b}$$

As before, a collision of two particles of equal size yields $\mathrm{m_{coll}} = ½\mathrm{m_{p1}}$, while a particle impacting a rigid wall yields $\mathrm{m_{coll}} = \mathrm{m_{p1}}$. If the two particles have different restitutions or if a single particle collides with a nonrigid wall, (11.34a) may be generalized as follows:

$$\mathrm{m_{p2}}\Delta\mathbf{v}_{\perp,2} = -\mathrm{m_{p1}}\Delta\mathbf{v}_{\perp,1} = (e_{\perp,1} + e_{\perp,2})\mathrm{m_{coll}}\mathbf{v}^{\mathrm{in}}_{\mathrm{coll},\perp}. \tag{11.35}$$

In this expression, the two restitution coefficients are associated with the two particles or with the particle and the wall, respectively.

If the surrounding fluid is a gas, fluid viscosity generally does not affect the interaction and the collision is often considered to occur in the "dry" limit. If the surrounding fluid is a liquid, the fluid viscosity may modify the restitution, and the collision is often considered to occur in the "wet" regime. In the following, a normal collision for a solid spherical particle(s) will be considered for an inviscid fluid (dry limit), followed by that for a viscous fluid (wet regime).

11.4.1 Normal Collisions for Spheres in an Inviscid Fluid

The perpendicular coefficient of restitution for a particle in an inviscid fluid can have several regimes, as shown in Figure 11.11. In the theoretical perfectly elastic regime, there are no energy losses, so the restitution coefficient is unity. This is a hypothetical limit, since some energy losses will always occur within the solid particle due to deformation and the associated strain. As such, the actual restitution coefficient is always less than unity (as shown in the figure), and this can be attributed to a combination of three factors: viscoelastic effects, plastic deformation effects, and surface adhesion effects. If these adhesion effects are strong, this can lead to immediate deposition. These three effects are discussed in the following for an inviscid surrounding fluid.

A *viscoelastic* collision is defined as one for which there are internal energy losses within the solid, but there is no permanent damage. In this case, the energy dissipation within the particles and/or wall results in a coefficient of restitution that is less than unity, and this value depends on the material properties. The coefficient of restitution for a viscoelastic solid collision in an inviscid fluid ($e_{\perp\mathrm{o}}$) is the most frequently reported experimental value in the literature and can be nearly a constant. This coefficient is related to the Young's modulus (which represents the amount of strain due to load) and the material's loss modulus (which represents the amount of strain energy that is converted into heat). Unfortunately, it is generally difficult to relate the loss modulus to the stiffness modulus, so there is not a robust theory for the viscoelastic restitution (Cross, 2000). In addition, this coefficient can have a weak dependence on impact velocity (Marshall, 2018). As such, empirical values based on

Table 11.1 Viscoelastic collision restitution and friction coefficients in "dry" conditions on smooth rigid walls for various particle materials based on experiments (McLaughlin, 1968; Foerster et al., 1994; Tsao and Koch, 1997; Richard and Quere, 2000; Joseph et al., 2001; Klaseboer et al., 2001; Gondret et al., 2002; Legendre et al., 2005; Brake et al., 2017; Sandeep et al., 2020).

Particle material	$e_{\perp o}$	$c_{surf,o}$
Steel	~0.97	~0.11
Glass	~0.98	~0.11
Aluminum 6061	~0.9	Not avail.
Nylon	~0.9	Not avail.
Cellulose acetate	~0.88	~0.22
Brass	~0.87	Not avail.
Titanium	~0.86	Not avail.
Rubber	~0.7	~0.65
Polyurethane	~0.65	Not avail.

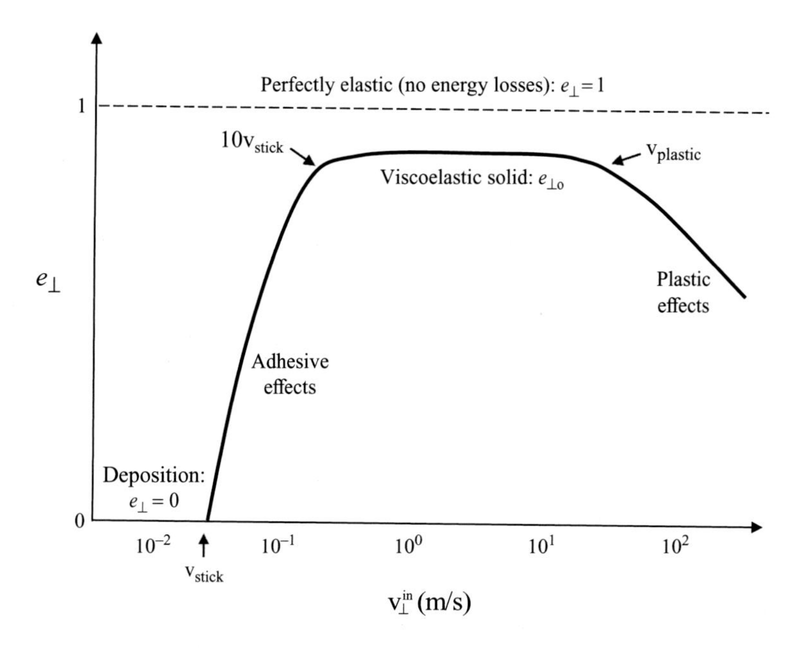

Figure 11.11 The four regimes of restitution (deposition, adhesion, viscoelastic rebound, and plastic reflection) as a function of impact velocity for collision in an inviscid fluid for a notional range of velocity values.

measurements are generally used as in Table 11.1. As noted in this table, softer materials generally have lower restitution coefficients. For example, the viscoelastic coefficient of rubber (which allows for high deformations) is much less than unity (indicating significant losses), while that of steel is nearly equal to unity. This nearly elastic response is the reason that the "Newton's cradle" toy, based on an array of suspended steel balls, is such a good demonstration of momentum conservation.

However, at high-impact velocities, the collision can enter the plastic regime with a rapid drop in the coefficient of restitution, as shown in Figures 11.11 and 11.12a. In this case, the collision results in permanent material changes (i.e., damage) to the particle or the wall. The plastic velocity ($v_{plastic}$) can be defined as the minimum impact velocity associated with the plastic regime. If the damage does not include fracture, this velocity and the viscoelastic restitution coefficient can be used to model the drop in restitution (McNamara and Falcon, 2005), as follows:

$$e_{\perp} \approx e_{\perp o}\left(\frac{v_{\perp}^{in}}{v_{plastic}}\right)^{-1/4}. \tag{11.36}$$

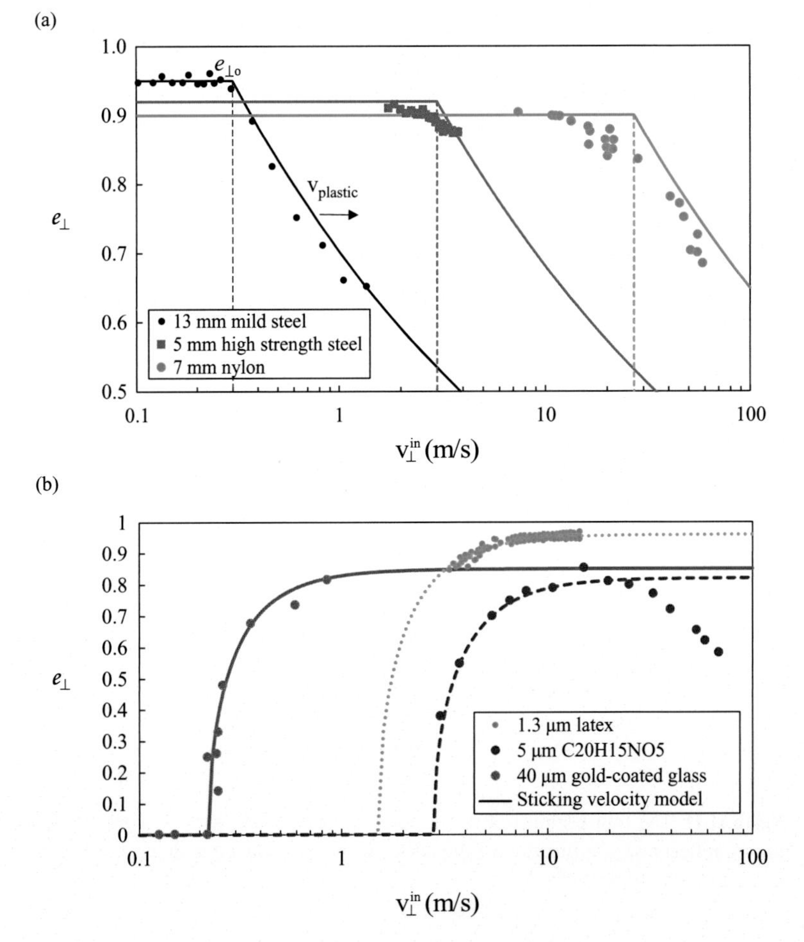

Figure 11.12 Normal wall collision restitution measurements (Labous et al., 1997; Thornton and Ning, 1998; McNamara and Falcon, 2005; Weir and Tallon, 2005; Güttler et al., 2013) and models for (a) millimeter-sized spheres showing an impact velocity transition into the plastic regime; and (b) μm-sized spheres showing an impact velocity transition from deposition to the adhesion regime, where ammonium fluorescein ($C_{20}H_{15}NO_5$) particles also enter the plastic regime at high speeds.

The RHS uses the product of the viscoelastic effects and plastic effects to approximately describe the overall restitution (Marshall, 2018). The plastic velocity can be approximately related to the material yield velocity (v_{yield}) and its viscoelastic restitution coefficient (Johnson, 1985) as follows:

$$v_{plastic} \approx 1.94 v_{yield} (e_{\perp o})^{-4}. \tag{11.37}$$

The yield velocity can, in turn, be theoretically obtained by equating the initial kinetic energy with the elastic work done up to the point of material yield (Wall et al., 1990), as follows:

$$v_{yield} = \left(\frac{\pi}{3\mathcal{E}}\right)^2 \sqrt{\frac{K_{yield}^5}{10\rho_p}}. \tag{11.38}$$

The RHS includes Young's modulus of elasticity ($\mathcal{E}$) and the yield stress (K_{yield}) of the material. As shown in Figure 11.12a, harder materials tend to have lower plastic velocities, which can be attributed to their higher moduli of elasticity per (11.38). For example, the plastic velocity for nylon is on the order of 30 m/s, compared to a range of 0.3 m/s to 3 m/s for mild to high-strength steel. Notably, (11.38) indicates that the yield and plastic velocities are not a function of particle size. As a result, even nanoparticles can exhibit reductions in restitution due to plastic deformation once $v > v_{plastic}$ (Takato et al., 2015).

In addition, surface attraction forces are also possible, especially for small particles. For a particle impacting a wall, these are termed *adhesive forces*, and for particles hitting each other, these are termed *cohesive forces*. There are various phenomenon that can cause such surface attraction, including van der Waals forces (due to molecular interactions between solids), chemical interactions, electrostatic charges, and liquid bridging (Marshall and Li, 2014). When these effects are strong enough, immediate deposition (no rebound) occurs, and the sticking velocity (v_{stick}) is defined as the maximum impact velocity for which no rebound will occur (Figure 11.11).

Surface attraction effects tend to increase as particles get smaller (due to higher surface-to-volume ratios) and as the impact velocity become smaller. For electrostatic and van der Waals forces, the sticking velocity (v_{stick}) for wall impacts is proportional to the interface energy (G_{stick}) as well as the particle modulus, density, and diameter (Wall et al., 1990), as follows:

$$v_{stick} = 3.28 \left(\frac{G_{stick}^5}{\mathcal{E}^2 d^5 \rho_p^3}\right)^{1/6}. \tag{11.39}$$

This relationship shows that the sticking velocity is proportional to d^{-5}, so adhesion effects are much more likely to occur for very small particles. This explains why the adhesion regimes are prevalent for μm-sized particles (as in Figure 11.12b), but are often absent for millimeter-sized particles (as in Figure 11.12a). The sticking velocity also becomes larger as the interface energy increases, which explains why the $C_{20}H_{15}NO_5$ particles of Figure 11.12b have the highest tendency to adhere.

However, G_{stick} is difficult to predict for realistic conditions since it can be a function of the chemistry and charge of both the particle and the surface (and any surface roughness). As such, G_{stick} is often obtained from measured values of v_{stick}.

For the impact velocities where adhesion effects are important but do not dominate ($v_{stick} \leq v \leq 10v_{stick}$), there will be a reduction in restitution that can be related to the sticking velocity and the viscoelastic restitution coefficient (Thornton and Ning, 1998), as follows:

$$e_{\perp} \approx e_{\perp o}\sqrt{1-\left(v_{stick}/v_{\perp}^{in}\right)^2}. \tag{11.40}$$

This model assumes the viscoelastic and adhesive effects combine directly (Marshall, 2018) and reasonably describes the adhesive regime, as shown in Figure 11.12b. Another hard-sphere approach to model this effect employs a cohesive impulse model (Kosinki and Hoffman, 2010).

In summary, the collision of a solid sphere with a wall or another particle in an inviscid fluid includes at least four regimes (deposition, adhesion, viscoelastic rebound, and plastic reflection), for which the restitution can be approximately modeled as follows:

$$e_{\perp} \approx 0 \qquad \textit{for}\, v \leq v_{stick}. \tag{11.41a}$$

$$e_{\perp} \approx e_{\perp o}\sqrt{1-\left(v_{stick}/v_{\perp}^{in}\right)^2} \qquad \textit{for}\, v_{stick} \leq v \leq 10v_{stick}. \tag{11.41b}$$

$$e_{\perp} \approx e_{\perp o} \qquad \textit{for}\, 10v_{stick} \leq v \leq v_{plastic}. \tag{11.41c}$$

$$e_{\perp} \approx e_{\perp o}\left(v_{\perp}^{in}/v_{plastic}\right)^{-1/4} \qquad \textit{for}\, v > v_{plastic}. \tag{11.41d}$$

This summary assumes no fracture and that the adhesion and the plastic regimes are far apart ($v_{stick} \ll v_{plastic}$). When these assumptions are not reasonable, nonlinear hard-sphere models can be employed or, better yet, a more physics-based soft-sphere model (Marshall, 2018). Fluid viscosity effects can also be considered in a hard-sphere approach, as follows.

11.4.2 Normal Collisions of Spheres in a Viscous Fluid

For particle collisions in a liquid, the particle restitution coefficient can be damped due to fluid viscosity, where energy is dissipated within the fluid (not just within the solid). As shown in Figure 11.13, this wet regime effect is generally more pronounced at lower-impact velocities, where viscous effects are stronger than inertial effects. Most of the fluid viscous effect occurs just before and just after the solid–solid contact when the gap is small compared to the particle radius so a thin film is formed. Just before contact, the decreasing gap causes film to be squeezed out as the particle moves toward the wall (or toward a target particle). Just after contact time, the reverse effect occurs, where the surrounding fluid is instead pulled in to a thin film to refill the gap as it opens. Theoretical studies to describe this effect generally employ lubrication theory

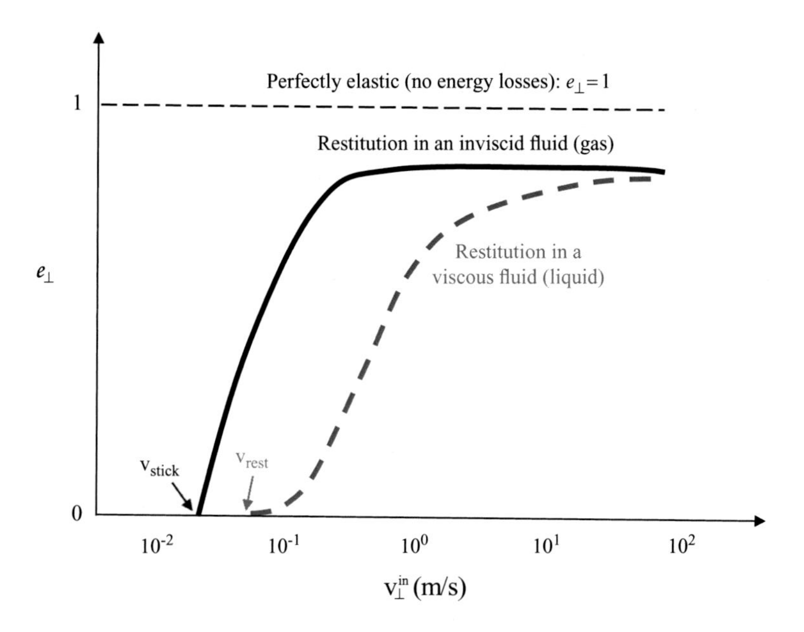

Figure 11.13 Notional schematic of perpendicular restitution for a sphere as a function of impact velocity comparing transitions in an inviscid fluid ($v_{rest} < v_{stick}$) to that in a viscous fluid ($v_{rest} > v_{stick}$).

(e.g., Davis et al., 1986) to determine the pressure distribution and fluid viscous losses in the thin film as it is being removed from the gap and then again as it is being refilled just after this contact. In both directions, the viscous losses reduce the particle momentum.

To characterize the wet regime (also called the *squeeze-film regime*), one may define the impact Stokes number ($St_\perp$) as the ratio of the particle inertial effects to the fluid viscous effects (Joseph et al., 2001). In particular, the impact Stokes number is defined as the ratio of the particle drag response time (τ_p of 5.11) to the drainage time scale (τ_{drain}), which is the time during which fluid is pushed out of the gap between the particle and the surface. For a spherical particle impacting a flat wall, this drainage time scale can be based on the normal impact velocity and particle diameter, yielding the following expression for the impact Stokes number:

$$St_\perp \equiv \frac{\tau_p}{\tau_{drainage}} = \frac{\left(\rho_p + c_\forall \rho_f\right) v_\perp^{in} d}{9\mu_f} \quad \textit{for wall collision.} \tag{11.42}$$

When this time scale ratio is small ($St_\perp \ll 1$), the particle inertia is weak and can quickly respond to the viscous forces and thus slow down before impacting. When this time scale ratio is high ($St_\perp \gg 1$), the particle inertia is high and will not be able to slow down significantly, so the effects of the surrounding viscosity effect will be weak. Note that the impact Stokes number of (11.42) includes the particle added-mass coefficient, which can be neglected when $\rho_p \gg \rho_f$ (Loth, 2000; Legendre et al., 2005).

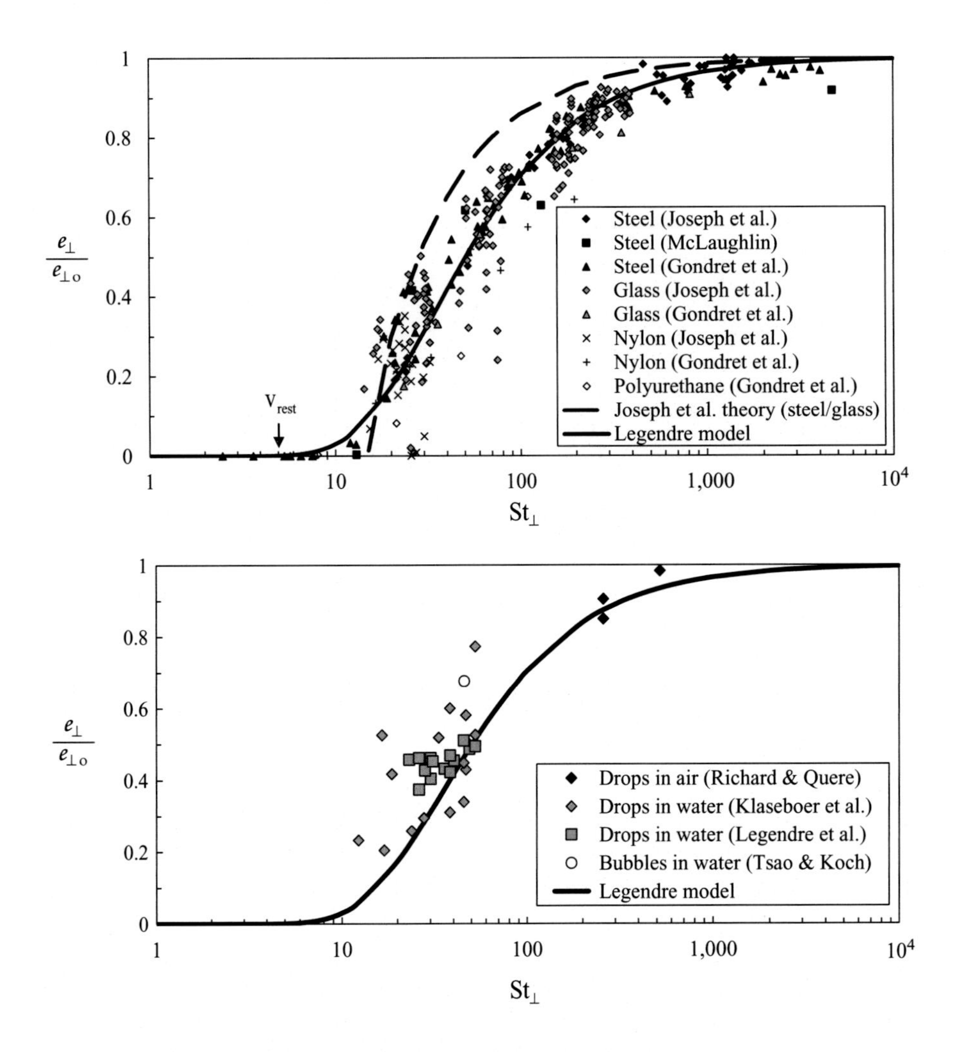

Figure 11.14 Viscous effects on normal wall collision restitution for (a) measurement and models for solid particles (McLaughlin, 1968; Joseph et al., 2001; Gondret et al., 2002; Legendre et al., 2005); and (b) fluid particles (Tsao and Koch, 1997; Richard and Quere, 2000; Klaseboer et al., 2001; Legendre et al., 2005), assuming $e_{\perp o} \approx 0.9$ for drops and $e_{\perp o} \approx 1$ for bubbles.

As shown in Figure 11.14a, experimental results for solid particles with very high particle inertias ($St_\perp > 2{,}000$) are consistent with the inviscid fluid (dry) limit of restitution, so there is no effect of surrounding fluid. As the impact Stokes number reduces, viscous effects become more important, and the restitution is reduced due to these losses. The experimental results are qualitatively consistent with the theoretical predictions using lubrication theory for rigid walls (Davis et al., 1986; Joseph et al., 2001). However, a simple model based on a mass-spring dissipative system (Legendre et al. (2005) that also provides reasonable correlation with the experimental data is given as follows:

$$\frac{e_{\perp}}{e_{\perp,\text{solid}}} = \exp\left(-\frac{35}{\text{St}_{\perp}}\right). \tag{11.43}$$

The value of 35 in this expression is empirical based on a best fit for solid spherical particles on smooth walls, as shown in Figure 11.14a, but this value may depend on particle roughness, density, shape, and elasticity. A hard-sphere model that incorporate these effects with higher fidelity is given by Marshall (2018).

At low particle inertias, there is no rebound at all since the surrounding fluid viscous effects dominate and dissipate all the kinetic energy. This zero-reflection limit for the impact Stokes number can be used to define a rest velocity (v_{rest}) as the maximum impact speed for which viscosity will damp out any rebound. As shown in Figure 11.14, rest generally occurs for $\text{St}_{\perp} < 5$ so

$$\text{v}_{\text{rest}} = \frac{45\mu_{\text{f}}}{\left(\rho_{\text{p}} + c_{\forall}\rho_{\text{f}}\right)\text{d}} \qquad \textit{for wall collision.} \tag{11.44}$$

Using this criteria for immediate deposition effectively assumes that $\text{v}_{\text{rest}} > \text{v}_{\text{stick}}$, which is generally reasonable for most collisions of solid particles in fluids (Marshall, 2018).

Experimental results for fluid particles (drops or bubbles) impacting a smooth solid surface are shown in Figure 11.14b. These results exhibit similar behavior, assuming $e_{\perp\text{o}} \approx 0.9$ for drops and $e_{\perp\text{o}} \approx 1$ for bubbles, to collapse the data. Figure 11.14a shows the data for solid particles. In addition, the restitution trends for fluid particles are again reasonably described by the Legendre model of (11.43). This indicates that the effects of internal particle recirculation and surface deformation for fluid particles do not dramatically alter the viscosity losses during drainage within the gap, at least for the conditions associated with the experiments.

The fluid viscosity effects can also be extended to the case of two particles colliding with each other in a liquid. In this case, the volume of the gap for a given gap distance will be larger due to the curvature of both particles, so the viscous losses will be reduced relative to that for a wall collision. This effect can be incorporated by generalizing the impact Stokes number definition to include the size ratio (s), defined as the ratio of the small to large particle diameters, as follows:

$$\text{St}_{\perp} \equiv \frac{\tau_{\text{p}}}{\tau_{\text{fluid}}} = \frac{2\text{v}_{\perp}^{\text{in}}\text{m}_{\text{coll}}(s+1)^2}{3\pi\mu_{\text{f}}\text{d}_{\text{small}}^2} \qquad \textit{for two particles with same density} \tag{11.45a}$$

$$s \equiv \frac{\text{d}_{\text{small}}}{\text{d}_{\text{large}}}. \tag{11.45b}$$

This impact Stokes number reverts to (11.42) for a wall collision ($s = 0$), and a similar correction can also be applied to (11.44) for the rest velocity of two particles colliding.

To summarize the effects for the wet regime where fluid viscosity effects dominate adhesion effects ($\text{v}_{\text{rest}} > \text{v}_{\text{stick}}$) and plastic deformation is neglected, the restitution can be modeled as follows:

$$e_\perp \approx 0 \qquad \textit{for}\ \mathrm{St}_\perp < 5. \tag{11.46a}$$

$$e_\perp = e_{\perp o}\exp(-\mathrm{St}_\perp/35) \qquad \textit{for}\ 5 < \mathrm{St}_\perp < 2{,}000. \tag{11.46b}$$

$$e_\perp \approx e_{\perp o} \qquad \textit{for}\ \mathrm{St}_\perp > 2{,}000. \tag{11.46c}$$

For particles in a gas, the rest velocity is generally relatively small ($v_{rest} \ll v_{stick}$) such that fluid viscous effects can generally be neglected and the outcomes of (11.41) are more appropriate. For example, the gold-coated glass particles in air of Figure 11.12b have a v_{stick} of 0.32 m/s with a v_{rest} of only 0.0041 m/s. As such, fluid viscosity effects are generally only important for particle collisions in liquids.

11.5 Oblique and Spin Collisions for Solid Spheres

For an oblique or a spin collision, the normal momentum change is still governed by the previously described relationships, but one must also consider the tangential momentum change. This change is related to the collision velocity at the contact surfaces (as shown in Figure 11.10 for a wall impact). Following the incoming decomposition of (11.30c), the outgoing collision tangential velocity (including the outgoing spin) can be defined as

$$\mathbf{v}^{out}_{coll,\parallel} \equiv \mathbf{v}^{out}_{\parallel} + r_p \boldsymbol{\Omega}^{out} \times \mathbf{n}_{coll}. \tag{11.47}$$

This outgoing component can also be related to the incoming component via a tangential restitution coefficient ($e_\parallel$) that reflects the degree of tangential momentum reversal, as follows:

$$\mathbf{v}^{out}_{coll,\parallel} = -e_\parallel \mathbf{v}^{in}_{coll,\parallel}. \tag{11.48}$$

This result can then be decomposed into the associated tangential translational and angular velocities changes of (11.28) as discussed by Crowe et al. (2011), as follows:

$$\Delta\mathbf{v}_\parallel \equiv \mathbf{v}^{out}_\parallel - \mathbf{v}^{in}_\parallel = -\frac{2}{7}(1+e_\parallel)\mathbf{v}^{in}_{coll,\parallel} \tag{11.49a}$$

$$\Delta\boldsymbol{\Omega} \equiv \boldsymbol{\Omega}^{out} - \boldsymbol{\Omega}^{in} = \frac{10}{7d}(1+e_\parallel)\mathbf{n}_{coll} \times \mathbf{v}^{in}_{coll,\parallel} \tag{11.49b}$$

for a particle – wall impact.

For a binary particle–particle collision with different masses, the jump relations for the momentum changes can be similarly obtained from the impulse equations and a tangential restitution coefficient. Using (11.31)–(11.34), these changes can then be decomposed into translational and angular velocity changes for the two particles as follows:

$$m_{p2}\Delta\mathbf{v}_{\parallel,2} = -m_{p1}\Delta\mathbf{v}_{\parallel,1} = \frac{2}{7}(1+e_\parallel)m_{coll}\mathbf{v}^{in}_{coll,\parallel} \tag{11.50a}$$

$$m_{p2}d_2\Delta\boldsymbol{\Omega}_2 = m_{p1}d_1\Delta\boldsymbol{\Omega}_1 = \frac{10}{7}(1+e_\parallel)m_{coll}\mathbf{n}_{coll} \times \mathbf{v}^{in}_{coll,\parallel} \tag{11.50b}$$

for two particles.

As such, all outgoing particle velocity components can be determined for a given set of incoming conditions once the normal restitution coefficient (discussed in the previous section) and parallel restitution coefficient (discussed in the following subsection) are specified.

11.5.1 Oblique Collisions of Spheres

For the hard-sphere model with an oblique collision ($\vartheta^{in} \neq 90°$), two different regimes can occur: *slide* and *roll*. The slide regime (significant slippage) dominates at near-grazing collisions ($\vartheta^{in} \sim 0°$), while the roll regime (significant traction) dominates at near-normal collisions ($\vartheta^{in} \sim 90°$). The resulting tangential restitution depends on which regime is controlling the tangential momentum. To determine this, we define the tangential velocity ratio (Ψ) for either the particle surface incoming or outgoing tangential velocity by normalizing with the *incoming* particle *normal* velocity:

$$\Psi^{in} \equiv \frac{v_{coll,\parallel}^{in}}{v_{coll,\perp}^{in}}. \tag{11.51a}$$

$$\Psi^{out} \equiv \frac{v_{coll,\parallel}^{out}}{v_{coll,\perp}^{in}} = -e_{\parallel}\Psi^{in}. \tag{11.51b}$$

Note that the RHS of (11.51b) stems from (11.48), while the RHS of (11.51a) reverts to $\tan^{-1}(\vartheta^{in})$ if there is no incoming particle spin, per (11.30c) and Figure 11.10. In the following, the roll and slide regimes are considered in terms of the incoming and outgoing tangential velocity ratios.

For the roll regime, the restitution coefficient is denoted as $e_{\parallel roll}$ and is bounded as $0 \geq e_{\parallel roll} \geq 1$ depending on how much of the surface tangential velocity is reversed. The roll coefficient has a typical experimentally-obtained value of 0.35 for impacts on hard, smooth surfaces. This coefficient can then be used to obtain the outgoing tangential velocity ratio of (11.51b) and the parallel collision velocity of (11.48) for the roll regime as follows:

$$\Psi_{roll}^{out} \equiv -e_{\parallel roll}\Psi^{in} \tag{11.52a}$$

$$v_{coll,\parallel}^{out} \equiv -e_{\parallel roll} v_{coll,\parallel}^{in} \tag{11.52b}$$

where $e_{\parallel roll} \approx 0.35$ *on hard smooth surfaces.*

Since the roll restitution coefficient is always positive, it is associated with a *reversal* of direction for the surface tangential velocity due to the collision. As an example, consider a particle with no spin moving at 2 m/s (downward and to the right) with a 60° impact angle relative to a wall that results in a roll outcome. The tangential incoming velocities for the particle centroid and for the particle surface are equal (as there is no spin) both with a value of 1 m/s (to the right). Based on (11.52b), the outgoing tangential surface velocity for the particle surface will be −0.35 m/s (to the left), indicating a direction reversal. Next, applying (11.49), the outgoing tangential translational velocity will be 0.614 m/s (to the right) and the outgoing angular surface velocity (based on Ωr) will be

0.964 m/s (clockwise). As such, the centroid tangential velocity will be preserved (not reversed) for the roll regime so long as there is no incoming spin.

For slide, the restitution coefficient is denoted as $e_{\|slide}$ but can be either positive or negative (i.e., the surface tangential velocity is not necessarily reversed). This is because the slide regime reduces the surface tangential velocity based on a coefficient of surface friction (c_{surf}) and the change in normal velocity as follows:

$$\mathbf{v}_{coll,\|slide}^{out} = \mathbf{v}_{coll,\|}^{in} - \frac{7}{2} c_{surf}\left(\mathbf{v}_{coll,\perp}^{in} - \mathbf{v}_{coll,\perp}^{out}\right). \tag{11.53}$$

The RHS indicates that the tangential friction impulse is proportional to the normal impulse. This is consistent with Coulomb's law of friction, whereby the tangential friction force for an object continuously sliding on a surface is equal to the product of the normal force and a friction factor.

The result of (11.53) can be used to obtain the tangential velocity ratio and restitution coefficient for slide as follows:

$$\Psi_{slide}^{out} \equiv \Psi^{in} - \frac{7}{2} c_{surf}(1 + e_{\perp}). \tag{11.54a}$$

$$e_{\|slide} \equiv \frac{7}{2} c_{surf}(1 + e_{\perp})\left(v_{coll,\perp}^{in}/v_{coll,\perp}^{out}\right) - 1. \tag{11.54b}$$

The surface friction coefficient for sliding collisions in dry conditions (surrounding fluid is a gas) is denoted by $c_{surf,o}$, and values for various surfaces are given in Table 11.1. For wet conditions (the surrounding fluid is a liquid or there is a liquid coating), the coefficient of friction can be dramatically reduced, especially as the liquid viscosity reduces. For example, steel particles have wet c_{surf} values on the order of 0.02–0.001, which is much less than their $c_{surf,o}$ value (for dry conditions) of about 0.11. As the wetting liquid viscosity becomes very small, c_{surf} tends to zero, so $e_{\|slide}$ tends to -1 based on (11.54a) and (11.52b), so the tangential momentum tends to be fully preserved after the collision. At the other extreme of high friction, the tangential restitution coefficient tends to a maximum based on the rolling value. Thus, the sliding restitution coefficient range is $-1 \leq e_{\|slide} \leq e_{\|roll}$.

Measured values of roll and slide behavior using the tangential velocity ratios of (11.51) are shown in Figure 11.15 for particles impacting a wall and for binary collisions of two particle impacting each other. The roll regime of (11.52a) is shown by a long-dashed line and is a reasonable representation of the experimental data for $\Psi^{in} < 1.05$ (consistent with $\vartheta^{in} > 46°$ for no incoming particle spin). The linear slope of this line is consistent with a constant value for $e_{\|roll}$. The predicted slide regime of (11.54a) is shown as the solid line in Figure 11.15, with a slope near unity, and reasonably represents the experimental data for $\Psi^{in} > 1.05$ for this data. This point where the roll and slide models intersect is termed the *critical tangential velocity ratio* and can more generally be determined from (11.52b) and (11.54a) as follows:

$$\Psi_{crit}^{in} = \frac{7 c_{surf}(1 + e_{\perp})}{2(1 + e_{\|roll})}. \tag{11.55}$$

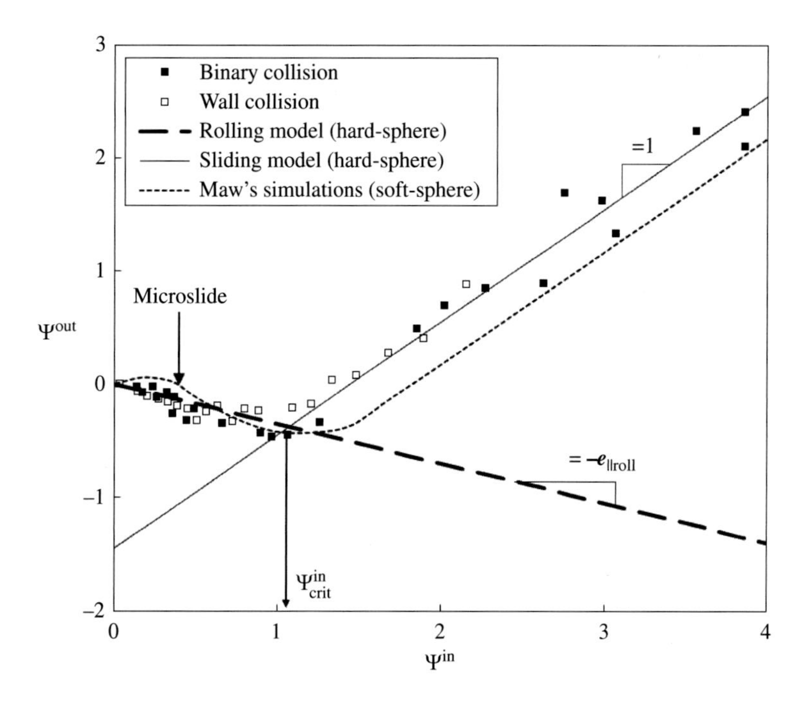

Figure 11.15 Measured velocity ratios for particle–particle and particle–wall interactions for spherical cellulose acetate particles (Foerster et al., 1994) and predictions from hard-sphere and soft-sphere models.

Consistent with this critical value and Figure 11.15, the roll versus slide outcome is determined by whichever gives the largest outgoing tangential velocity ratio and the smallest restitution, as follows:

$$\Psi^{out} = \max\left(\Psi^{out}_{roll}, \Psi^{out}_{slide}\right). \quad (11.56a)$$

$$e_{\|} = \min\left(e_{\|roll}, e_{\|slide}\right). \quad (11.56b)$$

Using these hard-sphere relations, the outgoing particle tangential velocity and spin values are then given by (11.49) for wall impact and by (11.50) for particle–particle impact.

Another approach is to use a soft-sphere model, where the time-evolving particle deformation and stresses though the collision (contact period) can be computed numerically based on material properties. An example soft-sphere simulation (Maw et al., 1976) shown in Figure 11.15 predicts an initial microslide region (which can sometimes occur at nearly normal collisions) as well as a more gradual slope change behavior near transition from roll to slide. However, the hard-sphere model is widely used when there are a large number of collisions due to its simplicity and computational efficiency and when the restitution coefficients are known. To demonstrate the robustness of the hard-sphere model, collisions for various particle materials and surfaces are shown in Figure 11.16. For dry collisions, the velocity ratios are

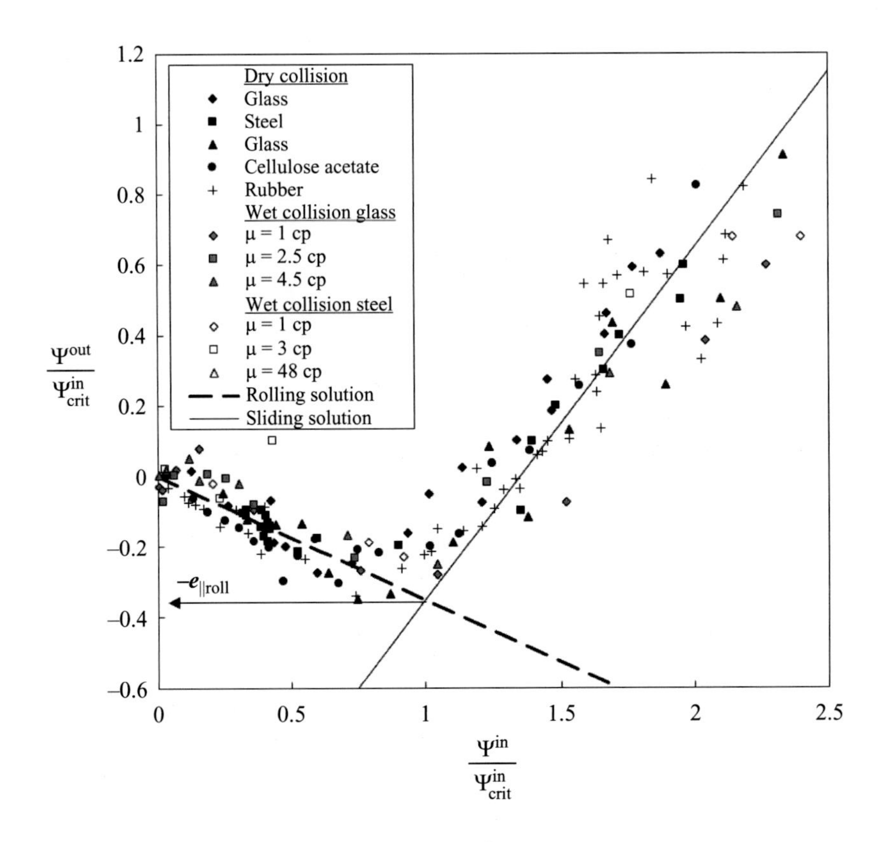

Figure 11.16 Oblique velocity ratio behavior for several particle materials in dry and wet collisions, where the latter includes various fluid viscosities (Maw et al., 1976; Foerster et al., 1994; Joseph and Hunt, 2004).

normalized by the critical value for each set of conditions, using $e_{\|roll} = 0.35$ and $c_{surf,o}$ from Table 11.1. For the wet conditions, the c_{surf} values were based on the measured critical tangential velocity ratio but can be also predicted in terms of fluid viscosity a semi-empirical relationship (Joseph and Hunt (2004). In general, the overall prediction of the collision velocity ratios (which determine the postcollision quantities) for dry and wet conditions is accurate to within experimental scatter for these collisions where the surfaces of the wall and particles are generally smooth. If the walls are rough and/or the particles are irregularly shaped, the collision outcome is generally stochastic, as discussed in the following.

11.5.2 Irregularly Shaped Particles and/or Walls

The influence of particle shape on particle reflection is illustrated in Figure 11.17, where the smooth spheres on a flat wall yield deterministic outcomes, while irregularly shaped particles and/or surfaces result in stochastic reflections. For the latter, the

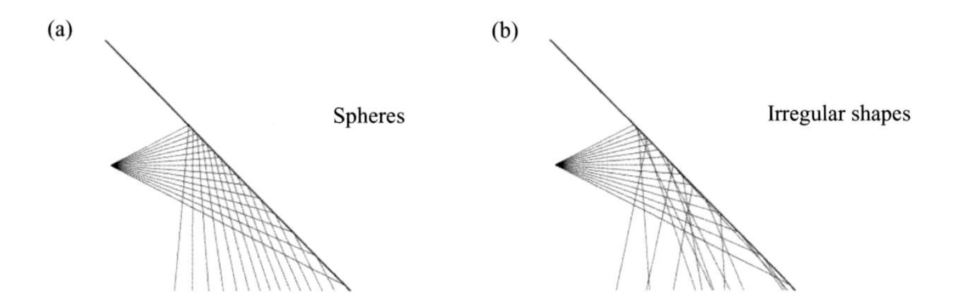

Figure 11.17 Schematic of particle trajectories for a range of incoming angles for (a) smooth spheres impacting a smooth wall with deterministic reflection, and (b) irregular particle shapes and or wall roughness yielding stochastic reflection.

outcome for a particle with a given incoming angle and velocity will have a mean restitution value about which there can be a significant variance.

To capture the effects of stochastic reflection due to irregular particle shapes, empirical models are typically employed. Often these are based on standardized test dusts to represent naturally occurring irregularly shaped particles. Of these test dusts, perhaps the most investigated is A4 (part of the Arizona Road Dust family), which has a broad size distribution with a peak probability at about 55 μm and is composed of about 73% quartz by volume, while the remainder primarily contains metal oxides (Connolly et al., 2019). Empirical models for particle restitution for A4 and similar test dusts are shown in Figure 11.18 as a range based on the mean value ± the standard deviation. For the normal restitution coefficient (Figure 11.18a), shallow impact angles ($\vartheta^{\text{in}} < 18°$) yield large scatter and the upper range can exceed unity due to conversion of the high tangential momentum to normal momentum (due to shape irregularity). On the other hand, near-normal impacts ($\vartheta^{\text{in}} > 72°$) have lower scatter and lower mean values for normal restitution (< 0.5) due to conversion to tangential momentum. The latter effect is also seen in Figure 11.18b, where high tangential restitution values are seen at near-normal impacts (Connolly et al., 2019) and are given as follows:

$$\overline{e_{\perp}} = 1 - 2.133\left(\vartheta^{\text{in}}/90^{\text{o}}\right) + 2.187\left(\vartheta^{\text{in}}/90^{\text{o}}\right)^2 - 0.729\left(\vartheta^{\text{in}}/90^{\text{o}}\right)^3. \tag{11.57a}$$

$$\overline{e_{\parallel}} = 0.8. \tag{11.57b}$$

$$\left(e_{\perp} - \overline{e_{\perp}}\right)_{\text{rms}} = 0.5\,\overline{e_{\perp}}. \tag{11.57c}$$

$$\left(e_{\parallel} - \overline{e_{\parallel}}\right)_{\text{rms}} = 0.2 + 0.4\left(\vartheta^{\text{in}}/90^{\text{o}}\right)^3. \tag{11.57d}$$

Notably, the mean restitutions are less than that for quartz spheres, and this can be attributed to energy losses from multiple contacts during the collision, the creation of high particle spin, and occasional particle fracture due to point impacts (Reagle et al., 2013; Sandeep et al., 2020). To employ such models as (11.57), random number generators can be sampled based on the corresponding mean and standard deviations for a given incoming angle.

(a)

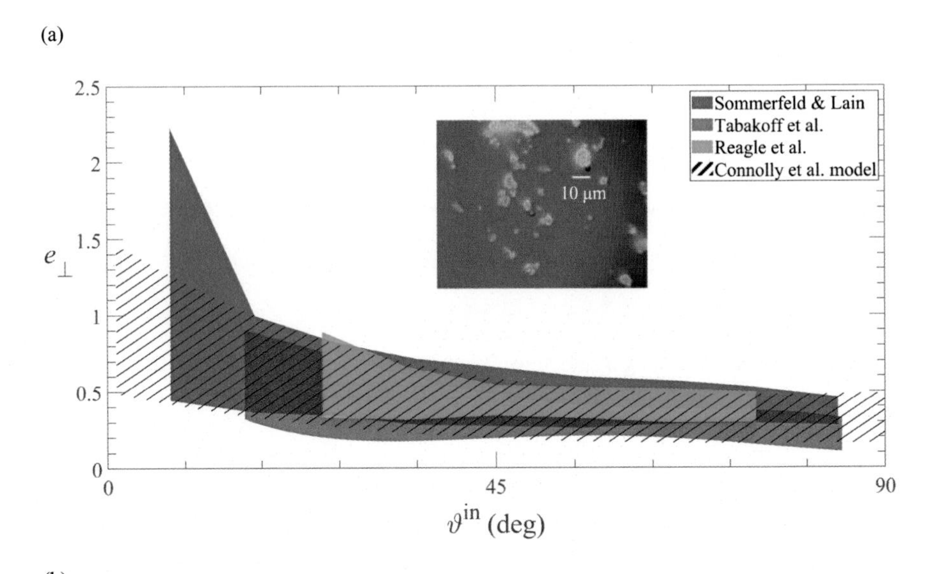

(b)

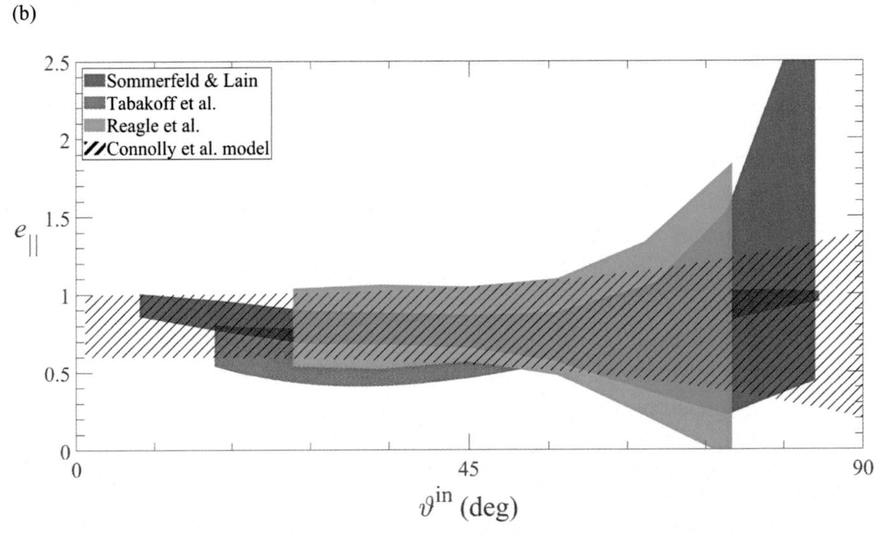

Figure 11.18 Empirical models of restitution showing the range for the mean value ± the standard deviation for test dust impacting an aluminum wall at velocities ranging from 20 m/s to 90 m/s (Tabakoff et al., 1996; Reagle et al., 2013; Sommerfeld and Lain, 2018; Connolly et al., 2019) for (a) normal coefficient of restitution (inset shows an image of A4 particles) and (b) tangential coefficient.

The impact of smooth particles onto walls with surface roughness also yields stochastic effects, as shown in Figure 11.17. However, the restitutions are instead obtained as a function of the local incoming angle to the wall, where this angle has a mean value based on a flat wall and a statistical variance based on the roughness geometry (Sommerfeld and Huber, 1999). The variance of local impact angles that the particle will "see" can be obtained by stochastically sampling local wall angles based on the distribution of roughness surface angles. A modified normal distribution function can be used for the roughness angle distribution, which can be based on a scanned surface geometry sampled at a wavelength equal to the particle diameter. Shallow impact angles can create a "shadow effect" that biases particles so they are more likely to see local surface inclinations toward the particle than away from the particle (Altmeppen et al., 2020). This sampling approach was extended to three-dimensional surfaces by Radenkovic and Simonin (2018) for a variety of surfaces with good predictions of particle outcomes.

As with particle shape irregularity, surface irregularity can reduce the mean restitutions. For example, surfaces with very small roughness heights (on the order 0.01% of the particle diameter) can reduce the normal impact restitution by 2–8% (Sandeep et al., 2020). This can be more exaggerated when adhesive effects are important; for example, nanoscale surface roughness can cause the sticking velocity micron-size particles to increase by as much as 50% (Li et al., 2020).

11.6 Collision Outcome Regimes for Impacting Fluid Particles

Fluid particles can coalesce, rebound, or breakup if they collide with other fluid particles. This can be important to overall cloud size distributions. For example, the growth of micron-sized cloud droplets into millimeter-sized raindrops is primarily attributed to coalescence via particle–particle collisions. Similarly, bubble size distributions are primarily governed by a balance between turbulent breakup and bubble coalescence. Breakup may also occur on wall surfaces due to drop splatter. The following subsection discusses the expected outcomes for two drops colliding in a gas, followed a discussion of bubbles and drops impacting each other in liquids. Finally, drop impacts on a wall are discussed.

11.6.1 Drop Collisions in a Surrounding Gas

Drop–drop binary collisions in a gas occur when two drops move toward each other in flight and come into direct contact. For two spherical drops of similar size, the outcomes can be classified by the five types shown in Figure 11.19. These include the following in order of increasing impact speeds for two drops of similar size:

- Slow coalescence (SC), where the drops have little deformation due to impact, but small instabilities and long contact times allow the interfaces to merge.
- Bounce (B), where the impact speeds allow for significant drop deformation, but the interaction time is too short for any coalescence.
- Fast coalescence (FC), where impact causes large deformation and instabilities that lead to merger, followed by stretching and dynamics, but finally staying as a single drop.
- Reflexive separation (RS), where impact leads to *temporary* merging (as in fast coalescence), but high stretching and inertia leads to separation with a residual filament.

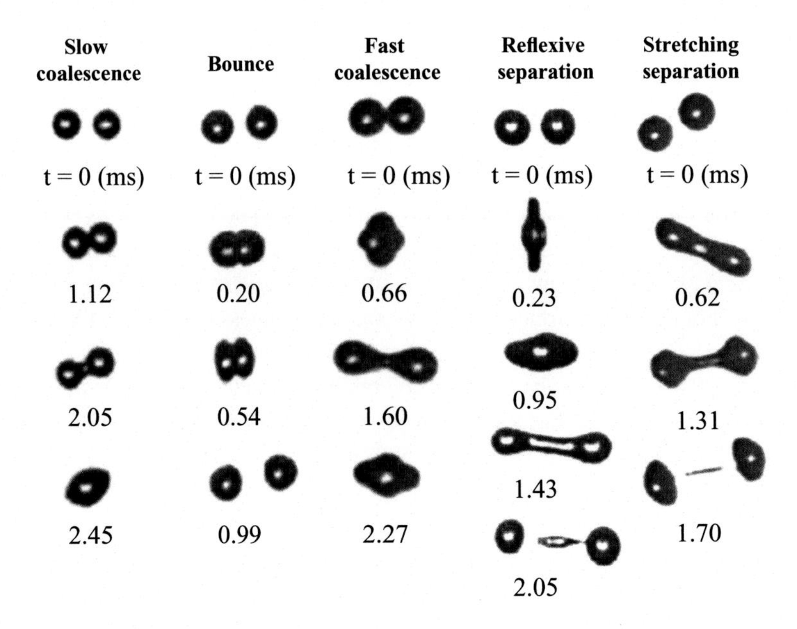

Figure 11.19 Regime classification of droplet–droplet outcomes based on images of a hydrocarbon liquid droplets colliding with increasing relative velocities (Qian and Law, 1997).

- Stretching separation (SS), where the drops impact at roughly grazing conditions, leading to temporary merging followed by stretching and then separation but with more complex shape modes, such that the residual filament is thinner and can lead to more satellite drops.

Spherical precollision droplet geometries generally occur when the aerodynamic Weber number, based on gas density and (4.41), is small. If this is not the case, additional outcome complexities can occur (Willis and Orme, 2000).

The hard-sphere restitution coefficients for drop–drop interaction can be modeled per §11.4 and §11.5. In particular, a drop bounce can be modeled using $e_{\perp o} \approx 0.9$ and $e_{\parallel slide} \approx -0.1$, while a merger outcome can be modeled based on the combined mass and momentum ($m_1 + m_2$ and $m_1\mathbf{v}_1 + m_2\mathbf{v}_1$). A temporary merger followed by separation can be modeled similar to that of a bounce, but a small fraction of the combined mass (and momentum) will generally be in the form of a filament, which results in one or more satellite drops. The number of satellite drops increases with the impact velocity (Brenn et al., 2001) and if the precollision drop shapes are nonspherical (Willis and Orme, 2000). Ignoring satellite drops, the outcome velocities can be determined once the type of collision outcome (merger or bounce) known.

To characterize the key parameters that control the outcome type, the aspects in Figure 11.20 are used. First, the collision Weber number is defined based on the collision velocity, the drop diameter, and the liquid–gas surface tension, as follows:

$$\mathrm{We}_{\mathrm{coll}} \equiv \frac{\rho_p v_{\mathrm{coll}}^2 d_{\mathrm{small}}}{\sigma}. \tag{11.58}$$

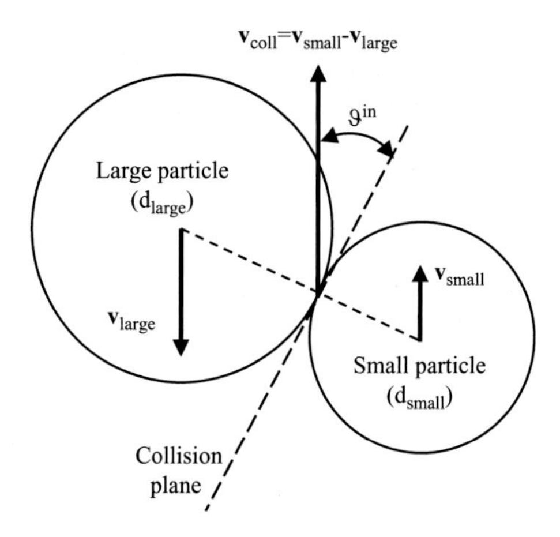

Figure 11.20 Collisional geometry between a small and a large drop whereby the impact angle (ϑ^{in}) is 90° for head-on collisions and approaches 0° for grazing collisions.

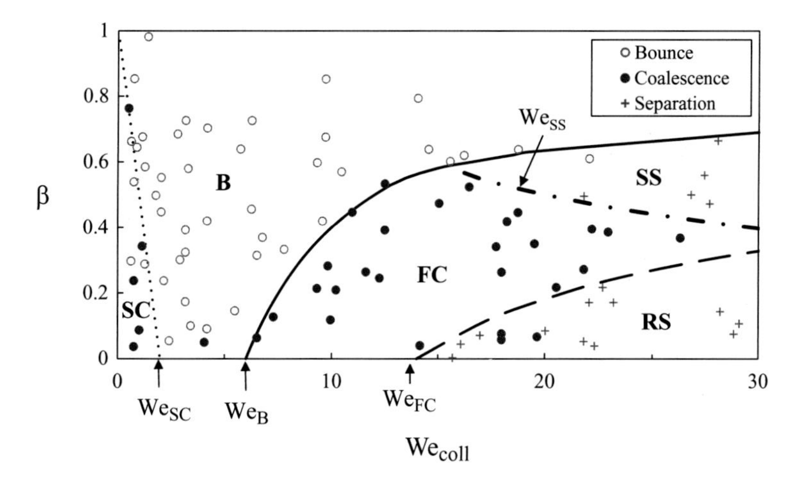

Figure 11.21 Stability nomogram of water droplet interaction in a 7.5 atm helium environment showing experimental outcomes as symbols (Qian and Law, 1997) and showing associated droplet-interaction outcome boundaries as lines.

Next, the incoming impact angle ϑ^{in} (based on the collision plane and the collision velocity) can be used to define the nondimensional collision impact parameter:

$$\beta \equiv \cos\left(\vartheta^{in}\right). \tag{11.59}$$

Thus, a head-on collision ($\vartheta^{in} = 90°$) corresponds to $\beta = 0$, while a near-grazing condition ($\vartheta^{in} = 0°$) corresponds to $\beta = 1$. Finally, the size ratio s of (11.45b) determines the mass ratio.

The impact Weber number and collision impact parameter can be used to construct a regime map of the outcomes for a given size ratio. For example, Figure 11.21 shows a regime map of outcomes using We_{coll} and β for equal drop sizes ($s = 1$). It can be seen that slow coalescence occurs when both We_{coll} and β are small (very slow head-on collisions). As, We_{coll} increases, bouncing next occurs, especially for high β (near-grazing conditions). For head-on collisions ($\beta = 0$), a further increase in Weber number leads to fast coalescence and then eventually reflexive separation. However, intermediate impact angles at high Weber numbers can also lead to stretching separation. As shown in Figure 11.21, the boundaries between these four regimes can be characterized by the three intercepts on the $\beta = 0$ axis: We_{SC}, We_B, and We_{FC}. In addition, the boundary between stretching separation and fast coalescence at $\beta = 0.5$ defines a fourth reference Weber number: We_{SS}. The following discusses theories for the boundaries and empirical models for the reference Weber numbers when droplet viscosity is important.

The first intercept (We_{SC}) represents the limit of slow coalescence with head-on bouncing. If the Weber number is very small, the drops will remain nearly spherical and the interfaces (if clean) will merge when collision contact occurs ($We_{coll} < We_{SC}$). However, higher Weber numbers (weaker surface tension effects) can cause the local droplet surfaces to deform and be flattened as they approach each other due to pressure buildup in the gas film trapped between the drops. As such, coalescence likelihood reduces for drop impact in high-pressure gases (Qian and Law, 1997) but is enhanced in vacuum conditions (Willis and Orme, 2000; Melean and Sigalotti, 2005). In particular, coalescence can be prevented when the gas film resonance frequency, which scales with the square root of pressure of (4.85), is high. In contrast, coalescence can be enhanced by fast liquid bridging, which occurs when the capillary frequency (which scales with $\sigma/d\mu_l$) is high. The ratio of these two frequencies can be used to model the head-on boundary, as follows (Krishnan and Loth, 2015):

$$We_{SC} \equiv \frac{f_{bridging}}{f_{resonance}} \approx 1.15 \frac{\sigma}{\mu_\ell} \sqrt{\frac{\rho_\ell}{p_g}} = \frac{1.15}{Oh_\ell} \sqrt{\frac{\sigma}{p_g d_{small}}}. \tag{11.60}$$

The RHS employs the Ohnesorge number of (4.54a) for the smaller drop and indicates that higher drop viscosity and higher gas pressure make bouncing more likely. For collisions that are not head on, experiments indicate that the boundary Weber number decreases monotonically with β (consistent with a reduced time for a liquid bridge to form) so that the slow coalescence boundary can be modeled as follows:

$$We_{SC/B} \approx We_{SC}(1 - \beta). \tag{11.61}$$

While this correlation was found to be generally reasonable, it does not account for effects of size ratio (s). However, it can incorporate the effects of an encapsulating liquid formed by condensation on the drops by using the Oh_ℓ of the outer liquid (Krishnan and Loth, 2015).

For the B/FC boundary (between bouncing and fast coalescence), the separating gas layer will remain intact if the interface deformations are small enough to yield a bounce outcome ($We_{coll} < We_B$). The boundary was obtained theoretically for inviscid conditions (Estrade et al., 1999) by comparing the kinetic energy associated with the collision and the surface tension energy for the outcome, where the result is expressed as follows:

$$\mathrm{We}_{\mathrm{B/FC,Estr}} = \mathrm{We}_{\mathrm{B,Estr}} \frac{s(1+s^2)}{2c_{\mathrm{B1}}\left[\cos\left(\sin^{-1}\beta\right)\right]}. \tag{11.62a}$$

$$c_{\mathrm{B1}} = 1 - 0.25(2 - c_{\mathrm{B2}})^2(1 + c_{\mathrm{B2}}) \qquad \textit{if}\; c_{\mathrm{B2}} > 1. \tag{11.62b}$$

$$c_{\mathrm{B1}} = 0.25c_{\mathrm{B2}}^2(3 - c_{\mathrm{B2}}) \qquad \textit{if}\; c_{\mathrm{B2}} \leq 1. \tag{11.62c}$$

$$c_{\mathrm{B2}} = (1 - \beta)(1 + s). \tag{11.62d}$$

In this expression, $\mathrm{We}_{\mathrm{B,Estr}}$ is the boundary for a head-on collision ($\beta = 0$) intercept and is typically given as 5 when droplet viscosity effects are weak. In general, the Estrade model gives reasonable predictions for $\beta > 0.5$, but droplet viscosity effects can stabilize the interfaces and prevent fast coalescence from occurring for $\beta < 0.5$. As shown in Figure 11.22a, measurements of the boundary head-on intercept ($\mathrm{We}_{\mathrm{B,o}}$) changes with droplet viscosity and peaks at $\mathrm{Oh}_\ell \sim 0.03$. This experimental trend for the intercept from Figure 11.22a can be modeled with a composite of two slopes and then combined with the Estrade model to describe the B/FC boundary as follows:

$$\mathrm{We}_{\mathrm{Bo}} = \min(5 + 800\mathrm{Oh}_\ell, 27.5 - 50\mathrm{Oh}_\ell). \tag{11.63a}$$

$$\mathrm{We}_{\mathrm{B/FC}} = \max\left(\mathrm{We}_{\mathrm{Bo}}, \mathrm{We}_{\mathrm{B/FC,Estr}}\right). \tag{11.63b}$$

In addition, $\mathrm{We}_{\mathrm{B,Estr}}$ can be adjusted for specific liquids (based on the shape factor) to further improve predictive performance (Sommerfeld & Pasternak, 2019).

The boundary between fast coalescence and reflexive separation was treated by the inviscid theory of Ashgriz and Poo (1989) based on the criterion that separation will occur if the internal reflexive kinetic energy of the two combined droplets is greater than 75% of the combined surface energy of the drops. This results in a set of equations that can be approximated with a single equation for convenience (Krishnan and Loth, 2015), as follows:

$$\mathrm{We}_{\mathrm{FC/RS}} = \frac{\mathrm{We}_{\mathrm{FC}}}{(1 - 3\beta)s}. \tag{11.64}$$

In general, $\mathrm{We}_{\mathrm{FC}}$ can be taken as 15 for the inviscid limit, but again droplet viscosity can cause damping, which causes the boundary to shift to higher Weber numbers. Based on the review of several experimental studies, Sommerfeld and Pasternak (2019) recommended the following empirical model for changes of the head-on intercept:

$$\mathrm{We}_{\mathrm{FC}} = \max\left(15 + 680\mathrm{Oh}_\ell, 9300\mathrm{Oh}_\ell^{1.7}\right). \tag{11.65}$$

This was found to be robust for water, alcohol, diesel fuels, and very high viscosity liquids.

The boundary between fast coalescence and stretching separation (FC/SS) was also obtained by inviscid theory (Brazier-Smith et al., 1972). This was based on comparing

(a)

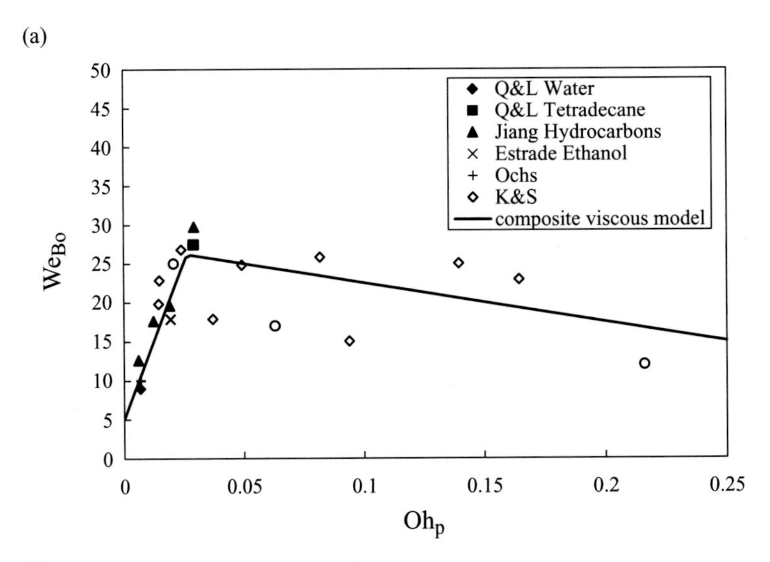

(b)

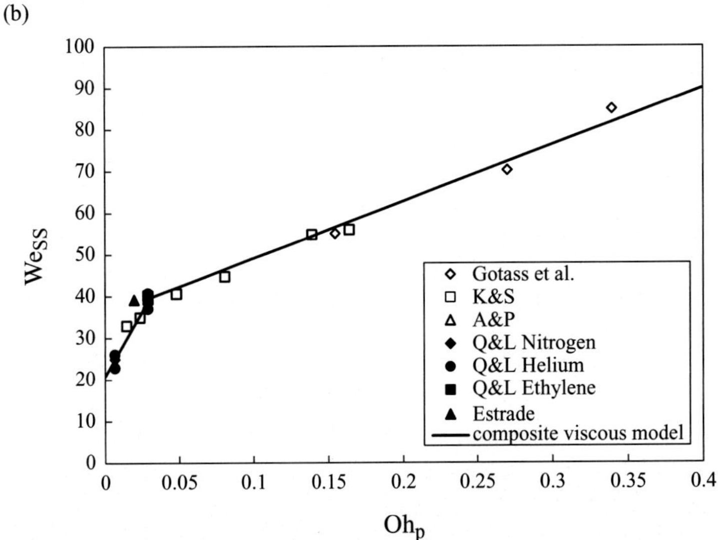

Figure 11.22 Influence of the drop Ohnesorge number on the horizontal regime intercepts for equal-sized drops ($s = 1$) for the boundaries between (a) bounce and fast coalescence and (b) fast coalescence and stretching separation. Experimental data from various sources (Ashgriz and Poo, 1989; Jiang et al., 1992; Ochs et al., 1995; Qian and Law, 1997; Estrade et al., 1999; Gotaas et al., 2007; Kuschel and Sommerfeld, 2013).

the rotational energy relative to the collision centroid (due to grazing angle conditions) with the necessary surface energy required to form two separate droplets and can be expressed as follows:

$$\mathrm{We}_{FC/SS} = \frac{\mathrm{We}_{SS}}{5.24\beta^2} \frac{\left[1 + s^2 - (1 + s^3)^{2/3}\right](1 + s^3)^{11/3}}{(1 + s)^2 s^5}. \tag{11.66}$$

For an inviscid interaction, Brazier-Smith et al. (1972) obtained $\mathrm{We}_{SS} = 25$, and this provided good results for low-viscosity liquids, such as water. Again, higher droplet viscosities cause the We_{SS} boundary to increase with the Ohnesorge number, as shown by the data in Figure 11.22b. Also shown in this figure is a two-part composite empirical model (Krishnan and Loth, 2015) that reflects the experimental change in slope seen at $\mathrm{Oh}_{\ell} \sim 0.03$:

$$\mathrm{We}_{SS} = \min(21 + 630\mathrm{Oh}_{\ell}, 36 + 135\mathrm{Oh}_{\ell}). \tag{11.67}$$

Sommerfeld and Pasternak (2019) also noted that liquid solutions can behave differently than pure fluids. It is important to note that the previous boundary regime models are only approximate, since there is a large amount of experimental scatter and outcome overlap (Krishnan and Loth, 2015; Sommerfeld and Pasternak, 2019). In addition, many issues are still unresolved, including effects of surfactants, temperature gradients, mass transfer, etc.

11.6.2 Drop and Bubble Collisions in a Liquid

Bubbles and immiscible drops impacting each other in a liquid can have similar outcomes as those seen for drops impacting each other in a gas. However, the collision Weber numbers are typically limited to values on the order of unity due to the reduced inertia of the particles and the higher viscosity of the surrounding fluid. As a result, collision outcomes tend to be confined to slow coalescence or bouncing due to reduced Weber numbers. For example, head-on clean bubble collisions are typified by $\mathrm{We}_{SC} \approx 0.2$ (Duineveld, 1995). However, We_{SC} can be increased due to contamination, since surfactants can prevent the bridging needed to rupture the interfaces and thus prevent coalescence (Walter and Blanch, 1986). As a result, bubbles in saltwater tend to have longer lifetimes than bubbles in freshwater since coalesce is prevented by the saltwater contaminants. Similarly, detergents, which include surfactants for cleaning, can prevent bubble coalescence, which can yield foams (or emulsions in the case of immiscible drops in liquids). Because of the complex nanoscale physics involved, models for coalescence in contaminated systems are often empirical (Tsouris and Tavlarides, 1994).

11.6.3 Drop Collisions on a Wall

For drops impacting normally on a wall, the outcomes again depend on the drop inertia, viscosity, and surface tension with respect to the gas (which specify We_{coll} and Oh_{∂}) but are also influenced by the surface tension with respect to the wall and the wall roughness (which specify surface wetting characteristics). As a result of these additional complexities, much is still being learned about drop–wall interactions (Josserand and Thoroddsen, 2016).

For a normal impact, there are at least six types of distinct morphological outcomes, as shown in Figure 11.23. These types include the following (in rough order of low-speed impact to high-speed impact):

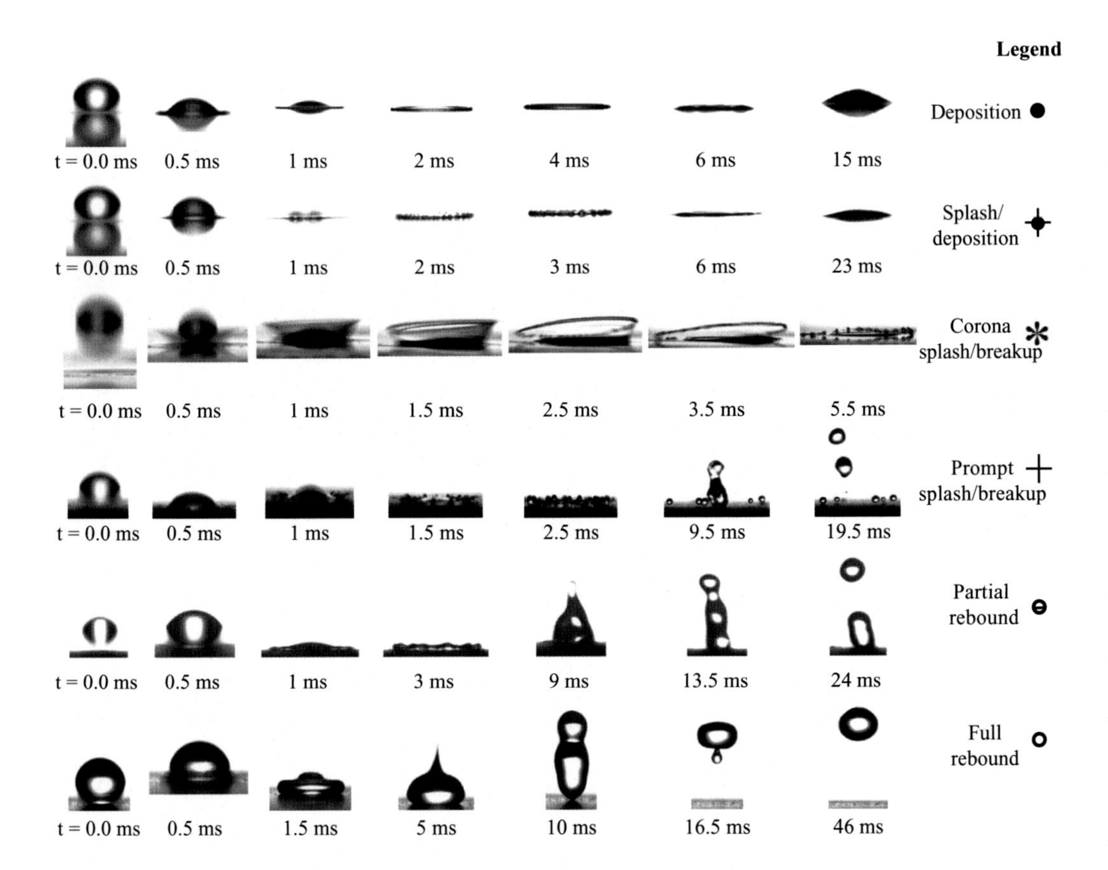

Figure 11.23 Observed outcomes of a drop–wall collision with a legend on right-hand side that defines symbols for a given outcome: + indicates a prompt splash, and o indicates a full rebound (Krishnan et al., 2023)

- Deposition, where the drop deforms during impact and stays attached to the surface during the entire process, yielding full deposition without any breakup.
- Beaded deposition, where instabilities form on the outer edge of the drop as it spreads that cause a beading of the outer rim at the outside edge, but the liquid stays attached to the surface and the eventual outcome is deposition of all the liquid in a single entity.
- Corona splash, where impact causes a rim of fluid to rebound upward, which breaks up due to instabilities into a ring of droplets that splash away from the surface.
- Prompt splash, where impact immediately generates breakup and droplets during the spreading phase and leaves some droplets on the surface while others repel upward.
- Partial rebound, where the rebound event causes part of the drop to detach and jet upward, while the remainder of the drop stays attached to the surface.
- Full rebound, where the drop bounces off the surface without leaving behind any liquid.

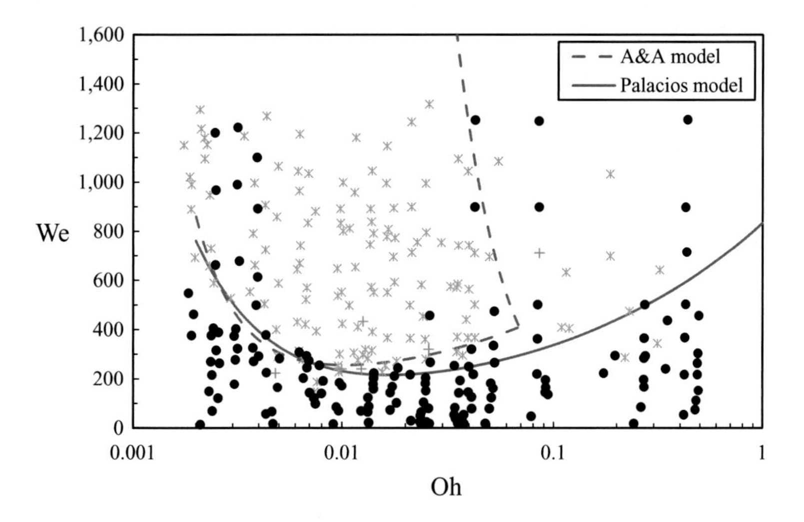

Figure 11.24 Experimental outcomes of full or beaded deposition (•) versus prompt or corona splash (X) for drop impact on high-wettability surfaces from data surveyed by A&A (Almohammadi and Amirfazli, 2019) along with the deposition/splashing boundaries of A&A and Palacios et al. (2010).

The first four outcomes generally occur on surfaces that are highly wettable (e.g., hydrophilic), while the last two occur on liquid-repellant surfaces (e.g., hydrophobic or superhydrophobic). In addition, droplet viscosity can dampen instabilities and modify the boundary Weber numbers. As such, the boundaries between outcomes are a function of the impact Weber number, the drop Ohnesorge number, and the surface wettability. For highly wettable surfaces (such as water drops on glass, metals, and many plastics), the two primary outcomes are deposition and splash, as shown in Figure 11.24. The figure shows that deposition (similar to slow coalescence) tends to occur at lower Weber numbers and always occurs for $\mathrm{We_{coll}} < 200$. It also shows that deposition is more likely for very low and very high Oh_ℓ. The minimum deposition Weber number tends to occur at about $\mathrm{Oh}_\ell \approx 0.025$. Also shown in Figure 11.24 are two empirical models for the splash–deposition boundary for highly wettable surfaces. There is significant uncertainty in the boundaries, but the model of Palacios et al. (2010) is perhaps the most robust and can be expressed as follows:

$$\mathrm{We_{coll,S/D}} = 0.4\mathrm{Oh}_\ell^{-6/5} + 834\mathrm{Oh}_\ell^{2/5}. \tag{11.68}$$

Models for intermediate and low wettability surfaces are dicussed by Krishnan et al. (2023) for normal impacts. Drops impacting at moderate oblique angles can be predicted to have a similar splash–deposition boundary if the collision Weber number of (11.68) is based on the normal velocity component. However, grazing impacts and higher Weber number impacts can yield complex outcomes with highly irregular shapes (Yeong et al., 2014).

Appendix

Table A.1 Properties of four fluids at Normal Temperature and Pressure (NTP), where NTP is a reference condition defined by $T_{ref} = 293\,K = 20°C$ and $p_{ref} = 101{,}320\,N/m^2$. Note that these liquid surface tensions are defined with respect to air while the diffusivities and Schmidt numbers for methane and ethanol are with respect to air and water. The data are based on King et al. (1965), White (2016), Schlichting and Gersten (2017), and Lide (2005).

Fluid	Dry air (gas)	Methane (gas)	Water (liquid)	Ethanol (liquid)
Density, ρ (kg/m^3)	1.205	0.669	998	789
Dynamic viscosity, μ (kg/m·s)	1.82×10^{-5}	1.10×10^{-5}	1.00×10^{-3}	9.67×10^{-4}
Kinematic viscosity, ν (m^2/s)	1.51×10^{-5}	1.65×10^{-5}	1.00×10^{-6}	1.23×10^{-6}
Surface tension, σ (N/m)	–	–	7.28×10^{-2}	2.28×10^{-2}
Speed of sound, a (m/s)	343	445	1,482	1,200
Specific heat, c_p (J/kg·K)	1,010	2,220	4,183	2,440
Gas constant, $\mathcal{R}_g$ (J/kg·K)	287	518	–	–
Mass diffusivity, Θ (m^2/s)	–	1.6×10^{-5}	–	1.12×10^{-9}
Schmidt number, Sc	–	1.06	–	1,100
Conductivity, $\mathcal{k}$ (J/m-K·s)	2.57×10^{-2}	3.35×10^{-2}	5.98×10^{-1}	1.68×10^{-1}
Prandtl number, Pr	0.717	0.729	7.01	14.04

References

Abrahamson, J. (1975) "Collision rates of small particles in a vigorously turbulent fluid," *Chemical Engineering Science*, Vol. 30, pp. 1371–1379.

Achenbach, E. (1974) "The effects of surface roughness and tunnel blockage on the flow past spheres," *Journal of Fluid Mechanics*, Vol. 65, Part 1, pp. 113–125.

Aerosty, J. (1962) "Sphere drag on a low-density flow," Report HE-150-192, University of California Berkeley, January.

Albertson, L. (1952) "Effect of shape on the fall velocity of gravel particles," in *Proceedings of the 5th Hydraulics Conference.* Studies in Engineering. Iowa City: University of Iowa.

Alger, G. R. and Simons, D. B. (1968) *Journal of Hydraulics Engineering Division, ASCE*, Vol. 94, pp. 721–737.

Aliseda, A., Cartellier, A., Hainaux, F., and Lasheras, J. C. (2002) "Effect of preferential concentration on the settling velocity of heavy particles in homogeneous isotropic turbulence," *Journal of Fluid Mechanics*, Vol. 468, pp. 77–105.

Aliseda, A. and Lasheras, J. C. (2011) "Preferential concentration and rise velocity reduction of bubbles immersed in a homogeneous and isotropic turbulent flow," *Physics of Fluids,* Vol. 23, 093301.

Almohammadi, H. and Amirfazli, A. (2019) "Droplet impact: viscosity and wettability effects on splashing," *Journal of Colloid and Interface Science*, Vol. 553, pp. 22–30.

Altmeppen, J., Sommerfeld, H., Koch, C., and Staudacher, S. "An analytical approach to estimate the effect of surface roughness on particle rebound," *Journal of Global Power and Propulsion Society*, Vol. 4, pp. 27–37.

Ambari, A. Gauthier-Manuel, B., and Guyon, E. (1984) "Wall effects on a sphere translating at constant velocity," *Journal of Fluid Mechanics*, Vol. 149, pp. 235–353.

Amin, A., Girolami, L., and Risso, F. (2021) "On the fluidization/sedimentation velocity of a homogeneous suspension in a low-inertia fluid," *Powder Technology*, Vol. 391, pp. 1–10.

Anfossi, D., Alessandrini, S., Trini Castelli, S., Ferrero, E., Oettl, D., and Degrazia, G. (2006) "Tracer dispersion simulation in low wind speed conditions with a new 2D Langevin equation system," *Atmospheric Environment*, Vol. 40, pp. 7234–7245.

Aoyama, S., Hayashi, K., Hosokawa, S., Lucas, D., and Tomiyama, A. (2017) "Lift force acting on single bubbles in linear shear flows," *International Journal of Multiphase Flow*, Vol. 96, pp. 113–122.

APSC (2020) Alyeska Pipeline Service Company, photograph with permission.

Aroesty, J. (1962) "Sphere drag in a low-density supersonic flow." Technical Report HE-150-192, University of California Berkeley.

Arolla, S. and Durbin, P. A. (2014) "LES of spatially developing turbulent boundary layer over a concave surface," *Journal of Turbulence*, Vol. 16. pp. 81–99.

Ashgriz, N. and Poo, J. Y. (1989) "Coalescence and separation in binary collisions of liquid drops," *Journal of Fluid Mechanics*, Vol. 221, pp. 183–204.

Asmolov, E. S. and McLaughlin, J. B. (1999) "The inertial lift of an oscillating sphere in a linear shear flow field," *International Journal of Multiphase Flow*, Vol. 25, pp. 739–751.

Auton, T. R., (1987) "The lift force on a spherical body in a rotational flow," *Journal of Fluid Mechanics,* Vol. 183, pp. 199–218.

Auton, T. R., Hunt, J. C. R., and Prud'Homme, M. (1988) "The force exerted on a body in inviscid unsteady non-uniform rotational flow," *Journal of Fluid Mechanics,* Vol. 197, pp. 241–257.

Aybers, N. M. and Tapucu, A. (1969) "The motion of gas bubbles rising through stagnant liquid," *Wärme Stoffübertrag.* Vol. 2, pp. 118–128.

Azuma, H. and Yoshihara, S. (1999) "Three-dimensional large-amplitude drop oscillations: experiments and theoretical analysis," *Journal of Fluid Mechanics,* Vol. 393, pp. 309–332.

Baba, J. and Komar, P. D. (1981) "Measurements and analysis of settling velocities of natural quartz sand grains," *Journal of Sedimentary Petrology,* Vol. 51, pp. 631–640.

Bagchi, P. and Balachandar, S. (2002a) "Effect of free rotation on motion of a solid sphere," *Physics of Fluids*, Vol. 14, No. 8, pp. 2719–2737.

Bagchi, P. and Balachandar, S. (2002b) "Shear versus vortex-induced lift on a rigid sphere at moderate Re," *Journal of Fluid Mechanics,* Vol. 473, pp. 379–388.

Bagchi, P. and Balachandar, S. (2002c) "Steady planar straining flow past a rigid sphere at moderate Reynolds number," *Journal of Fluid Mechanics,* Vol. 466, pp. 365–407.

Bagchi, P. and Balachandar, S. (2003) "Effect of turbulence on the drag and lift of a particle," *Physics of Fluids*, Vol. 15, No. 11, pp. 3496–3513.

Bagchi, P. and Balachandar, S. (2004) "Response of the wake of an isolated particle to an isotropic turbulent flow," *Journal of Fluid Mechanics,* Vol. 518, pp. 95–123.

Bailey, A. B. and Hiatt, J. (1972) "Sphere drag coefficients for a broad range of Mach and Reynolds numbers," *AIAA Journal*, Vol.10, pp 1436–1440.

Bailey, A. B. and Starr, R. F. (1976) "Sphere drag at transonic speeds and high Reynolds numbers," *AIAA Journal*, Vol. 14, No. 11, p. 1631.

Balachandar, S. and Eaton, J. K. (2010) "Turbulent dispersed multiphase flow," *Annual Review of Fluid Mechanics,* Vol. 42, pp. 111–133.

Balakumar, P. and Park. G. I. (2015) "DNS/LES simulations of separated flows at high Reynolds numbers," 45th AIAA Fluid Dynamics Conference, Dallas, AIAA 2015-2783.

Balaras, E., Benocci, C., and Piomelli, U. (1996) "Two-layer approximate boundary conditions for large-eddy simulations," *AIAA Journal*, Vol. 34, pp. 1111–1119.

Barbosa, C., Legendre, D., and Zenit, R. (2019) "Sliding motion of a bubble against an inclined wall from moderate to high bubble Reynolds number," *Physical Review Fluids*, Vol. 4, pp. 032201-1–032201-10.

Barkla, H. M. and Auchterlonie, L. J. (1971) "The Magnus or Robbins effect on rotating spheres," *Journal of Fluid Mechanics,* Vol. 47, pp. 437–448.

Barlow, J. B. and Domański, M. (2008) "Lift on stationary and rotating spheres under varying flow and surface conditions," *AIAA Journal*, Vol. 46, pp. 1932–1936.

Barua, S., Yoo, J.-W., Kolhar, P., Wakankar, A., Gokarn, Y. R., and Mitragotri, S. (2013) "Shape-induced enhancement of antibody specificity," *Proceedings of the National Academy of Sciences*, Vol. 110 pp. 3270–3275.

Basset, A. B. (1888) "On the motion of a sphere in a viscous liquid," *Philosophical Transactions of the Royal Society of London*, Vol. 179A, pp. 43–63.

Bataille, J., Lance, M., and Marie, J. L. (1991) "Some aspects of the modeling of bubbly flows," in G. F. Hewitt, F. Mayinger, and J. R. Riznic, eds., *Phase-Interface Phenomena in Multiphase Flow*. New York: Hemisphere Publishing Corporation, pp. 179–193.

Batchelor, G. K. (1967) *An Introduction to Fluid Dynamics.* Cambridge: Cambridge University Press.

Batchelor, G. K. (1972) "Sedimentation in a dilute dispersion of spheres," *Journal of Fluid Mechanics,* Vol. 52, pp. 245–268.

Beard, K. V. (1976) "Terminal velocity and shape of clouds and precipitation drops aloft," *Journal of Atmospheric Science*, Vol. 33, pp. 851–864.

Beard, K. V. and Prupacher, H. R. (1969) "A determination of the terminal velocity and drag of small water drops by means of a wind tunnel," *Journal of Atmospheric Science,* Vol. 26, pp. 1066–1071.

Bel Fdhila, R. and Duineveld, P. C. (1996) "The effect of surfactant on the rise of a spherical bubble at high Reynolds and Peclet numbers," *Physics of Fluids*, Vol. 8, pp. 310–321.

Benjamin, T. B. (1987) "Hamiltonian theory for motions of bubbles in an infinite liquid," *Journal of Fluid Mechanics*, Vol. 181, pp. 349–379.

Benson, C. M., Levin, D. A., Zhing, J., Gimelshein, S. F., and Montaser, A. (2004) "Kinetic model for simulation of aerosol droplets in high-temperature environments," *Journal of Thermophysics and Heat Transfer*, Vol. 18, pp. 122–134.

Bhaga, D. and Weber, M. E. (1981) "Bubbles in viscous liquids: shapes, wakes and velocities," *Journal of Fluid Mechanics*, Vol. 105, pp. 61–85.

Biesheuvel, A. and Spoelstra, S. (1989) "The added mass coefficient of a dispersion of spherical gas bubbles in liquid," *International Journal of Multiphase Flow*, Vol. 15, pp. 911–924.

Biesheuvel, A. and van Wijngaarden, L. (1984) "Two-phase flow equation for a dilute dispersion of gas bubbles in liquid," *Journal of Fluid Mechanics*, Vol. 148, pp. 301–318.

Bocksell, T. and Loth, E. (2001) "Discontinuous and continuous random walk models for particle diffusion in free-shear flows," *AIAA Journal*, Vol. 39, June, pp. 1086–1096.

Bocksell, T. and Loth, E. (2006) "Stochastic modeling of particle diffusion in a turbulent boundary layer," *International Journal of Multiphase Flow*, Vol. 32, pp. 1234–1253.

Boussinesq, J. (1903) *Théorie analytique de la chaleur*. Paris: L'Ecole Polytechnique, Vol. 2.

Boussinesq, J. (1997) *Théorie de l'écoulement tourbillonnant et tumultueux des liquides dans les lits rectilignes à grande section*. Paris: Gauthier–Villars.

Bragg, M. B. (1982) "A similarity analysis of the droplet trajectory equation," *AIAA Journal*, pp. 1681–1686.

Brake, M., Reu, P. L., and Aragon, D. S. (2017), "A comprehensive set of impact data for common aerospace metals," *ASME Journal of Computational Nonlinear Dynamics,* Vol. 12, No. 6, pp. 061011-1–061011-23.

Brazier-Smith, P. R., Jennings, S. G., and Latham, J. (1972) "The interaction of falling water drops: coalescence," *Proceedings of the Royal Society of London. Series A, Mathematical and Physical Sciences,* Vol. 326, No. 1566, pp. 393–408

Brenn, G., Valkovska, D. and Danov, K.D. (2001) "The formation of satellite droplets by unstable binary drop," *Physics of Fluids*, Vol. 13, pp. 2463–2477.

Brennen, C. E. (1995) *Cavitation and Bubble Dynamics.* New York, Oxford: Oxford University Press.

Brennen, C. E. (2005) *Fundamentals of Multiphase Flow.* Cambridge: Cambridge University Press.

Brenner, H. and Cox, R. G. (1963) "The resistance to a particle of arbitrary shape in translational motion at mall Reynolds numbers," *Journal of Fluid Mechanics,* Vol. 17, pp. 561–595.

Breuer, M., Peller, N., Rapp, Ch., and Manhart, M. (2009) ""Flow over periodic hills – numerical and experimental study in a wide range of Reynolds numbers," *Computers and Fluids*, Vol. 38, pp. 433–457.

Briggs, L. J. (1959) "Effects of spin and speed on the lateral deflection (curve) of a baseball and the Magnus effect for smooth spheres," *American Journal of Physics,* Vol. 27, pp. 589–596.

Brock, J. R. (1962) "On the theory of thermal forces acting on aerosol particles," *Journal of Colloid Science,* Vol. 17, pp. 768–780.

Brown, C. S., Dillon, S., Lahey, R. T., and Bolotnov, I. A. (2017) "Wall-resolved spectral cascade-transport turbulence model," *Nuclear Engineering and Design,* Vol. 320, pp. 309–324.

Brown, G. L. and Roshko, A. (1974) "On density effects and large structure in turbulent mixing layers," *Journal of Fluid Mechanics*, Vol. 64, pp. 775–816.

Brucato, A., Grusafi, F., and Montante, G. (1998) "Particle drag coefficients in turbulent fluids," *Chemical Engineering Science*, Vol. 53, pp. 3295–3314.

Brucker, C. (1999) "Structure and dynamics of the wake of bubbles and its relevance for bubble injection," *Physics of Fluids*, Vol. 11, pp. 1781–1796.

Builtjes, P. J. H. (1975) "Determination of the Eulerain longitudinal integral length scale in a turbulent boundary layer," *Applied Scienctific Research,* Vol. 31, pp. 397–399.

Butler, C. (2020) CBC News, Posted: April 14, 2020, 3:06 PM ET.

Burgers, J. M. (1942) "On the influence of the concentration of a suspension upon the sedimentation velocity," *Proceedings of the Koninklijke Nederlandse Akademie van Wetenschappen*, Vol. 45, pp. 126–128.

Byron, M. L., Tao, Y., Houghton, I. A. and Variano, E. A. (2019) "Slip velocity of large low-aspect-ratio cylinders in homogenous isotropic turbulence," *International Journal of Multiphase Flow*, Vol. 121, 103120.

Carlier, J. and Stanislas, M. (2005) "Experimental study of eddy structures in a turbulent boundary layer using particle image velocimetry," *Journal of Fluid Mechanics,* Vol. 535, pp. 143–188.

Carlson, D. J. and Hoglund, R. F. (1964) "Particle drag and heat transfer in rocket nozzles," *AIAA Journal,* Vol. 2, pp. 1980–1984.

Chen, R. C. and Wu, T. L. (1999) "The flow characteristics between two interactive spheres," ASME Fluids Engineering Conference, San Francisco, FEDSM99–7776, June.

Cherukat, P., McLaughlin, J. B., and Graham, A. L. (1994) "The inertial lift on a rigid sphere translating in a linear shear flow field," *International Journal of Multiphase Flow*, Vol. 20, No. 2, pp. 339–353.

Chester, D. (1993) *Volcanoes and Society.* London: Edward Arnold.

Chhabra, R. P. (1995) *Powder Technology.* Vol. 85, pp. 83–90.

Christiansen, E. B. and Barker, D. H. (1965) "The effect of shape and density on the free settling of particles at high Reynolds number," *AIChE Journal,* Vol. 11, pp. 145–151.

Chung, T. J. (2002) *Computational Fluid Dynamics.* Cambridge: Cambridge University Press.

Clamen, W. H. and Gauvin, A. (1969) "Effect of turbulence on the drag coefficients of spheres in a supercritical flow regime," *AIChE Journal,* Vol. 15, p. 184–189.

Clark, N. N., Gabriele, P., Shuker, S., and Turton, R. (1989) *Powder Technology*, Vol. 59, pp. 69–72.

Clift, R. and Gauvin, W. H. (1970) "The motion of particles in turbulent gas streams," *Proceedings of CHEMECA '70, Butterworth, Melbourne*, Vol. 1, pp. 14–28.

Clift, R., Grace, J. R., and Weber, M. E. (1978) *Bubbles, Drops and Particles*. New York: Academic Press.

Climent, E. and Maxey, M. R. (2003) "Numerical simulation of random suspensions at finite Reynolds numbers," *International Journal of Multiphase Flow*, Vol. 29, pp. 579–601.

Coimbra, C. F. M. and Kobayashi, M. H. (2002) "On the viscous motion of a small particle in a rotating cylinder," *Journal of Fluid Mechanics,* Vol. 469, pp. 257–286.

Connolly, B. J., Loth, E., and Smith, C. F. (2019) "Drag and bounce of irregular particles and test dust," *AIAA AIAA Propulsion and Energy 2019 Forum, Indianapolis, IN*, pp. 2019–4341.

Connolly, B. J., Loth, E., and Smith, C. F. (2020) "Shape and drag of irregular angular particles and test dust," *Powder Technology*, Vol. 363, pp. 275–285.

Coppen, S., Manno, V., and Rogers, C. B. (2001) "Turbulence characteristics along the path of a heavy particle," *Computers and Fluids*, Vol. 30, pp. 257–270.

Corrsin, S. (1963) "Turbulence: experimental methods," in C. Truesdell, ed., *Strömungsmechanik II [Fluid Dynamics II].* Handbuch der Physik [Encyclopedia of Physics]. Berlin, Heidelberg: Springer Berlin Heidelberg, Vol. 3, pp. 524–590.

Cox, R. G. (1969) "The deformation of a drop in a general time-dependent fluid flow," *Journal of Fluid Mechanics,* Vol. 37, pp. 601–623.

Cross, R. (2000) "The coefficient of restitution for collisions of happy balls, unhappy balls, and tennis balls," *American Journal of Physics*, Vol. 68, pp. 1025–1031.

Crow, S. C. (1970) "Stability theory for a pair of trailing vortices," *AIAA Journal*, Vol. 8, pp. 2172–2179.

Crowe, C. T. (2000) "REVIEW – numerical models for dilute gas-particle flows," *Journal of Fluid Engineering,* Vol. 104, pp. 297–303.

Crowe, C. T., Babcock, W. R., and Willoughby, P.G. (1972) "Drag coefficient for particles in rarefied low-Mach number flows," *Progress Heat Mass Transfer*, Vol. 6, pp. 419–431.

Crowe, C.T., Schwarzkopf, J. D., Sommerfeld, M., and Tsuji, Y. (2011) *Multiphase Flows with Droplets and Particles*. Boca Raton: CRC Press.

Crowe, C. T., Sommerfeld, M., and Tsuji, Y. (1998) *Multiphase Flows with Droplets and Particles*. Boca Raton: CRC Press.

Csanady, G. T. (1963) "Turbulent diffusion for heavy particles in the atmosphere," *Journal of Atmospheric Sciences*, Vol. 20, May, pp. 201–208.

Cuenot, B., Magnaudet, J., and Spennato, B. (1997) "Effects of slightly soluble surfactants on the flow around a spherical bubble," *Journal of Fluid Mechanics*, Vol. 339, pp. 25–53.

Cunningham, E. (1910) "On the velocity of steady fall of spherical particles through fluid medium," *Proceedings of the Royal Society of London, Series A: Mathematical and Physical Sciences*, Vol. 83, No. 563, pp. 302–323.

Czys, R. R. and Ochs, H. T. (1998) "The influence of charge on the coalescence of water drops in free fall," *Journal of the Atmospheric Sciences*, Vol. 45, November, pp. 3161–3168.

Dallavalle, J. M. (1948) *Micrometrics*. New York: Pitman Publishing.

Dandy, H. A. and Dwyer, D. S. (1990) "Some influences of particle shape on drag and heat transfer," *Physics of Fluids A*, Vol. 2, No. 12, pp. 2110–2118.

Darton R. C. and Harrison, D. (1974) "The rise of single gas bubbles in liquid fluidized beds," *Transactions of Institution Chemical Engineers*, Vol. 52, pp. 301–306.

Davies, J. M. (1949) "The aerodynamic of golf balls," *Journal of Applied Physics*, Vol. 20, pp. 821–828.

Davies, R. and Taylor, G. I. (1950) "The mechanics of large bubbles rising through extended liquids and though liquids in tubes," *Proceedings of the Royal Society of London*, Ser. A, Vol. 200, pp. 375–390.

Davis, R. H (1991) "Sedimentation of axisymmetric particles in a shear flows," *Physics of Fluids A*, Vol. 3, No. 9, pp. 2051–2060.

Davis, R. H., Serayssol, J. M., and Hinch, E. J. (1986) "The elastohydrodynamic collision of two spheres," *Journal of Fluid Mechanics*, Vol. 163, pp. 479–497.

DeCroix, D. S. (2003) "Visualizing chemical dispersion in populated regions," E-newsletter for Tecplot Users, No. 18.

Dennis, S. C. R., Singh, S. N., and Ingham, D. B. (1980) "The steady flow due to a rotating sphere at low and moderate Reynolds numbers," *Journal of Fluid Mechanics*, Vol. 101, pp. 257–279.

Derksen, J. J. (2020) "Particle-resolved simulations of liquid fluidization of rigid and flexible fibers" *Acta Mechanica*, Vol. 231, pp. 5193–5203.

De Vries, A. W. G., Biesheuvel, A., and van Wijngaarden, L. (2001) "Notes on the path and wake of a gas bubble rising in pure water," *International Journal of Multiphase Flow*, Vol. 28, No. 11, pp. 1823–1835

Di Bernardino, A., Monti, P., Leuzzi, G., et al. (2017) "Water-channel estimation of Eulerian and Lagrangian time scales of the turbulence in idealized two-dimensional urban canopies," *Boundary-Layer Meteorology*, Vol. 165, pp. 251–276.

Di Felice, R. (1994) "The voidage function for fluid-particle interaction systems," *International Journal of Multiphase Flow*, Vol. 20, pp. 153–159.

Dijkhuizen, W., Roghair, I., Van Sint Annaland, M., Kuipers, J. A. M. (2010) "DNS of gas bubbles behaviour using an improved 3D front tracking model – drag force on isolated bubbles and comparison with experiments" *Chemical Engineering Science*, Vol. 65, pp. 1415–1426.

Ding, J. and Gidaspow, D. (1990) "A bubbling fluidization model using kinetic theory of granular flow," *AIChE Journal*, Vol. 36, pp. 523–538.

Dorgan, A. and Loth, E. (2004) "Simulation of particles released near the wall in a turbulent boundary layer," *International Journal of Multiphase Flow*, Vol. 30, pp. 649–673.

Dorgan, A. and Loth, E. (2007) "Efficient calculation of the history force at finite Reynolds numbers," *International Journal of Multiphase Flow*, Vol. 33, August, pp. 833–848.

Dorgan, A. and Loth, E. (2009) "Dispersion of bubbles in a turbulent boundary layer," ASME Fluids Engineering Division Summer Conference (submitted).

Dorgan, A. J., Loth, E., Bocksell, T. L., and Yeung, P. K. (2005) "Boundary layer dispersion of near-wall injected particles of various inertias," *AIAA Journal*, Vol. 43, pp. 1537–1548

Drazin, P. G. and Reid, W. H. (1981) *Hydrodynamic Stability*. Cambridge: Cambridge University Press.

Dressel, M. (1985) "Dynamics shape factors for particle shape characterization," *Particle and Particle Systems Characterization,* Vol. 2, pp. 62–66.

Drew, D. A. (1983) "Mathematical modeling of two-phase flow," *Annual Review of Fluid Mechanics*, Vol. 15, pp. 261–291.

Drew, D. A. and Passman, S. L. (1998) "Theory of multi-component fluids," *Applied Mathematical Sciences*, Vol. 135. New York: Springer.

Drew, D. A. and Passman, S. L. (1999) *Theory of Multicomponent Fluids.* New York: Springer.

Druzhinin, O. A. and Elghobashi, S. (1998) "Direct numerical simulations of bubble-laden turbulent flows using the two-fluid formulation," *Physics of Fluids*, Vol. 10, No. 3, pp. 685–697.

Druzhinin, O. A. and Elghobashi, S. (1999) "On the decay rate of isotropic turbulence laden with micro-particles," *Physics of Fluids*, Vol. 11, No. 3, pp. 602–610.

Duineveld, P. C. (1995) "Rise velocity and shape of bubbles in pure water at high Reynolds number," *Journal of Fluid Mechanics*, Vol. 292, pp. 325–332.

Eaton, J. K (1999) "Local distortion of turbulence by dispersed particles," AIAA Fluid Dynamics Conference, Norfolk, VA, AIAA-99-3643, June.

Eaton, J. K. and Fessler, J. R. (1994) "Preferential concentration of particles by turbulence," *International Journal of Multiphase Flow*, Vol. 20, pp. 169–209.

Edge, R. M. and Grant, C. D. (1972) "Motion of drops in water contaminated with a surface-active agent," *Chemical Engineering Science*, Vol. 27, No. 9, pp. 1709–1721.

Elamworawutthikul, C. and Gould, R. D. (1999) "The flow structure around a suspended sphere at low Reynolds number in a turbulent freestream," 3rd ASME/JSME Joint Fluids Engineering Conference, July, FEDSM99–6996.

Elghobashi, S. (1994) "On predicting particle-laden turbulent flows," *Applied Scientific Research*, Vol. 52, pp. 309–329.

Elghobashi, S. and Abou-Arab, T. W. (1983) "A two-equation turbulence model for two-phase flows," *Physics of Fluids*, Vol. 26, pp. 931–938.

Elghobashi, S. and Truesdell, G. C. (1984) "Direct simulation of particle dispersion in a decaying isotropic turbulence," *Journal of Fluid Mechanics*, Vol. 242, pp. 655–700.

Ellingsen, K. and Risso, F. (2001) "On the rise of an ellipsoidal bubble in water: oscillatory paths and liquid-induced velocity," *Journal of Fluid Mechanics*, Vol. 440, pp. 235–268.

Endres, S. C., Ciacchi, L. C., and Madler, L. (2021) "A review of contact force models between nanoparticles in agglomerates, aggregates, and films," *Journal of Aerosol Science*, Vol. 153, 105719.

Epstein, P. S. (1929) "Zur theorie des radiometers," *Zeitschrift für Physik,* Vol. 54, pp. 537–563.

Esteban, L. B., Shrimpton, J., and Ganapathisubraman, B. (2019) "Study of the circularity effect on drag of disk-like particles," *International Journal of Multiphase Flow*, Vol. 110, January, pp. 189–197.

Estrade, J.-P., Carentz, H., Lavergne, G., and Biscos, Y. (1999) "Experimental investigation of dynamic binary collision of ethanol droplets – a model for droplet coalescence and bouncing," *International Journal of Heat and Fluid Flow,* Vol. 20, pp. 486–491.

Faeth, G. M., (1987) "Mixing, transport and combustion in sprays," *Progress in Energy and Combustion Science*, Vol. 13, pp. 293–345.

Fan, L.-S. and Tsuchiya, K. (1990), *Bubble Wake Dynamics in Liquids and Liquid-Solid Suspension.* Boston: Butterworth-Heinemann.

Fan, L.-S. and Tsuchiya, K. (1993) "Bubble flow in liquid–solid suspension," *Particulate Two-Phase Flow*, edited by M. C. Roco. Butterworth-Heinemann, Boston, chapter 23.

Faxen, H. (1922) "Resistance to the movement of a rigid sphere in a viscous fluid bounded by two parallel flat walls," *Annals of Physics,* Vol. 68, pp. 89–119.

Feldman, D. and Avila, M. (2018) "Overdamped large-eddy simulations of turbulent pipe flow up to Reτ=1500," *Journal of Physics: Conference Series*, Vol. 1001, 012016.

Felton, K. and Loth, E. (2001) "Spherical bubble motion in a turbulent boundary layer," *Physics of Fluids*, Vol. 13, No. 9, 2564–2577.

Feng, Z. C. and Leal, L. G. (1997) "Nonlinear bubble dynamics," *Annual Review of Fluid Mechanics,* Vol. 29, December, pp. 201–243.

Feng, Z.-G. and Michealidis, E. E. (2001) "Drag coefficients of viscous spheres at intermediate and high Reynolds numbers," *Journal of Fluids Engineering,* Vol. 123, December, pp. 841–849.

Ferry, J. and Balachandar, S. (2001) "A fast Eulerian method for disperse two-phase flow," *International Journal of Multiphase Flow,* Vol. 27, pp. 1199–1226.

Ferry, J. and Balachandar, S. (2002) "Equilibrium expansion for the Eulerian velocity of small particles," *Powder Technology*, Vol. 125, No. 2–3, pp. 131–139

Fessler, J. and Eaton, J. (1999) "Turbulence modification by particles in a backward-facing step flow," *Journal of Fluid Mechanics*, Vol. 394, pp. 97–117.

Fessler, J. R., Kulick, J. D., and Eaton, J. K. (1994) "Preferential concentration of heavy particles in a turbulent channel flow," *Physics of Fluids*, Vol. 6, pp. 3742–3749.

Field, S. B., Klaus, M., Moore, M. G. and Nori, F. (1997) "Chaotic dynamics of falling disks," *Nature,* Vol. 388, pp. 252–254.

Foerster, S. M., Louge, M. Y., Chang, H., and Khedidja, A. (1994) "Measurement of the collision properties of small spheres," *Physics of Fluids*, Vol. 6, pp. 1108–1115.

Fonseca, F. and Herrmann, H. J. (2004) "Sedimentation of oblate ellipsoids at low and moderate Reynolds numbers" *Physica A: Statistical Mechanics and Its Applications*, Vol. 342, pp. 447–461.

Ford, B. and Loth, E. (1998) "Forces on ellipsoidal bubbles in a turbulent free shear layer," *Physics of Fluids*, Vol. 10, No. 1, pp. 178–188.

Fornari, W., Ardekani, M. N., and Brandt, L. (2018) "Clustering and increased settling speed of oblate particles at finite Reynolds number," *Journal of Fluid Mechanics*, Vol. 848, pp. 696–721.

Forsberg, F., Goldberg, B. B., Liu, J.-B., Merton, D. A., Rawool, N. M. and Shi, W. T. (1999) "Tissue-specific US contrast agent for evaluation of hepatic and splenic parenchyma," *Radiology*, Vol. 210, pp. 125–132.

Fortes, A. F., Joseph, D. D., and Lundgren, T. S. (1987) "Nonlinear mechanics of fluidization of beds of spherical particles," *Journal of Fluid Mechanics*, Vol. 177, pp. 467–483.

Friedman, P. D. and Katz, J. (2002) "Mean rise rate of droplets in isotropic turbulence," *Physics of Fluids*, Vol. 14, pp. 3059–3073.

Gad-el-Hak, M. (1995) "Stokes' hypothesis for a Newtonian, isotropic fluid," *Journal of Fluids Engineering*, Vol. 117, No. 1, pp. 3–5.

Galdi, G., Vaidya, A., Pokorny, M., Joseph, D. D., and Feng, J. (2002) "Orientation of symmetric bodies falling in a second-order liquid at non-zero Reynolds number," *Mathematical Models and Methods in Applied Sciences*, Vol. 12, No. 11, pp. 1653–1690.

Ganser, G. H. (1993) "A rational approach to drag prediction of spherical and non-spherical particles," *Powder Technology*, Vol. 77, pp. 143–152.

Garg, K. and Nayar, S. K. (2004) "Photometric model of a raindrop," in *Proceedings of IEEE Conference on Computer Vision and Pattern Recognition*, Washington, June. 1063-6919/04.

Garnier, C., Lance, M., and Marie, J.L. (2002) "Measurement of local flow characteristics in buoyancy-driven bubbly flow at high void fraction," *Experimental Thermal and Fluid Science,* Vol. 26, pp. 811–815.

Garner, F. H. and Hammerton, D. (1954) "Circulation inside gas bubbles," *Chemical Engineering Science,* Vol. 3, No. 1, pp. 1–11.

Gauthier, D., Zerguerra, S., and Flamant, G. (1999) "Influence of the particle size distribution of powders on the velocities of minimum and complete fluidization," *Chemical Engineering Journal*, Vol. 74, pp. 181–196.

Gauthier, G., Gondret, P. and Rabaud, M. (1998) "Motions of anisotropic particles: application to visualization of three-dimensional flows," *Physics of Fluids*, Vol. 10, No. 9, pp. 2147–2154.

Gavze, E. and Shapiro, M. (1997) "Particles in a shear flow near a solid wall: effect of nonsphericity on forces and velocities," *International Journal of Multiphase Flow*, Vol. 23, No. 1, pp. 155–182.

Gibou, F. Hyde, D., and Fedkiw, R. (2019) "Sharp interface approaches and deep learning techniques for multiphase flows," *Journal of Computational Physics*, Vol. 380, pp. 442–463.

Gidaspow, D. (1994) *Multiphase Flow and Fluidization.* Boston: Academic Press.

Gloerfelt, X. and Cinnella, P. (2015) "Investigation of the flow dynamics in a channel constricted by periodic hills," 45th AIAA Fluid Dynamics Conference, Dallas hal-02156763.

Gogus, M., Ipecki, O. N., and Kokpinar, M. A. (2001) "Effect of particle shape on fall velocity of angular particles," *Journal of Hydraulic Engineering,* Vol. 127, No. 10, pp. 860–869.

Goldman, A. J., Cox, R. G., and Brenner, H. (1967) "Slow viscous motion of a sphere parallel to a plane wall," *Chemical Engineering Science*. Vol. 22, pp. 637–651.

Gondret, P., Lance, M., and Petit, L. (2002) "Bouncing motion of spherical particles in fluids," *Physics of Fluids*, Vol. 14, No. 2, pp. 643–652.

Good, G. H., Ireland, P. J., Bewley, G. P., Bodenschatz, E., Collins, L. R., and Warhaft, Z. (2014) "Settling regimes of inertial particles in isotropic turbulence," *Journal of Fluid Mechanics*, Vol. 759, R31–R3-12.

Gopalan, B., Malkiel, E., and Katz, J. (2008) "Experimental investigation of turbulent diffusion of slightly buoyant droplets in locally isotropic turbulence," *Physics of Fluids*, Vol. 20, 095102.

Gotaas, C., Havelka, P., Jakobsen, H.-A., Svendsen, H. F., Hase, M., Roth, N., and Weigand, B. (2007) "Effect of viscosity on droplet–droplet collision outcome: Experimental study and numerical comparison," *Physics of Fluids*, Vol. 19, 102106.

Grace, H. P. (1982) "Dispersion phenomena in high viscosity immiscible fluid systems and application of static mixers as dispersion devices in such systems," *Chemical Engineering Communications*, Vol. 14, pp. 225–277.

Grace, J. R. (1973) "Shapes and velocities of bubbles rising in infinite liquids," *Transactions of the Institute of Chemical Engineers*, Vol. 51, pp. 116–120.

Grace, J. R., Wairegi, T., and Nguyen, T. H. (1976) "Shapes and velocities of single drops and bubbles moving freely through immiscible liquids," *Transactions of Institute of Chemical Engineers*, Vol. 54, pp. 167–173.

Griffith, R. M. (1962) "The effect of surfactants on terminal velocity of drops and bubbles," *Chemical Engineering Science*, Vol. 17, pp. 1057–1070.

Groszmann, D. E., Fallon, T. M., and Rogers, C. B. (1999) "Decoupling the roles of inertia and gravity on the preferential concentration of particles," in *3rd ASME/JSME Joint Fluids Engineering Conference, FEDSM-99*, San Francisco. New York: ASME, pp. 83–87.

Güttler, C., Heißelmann, D., Blum, J., and Krijt, S. (2013) "Normal collisions of spheres: a literature survey on available experiments," arXiv preprint arXiv:1204.0001.

Haberman, W. L. and Morton, R. K. (1953) "Experimental investigation of drag and shape of air bubbles," David W. Taylor Model Basin Report, Report 802, NS 715-102, Published by Department of Navy, D.C.

Haberman, W. L. and Morton, R. K. (1954) "An experimental study of bubbles moving in liquids," *Transactions of the American Society of Civil Engineers*, Vol. 121, pp. 227–250.

Hagemans, F. (2020) "Colloidal synthesis, Transmission electron microscopy (TEM), NMR spectroscopy." https://colloid.nl/people/fabian-hagemans/.

Haider, A. and Levenspiel, O. (1989) "Drag coefficient and terminal velocity of spherical and non-spherical particles," *Powder Technology,* Vol. 58, pp. 63–70.

Hanna, S. R. (1981) "Lagrangian and Eulerian time-scale relations in the daytime boundary layer," *Journal of Applied Meteorology and Climatology*, Vol. 20, pp. 242–249.

Happel, J. and Brenner, H. (1973) *Low Reynolds Number Hydrodynamics*. Groningen: Noordhoff.

Hartunian, R. A. and Sears, W. R. (1957) "On instability of small gas bubbles moving uniformly in various liquids," *Journal of Fluid Mechanics*, Vol. 3, Part 1, pp. 27–47.

Head, M. R. (1982) *Flow Visualization II*. Washington: Hemisphere, pp. 399–403.

Heron, I., Davis, S., and Bretherton, F. (1975) "On the sedimentation of a sphere in a centrifuge," *Journal of Fluid Mechanics*, Vol. 68, pp. 209–234.

Herzhaft, B. and Guazzelli, E. (1999) "Experimental study of the sedimentation of dilute and semi-dilute suspensions of fibres," *Journal of Fluid Mechanics*, Vol. 384, pp. 133–158.

Hinch, E. J. and Acrivos, A. (1979) "Steady long slender droplets in two-dimensional straining motion," *Journal of Fluid Mechanics* Vol. 91 (3), pp. 404–414.

Hinch, E. J. and Acrivos, A. (1980) "Long slender drops in a simple shear flow," *Journal of Fluid Mechanics,* Vol. 98 (2), pp. 305–328.

Hinze, J. O. (1975), *Turbulence.* New York: McGraw-Hill.

Hirche, D., Birkholz, F., and Hinrichsen, O. (2019) "A hybrid Eulerian–Eulerian–Lagrangian model for gas–solid simulations," *Chemical Engineering Journal,* Vol. 377 (1), 119743.

Hirschfelder, J. O. and Bird, C. F. (1954) *Molecular Theory of Gases and Liquids*. New York: Wiley.

Hoerner, S. F. (1965) *Fluid-Dynamic Drag*. Self-published: Midland Park.

Homann, H., Bec, J., and Grauer, R. (2013) "Effect of turbulent fluctuations on the drag and lift forces on a towed sphere and its boundary layer," *Journal of Fluid Mechanics*, Vol. 721, pp. 155–179.

Hoque, M. M., Mitra, S., Sathe, M. J., Jose, J. B., and Evans, G. M. (2016) "Experimental investigation on modulation of homogeneous and isotropic turbulence in the presence of single particle using time-resolved PIV," *Chemical Engineering Science*, Vol. 153, pp. 308–329.

Hosokawa, S. and Tomiyama, A. (2004) "Turbulence modification in gas–liquid and solid–liquid dispersed two-phase pipe flows," *International Journal of Heat and Fluid Flow,* Vol. 25 pp. 489–498.

Hosokawa, S., and Tomiyama, A. (2013) "Bubble-induced pseudo turbulence in laminar pipe flows," *International Journal of Heat and Fluid Flow*, Vol. 40, pp. 97–105.

Houghton, H. G. (1950) "Spray nozzles," in J. H. Perry, ed., *Chemical Engineer's Handbook,* 3rd edition. New York: McGraw-Hill, p. 1170.

Hu, S. and Kintner, R. C. (1955) "The fall of single drops through water," *AIChE Journal*, Vol. 1, pp. 42–48.

Hu, Y. T. and Lips, A. (2003) "Transient and steady state three-dimensional drop shapes and dimensions under planar extensional flow," *Journal of Rheology*, Vol. 47, pp. 349–369.

Hwang, W. and Eaton, J. K. (2006) "Homogeneous and isotropic turbulence modulation by small heavy (St $\sim$ 50) particles," *Journal of Fluid Mechanics,* Vol. 564, pp. 361–393.

Ishii, M. (1975) *Thermo-Fluid Dynamic Theory of Two-Phase Flow*. Paris: Eyrolles.

Jayaweera, K. O. L. F. and Mason, B. J. (1965) "The behavior of freely falling cylinders and cones in viscous fluid," *Journal of Fluid Mechanics*. Vol. 22, pp. 709–720.

Jeffrey, G. B. (1922) "The motion of ellipsoidal particles immersed in a viscous fluid," *Proceedings of the Royal Society Service A*, Vol. 102, pp. 161–179.

Jeffrey, R. C. and Pearson, J. R. A. (1965) "Particle motion in a laminar vertical tube flow," *Journal of Fluid Mechanics*, Vol. 22, pp. 721–735.

Jenkins, J. T. and Savage, S. B. (1983) "A theory for the rapid flow of identical, smooth, nearly elastic particles," *Journal of Fluid Mechanics*, Vol. 130, pp. 186–202.

Jiang, Y. J., Umemura, A., and Law, C. K. (1992) "An experimental investigation on the collision behaviour of hydrocarbon droplets," *Journal of Fluid Mechanics,* Vol. 234, pp. 171–190.

Jimenez, J. A. and Madsen, O. S. (2003) "A simple formula to estimate settling velocity of natural sediments," *Journal of Waterway, Port, Coastal and Ocean Engineering*, Vol. 12, pp. 70–78.

Johansen, S. T., Anderson, N. M., and De Silva, S. R. (1990) "A two-phase model for particle local equilibrium applied to air classification of powders," *Powder Technology*, Vol. 63, pp. 121–132.

Johnson, K. L. (1985) *Contact Mechanics*. Cambridge: Cambridge University Press.

Johnson, P. L. (2020) "Predicting the impact of particle–particle collisions on turbophoresis with a reduced number of computational particles," *International Journal of Multiphase Flow*, Vol. 124, 103182.

Johnson, P. L., Bassenne, M., and Moin, P. (2020) "Turbophoresis of small inertial particles: theoretical considerations and application to wall-modelled large-eddy simulations," *Journal of Fluid Mechanics*, Vol. 883, p. A27.

Joseph, D. D. (2006) "Rise velocity of a spherical cap bubble," *Journal of Fluid Mechanics*, Vol. 488, pp. 213–223.

Joseph, D. D., Belanger, J., and Beavers, B. S. (1999) "Break-up of a liquid drop suddenly exposed to high-speed airstream," *International Journal of Multiphase Flow*, Vol. 25, pp. 1263–1303.

Joseph, G. G. and Hunt, M. L. (2004) "Oblique particle–wall collisions in a liquid," *Journal of Fluid Mechanics*, Vol. 510, pp. 71–93.

Joseph, G. G., Zenit, R., Hunt, M. L., and Rosenwinkel, A. M. (2001) "Particle–wall collisions in a viscous fluid," *Journal of Fluid Mechanics,* Vol. 433, pp. 329–346.

Josserand, C. and Thoroddsen, S. (2016) "Drop impact on a solid surface," *Annual Review of Fluid Mechanics*, 48, pp. 365–391.

JTJ (2013) "Advanced materials for plasma spray coating." Retrieved from www.amjtj.com/en/projekty/materialy-pro-plazmove-nastriky.

Kaftori, D., Hetsroni, G. and Banerjee, S. (1995), "Particle behavior in the turbulent boundary layer, Part I. motion, deposition and entrainment," *Physics of Fluids A*, Vol. 7, pp. 1095–1106.

Kameda, M. and Matsumoto, Y. (1996) "Shock waves in a liquid containing small gas bubbles," *Physics of Fluids*, Vol. 8, pp. 322–335.

Kane, R. S. and Pfeffer, R. (1973) "Heat transfer coefficients of dilute flowing gas-solids suspensions," NASA-CR-2266.

Kariyasaki, A. (1987) "Behavior of a single gas bubble in a liquid flow with a linear velocity profile," in *Proceedings of the 1987 ASME-JSME Thermal Engineering Joint Conference, ASME, New York.* Vol. 5, pp. 261–267.

Kawanisi, K. and Shiozaki, R. (2008) "Turbulent effects on the settling velocity of suspended sediment," *Journal of Hydraulic Engineering*, Vol. 134, pp. 261–266.

Keller, J. B. and Kolodner, I. I. (1956) "Damping of underwater explosion bubble oscillations," *Journal of Applied Physics*, Vol. 27, pp. 1152–1161.

Kendoush, A. A. (2003) "The virtual mass of a spherical-cap bubble," *Physics of Fluids*, Vol. 15, No. 9, pp. 2782–2785.

Khismatullin, D. B., Renardy, Y., and Cristini, V. (2003) "Inertia-induced breakup of highly viscous drops subjected to simple shear," *Physics of Fluids*, Vol. 15, No. 5, pp. 1351–1354.

Kim, I., Elghobashi, S. E., and Sirignano, W. (1993) "Three-dimensional flow over two spheres placed side by side," *Journal of Fluid Mechanics*, Vol. 246, pp. 465–488.

Kim, I., Elghobashi. S., and Sirignano, W. B. (1998) "On the equation for spherical-particle motion: effect of Reynolds and acceleration numbers," *Journal of Fluid Mechanics,* Vol. 367, pp. 221–253.

Kim, J., Choi, H., Park, H., and Yoo, J. Y. (2014) "Inverse Magnus effect on a rotating sphere: when and why," *Journal of Fluid Mechanics,* Vol. 754, pp. R2-1–R2-11.

King, C. J., Hsueh, L., and Mao, K. W. (1965) "Liquid phase diffusion of nonelectrolytes at high dilution," *Journal of Chemical and Engineering Data*, Vol. 10, No. 4, pp. 348–350.

Klaseboer, E., Chevaillier, J.-P., Mate, A., Masbernato, O., and Gourdon, C. (2001) "Model and experiments of a drop impinging on an immersed wall," *Physics of Fluids*, Vol. 13, No. 1, pp. 45–57.

Kleinstreuer, C. (2003) *Two-Phase Flow Theory and Applications*. New York: Routledge.

Koebe, M., Bothe, D., and Warnacke, H.-J. (2003) "Direct numerical simulation of air bubbles in water glycerol mixtures: shapes and velocity fields," *Proceedings of the 4th ASME–JSME Joint Fluids Engineering Conference,* FEDSM2003–45154, pp. 415–421.

Kojima, E., Akehata, T., and Shirai, T. (1968) "Rising velocity and shape of single air bubbles in highly viscous liquids," *Journal of Chemical Engineering of Japan*, Vol. 1, pp. 45–50.

Kok, J. B. W. (1993) "Dynamics of a pair of gas bubbles moving through liquid. Part I. Theory," *European Journal of Mechanics B: Fluids,* Vol. 12, pp. 515–540.

Kosinki, P. and Hoffman, A. C. (2010) "An extension of the hard-sphere particle–particle collision model to study agglomeration," *Chemical Engineering Science*, Vol. 65, pp. 3231–3239.

Kramer, O., de Moel, P., Baars, E., and van der Hoek, J. P. (2019) "Improvement of the Richardson–Zaki liquid–solid fluidisation model on the basis of hydraulics," *Powder Technology,* Vol. 343, February, pp. 465–478.

Krishnan, G. and Loth, E. (2015) "Effects of gas and droplet characteristics on drop–drop collision outcome regimes," *International Journal of Multiphase Flow*, Vol. 77, pp. 171–186.

Krishnan, G.H., Fletcher, K., and Loth, E. (2023) "Influence of Drop Viscosity and Surface Wettability on Impact Outcomes" Coatings Vol. 13, No. 5: 817.

Krogstadt, P. and Antonia, R. (1999) "Surface roughness effects in turbulent boundary layers," *Experiments in Fluids*, Vol. 27, pp. 450–460.

Kruis, F. E. and Kusters, K. A. (1997), "The collision rate of particle in turbulent flow," *Chemical Engineering Communications,* Vol. 158, pp. 201–230.

Kubota, M., Akehata, T., and Shirai, T. (1967) "The behavior of single air bubbles in liquids of small viscosity," *Kagaku Kogaku*, Vol. 31, pp. 1074–1080.

Kulick, J. D., Fessler, J. R., and Eaton, J. K. (1994) "Particle response and turbulence modification in fully developed channel flow," *Journal of Fluid Mechanics,* Vol. 277, pp. 109–134.

Kuo, K. K. (1986) *Principles of Combustion.* New York: John Wiley & Sons.

Kurose, R. and Komori, S. (1999) "Drag and lift forces on a rotating sphere on a linear shear flow," *Journal of Fluid Mechanics,* Vol. 384, pp. 183–206.

Kuschel, M. and Sommerfeld, M. (2013) "Investigation of droplet collisions for solutions with different solids content," *Experiments in Fluids,* Vol. 54, p. 1440–1447.

Kussin, J. and Sommerfeld, M. (2001) "Investigation of particle behavior and turbulence modification in particle laden channel flow," International Congress for Particle Technology, Session 12-046 Nuremberg, Germany, March 27–29.

Labous, L., Rosato, A. D., and Dave, R. N. (1997) "Measurements of collisional properties of spheres using high-speed video analysis," *Physical Review E*, Vol. 56, pp. 5717–5725.

Lackme, C. (1973) "Two regimes of a spray column in countercurrent flow," *National Heat Transfer Conference, Atlanta, Georgia, USA*, AIChE Symposium Series. CONF-730803-6, Vol. 70, p. 59.

Lai, R. Y. S. and Mockros, L. F. (1972) "The Stokes flow drag on prolate and oblate spheroids during axial translatory accelerations," *Journal of Fluid Mechanics,* Vol. 52, pp. 1–15.

Lamb, H. (1945) *Hydrodynamics.* New York: Dover.

Lance, M. and Bataille, J. (1991) "Turbulence in the liquid phase of a uniform bubbly air-water flow," *Journal of Fluid Mechanics*, Vol. 222, pp. 95–118.

Langmuir, I. and Blodgett, K. B. (1946) "A mathematical investigation of water droplet trajectories," *Army Air Force Tech. Report 5418*, Contract W-33-038-ac-9151.

Lasso, I. A. and Weidman, P. D. (1986) "Stokes drag on hollow cylinders and conglomerates," *Physics of Fluids*, Vol. 29, pp. 3921–3934.

Laufer, J. (1954) "The structure of turbulence in fully developed pipe flow," NASA Report 1174.

Launder, B. and Spalding, D. (1972) *Mathematical Models of Turbulence.* Waltham: Academic Press.

Lawrence, C. J. and Mei, R. (1995) "Long time behavior of the drag on a body in impulsive motion," *Journal of Fluid Mechanics*, Vol. 283, pp. 307–327.

Leal, L. G. (1980) "Particle motions in a viscous fluid," *Annual Review of Fluid Mechanics*, Vol. 12, pp. 435–476.

Leal, L. G. (2007) *Advanced Transport Phenomena: Fluid Mechanics and Convective Transport Processes*. Cambridge: Cambridge University Press.

Le Claire, B. P., Hamielec, A. E., and Pruppacher, H. R. (1970) "A numerical study of the drag on a sphere at low and intermediate Reynolds numbers," *Journal of the Atmospheric Sciences*, Vol. 27, pp. 308–315.

Lee, S. H., Heng, J., Zhou, D., et al. (2011) "Nano spray drying: A novel method for preparing protein nanoparticles for protein therapy," *International Journal of Pharmaceutics*, Vol. 403, pp. 192–200.

Lee, W. and Lee, J. (2020) "Experiment and modeling of lift force acting on single high Reynolds number bubbles rising in linear shear flow," *Experimental Thermal and Fluid Science*, Vol. 115, 110085.

Lefebvre, A. H., Wang, X. F., and Martin, C. A. (1988) "Spray characteristics of aerated-liquid pressure atomizers," *Journal of Propulsion and Power*, Vol. 4, No. 4, pp. 293–331.

Legendre, D., Daniel, C., and Guiraud, P. (2005) "Experimental study of a drop bouncing on a wall in a liquid," *Physics of Fluids*, Vol. 17, 097105.

Legendre, D. and Magnaudet, J. (1997) "A note on the lift force on a bubble or drop in a low Reynolds number shear flow," *Physics of Fluids*, Vol. 9, pp. 3572–3574.

Legendre, D. and Magnaudet, J. (1998) "Lift force on a bubble in a viscous linear shear flow," *Journal of Fluid Mechanics*, Vol. 368, pp. 81–126.

Legg, B. J. and Raupach, M. R. (1982) "Markov chain simulation of particle dispersion in inhomogeneous flow," *Boundary Layer Meteorology*, Vol. 24, pp. 3–13.

Leighton, T. G. (2004) "From sea to surgeries, from babbling brooks to baby scans: bubble acoustics at ISVR," *Proceedings of the Institute of Acoustics*, Vol. 26, Part 1, pp. 357–381.

Leith, D. (1987) "Drag on non-spherical objects," *Aerosol Science and Technology,* Vol. 6, pp. 153–161.

L'Esperance, D., Trolinger, J. D., Coimbra, C. F., and Rangel, R. H. (2006) "Particle response to low-Reynolds-number oscillation of a fluid in micro-gravity," *AIAA Journal,* Vol. 44, pp. 1060–1064.

Lessen, M., Sadler, G. S., and Liu, T. (1968) "Stability of pipe Poiseuille flow," *Physics of Fluids,* Vol. 11, pp. 1404–1409.

Letan, R. and Kehat, E. (1967) "The mechanics of a spray column," *AlChE Journal,* Vol. 13, pp. 443–449.

Levich, V. G. (1949) "Motion of a bubble at large Reynolds numbers," *Zhur Eksptl i Teoret Fiz,* Vol. 19, pp. 18–24.

Levich, V. G. (1962) *Physico Chemical Hydrodynamic.* Wilmington: Prentice Hall.

Li, X., Dong, M., Jiang, D., Li, S., and Shang, Y. (2020) "The effect of surface roughness on normal restitution coefficient, adhesion force and friction coefficient of the particle-wall collision" *Powder Technology,* Vol. 362, pp. 17–25.

Libbrecht, K. G. (2005) "The physics of snow crystals," *Reports on Progress in Physics,* Vol. 68, pp. 855–895.

Lide, D., ed. (2005) *CRC Handbook of Chemistry and Physics.* Boca Raton: CRC Press.

Liepmann, H. W. and Roshko, A. (1957) *Elements of Gasdynamics.* New York: John Wiley & Sons.

Lim, E. A., Coimbra, C. F. M., and Kobayashi, M. H. (2005) "Dynamics of suspended particles in eccentrically rotating flows," *Journal of Fluid Mechanics,* Vol. 535, pp. 101–110.

Lin, B., Yu, J., and Rics, S. A. (2000) "Direct measurements of constrained Brownian motion of an isolated sphere between two walls," *Physical Review E,* Vol. 62, pp. 3909–3918.

Lin, C. J., Perry, J. H., and Scholwater, W. R. (1970) "Simple shear flow round a rigid sphere: inertial effects and suspension rheology," *Journal of Fluid Mechanics,* Vol. 44, pp. 1–17.

List, R., Retsch, U. W., Byram, A. C., and Lozowski, E. P. (1973) "On the aerodynamics of spheroidal hailstone models," *Journal of the Atmospheric Sciences,* Vol. 30, No. 4, pp. 653–661,

List, R. and Schemenauer, R. S. (1971) "Free-fall behavior of planar snow crystals, conical graupel and small hail," *Journal of Atmospheric Sciences,* Vol. 28, pp. 110–115.

Loewenberg, M. (1993) "Stokes resistance, added mass, and Basset force for arbitrarily oriented finite-length cylinder," *Physics of Fluids A,* Vol. 5, pp. 765–767.

Loisy, A., Naso, A., and Spelt, P. D. M. (2017) "Buoyancy-driven bubbly flows: ordered and free rise at small and intermediate volume fraction," *Journal of Fluid Mechanics,* Vol. 816, pp. 94–141.

Loth, E. (2000) "Numerical approaches for motion of dispersed particles, bubbles, and droplets," *Progress in Energy and Combustion Sciences,* Vol. 26, pp. 161–223.

Loth, E. (2008a) "Review: quasi-steady shape and drag of deformable bubbles and drops," *International Journal of Multiphase Flow,* Vol. 34, pp. 523–546.

Loth, E. (2008b) "Review: lift of a spherical particle subject to vorticity and/or spin," *AIAA Journal,* Vol. 46, April, pp. 801–809.

Loth, E. (2008c) "Drag of non-spherical solid particles of regular and irregular shape," *Powder Technology,* Vol. 182, March, pp. 342–353.

Loth (2023) "Richardson-Zaki Exponents for Particles, Drops, and Bubbles" ASME IMECE, New Orleans, October, IMECE2023–109881.

Loth, E., Boris, J., and Emery, M. (1998) "Very large bubble cavitation in a temporally-evolving free shear layer," ASME Summer Fluids Engineering Meeting, Washington, June.

Loth, E., Daspit, J. T., Jeong, M., Nagata, T., and Nonomura, T. (2021) "Supersonic and hypersonic drag coefficients for a sphere" *AIAA Journal*, Vol. 59, pp. 3261–3274.

Loth, E. and Dorgan, A. J. (2009) "An equation of motion for particles of finite size and Reynolds number in a liquid," *Environmental Fluid Mechanics*, Vol. 9, pp. 187–214.

Loth, E., O'Brien, T. J., Syamlal, M., and Cantero, M. (2004) "Effective diameter for group motion of polydisperse particle mixtures," *Powder Technology*, Vol. 142, No. 2–3, pp. 209–218.

Loth, E. and Stedl, J. (1999) "Taylor and Lagrangian correlations in a turbulent free shear layer," *Experiments in Fluids*, Vol. 26, pp. 1–6.

Loth, E., Taebi-Rahni, M., and Tryggvason, G. (1997) "Deformable bubbles in a free shear layer," *International Journal of Multiphase Flow*, Vol. 23, No. 56, pp. 977–1001.

Lovalenti, P. M. and Brady, J. F. (1995) "The temporal behavior of the hydrodynamics force on a body in response to an abrupt change in velocity at small but finite Reynolds number," *Journal of Fluid Mechanics*, Vol. 293, pp. 35–46.

Loyalka, S. K. (1992) "Thermophoretic force on a single particle I. Numerical solution of the linearized Boltzmann equation," *Journal of Aerosol Science,* Vol. 23, pp. 291–300.

Lun, C. K. K., Savage, S. B., Jeffrey, D. J., and Chepurmly, N. (1984) "Kinetic theories for granular flow: inelastic particles in Couette flow and slightly inelastic particles in a general flowfield," *Journal of Fluid Mechanics*, Vol. 140, pp. 223–256.

Maccoll, J. H. (1928) "Aerodynamics of a spinning sphere," *Journal of the Royal Aeronautical Society*, Vol. 32, pp. 777–798.

Macrossan, N. (2004) "Scaling parameters in rarefied flow and the breakdown of the Navier-Stokes equations," Technical Report No: 2004/09, Department of Mechanical Engineering, University of Queensland, St Lucia 4072, Australia.

Madhav, G. V., and Chhabra, R. P. (1994) "Settling velocities of non-spherical particles in non-Newtonian polymer solutions," *Powder Technology*, Vol. 78, No. 1, pp. 77–83.

Maduta, R. and Jakirlic, S. (2012) "An eddy-resolving Reynolds stress transport model for unsteady flow computations," in S. Fu, W. Haasem, S-H. Peng, and D. Schwamborn, eds., *Progress in Hybrid RANS-LES Modelling.* Berlin, Heidelberg: Springer, pp. 77–89.

Maeda, M., Hishida, K., and Furutani, T. (1980) "Optical measurements of local gas and particle velocity in an upward flowing dilute gas–solids suspension," *Polyphase Flow Transport Technology, Century 2– Emerging Technology Conference, San Francisco*, pp. 211–216.

Magnaudet, J. (2003) "Small inertial effects on a spherical bubble, drop or particle moving near a wall in a time-dependent linear flow," *Journal of Fluid Mechanics*, Vol. 485, pp. 115–142.

Magnaudet, J. and Eames, I. (2000) "The motion of high-Reynolds-number bubbles in inhomogeneous flows," *Annual Review of Fluid Mechanics*, Vol. 32, pp. 659–708.

Mainardi, F. and Pironi, P. (1996) "The fractional Langevin equation Brownian motion revisited," *Extracta Mathematicae*, Vol. 11, pp. 140–154.

Mando, M. (2009) "Turbulence modulation by non-spherical particles," Ph.D. dissertation in mechanical engineering, Aalborg University.

Marchisio, D. and Fox, R. (2013) *Computational Models for Polydisperse Particulate and Multiphase Systems,* Cambridge Series in Chemical Engineering. Cambridge: Cambridge University Press.

Marie, J. L., Moursali, E., and Tran-Cong, S. (1997) "Similarity law and turbulence intensity profiles in a bubbly boundary layer at low void fractions," *International Journal of Multiphase Flow*, Vol. 23, No. 2, pp. 227–247.

Marshall, J. S. (2018) "Modeling and sensitivity analysis of particle impact with a wall with integrated damping mechanisms" *Powder Technology*, Vol. 339, pp. 17–24.

Marshall, J. S. and Li, S. (2014) *Adhesive Particle Flow*. Cambridge: Cambridge University Press.

Martinez-Bazan, C., Montanes, J. L. and Lasheras, J. C. (1999a) "On the breakup of an air bubble injected into a fully developed turbulent flow. Part 1. Breakup frequency," *Journal of Fluid Mechanics*, Vol. 401, pp. 157–182.

Martinez-Bazan, C., Monatnes, J. L., and Lasheras, J. C. (1999b) "On the breakup of an air bubble injected into a fully developed turbulent flow. Part 2. Size PDF of the resulting daughter bubbles," *Journal of Fluid Mechanics*, Vol. 401, pp. 183–207.

Martínez-Mercado, J., Palacios-Morales, C. A., and Zenit, R. (2007) "Measurement of pseudo-turbulence intensity in monodispersed bubbly liquids for 10 < Re <500," *Physics of Fluids*, Vol. 19, 103302.

Mason, J. (1978) "Physics of a raindrop," *Physics Bulletin,* Vol. 29, pp. 364–369.

Mathia, V., Huisman, S. G., Sun, C., Lohse, D., and Bourgoin, M. (2018) "Dispersion of air bubbles in isotropic turbulence," *Physics Review Letters*, Vol. 121, 054501.

Maude, A. D. and Whitmore, R. L. (1958) "A generalized theory of sedimentation," *British Journal of Applied Physics,* Vol. 9, pp. 477–482.

Maw, N., Barber, J. R., and Fawcett, J. N. (1976) "The oblique impact of elastic spheres," *Wear*, Vol. 38, pp. 101–114.

Maxey, M. R. (1987) "The gravitational settling of aerosol particles in homogeneous turbulence and random flow fields," *Journal of Fluid Mechanics*, Vol. 174, pp. 441–465.

Maxey, M. R. (1993) "Equation of motion for a small rigid sphere in a non-uniform or unsteady flow," *ASME/FED Gas–Solid Flows*, Vol. 166, pp. 57–62.

Maxey, M. R. and Patel, B. K. (2001) "Localized force representations for particles sedimenting in Stokes flow," *International Journal of Multiphase Flow*, Vol. 27, pp. 1603–1626.

Maxey, M. R., Patel, B. K., Chang, E. J., and Wang, L-P. (1997) "Simulations of dispersed turbulent multiphase flow," *Fluid Dynamics Research*, Vol. 20, pp. 143–156

Maxey, M. R. and Riley, J. J. (1983) "Equation of motion for a small rigid sphere in a non-uniform flow," *Physics of Fluids*, Vol. 26, No. 4, pp. 883–889.

McLaughlin, M. H. (1968) "An experimental study of particle–wall collision relating of flow of solid particles in a fluid," engineer's degree thesis, California Institute of Technology, Pasadena.

McLaughlin, J. B. (1991) "Inertial migration of a small sphere in linear shear flows," *Journal of Fluid Mechanics*, Vol. 224, pp. 261–274.

McLaughlin, J. B. (1996) "Numerical simulation of bubble motion in water," *Journal of Colloid and Interface Science*, Vol. 184, pp. 614–625.

McNamara, S. and Falcon, E. (2005) "Simulations of vibrated granular medium with impact-velocity-dependent restitution coefficient,'' *Physical Review E*, Vol. 71, 031302.

Mei, R. (1992) "An approximate expression for shear lift force on a spherical particle at a finite Reynolds number," *International Journal of Multiphase Flow*, Vol. 18, pp. 145–147.

Mei, R. (1994) "Effect of turbulence on the particle settling velocity in the nonlinear drag regime," *International Journal of Multiphase Flow*, Vol. 20, pp. 273–284.

Mei, R. and Adrian, R. J. (1992) "Flow past a sphere with an oscillation in the free-stream and unsteady drag at finite Reynolds number," *Journal of Fluid Mechanics*, Vol. 237, pp. 323–341.

Mei, R. and Klausner, J. F. (1992) "Unsteady force on a spherical bubble with finite Reynolds number with small fluctuations in the free-stream velocity," *Physics of Fluids A*, Vol. 4, No. 1, p. 63.

Mei, R. and Klausner, J. F. (1994) "Shear lift force on spherical bubbles," *International Journal of Heat and Fluid Flow*, Vol. 15, No. 1, pp. 62–65.

Mei, R., Klausner, J. F., and Lawrence, C. J. (1994) "A note on the history force on a spherical bubble at finite Reynolds number," *Physics of Fluids*, Vol. 6, No. 1, pp. 418–420.

Mei, R., Lawrence, C. J., and Adrian, R. J. (1991) "Unsteady drag on a sphere at finite Reynolds number with small fluctuations in the free-stream velocity," *Journal of Fluid Mechanics*. Vol. 233, pp. 613–631.

Melean, Y. and Sigalotti, L. (2005) "Coalescence of colliding van der Waals liquid drops," *International Journal of Heat and Mass Transfer*, Vol. 48, pp. 4041–4061.

Mena, S. E. and Curtis, J. S. (2020) "Experimental data for solid–liquid flows at intermediate and high Stokes numbers," *Journal of Fluid Mechanics*. Vol. 883, p. A24.

Mendelson, H. D. (1967) "The prediction of bubble terminal velocities from wave theory," *AIChE Journal*, Vol. 13, pp. 250–253.

Menter, F. R. (1994) "Two-equation eddy-viscosity turbulence models for engineering applications," *AIAA Journal*, Vol. 32, pp. 1598–1605.

Mercier, J., Lyrio, A., and Forslund, R. (1973) "Three-dimensional study of the nonrectilinear trajectory of air bubbles rising in water," *Journal of Applied Mechanics, Transactions ASME,* Vol. 40, Ser. E, No. 3, pp. 650-654.

Michaelides, E. (2006) *Particles, Bubbles, and Droplets: Their Motion, Hear and Mass Transfer*. Hackensack: World Scientific Publishing.

Millies, M. and Mewes, D. (1999) "Interfacial area density in bubbly flow," *Chemical Engineering and Processing*, Vol. 38, pp. 307–319.

Millikan, R. A. (1911) "The isolation of an ion: a precision measurement of its charge, and the correction of Stokes's law," *Physical Review*, Vol. 32, p. 349–397.

Millikan, R. A. (1923) "Coefficients of slip in gasses and the law of reflection of molecules from the surface of solids and liquids," *Physical Review*, Vol. 21, pp. 217–238.

Minnaert, M. (1933) "On musical air-bubbles and the sounds of running water," *Philosophical Magazine,* Vol. 16, pp. 235–248.

Mishin, G. I. (1997) "Experimental investigation of the flight of a sphere in weakly ionized air," AIAA Paper 1997-2298, June.

Miyahara, T. and Takahashi, T. (1985) "Drag coefficient of a single bubble rising through a quiescent liquid," *International Chemical Engineering*, Vol. 25, pp. 146–148.

Monchaux, R. and Dejoan, A. (2017) "Settling velocity and preferential concentration of heavy particles under two-way coupling effects in homogeneous turbulence" *Physical Review Fluids*, Vol. 2, 104302.

Monteith, J. L. and Unsworth, M. H. (1990) *Principles of Environmental Physics* London: Academic Press.

Moore, D. W. (1963) "The boundary layer on a spherical gas bubble," *Journal of Fluid Mechanics*, Vol. 16, pp. 161–176.

Moore, D. W. (1965) "The velocity rise of distorted gas bubbles in a liquid of small viscosity," *Journal of Fluid Mechanics*, Vol. 23, pp. 749–766.

Moorman, R. W. (1955) "Motion of a spherical particle in the accelerated portion of free-fall," Ph.D. dissertation, University of Iowa.

Mora, D. A., Obligado, M., Aliseda, A., and Cartellier, A. (2020), "The effect of Reλ and Rouse numbers on the settling of inertial particles in homogeneous isotropic turbulence," (preprint) American Institute of Physics.

Mudde, R. F. and Saito, T. (2001), "Hydrodynamical similarities between bubble column and bubbly pipe flow," *Journal of Fluid Mechanics.* Vol. 437, pp. 203–228.

Mugele, R. and Evans, H. D. (1951) "Droplet size distribution in sprays," *Industrial and Engineering Chemistry Research*, Vol. 43, pp. 1317–1324.

Nagata, T., Nonomura, T., Takahashi, S., and Fukuda, K. (2020) "Direct numerical simulation of subsonic, transonic and supersonic flow over an isolated sphere up to a Reynolds number of 1000," *Journal of Fluid Mechanics*, Vol. 904, October, pp. A36-1–A36-2.

Narciri, M. A. (1992) "Contribution a l'etude des forces exercees par un liquide sur une bulle de gaz: portance, masse ajoutee et interactions hydrodynamiques," Ph.D. thesis, L'ecole Centrale De Lyon, Lyons.

Nathan, A. M., Hopkins, J., Chong, L., and Kaczmarski, H. (2006) "The effect of spin on the flight of a baseball," International Sports Engineering Conference, Munich, July.

Neve, R. S. and Jaafar, F. B. (1982) "The effect of turbulence and surface roughness on the drag of spheres in thin jets," *Aeronautical Journal,* Vol. 86, pp. 331–336.

Nichols, R. H. and Nelson, C. C. (2003) "Application of hybrid RANS/LES turbulence models," 41st Aerospace Sciences Meeting and Exhibit, AIAA-2003-083, Reno.

Nielsen, T., Hebb, J., and Darling, S. L. (1999) "Large-scale CFB combustion demonstration project," 15th International Conference on Fluidized Bed Combustion, Savannah, GA, CONF-990534.

Niven, R. W., Lott, D. F., Ip, A. Y., Somaratne, K. D., and Kearney, M. (1994) "Development and use of an in vitro system to evaluate inhaler devices," *International Journal of Pharmaceutics*, Vol. 101, pp. 81–87.

Oberbeck, A. (1876) "Ueber Stationäre Flüssigkeitsbewegungen mit Berücksichtigung der inneren Reibung," *Journal für die reine und angewandte Mathematik,* Vol. 81, pp. 62–80.

Ochs III, H. T., Beard, K. V., Laird N. F., Holdridge, D. J., and Schaufelbergert, D. E. (1995) "Effects of relative humidity on the coalescence of small precipitation drops in free fall," *Journal of the Atmospheric Sciences*, Vol. 52, No. 21.

Odar, F. and Hamilton, W. S. (1964) "Forces on a sphere accelerating in a viscous fluid," *Journal of Fluid Mechanics.* Vol. 18, pp. 302–314.

O'Neill, M. E. (1964) "A slow motion of viscous liquid caused by a slowly moving solid sphere," *Mathematika*, Vol. 121, pp. 67–74.

Oran, E. S. and Boris, J. P. (1987) *Numerical Simulation of Reactive Flow.* New York: Elsevier.

Oseen, C. W. (1910) "Uber die Stokessche Formel und uber die verwandte Aufgabe in der Hydrodynamik," *Arkiv för Matematik, Astronomi och Fysik,* Vol. 6, No. 29, pp. 1–20.

Oseen, C. W. (1927) *Hydrodynamik*. Leipzig: Akademische Verlagsgesellschaft.

Oesterle, B. and Zaichik, L. I. (2004) "On Lagrangian time scales and particle dispersion modeling in equilibrium turbulent shear flows," *Physics in Fluid*, Vol. 16, pp. 2374–2384.

Ounis, H., Ahmadi, G., and McLaughlin, J. B. (1991) "Brownian diffusion of submicrometer particles in viscous sublayer," *Journal of Colloid and Interface Science*, Vol. 143, pp. 266–277.

Pal, S., Merkle, C. L., and Deutsch, S. (1988) "Bubble characteristics in a microbubble boundary layer," *Physics of Fluids,* Vol. 31, pp. 774–751.

Palacios, J., Gomez, P., Zanzi, C., Lopez, J., and Hernandez, J. (2010) "Experimental study on the splash/deposition limit in drop impact onto solid surfaces" 23rd Annual Conference on Liquid Atomization and Spray Systems, Brno, Czech Republic.

Pan, F. and Acrivos, A. (1968) "Shape of a drop or bubble at low Reynolds number," *Industrial and Engineering Chemistry Fundamentals*, Vol. 7, p. 227–232.

Parmar, M., Annamalai, S., Balachandar, S., and Prosperetti, A. (2018) "Differential formulation of the viscous history force on a particle for efficient and accurate computation," *Journal of Fluid Mechanics*, Vol. 844, pp. 970–993.

Parthasarathy, R. N. and Faeth, G. M. (1990) "Turbulence modulation in homogeneous dilute particle-laden flow," *Journal of Fluid Mechanics*, Vol. 220, pp. 485–537.

Park, W. C., Klauner, J. F., and Mei, R. (1995) "Unsteady forces on spherical bubbles," *Experiments in Fluids,* Vol. 19, pp. 167–172.

Pasquill, F. (1974) *Atmospheric Diffusion*. Chichester: Ellis Horwood.

Peng, C., Geneva, N., Guo, Z., and Wang, L.-P. (2018) "Overdamped large-eddy simulations of turbulent pipe flow up to Reτ = 1500," *Journal of Computational Physics*, Vol. 357, pp. 16–42.

Petersen, A. J., Baker, L., and Coletti, F. (2019) "Experimental study of inertial particles clustering and settling in homogeneous turbulence," *Journal of Fluid Mechanics*, Vol. 864, pp. 925–970.

Pettyjohn, E. S. and Christiansen, E. B. (1948) "Effect of particle shape on free settling rates of isometric particles," *Chemical Engineering Progress,* Vol. 4, pp. 157–172.

Phillips, W. F. (1975) *Physics of Fluids*, Vol. 18, pp. 1089–1093.

Pilch, M. and Erdman, C. A. (1987), "Use of break-up time data to predict the maximum size of stable fragment for acceleration induced breakup of a liquid drop," *International Journal of Multiphase Flow*, Vol. 13, pp. 741–757.

Piomelli, U. (1997) "Introduction to the modeling of turbulence: large eddy and direct simulation of turbulent flows," von Karman Institute for Fluid Dynamics, Lecture Series 1997-03, March.

Plesset, M. S. (1949) "The dynamics of cavitation bubbles," *Journal of Applied Mechanics,* Vol. 16, pp. 228–231.

Poe, G. G., and Acrivos, A (1975) "Closed-streamline flows past rotating single cylinders and spheres: inertia effects," *Journal of Fluid Mechanics*, Vol. 72, pp. 605–623.

Poelma, C., Westerweel, J., and Ooms, G. (2007) "Particle–fluid interactions in grid-generated turbulence," *Journal of Fluid Mechanics*, Vol. 589, pp. 315–351.

Poorte, R. E. G. and Biesheuvel, A. (2002) "Experiments on the motion of gas bubbles in turbulence generated by an active grid," *Journal of Fluid Mechanics*, Vol. 461, pp. 127–154.

Pope, S. B. (2000) *Turbulent Flows*. Cambridge: Cambridge University Press.

Prakash, R. S., Gadgil, H., and Raghunandan, B. N. (2014) "Breakup processes of pressure swirl spray in gaseous cross-flow" *International Journal of Multiphase Flow,* Vol. 66, pp. 79–91.

Prandtl, L. (1905) "über Flüssigkeitsbewegung bei sehr kleiner Reibung," *Verhandlungen des III. Internationalen Mathematiker Kongresses, Heidelberg, 8-13 August B. G. Teubner, Leipzig*, pp. 485–491.

Prosperetti. A. (1987) "The equation of bubble dynamics in a compressible liquid," *Physics of Fluids*, Vol. 30, p. 3626.

Prosperetti, A. (2007) "Averaged equations for multiphase flow," in Prosperetti, A. and Tryggvason, G., eds., *Computational Methods for Multiphase Flow.* Cambridge: Cambridge University Press.

Prosperetti, A. and Tryggvason, G., (2007) *Computational Methods for Multiphase Flow*. Cambridge: Cambridge University Press.

Proudman, I. (1969) "On the flow past a sphere at low Reynolds number," *Journal of Fluid Mechanics*, Vol. 37, pp. 759–760.

Proudman, I. and Pearson, J. R. A. (1957) "Expansions at small Reynolds number for the flow past a sphere and a cylinder," *Journal of Fluid Mechanics*, Vol. 2, pp. 237–262.

Putnam, A. (1961) "Integrable form of droplet drag coefficient," *American Rocket Society Journal*, Vol. 31, October, pp. 1467–1468.

Qian, J. and Law, C. K. (1997) "Regimes of coalescence and separation in droplet collision," *Journal of Fluid Mechanics*, Vol. 331, pp. 59–80.

Qiand, D. (2003) "Bubble motion, deformation, and breakup in stirred tanks," Ph.D. dissertation in chemical engineering, Clarkson University.

Qin, C., Loth, E., Li, P., Simon, T., and van de Ven, J. (2014) "Spray-cooling concept for wind-based compressed air energy storage," *Journal of Renewable and Sustainable Energy*, Vol. 6, 043125.

Radenkovic, D. and Simonon, O. (2018) "Stochastic modelling of three-dimensional particle rebound from isotropic rough wall surface," *International Journal of Multiphase Flow*, Vol. 109, pp. 35–50.

Ramirez, L. E. S, Lim, E. A., Coimbra, C. F. M., and Kobayashi, M. H. (2003), "On the dynamics of a spherical scaffold in rotating bioreactors," *Biotechnology and Bioengineering*, Vol. 84, pp. 382–389.

Rani, S. L. & Balachandar, S. (2003) "Evaluation of the equilibrium Eulerian approach for the evolution of particle concentration in isotropic turbulence," *International Journal of Multiphase Flow*, Vol. 29, pp. 1793–1816.

Ranz, W. E. and Marshall, W. R. (1952) "Evaporation from drops – part II," *Chemical Engineering Progress*, Vol. 48, pp. 141–146.

Rayleigh, L. (1917) "On the pressure developed in a liquid during the collapse of a spherical cavity," *Philosophical Magazine*, 34 (200): 94–98.

Raymond, F. and Rosant, J.-M. (2000) "A numerical and experimental study of the terminal velocity and shape of bubbles in viscous liquids," *Chemical Engineering Science*, Vol. 55, pp. 943–955.

Reagle, C. J., Delimont, J. M., Ng, W. F., Ekkad, S. V., and Rajendran, V. P. (2013) "Measuring the coefficient of restitution of high speed microparticle impacts using a PTV and CFD hybrid technique," *Measurement Science and Technology*, Vol. 24, No. 10, p. 105303.

Reeks, M. W. (1977) "On the dispersion of small particles suspended in an isotropic turbulent fluid," *Journal of Fluid Mechanics*, Vol. 83, Part 3, pp. 529–546.

Reeks, M. W. (2014) "Transport, mixing and agglomeration of particles in turbulent flows" *Journal of Physics: Conference Series,* Vol. 530, 012003.

Reichardt, T., Tryggvason, G., and Sommerfeld, M. (2017) "Effect of velocity fluctuations on the rise of buoyant bubbles," *Computers and Fluids*, Vol. 150, pp. 8–30.

Reichhardt, H. (1951) "Vollständige Darstellung der turbulenten Geschwindigkeitsverteilung in glatten Leitungen," *ZAMM*, Vol. 31, pp. 208– 219.

Reinhart, A. (1964) "Das Verhalten fallender Topfen," *Chemie Ingenieur Technik*, Vol. 36, pp. 740–746.

Revuelta, A., Rodríguez-Rodríguez, J., and Martínez-Bazán, C. (2006) "Bubble break-up in a straining flow at finite Reynolds numbers," *Journal of Fluid Mechanics*, Vol. 551, March 25, pp. 175–184.

Riboux, G., Risso, F., and Legendre, D. (2010) "Experimental characterization of the agitation generated by bubbles rising at high Reynolds number," *Journal of Fluid Mechanics*, Vol. 643, Part 3, pp. 509–539.

Richard, D. and Quere, D. (2000) "Bouncing water drops," *Europhysics Letters,* Vol. 50, pp. 769–775.

Richardson, J. F. and Zaki, W. N. (1954a) "Sedimentation and fluidization: part I," *Transactions of the Institution of Chemical Engineers,* Vol. 32, pp. 35–53.

Richardson, J. F. and Zaki, W. N. (1954b) "The sedimentation of a suspension of uniform spheres under conditions of viscous flow," *Chemical Engineering Science*, Vol. 8, pp. 65–73.

Richardson, L. F. (1922) *Weather Prediction by Numerical Process*. New York, Cambridge: Cambridge University Press.

Risso, F. and Fabre, J. (1998) "Oscillations and break-up of a bubble immersed in a turbulent field," *Journal of Fluid Mechanics*, Vol. 372, pp. 323–355.

Roessler, D. (1982) "Diesel particle mass concentration by optical techniques," *Applied Optics*, Vol. 21, No. 22, pp. 4077–4086.

Rogers, C. B. and Eaton, J. K. (1991) "The effect of small particles on fluid turbulence in a flat-plate, turbulent boundary layer in air," *Physics of Fluid A: Fluid Dynamics*, Vol. 3, pp. 928–937.

Rosa, B., Parishani, H., Ayalay, O., and Wang, L.-P. (2016) "Settling velocity of small inertial particles in homogeneous isotropic turbulence from high-resolution DNS," *International Journal of Multiphase Flow*, Vol. 83, pp. 217–231.

Rosin, P. and Rammler, E. (1933) "The laws governing the fineness of powdered coal," *Institute of Fuel*, Vol. 7, pp. 29–36.

Rowe, P. N. (1987) "A convenient empirical equation for estimation of the Richardson–Zaki exponent," *Chemical Engineering Science,* Vol. 42, p. 27952796.

Rubinow, S. I. and Keller, J. B. (1961) "The transverse force on spinning spheres moving in a viscous liquid," *Journal of Fluid Mechanics*, Vol. 11, No. 3, p. 447.

Rybalko, M., Loth, E., and Lankford, D. (2008) "Lagrangian sub-grid particle diffusion for LES/RANS flows," ASME Fluids Engineering Division Summer Meeting, FEDSM2008–55207, Jacksonville.

Ryskin, G. and Leal, L.G. (1984) "Numerical simulation of free-boundary problems in fluid mechanics. Part 3, bubble deformation in an axisymmetric straining flow," *Journal of Fluid Mechanics*, Vol. 148, No. 37, pp 37–43.

Sadhal, S. S. and Johnson, R. E. (1983) "Stokes flow past bubbles and drops partially coated with thin films. Part 1, stagnant cap of surfactant film – exact solution," *Journal of Fluid Mechanics*, Vol. 126, pp. 237–250.

Saffman, P. G. (1956) "On rise of small air bubbles in water," *Journal of Fluid Mechanics*, Vol. 1, pp. 249275.

Saffman, P. G. (1965) "The lift on a sphere in slow shear flow," *Journal of Fluid Mechanics*, Vol. 22, pp. 385–400.

Saffman, P. G. (1968) "The lift on a small sphere in slow shear flow," *Corrigendum*, Vol. 31, p. 624.

Saffman, P. G. and Turner, J. S. (1956) "On the collision of drops in turbulent clouds," *Journal of Fluid Mechanics*, Vol. 1, pp. 16–30.

Sanjeevi, S. K., Kuipers, J. A. M., and Padding, J.T. (2018) "Drag, lift and torque correlations for non-spherical particles from Stokes limit to high Reynolds numbers," *International Journal of Multiphase Flow*, Vol. 106, pp. 325–337.

Saito, S. (1913) "On the shape of the nearly spherical drop which falls through a viscous fluid," *Science Reports of Tohuku Imperial University, Sendai, Japan*, Vol. 2, pp. 179–185.

Salem, M. B. and Osterlé, B. (1998) "A shear flow around a spinning sphere: numerical study at moderate Reynolds numbers," *International Journal of Multiphase Flow*, Vol. 24, pp. 563–585.

Sandeep, C., Luo, L., and Senetakis, K. (2020) "Effect of grain size and surface roughness on the normal coefficient of restitution of single grains," *Materials*, Vol. 13, Article 814.

Sangani, A. S., Zhang, D. Z., and Prosperetti, A. (1991) "The added mass, Basset, and viscous drag coefficients in non-dilute bubbly liquids undergoing small-amplitude oscillatory motion," *Physics of Fluids A*, Vol. 3, pp. 2955–2970.

Sankaranarayanan, K., Shan, X., Kevrekidid, I. G., and Sundaresan, S. (2003) "Analysis of drag and virtual mass forces in bubble suspension using an implicit formulation of the Lattice Boltzmann Method," *Journal of Fluid Mechanics*, Vol. 452, pp. 61–966.

Sano, T. (1981) "Unsteady flow past a sphere at low Reynolds number," *Journal of Fluid Mechanics*, Vol. 112, pp. 443–441.

Sawatzki, O. (1970) "Stromungsfeld um eine rotierend Kugel," *Acta Mechanica,* Vol. 9, pp. 159–214.

Sawicki, G. S., Hubbard, M., and Stronge, W. (2003) "How to hit home runs: optimum baseball bat swing parameters for maximum range trajectories," *American Journal of Physics,* Vol. 71, pp. 1152–1162.

Schaaf, S. A. and Chambre, P. L. (1958) "The flow of rarefied gases," in H. W. Emmons, ed., *Fundamentals of Gas Dynamics.* Princeton: Princeton University Press, pp. 687–740.

Schiller, L. and Naumann, A. Z. (1933) "Über die grundlegenden Berechungen bei der Schwerkraftaufbereitung," *Verein Deutscher Ingenieure Ze*, Vol. 77, pp. 318–320.

Schlichting, H. and Gersten, G. (2017) *Boundary Layer Theory*, 9th edition, New York: Springer-Verlag.

Sene, K. J., Hunt, J. C. R., and Thomas, N. H. (1994) "The role of coherent structures in bubble transport by turbulent shear flows," *Journal of Fluid Mechanics*, Vol. 259, pp. 219–240.

Shardt, O. and Derksen, J. J. (2012) "Direct simulations of dense suspensions of non-spherical particles," *International Journal of Multiphase Flow*, Vol. 47, pp. 25–36.

Shew, W. L. and Pinton, J-F. (2006) "Dynamical model of bubble path instability," *Physical Review Letters*, Vol. 97, 144508.

Shi. P. and Rzehak, R. (2019) "Lift forces on solid spherical particles in unbounded flows," *Chemical Engineering Science*, Vol. 208, pp. 363–399.

Shi, P., Rzehak, R., Lucas, D., and Magnaudet, J. (2020) "Hydrodynamic forces on a clean spherical bubble translating in a wall-bounded linear shear flow," *Physical Review Fluids*, Vol. 5, pp. 1–31.

Shin, D. H., Sandberg, R. D., and Richardson, E. S. (2017) "Self-similarity of fluid residence time statistics in a turbulent round jet," *Journal of Fluid Mechanics*, Vol. 823, pp. 1–25.

Shirolkar, J. S., Coimbra, C. F. M., and Quirez McQuay, M. (1996) "Fundamental aspects of modeling turbulent particle dispersion in dilute flows," *Progress in Energy and Combustion Science*, Vol. 22, 115145.

Shyy, W., Thakur, S. S., Ouyang, H., Liu, J., and Blosch, E. (1997) *Computational Techniques for Complex Transport Phenomena.* Cambridge: Cambridge University Press.

Simonnet, M., Gentric, C., Olmos, E., and Midoux, N. (2007) "Experimental determination of the drag coefficient in a swarm of bubbles" *Chemical Engineering Science,* Vol. 62, pp. 858–866.

Simpkins, P. G. and Bales, E. L. (1972) "Water drop response to sudden accelerations," *Journal of Fluid Mechanics*, Vol. 55, pp. 629–639.

Singha, A. and Balachandar, R. (2011) “Structure of wake of a sharp-edged bluff body in a shallow channel flow,” *Journal of Fluids and Structures*, Vol. 27, pp. 233–249

Sirignano, W. A. (1993) “Fluid dynamics of sprays – 1992 Freeman Scholar Lecture,” *ASME Journal of Fluids Engineering*, Vol. 115, pp. 345–378.

Sirignano, W. A. (2010) *Fluid Dynamics and Transport of Droplets and Sprays.* New York: Cambridge University Press.

Sivier, S. A., Loth, E., Baum, J. D., and Lohner, R. (1994) “Unstructured adaptive remeshing finite element method for dusty shock flows,” *Shock Waves,* Vol. 4, No. 1, pp. 31–41.

Smith, D. A. and Cheung, K. F. (2003) “Settling characteristics of calcareous sand,” *Journal of Hydraulic Engineering*, June, pp. 479–483.

Smoluchowski, M. (1916) "Drei Vorträge über Diffusion, Brownsche Molekularbewegung und Koagulation von Kolloidteilchen," *Physikalische Zeitschrift* (in German), Vol. 17, pp. 557–571, 585–599.

Snyder, W. H. and Lumley, J. L. (1971) “Some measurements of particle velocity autocorrelation in approximately isotropic turbulence,” *Journal of Fluid Mechanics*, Vol. 48, pp. 41–71.

Sommerfeld, M. (2001) “Validation of a stochastic Lagrangian modeling approach for inter-particle collisions in homogeneous isotropic turbulence,” *International Journal of Multiphase Flow*, Vol. 27, pp. 1829–1858.

Sommerfeld, M. and Huber, N. (1999) “Experimental analysis and modeling of particle-wall collisions,” *International Journal of Multiphase Flow*, Vol. 25, pp. 1457–1489.

Sommerfeld, M., and Lain, S. (2018) “Stochastic modelling for capturing the behavior of irregular-shaped non-spherical particles in confined turbulent flows,” *Powder Technology*, Vol. 332, pp. 253–264.

Sommerfeld, M. and Pasternak (2019) “Advances in modeling binary droplet collision outcomes in sprays: a review of available knowledge,” *International Journal of Multiphase Flow*, Vol. 117, pp. 182–205.

Sone, W. and Aoki, K. (1983) “Forces on a spherical particle in a slightly rarefied gas,” *Progress in Aeronautics and Astronautics, Rarefied Gas Dynamics*, Vol. 51, pp. 417–433.

Soo, S. L. (1990) *Multiphase Fluid Dynamics.* Aldershot–Brookfield: Gower Technical.

Sosnick, A. and Seremeta, K. P. (2015) “Advantages and challenges of the spray-drying technology for the production of pure drug particles and drug-loaded polymeric carriers,” *Advances in Colloid and Interface Science*, Vol. 223, pp. 40–54

Spalart, P. R., Jou, W.-H., Strelets, M., and Allmaras, S. R. (1997) “Comments on the feasibility of LES for wings, and on a hybrid RANS/LES approach,” in United States Air Force Office of Scientific Research, *Advances in DNS/LES: Direct Numerical Simulation and Large Eddy Simulation*, pp. 137–148.

Spelt, P. D. M. and Biesheuvel, A. (1997) “On the motion of gas bubbles in homogeneous isotropic turbulence,” *Journal of Fluid Mechanics*, Vol. 336, pp. 221–244.

Springel, V. and Dullemond, C. P. (2011) “Numerische Strömungsmechanik,” chapter 9. www.ita.uni-heidelberg.de/~dullemond/lectures/num_fluid_2011/.

Squires, K. D. (2007) “Point particle models,” in Prosperetti, A. and Tryggvason, G., eds., *Computational Methods for Multiphase Flow.* Cambridge: Cambridge University Press, pp. 282–319.

Squires, K. D. and Eaton, J. K. (1990) “Particle response and turbulence modification in isotropic turbulence,” *Physics of Fluids A*, Vol. 2, pp. 1191–1203.

Sridhar, G. and Katz, J. (1995) “Drag and lift forces on microscopic bubbles entrained by a vortex,” *Physics of Fluids*, Vol. 7, No. 2, pp. 389–399.

Stadler, J. R. and Zurick, V. J. (1951) "Theoretical aerodynamic characteristics of bodies in free-molecule flow field," NACA TN 2423, July, pp. 12–53.

Stegeman, Y. V. (2002) "Time dependent behavior of droplets in elongational flows," Ph.D. thesis, Technische Universiteit Eindhoven.

Stokes, G. G. (1851) "On the effect of the inertial friction of fluids on the motion of pendulums," *Transactions of the Cambridge Philosophical Society*, Vol. 9 (part II), pp. 8–106

Stout, J. E., Arya, S. P., and Genikhovich, E. L. (1995) "Effect of nonlinear drag on the motion and settling of heavy particles," *Journal of Atmospheric Sciences*, Vol. 52, pp. 3836–3848.

Stringham, G. E., Simons, D. B., and Guy, H. P. (1969), "The behavior of large particles falling in quiescent liquids," professional paper, US Geological Survey, 562-C.

Sugiyama, K., Takagi, S., and Matsumoto, Y. (2001) "Multi-scale analysis of bubbly flows," *Computer Methods in Applied Mechanics and Engineering,* Vol. 191, pp. 689–704.

Sun, T.-Y. and Faeth, G. M. (1986) "Structure of turbulent bubbly jets," *International Journal of Multiphase Flow*, Vol. 12, pp. 115–126.

Sun, Z.-Q., Yang, X.-B., Wang, H.-D., Li, D.-L., Li. S.-Q., and Lu, Y. (2019) "Ceramic/resin composite powders with uniform resin layer synthesized from SiO2 spheres for 3D technology," *Journal of Inorganic Materials,* Vol. 34, No. 5, pp. 567–572.

Sundaram, S. and Collins, L. (1997) "Collision statistics in an isotropic particle-laden turbulent suspension, part 1: direct numerical simulations," *Journal of Fluid Mechanics,* Vol. 335, pp. 75–109.

Suslick, K. S. and Flannigan, D. (2005), "Plasma formation and temperature measurement during single-bubble cavitation," *Nature*, Vol. 434, March, pp. 52–55.

Suslick, K. S. and Price, G. J. (1999) "Applications of ultrasound to materials chemistry," *Annual Review of Materials Science*, Vol. 29, pp. 295–326.

Swamy, N. V. C., Gowda, B. H. L., and Lakshminath, V. R. (1979) "Auto-correlation measurements and integral time-scales in three-dimensional turbulent boundary layers," *Applied Scientific Research,* Vol. 35, pp. 265–316.

Syamlal, M., Rogers, W., and O'Brien, T. J. (1993) "MFIX documentation, theory guide," US Department of Energy Technical Note DOE/METC-94/1004.

Tabakoff, W., Hamed, A., and Murugan, D. M. (1996) "Effect of target materials on the particle restitution characteristics for turbomachinery application," *Journal of Propulsion and Power*, Vol. 12, No. 2, pp. 260–266.

Takagi, S. and Matsumoto, Y. (1999) "Numerical investigations of the lift force acting on bubbles and particles," 3rd JSME/ASME Joint Fluids Engineering Conference, San Francisco, FEDS99–7848, July.

Takata, S., Aoki, K., and Sone, Y. (1994) "Thermophoresis of a sphere with a uniform temperature: numerical analysis of the Boltzmann equation for hard-sphere molecules," *Progress in Astronautics and Aeronautics*, Vol. 159, pp. 626–639.

Takato, Y., Benson, M. E., and Sen, S. (2015) "Rich collision dynamics of soft and sticky crystalline nanoparticles: numerical experiments," *Physical Review E,* Vol. 92, 032403.

Takemura, F. (2004) "Migration velocities of spherical solid particles near a vertical wall for Reynolds number from 0.1 to 5," *Physics of Fluids*, Vol. 16, pp. 204–207.

Takemura, F. and Magnaudet, J. (2003) "The transverse force on clean and contaminated bubbles rising near a vertical wall at moderate Reynolds number," *Journal of Fluid Mechanics,* Vol. 495, pp. 234–253.

Takemura, F., Takagi, S., Magnaudet, J., and Matsumoto, Y. (2002) "Drag and lift forces on a bubble rising near a vertical wall in a viscous liquid," *Journal of Fluid Mechanics*, Vol. 461, pp. 277–300.

Talbot, L., Cheng, R. K., Schefer, R. W., and Willis, D.R. (1980) "Thermophoresis of particles in a heated boundary layer," *Journal of Fluid Mechanics,* Vol. 101, pp. 737–758.

Tanaka, T. and Eaton, J. K. (2010) "Sub-Kolmogorov resolution particle image velocimetry measurements of particle-laden forced turbulence," *Journal of Fluid Mechanics,* Vol. 643, pp. 177–206.

Tanaka, T., Yonemura, S., and Tsuji, Y. (1990) "Experiments of fluid forces on a rotating sphere and spheroid," in *Proceedings of the 2nd KSME-JSME Fluids Engineering Conference*, Vol. 1, pp. 366–369.

Tani, I. (1950) "Baseball's curved balls," *Kagaku*, Vol. 20, pp. 405–409.

Taylor, G. I. (1932) "The viscosity of a fluid containing small drops of another fluid," *Proceedings of the Royal Society A*, Vol. 138, pp. 41–48.

Taylor, G. I. (1934) "The formation of emulsions in definable fields of flow," *Proceedings of the Royal Society A*, Vol. 146, pp. 501–523.

Taylor, G. I. (1949) "The shape and acceleration of a drop in a high-speed air stream," for Advisory Council on Scientific Research and Technical Development, Ministry of Supply, AC 10647/Phys. C69.

Taylor, T. D. and Acrivos, A. (1964) "On the deformation and drag of a falling viscous drop at low Reynolds number," *Journal of Fluid Mechanics*, Vol. 18, pp. 466–476.

Ten Cate, A. and Sundaresan, S. (2006) "Analysis of unsteady forces in ordered arrays," *Journal of Fluid Mechanics*, Vol. 552, pp. 257–287.

Tennekes, H. and Lumley, J. L. (1972) *A First Course in Turbulence*. Cambridge: MIT Press.

Theofanous, T. G., Mitkin, V., and Chang, C. (2018) "Shock dispersal of dilute particle clouds," *Journal of Fluid Mechanics*, Vol. 841, February, pp. 732–745.

Theofanous, T. G. and Sullivan, J. (1982) "Turbulence in two-phase dispersed flow," *Journal of Fluid Mechanics*, Vol. 116, pp. 343–362.

Thompson, T. L. and Clark, N. N. (1991) "A holistic approach to particle drag prediction," *Powder Technology*, Vol. 67, pp. 57–66.

Thompson, P. (1972) *Compressible Fluid Dynamics*. New York: McGraw–Hill.

Thorpe, S. A. (1971) "Experiments on the instability of stratified shear flows: miscible fluids," *Journal of Fluid Mechanics*, Vol. 46, pp. 299–319.

Thornton, C. and Ning, Z. (1998) "A theoretical model for the stick/bounce behaviour of adhesive, elastic-plastic spheres," *Powder Technology*, Vol. 99, pp. 154–162.

Tomkins, M. R., Baldock, T. E., and Nielsen, P. (2005) "Hindered setting of sand grains," *Sedimentology*, Vol. 52, pp. 1425–1432.

Tomiyama, A. (1998) "Plenary lecture: struggle with computational bubble dynamics," International Conference on Multiphase Flow, Lyon, France, June.

Tomiyama, A., Celata, G. P., Hosokawa, S., and Yoshida, S. (2002a) "Terminal velocity of single bubbles in surface tension force dominant regime," *International Journal of Multiphase Flow*, Vol. 28, pp. 1497–1519.

Tomiyama, A., Tamai, H., Zun, I., and Hosokawa, S. (2002b) "Transverse migration of single bubbles in simple shear layers," *Chemical Engineering Science*, Vol. 57, pp. 1849–1858.

Torobin, L. B. and Gauvin, W. H. (1960) *Canadian Journal of Chemical Engineering,* Vol. 38, pp. 142–153.

Tran-Cong, S., Gay, M., and Michaelides, E. E. (2004) "Drag coefficients of irregularly shaped particles," *Powder Technology*, Vol. 139, pp. 21–32.

Tri, B. D., Oesterle, B., and Deneu, F. (1990) "Premiers resultants sur la portance d'une sphere en rotation aux nombres de Reynolds intermediaies," *Comptes rendus de l'Académie des Sciences, Ser. II: Mec., Phys., Chim., Sci. Terre Universe*, Vol. 311, pp. 27–31.

Tryggvason, G., Scardovelli, R., and Zaleski, S. (2011) *Direct Numerical Simulations of Gas–Liquid Multiphase Flows*. Cambridge, New York: Cambridge University Press.

Tsao, H. K. and Koch, D. T (1997) "Observations of high Reynolds number bubbles interacting with a rigid wall," *Physics of Fluids*, Vol. 9, No. 1, pp. 44–56.

Tsouris, C. and Tavlarides, L. L. (1994) "Breakage and coalescence models for drops in turbulent dispersions," *AIChE Journal*, Vol. 40, pp. 395–406.

Tsuge, H. and Hibino, S. I. (1977) "The onset of oscillatory motion of single gas bubbles rising in various liquids," *Journal of Chemical Engineering of Japan*. Vol. 10, pp. 66–68.

Tsuji, Y. and Morikawa, Y. (1982) "LDV measurements of an air-solid two-phase flow in a horizontal pipe," *Journal of Fluid Mechanics*, Vol. 120, pp. 385–409.

Tsuji, Y., Morikawa, Y., and Mizuno, O. (1985) "Experimental measurements of the Magnus force on a rotating sphere at low Reynolds numbers," *Journal of Fluids Engineering,* Vol. 107, No. 9, pp. 484–488.

Tsuji, Y., Morikawa, Y., and Shiomi, H. (1984) "LDV measurements of an air–solid two-phase flow in a vertical pipe," *Journal of Fluid Mechanics,* Vol. 139, pp. 417–434.

Tunstall, E. B. and Houghton, G. (1968) "Retardation of falling spheres by hydrodynamic oscillations," *Chemical Engineering Science*, Vol. 23, No. 9, pp. 1067–1081.

Turney, M. A., Cheung, M. K., McCarthy, M. J., and Powell, R. L. (1995) "Magnetic resonance imaging study of sedimenting suspensions of noncolloidal spheres," *Physics of Fluids*, Vol. 7, pp. 904–911.

Urbin, G. and Knight, D. (2001) "Large eddy simulation of a supersonic boundary layer using an unstructured grid," *AIAA Journal*, Vol. 39, No. 7, pp. 1288–1295.

Vanderwel, C. and Tavoularis, S. (2014) "Measurements of turbulent diffusion in uniformly sheared flow," *Journal of Fluid Mechanics*, Vol. 754, pp. 488–514.

Van Donkelaar, A. (2010) "Global satellite-derived map of PM2.5 averaged over 2001–2006." www.nasa.gov/topics/earth/features/health-sapping.html.

Van Driest, E. R. (1956) "On turbulent flow near a wall" *Journal of Aeronautical Sciences,* Vol. 23, pp. 1007–1012.

Van Dyke, M. (1982) *An Album of Fluid Motion.* Stanford: Parabolic Press.

Van Wachem, B., Curran, T., and Evrard, F. (2020) "Fully correlated stochastic inter–particle collision model for Euler–Lagrange gas–solid flows," *Flow, Turbulence and Combustion,* Vol. 105, pp. 935–963.

Van Wijngaarden, T. (1976) "Hydrodynamic interaction between gas bubbles in liquid," *Journal of Fluid Mechanics*, Vol. 77, pp. 27–44.

Van Wijngaarden, T. (1998) "On pseudo turbulence," *Theoretical and Computational Fluid Dynamics*, Vol. 10, pp. 449–458.

Varaksin, A. Y., Polezhaev, Y. V., and Polyakov, A. F. (2000) "Effect of particle concentration on fluctuating velocity of the disperse phase for turbulent pipe flow," *International Journal of Heat and Fluid Flow*, Vol. 21, pp. 562–567.

Vasseur, P. and Cox, R. G. (1977) "The lateral migration of spherical particles sedimenting in a stagnant bounded fluid," *Journal of Fluid Mechanics*, Vol. 80, pp. 561–591.

Venerus, D. C. and Simavilla, D. N. (2015) "Tears of wine: new insights on an old phenomenon," *Scientific Reports*, Vol. 5, 16162.

Wakaba, L. V. and Balachandar, S. (2005) "History force on a sphere in a weak linear shear flow," *International Journal of Multiphase Flow*, Vol. 31, pp. 996–1014.

Waldmann, L. (1961) "On the motion of spherical particles in nonhomogeneous gases," in L. Talbot, ed., *Rarefied Gas Dynamics*. New York: Academic Press, pp. 323–344.

Wall, S., John, W., Wang, H. C., and Goren, S. L., (1990) "Measurements of kinetic energy loss for particles impacting surfaces," *Aerosol Science and Technology*, Vol. 12, pp. 926–946.

Wallis, G. B. (1969) *One-Dimensional Two-Phase Flow*. New York: McGraw-Hill.

Wallis. G. B. (1974) "The terminal speed of single drops in an infinite medium," *International Journal of Multiphase Flow*, Vol. 1, pp. 491–511.

Walsh, D. E. (1988) "A study of factors suspected of influencing the settling velocity of fine gold particles," University of Alaska Mineral Industry Research Laboratory.

Walter, J. F. and Blanch, H. W. (1986) "Bubble break-up in gas-liquid bioreactors: break-up in turbulent flows," *Chemical Engineering Journal*, Vol. 321, pp. B7–B17.

Wang, B., Clemens, N. T., Varghese, P. L., and Barlow, R. S. (2008) "Turbulent time scales in a nonpremixed turbulent jet flame by using high-repetition rate thermometry," *Combustion and Flame*, Vol. 152, pp. 317–335.

Wang, B. and Manhart, M. (2012) "Two-phase micro- and macro-time scales in particle-laden turbulent channel flows," *Acta Mechanica Sinica*, Vol. 28, pp. 595–604.

Wang, L.-P. and Maxey, M. R. (1993) "Settling velocity and concentration distribution of heavy particles in homogeneous, isotropic turbulence," *Journal of Fluid Mechanics*. Vol. 256, pp. 27–68.

Wang, L.-P. and Stock, D. E. (1993) "Dispersion of heavy particles by turbulent motion," *Journal of the Atmospheric Sciences*, Vol. 50, No. 13, pp. 1897–1913.

Warnica, W. D., Renksizbulut, M., and Strong, A. B. (1995) "Drag coefficients of spherical liquid droplet. Part II turbulent gaseous fields," *Experiments in Fluids*, Vol. 18, pp. 265–276.

Watts, R. G. and Ferrer, R. (1987) "The lateral force on a spinning sphere: aerodynamics of a curveball," *American Journal of Physics*. Vol. 55, pp. 40–44.

Wegener, P. P. and Ashkenas, H. (2006) "Wind tunnel measurements of sphere drag at supersonic speeds and low Reynolds numbers," *Journal of Fluid Mechanics*, Vol. 10, No. 4, pp. 550–560.

Wegener, P. P., Sundell, R. E., and Parlange, J.-Y. (1971) "Spherical-cap bubbles rising in liquids," *Z. Flugwissenschaften*, Vol. 19, pp. 347–352.

Weir, G. and Tallon, S. (2005) "The coefficient of restitution for normal incident, low velocityparticle impacts," *Chemical Engineering Science*, Vol. 60, pp. 3637–3647.

Welleck, R. M., Agrawal, A. K., and Skelland, A. H. P. (1966) "Shape of liquid drops moving in liquid media," *AIChE Journal*, Vol. 12, pp. 854–862.

Wells, M. R. and Stock, D. E. (1983) "The effects of crossing trajectories on the dispersion of particles in a turbulent flow," *Journal of Fluid Mechanics*, Vol. 136, pp. 31–62.

Wen, C. Y. and Yu, Y. H. (1966) "Mechanics of fluidization," *Chemical Engineering Progress Symposium Series*, Vol. 62, pp. 100–111.

Wen, F., Kamalu, N., Chung, J. N., Crowe, C. T., and Trout, T. R. (1992) "Particle dispersion by vortex structures in plane mixing layers," *Journal of Fluids Engineering*, Vol. 114, pp. 657–666.

Werlé, H. (1980) "Transition and separation: visualizations in the ONERA water tunnel," *La Recherche aérospatial*, Vol. 1980–5, pp. 35–49.

White, F. M. (2016) *Viscous Fluid Flow*. New York: McGraw-Hill.

Wilcox, D. C. (2006) *Turbulence Modeling for CFD*, 3rd edition. La Canada: DCW Industries.

Williams, F. A. (1965) *Combustion Theory*. Reading: Addison-Wesley.

Williams, J. J. E. and Crane, R. I. (1983) "Particle collision rate in turbulent flow," *International Journal of Multiphase Flow*, Vol. 9, pp. 421–435.

Willis, K. D. and Orme, M. E. (2000) "Experiments on the dynamics of droplet collisions in a vacuum," *Experiments in Fluids*, Vol. 29, pp. 347–358.

Willmarth, W. W., Hawk, N. E., and Harvey, R. L. (1964) "Steady and unsteady motions and wakes of freely falling disks," *Physics of Fluids*, Vol. 7, pp. 197–208.

Winnikow, S. and Chao, B. T. (1966) "Droplet motion in purified systems," *Physics of Fluids*, Vol. 9, pp. 50–61.

Wolfrum, B., Mettin, R., Kurz, T., and Lauterborn, W. (2003) "Cavitation induced cell detachment and membrane permeabilization," 2003 IEEE Ultrasonics Symposium-837.

Wu, J.-S. and Faeth, G. M. (1994) "Sphere wakes at moderate Reynolds numbers in a turbulent environment," *AIAA Journal*, Vol. 32, No. 2, pp. 535–554.

Wu, M. and Gharib, M. (2002) "Experimental studies on the shape and path of small air bubbles rising in clean water," *Physics of Fluids*, Vol. 14, No. 49, 10.1063/1.1485767.

Wu, W., and Wang, S. S. (2006) "Formulas for sediment porosity and settling velocity," *Journal of Hydraulic Engineering*, Vol. 132, pp. 858–862.

Wu, X. and Adrian, R. J. (2012) "Direct numerical simulation of a 30R long turbulent pipe flow at R+ = 685: large- and very large-scale motions," *Journal of Fluid Mechanics*, Vol. 698, pp. 235–281.

Wygnanski, I. and Fiedler, H. (1969) "Some measurements in the self-preserving jet," *Journal of Fluid Mechanics*, Vol. 38, pp. 577–612.

Wygnanski, I. and Fiedler, H. (1970) "The two-dimensional mixing region," *Journal of Fluid Mechanics*, Vol. 41, pp. 327–361.

Xie, H.-Y. and Zhang, D.-W. (2001) "Stokes shape factor and its application in the measurement of sphericity of non-spherical particles," *Powder Technology*, Vol. 114, pp. 102–105.

Yamamato, K. and Ishihara, Y. (1988) "Thermophoresis of a spherical particle in a rarefied gas of a transition regime," *Physics of Fluids*, Vol. 31, pp. 3618–3624.

Yan, X., Jia, Y., Wang, L., and Cao, Y. (2017) "Drag coefficient fluctuations of a single bubble rising in water," *Chemical Engineering Journal*, Vol. 316, pp. 553–562.

Yang, X., Muhlassen, M.-P., and Frohlich, J. (2021) "Efficient simulation of bubble dispersion and resulting interaction," *Experimental and Computational Multiphase Flow*, Vol. 3, No. 3, pp. 152–170.

Yang, T. S. and Shy, S. S. (2003) "The settling velocity of heavy particles in an aqueous near-isotropic turbulence," *Physics of Fluids*, Vol. 15, No. 4, pp. 868–880.

Yang, T. S. and Shy, S. S. (2005) "Two-way interaction between solid particles and homogenous air turbulence: particle settling rate and turbulence modification measurements," *Journal of Fluid Mechanics*, Vol. 526, pp. 171–216.

Yeong, Y., Burton, J., and Loth, E. (2014) "Drop impact and rebound dynamics on an inclined superhydrophobic surface," *Langmuir*, Vol. 30, pp. 12027–12038.

Yih, C. (1969) *Fluid Mechanics*. New York: McGraw-Hill.

Yoon, J. (2001) "Jet engine types," www.aerospaceweb.org/question/propulsion/q0033.shtml.

Yoshizawa, A. and Horiuti, K. (1985) "A statistically-derived subgrid-scale kinetic energy model for the large-eddy simulation of turbulent flows," *Journal of the Physical Society*, Vol. 54, pp. 2834–2839.

Young, J. and Leeming, A. (1997) "A theory of particle deposition in a turbulent pipe flow," *Journal of Fluid Mechanics*, Vol. 340, pp. 129–159.

Young, J. B. and Hanratty, T. J. (1991) "Trapping of solid particles at a wall in a turbulent flow," *AIChE Journal,* Vol. 37, No. 10, pp. 1529–1536.

Yuan, C., Fox, R. O. (2011) "Conditional quadrature method of moments for kinetic equations," *Journal of Computational Physics*, Vol. 230, pp. 8216–8246.

Yuan, H. and Prosperetti, A. (1994) "On the in-line motion of two spherical bubbles in a viscous fluid," *Journal of Fluid Mechanics*, Vol. 278, pp. 325–349.

Zakharov, L. V., Ovchinnikov, A. A., and Nikolayev, N.A. (1993) "Modeling of the effect of turbulent two-phase flow friction decrease under the influence of dispersed phase elements," *International Journal of Heat and Mass Transfer*, Vol. 36, pp. 1981–1991.

Zayas, G., Chiang, M. C., Wong, E., et al. (2012) "Cough aerosol in healthy participants: fundamental knowledge to optimize droplet-spread infectious respiratory disease management," *BMC Pulmonary Medicine*, Vol. 12, Article 11.

Zeng, L., Najjar, F., Balachandar, S., and Fisher, P. (2009) "Forces on a finite-sized particle located close to a wall in a linear shear flow," *Physics of Fluids*, Vol. 21, 033302.

Zenit, R., Koch, D. L., and Sangani, A. S. (2001) "Measurements of the average properties of a suspension of bubbles rising in a vertical channel," *Journal of Fluid Mechanics,* Vol. 429, pp. 307–342.

Zhang, D. Z. and Prosperetti, A. (1994) "Averaged equations for inviscid dispersed two-phase flow," *Journal of Fluid Mechanics*, Vol. 267, pp. 185–219.

Zhang, Z. (2019) "Micro-bubble dynamics in turbulent flow," Ph.D. dissertation, University of Toulouse, Institut National Polytechnique de Toulouse.

Zhao, B., Chen, C., and Tan, Z. (2009) "Modeling of ultrafine particle dispersion in indoor environments with an improved drift flux model" Aerosol *Science*, Vol. 40, pp. 29–43.

Zhao, L., Andersson, H. I., and Gillissen, J. J. J. (2013) "Interphasial energy transfer and particle dissipation in particle-laden wall turbulence," *Journal of Fluid Mechanics*, Vol. 715, pp. 32–59.

Zheng, F. (2002) "Thermophoresis of spherical and non-spherical particles: a review of theories and experiments," *Advances in Colloid and Interface Science*, Vol. 97, pp. 255–278.

Zhong, H., Chen, S., and Lee, C. (2011) "Experimental study of freely falling thin disks: transition from planar zigzag to spiral," *Physics of Fluids*, Vol. 23, 011702; https://doi.org/10.1063/1.3541844.

Ziegenhein, T., Tomiyama, A., and Lucas, D. (2018) "A new measuring concept to determine the lift force for distorted bubbles in low Morton number system: results for air/water" *International Journal of Multiphase Flow*, Vol. 108, pp. 11–24.

Zuber, N. (1964) "On the dispersed two-phase flow in the laminar flow regime," *Chemical Engineering Science*, Vol. 19, pp. 897–917.

Zun, I., Kljenek, I., and Serizaw, A. (1992) "Bubble coalescence and transition from wall void peaking to core void peaking in turbulent bubbly flow," in O. C. Jones and Michiyoshi, I., eds., *Dynamics of Two-Phase Flows.* Boca Raton: CRC Press, pp. 233–239.

Index

Printed in the United States
by Baker & Taylor Publisher Services